ENGINEERING
STATISTICS

ENGINEERING STATISTICS

SECOND EDITION

ALBERT H. BOWKER

*Chancellor of the University of California
at Berkeley*

GERALD J. LIEBERMAN

*Professor of Statistics and Operations Research
Stanford University*

PRENTICE-HALL, INC.

Englewood Cliffs, New Jersey

Library of Congress Cataloging in Publication Data

Bowker, Albert Hosmer
  Engineering statistics.

  1. Engineering — Statistical methods. I. Lieberman,
Gerald J., joint author. II. Title.
TA153.B67 1972   620'.001'5195   72-39106
ISBN 0-13-279455-1

20 19 18 17 16 15 14 13 12 11

ISBN: 0-13-279455-1

Printed in the United States of America

Prentice-Hall International, Inc., *London*
Prentice-Hall of Australia, Pty. Ltd., *Sydney*
Prentice-Hall of Canada, Ltd., *Toronto*
Prentice-Hall of India Private Limited, *New Delhi*
Prentice-Hall of Japan, Inc., *Tokyo*

CONTENTS

Chapter 1   Histograms and Empirical Distributions   **1**

1.1   Introduction   1
1.2   Empirical Distributions   2
1.3   Measures of Central Tendency   7
1.4   Measures of Variation   8
* 1.5   Computation of the Mean and Standard Deviation of the Data from the Frequency Table   9

Chapter 2   Random Variables and Probability Distributions   **13**

2.1   Introduction   13
2.2   Set of All Possible Outcomes of the Experiment—the Sample Space   14
2.3   Events   18
2.4   Random Variables   19
2.5   Probability   24
2.6   Cumulative Distribution Functions   27
2.7   Discrete Probability Distributions   33
2.8   Continuous Random Variables and Density Functions   37
2.9   Expectation   43
2.10   Moments   47
2.11   Some Properties of Functions of Random Variables   51
2.12   Bivariate Probability Distributions   52
2.13   Conditional Probability and Independent Events   57
2.14   Independent Random Variables and a Random Sample   60
2.15   Conditional Probability Distributions   64

Chapter 3   The Normal Distribution   **75**

3.1   Definitions   75
3.2   The Mean and Variance of the Normal Distribution   76
    *3.2.1.   Evaluation of the Mean and Variance   78
3.3   Tables of the Normal Integral   79
3.4   Combinations of Normally Distributed Variables   82
3.5   The Standardized Normal Random Variable   84

3.6	The Distribution of the Sample Mean	84
3.7	Tolerances	86
* **3.8**	Tolerances in "Complex Items"	93
3.9	The Central Limit Theorem	99

Chapter 4   Other Probability Distributions   **106**

4.1	Introduction		106
4.2	The Chi-Square Distribution		107
	4.2.1.	The Chi-Square Random Variable	107
	4.2.2	The Addition Theorem	110
	4.2.3	The Distribution of the Sample Variance, S^2	111
4.3	The t Distribution		114
	4.3.1	The t Random Variable	114
	4.3.2	The Distribution of $(\overline{X} - \mu)\sqrt{n}/S$	117
	4.3.3	The Distribution of the Difference Between Two Sample Means	118
4.4	The F Distribution		120
	4.4.1	The F Random Variable	120
	4.4.2	The Distribution of the Ratio of Two Sample Variances	122
4.5	The Binomial Distribution		123
	4.5.1	The Binomial Random Variable	123
	* **4.5.2**	Tables of the Binomial Probability Distribution	126
	* **4.5.3**	The Normal Approximation to the Binomial	127
	* **4.5.4**	The Arc Sine Transformation	128
	* **4.5.5**	The Poisson Approximation to the Binomial	128

Chapter 5   Decision Making   **134**

5.1	Introduction		134
5.2	Decision Making Without Experimentation		134
	5.2.1	Action Space	135
	5.2.2	States of Nature	136
	5.2.3	Loss Function	137
	5.2.4	Criteria for Choosing Among Actions	138
	5.2.5	Bayes Principle	139
	5.2.6	Evaluation of the Loss Function for the Rockwell Hardness Example	141
	5.2.7	Further Examples	145
5.3	Decision Making With Experimentation		148
	5.3.1	Decision Procedures	148
	5.3.2	Risk Function	150
	5.3.3	Bayes Decision Procedures	152
	5.3.4	Calculation of the Posterior Distribution	155
		* **5.3.4.1** Derivation of Posterior Distribution	158
	5.3.5	Further Examples	158
	5.3.6	Assessment of the Bayesian Approach	161
5.4	Significance Tests		163
	5.4.1	The Operating Characteristic Curve	165
	5.4.2	Comparison of the OC Curve with the Risk Function	167
	5.4.3	Comparison of OC Curves	168
	5.4.4	Tests of Hypotheses	172
	5.4.5	One- and Two-Sided Procedures	174

Chapter 6   Tests of Hypotheses about a Single Parameter   **183**

6.1   Test of the Hypothesis that the Mean of a Normal Distribution
Has a Specified Value when the Standard Deviation Is Known   183
 6.1.1   Choice of an OC Curve   183
 6.1.2   Tables and Charts for Determining Decision Rules   184
  6.1.2.1   Tables and Charts for Two-Sided
   Procedures   185
  6.1.2.2   Summary for Two-Sided Procedures Using
   Tables and Charts   188
  6.1.2.3   Tables and Charts for One-Sided Procedures   189
  6.1.2.4   Summary for One-Sided Procedures Using
   Tables and Charts   191
  6.1.2.5   Tables and Charts for OC Curves   192
 6.1.3   Analytical Determination of Decision Rules   192
  6.1.3.1   Acceptance Regions and Sample Sizes   192
  6.1.3.2   The OC Curve   196
 6.1.4   Example   197
6.2.   Test of the Hypothesis that the Mean of a Normal Distribution
Has a Specified Value when the Standard Deviation is Unknown   198
 6.2.1   Choice of an OC Curve   198
 6.2.2   Tables and Charts for Carrying Out *t* Tests   199
  6.2.2.1   Tables and Charts for Two-Sided Procedures   200
  6.2.2.2   Summary for Two-Sided Procedures   200
  6.2.2.3   Tables and Charts for One-Sided Procedures   202
  6.2.2.4   Summary for One-Sided Procedures   204
  6.2.2.5   Tables and Charts for OC Curves   205
 6.2.3.   Examples of *t* Tests   206
6.3   Test of the Hypothesis that the Standard Deviation of a Normal
Distribution Has a Specified Value   207
 6.3.1   Choice of an OC Curve   207
 6.3.2   Charts and Tables to Design Tests of Dispersion   208
  6.3.2.1   Tables and Charts for Two-Sided Procedures   208
  6.3.2.2   Summary for Two-Sided Procedures Using
   Tables and Charts   209
  6.3.2.3   Tables and Charts for One-Sided Procedures   210
  6.3.2.4   Summary for One-Sided Procedures Using
   Tables and Charts   213
  6.3.2.5   Tables and Charts for OC Curves   214
 6.3.3   Analytical Treatment for Chi-Square Tests   215
 6.3.4   Example   216

Chapter 7   Tests of Hypotheses about Two Parameters   **225**

7.1   Test of the Hypothesis that the Means of Two Normal Distributions Are Equal when Both Standard Deviations Are Known   225
 7.1.1   Choice of an OC Curve   225
 7.1.2   Tables and Charts for Determining Decision Rules   226
  7.1.2.1   Tables and Charts for Two-Sided Procedures   226
  7.1.2.2   Summary for Two-Sided Procedures Using
   Tables and Charts   228

	7.1.2.3	Summary for One-Sided Procedures Using Tables and Charts	229
	7.1.2.4	Tables and Charts for OC Curves	230
7.1.3	Analytical Determination of Decision Rules		230
	7.1.3.1.	Acceptance Regions and Sample Sizes	230
	7.1.3.2	The OC Curve	232
7.1.4	Example		234
7.2	Test of the Hypothesis that the Means of Two Normal Distributions Are Equal, Assuming that the Standard Deviations Are Unknown but Equal		235
7.2.1	Choice of an OC Curve		235
7.2.2	Tables and Charts for Carrying out Two-Sample *t* Tests		235
	7.2.2.1	Tables and Charts for Two-Sided Procedures	236
	7.2.2.2	Summary for Two-Sided Procedures Using Tables and Charts	237
	7.2.2.3	Summary for One-Sided Procedures Using Tables and Charts	237
	7.2.2.4	Tables and Charts for OC Curves	238
7.2.3	Example		239
7.3	Test of the Hypothesis that the Means of Two Normal Distributions Are Equal, Assuming that the Standard Deviations Are Unknown and Not Necessarily Equal		240
7.3.1	Test Procedure		240
7.3.2	Example		241
7.4	Test for Equality of Means when the Observations Are Paired		242
7.4.1	Test Procedure		242
7.4.2	Example		244
7.5	Non-parametric Tests		246
7.5.1	The Sign Test		246
7.5.2	The Wilcoxon Signed Rank Test		249
7.5.3	Wilcoxon Test for Two Independent Samples		251
7.6	Test of the Hypothesis that the Standard Deviations of Two Normal Distributions Are Equal		254
7.6.1	Choice of an OC Curve		254
7.6.2	Charts and Tables for Carrying out *F* Tests		255
	7.6.2.1	Tables and Charts for Two-Sided Procedures	255
	7.6.2.2	Summary for Two-Sided Procedures Using Tables and Charts	256
	7.6.2.3	Tables and Charts for One-Sided Procedures	257
	7.6.2.4	Summary for One-Sided Procedures Using Tables and Charts	258
	7.6.2.5	Tables and Charts for OC Curves	258
*7.6.3	Analytical Treatment for Tests		260
7.6.4	Example		262
7.7	Cochran's Test for the Homogeneity of Variances		263

Chapter 8   Estimation 279

8.1	Introduction	279
8.2	Point Estimation	279
8.2.1	Comparison of Estimators	280

8.2.2	Unbiased Estimators	283
8.2.3	Consistent Estimators	284
8.2.4	Efficient Unbiased Estimators	285
8.2.5	Estimation by the Method of Maximum Likelihood	286
8.2.6	Estimation by the Method of Moments	290
8.2.7	Estimation by the Method of Bayes	292
8.3	Confidence Interval Estimation	294
8.3.1	Confidence Interval for the Mean of a Normal Distribution when the Standard Deviation Is Known	295
8.3.2	Confidence Interval for the Mean of a Normal Distribution when the Standard Deviation Is Unknown	296
8.3.3	Confidence Interval for the Standard Deviation of a Normal Distribution	297
8.3.4	Confidence Interval for the Difference between the Means of Two Normal Distributions when the Standard Deviations Are Both Known	298
8.3.5	Confidence Interval for the Difference between the Means of Two Normal Distributions where the Standard Deviations Are Both Unknown but Equal	300
8.3.6	Confidence Interval for the Ratio of Standard Deviations of Two Normal Distributions	301
8.3.7	A Table of Point Estimates and Interval Estimates	302
8.3.8	Approximate Confidence Intervals	302
8.3.9	Simultaneous Confidence Intervals	304
8.3.10	Bayesian Intervals	308
8.4	Statistical Tolerance Limits	309
8.4.1	Statistical Tolerance Limits Based on the Normal Distribution	309
8.4.2	One-Sided Statistical Tolerance Limits Based on the Normal Distribution	310
8.4.3	Distribution-Free Tolerance Limits	310

Chapter 9   Fitting Straight Lines   325

9.1	Introduction	325
9.2	Types of Linear Relationships	328
9.3	Least Squares Estimators of the Slope and Intercept	331
9.3.1	Formulation of the Problem and Results	331
***9.3.2**	Theory	333
9.4	Confidence Interval Estimators of the Slope and Intercept	336
9.4.1	Formulation of the Problem and Results	336
***9.4.2**	Theory	337
9.5	Point Estimators and Confidence Interval Estimators of the Average Value of Y for a Given x	338
9.5.1	Formulation of the Problem and Results	338
***9.5.2**	Theory	339
9.6	Point Estimators and Interval Estimators of the Independent Variable x Associated with an Observation on the Dependent Variable Y	340
9.7	Prediction Interval for a Future Observation on the Dependent Variable	342

9.7.1   Formulation of the Problem and Results   342
*****9.7.2**   Theory   343
9.8   Tests of Hypotheses about the Slope and Intercept   345
9.9   Estimation of the Slope B when A Is Known to be Zero   346
9.10   Ascertaining Linearity   349
9.11   Transformation to a Straight Line   351
9.12   Work Sheets for Fitting Straight Lines   352
9.13   Illustrative Examples   352
9.14   Correlation   362

Chapter 10   Analysis of Variance   377

10.1   Introduction   377
10.2   Model for the One-Way Classification   377
    10.2.1   Fixed Effects Model   377
    10.2.2   Random Effects Model   380
    10.2.3   Further Examples of Fixed Effects and of the Random Effects Models   380
    10.2.4   Computational Procedure, One-Way Classification   381
    10.2.5   The Analysis of Variance Procedure   383
        10.2.5.1   A Heuristic Justification   383
        *****10.2.5.2**   The Partition Theorem   383
    10.2.6   Analysis of the Fixed Effects Model, One-Way Classification   385
    10.2.7   The OC Curve of the Analysis of Variance for the Fixed Effects Model   389
    10.2.8   Example Using the Fixed Effects Model   395
    10.2.9   Analysis of the Random Effects Model   396
    10.2.10   The OC Curve for the Random Effects Model   397
    10.2.11   Example Using the Random Effects Model   402
    10.2.12   Randomization Tests in the Analysis of Variance   403
10.3   Two-Way Analysis of Variance, One Observation per Combination   405
    10.3.1   Fixed Effects Model   405
    10.3.2   Random Effects Model   408
    10.3.3   Mixed Fixed Effects and Random Effects Model   409
    10.3.4   Computational Procedure, Two-Way Classification, One Observation per Combination   409
    10.3.5   Analysis of the Fixed Effects Model, Two-Way Classification, One Observation per Combination   411
    10.3.6   The OC Curve of the Analysis of Variance for the Fixed Effects Model, Two-Way Classification, One Observation per Combination   413
    10.3.7   Example Using the Fixed Effects Model   414
    10.3.8   Analysis of the Random Effects Model, Two-Way Classification, One Observation per Combination   416
    10.3.9   The OC Curve for the Random Effects Model, Two-Way Classification   417
    10.3.10   Example Using the Random Effects Model   418
    10.3.11   Analysis of the Mixed Effects Model, Two-Way Classification, One Observation per Combination   418

	10.3.12	The OC Curve of the Analysis of Variance for the Mixed Effects Model, Two-Way Classification, One Observation per Cell	420
	10.3.13	Example Using the Mixed Effects Model	420
10.4		Two-Way Analysis of Variance, *n* Observations per Combination	421
	10.4.1	Description of the Various Models	421
	10.4.2	Computational Procedure, Two-Way Classification, *n* Observations per Combination	423
	10.4.3	Analysis of the Fixed Effects Model, Two-Way Classification, *n* Observations per Combination	424
	10.4.4	The OC Curve of the Analysis of Variance for the Fixed Effects Model, Two-Way Classification, *n* Observations per Cell	427
	10.4.5	Example Using the Fixed Effects Model, Two-Way Classification, Three Observations per Combination	428
	10.4.6	Analysis of the Random Effects Model, Two-Way Classification, *n* Observations per Combination	430
	10.4.7	The OC Curve of the Random Effects Model, Two-Way Classification, *n* Observations per Combination	431
	10.4.8	Example Using the Random Effects Model	432
	10.4.9	Analysis of the Mixed Effects Model, Two-Way Classification, *n* Observations per Cell	432
	10.4.10	The OC Curve of the Analysis of Variance for the Mixed Effects Model, Two-Way Classification, *n* Observations per Cell	434
	10.4.11	Example Using the Mixed Effects Model	434
10.5		Summary of Models and Tests	436

Chapter 11   Some Further Techniques of Data Analysis   452

11.1		Introduction	452
11.2		Qualitative Techniques for Determining the Form of a Distribution	452
11.3		Quantitative Techniques for Determining the Form of a Distribution	454
	11.3.1	The Kolmogorov-Smirnov Test	454
	11.3.2	The Chi-Square Goodness of Fit Test	458
11.4		Chi-Square Tests	460
	11.4.1	The Hypothesis Completely Specifies the Theoretical Frequency	461
	11.4.2	Dichotomous Data	461
	11.4.3	Test of Independence in a Two-Way Table	463
	11.4.4	Computing Form for Test of Independence in a 2×2 Table	464
11.5		Comparison of Two Percentages	465
11.6		Confidence Intervals for Proportion	466

Chapter 12   Statistical Quality Control: Control Charts   472

12.1	Introduction	472
12.2	An Overview of Control Charts	473

12.3	Control Chart for Variables: $\overline{X}$-Charts	474
12.3.1	Statistical Concepts	474
12.3.2	Estimate of $\overline{X}'$	475
12.3.3	Estimate of σ' by $\overline{\sigma}$	476
12.3.4	Estimate of σ' by $\overline{R}$	476
12.3.5.	Starting a Control Chart for $\overline{X}$	478
12.3.6	Relation between Natural Tolerance Limits and Specification Limits	479
12.3.7	Interpretation of Control Charts for $\overline{X}$	480
12.4	R Charts and σ *Charts*	481
12.4.1	Statistical Concepts	481
12.4.2	Setting up a Control Chart for R or σ	483
12.5	Example of $\overline{X}$ and R Chart	483
12.6	Control Chart For Fraction Defective	485
12.6.1	Relation between Control Charts Based on Variables Data and Charts Based on Attributes Data	485
12.6.2	Statistical Theory	486
12.6.3	Starting the Control Chart	487
12.6.4	Continuing the p Chart	488
12.6.5	Example	489
12.7	Control Charts For Defects	489
12.7.1	Difference between a Defect and a Defective Item	489
12.7.2	Statistical Theory	490
12.7.3	Starting and Continuing the c Chart	490
12.7.4	Example	491
12.8	Further Developments on Control Charts	492
12.8.1	The Signed Sequential Rank Control Chart	493
12.8.2	The Cumulative Sum Control Chart	495

Chapter 13   Sampling Inspection by Attributes   503

13.1	The Problem of Sampling Inspection	503
13.1.1	Introduction	503
13.1.2	Drawing the Sample	504
13.2	Lot-by-Lot Sampling Inspection by Attributes	505
13.2.1	Single Sampling Plans	505
13.2.1.1	Single Sampling	505
13.2.1.2	Choosing a Sampling Plan	507
13.2.1.3	Calculation of OC Curves for Single Sampling Plans	507
13.2.1.4	Example	508
13.2.2	Double Sampling Plans	508
13.2.2.1	Double Sampling	508
*****13.2.2.2**	OC Curves for Double Sampling Plans	509
*****13.2.2.3**	Example	509
13.2.3	Multiple Sampling Plans	510
13.2.4	Classification of Sampling Plans	511
13.2.4.1	Classification By AQL	511
13.2.4.2	Classification By LTPD	512
13.2.4.3	Classification By Point of Control	512
13.2.4.4	Classification By AOQL	512

13.2.5		Dodge-Romig Tables	513
	13.2.5.1	Single Sampling Lot Tolerance Tables	514
	13.2.5.2	Double Sampling Lot Tolerance Tables	514
	13.2.5.3	Single Sampling AOQL Tables	518
	13.2.5.4	Double Sampling AOQL Tables	518
13.2.6		Military Standard 105D	518
	13.2.6.1	History	518
	13.2.6.2	Classification of Defects	522
	13.2.6.3	Acceptable Quality Levels	522
	13.2.6.4	Normal, Tightened, and Reduced Inspection	523
	13.2.6.5	Sampling Plans	525
	13.2.6.6	Summary of the Procedure To Be Followed in the Selection of a Sampling Plan from MIL-STD-105D	527
13.2.7		Designing Your Own Attribute Plan	527
	13.2.7.1	Computing the OC Curve of a Single Sampling Plan	527
	13.2.7.2	Finding a Sampling Plan Whose OC Curve Passes through Two Points	537
	13.2.7.3	Design of Item-by-Item Sequential Plans	537
13.2.8		A Bayesian Approach to Sampling Inspection	545
	13.2.8.1	Economic Structure	545
	13.2.8.2	A Decision Analysis Model	546
	13.2.8.3	Bayes Procedures	547
13.3		Continuous Sampling Inspection	550
13.3.1		Introduction	550
13.3.2		Dodge Continuous Sampling Plans	551
13.3.3		Multi-Level Sampling Plans	552
13.3.4		The Dodge CSP-1 Plan without Control	556
13.3.5		Wald-Wolfowitz Continuous Sampling Plans	557
13.3.6		Girshick Continuous Sampling Plan	558
13.3.7		Plans Which Provide for Termination of Production	559

Chapter 14   Lot-by-Lot Sampling Inspection by Variables   **565**

14.1		Introduction	565
14.2		General Inspection Criteria	566
14.3		Estimates of the Percent Defective	568
	14.3.1	Estimate of the Percent Defective when the Standard Deviation Is Unknown but Estimated by the Sample Standard Deviation	568
	14.3.2	Estimate of the Percent Defective when the Standard Deviation Is Unknown but Estimated by the Average Range	569
	14.3.3	Estimate of the Percent Defective when the Standard Deviation Is Known	571
14.4		Comparison of Variables Procedures with M and k	574
14.5		The Military Standard for Inspection by Variables, MIL-STD-414	575
	14.5.1	Introduction	575
	14.5.2	Section A — General Description of Sampling Plans	576
	14.5.3	Section B — Variability Unknown, Standard Deviation Method	585

14.5.4 Section C — Variability Unknown, Range Method   585
14.5.5 Section D — Variability Known   590
14.5.6 Example Using MIL-STD-414   591

Appendix   **599**

Answers to Selected Problems   **618**

Index   **623**

PREFACE

This second edition of *Engineering Statistics* represents a major revision of the original text. The material on probability theory (Chapter 2) has been completely rewritten. A chapter on decision analysis (Chapter 5) has been added and its contents integrated into the development of testing hypotheses and estimation. The material on point estimation has been revised (Chapter 8) and results on approximate and simultaneous confidence interval estimation have been included. Other important developments over the past several years have also been added. Yet, the spirit of the original text remains. As stated in the Preface to the first edition: "The book attempts to present statistics as a science, rather than an art, for making decisions .... The book is not intended to be a cookbook or a treatise in mathematical statistics." It is still our aim to provide a logical presentation of the statistical techniques used in engineering and the physical sciences, requiring only an elementary knowledge of the calculus as a prerequisite.

Although the number of pages in this edition has been increased slightly, the material contained in the first ten chapters has been used at Stanford in a one-quarter course containing approximately 40 periods (actually, only half of Chapter 10 was covered). The contents of the entire book can easily be covered in a two-quarter (or two-semester) course.

We are grateful to many people for their help in the preparation of this second edition—too many to mention them all. Still, there are some who deserve special thanks. In particular, we wish to express our gratitude to Professors Samuel Eilon, Frederick S. Hillier (and Ann Hillier), Steven Mathewson, and Howard M. Taylor for their many helpful comments and suggestions. Many students helped to debug the second edition, but we specifically acknowledge Chris Albright, Hull Belmore, Peter Bryant, Malcolm Hudson, and Marion Reynolds for their valuable aid. We are also indebted to the Office Of Naval Research for supporting much of the research that is included in the material in the chapters on quality control and sampling inspection.

Finally, one of the authors (GJL) spent the 1969–70 academic year on sabbatical leave at Imperial College in London where he made his contribution to the revision. He is particularly grateful to Helen for her understanding and invaluable assistance, and to Janet, Joanne, Michael, and Diana for their indulgence.

A.H.B.

G.J.L.

PREFACE TO
THE FIRST EDITION

This book is intended as a text for a first course in statistics for students in engineering and the physical sciences. At Stanford it is the basis for a one-quarter course entitled "Statistical Methods in Engineering and the Physical Sciences," which takes up the material in the first ten chapters. This course meets for approximately forty periods (four times a week) for one quarter and would be roughly the equivalent, in classroom hours, of a semester course meeting three times a week. A course limited to the first ten chapters concentrates primarily on basic probability and distribution theory and experimental statistics. Additional topics in experimental statistics, and the important subjects of quality control and acceptance sampling, could be incorporated in a two-quarter or two-semester course. This would include the material in Chapters XI, XII, and XIII. Chapter XIII, in particular, is a very long and detailed presentation of sampling inspection, including the basic tables now in practical use for both attribute and variables inspection.

The book attempts to present statistics as a science, rather than an art, for making decisions. After a thorough grounding in the necessary fundamentals of probability and distribution theory, the text turns to a discussion of the statistical techniques commonly used by engineers and physical scientists. Ideally in statistical work, decision procedures should be specified before the data are available and the experiment is run. The book emphasizes the notion of operating characteristic curves and the considerations in the choice of a particular experimental procedure and sample size. Each technique is illustrated by fully worked examples drawn from engineering and science. A complete set of tables, both for the design of the experiment and the final test of significance, is included.

The book is not intended to be a cookbook or a treatise in mathematical statistics. It is our hope to provide a logical presentation of the statistical techniques used in engineering and physical sciences. The only prerequisite for the book is an elementary knowledge of calculus. Sections which can be deleted without losing the continuity of the text are starred. As a rule these are sections which have detailed or difficult mathematical arguments.

We are grateful to Professors E. L. Grant and W. G. Ireson, editors of the *Handbook of Industrial Engineering and Management*, for many helpful comments and for their permission to reprint much of the material from our chapter in that handbook. We are indebted to many members of the Statistics Department and the Applied Mathematics and Statistics Laboratory at Stanford, and to several classes of students who have helped to eradicate some of the errors and clarify some of the explanations. Particular mention should be made of Mrs. Peggy Hathaway, who typed the manuscript, of Miss Barbara Bonesteele, who undertook the tedious chore of proofreading, of Professor Donald Guthrie and Mr. Peter Zehna, who assisted in the course and helped check much of the material, of Mrs. Gladys Garabedian for invaluable assistance in the preparation of the charts, of Miss Shirley Smith for her general assistance, and of Professor Herbert Solomon, without whose aid this book might never have been completed.

Finally, we are indebted to Professor Sir Ronald A. Fisher, Cambridge, and to Dr. Frank Yates, Rothanstad, and to Messrs. Oliver and Boyd Ltd., Edinburgh, for permission to abridge Tables III, IV, and V from their book, *Statistical Tables for Biological, Agricultural, and Medical Research*.

A.H.B.

G.J.L.

1

HISTOGRAMS AND
EMPIRICAL DISTRIBUTIONS

1.1 Introduction

The science of statistics deals with making decisions based on observed data in the face of uncertainty. However, the popular conception of statistics is that it involves large masses of data and concerns itself with percentages, averages, or presentation of data in tables or charts. These, in fact, represent only a very small part of the field today and are of less interest to engineers than are other aspects of statistics, e.g., quality control, sampling inspection, decision analysis, and the design and analysis of experiments. The latter are treated in this book.

Most scientific investigations, whether concerned with the effect of a new drug on polio, methods of allaying traffic congestion in a big city, consumer reaction to a new product, or the quality of manufactured products, depend on observations, even if they are of a rudimentary sort. In scientific and industrial experimentation, these observations are taken to study the effect of variation of certain factors or the relation between certain factors. One may wish to study the quality of a raw material from a new supplier, the relation between tensile strength and hardness for a particular alloy, or the optimum combination of conditions for a manufacturing process. Ultimately, these observations are to be used for making decisions, and the main subject matter of this book will deal with providing procedures for making decisions with preassigned risks on the basis of the limited information in samples. These procedures will be illustrated by examples.

The raw material of statistics consists of data or observations. Usually these observations are physical measurements: length, decimeter temperature, or resistance, and may be any number on a continuous scale. But these continuous measurements are not the only kind of observations. In the inspection of manufactured products, items are often classified as defective or non-defective and the observation for each item is not a number but simply the category into which it falls. In studying the grading system in a statistics course, each observation may be a letter, say, A or B or C, etc.— the grades given to each student. Whether the observations are numbers, letters, categories, or whatever, the same general methods of statistical analysis are applicable. In most of the book, the concern is with small numbers of observations, but before turning to this topic a few of the techniques useful in arranging, presenting, and summarizing large numbers of observations will be discussed. With the advent of high-speed electronic computing machines, enormous quantities of data can be stored, and then made available, thereby emphasizing the need for meaningful techniques of data presentation.

1.2 Empirical Distributions

A basic notion of statistics is the notion of variation. There is no real meaning in speaking of the life of an incandescent lamp produced under certain conditions; some bulbs may last many times as long as others. Consider, for example, the data in Table 1.1 on the lifetimes in hours of 417 forty-watt, 110-volt, internally frosted incandescent lamps taken from forced life tests.[1] Many factors including variations in raw materials, workmanship, function of automatic machinery, and, in fact, test conditions may account for these differences; some of these factors may be controlled carefully, but some pattern of variation is inherent in all observational data. Perhaps no one would expect all light bulbs to have exactly the same life, but even the results of the most easily controlled experiments, such as those designed to measure the velocity of light, exhibit variation from experimental error.

Returning to the life of bulbs, the data in Table 1.1 are listed in the order in which the life tests were made; arranged serially in this way the numbers present a meaningless jumble. The best way to arrange the data depends, of course, on the question to be answered. In a study of the life of light bulbs, there are many questions which might arise. How long do the bulbs last on the average? What fraction of light bulbs manufactured in this way can be expected to last 1,000 hours? What value of life will 95% of the bulbs outlast? What range of life can we expect? What interval around the average will include 50% of the observations? A careful investigation of the table will reveal that the life varies from a low of 225 hours to a maximum of 1,690, but the other questions cannot be answered from this table.

The first step in analyzing data of this sort is to arrange them into a fre-

---

[1] D. J. Davis, "An Analysis of Some Failure Data," *Journal of the American Statistical Association*, Vol. 47, 1952.

quency table, grouping adjacent observations into classes which are usually called class intervals or cells. If this is done by hand, the class intervals are listed on a ruled sheet and a tally mark for each observation is placed opposite the appropriate interval. A convenient way to keep track is to make the fifth tally mark diagonally through the preceding four. A worksheet is presented in Table 1.2; the completed frequency table is given in Table 1.3. This table indicates that there is one lamp with a life between 201 and 300 hours, none between 301–400 and 401–500, three with lives between 501 and 600, etc.

The frequency table is often displayed in graphical form by drawing a series of rectangles, each with the class interval as base and height equal to the number of items in the interval (see Fig. 1.1). Charts of this form are called histograms. Both the histogram and the frequency table provide at a glance much more information than the original data. There is a high concentration of lives between 900 and 1,300 hours, and all but a handful lie between 600 and 1,500 hours.

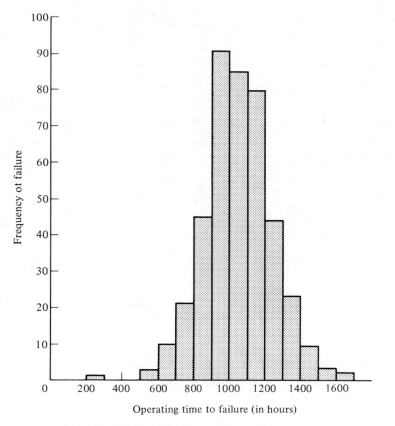

Fig. 1.1. *Life length histogram for incandescent lamps.*

Table 1.1. Life in Hours of 417 Incandescent Lamps

Date	Item Lifetimes										Average of Subgroup
1- 2-47	1,067	919	1,196	785	1,126	936	918	1,156	920	948	997
1- 9-47	855	1,092	1,162	1,170	929	950	905	972	1,035	1,045	1,012
1-16-47	1,157	1,195	1,195	1,340	1,122	938	970	1,237	956	1,102	1,121
1-23-47	1,022	978	832	1,009	1,157	1,151	1,009	765	958	902	978
1-30-47	923	1,333	811	1,217	1,085	896	958	1,311	1,037	702	1,027
2- 6-47	521	933	928	1,153	946	858	1,071	1,069	830	1,063	937
2-13-47	930	807	954	1,063	1,002	909	1,077	1,021	1,062	1,157	998
2-20-47	999	932	1,035	944	1,049	940	1,122	1,115	833	1,320	1,029
2-27-47	901	1,324	818	1,250	1,203	1,078	890	1,303	1,011	1,102	1,088
3- 6-47	996	780	900	1,106	704	621	854	1,178	1,138	951	923
3-13-47	1,187	1,067	1,118	1,037	958	760	1,101	949	992	966	1,014
3-20-47	824	653	980	935	878	934	910	1,058	730	980	888
3-27-47	844	814	1,103	1,000	788	1,143	935	1,069	1,170	1,067	993
4- 3-47	1,037	1,151	863	990	1,035	1,112	931	970	932	904	993
4-10-47	1,026	1,147	883	867	990	1,258	1,192	922	1.150	1,091	1,053
4-17-47	1,039	1,083	1,040	1,289	699	1,083	880	1,029	658	912	971
4-23-47	1,023	984	856	924	801	1,122	1,292	1,116	880	1,173	1,017
5- 1-47	1,134	932	938	1,078	1,180	1,106	1,184	954	824	529	986
5- 8-47	998	996	1,133	765	775	1,105	1,081	1,171	705	1,425	1,015
5-15-47	610	916	1,001	895	709	860	1,110	1,149	972	1,002	922
5-22-47	990	1,141	1,127	1,181	856	716	1,308	943	1,272	917	1,045
5-29-47	1,069	976	1,187	1,107	1,230	836	1,034	1,248	1,061	1,550	1,130
6- 5-47	1,240	932	1,165	1,303	1,085	813	1,340	1,137	773	787	1,058
6-12-47	1,438	1,009	1,002	1,061	1,277	892	900	1,384	1,148		1,123
6-19-47	1,117	1,225	1,176	709	1,485	1,225	1,011	1,028	1,227	1,277	1,148
6-26-47	1,222	912	885	1,562	1,118	1,197	976	1,080	924	1,233	1,111
7- 3-47	1,135	623	983	883	1,088	1,029	1,201	898	970	1,058	987
7-10-47	1,160	831	1,023	1,354	1,218	1,121	1,172	1,169	1,113	1,308	1,147
7-17-47	1,166	1,470	1,635	1,141	1,555	1,054	1,461	1,057	1,228	1,187	1,295
8- 7-47	1,016	744	1,197	1,122	666	1,022	964	1,085	612	1,003	943
8-14-47	1,235	942	1,055	893	1,235	1,056	968	1,056	1,014	1,096	1,055
8-21-47	1,013	889	1,430	926	1,297	1,033	1,024	1,103	1,385		1,122
8-28-47	1,077	813	1,121	960	1,156	1,033	1,255	225	525	675	884
9- 4-47	1,211	995	924	732	935	1,173	1,024	1,254	1,014		1,029
9-11-47	798	1,080	862	1,220	1,024	1,170	1,120	898	918	1,086	1,018
9-18-47	1,028	1,122	872	826	1,337	965	1,297	1,096	1,068	943	1,055
9-25-47	1,490	918	609	985	1,233	985	985	1,075	1,240	985	1,051
10- 2-47	1,105	1,243	1,204	1,203	1,310	1,262	1,234	1,104	1,303	1,185	1,215
10- 9-47	759	1,404	944	1,343	932	1,055	1,381	816	1.067	1,252	1,095
10-16-47	1,248	1,324	1,000	984	1,220	972	1,022	956	1,093	1,358	1,118
10-23-47	1,024	1,240	1,157	1,415	1,385	824	1,690	1,302	1,233	1,331	1,260
10-30-47	1,109	827	1,209	1,202	1,229	1,079	1,176	1,173	769	905	1,068

Questions about life of bulbs can be answered much more precisely if, instead of presenting the frequency for a given class interval, the number that will achieve a specified value is presented. This is particularly useful with life data; a common question is: how many bulbs will fail by 1,000 hours? From the frequency table, the number of bulbs with lives that fail

Table 1.2. Tally Sheet for Length of Life of Incandescent Lamps

Class Interval	Frequency
201- 300	|
301- 400	
401- 500	
501- 600	|||
601- 700	卌 卌
701- 800	卌 卌 卌 卌 |
801- 900	卌 卌 卌 卌 卌 卌 卌 卌 卌
901-1,000	卌 卌 卌 卌 卌 卌 卌 卌 卌 卌 卌 卌 卌 卌 卌 卌 卌 卌 |
1,001-1,100	卌 卌 卌 卌 卌 卌 卌 卌 卌 卌 卌 卌 卌 卌 卌 卌 卌
1,101-1,200	卌 卌 卌 卌 卌 卌 卌 卌 卌 卌 卌 卌 卌 卌 卌 卌
1,201-1,300	卌 卌 卌 卌 卌 卌 卌 卌 ||||
1,301-1,400	卌 卌 卌 卌 |||
1,401-1,500	卌 ||||
1,501-1,600	|||
1,601-1,700	||

**Table 1.3. Frequency Table for Length
of Life of Incandescent Lamps**

Class Interval (100 hr)	Frequency f
201- 300	1
301- 400	—
401- 500	—
501- 600	3
601- 700	10
701- 800	21
801- 900	45
901-1,000	91
1,001-1,100	85
1,101-1,200	80
1,201-1-300	44
1,301-1,400	23
1,401-1,500	9
1,501-1,600	3
1,601-1,700	2
	417

before 1,000 hours can be calculated by adding the frequencies in cells below 1,000. The result is $1 + 3 + 10 + 21 + 45 + 91 = 171$. If this is done for every cell in the table, a cumulative frequency table is obtained as in Table 1.4. If each of the cumulative frequencies is divided by the total number of observations, the fraction cumulative frequency distribution is obtained. A plot of this cumulative frequency function is presented in Fig. 1.2.

From this curve it is easily seen that 90% of the bulbs will exceed 820, or that half of them will last 1,040 hours. From such a curve replacement of bulbs can be scheduled so as to minimize total cost, or the future demand for bulbs can be forecast.

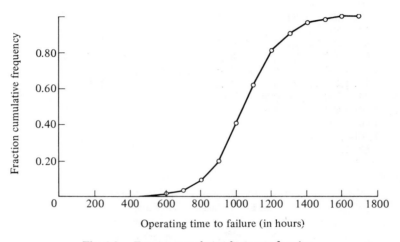

Operating time to failure (in hours)

Fig. 1.2. *Fraction cumulative frequency function.*

Table 1.4. **Cumulative Frequency Table**

Class Interval (100 hr)	Frequency f	Cumulative Frequency	Fraction Cumulative Frequency
201- 300	1	1	0.002
301- 400	—	1	0.002
401- 500	—	1	0.002
501- 600	3	4	0.009
601- 700	10	14	0.034
701- 800	21	35	0.084
801- 900	45	80	0.192
901-1,000	91	171	0.410
1,001-1,100	85	256	0.614
1,101-1,200	80	336	0.806
1,201-1,300	44	380	0.913
1,301-1,400	23	403	0.966
1,401-1,500	9	412	0.988
1,501-1,600	3	415	0.995
1,601-1,700	2	417	1.000

1.3 Measures of Central Tendency

The preceding sections have been concerned with presenting a mass of data in useful form. Often this is not enough and it is necessary to replace a collection of data by a single number. Someone may ask, "How long do these light bulbs really last? How long do they last on the average? Can you give me some idea about length of life in one number?" The questioner may not be satisfied with the evasive answer that it all depends, and that some bulbs last longer than others. Reluctantly, the statistician must agree that there are certain questions which can be answered by a single number.

The most common and most useful single descriptive measure is the mean or arithmetic average of the data. If the observations are denoted by $x_1, x_2, x_3, \ldots, x_N$, the arithmetic mean of the data is

$$\bar{x} = \frac{\sum_{i=1}^{N} x_i}{N}.$$

In order to avoid confusion later, the reader is reminded that a large mass of data is being considered. All the data are available, if desired, i.e., the N observations can be produced. However, a measure of its central tendency is sought which "describes" these N recorded observations. This is to be contrasted with the situation to be studied later where all the data *is not* available, and only a "random sample" of size n can be drawn (n usually small relative to N). In the former instance, the terminology "mean of the data" is used whereas in the latter instance, the terminology "sample mean" will be used.

If the data as grouped in a histogram present a well-behaved picture which rises and falls smoothly, the mean of the data is a reasonably typical value in the sense that it will occur where the observations cluster. For the light bulb data,

$$\bar{x} = \frac{435,921}{417} = 1,045,$$

which is a value close to the intuitive center of the frequency curve.

A word of caution must be introduced. Most data in engineering and science follow reasonably smooth curves, but occasionally one may observe a pathological frequency function—for example, one which might look like Fig. 1.3.

The mean of such a set of data is still the middle in some sense, but does not represent a point of high concentration of the data. Actually, if each observation is considered to have an associated unit mass, the mean is simply the center of gravity.

Another common measure of central tendency is the median of the data, defined as the middle observation when the numbers are arranged in order

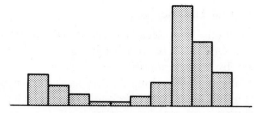

Fig. 1.3. *Histogram of a pathological frequency function.*

of magnitude. If there are an odd number, the median is uniquely defined; if there are an even number, the average of the neighboring numbers is usually taken; i.e., the median of the numbers 7, 11, 21, 24, 31, 92, 1,017 is the middle one, or 24. The median of the numbers 7, 11, 21, 24, 31, 92 is (21 + 24)/2 or 22.5. The median is easy to calculate if the data are arranged properly or can be sorted, say, on punch cards. If a cumulative frequency function is given, the median can be read off easily, at least approximately.

A third measure of central tendency sometimes used is the mode of the data, defined as the observation of maximum frequency. This measure is defined for grouped data and is taken to be the midpoint of this class interval with maximum frequency, 950 for the light bulb data given in Table 1.3.

This text does not make much use of the median and the mode as they do not play an important role in the field of analytical statistics. However, they are so widely used in descriptive work that it seems inappropriate to follow one's inclination to write a statistics book without mentioning them.

1.4 Measures of Variation

Two methods of summarizing data have been presented: one which groups the data but still exhibits the patterns of variation, and one which replaces the data by a single measure of central tendency. A compromise is to characterize data by two statistics: one a measure of central tendency, and one a measure of variation. One simple measure of variation is the range of the data, the largest minus the smallest. Another more common measure is the standard deviation of the data, s, defined as the square root of the average squared deviation from the mean; i.e.,

$$s = \sqrt{\frac{\sum_{i=1}^{N} (x_i - \bar{x})^2}{N - 1}}.$$

Again, the aforementioned comment about contrasting the current situation where all the data are available, if desired, with the situation to be studied later where only a "random sample" can be drawn is applicable. In the former instance the terminology "standard deviation of the data" is used whereas in the latter instance, the terminology "sample standard deviation" will be used. Some writers divide by N instead of $N - 1$ in defining s. For N large it doesn't make any appreciable numerical difference. In order to use essentially the same definition for the sample standard deviation, where the sample size is small, there are important reasons for dividing by $N - 1$ rather than N.

A detailed discussion of the relation between s and the pattern of variation will not be presented here because this can be illustrated very quickly after the normal distribution is introduced. Let it suffice to say that small values of s are associated with a high concentration of the observations around the mean and that this is, strictly speaking, valid only for reasonably symmetric bell-shaped histograms. The variance of the data is really the moment of inertia of the observations.

The algebraic identity

$$\sum_{i=1}^{N} (x_i - \bar{x})^2 = \sum_{i=1}^{N} x_i^2 - N\bar{x}^2 = \sum_{i=1}^{N} x_i^2 - \frac{\left(\sum_{i=1}^{N} x_i\right)^2}{N}$$

provides a useful computation formula for s. It is not efficient to subtract the mean from numbers and square the difference, but much better to calculate the sum and sum of squares of the observations. For the life data,

$$\sum_{i=1}^{417} x_i = 435{,}921 \quad \text{and} \quad \sum_{i=1}^{417} x_i^2 = 470{,}808{,}333.$$

Hence, $\quad s^2 = \dfrac{470{,}808{,}333 - \dfrac{190{,}027{,}118{,}241}{417}}{416} = 36{,}316.85,$

and $\quad s = 190.57.$

*1.5 Computation of the Mean and Standard Deviation of the Data from the Frequency Table

Besides providing a useful tabulation of a large quantity of data, the frequency table may be used to compute the mean and standard deviation. There is some approximation involved in the method; essentially it replaces each observation by the midpoint of the class interval in which it falls, but the error is usually negligible. The computing method is illustrated in Table 1.5; it involves a new scale with an arbitrary origin and the width of the class interval taken as unity. The mean and standard deviation are computed in terms of the new scale and then transformed to the original units. For the life data, the mean computed from grouped data is 1,048 compared with 1,045 from raw data; the standard deviations are 190.7 and 190.6.

Table 1.5. Computation of Mean and Standard Deviation from Grouped Data

Class Interval (100 hr)	Frequency f	Midpoint of Class Interval	Deviation from Arbitrary Origin d	fd	fd^2
201- 300	1	250	−8	−8	64
301- 400	—	350	−7	0	0
401- 500	—	450	−6	0	0
501- 600	3	550	−5	−15	75
601- 700	10	650	−4	−40	160
701- 800	21	750	−3	−63	189
801- 900	45	850	−2	−90	180
901-1,000	91	950	−1	−91	91
1,001-1,100	85	1,050	0	0	0
1,101-1,200	80	1,150	1	80	80
1,201-1,300	44	1,250	2	88	176
1,301-1,400	23	1,350	3	69	207
1,401-1,500	9	1,450	4	36	144
1,501-1,600	3	1,550	5	15	75
1,601-1,700	2	1,650	6	12	72
TOTALS	417			−7	1,513

Mean:

$\bar{x}$ in class interval from arbitrary origin of $1,050 = \dfrac{\sum fd}{N} = \dfrac{-7}{417} = -0.0167$.

$\bar{x}$ in original units = arbitrary origin $+ \dfrac{\sum fd}{N}$ (class interval)

$$= 1,050 - (0.0167)(100) = 1,048.$$

Standard Deviation:

$$s \text{ class interval units} = \sqrt{\dfrac{\sum fd^2 - \dfrac{(\sum fd)^2}{N}}{N-1}}$$

$$= \sqrt{\dfrac{1,513 - \dfrac{49}{417}}{416}} = 1.907.$$

s in original units = s in class interval units $\times$ 100 = 190.7.

Note that the midpoint of the class interval 901–1,000 is taken as 950. This assumes that a bulb which expires with a life of, say, 900.2 hours is recorded as 901; the test rack is checked every hour. If the data were measurements of some dimension, the interpretation would be different. It might be that 901 actually represented not any number between 900 and 901 but any number which would round to 901, say, any number between 900.5 and 901.5. In this case the class interval 901 to 1,000 would actually represent 900.5 to 1,000.5 and its midpoint would be 950.5. The precise definition of

endpoints of class intervals affects both computation and graphical presenta-
tions and deserves consideration when class intervals are chosen.

PROBLEMS

1. For the data shown in the table (a) compute a frequency table with class intervals of 0.02 starting with 6.60–6.61, 6.62–6.63, etc., (b) draw a histogram, and
(c) plot a cumulative frequency function.

Diameters of Rivet Heads in Hundredths of an Inch

6.72	6.77	6.82	6.70	6.78	6.70	6.62
6.75	6.66	6.66	6.64	6.76	6.73	6.80
6.72	6.76	6.76	6.68	6.66	6.62	6.72
6.76	6.70	6.78	6.76	6.67	6.70	6.72
6.74	6.81	6.79	6.78	6.66	6.76	6.72
6.74	6.70	6.78	6.76	6.70	6.76	6.76
6.67	6.62	6.68	6.74	6.74	6.81	6.66
6.68	6.72	6.74	6.64	6.79	6.72	6.82
6.80	6.74	6.73	6.81	6.77	6.60	6.72
6.68	6.78	6.76	6.74	6.70	6.64	6.78
6.72	6.71	6.64	6.70	6.70	6.75	
6.67	6.72	6.76	6.64	6.69	6.73	
6.67	6.66	6.84	6.73	6.66	6.66	
6.62	6.72	6.80	6.72	6.76	6.72	

2. From the raw data of Problem 1, compute the mean and standard deviation of
these data. Compute the mean and standard deviation of the grouped data of
Problem 1.

3. In an experiment measuring the percent shrinkage on drying, 50 plastic clay
test specimens produced the following results:

19.3	20.5	17.9	17.3	17.1
15.8	16.9	17.1	19.5	22.5
20.7	18.5	22.5	19.1	17.9
18.4	18.7	18.8	17.5	17.5
14.9	12.3	19.4	16.8	19.3
17.3	19.5	17.4	16.3	18.8
21.3	23.4	18.5	19.0	19.0
16.1	18.8	17.5	18.2	17.4
18.6	18.3	16.5	17.4	17.4
20.5	16.9	17.5	18.2	22.5

(a) Group these percentages into a frequency table with class intervals of 1 % starting
with 12.0–12.9. (b) Draw a histogram and (c) plot a cumulative frequency function.

4. In Problem 3, compute the mean and standard deviation of the raw data and
of the grouped data.

5. Take the first 50 observations (from 1–2–47 to 1–30–47) of Table 1.1 and
(a) compute a frequency table with class intervals of 100 hours starting with
201–300, (b) draw a histogram, and (c) plot a cumulative frequency function.

6. From the raw data of Problem 5, compute the mean and standard deviation of these data. Compute the mean and standard deviation of the grouped data of Problem 5.

7. Repeat Problem 5 using the second 50 observations (from 2–6–47 to 3–6–47).

8. From the raw data of Problem 7, compute the mean and standard deviation of these data. Compute the mean and standard deviation of the grouped data of Problem 7.

9. Repeat Problem 5 using the third 50 observations (from 3–13–47 to 4–10–47).

10. From the raw data of Problem 9, compute the mean and standard deviation of these data. Compute the mean and standard deviation of the grouped data of Problem 9.

11. Repeat Problem 5 using the fourth 50 observations (from 4–17–47 to 5–15–47).

12. From the raw data of Problem 11, compute the mean and standard deviation of these data. Compute the mean and standard deviation of the grouped data of Problem 11.

13. Show that $\sum_{i=1}^{N} (x_i - \bar{x}) = 0$.

14. Prove the algebraic identity $\sum_{i=1}^{N} (x_i - \bar{x})^2 = \sum_{i=1}^{N} x_i^2 - N\bar{x}^2$.

15. If all the observations in Table 1.1 are changed by subtracting 500 hours from each, what effect will this have on the mean and standard deviation of the data?

16. If all the observations in Table 1.1 are changed by dividing each by 100 hours, what effect will this have on the mean and standard deviation of the data?

2

RANDOM VARIABLES AND PROBABILITY DISTRIBUTIONS

2.1 Introduction

Chapter 1 dealt with the aspect of statistics related to the processing of data. It was assumed that a large amount of data was amassed and had to be processed. In this chapter and the ensuing chapters, that aspect of statistics dealing with making decisions in the face of uncertainties will be considered.

In most engineering problems decisions must be made on the basis of experimentation. An experiment specifies exactly what tests or trials are to be performed and what is to be observed. These tests, leading to outcomes or observations, usually are repeated several times under uniform or constant conditions. Even though great care is taken to keep the conditions of the experiment as uniform as possible, the individual observations exhibit an intrinsic variability that cannot be eliminated. For example, steel ingots may be produced and placed in a Rockwell hardness testing machine[1] for the purpose of measuring the hardness of a particular type of steel. Even if the ingots are taken from the same batch and the experiment performed under similar conditions, the readings for the different ingots will generally differ. This inherent variability is often referred to as the experimental error, which is a convenient name for a source of variation that eludes control. Thus, in all types of repeated experiments performed under "controlled" conditions, the outcomes of the individual repetitions vary; hence, the results of any given repetition usually cannot be predicted exactly.

---

[1] The principle of a Rockwell hardness testing machine consists of impressing a hardened steel or diamond point into the surface to be tested, and measuring the depth of the penetration—the resistance of the metal indicating the degree of hardness. The depth measurement is recorded on a dial, the scale of which depends on the amount of pressure used in making the indentation.

Rather than ignore this variability, or treat it qualitatively, it can be incorporated into a "mathematical model" of the physical phenomenon being studied. Such a model is a mathematical description of the process and is generally of a simplified nature. This formulation can then be used to characterize the physical process and be used for subsequent analysis. However, it must be noted that this "mathematical model" may not be representative of the true physical phenomenon since, by necessity, it may have been an oversimplification of the real process.

The remaining sections will deal with concepts which provide the basis for including "inherent variability" in the mathematical model.

2.2 Set of All Possible Outcomes of the Experiment—The Sample Space

Although the results of any given experiment cannot be predicted exactly, it is possible to characterize the set of all possible outcomes of the experiment. This set is often called the sample space and is denoted by Ω. Elements of the sample space which represent the possible outcomes of the experiment are denoted by ω. For example, suppose a single steel ingot is to be placed in the Rockwell hardness testing machine. The Rockwell hardness of this ingot on the B scale must fall between 0 and the upper limit of the scale U. Prior to doing the experiment, the value of the actual observation to be obtained will be unknown, but will lie within the interval 0 and U. Thus, the sample space Ω consists of all points ω lying in the interval 0 and U. This can be written in set notation as

$$\Omega = \{\omega \,|\, 0 \leq \omega \leq U\}.^{1}$$

If two steel ingots are measured, each ingot can have a hardness between 0 and U. Denote the Rockwell hardness of the first ingot by x_1 and the Rockwell hardness of the second ingot by x_2. Thus, the set Ω consists of all points $\omega = (x_1, x_2)$ such that

$$0 \leq x_1 \leq U \quad \text{and} \quad 0 \leq x_2 \leq U.$$

The sample space can be written, again, in set notation as

$$\Omega = \{\omega = (x_1, x_2) \,|\, 0 \leq x_1, x_2 \leq U\}.$$

This sample space can also be represented graphically by the shaded area in Fig. 2.1.

If $U = 70$, one possible outcome of the experiment is $\omega = (x_1, x_2) = (43, 56)$. This would result if the Rockwell hardness of the first specimen were 43 and the second specimen 56. Thus (43, 56) is a single point in Ω. Similarly, if a sample of n ingots are measured and x_i represents the Rockwell hardness of the ith ingot, the set of all possible outcomes of the experiment

---

[1] This notation is read as "Ω is the set of all ω, where ω is a real number between zero and U inclusive."

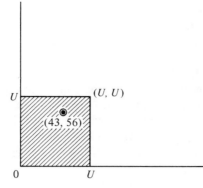

Fig. 2.1. *The sample space for the measurement of two steel ingots.*

Ω consists of all points $\omega = (x_1, x_2, \ldots, x_n)$ such that

$$0 \leq x_1 \leq U, \qquad 0 \leq x_2 \leq U, \ldots, \qquad 0 \leq x_n \leq U.$$

In set notation this sample space can be written as

$$\Omega = \{\omega = (x_1, x_2, \ldots, x_n) \,|\, 0 \leq x_1, x_2, \ldots, x_n \leq U\}.$$

A second example deals with the life of transistors. From a group of transistors produced under similar conditions a single unit is chosen, placed under test in an environment similar to its designed use, and then tested to failure. Since the life of the transistor can be any non-negative number, the sample space Ω consists of all points ω lying in the interval 0 and ∞, i.e.,

$$\Omega = \{\omega \,|\, 0 \leq \omega \leq \infty\}.[1]$$

If instead of testing one transistor, the experiment consists of choosing two of the above transistors, placing them into a test bank, and testing them to failure, the sample space Ω is given by

$$\Omega = \{\omega = (x_1, x_2) \,|\, 0 \leq x_1, x_2 \leq \infty\},$$

where x_1 denotes the time to failure of the transistor designated as number 1 and x_2 denotes the time to failure of the transistor designated as number 2. Similarly, if a sample of 5 transistors is placed into the test bank and tested to failure, the sample space Ω is given by

$$\Omega = \{\omega = (x_1, x_2, \ldots, x_5) \,|\, 0 \leq x_1, x_2, \ldots, x_5 \leq \infty\},$$

where x_i denotes the time to failure of the transistor designated as number i.

---

[1] From a practical point of view it must be recognized that the life of a transistor cannot be infinite nor are the measuring instruments able to record continuous values (similarly for the Rockwell hardness machine). However, as indicated in Sec. 2.1, the mathematical model is only an approximation of the physical phenomenon being studied. In the examples, these approximations generally will not distort the results.

One possible outcome of the experiment is $\omega = (x_1, x_2, x_3, x_4, x_5) = $ (632, 554, 681, 695, 408). Assuming that the time units are in hours, this point would be obtained if the time to failure of the first transistor were 632 hours, the second transistor 554 hours, the third 681 hours, the fourth 695 hours, and the fifth 408 hours.

A third example concerns the tossing of a die. The experiment consists of tossing a die and observing the upturned face. Prior to the toss, the value of the upturned face cannot be predicted with certainty; but the set of all possible outcomes of the experiment, Ω, can be recorded. The upturned face can be 1, 2, 3, 4, 5, or 6. Hence, Ω consists of these 6 values. In set notation this sample space can be written as

$$\Omega = \{1, 2, 3, 4, 5, 6\}.$$

If two dice are tossed, each die can take on a number from 1 to 6 inclusive. Denote the value of the upturned face of die number 1 by x_1 and the value of the upturned face of die number 2 by x_2. The sample space Ω consists of all points $\omega = (x_1, x_2)$ such that x_1 and x_2 are integers that lie between 1 and 6 inclusive.

An enumeration of these 36 points in the sample space leads to

$$\Omega = \begin{cases} (1, 1); (1, 2); (1, 3); (1, 4); (1, 5); (1, 6) \\ (2, 1); (2, 2); (2, 3); (2, 4); (2, 5); (2, 6) \\ (3, 1); (3, 2); (3, 3); (3, 4); (3, 5); (3, 6) \\ (4, 1); (4, 2); (4, 3); (4, 4); (4, 5); (4, 6) \\ (5, 1); (5, 2); (5, 3); (5, 4); (5, 5); (5, 6) \\ (6, 1); (6, 2); (6, 3); (6, 4); (6, 5); (6, 6) \end{cases}.$$

Any one of the 36 points above can be obtained as the result of experimentation: $\omega = (x_1, x_2) = (3, 4)$ means that the result of the first toss is a 3 and the result of the second toss is a 4. If a single die is tossed twice, the sample space is the same as that above where x_1 now denotes the value of the upturned face of the first toss and x_2 denotes the value of the upturned face of the second toss. If the same die is tossed n times, the sample space Ω consists of all points $\omega = (x_1, x_2, \ldots, x_n)$ such that x_i, the value of the upturned face of the ith toss, lies between 1 and 6 inclusive.

A fourth example concerns the number of α particles reaching a counter placed at a known distance from the source during a 10-second interval. The number of such particles is a non-negative integer. Hence, the sample space can be represented as

$$\Omega = \{0, 1, 2, \ldots\}.[1]$$

---

[1] Although the possible outcomes of the experiment are depicted as being unbounded, this is clearly a mathematical approximation since the number of α particles of a radioactive substance must be finite.

The final example concerns tossing a coin and is given to illustrate the fact that the sample space need not consist of a set of numbers. The experiment consists of tossing a coin and observing the upturned face. The face can be either a head or a tail (ruling out the possibility of standing on edge) and hence, Ω consists of two points, a head and a tail. If the coin is tossed twice, the first toss can be either a head or a tail and similarly for the second toss. If x_1 denotes the outcome of the first toss and x_2 denotes the outcome of the second toss, the sample space Ω consists of all points $\omega = (x_1, x_2)$ such that x_1 represents a head or a tail and x_2 represents a head or a tail. An enumeration of the four points of Ω is given by $\omega = (x_1, x_2) = (H, H); (H, T); (T, H);$ and (T, T). Thus, $\omega = (x_1, x_2) = (H, H)$ means that the result of the first toss is a head and the result of the second toss is a head. This example is a special case of a more general experiment in which the outcome can be characterized by one of two values. Instead of flipping a single coin, the experiment can consist of observing the sex of a newborn child. The sample space consists of the two points, male and female. Another type of experiment can consist of measuring whether an item is defective or non-defective. The sample space consists of the two points, defective and non-defective.

The experiments described in the illustrations above were actually much more complex than indicated. For example, in addition to measuring the Rockwell hardness of a steel ingot, the outcome of the experiment consisted of wear on the testing equipment, stress set up in the floor, energy consumed by the operator, generation of heat, etc. There are many factors resulting from the experiment which could be included in the set of all possible outcomes of the experiment but which are unimportant from the point of view of making decisions based on the outcome of the experiment. Hence, these irrelevant factors will be excluded from the formal enumeration of the sample space, and Ω will be the set of all possible outcomes *of the relevant factors* of the experiment.

In all of the previous examples the experiment was viewed in the context of specifying exactly what tests or trials were to be performed and what was to be observed beforehand. The sample space related to the experiment was then constructed. For example, the tossing of a die twice was considered as a single experiment and the sample space Ω consisting of 36 points was obtained. It is natural to question whether or not this experiment can be treated as two simple experiments, the first experiment consisting of tossing the die for the first time and the second experiment consisting of tossing the die for the second time. This will result in sample spaces Ω_1 and Ω_2 associated with these two simple experiments. In principle, the overall sample space Ω can be constructed from the sample spaces Ω_1 and Ω_2. However, this may sometimes be cumbersome and somewhat misleading so that for the purposes of this text, only sample spaces constructed from the overall experiment will be considered.

2.3 Events

The sample space is an important notion associated with a particular experiment. Another important concept is an event. An event is simply any set of outcomes in the sample space of the experiment. Thus, an event E is any subset of Ω, including such subsets as a single point ω, a collection of such single points, and even Ω itself. Some examples of events follow. In the Rockwell hardness experiment where a single item was tested, the set E_1, determined by

$$E_1 = \{\omega \mid 0 \leq \omega \leq 30\},$$

is the event which indicates obtaining a Rockwell hardness not exceeding 30.

In the transistor experiment where two transistors were tested to failure, the set E_2, determined by

$$E_2 = \{\omega = (x_1, x_2) \mid 400 < x_1, x_2 \leq \infty\},$$

is the event which denotes that the time to failure of each transistor will exceed 400 hours. This event is shown by the shaded area in Fig. 2.2. In this same figure the sample space Ω is the entire quadrant.

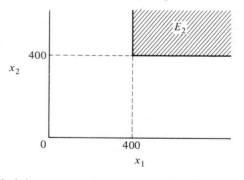

Fig. 2.2.   *Shaded area represents* $E_2 = \{\omega = (x_1, x_2) \mid 400 < x_1, x_2 \leq \infty\}$.

It should be noted that the set consisting of those average times to failure of the two transistors exceeding 400 hours is *not* an event in the sample space. However, it will be seen in Sec. 2.6 that an event in the sample space which has this propetry can be constructed, i.e., each ω in this event leads to an average time to failure of the two transistors exceeding 400 hours.

In the dice experiment where two dice are thrown, the event $E_3 = \{6, 6\}$ is the event which represents an upturned face showing a 6 for each die. In most games involving tossing two dice, the sum of the upturned faces is of interest. Again, it must be noted that sets involving sums of the upturned faces are not events in the sample space, e.g., the number $12 = 6 + 6$ does not even appear in the sample space.

Finally, in the particle emission experiment, the event $E_4 = \{0, 1\}$ is the event which represents no more than one α particle reaching the counter during the 10-second test interval.

Before concluding this section, some definitions related to events will be given. Events can be combined in order to obtain new events. The union (or addition) of two events E and F, denoted by $E \cup F$ or $E + F$, is defined to be that event which contains elements which are in either E or F or both, i.e.,

$$E \cup F = \{\omega \,|\, \omega \text{ is in } E \text{ or } F \text{ or both}\}.$$

Thus, in the transistor example if E is the event which denotes that the time to failure of each of the two transistors will exceed 400 hours and F is the event which denotes that the time to failure of each of the two transistors will exceed 500 hours, then the event $E \cup F = E$.

The intersection of two events E and F, denoted by $E \cap F$ or EF, is defined to be that event which contains elements which are in both E and F, i.e.,

$$E \cap F = \{\omega \,|\, \omega \text{ is in both } E \text{ and } F\}.$$

In the above example $E \cap F = F$.

Finally, two events E and F are said to be disjoint or mutually exclusive events if their intersection is empty (does not contain any elements), i.e., these events cannot occur at the same time. In the example, E and F are not mutually exclusive. However, if G is the event which denotes that the time to failure of each of the two transistors will be less than 200 hours, then E and G (and F and G) are disjoint events.

2.4 Random Variables

Experiments are usually performed in order to obtain information on which to base decisions. Sometimes such decisions can be made directly by referring to the occurrence or non-occurrence of an event in the sample space. For example, suppose the transistors alluded to in Sec. 2.2 are to be used in the assembly of radio equipment which is part of the gear going into a spacecraft. The radio equipment is to operate for approximately 200 hours. The contractor has arranged with the supplier of these transistors for a single unit to be tested. If the time to failure of this transistor exceeds 400 hours, he will purchase the entire production run. Otherwise, he will not accept the transistors. In this example, the contractor is partitioning the sample space into two disjoint events A and B, where A is the event which denotes that the time to failure will exceed 400 hours and B is the event which denotes that the time to failure will be less than or equal to 400 hours. If the actual outcome of the experiment lies in A, the transistors will be purchased. If the actual outcome of the experiment lies in B, the transistors will not be purchased.

On the other hand, it is conceivable that the contractor may not want to make his decision by referring to the occurrence or non-occurrence of an event in the sample space. For example, suppose the contractor has arranged with the supplier for two transistors to be tested. Suppose further that they have agreed that if the average time to failure of the two tested transistors exceeds 400 hours, the contractor will purchase the entire production run. Otherwise, he will not accept the transistors. It has already been pointed out in Sec. 2.3 that the set which represents the average time to failure of the two transistors exceeding 400 hours is *not* an event in the sample space (nor is the average time to failure a point ω in the sample space) so that the contractor cannot base his decision on *direct* observation of the outcome of the experiment. Instead, he makes his decision by using a rule which assigns to each point ω in the sample space a numerical value obtained by averaging the times to failure for the two transistors, x_1 and x_2, respectively, associated with that point ω, i.e. for every $\omega = (x_1, x_2)$ in Ω, $\bar{X}(\omega) = (x_1 + x_2)/2$ is calculated. This resulting set of values is then partitioned into two groupings: those values of the average which are greater than 400 hours and those values of the average which are less than or equal to 400 hours. If the actual value of the average obtained from the particular experiment performed lies in the partition having values greater than 400 hours, the entire production run of transistors is purchased. Otherwise, the transistors are not accepted. Equivalently, if the actual outcome of the experiment ω were such that the rule which averaged the times to failure of the two transistors led to a value exceeding 400 hours, the production run is purchased.

The rule, which in the example assigned to each point in the sample space a numerical value (obtained by averaging the times to failure for the two transistors), is evidently of importance, and hence is given the special name of "random variable." Since a "rule" is nothing more than a function in a mathematical sense, a formal definition can be given as follows:

Let Ω denote the sample space associated with an experiment. A random variable X is a numerical valued function which assigns a real number $X(\omega)$ = x to every point ω in the sample space Ω.

Throughout the text, random variables are denoted by capital letters and the values the random variable takes on are denoted by lower-case letters. In the particular example of testing *two* transistors to failure, the random variable $\bar{X}$, the average time to failure, was chosen for decision-making purposes, i.e., the function that assigns to each point $\omega = (x_1, x_2)$ in Ω a value $\bar{X}(\omega) = (x_1 + x_2)/2$.[1] When $\omega = (0, 0)$, corresponding to the possible outcome of the experiment where both transistors fail immediately, $\bar{X}(\omega)$

---

[1] Note that $\bar{X}$ is the function, and perhaps the notation should really show the function as $\bar{X}(\cdot)$ where $(\cdot)$ is used to indicate that the function is defined over some domain. When an element from this domain is substituted for the $\cdot$, then the function takes on a particular value.

$= \bar{X}(0, 0) = (0 + 0)/2 = 0$. Similarly, when $\omega = (636, 204)$, corresponding to the possible outcome of the experiment where the first transistor fails at 636 hours and the second transistor fails at 204 hours, $\bar{X}(\omega) = \bar{X}(636, 204) = (636 + 204)/2 = 420$. For each ω in the sample space a similar calculation can be made, and it becomes evident that the values the random variable $\bar{X}$ takes on are the set of values $\bar{x}$ where $0 \leq \bar{x} \leq \infty$. As indicated previously, this range of values that the random variable takes on is then partitioned into two sets, those values above 400 and those less than or equal to 400. If the *actual* outcome of the experiment had been the point $\omega = (636, 204)$, then $\bar{x} = 420$ is the *actual* value that the random variable $\bar{X}$ takes on. Since this lies in the partition where the values exceed 400 hours, the contractor would then purchase the transistors.

Random variables are very important in the study of probability theory. It has already been suggested that experiments are extremely complex and that in this text the sample space will contain the set of all possible outcomes *of the relevant factors* of the experiment. However, it is often difficult to differentiate between relevant and irrelevant factors at the time the experiment is performed, and it is wise to record as much information as possible. For example, rather than record the times to failure of the two transistors, it may be useful to record the hour and date that each transistor is placed on test and the hour and date that each fails. Subsequently, if only the average time to failure is important for decision-making purposes, the appropriate random variable can be chosen. This random variable is somewhat more complicated than the previous one because the function must first find the times to failure associated with each ω before averaging. However, if it were suspected that a power surge occurred during the experiment, the exact time of the failures rather than the time to failure would be very important data. The choice of the random variable then enables the experimenter to choose those factors that appear to be important, while discarding those factors which appear irrelevant and often difficult to characterize, although these latter factors have been duly recorded in the "sample space" for possible review at a later date. Another use for random variables occurs when experimental data are not recorded as numerical data. This might occur when items are classified only as defective or non-defective, data recorded in graphical form, etc. Here again, the use of random variables enables the decision maker to quantify his data.

There are a very large number of random variables that may be associated with a particular experiment. In the experiment where two transistors are tested to failure, the average time to failure, $\bar{X}$, was shown to be a random variable. Another example of a random variable, denoted by S^2, is described as follows: For every $\omega = (x_1, x_2)$ in Ω, the function S^2 takes on values

$$S^2(\omega) = S^2(x_1, x_2) = (x_1 - \bar{x})^2 + (x_2 - \bar{x})^2,$$

that is, the random variable measures the sum of the squares of the deviation of each coordinate of the point (x_1, x_2) from its average coordinate. Thus, for $\omega = (0, 0)$, which corresponds to the possible outcome of the experiment where both transistors fail immediately, $S^2(\omega) = S^2(0, 0) = (0 - 0)^2 + (0 - 0)^2 = 0$. Similarly, when $\omega = (636, 204)$, which corresponds to the possible outcome of the experiment where the first transistor fails at 636 hours and the second transistor fails at 204 hours, $S^2(\omega) = S^2(636, 204) = (636 - 420)^2 + (204 - 420)^2 = 93,312$. For each ω in the sample space a similar calculation can be made, and it becomes evident that the values the random variable S^2 takes on are the set of values s^2 where $0 \leq s^2 \leq \infty$.

A third example of a random variable, denoted by X_1, is described as follows: For every $\omega = (x_1, x_2)$ in Ω, the function disregards the x_2 coordinate of the point (x_1, x_2) and transforms the x_1 coordinate into itself. Thus, for $\omega = (0, 0)$, $X_1(\omega) = X_1(0, 0) = 0$ and for $\omega = (636, 204)$, $X_1(\omega) = X_1(636, 204) = 636$. Similar calculations can be made for each ω in the sample space. This random variable represents the time to failure of the transistor labeled number 1, and the values it takes on is the set of non-negative numbers x_1 where $0 \leq x_1 \leq \infty$. The random variable X_2, which represents the time to failure of the transistor labeled number 2, can be defined in a similar manner, i.e., for every ω in Ω, the function X_2 disregards the x_1 coordinate of the point (x_1, x_2) and transforms the x_2 coordinate into itself. It is worth noting that if the experiment consists of testing n transistors to failure, a sequence of random variables $X_1, X_2, \ldots, X_n$ can be generated in exactly the same manner as described for the case of two. The random variable X_i represents the time to failure of the ith transistor (labeled number i) and is defined as that function which disregards the other $n - 1$ coordinates of the point $(x_1, x_2, \ldots, x_n)$ and transforms the x_i coordinate into itself. This sequence of random variables associated with an experiment is often called a "sample" and plays an important role in statistics. Another interesting comment about such random variables is that the function, say X_1, cannot be represented as a simple mathematical expression and, instead, must be described in words. Many random variables have this property.

A final example of a random variable associated with the experiment of placing two transistors on test will be denoted by X_{max} and can be described as follows: For every ω in Ω, the function X_{max} selects the larger of the two coordinates of the point (x_1, x_2), i.e., the function X_{max} takes on the values

$$X_{max}(\omega) = X_{max}(x_1, x_2) = \max(x_1, x_2).$$

Thus, for $\omega = (0, 0)$, $X_{max}(\omega) = X_{max}(0, 0) = \max(0, 0) = 0$ and for $\omega = (636, 204)$, $X_{max}(\omega) = X_{max}(636, 204) = \max(636, 204) = 636$. Similar calculations can be made for each point ω in the sample space Ω, and again the possible values that the random variable X_{max} can take on is the set of non-negative numbers x_{max}, where $0 \leq x_{max} \leq \infty$. Although the values that the random variable takes on have always been the set of non-negative numbers in all the examples presented, it is not difficult to find examples of random variables where this does not occur. For example, the random variable $(X_1 - X_2)$, which represents the difference in the times to failure of the two transistors on test, takes on values which are the set of all numbers $(x_1 - x_2)$ where $-\infty \leq (x_1 - x_2) \leq \infty$. Note that this random variable is described by the function $(X_1 - X_2)$, which takes on values

$$(X_1 - X_2)(\omega) = (X_1 - X_2)(x_1, x_2) = (x_1 - x_2).$$

It is evident that there are an infinite number of random variables that can be obtained from this simple experiment. In fact, any function of random variables is itself a random variable since a function of a function must also be a function. Thus, the random variable $(X_1 - X_2)$, which represents the difference in the times to failure of the two transistors, was shown to be a random variable by referring to the definition of a random variable (and the sample space). Alternatively, it can be shown to be a random variable by recognizing it as a function of the random variables X_1 and X_2. Similarly, $\bar{X}$, S^2, and X_{max} can be shown to be random variables by noting that they, too, are functions of the random variables X_1 and X_2. Although there are a large number of random variables associated with a simple experiment, seldom are more than one of interest for decision-making purposes.

Often random variables encountered in practice can be classified as either continuous or discrete random variables. *A random variable is said to be continuous if the possible values that it can take on is a continuum of values.* All the random variables described so far in this section have been continuous random variables. *A random variable is said to be discrete if the number of possible values that the random variable can take on is finite or countably infinite.*[1] An example of a discrete random variable is obtained by considering the experiment of tossing two dice. A discrete random variable T is the sum of the upturned faces. For each of the 36 values of $\omega = (x_1, x_2)$ in the sample space where x_1 and x_2 are integers that lie between 1 and 6 inclusive, the function T adds the coordinates of each such point. Thus, when $\omega = (1, 1)$, which corresponds to the possible outcome of the experiment where each die results in an upturned face of 1, $T(\omega) = T(1, 1) = 1 + 1 = 2$. For each of the other 35 points in Ω, a similar calculation can be made, and the resulting set of values that the random variable T can take on is {2, 3, 4, 5, 6, 7, 8, 9, 10, 11, 12}, which contains 11 elements (finite number).

Another example of a discrete random variable is obtained by considering

---

[1] A "countably infinite" set of values is a set whose elements can be put into one-to-one correspondence with the positive integers. For example, the set of positive even integers is countably infinite; 2 is paired with 1, 4 is paired with 2, 6 is paired with 3, etc. The set of all real numbers between zero and one is not countably infinite since there are "too many" numbers to be paired with the positive integers.

the experiment of measuring the number of α particles emitted by a radio-active substance. A discrete random variable X is just the number of α particles reaching the counter. For each ω in Ω, the function X transforms ω into itself. Thus, the set of all possible values that the random variable takes on is given by $X(0) = 0$, $X(1) = 1$, $X(2) = 2, \ldots$ and is countably infinite. Furthermore, it is evident that the random variable X is just the identity function. In general, the sample space and the set of all possible values that the random variable takes on coincide when the random variable is the identity function. This random variable is frequently encountered in practice because the experimenter often simplifies the sample space so that it coincides with the set of all possible values that the random variable takes on, i.e., his description of the set of all possible outcomes of the experiment implies that the random variable to be chosen for decision-making purposes is the identity function. As was previously indicated, this possible oversim-plification of the sample space, with the corresponding inadequate recording of appropriate data, could lead to serious difficulties.

A final example of a discrete variable is concerned with a coin. The sample space consists of two points, a head and a tail. Define a random variable X by assigning it the value 0 if ω is heads and by assigning it the value 1 if ω is tails, i.e., $X(\text{heads}) = 0$ and $X(\text{tails}) = 1$.

In summary then, a random variable X is a numerical valued function which assigns a real number $X(\omega) = x$ to every point ω in Ω. This can be represented pictorially as is shown in Fig. 2.3.

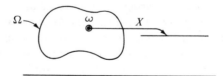

Fig. 2.3.   *Transformation of the sample space into the real line or a subset of the real line by means of the random variable X.*

2.5 Probability

The term "probability" has several meanings. It is used in everyday speech in a qualitative manner, and almost everyone has some feeling for the con-cept. Although a rigorous treatment of probability is desirable, the mathe-matical sophistication required is beyond the scope of this text. Instead, some basic properties of probability will be described and shown to conform to one's intuitive notions about probability theory.

Some of the material presented earlier can be summarized briefly as follows: Generally the results of a given experiment cannot be predicted exactly because of the inherent variability associated with the experiment. However, it is possible to characterize the sample space Ω which contains elements ω that represent the possible outcomes of the experiment. An event E is a set of outcomes in the sample space. *Probability may be defined as numbers associated with these events* and is denoted by P$\{E\}$. These numbers are a measure of the inherent variability alluded to and are related to the occurrence of the event E when the experiment is performed. For example, in the Rockwell hardness experiment where a single item was tested and E_1 represents obtaining a Rockwell hardness not exceeding 30, i.e.,

$$E_1 = \{\omega \,|\, 0 \leq \omega \leq 30\},$$

then P$\{E_1\}$ is the number associated with this event and will be read as the "probability of the occurrence of this event" or the "probability that the Rockwell hardness (the outcome of the experiment) did not exceed 30." Similarly, in the testing of two transistors where E_2 denotes the event that the time to failure of each transistor will exceed 400 hours, P$\{E_2\}$ is the number associated with this event and is read as "the probability that the time to failure of each transistor will exceed 400 hours." In the dice experiment where two dice are thrown and E_3 is the event which represents an upturned face showing a 6 for each die, P$\{E_3\}$ is the number associated with this event and is read as "the probability that both upturned faces show a 6."

If P$\{E\}$ were known for all events E in the sample space, this information would be very useful for decision-making purposes. Unfortunately, obtaining these numerical values is usually quite difficult. Additional assumptions about the mathematical model of the physical phenomenon may be required, or occasionally further experimentation may have to be performed, in order to obtain approximate values of these probabilities. However, even without describing how to obtain these numbers, their existence can be postulated. Furthermore, properties of these probabilities that appear to be reasonable can also be postulated.

Let Ω *denote the sample space associated with an experiment. For each event E in this sample space, let* P$\{E\}$ *be the number associated with this event.* P$\{E\}$ *is known as the probability of the occurrence of the event and satisfies the following reasonable conditions:*

1. If E is any event in the sample space, then $0 \leq$ P$\{E\} \leq 1$. This implies that probabilities are numbers that fall between zero and one. If 80 is the upper value on the Rockwell B scale, and E is the event of obtaining a Rockwell hardness falling between 20 and 80, then P$\{E\}$ is a nonnegative number which is less than or equal to one.
2. If N is an event which cannot occur, then P$\{N\} = 0$. If N is the event of obtaining a negative reading on the Rockwell tester, implying a negative hardness, then P$\{N\} = 0$.

3. If the event E is the entire sample space, i.e., $E = \Omega$, then $P\{\Omega\} = 1$. If 80 is the upper value on the Rockwell B scale, then the probability of obtaining a Rockwell hardness between zero and eighty is one.
4. If E and F are disjoint events in the sample space, then $P\{E \cup F\} = P\{E\} + P\{F\}$. Suppose the experiment consists of testing a single specimen on a Rockwell hardness machine which has 80 as the upper value on the B scale. If E is the event of obtaining a Rockwell hardness less than 20 and F is the event of obtaining a Rockwell hardness greater than 40, then $E \cup F$ is the event of obtaining a Rockwell hardness less than 20 or greater than 40. Furthermore,

$$P\{E \cup F\} = P\{E\} + P\{F\}.$$

These conditions appear to be reasonable because they tend to conform to the everyday notions about the meaning of probability. However, as indicated earlier, they do not provide a method for either calculating the value of a probability, approximating the value of a probability, or interpreting the meaning of a probability statement. Whereas calculating the value of a probability may require additional assumptions which may or may not be possible, approximating the value of a probability and interpreting the meaning of a probability statement may be accomplished through the "frequency interpretation" of probability.

Suppose an experiment is repeated n times. Denote by m the number of (successful) occurrences of an event E in these n repetitions. Consider the relative frequency of the occurrence of event E, i.e., m/n. Then $P\{E\}$ can be interpreted as

$$P\{E\} = \lim_{n \to \infty} (m/n)$$

provided the limit exists, and m/n can be used to approximate the value of $P\{E\}$.

Note that the ratio m/n fulfills the four conditions that probabilities satisfy, i.e.,

1. The number of occurrences of an event in n repeated trials must lie between 0 and n so that $0 \leq m/n \leq 1$.
2. If an event cannot occur, m must equal zero and $m/n = 0$.
3. If the event must occur on each trial, then $m = n$ and $m/n = 1$.
4. If m_1 is the number of (successful) occurrences of an event E in n trials, and m_2 is the number of (successful) occurrences of an event F in these same n trials, and E and F are disjoint, then the number of occurrences of the event E or F in the n trials is $m_1 + m_2$. Therefore, the relative frequency of occurrences of the event E or F in the n trials, $(m_1 + m_2)/n$, is equal to the sum of the relative frequency of the occur-

rence of the event E, m_1/n, and the relative frequency of the occurrence of the event F, m_2/n, i.e.,

$$(m_1 + m_2)/n = m_1/n + m_2/n.$$

The relative frequency m/n, then, which can be used to approximate the true value of $P\{E\}$, satisfies the properties imposed on $P\{E\}$. Furthermore, if these conditions are satisfied by the relative frequency m/n, it seems reasonable to expect them to be satisfied by the limiting value of the relative frequency, i.e., $\lim_{n \to \infty} (m/n)$. It is this limiting value that is suggested to be the interpretation of the term $P\{E\}$. Unfortunately, however, this interpretation of $P\{E\}$ does not provide a method for evaluating $P\{E\}$ since it does not specify how large n must be. Furthermore, even approximating $P\{E\}$ by m/n for finite n has the serious disadvantage of requiring experimentation just for that purpose. As shall be seen subsequently, evaluating exact probabilities of events in the sample space may not be necessary. *What is important to the development of the mathematical model is postulating the existence of the numbers* $P\{E\}$ *and the properties that these numbers satisfy.*

If the evaluation of the numbers $P\{E\}$ is desirable, they can be approximated experimentally by the relative frequency, or they may be evaluated exactly by imposing additional assumptions on the model. Probabilities obtained by the latter method generally reflect the attitude of the model builder. Furthermore, they are often not verifiable experimentally, in which case they are referred to as personal or subjective probabilities.

2.6 Cumulative Distribution Functions

It has been shown that postulating the existence of probabilities for events occurring in the sample space, and the properties that these probabilities satisfy, is important for the development of a mathematical model of the physical phenomenon under consideration. It has also been pointed out that explicit reference to events in the sample space is often inadequate for decision-making purposes. Instead, decisions are usually dependent on a random variable of interest and the value that this random variable takes on. Hence, the relationship between random variables, the values random variables take on, and probabilities of events in the sample space require further study. This necessitates the introduction of some additional notation. An event of particular importance in the sample space is the event E_b^X, which contains those outcomes of the experiment having the following property: For a given random variable X and real number b, E_b^X contains all ω such that $X(\omega)$ is less than or equal to b, i.e.,

$$E_b^X = \{\omega \mid X(\omega) \leq b\}.$$

An alternative shortened version of this notation, which suppresses the

ω, and which will be used interchangeably in the subsequent development, is given by

$$E_b^X = \{X \leq b\}.$$

Thus, an event in the sample space of this form is seen to depend on the choice of the random variable of interest and the real number b. For example, in the experiment of tossing two dice, suppose the random variable of interest is the sum of the upturned faces, denoted by T. Arbitrarily choose $b = 3.5$. It then follows that

$$E_{3.5}^T = \{T \leq 3.5\} = \{\omega \mid T(\omega) \leq 3.5\} = \{\omega = (1, 1); \omega = (1, 2); \omega = (2, 1)\}$$

since only the outcomes $\omega = (1, 1); \omega = (1, 2); \omega = (2, 1)$ lead to the random variable T taking on values less than or equal to 3.5. Note that changing either the random variable or the value of b may change the event E_b^X in the sample space. If T remains as the sum of the upturned faces and b becomes 2, then $E_2^T = \{T \leq 2\} = \{\omega \mid T(\omega) \leq 2\} = \{\omega = (1, 1)\}$. On the other hand, if $b = 2.5$, then $E_{2.5}^T = \{T \leq 2.5\} = \{\omega \mid T(\omega) \leq 2.5\} = \{\omega = (1, 1)\}$ so that $E_2^T = E_{2.5}^T$. In the experiment which tests two transistors to failure, the contractor was interested in the random variable $\bar{X}$, the average time to failure. Furthermore, an agreement had been reached *not* to purchase the production run if the average time to failure of the two tested transistors was less than or equal to 400 hours. Thus, the transistors would not be purchased if the actual outcome of the experiment led to the random variable $\bar{X}$ taking on a value less than or equal to 400, i.e., if the actual outcome ω were a point in the set $E_{400}^{\bar{X}}$, where

$$E_{400}^{\bar{X}} = \{\bar{X} \leq 400\} = \{\omega \mid \bar{X}(\omega) \leq 400\}.$$

The event $E_{400}^{\bar{X}}$ is shown by the shaded area in Fig. 2.4. Since the events $E_b^X = \{X \leq b\}$ are events appearing in the sample space Ω, there exist probabilities $P\{E_b^X\}$ attached to them. Because these events are of a particular form, which is dependent upon the random variable X and the real number b, these probabilities are given the special name of cumulative distribution

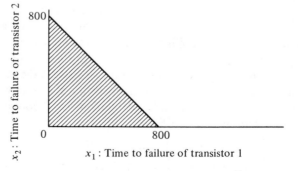

Fig. 2.4. *The shaded area is the event* $E_{400}^{\bar{X}}$.

function (CDF). *The cumulative distribution function $F_X(b)$[1] of the random variable X is defined to be*

$$F_X(b) = P\{E_b^X\} = P\{\omega \mid X(\omega) \leq b\} = P\{X \leq b\},$$

for all real values b, $-\infty \leq b \leq \infty$. Thus, a cumulative distribution function is associated with every random variable. Furthermore, this CDF is not an arbitrary function, but is induced by probabilities associated with events of the form E_b^X appearing in the sample space. Since such probabilities have been postulated to exist, the CDF can then be determined. For example, returning to the experiment where two dice are thrown, suppose it is postulated that each of the 36 possible outcomes of the experiment occurs with equal probability.[2] The random variable of interest is chosen to be T, the sum of the upturned faces. The events E_b^T can be constructed as follows:

for b such that $-\infty \leq b < 2$, E_b^T is the event which cannot occur

for b such that $\quad 2 \leq b < 3$, $E_b^T = \{\omega = (1, 1)\}$

for b such that $\quad 3 \leq b < 4$, $E_b^T = \{\omega = (1, 1); \omega = (1, 2);$
$$\omega = (2, 1)\}$$

for b such that $\quad 4 \leq b < 5$, $E_b^T = \{\omega = (1, 1); \omega = (1, 2);$
$$\omega = (2, 1); \omega = (2, 2);$$
$$\omega = (1, 3); \omega = (3, 1)\}$$

$$\vdots$$

for b such that $11 \leq b < 12$, $E_b^T = \{$all 36 possible outcomes except
for $\omega = (6, 6)\}$

for b such that $12 \leq b \leq \infty$, $E_b^T = \{$entire sample space containing 36 points$\}$.

Using the properties that probabilities are assumed to satisfy, the CDF can be obtained and is shown in Fig. 2.5.

A particular point on the graph in Fig. 2.5 is easily constructed. Choose $b = 3.5$, then $F_T(3.5) = P\{E_{3.5}^T\} = P\{T \leq 3.5\} = P\{\omega = (1, 1); \omega = (1, 2);$ $\omega = (2, 1)\}$. Using property (4) of Sec. 2.5 for disjoint events, it is evident that

$$P\{\omega = (1, 1); \omega = (1, 2); \omega = (2, 1)\}$$
$$= P\{\omega = (1, 1)\} + P\{\omega = (1, 2)\} + P\{\omega = (2, 1)\}.$$

Since it has been assumed that the probability of occurrence of the event

---

[1] When the choice of the random variable X is obvious, an alternative notation that suppresses it, $F(b)$, will be used interchangeably.

[2] It can be shown that this assumption is sufficient for determining the probabilities of all events occurring in the sample space.

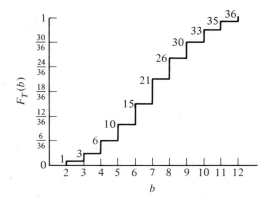

Fig. 2.5. *Cumulative distribution function of the random variable which is the sum of the upturned faces of two fair dice.*

consisting of a single point ω for each of the 36 points ω is the same (and hence equal to 1/36), it follows that

$$F_T(3.5) = \tfrac{1}{36} + \tfrac{1}{36} + \tfrac{1}{36} = \tfrac{1}{12}.$$

Another example concerns the experiment which tests two transistors to failure, and the random variable of interest is $\bar{X}$, the average time to failure. If $b = 400$, the CDF evaluated at this point is given by

$$F_{\bar{X}}(400) = P\{E_{400}^{\bar{X}}\} = P\{\bar{X} \le 400\},$$

which is just the probability that the outcome of the experiment falls within the event shown by the shaded area in Fig. 2.4. The CDF evaluated for other values of b can be obtained by evaluating the probability that the outcome of the experiment falls within events similar to the shaded area shown in Fig. 2.4, but bounded by lines running parallel (dependent on the value of b chosen). Since probabilities are defined for all events over the sample space, the CDF can be evaluated for this random variable.

In the same experiment, if the random variable chosen is X_1, the time to failure of the transistor labeled as number 1, the cumulative distribution function can be evaluated by referring to the events $E_b^{X_1}$. These events are of the form

$$E_b^{X_1} = \{\omega \mid X_1(\omega) \le b\} = \{X_1 \le b\}.$$

The CDF is given by

$$F_{X_1}(b) = P\{E_b^{X_1}\} = P\{\omega \mid X_1(\omega) \le b\} = P\{X_1 \le b\},$$

and presumably can be evaluated if the probabilities of events defined over the sample space are known.

Throughout the discussion of the cumulative distribution function, it has been made clear that the CDF of a random variable is evaluated by referring to events in the sample space of the experiment; i.e.,

$$F_X(b) = P\{E_b^x\} = P\{\omega \mid X(\omega) \leq b\} = P\{X \leq b\}.$$

Although the right-hand side of this last equality is just the probability of an event occurring in the sample space, it is usually read as "the probability that the random variable X takes on a value less than or equal to b." This is "loose" terminology because probability has been defined only for events and not for random variables. However, this terminology is used throughout the text with the caution that the reader interpret the statement properly, i.e., the probability that a random variable X is less than or equal to b is just the probability of the occurrence of the event E_b^x which contains all ω such that $X(\omega)$ is less than or equal to b.

In addition to being useful for evaluating probabilities of events of the form E_b^x, the CDF is also useful for evaluating probabilities of events in the sample space of the form

$$\{\omega \mid a < X(\omega) \leq b\}.$$

The probability of the occurrence of such an event is denoted by

$$P\{a < X \leq b\}$$

and is read as "the probability that the random variable X takes on a value greater than a but less than or equal to b. This probability can be evaluated in terms of the CDF by noting that the event

$$\{\omega \mid -\infty \leq X(\omega) \leq b\}$$

can be decomposed into two disjoint sets as follows:

$$\{\omega \mid -\infty \leq X(\omega) \leq b\} = \{\omega \mid -\infty \leq X(\omega) \leq a\} \cup \{\omega \mid a < X(\omega) \leq b\}.$$

Using property (4) of Sec. 2.5 for disjoint events, the equality

$$P\{\omega \mid -\infty \leq X(\omega) \leq b\} = P\{\omega \mid -\infty \leq X(\omega) \leq a\} + P\{\omega \mid a < X(\omega) \leq b\}$$

or
$$F_X(b) = F_X(a) + P\{a < X \leq b\}$$

is obtained. Hence, $P\{a < X \leq b\}$ can be evaluated in terms of the CDF, i.e.,

$$P\{a < X \leq b\} = F_X(b) - F_X(a).$$

The cumulative distribution function plays an important role in formulating a mathematical model since a CDF is associated with every random variable. In addition to being induced by probabilities associated with events appearing in the sample space, the CDF $F_X(b)$ has the following properties:

1. It is a numerical valued function defined for all b, $-\infty \leq b \leq \infty$.
2. $F_X(b)$ is a non-decreasing function of b.
3. $F_X(+\infty) = \lim_{b \to \infty} F_X(b) = 1$.
4. $F_X(-\infty) = \lim_{b \to -\infty} F_X(b) = 0$.

The CDF has a simple frequency interpretation. Suppose an experiment is performed, a random variable X chosen, and the value the random variable takes on, x, computed. Suppose further that the experiment is *repeated n* times and the values the random variable X takes on are denoted by $x_1, x_2, \ldots, x_n$, respectively. Order these values that the random variable takes on by letting $x_{(1)}$ be the smallest, $x_{(2)}$ the second smallest, $\ldots$, and $x_{(n)}$ the largest. Plot the "Sample Cumulative Distribution Function" $F_n(x)$, which is shown in Fig. 2.6.[1] Note that the sample CDF changes by $1/n$ units

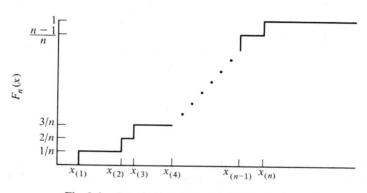

Fig. 2.6.   *Sample cumulative distribution function.*

at the values that the random variable takes on. $F_n(x)$ is just the fraction of values that the random variable takes on which are less than or equal to x. *The sample CDF, $F_n(x)$, can be interpreted to be an approximation to the CDF of the random variable, $F(x)$.* In fact, it can be shown that as the number of repetitions, n, of the experiment gets large, the sample CDF approaches the CDF of the random variable, i.e., for every x

$$\lim_{n \to \infty} F_n(x) = F(x).$$

Since decisions are usually based on the outcome of random variables, the decision maker need only be concerned with the cumulative distribution function associated with the random variable of interest. Rarely is there need to be concerned directly with events defined over the sample space and their probabilities of occurrence. The development of the mathematical model generally proceeds as follows: an experiment is to be performed and a random variable (or random variables) chosen. Assumptions are then made about the form of the CDF associated with the random variable, assumptions which can often be justified on theoretical grounds. For example, in the

---

[1] When only grouped data are available the sample CDF is equivalent to the cumulative frequency function described in Sec. 1.2.

experiment which consisted of tossing two dice, let X_1, the upturned face of the die designated as number 1, be the random variable of interest. Suppose the experimenter is willing to assume that the cumulative distribution function of this random variable is given by the graph in Fig. 2.7. This assumption is

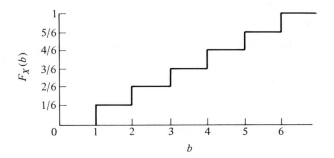

Fig. 2.7. *The CDF of the random variable which is the upturned face of a die.*

equivalent to assuming that the die is a "fair" die, an assumption that real world dice players are usually willing to make. If similar assumptions are made about X_2, the upturned face of the die designated as number 2, and some assumptions made about the "degree of association" between X_1 and X_2, the cumulative distribution function of the random variable $T = X_1 + X_2$, the sum of the upturned faces can be obtained. It is evident that assumptions made about the CDF cannot be arbitrary because they imply assumptions about probabilities associated with events in the sample space. The experimenter must be cautioned that any assumptions made about the CDF of a random variable should be justified to his satisfaction on theoretical grounds (see Sec. 3.9) or by empirical evidence. Hence, whereas very little subsequent mention will be made of probabilities associated with events in the sample space, and cumulative distribution functions of important random variables will be studied in detail, readers should continue to be aware of the close relationship between these two concepts.

2.7 Discrete Probability Distributions

A discrete random variable X has been defined as one in which the number of possible values that the random variable can take on is finite or countably infinite. Denote the possible values by $x_1, x_2, \ldots, x_m, \ldots$ (for the finite case, the sequence terminates at m) and assume that $x_1 < x_2 < x_3 < \ldots$. Recall that the CDF for a random variable X is defined to be

$$F_X(b) = P\{\omega \,|\, X(\omega) \leq b\} = P\{X \leq b\}.$$

Since X is also discrete, the event $\{\omega \,|\, X(\omega) \leq b\}$ can be decomposed into disjoint events and expressed as

$$\{\omega \mid X(\omega) \leq b\} = \{\omega \mid X(\omega) = x_1\} \cup \{\omega \mid X(\omega) = x_2\} \cup$$
$$\cdots \cup \{\omega \mid X(\omega) = x_{[b]}\},$$

where $x_{[b]}$ is the largest value of the x's less than or equal to b. The CDF may then be written as a sum of terms, i.e.,

$$F_X(b) = P\{X = x_1\} + P\{X = x_2\} + \cdots + P\{X = x_{[b]}\},$$

where, for example, $P\{X = x_1\}$ is the probability of the occurrence of the event containing those ω in the sample space such that $X(\omega)$ equals x_1. This representation for the CDF can also be expressed as

$$F_X(b) = \sum_{\text{all } k \leq b} P\{X = k\},$$

where k is an index which is allowed to range over all possible values that the random variable takes on, i.e., over all the x's. The collection of terms, $P\{X = k\}$, is called the probability distribution, the probability function, or the probability density of the discrete random variable X. An alternative notation for $P\{X = k\}$ is given by

$$P\{X = k\} = P_X(k),$$

or when there is no ambiguity about which random variable is being discussed, simply by

$$P\{X = k\} = P(k).$$

Thus, the CDF (or the probability distribution) of a discrete random variable is characterized by the pairs $[k, P(k)]$, with k ranging over all the values the random variable takes on. Furthermore, the $P(k)$ must satisfy only the two conditions

(i) $P(k) \geq 0$ for all k,

(ii) $\sum_{\text{all } k} P(k) = 1$.

The CDF for a discrete random variable is a step function which jumps at the values $x_1, x_2, \ldots$, and is constant between adjacent jump points. The magnitude of the jump at the point $x = k$ is just $P(k)$ as shown in Fig. 2.8. The probability distribution of a discrete random variable is often graphed since this is a simple representation for the pairs $[k, P(k)]$. This is shown in Fig. 2.9.

The cumulative distribution functions of some simple examples of discrete random variables have already been presented in Sec. 2.6. Figure 2.7 is a graph of a CDF for a discrete random variable which is the upturned face of a die. Recall that postulating this particular CDF was equivalent to assuming that the die was "fair," or alternatively, $P(k) = 1/6$ for $k = 1, 2, 3, 4, 5, 6$.

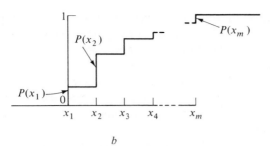

b

Fig. 2.8. *Cumulative distribution function of a discrete random variable which takes on m values.*

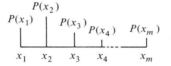

Fig. 2.9. *Probability distribution of a discrete random variable which takes on m values.*

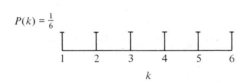

k

Fig. 2.10. *Probability distribution of the random variable which is the upturned face of a fair die.*

The probability distribution for this random variable is shown in Fig. 2.10. An analogy can be drawn between a discrete probability distribution and a mass of unit weight distributed over a weightless bar. The unit mass is concentrated only at values that the random variable can take on. This is depicted in Fig. 2.11 where each circle represents a mass of weight 1/6.

If the probability distribution is as shown above, i.e., the die is fair, the actual outcomes of a random variable should follow the laws of probability. In the long run, each number should appear about 1/6 of the time. However,

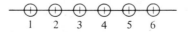

Fig. 2.11. *Mass analogy to the probability distribution of the face of a fair die.*

if the fairness of the die is in question, and its probability distribution unknown, as is the case in most statistical problems, the outcomes of the random variable will be used to make decisions about the unknown probability distribution.

The random variable T, the sum of the upturned faces, in the experiment concerning the tossing of two dice is another example of a discrete random variable presented earlier. In Sec. 2.6 some assumptions about the sample space were made and a CDF for this random variable was obtained and presented in Fig. 2.5. The probability distribution for this random variable is shown in Fig. 2.12.

Figure 2.13 is obtained by using the analogy of a body having unit mass.

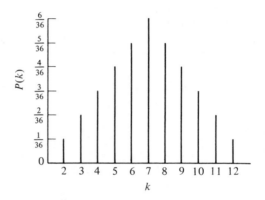

Fig. 2.12. *Probability distribution of the random variable which is the sum of upturned faces of two dice.*

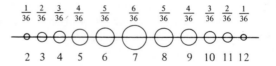

Fig. 2.13. *Mass analogy to the probability distribution of the sum of two fair dice.*

The mass analogy to a probability distribution is a useful concept enabling one to use certain well-known concepts belonging to the field of mechanics to illustrate definitions and properties of probability theory.

The final example of the CDF for a discrete random variable pertains to the experiment of measuring the number of α particles emitted by a radioactive substance. Let X be the random variable which is the number of α particles reaching the counter during a 10-second interval. It can be shown that the simplest physical assumptions lead to a CDF of the form

$$F(b) = \sum_{k=0}^{[b]} \frac{\lambda^k e^{-\lambda}}{k!}$$

where λ is a positive constant (often called a parameter of the CDF), $[b]$ is the greatest integer less than or equal to b, and $k! = k(k - 1)(k - 2) \cdots 3 \cdot 2 \cdot 1$ (with $0!$ defined equal to 1). The corresponding probability distribution is given by

$$P(k) = \frac{\lambda^k e^{-\lambda}}{k!}, \qquad \text{for } k = 0, 1, 2, \ldots .$$

A random variable possessing these properties is called a Poisson random variable and the corresponding probability distribution is called a Poisson distribution. It should be noted that $P(k) \geq 0$ and $\sum_{k=0}^{\infty} P(k) = 1$ so that the Poisson distribution is, indeed, a probability distribution.

2.8 Continuous Random Variables and Density Functions

A continuous random variable X has been defined as one in which all the possible values that it can take on form a continuum of values. The cumulative distribution function for such random variables can usually be expressed as

$$F_X(b) = P\{\omega \,|\, X(\omega) \leq b\} = P\{X \leq b\} = \int_{-\infty}^{b} f_X(z)\, dz,$$

where $f_X(z)$ is called the density function, probability density, or probability distribution of the continuous random variable X.

The density function $f_X(z)$ plays an important role in probability and statistics because many random variables of practical import are continuous, and possess densities. Before studying the properties that density functions possess, the notation should be clearly understood. The subscript X denotes the random variable whose density function is under consideration. When there is no ambiguity about which random variable is being discussed, $f_X(z)$ may be written simply as $f(z)$. The symbol z in $f_X(z)$ is just a dummy variable and represents the value at which the density is evaluated, or it represents the variable of integration when the CDF is being evaluated at the point b, i.e.,

$$F_X(b) = \int_{-\infty}^{b} f_X(z)\, dz.$$

In fact, the right-hand side of this equality may be viewed an alternate way

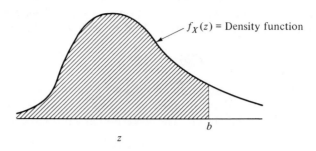

Fig. 2.14.   *A typical density function of a random variable X.*

for computing the probability

$$P\{X \leq b\}$$

when the random variable is continuous and when the density function is known. This is illustrated in Fig. 2.14 where the shaded area under the density function corresponds to $F_X(b)$. Although this simple computational device will always be used when appropriate and the probability being calculated will usually be referred to as the "probability that the random variable X takes on values less than or equal to b," it must be emphasized that what is actually being computed is the probability of the occurrence of the event containing those ω in the sample space such that $X(\omega)$ is less than or equal to b. In a similar manner, the density, when known, can be used to compute other "probabilities" such as

$$P\{a < X \leq b\} = F_X(b) - F_X(a) = \int_a^b f_X(z)\,dz.$$

Thus, the CDF is readily obtainable from a knowledge of the density function. It is also evident that the density function can be obtained from a knowledge of the CDF, i.e.,

$$f_X(z) = \frac{dF_X(z)}{dz}.$$

This follows immediately by noting that

$$\frac{dF_X(z)}{dz} = \frac{d}{dz} \int_{-\infty}^z f_X(y)\,dy = f_X(z).$$

An analogy can be drawn between a density function and a mass of unit weight distributed over a bar similar to that drawn between a probability distribution of a discrete random variable and a mass of unit weight. The mass is distributed along the bar of unit weight according to some curve as illustrated in Fig. 2.15. The curve which gives this distribution of mass is just the density function. Thus, the amount of mass along the bar greater

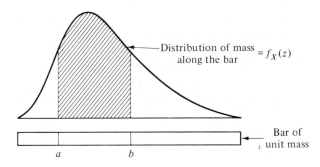

Fig. 2.15. *Analogy between a bar of unit mass and a density function.*

than a but less than or equal to b is given by the shaded area in Fig. 2.15 and is equivalent to

$$P\{a < X \leq b\} = F(b) - F(a) = \int_a^b f_X(z)\, dz.$$

There are several remarks that will be made pertaining to the CDF and density function of a continuous random variable.

1. For any arbitrary x_0 the probability of the event E_0 in the sample space which contains those ω such that $X(\omega) = x_0$, i.e., $P\{X = x_0\}$, has not been defined in terms of the density. As a consequence of expressing "probability" as an area under the density function, $P\{X = x_0\}$ will be taken to be zero. Although this result may be intuitively unappealing, a little thought will reveal its reasonableness. The event E_0 is not being viewed as an impossible event (an event which cannot occur). Instead, E_0 is an event in the sample space which may occur but has a probability of zero assigned to it. Since the values that a continuous random variable may take on form a continuum, choosing the *precise* outcome x_0 of the random variable would be quite surprising; hence, the reasonableness of $P\{X = x_0\} = 0$. Still, the random variable X must take on *some* value as a result of the experiment. As a consequence of taking $P\{X = x_0\} = 0$, the following probabilities are equivalent when X is a continuous random variable,

$$P\{a < X \leq b\} = P\{a \leq X \leq b\} = P\{a < X < b\} = P\{a \leq X < b\}.$$

2. In expressing the CDF of a continuous random variable in terms of the density function, i.e.,

$$F_X(b) = \int_{-\infty}^b f_X(z)\, dz,$$

the notation assumes that $f_X(z)$ is defined over the entire real line. Because of physical considerations some random variables cannot take on values over this entire region and may be constrained to take on values over only a part of the entire real line, e.g., the time to failure of a transistor must be non-negative. This is resolved easily by defining $f_X(z)$ to be zero outside of the constrained region. For the transistor example, $f_X(z)$ is generally defined to be zero for all z less than zero.

3. Since many important random variables are continuous and possess densities, it is often necessary to verify that the functions are indeed densities. Any function $f_X(z)$ can be viewed as a density if it possesses the following two properties:

   (i) $f_X(z) \geqq 0$
   (ii) $\int_{-\infty}^{\infty} f_X(z)\, dz = 1.$

   If (ii) is not satisfied, i.e., $\int_{-\infty}^{\infty} f_X(z)\, dz = c > 0$, then $f_X(z)/c$ becomes a density function.

4. A typical CDF for a random variable whose density is zero below b_0 is shown in Fig. 2.16. If X is a continuous random variable, then the CDF, $F_X(b)$, is a continuous function for all b.

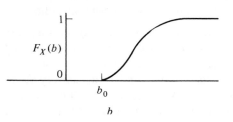

Fig. 2.16.   *The CDF of a continuous random variable.*

5. The density function, $f_X(z)$, for a continuous random variable can be given a probability interpretation. For small Δz, $P\{z \leqq X \leqq z + \Delta z\} \simeq f_X(z)\, \Delta z$ so that $f_X(z)\, \Delta z$ is approximately the probability of the random variable X falling into the small interval $(z, z + \Delta z)$. Thus, $f_X(z)$ will be a measure of the frequency with which a random variable X will tend to assume a value in a small interval near z.

Some examples of continuous random variables, their densities, and their CDF's follow. In a Rockwell hardness experiment a single steel ingot is placed in the Rockwell hardness testing machine and measured on the B scale. The random variable, X, of interest is the Rockwell hardness of the specimen. This is a continuous random variable and it will be assumed that its density is given by

$$f_X(z) = \begin{cases} 0, & \text{for} & z < 50 \\ 1/20, & \text{for } 50 \leqq z \leqq 70 \\ 0, & \text{for} & z > 70. \end{cases}$$

This density function is known as a rectangular density, and the random variable is said to have a rectangular distribution. The density function shown in Fig. 2.17 is always non-negative and its total area is clearly one. The

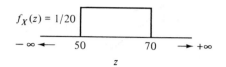

Fig. 2.17. *The density function of a rectangularly distributed random variable.*

cumulative distribution function is given by

$$F_X(b) = P\{X \leq b\} = \int_{-\infty}^{b} f_X(z)\, dz$$

$$= \begin{cases} 0, & \text{for} \quad b < 50 \\ \int_{50}^{b} \frac{1}{20}\, dx = \frac{1}{20}(b-50), & \text{for } 50 \leq b \leq 70 \\ 1, & \text{for} \quad b > 70. \end{cases}$$

Thus, for example,

$$P(X \leq 60) = F_X(60) = \tfrac{1}{20}(60 - 50) = \tfrac{1}{2}.$$

The complete CDF is shown in Fig. 2.18.

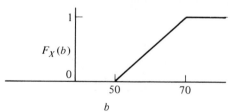

Fig. 2.18. *The CDF of a rectangular distribution.*

A second example deals with the experiment which tests two transistors to failure. Let X_1 be the continuous random variable which measures the time to failure of the transistor labeled as number 1. There appears to be some empirical evidence that the density function for such a random variable is of the following functional form,

$$f_{X_1}(z) = \begin{cases} 0, & \text{for } z < 0 \\ \frac{1}{\theta} e^{-z/\theta}, & \text{for } z \geq 0 \end{cases}$$

where $\theta > 0$ is called the parameter of the density function. A random variable having this density is said to have an exponential (probability) distribution. For this type of transistor, it will be assumed that $\theta = 1,000$ hours. This density function is shown in Fig. 2.19.

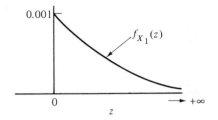

Fig. 2.19. *The exponential density function.*

The exponential density function is always non-negative and its total area is easily verified to be one. The cumulative distribution function is given by

$$F_{X_1}(b) = P\{X_1 \leq b\} = \int_{-\infty}^{b} f_{X_1}(z)\, dz$$

$$= \begin{cases} 0, & \text{for } b < 0 \\ \int_{0}^{b} \frac{1}{1000} e^{-z/1000}\, dz = -e^{-z/1000} \big|_{0}^{b} = 1 - e^{-b/1000}, & \text{for } b \geq 0. \end{cases}$$

Thus, the probability that X_1, the life of the transistor labeled as number 1, fails before 400 hours is given by

$$F_{X_1}(400) = P\{X_1 < 400\} = 1 - 1/e^{0.4} = 0.330.$$

The complete CDF is shown in Fig. 2.20.

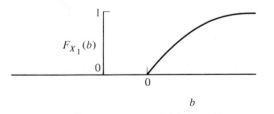

Fig. 2.20. *The CDF of the exponential distribution.*

The assumption that the random variable X_1 has an exponential distribution leads to restrictive physical implications. It can be shown that exponentially distributed random variables do not "age" with time, i.e., the probability that a transistor will survive to 800 hours given that it has already survived to 700 hours (i.e., survive an additional 100 hours) is the same as the probability that a "new" transistor will survive to 100 hours. It then follows that components whose measurable characteristics have exponential distributions should never be replaced until they fail, assuming that they are being replaced by components having identical exponential distributions. In practice, the time to failure of many electronic components is assumed to have an exponential distribution even though it is known that these components must "age" in time. Hopefully, the "aging process" will be negligible over the time span that the component is supposed to operate so that the exponential density will be a good mathematical approximation to the true distribution of the time to failure.

2.9 Expectation

Instead of specifying the complete cumulative distribution function (or probability distribution, or density function) of a random variable, it is often necessary to describe this function by a "typical quantity." A quantity which is quite suggestive is the expected value (or expectation) of the random variable.

Denote by X the random variable. If X is discrete, let the corresponding probability distribution of X be represented by $P\{X = k\} = P_X(k)$ for all k corresponding to the possible values that the random variable can take on. If X is continuous, let $f_X(z)$ denote the density function. The expected value of X is defined as

$$E(X) = \begin{cases} \displaystyle\sum_{\text{all } k} k P_X(k), & \text{if } X \text{ is a discrete random variable} \\[2ex] \displaystyle\int_{-\infty}^{\infty} z f_X(z)\, dz, & \text{if } X \text{ is a continuous random variable.} \end{cases}$$

The symbol $E(X)$ is read as "the expected value of X."

For a discrete random variable, $E(X)$ is obtained by first multiplying every value that the random variable can take on by the probability that it takes on this value, i.e., $k P_X(k)$, and then adding all of these terms together. The expected values of the discrete random variables presented as examples in Sec. 2.7 will be calculated. For the random variable X, which is the upturned face of a single die, $P_X(k) = \frac{1}{6}$ for $k = 1, 2, 3, 4, 5$, and 6. Hence, the expected value of this random variable is then given by

$$\begin{aligned} E(X) &= \sum_{k=1}^{6} k P_X(k) = \sum_{k=1}^{6} k(\tfrac{1}{6}) \\ &= (1)(\tfrac{1}{6}) + (2)(\tfrac{1}{6}) + (3)(\tfrac{1}{6}) + (4)(\tfrac{1}{6}) + (5)(\tfrac{1}{6}) + (6)(\tfrac{1}{6}) \\ &= 3\tfrac{1}{2}. \end{aligned}$$

It is important to note that this example illustrates that $E(X)$ need not be one of the possible values that the random variable takes on.

For the random variable T, which is the sum of the upturned faces when two dice are tossed, $k = 2, 3, 4, 5, 6, 7, 8, 9, 10, 11, 12$ and the corresponding $P_X(k) = \frac{1}{36}, \frac{2}{36}, \frac{3}{36}, \frac{4}{36}, \frac{5}{36}, \frac{6}{36}, \frac{5}{36}, \frac{4}{36}, \frac{3}{36}, \frac{2}{36}$, and $\frac{1}{36}$, respectively. The expected value is then given by

$$E(X) = \sum_{k=2}^{12} kP_X(k)$$

$$= (2)(\tfrac{1}{36}) + (3)(\tfrac{2}{36}) + (4)(\tfrac{3}{36}) + (5)(\tfrac{4}{36}) + (6)(\tfrac{5}{36}) + (7)(\tfrac{6}{36})$$
$$+ (8)(\tfrac{5}{36}) + (9)(\tfrac{4}{36}) + (10)(\tfrac{3}{36}) + (11)(\tfrac{2}{36}) + (12)(\tfrac{1}{36}) = 7.$$

For the random variable X, which is the number of α particles reaching the counter during a 10-second interval, $P_X(k) = \lambda^k e^{-\lambda}/k!$ for $k = 0, 1, 2,$ .... The expected value of this Poisson random variable is then given by

$$E(X) = \sum_{k=0}^{\infty} kP_X(k) = \sum_{k=0}^{\infty} \frac{k\lambda^k e^{-\lambda}}{k!} = \sum_{k=1}^{\infty} \frac{\lambda^k e^{-\lambda}}{(k-1)!}.$$

Letting $j = (k-1)$, $E(X)$ becomes

$$E(X) = \sum_{j=0}^{\infty} \frac{\lambda^{j+1} e^{-\lambda}}{j!} = \lambda \sum_{j=0}^{\infty} \frac{\lambda^j e^{-\lambda}}{j!} = \lambda \quad \left(\text{since } \sum_{j=0}^{\infty} \frac{\lambda^j e^{-\lambda}}{j!} = 1\right).$$

For a continuous random variable, the expectation is obtained by integrating between minus and plus infinity the product $zf_X(z)$. The expected value of the continuous random variables presented as examples in Sec. 2.8 will be calculated. For the random variable X, which is the Rockwell hardness of a specimen, the density function is rectangular, and given by

$$f_X(z) = \begin{cases} 0, & \text{for} \quad z < 50 \\ \frac{1}{20}, & \text{for } 50 \leq z \leq 70 \\ 0, & \text{for} \quad z > 70. \end{cases}$$

The expected value of this random variable is then given by

$$E(X) = \int_{-\infty}^{\infty} zf_X(z)\, dz = \int_{-\infty}^{50} zf_X(z)\, dz + \int_{50}^{70} zf_X(z)\, dz + \int_{70}^{\infty} zf_X(z)\, dz$$

$$= 0 + \int_{50}^{70} z(\tfrac{1}{20})\, dz + 0 = 60.$$

For the random variable X_1, the time to failure of the transistor labeled as number 1, the density function was assumed to be exponential, i.e.,

$$f_{X_1}(z) = \begin{cases} 0, & \text{for } z < 0 \\ \dfrac{1}{\theta} e^{-z/\theta}, & \text{for } z \geq 0. \end{cases}$$

The expected value of this random variable is then given by

$$E(X_1) = \int_{-\infty}^{\infty} zf_{X_1}(z)\, dz = \int_{-\infty}^{0} zf_{X_1}(z)\, dz + \int_{0}^{\infty} zf_{X_1}(z)\, dz$$

$$= 0 + \int_{0}^{\infty} \frac{z}{\theta} e^{-z/\theta}\, dz,$$

which can be easily evaluated by integration by parts or from a standard table of integrals, and seen to equal θ. In the example, the parameter θ was chosen to be equal to 1,000 hours so that the expected value of the time to failure is also 1,000 hours.

The expectation is quite important because it follows the customary notion of an "average." Suppose an experiment is performed consisting of testing n units. Choose as random variables of interest the sequence $X_1, X_2, \ldots, X_n$, where X_i represents the measurable characteristic associated with the ith unit. Assume that the probability distribution associated with each random variable is the same, i.e., the random variables are identically distributed. Assume further that these n random variables are independent, i.e., knowledge of the outcomes of one or more of these random variables does not affect the probability distribution (and hence the outcomes) of the remaining random variables.[1] Thus, the "sample of $X_1, X_2, \ldots, X_n$" forms a sequence of independent identically distributed random variables. Consider a new random variable $\bar{X}$, the arithmetic mean, which is formed by averaging the sequence of random variables, i.e.,

$$\bar{X} = \frac{X_1 + X_2 + \cdots + X_n}{n}.$$

This random variable is also referred to as the sample mean. The relationship between this sample mean, $\bar{X}$, and the expected value of any of the X_i (each X_i has the same expected value since each X_i has the same probability distribution) is that $\bar{X}$ *tends to* $E(X_i)$ *as n becomes large.*[2] For example, suppose n transistors are placed on test. Let $X_1, X_2, \ldots, X_n$, the times to failure, be independent, identically exponentially distributed random variables with parameter θ equal to 1,000 hours. The expected time to failure for each transistor has been shown to be 1,000 hours. If $x_1, x_2, \ldots, x_n$ denote the observed times to failures for each of the n transistors, then their arithmetic mean, $\bar{x} = (x_1 + x_2 + \cdots + x_n)/n$, will be close to 1,000 provided that n is sufficiently large. Alternatively, if the expected value of the time to failure, θ, were unknown, the value that the random variable $\bar{X}$ takes on will be a good approximation to, or estimate of, θ provided that n is sufficiently large.

It is unnecessary to confine the discussion about expectation solely to the expectation of the random variable X. Let $h(X)$ be an arbitrary function of the random variable X. The expected value of $h(X)$ is defined to be

$$E[h(X)] = \begin{cases} \sum_{\text{all } k} h(k)P_X(k), & \text{if } X \text{ is a discrete random variable} \\ \int_{-\infty}^{\infty} h(z)f_X(z)\, dz, & \text{if } X \text{ is a continuous random variable.} \end{cases}$$

---

[1] A more quantitative definition of independent random variables will be given in Sec. 2.14.

[2] A formal statement of this property is known as the Law of Large Numbers.

Thus, the expectation of the function of X is expressed in terms of the distribution of the random variable X. However, $h(X)$ is, itself, a random variable and, hence, has a CDF associated with it. Denote this random variable by Y, i.e., $Y = h(X)$. The CDF of Y can be obtained as follows:

$$F_Y(b) = P\{Y \leq b\} = P\{X \text{ takes on such values that } h(X) \leq b\}.$$

The right-hand side of this last expression can be evaluated if the CDF of the random variable X is known. The probability distribution or density function, whichever is applicable, can then be found.[1] Let $Q_Y(j)$ be the probability distribution of Y if Y is discrete, and let $G_Y(w)$ be the density function of Y if Y is continuous. Using the definition of the expectation of a random variable presented earlier,

$$E(Y) = \begin{cases} \sum_{\text{all } j} jQ_Y(j), & \text{if } Y \text{ is a discrete random variable} \\ \int_{-\infty}^{\infty} wG_Y(w)\, dw, & \text{if } Y \text{ is a continuous random variable.} \end{cases}$$

A natural question to raise is whether

$$E[h(X)] = E(Y).$$

The answer is presented by the Theorem of the Unconscious Statistician[2] which states that they are, indeed, equal. Thus, the expectation of a function of a random variable, $E[h(X)]$, can be obtained by using the distribution of the original random variable X or by using the distribution of the transformed variable Y. As an example, consider the tossing of a die experiment. Let X denote the upturned face. Define $h(X)$ as follows:

$h(X)$ takes on the value -1 when X takes on the values 1, 2, or 3.
$h(X)$ takes on the value $+1$ when X takes on the values 4, 5, or 6.

Using the definition of $E[h(X)]$, this expectation is found to be

$$E[h(X)] = \sum_{k=1}^{6} h(k)\tfrac{1}{6} = -1(\tfrac{1}{6}) - 1(\tfrac{1}{6}) - 1(\tfrac{1}{6}) + 1(\tfrac{1}{6}) + 1(\tfrac{1}{6}) + 1(\tfrac{1}{6}) = 0.$$

In order to evaluate this expectation by using the distribution of $Y = h(X)$, it is necessary to find the probability distribution of Y. The random variable takes on two possible values, -1 and $+1$. Since Y takes on the value -1 whenever X takes on the values 1, 2, or 3, $Q_Y(-1) = P\{X \leq 3\} = \tfrac{1}{2}$. Similarly, since Y takes on the value of $+1$ whenever X takes on the values 4, 5, or 6, $Q_Y(+1) = P\{X \geq 4\} = \tfrac{1}{2}$. Thus,

$$E[Y] = \sum_{\text{all } j} jQ_Y(j) = -1(\tfrac{1}{2}) + 1(\tfrac{1}{2}) = 0,$$

---

[1] If X and Y are discrete, the probability distribution of Y can also be obtained from
$$P_Y(k) = P\{Y = k\} = P\{X \text{ takes on such values that } h(X) = k\}.$$
If Y has a density function, it can be obtained by differentiating $F_Y(b)$.

[2] This peculiar name arises from the fact that the statistician often uses these identities interchangeably, and is oblivious to the fact that this theorem requires proof.

thereby verifying the conclusions of the Theorem of the Unconscious Statistician. It is worth remarking that the random variable Y corresponds to the payoff in the game between two players with the rules that if the upturned face of the die results in a 1, 2, or 3, player I loses one dollar and if the upturned face of the die results in a 4, 5, or 6, he wins one dollar. An expected value of the payoff equal to zero shows that this is a fair game.

2.10 Moments

In elementary mechanics moments are associated with the physical properties of bodies of mass. The first moment about the origin is related to the center of gravity, and the second moment about the center of gravity is known as the moment of inertia. Similarly, the concept of moments is extremely important for distributions. The analogy between bodies of unit mass and distributions leads to the usefulness of the analogy between the moment properties. Just as bodies are completely characterized by their moments, distribution functions are completely characterized by their moments. The moments of a distribution are expectations of particular functions of the attached random variable. Such expectations, $E[h(X)]$, were defined in the previous section.

The first moment about the origin is obtained by letting $h(X) = X$, and, therefore, coincides with the expected value of the random variable itself. It is often called the mean of the distribution or simply the mean, and is denoted by the symbol μ.[1] Thus, the mean of the distribution is given by

$$E(X) = \mu = \begin{cases} \displaystyle\sum_{\text{all } k} k P_X(k), & \text{if } X \text{ is a discrete random variable} \\ \displaystyle\int_{-\infty}^{\infty} z f_X(z)\, dz, & \text{if } X \text{ is a continuous random variable.} \end{cases}$$

The mean, μ, is also the center of gravity of the probability mass and can be interpreted as the point where the fulcrum is placed in order to balance the mass.[2]

The second moment about the origin, obtained by letting $h(X) = X^2$, is given by

$$E(X^2) = \begin{cases} \displaystyle\sum_{\text{all } k} k^2 P_X(k), & \text{if } X \text{ is a discrete random variable} \\ \displaystyle\int_{-\infty}^{\infty} z^2 f_X(z)\, dz, & \text{if } X \text{ is a continuous random variable.} \end{cases}$$

---

[1] The mean of the distribution, μ, should never be confused with the sample mean, $\bar{X}$, which is simply the arithmetic average. The former is a fixed constant associated with the probability distribution whereas the latter is a random variable. Throughout this text the terminology "sample mean" is used whenever referring to this random variable. The terminology "mean" refers only to the mean of the distribution.

[2] The total probability mass is one. Hence, it is unnecessary to divide this expression by the total mass as is usually done in mechanics to obtain the center of gravity.

The second moment about the mean, obtained by letting $h(X) = (X - \mu)^2$, is given by

$$E(X - \mu)^2 = \begin{cases} \sum_{\text{all } k} (k - \mu)^2 P_X(k), & \text{if } X \text{ is a discrete random variable} \\ \int_{-\infty}^{\infty} (z - \mu)^2 f_X(z) \, dz, & \text{if } X \text{ is a continuous random variable.} \end{cases}$$

This second moment about the mean is also known as the variance of the distribution, or simply as the variance, and is denoted by the symbol σ_X^2 or by σ^2 when there is no ambiguity. The square root of the variance is called the standard deviation and is denoted by σ. Note that the variance is analogous to the moment of inertia in mechanics.

The parallel axis theorem of elementary mechanics is, of course, applicable here. The second moment about the mean is equal to the second moment about any other point minus the square of the distance between the mean and this point, i.e.,

$$E(X - \mu)^2 = E(X - a)^2 - (\mu - a)^2.$$

In particular, the second moment about the origin can be written as

$$E(X^2) = E(X - \mu)^2 + \mu^2 = \sigma^2 + \mu^2.$$

This theorem is easily proven by reverting to the definitions. It is presented here for the case of a continuous random variable, and the proof for the discrete case is left to the reader. Using the definition of the variance, the following result is obtained:

$$E(X - \mu)^2 = \int_{-\infty}^{\infty} (z - \mu)^2 f_X(z) \, dz = \int_{-\infty}^{\infty} [(z - a) - (\mu - a)]^2 f_X(z) \, dz.$$

Squaring the terms in brackets,

$$E(X - \mu)^2 = \int_{-\infty}^{\infty} (z - a)^2 f_X(z) \, dz - 2(\mu - a) \int_{-\infty}^{\infty} (z - a) f_X(z) \, dz$$
$$+ (\mu - a)^2 \int_{-\infty}^{\infty} f_X(z) \, dz.$$

Since

$$\int_{-\infty}^{\infty} f_X(z) \, dz = 1, \quad \int_{-\infty}^{\infty} (z - a)^2 f_X(z) \, dz = E(X - a)^2, \quad \text{and}$$

$$\int_{-\infty}^{\infty} (z - a) f_X(z) \, dz = \int_{-\infty}^{\infty} z f_X(z) \, dz - a \int_{-\infty}^{\infty} f_X(z) \, dz = \mu - a,$$

the variance can be evaluated as

$$E(X - \mu)^2 = E(X - a)^2 - 2(\mu - a)^2 + (\mu - a)^2$$
$$= E(X - a)^2 - (\mu - a)^2.$$

Returning to the examples of Sec. 2.7, the first moment about the origin for the single die example has already been calculated and found equal to

$3\frac{1}{2}$. The second moment about the origin is given by

$$E(X^2) = \sum_{\text{all } k} k^2 P_X(k)$$

$$= (1)^2 \tfrac{1}{6} + (2)^2 \tfrac{1}{6} + (3)^2 \tfrac{1}{6} + (4)^2 \tfrac{1}{6} + (5)^2 \tfrac{1}{6} + (6)^2 \tfrac{1}{6} = 15\tfrac{1}{6}.$$

Hence, the variance is given by

$$\sigma^2 = E(X - \mu)^2 = E(X^2) - \mu^2 = 15\tfrac{1}{6} - (3\tfrac{1}{2})^2 = \tfrac{35}{12},$$

and the standard deviation by $\sigma = \sqrt{\tfrac{35}{12}}$.

For the two dice example, the mean of the distribution was found equal to 7. The second moment about the origin is given by

$$E(X^2) = \sum_{\text{all } k} k^2 P_X(k) = (2)^2 \tfrac{1}{36} + (3)^2 \tfrac{2}{36} + (4)^2 \tfrac{3}{36} + (5)^2 \tfrac{4}{36}$$

$$+ (6)^2 \tfrac{5}{36} + (7)^2 \tfrac{6}{36} + (8)^2 \tfrac{5}{36} + (9)^2 \tfrac{4}{36} + (10)^2 \tfrac{3}{36}$$

$$+ (11)^2 \tfrac{2}{36} + (12)^2 \tfrac{1}{36}$$

$$= 54\tfrac{5}{6}.$$

For the particle counter example, the mean (of the Poisson distribution) is λ. The second moment about the origin is given by

$$E(X^2) = \sum_{\text{all } k} k^2 P_X(k) = \sum_{k=0}^{\infty} \frac{k^2 \lambda^k e^{-\lambda}}{k!} = \sum_{k=1}^{\infty} \frac{k \lambda^k e^{-\lambda}}{(k-1)!}.$$

Letting $j = k - 1$, $E(X^2)$ becomes

$$E(X^2) = \sum_{j=0}^{\infty} (j+1) \frac{\lambda^{j+1} e^{-\lambda}}{j!} = \lambda \sum_{j=0}^{\infty} j \frac{\lambda^j e^{-\lambda}}{j!} + \lambda \sum_{j=0}^{\infty} \frac{\lambda^j e^{-\lambda}}{j!}$$

$$= \lambda^2 + \lambda \left(\text{since } \sum_{j=0}^{\infty} j \frac{\lambda^j e^{-\lambda}}{j!} = E(X) = \lambda \right).$$

The variance is given by

$$\sigma^2 = E(X^2) - \mu^2 = \lambda^2 + \lambda - \lambda^2 = \lambda,$$

so that the Poisson distribution possesses the interesting property of having the mean and variance equal.

The first moments about the origin for the examples of Sec. 2.8 involving continuous random variables were also calculated in the previous section. For the Rockwell hardness example, the mean of the distribution was found to be 60. The second moment about the origin is given by

$$E(X^2) = \int_{-\infty}^{\infty} z^2 f_X(z) \, dz = \int_{50}^{70} z^2 \left(\frac{1}{20} \right) dz = \frac{1}{20} \frac{z^3}{3} \Big|_{50}^{70}$$

$$= \frac{1}{60}(343{,}000 - 125{,}000) = 3{,}633\tfrac{1}{3}.$$

The variance is given by

$$\sigma^2 = E(X^2) - \mu^2 = 3633\tfrac{1}{3} - (60)^2 = 33\tfrac{1}{3}.$$

For the transistor example, the mean of the distribution was found to be

1,000 hours (in general the mean is the parameter θ). The second moment about the origin is given by

$$E(X^2) = \int_{-\infty}^{\infty} z^2 f_X(z)\, dz = \int_0^{\infty} \frac{z^2}{1,000} e^{-z/1,000}\, dz.$$

Integrating by parts leads to

$$E(X^2) = 2(1,000)^2,$$

so that the variance becomes

$$\sigma^2 = E(X^2) - \mu^2 = 2(1,000)^2 - (1,000)^2 = (1,000)^2.$$

For any general value of the parameter θ, the variance for an exponential distribution is simply θ^2, and the standard deviation is θ. Note that the exponential distribution possesses the interesting property of having the mean and standard deviation equal.

In calculating the variance of the distribution in all the above examples use was made of the parallel axis theorem. This was done for ease of computation, and the alternate form, i.e., calculating $E(X - \mu)^2$ directly, would have been used if it led to simpler calculations.

Higher moments may be calculated in a manner similar to that shown for the first two. In general, the jth moment about the origin, $j = 1, 2, \ldots$, obtained by letting $h(X) = X^j$, is given by

$$E(X^j) = \begin{cases} \sum_{\text{all } k} k^j P_X(k), & \text{if } X \text{ is a discrete random variable} \\ \int_{-\infty}^{\infty} z^j f_X(z)\, dz, & \text{if } X \text{ is a continuous random variable.} \end{cases}$$

Similarly, the jth moment about the mean, $j = 1, 2, \ldots$, obtained by letting $h(X) = (X - \mu)^j$, is given by

$$E(X - \mu)^j = \begin{cases} \sum_{\text{all } k} (k - \mu)^j P_X(k), & \text{if } X \text{ is a discrete random variable} \\ \int_{-\infty}^{\infty} (z - \mu)^j f_X(z)\, dz, & \text{if } X \text{ is a continuous random variable.} \end{cases}$$

If some moment of a distribution fails to be finite, then the moment is said not to exist. Furthermore, all moments of higher order will also not exist.

From the discussion in this section, it is evident that if the distribution is known, all its moments can be computed. Similarly, if all the moments of a distribution are known, the distribution is completely characterized. Unfortunately, in practice, a knowledge of moments beyond the second is rare. Hence, a reasonable question to formulate is, "To what extent do the first two moments, μ and σ^2, characterize the distribution of the random variable?" The answer, known as Tchebycheff's Inequality, is as follows: For any positive number k, the probability that the random variable, X, lies in the interval $(\mu - k\sigma, \mu + k\sigma)$ is larger than $1 - (1/k^2)$, i.e.,

$$P\{\mu - k\sigma \leq X \leq \mu + k\sigma\} > 1 - \frac{1}{k^2}$$

where X is a random variable having *any* distribution with mean μ and variance σ^2. For example, if $k = 2$, Tchebycheff's inequality indicates that

$$P\{\mu - 2\sigma \leq X \leq \mu + 2\sigma\} > \tfrac{3}{4};$$

if $k = 3$, the inequality indicates that

$$P\{\mu - 3\sigma \leq X \leq \mu + 3\sigma\} > \tfrac{8}{9}.$$

2.11 Some Properties of Functions of Random Variables

In Sec. 2.9, the expected value of the function $h(X)$ was defined as

$$E[h(X)] = \begin{cases} \sum_{\text{all } k} h(k)P_X(k), & \text{if } X \text{ is a discrete random variable} \\ \int_{-\infty}^{\infty} h(z)f_X(z)\,dz, & \text{if } X \text{ is a continuous random variable.} \end{cases}$$

Some useful results about particular functions follow:

1. Let $h(X)$ be equal to aX, where a is a constant.

$$E(aX) = \begin{cases} \sum_{\text{all } k} akP_X(k), & \text{if } X \text{ is discrete} \\ \int_{-\infty}^{\infty} azf_X(z)\,dz, & \text{if } X \text{ is continuous} \end{cases} = aE(X) = a\mu.$$

Thus, the expected value of a constant times a random variable is equal to the constant times the expected value of the random variable.

2. Let $h(X)$ be equal to $(aX - a\mu)^2$, where a is a constant.

$$E(aX - a\mu)^2 = \sigma_{aX}^2 = \begin{cases} \sum_{\text{all } k} (ak - a\mu)^2 P_X(k), & \text{if } X \text{ is discrete} \\ \int_{-\infty}^{\infty} (az - a\mu)^2 f_X(z)\,dz, & \text{if } X \text{ is continuous} \end{cases}$$

$$= a^2\sigma_X^2.$$

Thus, the variance of a constant times a random variable is equal to the square of the constant times the variance of the random variable.

3. Let $h(X)$ be a constant a.

$$E(a) = \begin{cases} \sum_{\text{all } k} aP_X(k), & \text{if } X \text{ is discrete} \\ \int_{-\infty}^{\infty} af_X(z)\,dz, & \text{if } X \text{ is continuous} \end{cases} = a.$$

Thus, the expected value of a constant is equal to the constant.

4. Finally, let $h(X)$ be $(a - a)^2 = 0$.

$$E(a - a)^2 = \sigma_a^2 = \begin{cases} \sum_{\text{all } k} (a - a)^2 P_X(k), & \text{if } X \text{ is discrete} \\ \int_{-\infty}^{\infty} (a - a)^2 f_X(z)\, dz, & \text{if } X \text{ is continuous} \end{cases} = 0.$$

Thus, the variance of a constant is zero.

2.12 Bivariate Probability Distributions

An experiment can generate many random variables of importance. An experimenter, or decision maker, may be interested in the *simultaneous* performances of two or more of these random variables. Therefore, it is important to look at their "joint probability distribution." For example, the times to failure of the two transistors in the transistor experiment are important random variables for decision making purposes. In the dice tossing experiment, the upturned faces play important roles in the game of craps. In an experiment where both the Rockwell hardness and abrasion loss of the same specimen are measured, the random variables, Rockwell hardness and abrasion loss, considered jointly, may be useful for predictive purposes. The first two examples are concerned with random variables that are each associated with different "specimens." This is not a requirement as evidenced by the last example where the same "specimen" is used to obtain the Rockwell hardness and the abrasion loss.

The "joint probability distribution" for the case of two random variables will now be studied in detail. Let X_1 and X_2 be random variables defined on a sample space Ω. Then X_1 and X_2 will be called a bivariate random variable. Denote by the symbol $E_{b_1, b_2}^{X_1, X_2}$ the event consisting of the set of possible outcomes ω of the experiment such that $X_1(\omega) \leq b_1$ *and* $X_2(\omega) \leq b_2$, i.e.,

$$E_{b_1, b_2}^{X_1, X_2} = \{\omega \mid X_1(\omega) \leq b_1, X_2(\omega) \leq b_2\}, \quad \text{or equivalently,}$$

$$E_{b_1, b_2}^{X_1, X_2} = \{X_1 \leq b_1, X_2 \leq b_2\}.$$

The probability of this event is then well defined and given by

$$P\{E_{b_1, b_2}^{X_1, X_2}\} = P\{X_1 \leq b_1, X_2 \leq b_2\}.$$

This leads to the definition of a joint cumulative distribution function which is analogous to the CDF for a single random variable.

The joint cumulative distribution function $F_{X_1 X_2}(b_1, b_2)$ *of the bivariate random variable* X_1, X_2 *is given by*

$$F_{X_1 X_2}(b_1, b_2) = P\{E_{b_1, b_2}^{X_1, X_2}\} = P\{X_1 \leq b_1, X_2 \leq b_2\},$$

for all real values b_1, b_2, $-\infty \leq b_1, b_2 \leq \infty$. Thus, a joint CDF is associated with every bivariate random variable. Furthermore, this joint CDF is not an arbitrary function, but is induced by probabilities associated with events of the form $E_{b_1, b_2}^{X_1, X_2}$ appearing in the sample space.

As an example, consider the experiment where two dice are tossed. Assume that each of the 36 possible outcomes of the experiment have equal probability of $\frac{1}{36}$ assigned to them. Let X_1 and X_2 be the upturned faces of the dice designated as number 1 and number 2, respectively. The joint CDF can be evaluated at a particular point, say, $b_1 = 1.5$, $b_2 = 3$ as follows:

$$F_{X_1, X_2}(1.5, 3) = P\{X_1 \leq 1.5, X_2 \leq 3\}.$$

The set of ω in the sample space having the property that $X_1 \leq 1.5$ and $X_2 \leq 3$ form the set $E_{1.5, 3}^{X_1, X_2}$, which is given by

$$E_{1.5, 3}^{X_1, X_2} = \{(1, 1), (1, 2), (1, 3)\}.$$

Hence, $\qquad F_{X_1, X_2}(1.5, 3) = P\{E_{1.5, 3}^{X_1, X_2}\} = \frac{3}{36} = \frac{1}{12}.$

In a similar manner the entire joint CDF can be obtained.

In general, the joint cumulative distribution function has the following properties:

1. It is a numerical valued function defined for all b_1 and b_2, $-\infty \leq b_1$, $b_2 \leq \infty$.
2. $F_{X_1 X_2}(b_1, \infty) = P\{X_1 \leq b_1, X_2 \leq \infty\} = P\{X_1 \leq b_1\} = F_{X_1}(b_1).$ The function $F_{X_1}(b_1)$ is the CDF for the individual random variable X_1.
3. $F_{X_1 X_2}(\infty, b_2) = P\{X_1 \leq \infty, X_2 \leq b_2\} = P\{X_2 \leq b_2\} = F_{X_2}(b_2).$ The function $F_{X_2}(b_2)$ is the CDF for the individual random variable X_2.
4. $F_{X_1 X_2}(b_1, -\infty) = P\{X_1 \leq b_1, X_2 \leq -\infty\} = 0$ and $F_{X_1 X_2}(-\infty, b_2) = P\{X_1 \leq -\infty, X_2 \leq b_2\} = 0.$
5. $F_{X_1 X_2}(b_1 + \Delta_1, b_2 + \Delta_2) - F_{X_1 X_2}(b_1 + \Delta_1, b_2) - F_{X_1 X_2}(b_1, b_2 + \Delta_2) + F_{X_1 X_2}(b_1, b_2) \geq 0$ for every $\Delta_1, \Delta_2 \geq 0$ and b_1, b_2.

Just as a single random variable has been classified as discrete or continuous, a bivariate random variable can be similarly categorized. A bivariate random variable X_1, X_2 is called discrete if both X_1 and X_2 are discrete random variables. Similarly, a bivariate random variable X_1, X_2 is called continuous if both X_1 and X_2 are continuous random variables. It must be noted that bivariate random variables can be neither discrete nor continuous. However, many bivariate random variables of importance are either discrete or continuous.

If the bivariate random variable X_1, X_2 is discrete, the joint cumulative distribution function can be expressed as

$$F_{X_1 X_2}(b_1, b_2) = \sum_{\text{all } j \leq b_1} \sum_{\text{all } k \leq b_2} P\{\omega \mid X_1(\omega) = j, X_2(\omega) = k\}$$

$$= \sum_{\text{all } j \leq b_1} \sum_{\text{all } k \leq b_2} P_{X_1 X_2}(j, k).$$

The terms $P_{X_1 X_2}(j, k) = P\{X_1 = j, X_2 = k\}$ (or equivalently, $P\{\omega \mid X_1(\omega) = j,$

$X_2(\omega) = k\}$) are called the joint probability distribution of the discrete bivariate random variable X_1, X_2. In the dice tossing example, $P_{X_1X_2}(j, k)$ $= \frac{1}{36}$ for all j, k which are integers between 1 and 6 inclusive.

It is important to note the relationship between the joint probability distribution $P_{X_1X_2}(j, k)$ of the discrete bivariate random variable X_1, X_2 and the probability distributions, $P_{X_1}(j)$ and $P_{X_2}(k)$, of the individual discrete random variables X_1 and X_2, respectively.[1] Since

$$\sum_{\text{all } j \leq b_1} \sum_{\text{all } k} P_{X_1X_2}(j, k) = F_{X_1X_2}(b_1, \infty) = F_{X_1}(b_1) = \sum_{\text{all } j \leq b_1} P_{X_1}(j),$$

it follows that $P_{X_1}(j) = \sum_{\text{all } k} P_{X_1X_2}(j, k)$. $P_{X_1}(j)$ is called the marginal probability distribution of the discrete random variable X_1. Similarly, $P_{X_2}(k) = \sum_{\text{all } j} P_{X_1X_2}(j, k)$ is called the marginal probability distribution of the discrete random variable X_2.

If the joint cumulative distribution function of a continuous bivariate random variable X_1, X_2 can be expressed as

$$F_{X_1X_2}(b_1, b_2) = \int_{-\infty}^{b_1} \int_{-\infty}^{b_2} f_{X_1X_2}(u, v) \, du \, dv,[2]$$

then $f_{X_1X_2}(u, v)$ is called the joint density function of the continuous bivariate random variable X_1, X_2. This joint density function can be interpreted as a three-dimensional surface of total unit volume where the volume within this surface over regions in the u, v plane corresponds to "probabilities." Thus, if the joint density function is known, it can be used to compute such quantities as

$$P\{a_1 \leq X_1 \leq b_1, a_2 \leq X_2 \leq b_2\} = \int_{a_1}^{b_1} \int_{a_2}^{b_2} f_{X_1X_2}(u, v) \, du \, dv.$$

The times to failure for the two transistors in the transistor experiment are an example of a continuous bivariate random variable. If some assumptions are made about the form of its joint density function, then the CDF can be utilized to aid in making decisions.

As an example, assume that the joint density function for the times to failure of the two transistors is given by

$$f_{X_1X_2}(u, v) = \begin{cases} \dfrac{1}{\theta_1\theta_2} e^{-(u/\theta_1 + v/\theta_2)}, & \text{for } u, v \geq 0 \\ 0, & \text{otherwise.} \end{cases}$$

---

[1] Note that the notation implies that $P_{X_1}(j)$, $P_{X_2}(k)$, and $P_{X_1X_2}(j, k)$ are generally *different* functions.

[2] As in the case for a single random variable, the notation assumes that $f_{X_1X_2}(u, v)$ is defined over the entire plane. If either X_1 or X_2 or both are constrained to take on values over only a part of the plane, then $f_{X_1X_2}(u, v)$ is defined to be zero outside of the constrained region.

The joint CDF is then easily obtained, i.e.,

$$F_{X_1 X_2}(b_1, b_2) = \begin{cases} (1 - e^{-b_1/\theta_1})(1 - e^{-b_2/\theta_2}), & \text{for } b_1, b_2 \geqq 0 \\ 0, & \text{otherwise.} \end{cases}$$

Although assumptions about the form of the joint density should be based on physical conditions, the only mathematical constraints that $f_{X_1 X_2}(u, v)$ must satisfy are the following two conditions:

1. $f_{X_1 X_2}(u, v) \geqq 0$
2. $\int_{-\infty}^{\infty} \int_{-\infty}^{\infty} f_{X_1 X_2}(u, v) \, du \, dv = 1$.

It is important to note the relationship between the joint density function $f_{X_1 X_2}(u, v)$ of the bivariate random variable X_1, X_2 and the density functions, $f_{X_1}(u)$ and $f_{X_2}(v)$, of the individual continuous random variables X_1 and X_2, respectively.[1] Since

$$\int_{-\infty}^{b_1} \int_{-\infty}^{\infty} f_{X_1 X_2}(u, v) \, dv \, du = F_{X_1 X_2}(b_1, \infty) = F_{X_1}(b_1)$$

$$= \int_{-\infty}^{b_1} f_{X_1}(u) \, du,$$

it follows that

$$f_{X_1}(u) = \int_{-\infty}^{\infty} f_{X_1 X_2}(u, v) \, dv.$$

$f_{X_1}(u)$ is called the marginal density function of the continuous random variable X_1. Similarly,

$$f_{X_2}(v) = \int_{-\infty}^{\infty} f_{X_1 X_2}(u, v) \, du$$

is called the marginal density function of the continuous random variable X_2. For the transistor example just presented,

$$f_{X_1}(u) = \int_{-\infty}^{\infty} \frac{1}{\theta_1 \theta_2} e^{-(u/\theta_1 + v/\theta_2)} \, dv = \frac{1}{\theta_1} e^{-u/\theta_1}$$

and

$$f_{X_2}(v) = \int_{-\infty}^{\infty} \frac{1}{\theta_1 \theta_2} e^{-(u/\theta_1 + v/\theta_2)} \, du = \frac{1}{\theta_2} e^{-v/\theta_2},$$

so that both the marginal density of X_1 and the marginal density of X_2 are exponential.

Very frequently, the distribution of a function of a bivariate random variable is of importance, e.g., in the transistor experiment, the decision maker is interested in the distribution of the sample mean $\bar{X} = (X_1 + X_2)/2$. Let h be a random variable which is a function of a bivariate random variable

---

[1] Again, such notation implies that $f_{X_1}(u)$, $f_{X_2}(v)$, $f_{X_1 X_2}(u, v)$ are generally *different* functions.

X_1, X_2. If X_1, X_2 is a continuous bivariate random variable having a joint density $f_{X_1 X_2}(u, v)$, the CDF of $Y = h(X_1, X_2)$ can be obtained from

$$F_Y(b) = P\{Y \le b\} = \iint_{R_b} f_{X_1 X_2}(u, v)\, du\, dv$$

where the region of integration R_b is the region in the u, v space such that $h(u, v) \le b$. A similar expression holds for a discrete bivariate random variable with the integrals being replaced by summations. If $F_Y(b)$ is known, the expectation of Y can easily be obtained. Although this can be accomplished in principle, evaluating the CDF of Y is often difficult. Hence, alternative techniques for obtaining the expectation $E(Y)$ are desirable.

The expectation of a function of a bivariate random variable can be defined as follows:

Let $h(X_1, X_2)$ *be a function of the bivariate random variable* X_1, X_2. *Then,*

$$E[h(X_1, X_2)] = \begin{cases} \displaystyle\sum_{\text{all } j}\sum_{\text{all } k} h(j, k) P_{X_1 X_2}(j, k), & \text{if } X_1, X_2 \text{ is a} \\ & \text{discrete bivariate} \\ & \text{random variable} \\ \displaystyle\int_{-\infty}^{\infty}\int_{-\infty}^{\infty} h(u, v) f_{X_1 X_2}(u, v)\, du\, dv, & \text{if } X_1, X_2 \text{ is a} \\ & \text{continuous bivariate} \\ & \text{random variable.} \end{cases}$$

Furthermore, by an extension of the Theorem of the Unconscious Statistician (Sec. 2.9), the two methods for computing the expectation yield the same results, i.e.,

$$E[Y] = E[h(X_1, X_2)].$$

The definition of $E[h(X_1, X_2)]$ leads to several useful results. These will be given for continuous bivariate random variables, but similar results hold for discrete bivariate random variables.

1. Let $h(X_1, X_2)$ be equal to X_1.

$$E(X_1) = \int_{-\infty}^{\infty}\int_{-\infty}^{\infty} u f_{X_1 X_2}(u, v)\, du\, dv = \int_{-\infty}^{\infty} u \left[\int_{-\infty}^{\infty} f_{X_1 X_2}(u, v)\, dv \right] du$$

$$= \int_{-\infty}^{\infty} u f_{X_1}(u)\, du.$$

This is just the expectation of the single random variable X_1 taken with respect to its marginal density function.

2. Let $h(X_1, X_2)$ be equal to $(X_1 - E(X_1))^2$.

$$E(X_1 - E(X_1))^2 = \int_{-\infty}^{\infty}\int_{-\infty}^{\infty} (u - E(X_1))^2 f_{X_1 X_2}(u, v)\, du\, dv$$

$$= \int_{-\infty}^{\infty} (u - E(X_1))^2 \left[\int_{-\infty}^{\infty} f_{X_1 X_2}(u, v)\, dv \right] du$$

$$= \int_{-\infty}^{\infty} (u - E(X_1))^2 f_{X_1}(u)\, du.$$

This is just the variance of the single random variable X_1 taken with respect to its marginal density function.

3. Let $h(X_1, X_2)$ be equal to $a_1X_1 + a_2X_2$, where a_1 and a_2 are constants.

$$E(a_1X_1 + a_2X_2) = \int_{-\infty}^{\infty} \int_{-\infty}^{\infty} (a_1u + a_2v)f_{X_1X_2}(u, v) \, du \, dv$$

$$= a_1 \int_{-\infty}^{\infty} uf_{X_1}(u) \, du + a_2 \int_{-\infty}^{\infty} vf_{X_2}(v) \, dv$$

$$= a_1E(X_1) + a_2E(X_2).$$

Thus, the expectation of a linear combination of two random variables is just the sum of the expectations of the individual random variables, with each expectation multiplied by the appropriate coefficient.

Applying this last result to the transistor example, $E(\bar{X})$, the expectation of the sample average time to failure, is easily obtained. Note that $\bar{X} = (X_1 + X_2)/2 = (\frac{1}{2})X_1 + (\frac{1}{2})X_2$ so that $a_1 = a_2 = \frac{1}{2}$. Furthermore, since the marginal densities of X_1 and X_2 are exponential with parameters θ_1 and θ_2, respectively, $E(X_1) = \theta_1$ and $E(X_2) = \theta_2$. Therefore, $E(\bar{X}) = (\theta_1 + \theta_2)/2$. It is interesting to observe that if $\theta_1 = \theta_2 = \theta$, i.e., the times to failure have *identical* distributions, then $E(\bar{X}) = \theta$. Thus, it is evident that if random variables have identical probability distributions each with expectation μ, then $E(\bar{X}) = \mu$. The result holds for any (finite) number of random variables.

Although all the results in this section have been based on a study of the bivariate random variable X_1, X_2, they are all easily extended for the multivariate random variable $X_1, X_2, \ldots, X_n$.

2.13 Conditional Probability and Independent Events

In the previous sections of this chapter it has been implied that the "information" derived from an experiment appears all at once rather than sequentially in time. Frequently, partial information about the outcome of the experiment becomes known, and the decision maker may want to utilize this information immediately. Alternatively, there may be several "related" random variables of interest, and information about the outcomes of some may be useful for predicting the outcomes of those remaining. The transistor experiment may be viewed as an example of the former. The contractor has agreed to purchase the entire production run if the average time to failure of two tested transistors exceeds 400 hours. Assume that the transistors are to be tested sequentially, i.e., the second transistor is placed on test after the first fails. After 600 hours of testing it is noted that the first transistor still has not failed. The contractor may now wish to make preparations for the delivery of the entire production run rather than wait for the completion of the experiment because it is "likely" that the average time to failure of both transistors will exceed 400 hours, given that the first transistor has survived at least 600 hours. An example of the latter "related" random variable case is obtained by the tossing of a single die. Let X_1 be the discrete random variable which takes on the value 0 if the upturned face is even and 1 if the upturned face is odd. Let X_2 be the discrete random variable which denotes the upturned face (identity function). Certainly, given the information that X_1 is zero (or one) very much affects the probability distribution of the random variable X_2. In order to be able to utilize such information as just described, a study of conditional probability defined for events in the original sample space will first be made.

Let E_1 and E_2 be events in the sample space Ω. Denote the conditional probability of the occurrence of the event E_2, given that the event E_1 has occurred, by the symbol $P\{E_2 | E_1\}$. In the die example let E_2 be the event in the sample space consisting of the points 2, 3, and 4, i.e., $E_2 = \{2, 3, 4\}$, and let E_1 be the event consisting of the three points 2, 4, and 6, i.e., $E_1 = \{2, 4, 6\}$. Then $P\{E_2 | E_1\}$ is the conditional probability that the upturned face is 2, 3, or 4, given that the upturned face is even.

In order to evaluate such expressions, the following definition of conditional probability is made:

Let E_1 and E_2 be events in the sample space Ω. Then

$$P\{E_2 | E_1\} = \frac{P\{E_1 \cap E_2\}}{P\{E_1\}},$$

provided $P\{E_1\} > 0$.

Thus, in the above example, if each ω in the sample space is assumed to have a probability of $\frac{1}{6}$ associated with it (a fair die is tossed), then $\{E_1 \cap E_2\} = \{2, 4\}$ and

$$P\{E_2 | E_1\} = \frac{P\{2, 4\}}{P\{2, 4, 6\}} = \frac{2/6}{3/6} = \frac{2}{3}.$$

In the transistor example, choose E_2 as the event in the sample space consisting of those possible outcomes of the experiment corresponding to failure times of the second transistor exceeding 200 hours and any failure time for the first transistor, i.e.,

$$E_2 = \{\omega = (x_1, x_2) | 0 \leq x_1 \leq \infty, 200 < x_2 \leq \infty\}.$$

Let E_1 denote the event in the sample space consisting of those possible outcomes of the experiment corresponding to failure times of the first transistor greater than or equal to 600 hours, and any failure time for the second transistor, i.e.,

$$E_1 = \{\omega = (x_1, x_2) | 600 \leq x_1 \leq \infty, 0 \leq x_2 \leq \infty\}.$$

Then $P\{E_2 | E_1\}$, the conditional probability that the time to failure of the second transistor exceeds 200 hours, given that the time to failure of the first transistor is greater than or equal to 600 hours, is given by

$$P\{E_2 \mid E_1\} = \frac{P\{E_1 \cap E_2\}}{P\{E_1\}}$$

$$= \frac{P\{\omega = (x_1, x_2) \mid 600 \leq x_1 \leq \infty, 200 < x_2 \leq \infty\}}{P\{\omega = (x_1, x_2) \mid 600 \leq x_1 \leq \infty, 0 \leq x_2 \leq \infty\}}.$$

Although the terminology for conditional probability implies that it is a probability, it must be emphasized that the expression for conditional probability is simply a definition. Fortunately, this definition is well motivated in that (1) it conforms to the frequency interpretation of probability, that is, the conditional relative frequency of E_2, given E_1, is approximately $P\{E_1 \cap E_2\}/P\{E_1\}$, and (2) it satisfies the four conditions that regular probabilities satisfy, described in Sec. 2.5.

The computation of conditional probability is actually equivalent to calculating probabilities of events defined in a "reduced" sample space, i.e., $P\{E_2 \mid E_1\}$ is essentially the probability of the occurrence of the event E_2 relative to the "reduced" sample space E_1. In the die example, the "reduced" sample space is the event E_1 which consists of the even upturned faces. All events to be considered must be subsets of E_1 so that only that part of E_2 which is a subset of E_1 is to be considered, i.e., the points 2 and 4 in the example. The probabilities associated with these subsets of E_1 are obtained by dividing the original probabilities defined for these subsets by $P\{E_1\}$.

The very thought of finding the conditional probability of the occurrence of an event E_2 given that an event E_1 occurred implies that the occurrence of E_1 provides *useful* information about the occurrence of E_2. Sometimes this is true, but sometimes it is false. This leads to the concept of independent events which is defined as follows:

Two events E_1 and E_2 are said to be independent if $P\{E_1 \cap E_2\} = P\{E_1\}P\{E_2\}$.

Note that, if $P\{E_1\} > 0$, this definition implies that $P\{E_2 \mid E_1\} = P\{E_2\}$ so that this last expression can be taken as an alternative definition. In the die example, $P\{E_2 \mid E_1\}$ was shown to equal $\frac{2}{3}$. $P\{E_2\}$ is simply $\frac{1}{2}$ so that E_1 and E_2 are not independent events.

On the other hand if the event E_3 is taken to be that event consisting of the points 3 and 4, i.e., $E_3 = \{3, 4\}$, then

$$P\{E_3 \cap E_1\} = P\{4\} = 1/6.$$

Since $P\{E_3\} = \frac{1}{3}$ and $P\{E_1\} = \frac{1}{2}$, then $P\{E_3 \cap E_1\} = P\{E_3\}P\{E_1\} = \frac{1}{6}$ (or equivalently, $P\{E_3 \mid E_1\} = P\{E_3\} = \frac{1}{3}$) so that the events E_1 and E_3 are independent. Thus, it is seen that knowledge of the occurrence of the event E_1 does affect the probability of the occurrence of the event E_2 whereas it does not affect the probability of the occurrence of the event E_3. Hence, in a given sample space associated with an experiment, some types of events may be independent while others may be dependent.

The above definition of independence is not particularly useful for judging whether events are independent in the absence of complete knowledge of the probabilities associated with events in the sample space. Instead assumptions about the independence or dependence of events are usually made by viewing the physical properties associated with the performance of the experiment. For example, in testing two transistors sequentially, the performance of the first transistor should not be expected to affect the performance of the second transistor. Thus, an event defined only by the performance of transistor 1, i.e., the event E_1 in the example, may be reasonably assumed to be independent of an event defined only by the performance of transistor 2, i.e., the event E_2 in the example. This is to be contrasted with the situation where a single specimen is tested for both its Rockwell hardness and abrasion loss. A typical sample space is shown by the shaded area in Fig. 2.21. These two

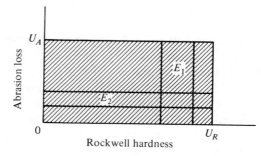

Fig. 2.21. *A typical sample space where a single specimen is tested for both its Rockwell hardness and abrasion loss.*

physical characteristics may very well be related so that knowledge about the occurrence of an event pertaining only to the Rockwell hardness, say, E_1, may affect the probability of occurrence of an event pertaining only to the abrasion loss, say, E_2. Hence, such events could not be assumed to be independent.

The definition of independent events can be extended to n events as follows: *The n events $E_1, E_2, \ldots, E_n$ are said to be independent if for every subset of m events $E_{j_1}, E_{j_2}, \ldots, E_{j_m}$,*

$$P\{E_{j_1} E_{j_2} \ldots E_{j_m}\} = P\{E_{j_1}\} P\{E_{j_2}\} \ldots P\{E_{j_m}\}.$$

Thus, if $n = 3$, this definition implies that E_1, E_2, and E_3 are independent if they are pairwise independent *and* if $P\{E_1 E_2 E_3\} = P\{E_1\} P\{E_2\} P\{E_3\}$.

2.14 Independent Random Variables and A Random Sample

In Sec. 2.13, a definition of independence of two events was introduced: Two events E_1 and E_2 were said to be independent if $P\{E_1 \cap E_2\} = P\{E_1\} P\{E_2\}$. Note that in Sec. 2.12, the joint cumulative distribution function $F_{X_1 X_2}(b_1, b_2)$ for the bivariate random variable X_1, X_2 was defined, i.e.,

$$F_{X_1X_2}(b_1, b_2) = P\{\omega \,|\, X_1(\omega) \leq b_1, X_2(\omega) \leq b_2\} = P\{X_1 \leq b_1, X_2 \leq b_2\}.$$

Denote by E_1 the event containing those ω such that $X_1(\omega) \leq b_1$ and $X_2(\omega)$ takes on any value, i.e.,

$$E_1 = \{\omega \,|\, X_1(\omega) \leq b_1, X_2(\omega) \leq \infty\} = \{X_1 \leq b_1\}.$$

Denote by E_2 the event containing those ω such that $X_2(\omega) \leq b_2$ and $X_1(\omega)$ takes on any value, i.e.,

$$E_2 = \{\omega \,|\, X_1(\omega) \leq \infty, X_2(\omega) \leq b_2\} = \{X_2 \leq b_2\}.$$

If E_1 and E_2 are independent, then $P\{E_1 \cap E_2\} = P\{E_1\}P\{E_2\}$ implies that $P\{X_1 \leq b_1, X_2 \leq b_2\} = P\{X_1 \leq b_1\}P\{X_2 \leq b_2\}$, or alternatively,

$$F_{X_1X_2}(b_1, b_2) = F_{X_1}(b_1)F_{X_2}(b_2).$$

This last result motivates the definition of independence of two random variables:

The random variables X_1 and X_2 are said to be independent if

$$F_{X_1X_2}(b_1, b_2) = F_{X_1}(b_1)F_{X_2}(b_2),$$

for all values of b_1 and b_2.

It is most important to emphasize that this definition requires that the joint CDF of independent random variables factor into the individual CDF's for *all* values of b_1 and b_2. It was noted previously (Sec. 2.13) that some events in the sample space may be independent while others are dependent. However, if random variables X_1 and X_2 are to be independent, then an event of the form $\{X_1 \leq b_1\}$ must be independent of an event of the form $\{X_2 \leq b_2\}$ for *all* values of b_1 and b_2.

The aforementioned definition of independence for random variables has special implications for discrete bivariate random variables and continuous bivariate random variables. In particular, if X_1, X_2 is a discrete bivariate random variable, and X_1 and X_2 are independent, then their joint probability distribution factors into the product of their individual (marginal) probability functions, i.e.,

$$P_{X_1X_2}(j, k) = P_{X_1}(j)P_{X_2}(k),$$

for all possible values of j and k.

Note that this property implies that if X_1 and X_2 are independent, then an event in the original sample space of the form $\{X_1 = j\}$ is independent of an event of the form $\{X_2 = k\}$ for all j and k. Alternatively, if the original sample space can be represented by points $\omega = (j, k)$, and X_1 and X_2 are random variables such that $X_1(\omega) = j$ and $X_2(\omega) = k$, then the independence of X_1 and X_2 implies that the probabilities associated with all the events consisting of single points $\omega = (j, k)$ factors into two components, depending only on j and k, respectively. The experiment of tossing two "fair" dice is such an example.

If X_1, X_2 is a continuous bivariate random variable, and X_1 and X_2 are independent, then the joint density function factors into the product of their individual (marginal) density functions, i.e.,

$$f_{X_1 X_2}(u, v) = f_{X_1}(u) f_{X_2}(v).$$

For the transistor example in which the joint density function for the times to failure of the two transistors is given by

$$f_{X_1 X_2}(u, v) = \begin{cases} \dfrac{1}{\theta_1 \theta_2} e^{-(u/\theta_1 + v/\theta_2)}, & \text{for } u, v \geq 0 \\ 0, & \text{otherwise,} \end{cases}$$

the joint density factors into the marginal densities of X_1 and X_2, i.e.,

$$\frac{1}{\theta_1 \theta_2} e^{-(u/\theta_1 + v/\theta_2)} = \left(\frac{1}{\theta_1} e^{-u/\theta_1} \right) \left(\frac{1}{\theta_2} e^{-v/\theta_2} \right).$$

Hence, X_1 and X_2 are independent random variables.

Unfortunately, this definition of independence is not particularly useful for judging whether or not two random variables are independent by observing their outcomes. Yet the property of independence plays an important role in decision making. Hence, assumptions about the independence of random variables are usually made by viewing the physical properties of the experiment. The definition of independence implies that the probability of occurrence of a certain type of event determined by the bivariate random variable X_1, X_2 can be factored into two components, namely, the probability of occurrence of an event determined solely by X_1 and the probability of occurrence of an event determined solely by X_2, i.e., these latter two events are independent. Alternatively, random variables may be considered to be independent if knowledge of the outcome of one does not affect the distribution of the other. For example, in testing two transistors sequentially, it is reasonable to expect events of the form $\{X_1 \leq b_1\}$ and $\{X_2 \leq b_2\}$ to be independent for all b_1 and b_2 (the performance of the first transistor should not be expected to affect the distribution of the second transistor) so that X_1 and X_2 may be assumed to be independent random variables. If the transistors are tested simultaneously in a "bank," the assumption of independence may or may not be reasonable. If a failure of one transistor causes a power surge in the bank, thereby affecting the life of the second transistor, an independence assumption may be unwarranted.

The definition of independence of two random variables is easily extended to a definition for any finite number of random variables.

Let $F_{X_1 X_2 \ldots X_n}(b_1, b_2, \ldots, b_n)$ be the joint CDF for the multivariate random variable X_1, X_2, ..., X_n, and let $F_{X_i}(b_i)$ be the CDF for the random variable X_i.

A sequence of n random variables X_1, X_2, ..., X_n are said to be independent if

$$F_{X_1 X_2 \ldots X_n}(b_1, b_2, \ldots, b_n) = F_{X_1}(b_1) F_{X_2}(b_2) \ldots F_{X_n}(b_n),$$

for all values of $b_1, b_2, \ldots, b_n$.

It has already been shown that if $h(X_1, X_2) = a_1 X_1 + a_2 X_2$, where a_1 and a_2 are constants, then $E[a_1 X_1 + a_2 X_2] = a_1 E(X_1) + a_2 E(X_2)$. A similar result holds for the variance of a linear combination of *independent* random variables. In particular, if X_1 and X_2 are independent random variables, and if $h(X_1, X_2) = [a_1 X_1 + a_2 X_2 - (a_1 E(X_1) + a_2 E(X_2))]^2$, then $E[h(X_1, X_2)] =$ Variance of $(a_1 X_1 + a_2 X_2) = a_1^2$ Variance $X_1 + a_2^2$ Variance X_2.

The term "sample" was introduced in Secs. 2.4 and 2.9 and used to denote a sequence of identically distributed random variables. Now that the concept of independent random variables has been introduced, the term random sample can be defined:

The random variables $X_1, X_2, \ldots, X_n$ *form a random sample of size n if they are independent and identically distributed.*

As an example, once again consider the transistor experiment. The random variables X_1 and X_2, the times to failure of transistors labeled 1 and 2, respectively, form a random sample provided that X_1 and X_2 are independent and have the same probability distribution, e.g., X_1 and X_2 have exponential distributions with the *same* parameter θ. If the parameter differs, or the distribution differs, or if X_1 and X_2 are not independent, they do not constitute a random sample.

If $X_1, X_2, \ldots, X_n$ form a random sample of n discrete random variables, they are independent and identically distributed so that the joint probability function can be expressed as

$$
\begin{aligned}
\mathrm{P}\{X_1 = k_1, X_2 = k_2, \ldots, X_n = k_n\} &= P_{X_1 X_2 \ldots X_n}(k_1, k_2, \ldots, k_n) \\
&= \mathrm{P}\{X_1 = k_1\} \mathrm{P}\{X_2 = k_2\} \ldots \mathrm{P}\{X_n = k_n\} \\
&= \mathrm{P}(k_1) \mathrm{P}(k_2) \ldots \mathrm{P}(k_n),
\end{aligned}
$$

where the marginal probability functions have the identical functional form. Similarly, if $X_1, X_2, \ldots, X_n$ form a random sample of n continuous random variables, they are independent and identically distributed so that the joint density function can be expressed as

$$f_{X_1, X_2, \ldots, X_n}(z_1, z_2, \ldots, z_n) = f(z_1) f(z_2) \ldots f(z_n),$$

where the marginal density functions have the identical functional form. For example, if $X_1, X_2, \ldots, X_n$ is a random sample of n exponentially distributed random variables with parameter θ, the joint density is given by

$$
f_{X_1, X_2, \ldots, X_n}(z_1, z_2, \ldots, z_n)
$$
$$
= \begin{cases} \left(\dfrac{1}{\theta} e^{-z_1/\theta}\right)\left(\dfrac{1}{\theta} e^{-z_2/\theta}\right) \cdots \left(\dfrac{1}{\theta} e^{-z_n/\theta}\right) = \dfrac{1}{\theta^n} e^{-\sum\limits_{i=1}^{n} z_i/\theta}, \\[2ex] \qquad\qquad\qquad\qquad \text{for } z_1, z_2, \ldots, z_n \geqq 0 \\[1ex] 0, \qquad\qquad\qquad\qquad \text{otherwise.} \end{cases}
$$

2.15 Conditional Probability Distributions

Section 2.13 was concerned with the concept of conditional probability of events defined in the sample space. The study of conditional probability was motivated by the occasional presence of partial information about the outcome of the experiment (such as noting that the first transistor tested still had not failed after 600 hours) or by the presence of several "related" random variables of interest and the availability of information about one or more of them (such as noting that the upturned face of a die is even before placing a wager on the number appearing on the upturned face). For each of these situations the decision maker is interested in the probability distribution of a random variable X_2, given information about the random variable X_1.

Specifically, suppose that X_1, X_2 is a bivariate discrete random variable having joint probability distribution $P_{X_1 X_2}(j, k)$ and marginal probability distributions $P_{X_1}(j)$ and $P_{X_2}(k)$. Using the analogy with conditional probability, with the event

$$\{\omega \mid X_1(\omega) = j, X_2(\omega) = \text{anything}\} = \{X_1 = j\}$$

playing the role of E_1, and the event

$$\{\omega \mid X_1(\omega) = \text{anything}, X_2(\omega) = k\} = \{X_2 = k\}$$

playing the role of E_2, the conditional probability distribution is defined as follows:

Denote the conditional probability distribution of X_2, given X_1, by $P_{X_2 \mid X_1 = j}(k)$. Then

$$P_{X_2 \mid X_1 = j}(k) = P\{X_2 = k \mid X_1 = j\} = \frac{P_{X_1 X_2}(j, k)}{P_{X_1}(j)},$$

if $P_{X_1}(j) > 0$.

Similarly, denote the conditional probability distribution of X_1, given X_2, by $P_{X_1 \mid X_2 = k}(j)$. Then

$$P_{X_1 \mid X_2 = k}(j) = P\{X_1 = j \mid X_2 = k\} = \frac{P_{X_1 X_2}(j, k)}{P_{X_2}(k)},$$

if $P_{X_2}(k) > 0$.

For example, suppose a single die is tossed. Let X_1 denote the random variable which takes on the value 0 if the upturned face is even and 1 if the upturned face is odd. Let X_2 be the random variable which denotes the upturned face. Suppose that it is known that $X_1 = 0$. Then

$$P_{X_2 \mid X_1 = 0}(k) = \begin{cases} \dfrac{0}{1/2} = 0, & \text{for } k = 1, 3, 5 \\[2mm] \dfrac{1/6}{1/2} = \dfrac{1}{3}, & \text{for } k = 2, 4, 6. \end{cases}$$

It should be noted that for fixed $X_1 = j$, the conditional probability distribution, $P_{X_2|X_1=j}(k)$ satisfies the two conditions required for discrete probability distributions, i.e.,

$$P_{X_2|X_1=j}(k) \geq 0 \qquad \text{and} \qquad \sum_{\text{all } k} P_{X_2|X_1=j}(k) = 1.$$

The analogy between conditional probability and the conditional density function for a continuous bivariate random variable is not so straightforward. However, this approach will be followed initially. Suppose that X_1, X_2 is a bivariate continuous random variable having joint density $f_{X_1X_2}(u, v)$ and marginal densities $f_{X_1}(u)$ and $f_{X_2}(v)$. Using the analogy with conditional probability, with the event

$$\{\omega \,|\, b_1 < X_1(\omega) \leq b_1 + h, X_2(\omega) \leq \infty\} = \{b_1 < X_1 \leq b_1 + h\}$$

playing the role of E_1, and the event

$$\{\omega \,|\, X_1(\omega) \leq \infty, X_2(\omega) \leq b_2\} = \{X_2 \leq b_2\}$$

playing the role of E_2, the conditional probability of event E_2, given E_1, is given by

$$P\{X_2 \leq b_2 \,|\, b_1 < X_1 \leq b_1 + h\} = \frac{\int_{b_1}^{b_1+h} \int_{-\infty}^{b_2} f_{X_1X_2}(u, v) \, du \, dv}{\int_{b_1}^{b_1+h} f_{X_1}(u) \, du}.$$

For any finite h, this is a well defined probability. However, a definition of conditional density is desired where the conditioning is on a value that X_1 takes on rather than X_1 lying in an interval, i.e., when $h = 0$. This cannot be done directly since the probability that a continuous random variable equals any constant is zero. Instead, the numerator and denominator in the above expression will be divided by h, and then, as $h \longrightarrow 0$, the right-hand side tends to

$$\frac{\int_{-\infty}^{b_2} f_{X_1X_2}(b_1, v) \, dv}{f_{X_1}(b_1)}.$$

Hence, the following definition is presented:

Denote the conditional density function of X_2, *given that* $X_1 = x_1$, *by* $f_{X_2|X_1=x_1}(v)$. *Then*

$$f_{X_2|X_1=x_1}(v) = \frac{f_{X_1X_2}(x_1, v)}{f_{X_1}(x_1)},$$

if $f_{X_1}(x_1) > 0$.

Similarly, denote the conditional density function of X_1, *given that* $X_2 = x_2$, *by* $f_{X_1|X_2=x_2}(u)$. *Then*

$$f_{X_1|X_2=x_2}(u) = \frac{f_{X_1X_2}(u, x_2)}{f_{X_2}(x_2)},$$

if $f_{X_2}(x_2) > 0$.

Note that for fixed $X_1 = x_1$, the conditional density function $f_{X_2|X_1=x_1}(v)$ satisfies the two conditions required for densities, i.e.,

$$f_{X_2|X_1=x_1}(v) \geqq 0 \quad \text{and} \quad \int_{-\infty}^{\infty} f_{X_2|X_1=x_1}(v)\, dv = 1.$$

The definitions of conditional probability distributions and conditional densities lead to an interesting interpretation of independent random variables. If X_1 and X_2 are independent random variables, then the joint probability distribution or the joint density function factors into the marginals. This leads to

$$P_{X_2|X_1=j}(k) = P_{X_2}(k), \quad \text{if } X_1, X_2 \text{ are discrete,}$$

and
$$f_{X_2|X_1=x_1}(v) = f_{X_2}(v), \quad \text{if } X_1, X_2 \text{ are continuous.}$$

Hence, if X_1 and X_2 are independent discrete random variables, the conditional probability distribution of X_2, given $X_1 = j$, is equal to the marginal probability distribution of X_2 so that information about the value that X_1 takes on doesn't affect the probability distribution of X_2. A similar statement can be made for independent continuous random variables.

PROBLEMS

1. A sample space Ω consists of three points ω_1, ω_2, and ω_3. The probabilities assigned to these points are $\frac{1}{6}, \frac{1}{3}, \frac{1}{2}$, respectively. A random variable X is defined as

$$X(\omega_1) = 0, \quad X(\omega_2) = 0, \quad \text{and} \quad X(\omega_3) = 1.$$

(a) Find the probability distribution of X, i.e., $P_X(k)$.
(b) Find the CDF of X, i.e., $F_X(b)$.

2. A die has its numbers removed and two of its sides painted black, two of its sides painted red, one of its sides painted yellow, and one of its sides painted green. The die is tossed once. Describe the sample space.

3. In Problem 2, probabilities over the "point" events are generated in accordance with the probabilities associated with a "fair" die, i.e., P{black} = $\frac{2}{6}$, P{red} = $\frac{2}{6}$, P{yellow} = $\frac{1}{6}$, and P{green} = $\frac{1}{6}$. A random variable X is chosen which assigns the number 0 to black, 1 to red, 1 to yellow, and 2 to green. (a) Find the probability distribution of X, i.e., $P_X(k)$. (b) Find the CDF of X, i.e., $F_X(b)$.

4. Three "fair" dice are tossed. Describe the sample space.

5. In Problem 4, consider the random variable which is the total number of ones that appear. (a) Find the probability distribution of X, i.e., $P_X(k)$. (b) Sketch the CDF of X.

6. During the course of a day, a machine produces three items, each of whose quality, defective or non-defective, is determined at the end of the day. Describe the sample space generated by a day's production.

7. In Problem 6, consider the random variable X which is the number of defective units. Assume that each point in the sample space has equal probability. (a) Find the probability distribution of X, i.e., $P_X(k)$. (b) Sketch the CDF of X.

8. In Problem 6, assume that a non-defective item yields a profit of $1,000 whereas a defective item results in a loss of $250. Let the random variable Y denote the total profit for the day. Assuming that each point in the sample space has equal probability, find the probability distribution for the random variable Y, i.e., $P_Y(k)$.

9. An airplane wing is assembled with a large number of rivets. A single unit is inspected and the number of defective rivets is the factor of importance. Describe the sample space.

10. (a) The shear strengths of two spot welds are to be measured. Assuming that the upper limit is given by U, describe the sample space. Denote by X_{max} the larger of the two shear strengths. Find the possible values that this random variable can take on. (b) An experiment is performed which measures the fraction of sulphur dioxide in a specimen of air. Describe the sample space. Let the random variable X denote the fraction of sulphur dioxide present. Describe the range of this random variable.

11. Assume that the daily demand for gasoline in a gas station is bounded by 1,000 gallons. A measurement is recorded. (a) Describe the sample space. (b) Each gallon sold results in a profit of 6 cents whereas each unsold gallon results in a loss of $\frac{1}{2}$ cent (due to storage costs). If X denotes the random variable which is the demand and Y the random variable profit, describe the random variable Y in terms of X.

12. Suppose that a "fair" coin has the number r_1 inscribed on one side and the number r_2 inscribed on the other. Let X be the random variable which takes on the number (r_1 or r_2) that turns up when the coin is flipped. (a) Determine the numbers r_1 and r_2 so that $E(X) = 0$ and the standard deviation of X is 2. (b) Sketch the cumulative distribution function of X.

13. Let X be the random variable, the toss of a loaded die. The probability distribution for X is given by

$$P_X(1) = P_X(2) = \tfrac{1}{6}; \qquad P_X(3) = \tfrac{1}{12}; \qquad P_X(4) = P_X(5) = \tfrac{1}{4}; \qquad P_X(6) = d.$$

(a) Find the value of d. (b) Evaluate the CDF at $b = 3.6$, i.e., $F_X(3.6)$. (c) Find $P\{3 \le X < 5\}$.

14. For the experiment described in Problem 9, the number of defective rivets X is a random variable whose probability distribution is closely approximated by the Poisson distribution with parameter $\lambda = 3$. Find the probability that there are no defective rivets on an airplane wing, i.e., $P_X(0)$.

15. A port is able to accommodate 4 ships of a given type for overnight accommodation. The port fees result in a profit of $1,000 per housed ship. Let X denote the random variable which is the number of ships seeking berths per night, and assume that $P_X(k) = \tfrac{1}{6}$ for $k = 0, 1, 2, 3, 4, 5$. (a) If Y is the random variable which denotes the nightly profit, describe the random variable Y in terms of the random variable X. (b) Find the probability distribution for Y, i.e., $P_Y(k)$.

16. In a mandrel bend test a $\frac{3}{8}''$ section across a weld in large diameter steel pipe is bent around a mandrel of known diameter. This stresses the weld area, and any length of weld crack that develops is recorded. Assume that the length, in inches, of an observed weld crack is a random variable X having density function given by

$$f_X(z) = \begin{cases} 0, & \text{for } z < 0 \\ \dfrac{1}{0.04} e^{-z/0.04}, & \text{for } z \geq 0. \end{cases}$$

(a) Find the probability of observing a crack exceeding $\frac{1}{2}$ inch. (b) Find the CDF, i.e., $F_X(b)$.

17. Suppose the life, L, in hours of a certain type tube has a density function given by

$$f_L(z) = \begin{cases} 0, & \text{for } z < 1,000 \\ a/z^2, & \text{for } z \geq 1,000. \end{cases}$$

(a) Find the value of a. (b) Determine the expression for the CDF. (c) What is the probability that a tube will last at least 1,500 hours?

18. Let X be a random variable whose density function is given by

$$f_X(z) = \begin{cases} 0, & \text{for } z < 10 \\ e^{-(z-10)}, & \text{for } z \geq 10. \end{cases}$$

(a) Find the number c such that X is equally likely to be greater than or less than c. (b) Find the number d such that the probability that X will exceed d is equal to 0.05.

19. The density function of coded measurements of pitch diameter of threads of a fitting is given by

$$f_D(z) = \begin{cases} 0, & \text{for } z < 0 \\ 1/(1 + z)^2, & \text{for } z \geq 0. \end{cases}$$

(a) Find the probability that D exceeds 2. (b) What is the expression for the CDF?

20. The density function of the random variable X, the shear strength of test spot welds, is given by

$$f_X(z) = \begin{cases} z/250,000, & \text{for } 0 \leq z \leq 500 \\ (1,000 - z)/250,000, & \text{for } 500 < z \leq 1,000 \\ 0, & \text{elsewhere.} \end{cases}$$

Find the number a such that $P\{X < a\} = 0.50$ and the number b such that $P\{X < b\} = 0.90$.

21. The density function of the random variable X, the finished diameter on armored electric cable, is given by

$$f_X(z) = \begin{cases} (z - 0.70)/a, & \text{for } 0.70 \leq z \leq 0.75 \\ (0.80 - z)/a, & \text{for } 0.75 \leq z \leq 0.80 \\ 0, & \text{elsewhere.} \end{cases}$$

Find the number a, $F_X(0.79)$, and $F_X(0.85)$.

22. Let X be the random variable which denotes the life, in hours, of an electric light bulb. The cumulative distribution function is given by

$$F_X(b) = \begin{cases} (1 - k/b), & \text{for } b \geq 1,000 \\ 0, & \text{for } b < 1,000. \end{cases}$$

(a) Find the value of k. (b) What is the density function for the random variable X? (c) What is the probability of a bulb lasting more than 1,500 hours?

23. Let X be a random variable with CDF given by

$$F_X(b) = \begin{cases} 0, & \text{for } b < 0 \\ (b/\theta)^2, & \text{for } 0 \leq b \leq \theta \\ 1, & \text{for } b > \theta, \end{cases}$$

where θ is an unspecified constant. (a) What value(s) of θ makes $F_X(b)$ a bona fide CDF? (b) Find $P\{X > 1\}$. (c) Find the density function of X.

24. The random variable X has probability density function given by

$$f_X(z) = \begin{cases} Kz(1 - z), & \text{for } 0 \leq z \leq 1 \\ 0, & \text{otherwise.} \end{cases}$$

(a) Find the value of K which will make $f_X(z)$ a true density function. (b) Find the CDF.

25. The random variable X has density function given by

$$f_X(z) = \begin{cases} K(z/\theta)^2, & \text{for } 0 \leq z \leq \theta \\ 0, & \text{otherwise.} \end{cases}$$

(a) Determine K so that $f_X(z)$ is a true density. (b) Find the CDF.

26. A random variable X, which represents the weight (in ounces) of an article, has density function given by $f_X(z)$,

$$f_X(z) = \begin{cases} (z - 8), & \text{for } 8 \leq z \leq 9 \\ (10 - z), & \text{for } 9 < z \leq 10 \\ 0, & \text{otherwise.} \end{cases}$$

(a) Calculate the mean and variance of the random variable X. (b) The manufacturer sells the article for a fixed price of $2.00. He guarantees to refund the purchase money to any customer who finds the weight of his article to be less than 8.25 oz. His cost of production is related to the weight of the article by the relation $(0.05)X + 0.30$. Express the random variable profit P in terms of the random variable X, i.e., find the function $h(X)$ such that $P = h(X)$. (c) Find the expected profit per article.

27. A machine makes a product which is screened (inspected 100%) before being shipped. The measuring instrument is such that it is difficult to read between 1 and $1\frac{1}{3}$ (coded data). After the screening process takes place, the measured dimension has density

$$f_X(z) = \begin{cases} kz^2, & \text{for } 0 \leq z \leq 1 \\ 1, & \text{for } 1 < z \leq 1\frac{1}{3} \\ 0, & \text{otherwise.} \end{cases}$$

(a) Find the value of k. (b) What fraction of the items will fall outside the twilight zone (fall between 0 and 1)? (c) Find the mean and variance of this random variable.

28. Prove the parallel axis theorem for a discrete random variable.

29. Find $E(X)$, $E(X^2)$, and $E(X - \mu)^2$ for the random variable described in Problem 1.

30. Find $E(X)$, $E(X^2)$, and $E(X - \mu)^2$ for the random variable described in Problem 3.

31. Find the mean and variance of the random variable described in Problem 5.

32. Find the mean and variance of the random variable described in Problem 7.

33. (a) Find the mean and variance of the random variable described in Problem 8. (b) Express the random variable Y (total profit) in terms of the random variable X (number of defective units) of Problem 7, i.e., find the function $h(X)$ such that $Y = h(X)$. (c) Find $E[h(X)]$ using the probability distribution found in Problem 7, i.e., $P_X(k)$.

34. Find the mean and variance of the random variable described in Problem 13.

35. (a) Find the mean and variance of the random variable X described in Problem 15. (b) Show that $E(Y) = E[h(X)]$ in Problem 15 where the function h is the solution of part (a) of Problem 15.

36. Find the mean and variance of the random variable described in Problem 16.

37. Show that the mean and variance of the random variable described in Problem 17 is not finite.

38. Find the mean and variance of the random variable described in Problem 18.

39. Find the mean and variance of the random variable described in Problem 19.

40. Find the mean and variance of the random variable described in Problem 20.

41. Find the mean and variance of the random variable described in Problem 21.

42. Find the mean and variance of the random variable described in Problem 22.

43. Find the mean and variance of the random variable described in Problem 23.

44. Find the mean and variance of the random variable described in Problem 24.

45. Find the mean and variance of the random variable described in Problem 25.

46. Using the results of Problem 29, find $E(2X)$ and the variance of $2X$.

47. Using the results of Problem 16, find $E(3X)$.

48. If X is a positive continuous random variable, i.e., the density function $f_X(z) = 0$ for $z < 0$, prove that

$$E(X) = \int_0^\infty [(1 - F_X(b)] \, db,$$

where $F_X(b)$ is the CDF of the random variable X.

49. Suppose X is a random variable with continuous density function $f_X(z)$ where $f_X(z)$ is symmetric about some number a, i.e., $f_X(z + a) = f(a - z)$. Prove that $E(X) = a$.

50. The CDF of the random variable X given in Problem 16 is given by

$$F_X(b) = \begin{cases} 0, & \text{for } b < 0 \\ 1 - e^{-b/0.04}, & \text{for } b \geq 0. \end{cases}$$

Consider the random variable $Y = \ln X$. Find the density function of the random variable Y. *Hint:* Note that the CDF of Y is given by

$$F_Y(b) = P\{Y \leq b\} = P\{\ln X \leq b\} = P\{X \leq e^b\}.$$

51. Suppose that X is a continuous random variable with CDF $F_X(b)$. Let $Y = F_X(X)$. Show that Y has a uniform density function, i.e.,

$$f_Y(z) = \begin{cases} 1, & \text{if } 0 \leq z \leq 1 \\ 0, & \text{otherwise.} \end{cases}$$

Hint: Note that

$$F_Y(d) = P\{Y \le d\} = P\{F_X(X) \le d\} = P\{X \le F_X^{-1}(d)\}.$$

52. Suppose that X is a continuous random variable with CDF $F_X(b)$. Let $Y = (X - k)/a$, where $a > 0$ and k are constants. Show that the CDF of Y is given by

$$F_Y(d) = F_X(ad + k).$$

53. A random variable X is said to have a Weibull distribution if the CDF is given by

$$F_X(b) = \begin{cases} 0, & \text{for } b < 0 \\ 1 - e^{-(b/\alpha)^\beta}, & \text{for } b \ge 0, \text{ where } \alpha, \beta > 0. \end{cases}$$

(a) Show that this class of distributions includes the exponential. (b) Let $Y = X^\beta$. Find the CDF of Y, i.e., $F_Y(d)$. *Hint:* $F_Y(d) = P\{Y \le d\} = P\{X^\beta \le d\} = P\{X \le d^{1/\beta}\}$.

54. Unmanned space vehicles are launched until the first successful launching has taken place. If this does not occur within 5 attempts, the experiment is halted and the equipment inspected. Suppose that there is a constant probability of 0.8 of having a successful launching and that successive attempts are independent. Assume that the cost of the first launching is K dollars while subsequent launchings cost $K/3$ dollars. Whenever a successful launching takes place, a certain amount of information is obtained which may be expressed as financial gain of, say, C dollars. If the five attempts all result in failures, no gain is obtained. (a) Let N denote the number of trials until a success is obtained. Find the probability distribution of N, i.e.,

$$P\{N = 1\}, \quad P\{N = 2\}, \quad P\{N = 3\}, \quad P\{N = 4\}, \quad P\{N = 5\}, \quad P\{N > 5\}.$$

Note that $P\{N > 5\}$ is the probability that the experiment is halted without a successful launching. (b) If T is the random variable, the net cost of this experiment, find the probability distribution of T. (c) Find $E(T)$.

55. Let X_1 and X_2 be a random sample of size 2 with CDF $F(b)$ and density $f(z)$. Let $Y = \max\{X_1, X_2\}$. Show that the CDF of the random variable Y is given by

$$F_Y(d) = [F(d)]^2$$

and the corresponding density by

$$f_Y(z) = 2f(z)F(z).$$

Hint: $F_Y(d) = P\{Y \le d\} = P\{X_1 \le d, X_2 \le d\}$.

56. If X_1, X_2 is a bivariate random variable, show that

$$P\{a_1 < X_1 \le b_1, a_2 < X_2 \le b_2\} = F_{X_1X_2}(b_1, b_2) - F_{X_1X_2}(b_1, a_2)$$
$$- F_{X_1X_2}(a_1, b_2) + F_{X_1X_2}(a_1, a_2).$$

57. If $h(X_1, X_2) = [X_1 - E(X_1)][X_2 - E(X_2)]$, then $E[h(X_1, X_2)] = \sigma_{X_1X_2}$ is called the covariance of the random variable X_1, X_2. Show that if X_1 and X_2 are independent random variables, then $\sigma_{X_1X_2}$ is zero.

58. If $h(X_1, X_2) = (X_1 + X_2 - [E(X_1) + E(X_2)])^2$, show that

$$E[h(X_1, X_2)] = \text{Variance of } (X_1 + X_2)$$
$$= \text{Variance } X_1 + \text{Variance } X_2 + 2 \text{ Covariance } X_1 X_2.$$

59. The correlation coefficient between X_1 and X_2 is defined to be

$$\rho = \frac{E[X_1 - E(X_1)][X_2 - E(X_2)]}{\sqrt{E[X_1 - E(X_1)]^2 E[X_2 - E(X_2)]^2}} = \frac{\sigma_{X_1 X_2}}{\sigma_{X_1} \sigma_{X_2}}$$

Show that if X_1 and X_2 are independent random variables, then $\rho = 0$.

60. Consider the random variable which is the performance of launching a satellite: If there is a successful launching, the outcome is denoted by a 1. If there is an unsuccessful launching, the outcome is denoted by a 0. The probability of a successful launching is given by p. (a) If the variance of this random variable is known to equal $\frac{1}{4}$, what is the value of p? (b) If a sample of two satellites is launched, the number of successes is given by the random variable $X_1 + X_2$ where X_1 is the performance of the first satellite and X_2 the performance of the second. As noted above, X_1 and X_2 each take on the possible values of 1 or 0, depending on whether the launching is successful or not. What is the expected value of the number of successes (leave answer in terms of p)?

61. In Problem 1, another random variable Y is defined as follows:

$$Y(\omega_1) = 1, \qquad Y(\omega_2) = 2, \qquad \text{and} \qquad Y(\omega_3) = 3.$$

(a) Find the joint probability distribution of X, Y. (b) Find the marginal probability distribution of Y. (c) Find $E(Y)$.

62. In Problem 15, a second port is available to handle the overflow, if any. Denote by Z the number of ships seeking berths in the second port (which occurs only if the first port is filled). (a) Find the joint probability distribution of X, Z. (b) Find the marginal probability distribution of Z. (c) Find $E(Z)$.

63. A product is classified according to the number of defects it contains and the factory that produces it. Let X_1 and X_2 be the random variables which represent the number of defects per unit (taking on possible values of 0, 1, 2, or 3) and the factory number (taking on possible values 1 or 2), respectively. The entries in the table represent the joint probability distribution, e.g., $P_{X_1 X_2}(0, 1) = \frac{1}{8}$.

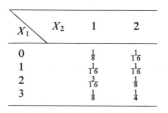

X_1 \\ X_2	1	2
0	$\frac{1}{8}$	$\frac{1}{16}$
1	$\frac{1}{16}$	$\frac{1}{16}$
2	$\frac{3}{16}$	$\frac{1}{8}$
3	$\frac{1}{8}$	$\frac{1}{4}$

(a) Find the marginal probability distributions of X_1 and X_2. (b) Find $E(X_1)$, $E(X_2)$, Variance X_1, and Variance X_2.

64. Two vacuum tubes are tested to failure. Let X_L be the time to failure of the smaller and let X_U be the time to failure for the larger. The joint density function is given by

$$f_{X_L X_U}(l, u) = \begin{cases} \dfrac{2}{(100)^2} e^{-(l/100 + u/100)}, & \text{for } 0 < l < u \\ 0, & \text{otherwise.} \end{cases}$$

Find the marginal densities of X_L and X_U.

65. The dimensions of a rectangular table are recorded. Let X and Y denote the measurement errors whose joint distribution is given by

$$f_{XY}(u, v) = \frac{1}{2\pi} e^{-[u^2 + v^2]/2}, \qquad -\infty \leq u, v \leq \infty.$$

(a) Find the marginal densities of X and Y. (b) Find $E(X)$ and Variance X.

66. A fuel element of diameter D is to be placed into a tube of diameter T. The joint density is given by

$$f_{DT}(u, v) = \begin{cases} 100, & \text{for } 1.95 \leq u \leq 2.05 \\ & \qquad 2.00 \leq v \leq 2.10 \\ 0, & \text{otherwise.} \end{cases}$$

(a) Find the marginal densities of D and T. (b) Find $E(D)$, $E(T)$, Variance D, and Variance T.

67. Suppose that the Rockwell hardness X and abrasion loss Y of a specimen (coded data) have a joint density given by

$$f_{XY}(u, v) = \begin{cases} (u + v), & \text{for } 0 \leq u, v \leq 1 \\ 0, & \text{otherwise.} \end{cases}$$

(a) Find the marginal densities of X and Y. (b) Find $E(X)$ and Variance X.

68. Show that the definition of conditional probability satisfies the four conditions that regular probabilities satisfy as described in Sec. 2.5.

69. Let E_1, E_2, E_3 denote events in the sample space. Show

$$P\{E_1 \cap E_2 \cap E_3\} = P\{E_1 \mid E_2 \cap E_3\} P\{E_2 \mid E_3\} P\{E_3\}.$$

70. Let $E_1, E_2, \ldots, E_m$ be mutually exclusive events in the sample space Ω such that one must occur, i.e., $E_1 \cup E_2 \cup \cdots \cup E_m = \Omega$. Let B be any event in Ω. (Thus, B can occur only together with some event E_i.) Therefore, $B = BE_1 \cup BE_2 \cup \cdots \cup BE_m$. Show that

$$P\{B\} = \sum_{i=1}^{m} P\{B \mid E_i\} P\{E_i\}.$$

71. Let $E_1, E_2, \ldots, E_m$ be mutually exclusive events in the sample space Ω such that one must occur. Let B be any event in Ω. Using the results of Problem 70, prove *Bayes' formula*, i.e.,

$$P\{E_i \mid B\} = \frac{P\{B \mid E_i\} P\{E_i\}}{P\{B \mid E_1\} P\{E_1\} + P\{B \mid E_2\} P\{E_2\} + \cdots + P\{B \mid E_m\} P\{E_m\}}.$$

72. In Problem 61, determine if the events $E_1 = \{\omega : X(\omega) \leq 0\}$ and $E_2 = \{\omega : Y(\omega) \leq 1\}$ are independent events. Are the events $E_3 = \{\omega : X(\omega) \leq 1\}$ and E_2 independent?

73. In Problem 61, are X and Y independent random variables?

74. In Problem 62, are X and Z independent random variables?

75. In Problem 63, are X_1 and X_2 independent random variables?

76. In Problem 64, are X_L and X_U independent random variables?

77. In Problem 65, are X and Y independent random variables?

78. In Problem 66, are D and T independent random variables? Find the expected value and variance of the clearance between the fuel element and the tube.

79. In Problem 67, are X and Y independent random variables?

80. Consider two *independent* random variables X_1 and X_2 each of which can take on only the values $-1, 0, +1$. Suppose that $P\{X_1 = -1\} = P\{X_1 = +1\} = \frac{1}{4}$ and that $P\{X_2 = -1\} = P\{X_2 = 0\} = \frac{1}{3}$. (a) Find the means and variances of X_1 and X_2. (b) If $Y = 4X_1 + 3X_2$, find $E(Y)$. (c) Find the joint probability distribution of X_1, X_2.

81. In Problem 61, find the conditional probability distribution of X, given $Y = 1$.

82. In Problem 62, find the conditional probability distribution of Z, given $X = 4$.

83. In Problem 63, find the conditional probability distribution of X_1, given $X_2 = 1$.

84. In Problem 64, find the conditional density of X_U, given $X_L = 60$.

85. In Problem 65, find the conditional density of X, given $Y = 0.1$.

86. In Problem 66, find the conditional density of D, given $T = 2.00$.

87. In Problem 67, find the conditional density of X, given $Y = \frac{1}{2}$.

3

THE NORMAL DISTRIBUTION

3.1 Definitions

In Chapter 1, histograms were drawn of experimental data. These histograms represented the outcomes of some random variables. One such random variable often encountered in practice is the continuous random variable which has a normal distribution. *A continuous random variable whose density function can be expressed as*

$$f_X(z) = \frac{1}{\sqrt{2\pi}D}\, e^{-(z-C)^2/2D^2}, \qquad -\infty < z < \infty,$$

where $D > 0$ and C and D are parameters (constants) is said to be a normally distributed random variable, or alternatively, is said to have a normal distribution.

The range of this density function is $-\infty$ to $+\infty$. This may appear to restrict the usefulness of this distribution. For example, measurement errors are often assumed to be normally distributed. However, it is evident that such errors are bounded. Measurements themselves are often assumed to be normally distributed yet are non-negative by the nature of the physical situation. In order to reconcile these apparent contradictions, it must be pointed out that the assumption of a random variable having a normal distribution is simply an assumption regarding the form of a mathematical model which, at best, is just an approximation to a real situation. In the examples above it is tacitly assumed that the probability of getting large measurement errors is very small and the probability of getting negative measurements is also very small. Although the physical phenomena require these probabilities to be zero, using the mathematical model approximates the real situation by assigning small probability to these events. This approximation is similar to the approximation made in assuming that random variables are continuous when measuring devices are all discrete.

The probability that a normally distributed random variable X is less than or equal to b is given by

$$F_X(b) = P\{X \le b\} = \int_{-\infty}^{b} \frac{1}{\sqrt{2\pi}D} e^{-(z-C)^2/2D^2} \, dz.$$

This density function cannot be integrated directly. However, the probability that X is less than or equal to b can be represented by the shaded area in Fig. 3.1, and its magnitude is determined with the aid of tables as shown in Sec. 3.3. A diagram of the cumulative distribution function is given in Fig. 3.2. Note that the normal density function is dependent only on the two parameters C and D. If C and D are both specified, the aforementioned probabilities can be obtained (using the method discussed in Sec. 3.3).

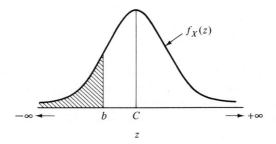

Fig. 3.1. *The density function of the normal distribution.*

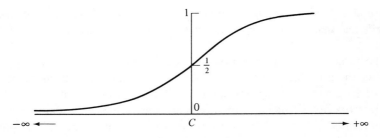

Fig. 3.2. *The CDF of the normal distribution.*

3.2 The Mean and Variance of the Normal Distribution

The constants C and D are related to the moments of the normal distribution. The expected value of X is equal to C, i.e.,

$$\mu = E(X) = \int_{-\infty}^{\infty} z \frac{1}{\sqrt{2\pi}D} e^{-(z-C)^2/2D^2} \, dz = C$$

and the variance of X is equal to D^2, i.e.,

$$\sigma^2 = E(X - \mu)^2 = \int_{-\infty}^{\infty} (z - \mu)^2 \frac{1}{\sqrt{2\pi}D} e^{-(z-\mu)^2/2D^2} \, dz = D^2.[1]$$

---

[1] The proofs of these statements appear in Sec. 3.2.1.

Thus, the two parameters which completely specify the normal distribution are its mean and its variance.

The density function of a normally distributed random variable can then be written as

$$f_X(z) = \frac{1}{\sqrt{2\pi}\sigma} e^{-(z-\mu)^2/2\sigma^2},$$

where μ and σ^2 are the mean and the variance, respectively. If

$$f_X(z) = \frac{1}{\sqrt{2\pi}} e^{-z^2/2},$$

the random variable X has a normal distribution with mean equal to 0 and variance equal to 1. Similarly, if

$$f_X(z) = \frac{1}{\sqrt{2\pi}(14.3)} e^{-(z-3.6)^2/2(14.3)^2},$$

X has a normal distribution with mean 3.6 and standard deviation 14.3. In other words, any random variable X whose density function has the form

$$f_X(z) = \frac{1}{\sqrt{2\pi}\sigma} e^{-(z-\mu)^2/2\sigma^2}$$

is normally distributed with a mean equal to the value of the constant subtracted from z in the exponent of e, i.e., μ, and variance equal to the value of the constant σ^2 in the denominator of the exponent of e.

Mean values can be represented on the graphs of the density functions as shown in Fig. 3.3. It is evident that the normal density function is symmetric about its mean.

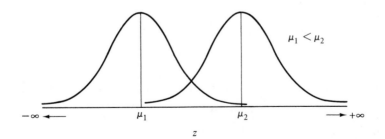

Fig. 3.3. *Comparison of two normal distributions with different means.*

The variance cannot be represented as easily, but it does have some intuitive meaning. It is a measure of the variability or dispersion of the random variable. The larger the variance, the larger is the variability. This can be shown graphically by Fig. 3.4. The standard deviation, σ, can be used to locate the points of inflection of the density function. The points of inflection appear at $\mu - \sigma$ and $\mu + \sigma$.

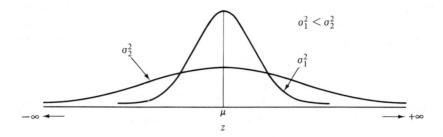

Fig. 3.4. *Comparison of two normal distributions with different variances.*

*3.2.1 EVALUATION OF THE MEAN AND VARIANCE OF THE NORMAL DISTRIBUTION

This section will contain the proofs of the results that $E(X) = C$ and $E(X - \mu)^2 = D^2$. It will be shown first that

$$\mu = E(X) = C.$$

From the basic definition of the mean,

$$\mu = E(X) = \int_{-\infty}^{\infty} z \frac{1}{\sqrt{2\pi}D} e^{-(z-C)^2/2D^2} \, dz.$$

Let $(z - C)/D = v$ so that $dz = D \, dv$. Then

$$\mu = \int_{-\infty}^{\infty} \left(\frac{vD + C}{\sqrt{2\pi}} \right) e^{-v^2/2} \, dv = \frac{D}{\sqrt{2\pi}} \int_{-\infty}^{\infty} v e^{-v^2/2} \, dv + \frac{C}{\sqrt{2\pi}} \int_{-\infty}^{\infty} e^{-v^2/2} \, dv.$$

But

$$\frac{D}{\sqrt{2\pi}} \int_{-\infty}^{\infty} v e^{-v^2/2} \, dv = \frac{-D}{\sqrt{2\pi}} e^{-v^2/2} \Big|_{-\infty}^{\infty} = 0,$$

and it is well known that

$$\int_{-\infty}^{\infty} \frac{1}{\sqrt{2\pi}} e^{-v^2/2} \, dv = 1.$$

Hence, $$\mu = C. \qquad \text{Q.E.D.}$$

The second result concerns the variance, i.e., it will be shown that

$$\sigma^2 = E(X - \mu)^2 = D^2.$$

From the basic definition of the variance

$$\sigma^2 = E(X - \mu)^2 = \int_{-\infty}^{\infty} \frac{(z - \mu)^2}{\sqrt{2\pi}D} e^{-(z-\mu)^2/2D^2} \, dz.$$

Let $(z - \mu)/D = v$ so that $dz = D \, dv$. Then

$$\sigma^2 = \int_{-\infty}^{\infty} \frac{D^2 v^2}{\sqrt{2\pi}} e^{-v^2/2} \, dv = D^2 \int_{-\infty}^{\infty} \frac{1}{\sqrt{2\pi}} v^2 e^{-v^2/2} \, dv.$$

Integrating by parts, it can be shown that

$$\int_{-\infty}^{\infty} \frac{1}{\sqrt{2\pi}} v^2 e^{-v^2/2} \, dv = 1.$$

Hence, $\qquad\qquad\qquad\qquad \sigma^2 = D^2. \qquad\qquad\qquad\qquad$ Q.E.D.

3.3 Tables of the Normal Integral

Given values of the mean and the variance of a random variable X which has a normal distribution, it is only a matter of calculus to find such probabilities as

$$P\{X > b\} = \int_{b}^{\infty} \frac{1}{\sqrt{2\pi}\,\sigma} e^{-(z-\mu)^2/2\sigma^2} \, dz$$

or

$$P\{X \leq b\} = 1 - P\{X > b\} = \int_{-\infty}^{b} \frac{1}{\sqrt{2\pi}\,\sigma} e^{-(z-\mu)^2/2\sigma^2} \, dz.$$

However, this is not an elementary function, and hence its integral cannot be written down in simple form. If it were tabulated in the form above, a separate table would be required for each pair μ and σ chosen. This is evidently an impossible task, but fortunately a simple transformation can be made to reduce the problem to the computation of a single table. Let

$$v = \frac{z - \mu}{\sigma}; \qquad dz = \sigma \, dv.$$

The probability that X is greater than b can then be written as

$$P\{X > b\} = \int_{(b-\mu)/\sigma}^{\infty} \frac{1}{\sqrt{2\pi}\,\sigma} e^{-v^2/2}(\sigma \, dv)$$

$$= \int_{(b-\mu)/\sigma}^{\infty} \frac{1}{\sqrt{2\pi}} e^{-v^2/2} \, dv = \int_{(b-\mu)/\sigma}^{\infty} f(v) \, dv.$$

Note that $f(v)$ has the form of a normal density function with mean 0 and variance 1. *A random variable N having such a density function is known as a standardized normal random variable.* The probability that X is greater than b can always be written in the form of an integral of the normal density with mean 0 and variance 1, where the lower limit of integration depends on b, μ, and σ. Therefore, a tabulation of the normal integral for a normally distributed random variable having mean 0 and variance 1 is sufficient. Appendix Table 1 is such a table. It presents the area under the density function of a standardized normal random variable N from K_α to plus infinity, i.e., the shaded area in Fig. 3.5. K_α is known as the upper α percentage point or the normal deviate corresponding to α and is defined by

$$P\{N > K_\alpha\} = \int_{K_\alpha}^{\infty} \frac{1}{\sqrt{2\pi}} e^{-v^2/2} \, dv = \alpha = \text{shaded area.}$$

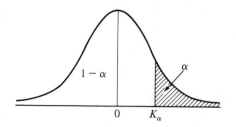

Fig. 3.5. *The standardized normal distribution.*

Thus,
$$P\{X > b\} = \int_b^\infty \frac{1}{\sqrt{2\pi}\sigma} e^{-(z-\mu)^2/2\sigma^2} \, dz$$

$$= P\{N > K_\alpha\} = \int_{K_\alpha}^\infty \frac{1}{\sqrt{2\pi}} e^{-v^2/2} \, dv = \alpha$$

where $K_\alpha = (b - \mu)/\sigma$. This probability can be represented by the two equal shaded areas in Fig. 3.6. In other words, in order to find this probability, subtract the mean from b and divide the result by the standard deviation, i.e., compute $K_\alpha = (b - \mu)/\sigma$. Enter Appendix Table 1 with this value and read out the probability. For example, let $\mu = 2$, $\sigma^2 = 9$, and $b = 8$.

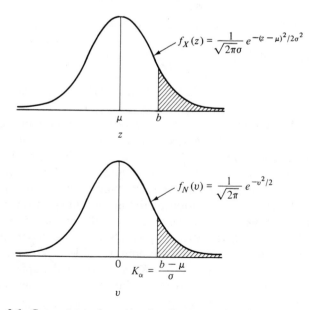

Fig. 3.6. *Comparison of areas under the regular normal density and the standardized normal density.*

The probability that X is greater than 8 is given by

$$P\{X > 8\} = \int_8^\infty \frac{1}{\sqrt{2\pi(3)}} e^{-(z-2)^2/2(9)} \, dz$$

$$= \int_{(8-2)/3=2}^\infty \frac{1}{\sqrt{2\pi}} e^{-v^2/2} \, dv = 0.0228$$

This result is obtained by entering Appendix Table 1 with $K_\alpha = 2$. Summarizing then, the probability that X is greater than 8 when the mean of X is 2 and the variance of X is 9 equals 0.0228 and is equivalent to the probability that N is greater than 2, where N is the standardized normal random variable. The value 0.0228 is the value of α corresponding to $K_\alpha = 2$.

A probability of the form $P\{X \leq b\}$ can be obtained by using the relation

$$P\{X \leq b\} = 1 - P\{X > b\}.$$

Note that Appendix Table 1 does not contain negative values of K_α. Such values of K_α can be obtained by recognizing that the standardized normal distribution is symmetric about 0. It is easily seen that $P\{N < -K_\alpha\} = P\{N > K_\alpha\}$. This result can be verified by referring to Fig. 3.7. Thus,

$$P\{N < -2\} = P\{N > 2\} = 0.0228.$$

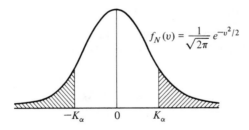

Fig. 3.7. *The standardized normal distribution.*

A probability of the form $P\{N \geq -K_\alpha\}$ is also easy to obtain.

$$P\{N \geq -K_\alpha\} = 1 - P\{N < -K_\alpha\} = 1 - P\{N > K_\alpha\}.$$

From this result

$$P\{N \geq -2\} = (1 - 0.0228)$$
$$= 0.9772.$$

A probability of the form $P\{a < X \leq b\}$ can also be obtained by a simple extension of the ideas above. This probability is represented by the shaded area in Fig. 3.8 and it can be evaluated by referring to the probabilities $P\{X > a\}$ and $P\{X > b\}$ shown in Fig. 3.9. A probability of the form

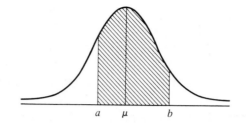

Fig. 3.8. *Probability of the random variable falling in the finite interval* [a, b].

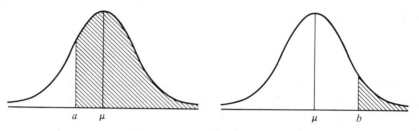

Fig. 3.9. (a) *Probability that* $X >$ a. (b) *Probability that* $X >$ b.

$P\{a < X \leq b\}$ can be expressed as

$$P\{a < X \leq b\} = P\{X > a\} - P\{X > b\}.$$

Both of the probabilities on the right-hand side of the equality can be found by entering Appendix Table 1. For example, let $\mu = 2$, $\sigma^2 = 9$, $b = 8$, and $a = 5$.

$$P\{5 < X \leq 8\} = \int_5^8 \frac{1}{\sqrt{2\pi}(3)} e^{-(z-2)^2/2(9)} \, dz = P\{X > 5\} - P\{X > 8\}$$

$$= \int_{(5-2)/3=1}^{\infty} \frac{1}{\sqrt{2\pi}} e^{-v^2/2} \, dv - \int_{(8-2)/3=2}^{\infty} \frac{1}{\sqrt{2\pi}} e^{-v^2/2} \, dv$$

$$= 0.1587 - 0.0228 = 0.1359.$$

3.4 Combinations of Normally Distributed Random Variables

In the previous chapter, some basic concepts about functions of several random variables were presented. In particular, the expectation of a function, $h(X_1, X_2)$, of two random variables was defined. Furthermore, the cumulative distribution function of such a function was alluded to. The distribution of functions of several random variables is essential to the study of statistics. For example, decisions are often based on the outcome of the random variable, the sample mean. In fact, the distribution of linear combinations of random variables, of which the sample mean is a special case, plays an important role. It is the purpose of this section to present the distribution of linear combinations of normally distributed random variables. These results on linear combinations of normal (LCN) random variables will be referred to as LCN in subsequent discussions.

LCN: An experiment is performed. Let $X_1, X_2, \ldots, X_n$ be independent random variables having means $\mu_1, \mu_2, \ldots, \mu_t$ and variances $\sigma_1^2, \sigma_2^2, \ldots, \sigma_n^2$, respectively.[1] Let $a_1, a_2, \ldots, a_n$ be constants. Denote by Y the random variable which is the linear combination of the X's, i.e.,

$$Y = a_1 X_1 + a_2 X_2 + \cdots + a_n X_n.$$

Then Y possesses these three properties:

1.
$$\begin{aligned} E(Y) = \mu_Y &= E(a_1 X_1 + a_2 X_2 + \cdots + a_n X_n) \\ &= a_1 \mu_1 + a_2 \mu_2 + \cdots + a_n \mu_n. \end{aligned}$$

This statement says that the expected value of a linear combination of random variables is equal to the linear combination of the expected values. Furthermore this result is valid even if the X's are dependent random variables. The case for $n = 2$ was obtained in Sec. 2.12. The result for general n is a simple extension and is obtained by letting $h(X_1, X_2, \ldots, X_n) = a_1 X_1 + a_2 X_2 + \cdots + a_n X_n$ and then finding its expectation.

2.
$$\begin{aligned} E(Y - \mu_Y)^2 = \sigma_Y^2 &= \text{Variance } (a_1 X_1 + a_2 X_2 + \cdots + a_n X_n) \\ &= a_1^2 \sigma_1^2 + a_2^2 \sigma_2^2 + \cdots + a_n^2 \sigma_n^2. \end{aligned}$$

This statement says that the variance of a linear combination of independent random variables is equal to the sum of the products of variances and squared constants. The case for $n = 2$ was discussed in Sec. 2.14. Again, the result for general n is a simple extension.

3. If $X_1, X_2, \ldots, X_n$ are *normally distributed* random variables, then Y is also normally distributed with mean μ_Y and variance σ_Y^2.[2] In principal, this result can be obtained by the method presented in Sec. 2.12, i.e., the CDF of the random variable Y is given by

$$\begin{aligned} F_Y(b) &= \iint_{R_b} \cdots \int f_{X_1 X_2 \ldots X_n}(z_1, z_2, \ldots, z_n) \, dz_1 \, dz_2 \ldots dz_n \\ &= \iint_{R_b} \cdots \int \frac{1}{(2\pi)^{n/2} \sigma_1 \sigma_2 \ldots \sigma_n} e^{-(1/2)\{[(z_1 - \mu_1)/\sigma_1]^2 + [(z_2 - \mu_2)/\sigma_2]^2 + \ldots + [(z_n - \mu_n)/\sigma_n]^2\}} \end{aligned}$$

$$dz_1 \, dz_2 \ldots dz_n$$

where the region of integration R_b is the region in the $z_1, z_2, \ldots, z_n$ space

---

[1] A constant can also be considered as a degenerate random variable having mean equal to the constant and variance equal to zero.

[2] A constant can also be considered as a degenerate normally distributed random variable with mean equal to the constant and variance equal to zero.

such that $a_1z_1 + a_2z_2 + \cdots + a_nz_n \leq b$. However, this method is not simple so that this result is usually proven by techniques which are not included in this text.

The ensuing sections will deal with examples of applications of LCN.

3.5 The Standardized Normal Random Variable

In Sec. 3.3, the standardized normal random variable, i.e., that random variable having mean 0 and variance 1, was obtained by making a transformation of the normal random variable X having mean μ and variance σ^2. As the first example, this result will be obtained again. Consider the random variable

$$Y = \frac{X - \mu}{\sigma} = \left(\frac{1}{\sigma}\right)X - (1)\frac{\mu}{\sigma}$$

and let $a_1 = 1/\sigma$ and $a_2 = -1$. (μ/σ will be considered as a degenerate random variable with mean μ/σ and variance 0.) From the results on linear combinations of random variables

$$E(Y) = \left(\frac{1}{\sigma}\right)\mu - (1)\frac{\mu}{\sigma} = 0;$$

$$\sigma_Y^2 = \left(\frac{1}{\sigma^2}\right)\sigma^2 - 0 = 1.$$

Furthermore, since X and μ/σ are normally distributed (μ/σ being considered a normally distributed random variable with mean equal to μ/σ and variance 0), Y is also normally distributed with mean 0 and variance 1. Thus, if the difference, obtained by subtracting the mean of a normally distributed random variable from the random variable itself, is divided by the standard deviation of the variable [i.e., $Y = (X - \mu)/\sigma$], the resulting random variable is normally distributed with zero mean and unit standard deviation.

3.6 The Distribution of the Sample Mean

Another example of an application of the results of Sec. 3.4 is for the case of the sample mean $\bar{X}$. Let

$$Y = \bar{X} = \frac{X_1 + X_2 + \cdots + X_n}{n} = \frac{1}{n}X_1 + \frac{1}{n}X_2 + \cdots + \frac{1}{n}X_n$$

where $X_1, X_2, \ldots, X_n$ are a random sample of n observations (X's are independent and identically distributed). Call the common mean μ and the common variance σ^2. The a's are all $1/n$ so that

$$E(Y) = \left(\frac{1}{n}\right)\mu + \left(\frac{1}{n}\right)\mu + \cdots + \left(\frac{1}{n}\right)\mu = \mu;$$

$$\sigma_Y^2 = \left(\frac{1}{n}\right)^2\sigma^2 + \left(\frac{1}{n}\right)^2\sigma^2 + \cdots + \left(\frac{1}{n}\right)^2\sigma^2 = \frac{\sigma^2}{n}.$$

Furthermore, if $X_1, X_2, \ldots, X_n$ are normally distributed with mean μ and variance σ^2, $\bar{X}$ is normally distributed with mean μ and variance σ^2/n.

In earlier sections, it was shown that by subtracting the mean from a normally distributed random variable and then dividing by its standard deviation, the resulting random variable is normally distributed with mean 0 and variance 1. Therefore, $(\bar{X} - \mu)/(\sigma/\sqrt{n}) = (\bar{X} - \mu)\sqrt{n}/\sigma$ is normally distributed with mean 0 and variance 1, provided the X's are all independently normally distributed with common mean and variance.

As pointed out in Chapter 2, the sample mean is a random variable. It has now been shown that the expected value of this random variable is the same as the expected value of the individual random variables.[1] The variance, however, is reduced to σ^2/n. Furthermore, if the original n random variables are normally distributed, the sample mean has a normal distribution. A comparison of the distribution of the sample mean $\bar{X}$ and one of the original normally distributed random variables X is given in Fig. 3.10.

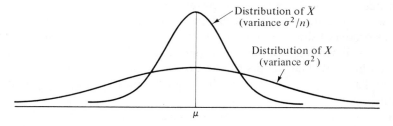

Fig. 3.10. *Comparison of the probability distribution of X and the probability distribution of $\bar{X}$.*

There is a rather important concept involved in discussing the distribution of the sample mean. Whereas the outcome of n random variables (the individual observations) are contained in the computed sample mean, this value may be considered as the outcome of a single random variable, the sample mean. For example, two experiments are to be performed. Let $X_1, X_2, \ldots,$ X_5 be 5 independent normally distributed random variables, each with mean 0 and variance 1. The distribution of the sample mean is then also normal with mean 0 but with variance $\frac{1}{5}$. Suppose the outcome of the X's is observed and the value taken on by the random variable $\bar{X}$ is to be computed. The second experiment involves a random variable Y which is normally distributed with mean 0 and variance $\frac{1}{5}$. A *single* observation on this random variable is to be taken. An announcement is made that the outcome of one of the experiments is 0.25, but no mention is made of which experiment led to this result. The problem is to decide whether the experiment resulting in the calculation of the value the random variable $\bar{X}$ takes on, or

---

[1] This result for $n = 2$ was also obtained in Sec. 2.12.

the experiment involving Y, belongs to the 0.25. The answer is, of course, that the experiments were equivalent and that it is impossible to differentiate between the two. The distribution of $\bar{X}$ is identical with the distribution of Y. Thus, an observation on the sample mean may be considered as the outcome of an experiment (a sample of size one) where the random variable has the distribution of the sample mean. Using this probability distribution, probabilities involving the sample mean may be computed.

3.7 Tolerances

A third example using the results on linear combinations of random variables deals with tolerances. Design specifications on a dimension are usually given as a nominal value plus or minus some increment, e.g., 1.530 $\pm$ 0.003. If asked exactly what these limits mean, the designer will usually answer that *all* parts produced should fall within these limits. If he is pressed further, the designer will admit that he really doesn't expect all, but *almost all*, the items to fall within these dimensions. If this statement is questioned, the answer will finally result in stating that no more than a given fraction of the items produced should fall outside these limits. In other words, he recognizes that this dimension is a random variable. Hopefully, he has set the specification limits so that the distribution of this random variable allows for no more than the given fraction to fall outside the lower and upper specification limits.

It is well to distinguish between specification limits and natural tolerance limits. The specification limits are limits that are set somewhat arbitrarily, say, by the designer, without regard to what the process can actually achieve. The natural tolerance limits are the actual capabilities of the process, and can be defined *as the limits within which all but a given allowable fraction α of the items produced will fall.* The magnitude of α is usually chosen by mutual agreement. If it can be assumed that the dimension is a normally distributed random variable, a good design will usually have a mean value coincident with the nominal value and a standard deviation which will permit only the small allowable fraction α of the items produced to fall outside of the specification limits. In other words, the natural tolerances will coincide with the design specifications. In any event, in order for a process to be acceptable, the natural tolerances must fall *within* the specification limits, which insures that the specification limits will include *at least* the fraction $1 - \alpha$ of the items produced. For example, if the allowable fraction of items falling outside the natural tolerance limits is 27 in 10,000, and the specifications are 1.530 $\pm$ 0.003, the process whose dimension is normally distributed with mean 1.530 and standard deviation 0.001 will have natural tolerances which will coincide with the specification limits. This is evident from Fig. 3.11. If the standard deviation exceeds 0.001,

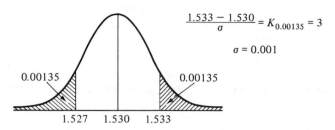

$$\frac{1.533 - 1.530}{\sigma} = K_{0.00135} = 3$$

$$\sigma = 0.001$$

0.00135   0.00135

1.527   1.530   1.533

Fig. 3.11. *The distribution of a dimension whose natural tolerances coincide with the specifications.*

the natural tolerance limits will fall outside the specification limits, thereby resulting in an unacceptable process. If the standard deviation is less than 0.001, the natural tolerance limits will fall within the specification limits, thereby resulting in an acceptable, but possibly costly, process.

In other words, given symmetrical specifications and an allowable fraction of defectives α, the maximum allowable standard deviation (if the dimension of the product is centered at the nominal dimension) enabling the natural tolerances to fall within the specification limits is obtained by finding the normal deviate (percentage point) $K_{\alpha/2}$ corresponding to the fraction above the upper limit, and dividing this number into the difference between the upper specification limit U and the nominal dimension, i.e., $(U - \mu)/K_{\alpha/2}$.

Given the actual standard deviation σ, it is a simple problem to solve the inverse problem, i.e., to find out what fraction of the items will fall outside the specification limits.

It is often the case that a dimension of an assembled product is the sum of the dimensions of several parts. An electrical resistance may be the sum of several electrical resistances. A weight may be the sum of a number of individual weights. The distribution of the individual components may be known, and what is of interest is the distribution of the sum $Y = X_1 + X_2 + \cdots + X_n$, where $X_1, X_2, \ldots, X_n$ are independent random variables (measurable characteristics of the components making up the finished product) having means $\mu_1, \mu_2, \ldots, \mu_n$ and variances $\sigma_1^2, \sigma_2^2, \ldots, \sigma_n^2$, respectively. Referring to the results about linear combinations of independent random variables (LCN of Sec. 3.4), it is evident that all the a's are equal to 1, so that the expected value of Y is given by

$$\mu_Y = E(Y) = \mu_1 + \mu_2 + \cdots + \mu_n$$

and the variance of Y is given by

$$\sigma_Y^2 = \sigma_1^2 + \sigma_2^2 + \cdots + \sigma_n^2.$$

Furthermore, if all the X's are normally distributed random variables, Y is also normally distributed. Hence, if all the μ's and σ^2's are known, the fraction falling outside the specification limits can be obtained.

Another problem of interest is the assigning of specification limits to individual components, given the specification limits of final assembly. As an example, consider the problem of assigning individual specification limits to a component shown in Fig. 3.12 such that the over-all dimension is 2.020 $\pm$ 0.030 inches. It will be assumed that the five components are independent normal random variables having means as shown in Fig. 3.12 and *common*

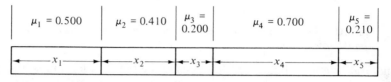

Fig. 3.12. *Assembly consisting of five components.*

variance σ^2. It will also be assumed that the natural tolerance limits are defined such that $\alpha = 0.0027$ for the final assembly as well as for the individual components, i.e., the allowable fraction falling outside the natural tolerance limits of the final assembly, as well as the allowable fraction falling outside the natural tolerance limits of the individual components, is 0.0027.[1] The individual specification limits are to be set so that if the natural tolerance limits of the individual components fall within these limits, the natural tolerance limits of the final assembly will fall within the specification limits of the final assembly. It then follows that the fraction falling outside the specification limits of the final assembly will not exceed 0.0027. Let $Y = X_1 + X_2 + X_3 + X_4 + X_5$. The expected value of Y is given by

$$E(Y) = 0.500 + 0.410 + 0.200 + 0.700 + 0.210 = 2.020$$

and the variance of Y is given by

$$\sigma_Y^2 = 5\sigma^2.$$

The most conservative acceptable situation results when the natural tolerance limits of the final assembly coincide with the given specification limits for the final assembly. If this occurs, the standard deviation of Y will have its maximum permissible value since Y is normally distributed and $E(Y)$ is equal to the nominal value of the final assembly. Denote this maximum permissible value of the standard deviation by σ_{YM}. Thus, if $\sigma_Y > \sigma_{YM}$, the natural tolerance limits of the final assembly will fall outside the specification limits of the final assembly; if $\sigma_Y < \sigma_{YM}$, the natural tolerance limits will fall within the specification limits.

Figure 3.13 illustrates this case of natural tolerance limits coinciding with specification limits so that the normal deviate (percentage point) corre-

---

[1] The implications of this assumption will be discussed later.

sponding to an area of 0.00135 is given by $K_{0.00135} = 3$ and must be equal to $(2.050 - 2.020)/\sigma_{YM}$. Solving the equation $3 = (2.050 - 2.020)/\sigma_{YM}$ for σ_{YM}, it is evident that σ_{YM} equals 0.01. Hence, $\sigma_{YM}^2 = 0.0001$ is the maximum permissible variance for the final assembly which will allow the natural tolerances to fall within the design specifications.

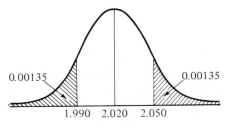

Fig. 3.13. *The distribution of Y.*

Thus, if the variance of Y is less than or equal to 0.0001, the natural tolerance of the over-all process will fall within the given specification limits.

Using the relation $\sigma_Y^2 = 5\sigma^2$, denote by σ_M^2 the value of σ^2 corresponding to $\sigma_Y^2 = \sigma_{YM}^2$. It then follows that $\sigma_M^2 = 0.0001/5 = 0.00002$ is the maximum permissible variance for the individual components enabling the natural tolerance limits to fall within the specification limits. The corresponding standard deviation is given by $\sigma_M = 0.0045$. Therefore, if the standard deviation of the individual components is less than or equal to 0.0045, the natural tolerances of the final assembly will fall within the given specification limits. Since natural tolerances for the individual components correspond to an α of 0.0027, the limits are given by $\mu \pm 3\sigma$. Thus, the natural tolerance limits of the individual components, when their standard deviations are at their maximum permissible values σ_M, are equal to $\mu \pm 3(0.0045)$ and will contain all but 0.0027 of the items produced. The desired specification limits for the individual components can now be produced. These limits should contain their natural tolerances, and furthermore, when these component natural tolerances fall within their specification limits, they must insure the inclusion of the natural tolerances of the final assembly within the over-all specification limits. Such specification limits are:

Component 1	0.500 ± 0.0135
Component 2	0.410 ± 0.0135
Component 3	0.200 ± 0.0135
Component 4	0.700 ± 0.0135
Component 5	0.210 ± 0.0135

The choice of the value α for the allowable fraction of defectives defining the natural tolerance limits is irrelevant to the solution of the problem, provided that the *same* value pertains to the individual components as well as the over-all unit. This is most easily seen by choosing a fraction $\alpha > 0$ which pertains to both sets of limits. The natural tolerances of the final assembly should be such that no more than α of the values will fall outside of the specification limits of 2.020 ± 0.030. Similarly, the natural tolerances of the individual components should be such that no more than α of the values will fall outside the determined specification limits.

As before, let

$$Y = X_1 + X_2 + X_3 + X_4 + X_5$$

so that the expected value of Y is given by

$$E(Y) = 0.500 + 0.410 + 0.200 + 0.700 + 0.210 = 2.020$$

and the variance of Y is given by

$$\sigma_Y^2 = 5\sigma^2.$$

As noted previously, σ_{YM} represents the maximum permissible value of the standard deviation, and corresponds to the standard deviation of the final assembly when the tolerance limits of the final assembly coincide with the specification limits. Figure 3.14 illustrates this case of the natural tolerance

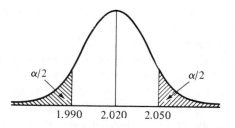

Fig. 3.14. *The distribution of the final assembly.*

limits coinciding with the specification limits so that the normal deviate (percentage point) corresponding to an area of $\alpha/2$ is given by $K_{\alpha/2}$ and must equal $K_{\alpha/2} = (2.050 - 2.020)/\sigma_{YM}$. Solving this equation for σ_{YM}, it is evident that $\sigma_{YM} = 0.030/K_{\alpha/2}$. Hence, $\sigma_{YM}^2 = (0.030/K_{\alpha/2})^2$ is the maximum permissible variance for the final assembly allowing the natural tolerances to fall within the specification limits. Using the relation $\sigma_Y^2 = 5\sigma^2$, denote by σ_M^2 the value of σ^2 corresponding to $\sigma_Y^2 = \sigma_{YM}^2$. It follows that $\sigma_M^2 = (0.030)^2/5K_{\alpha/2}^2$ is the maximum permissible variance for the individual components enabling the natural tolerances of the final assembly to fall within the given specification limits. The corresponding standard deviation is given by $\sigma_M = 0.030/\sqrt{5}\,K_{\alpha/2}$.

The natural tolerances of the individual components, when their standard deviations are at their maximum permissible values σ_M, are then given by $(\mu \pm K_{\alpha/2}\sigma) = (\mu \pm 0.030/\sqrt{5})$, and they will contain all but α of the items produced. The desired specification limits for the individual components can now be produced. These limits should contain their natural tolerances, and furthermore, when these component natural tolerances fall within their specification limits, they must insure the inclusion of the natural tolerances of the final assembly within the over-all specification limits. Such specification limits are:

Component 1	0.500 ± 0.0135
Component 2	0.410 ± 0.0135
Component 3	0.200 ± 0.0135
Component 4	0.700 ± 0.0135
Component 5	0.210 ± 0.0135

Note that the results are the same as those obtained previously. This observation has an interesting implication about tolerances. Consider the range of the natural tolerances of the final assembly $(T_{max} - T_{min})_{sum}$. Then let $(T_{max} - T_{min})_i$ represent the range of the natural tolerance of the ith component. If the same but unknown value of α pertains to the over-all assembly, as well as to the individual components, a relationship between the $(T_{max} - T_{min})_{sum}$ and the $(T_{max} - T_{min})_i$ exists. This is found by recognizing that

$$(T_{max} - T_{min})_{sum} = 2K_{\alpha/2}\sigma_{sum}$$

and

$$(T_{max} - T_{min})_i = 2K_{\alpha/2}\sigma_i$$

for all i.

Using the equality

$$\sigma_{sum} = \sqrt{\sigma_1^2 + \sigma_2^2 + \cdots + \sigma_n^2}$$

and then expressing the standard deviations in terms of the appropriate $(T_{max} - T_{min})$, the equation

$$(T_{max} - T_{min})_{sum}$$
$$= \sqrt{(T_{max} - T_{min})_1^2 + (T_{max} - T_{min})_2^2 + \cdots + T_{max} - T_{min})_n^2}$$

is obtained. Applying this result in the previous example for the situation where the natural tolerance limits of the final assembly coincide with the given specification limits leads to

$$0.060 = \sqrt{5(T_{max} - T_{min})^2}$$

so that $(T_{max} - T_{min})$ for each component is 0.0270. The value 0.0270 corresponds to the range of the tolerance limits for the case where the natural tolerance limits of the individual components coincide with their specification limits. Thus, the range of the specification limits for the individual components must equal 0.0270, which is the result obtained previously by the other methods.

Another application of the results on linear combinations of independent random variables is to the problem of the clearance between components. Consider the following example: The mean external diameter of a shaft is $\mu_S = 1.048$ inches and the standard deviation is $\sigma_S = 0.0013$ inch. The mean inside diameter of the mating bearing is $\mu_B = 1.059$ inches and the standard deviation is $\sigma_B = 0.0017$ inch. Assuming that both the external shaft diameters and the internal bearing diameters are normally distributed and independent, what is the minimum clearance that can be expected to occur by random matching of shafts and bearings in assembly? The minimum clearance will be defined as that value L such that P{clearance $< L$} $= 0.00135.$[1] In other words, L is defined as the natural tolerance, and is compared with the design specification. Consider the diagram in Figure 3.15.

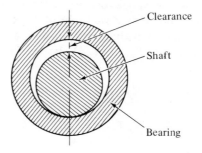

Fig. 3.15. *Shaft and mating bearing.*

It is evident that what is of interest is the difference between the two random variables, the diameter of the bearing, B, and the diameter of the shaft, S. This is the clearance. Let the clearance be denoted by Y so that $Y = (B - S)$. Using the results on linear combinations of random variables with $a_1 = 1$ and $a_2 = -1$, the expected value of the clearance is given by

$$E(Y) = \mu_B - \mu_S = 0.011$$

and the variance of the clearance is given by

$$\sigma_Y^2 = (1)^2\sigma_B^2 + (-1)^2\sigma_S^2 = \sigma_B^2 + \sigma_S^2 = (0.0017)^2 + (0.0013)^2 = 0.00000458.$$

Furthermore, the clearance is normally distributed.

Thus, the distribution of clearance is as shown in Fig. 3.16. Consequently,

---

[1] The choice of the value 0.00135 was arbitrary. The same method of analysis would be used if any other probability were assigned with the final result depending on the particular choice.

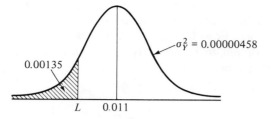

Fig. 3.16. *Distribution of clearance.*

from Appendix Table 1, the normal deviate (percentage point) corresponding to 0.99865 (the area above L) is found to be $K_{0.99865} = -3$ and must be equal to

$$-3 = K_{0.99865} = \frac{L - \mu}{\sigma_Y} = \frac{L - 0.011}{0.00214}.$$

Solving this equation for L, it is evident that $L = 0.011 - 0.00642 = 0.00458$.

Interference occurs when the shaft diameter exceeds the bearing diameter. Using the same distributions for the shaft and bearing, the probability of interference is given by

$$P\{Y < 0\}$$

where Y is a normally distributed random variable (the clearance) with mean equal to 0.011 and standard deviation 0.00214.

$$P\{Y < 0\} = P\left\{N < \frac{0 - 0.011}{0.00214} = -5.1\right\}$$

where N is normally distributed with mean 0 and standard deviation 1, i.e., it is the standardized normal random variable.

Since -5.1 is a negative value, the identity

$$P\{N < -5\} = P\{N > 5\}$$

is used. Referring to Appendix Table 1, the value 5 exceeds the range of the table so that

$$P\{N > 5\} = 0$$

for all practical purposes. Therefore, the probability of interference is negligible.

*3.8 Tolerances in "Complex Items"

In the previous section, tolerances for components which were the result of direct addition or subtraction were considered. However, similar problems exist for more complex assemblies, where final assembly is a nonlinear function of the components. This situation is handled by approximating this function by a linear function in the region of interest. This is usually accomplished by expanding the function into a Taylor series around the nominal dimensions. For example, in the electrical circuit of Fig. 3.17,[1] the voltage E_0 is required to be 60 volts $\pm 3\%$ with components having the following specifications:

$$E_1 = 40 \pm 0.5 \text{ volts};$$

$$N = 2 \text{ to } 1 \pm 1\%;$$

$$K = 3 \pm 2\%.$$

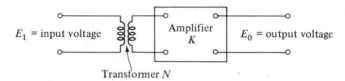

Fig. 3.17. *The output voltage* E_0 *of this circuit is a function of the product of the components.*

The output voltage is given by $E_0 = E_1 NK$. If E_0 is expanded into a Taylor series around the nominal dimensions E_1^*, N^*, and K^*, E_0 can be expressed as

$$E_0 = E_1^* N^* K^* + (E_1 - E_1^*)\frac{\partial E_0}{\partial E_1}\bigg|_{E_1^*, N^*, K^*} + (N - N^*)\frac{\partial E_0}{\partial N}\bigg|_{E_1^*, N^*, K^*}$$

$$+ (K - K^*)\frac{\partial E_0}{\partial K}\bigg|_{E_1^*, N^*, K^*} + \text{terms of higher order.}$$

Neglecting the terms of higher order,

$$E_0 \cong E_1^* N^* K^* + (E_1 - E_1^*)N^* K^* + (N - N^*)E_1^* K^*$$
$$+ (K - K^*)E_1^* N^*.$$

Furthermore, if these components are normally distributed independent random variables with means at the nominal values, then from the results on linear combinations of random variables, LCN, E_0 is approximately normally distributed with mean, $E(E_0)$, approximately equal to $E_1^* N^* K^*$; and variance, $\sigma^2(E_0)$, approximately equal to

$$(N^* K^*)^2 \sigma_{E_1}^2 + (E_1^* K^*)^2 \sigma_N^2 + (E_1^* N^*)^2 \sigma_K^2.$$

The symbols $\sigma_{E_1}^2$, σ_N^2, and σ_K^2 represent the variances of the input voltage, transformer, and amplifier, respectively. Putting the numerical constants of the example into the expression for the mean, $E(E_0) \cong 40 \times 3 \times \frac{1}{2} =$

[1] This example is taken from "How to Evaluate Assembly Tolerances," by R. H. Johnson, *Product Engineering*, Jan., 1953.

60; the variance can be expressed as

$$\sigma_{E_0}^2 \cong (\tfrac{1}{2} \times 3)^2 \sigma_{E_1}^2 + (40 \times 3)^2 \sigma_N^2 + (40 \times \tfrac{1}{2})^2 \sigma_K^2$$
$$= \tfrac{9}{4}\sigma_{E_1}^2 + (14{,}400)\sigma_N^2 + 400\sigma_K^2.$$

Assume that each component can be centered at its nominal values and consider the case of the most conservative acceptable situation, i.e., the magnitude of the variance of each component is such that the natural tolerance limits *coincide* with the specification limits. Furthermore, choose α to be equal to 0.0027. Thus, for the input voltage all but 0.0027 of the values will fall within

$$40 \pm 0.5 = (39.5, 40.5).$$

The normal deviate (percentage point) corresponding to 0.00135 is given by $K_{0.00135} = 3$ and must equal $(40.5 - 40)/\sigma_{E_1}$. This expression is solved for σ_{E_1}, i.e., $\sigma_{E_1} = 0.5/3$. The standard deviation σ_{E_1} equal to 0.5/3 can be interpreted as the largest value of the standard deviation of the input voltage allowing the natural tolerances to fall within the design specifications. Similarly, for the transformer all but 0.0027 of the values will fall within

$$\tfrac{1}{2} \pm 1\% = \tfrac{1}{2} \pm 0.005 = (0.495, 0.505)$$

so that

$$3 = \frac{0.505 - 0.500}{\sigma_N}, \quad \text{or} \quad \sigma_N = \frac{0.005}{3}.$$

Finally, for the amplifier all but 0.0027 of the values will fall within

$$3 \pm 2\% = 3 \pm 0.06 = (2.94, 3.06)$$

so that

$$3 = \frac{3.06 - 3.00}{\sigma_K}, \quad \text{or} \quad \sigma_K = 0.02.$$

Standard deviations σ_N equal to 0.005/3 and σ_K equal to 0.02 can be interpreted as the largest values of the standard deviation of the transformer and amplifier, respectively, allowing the natural tolerances of these components to fall within the design specifications. Hence, the largest value of the variance of the output voltage, subject to the constraint that the natural tolerances of the individual components fall within their specifications, is given by

$$\sigma_{E_0}^2 = \left(\frac{9}{4}\right)\left(\frac{0.25}{9}\right) + (14{,}400)\left(\frac{0.000025}{9}\right) + (400)(0.0004)$$
$$= 0.0625 + 0.040 + 0.16 = 0.2625,$$

so that the corresponding largest standard deviation is equal to 0.512. If the natural tolerances of the individual components all fall within their specification limits, the natural tolerances for the output voltage will not exceed

$$60 \pm 3(0.512) = 60 \pm 1.536 = 60 \pm 2.56\%,$$

which is within the design specification of $60 \pm 3\%$. Thus, the natural tolerances of the output voltage will fall within the design specifications, provided that the natural tolerances of the individual components fall within their specification limits.

A little thought will reveal that $\alpha = 0.0027$ and the corresponding $K_{0.00135} = 3$ really play no role in this problem, provided that it is assumed that the allowable fraction of defectives defining the natural tolerance limits is the same for all the components as well as E_0. If $K_{\alpha/2}$ denotes the percentage point corresponding to any allowable fraction of defectives, α, ($\alpha/2$ falling outside of each limit) the maximum standard deviations of the components allowing the natural tolerances to fall within the design specifications are given by

$$\sigma_{E_1} = \frac{0.5}{K_{\alpha/2}}; \qquad \sigma_N = \frac{0.005}{K_{\alpha/2}}; \qquad \sigma_K = \frac{0.06}{K_{\alpha/2}}.$$

The largest value of the variance of the output voltage, subject to the constraint that the natural tolerances of the individual components fall within their specifications, is given by

$$\sigma_{E_0}^2 = \left(\frac{9}{4}\right)\left(\frac{0.25}{(K_{\alpha/2})^2}\right) + (14,400)\left(\frac{0.000025}{(K_{\alpha/2})^2}\right) + (400)\left(\frac{0.0036}{(K_{\alpha/2})^2}\right)$$

$$= \frac{0.5625}{K_{\alpha/2}^2} + \frac{0.3600}{K_{\alpha/2}^2} + \frac{1.44}{K_{\alpha/2}^2} = \frac{2.3625}{K_{\alpha/2}^2},$$

so that the corresponding largest standard deviation is equal to $1.536/K_{\alpha/2}$. If the natural tolerances of the individual components all fall within their specification limits, the natural tolerances for the output voltage will not exceed

$$60 \pm K_{\alpha/2}\frac{1.536}{K_{\alpha/2}} = 60 \pm 1.536 = 60 \pm 2.56\%,$$

which is exactly the same result as obtained previously.

The function $E_0 = E_1 NK$ was expanded into a Taylor series, with higher order terms neglected. Other functions of n variables $g(X_1, X_2, \ldots, X_n)$ can be expanded into a Taylor series about a particular point, say $X_1^*, X_2^*, \ldots, X_n^*$, i.e.,

$$g(X_1, X_2, \ldots, X_n) = g(X_1^*, X_2^*, \ldots, X_n^*) + (X_1 - X_1^*)\frac{\partial g}{\partial X_1}\Big|_{X_1^*, X_2^*, \ldots, X_n^*} + \cdots$$

$$+ (X_n - X_n^*)\frac{\partial g}{\partial X_n}\Big|_{X_1^*, X_2^*, \ldots, X_n^*}$$

$$+ \text{terms of higher order.}$$

Neglecting terms of higher order and assuming that the random variables are independent with means at $X_1^*, X_2^*, \ldots, X_n^*$, respectively, the expected value of the function is approximately given by

$$E[g(X_1, X_2, \ldots, X_n)] \cong g(X_1^*, X_2^*, \ldots, X_n^*)$$

and the variance of the function is approximately given by

$$\sigma^2_{g(X_1, X_2, \ldots, X_n)} \cong \sigma^2_{X_1}\left[\frac{\partial g}{\partial X_1}\bigg|_{X_1^*, X_2^*, \ldots, X_n^*}\right]^2 + \cdots + \sigma^2_{X_n}\left[\frac{\partial g}{\partial X_n}\bigg|_{X_1^*, X_2^*, \ldots, X_n^*}\right]^2.$$

Furthermore, if the X's are normally distributed, $g(X_1, X_2, \ldots, X_n)$ is approximately normally distributed.

As a final example,[1] suppose that a phase-shifting synchro is added to the output of the electrical circuit in the previous example as shown in Fig. 3.18. The new output voltage is given by

$$E_0 = E_2 \cos \theta + E_1 NK \sin \theta$$

where θ refers to the angular relationship of the synchro inputs

$$\frac{\partial E_0}{\partial E_1} = NK \sin \theta$$

$$\frac{\partial E_0}{\partial E_2} = \cos \theta$$

$$\frac{\partial E_0}{\partial N} = E_1 K \sin \theta$$

$$\frac{\partial E_0}{\partial K} = E_1 N \sin \theta$$

$$\frac{\partial E_0}{\partial \theta} = -E_2 \sin \theta + E_1 NK \cos \theta.$$

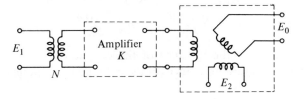

Fig. 3.18. *Adding a phase-shifting synchro to the circuit of Fig. 3.17 causes the output voltage to become a more complicated function of the components.*

Assigning typical values and specifications to the components

$$E_1: 40 \pm 0.5 \text{ volts}$$
$$N: 2 \text{ to } 1 \pm 1\%$$
$$K: 3 \pm 2\%$$
$$\theta: 60 \pm \tfrac{1}{4} \text{ degrees (0.00436 radian)}$$
$$E_2: 40 \pm 0.4 \text{ volts}$$

---

[1] This example is taken from "How to Evaluate Assembly Tolerances," by R. H. Johnson, *Product Engineering*, Jan., 1953.

and assuming that the components are independently distributed normal random variables with means at the nominal values (points about which the function is expanded), it follows that E_0 is approximately normally distributed with mean

$$E(E_0) \cong E_2^* \cos \theta^* + E_1^* N^* K^* \sin \theta^*$$

$$= 40 \times \frac{1}{2} + 40 \times \frac{1}{2} \times 3 \times \frac{\sqrt{3}}{2} = 72 \text{ volts}$$

and variance

$$\sigma_{E_0}^2 \cong \sigma_{E_1}^2 (N^* K^* \sin \theta^*)^2 + \sigma_{E_2}^2 (\cos \theta^*)^2 + \sigma_N^2 (E_1^* K^* \sin \theta^*)^2$$
$$+ \sigma_K^2 (E_1^* N^* \sin \theta^*)^2 + (E_1^* N^* K^* \cos \theta^* - E_2 \sin \theta^*)^2 \sigma_\theta^2$$

$$= \sigma_{E_1}^2 \left(\frac{1}{2} \times 3\frac{\sqrt{3}}{2} \right)^2 + \sigma_{E_2}^2 \left(\frac{1}{2} \right)^2 + \sigma_N^2 \left(40 \times 3\frac{\sqrt{3}}{2} \right)^2$$

$$+ \sigma_K^2 \left(40 \times \frac{1}{2} \times \frac{\sqrt{3}}{2} \right)^2 + \sigma_\theta^2 \left(40 \times \frac{1}{2} \times 3 \times \frac{1}{2} - 40\frac{\sqrt{3}}{2} \right)^2$$

$$= \left(\frac{27}{16} \right) \sigma_{E_1}^2 + \left(\frac{1}{4} \right) \sigma_{E_2}^2 + (10{,}800) \sigma_N^2 + (300) \sigma_K^2 + (21.54) \sigma_\theta^2.$$

If $K_{\alpha/2}$ denotes the percentage point corresponding to any allowable fraction of defectives, α, ($\alpha/2$ falling outside of each limit) the maximum standard deviations of the components allowing the natural tolerances to fall within the design specifications are given by

$$\sigma_{E_1} = \frac{0.5}{K_{\alpha/2}}; \qquad \sigma_N = \frac{0.005}{K_{\alpha/2}}; \qquad \sigma_K = \frac{0.06}{K_{\alpha/2}};$$

$$\sigma_\theta = \frac{0.00436}{K_{\alpha/2}}; \qquad \sigma_{E_2} = \frac{0.4}{K_{\alpha/2}}.$$

The largest value of the variance of the output voltage, subject to the constraint that the natural tolerances of the individual components fall within their specifications, is given by

$$\sigma_{E_0}^2 = \frac{27}{16} \frac{0.25}{K_{\alpha/2}^2} + \frac{1}{4} \frac{0.16}{K_{\alpha/2}^2} + \frac{(10{,}800)(0.000025)}{K_{\alpha/2}^2}$$

$$+ (300) \frac{(0.0036)}{K_{\alpha/2}^2} + \frac{(21.54)(0.00001901)}{K_{\alpha/2}^2}$$

$$= \frac{1.81}{K_{\alpha/2}^2},$$

so that

$$\sigma_{E_0} = \frac{1.35}{K_{\alpha/2}}.$$

Hence, the design specifications on E_0 should not be less than

$$72 \pm 1.35 \text{ volts}.$$

3.9 The Central Limit Theorem

In many of the previous examples, it was assumed, somewhat arbitrarily, that the random variables were normally distributed. Although in actual practice many random variables are normally distributed, many do not resemble a normal distribution at all, and such assumptions can lead to difficulties. Fortunately, however, there are certain natural factors which, in some cases, tend to make the random variable have a normal distribution. This can be best explained by stating the central limit theorem.

Central Limit Theorem. Let the random variable $Y = X_1 + X_2 + \cdots + X_n$ where $X_1, X_2, \ldots, X_n$ are identically distributed independent random variables, each having mean μ and finite variance σ^2. Then the distribution of $N = (Y - n\mu)/\sqrt{n}\,\sigma$ approaches the normal distribution with mean 0 and variance 1 as n becomes infinite.

This theorem implies that the sum of a large number of random variables will be approximately normally distributed regardless of the distribution of the individual random variables. Furthermore, note that

$$\frac{Y - n\mu}{\sqrt{n}\,\sigma} = \left(\frac{\bar{X} - \mu}{\sigma}\right)\sqrt{n}$$

so that this theorem also implies that the mean of n identically distributed independent random variables, i.e., the mean of a random sample, will be approximately normally distributed regardless of the distribution of the individual variables.

Under certain conditions the Central Limit Theorem is also valid when the random variables are not identically distributed. Loosely speaking, this generalization holds whenever the individual random variables make only a "small" contribution relative to the entire sum. In this situation, $n\mu$ is replaced by $\sum_{i=1}^{n} \mu_i$ and $\sqrt{n}\,\sigma$ is replaced by $\sqrt{\sum_{i=1}^{n} \sigma_i^2}$ where μ_i and σ_i denote the mean and standard deviation of the random variable X_i.

The magnitude of n that is required before the results of the theorem begin to hold depends to a great extent on the shape of the distribution of the original random variables. Figure 3.19 presents some empirical evidence of the validity of the Central Limit Theorem. Repeated random samples of size four were drawn from rectangular and triangular distributions, and the outcomes of the random variable, the sample mean, were recorded. Frequency histograms were constructed, with points plotted instead of bar graphs. Smooth curves were drawn through the points, resulting in near bell-shaped curves characteristic of the normal distribution. The experiment demonstrated that even when the underlying distribution is a rectangular or a triangular distribution, the distribution of $\bar{X}$ values from samples of four is approximately normal.

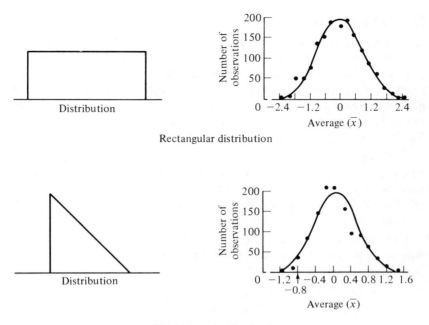

Right triangular distribution

Fig. 3.19. *Distribution of the sample average from rectangular and triangular distributions.*

From Economic Control of Quality of Manufactured Product, *by W. A. Shewhart, copyright © 1931, by Litton Educational Publishing Inc. Published by Van Nostrand Reinhold Company.*

The Central Limit Theorem, then, is the key to the importance of the normal distribution in statistics.

In the theory of errors, which deals with the distribution of errors of measurements, errors can usually be assumed to have a normal distribution, since they are usually composed of the sum of many small independent components.

Most of the procedures for estimating parameters or testing hypotheses are dependent on the sample mean. If the individual random variables going into the sample are normally distributed, then the sample mean is certainly normally distributed, and the distribution theory for normally distributed random variables is relevant. If the individual random variables going into the sample are not normally distributed, but their number is sufficiently large, then the Central Limit Theorem indicates that the sample mean tends toward normality, and the distribution theory for normally distributed random variables is relevant. If the individual random variables going into the sample have an unknown distribution, and their number is not sufficiently large, the distribution of the sample mean *cannot* be assumed to be normal.

In such cases, non-parametric techniques should be employed. Such techniques are presented in Chapter 7.

PROBLEMS

1. If X is normally distributed with mean 14 and variance 4, find (a) $P\{8 < X < 12\}$, (b) $P\{X < 20\}$, (c) $P\{X > 5\}$.

2. Find the number k such that for a normally distributed random variable X having mean μ and variance σ^2,

$$P\{\mu - k\sigma < X < \mu + k\sigma\} = 0.683, 0.90, 0.95, 0.99, 0.995.$$

Find k such that $P\{X < \mu + k\sigma\} = 0.841, 0.95, 0.975, 0.99$.

3. The life of army shoes is normally distributed with mean 12 months and standard deviation 2 months. If 10,000 pairs are issued, how many pairs would be expected to need replacement after 15 months?

4. The finished diameter on armored electric cable is normally distributed with mean 0.775 and standard deviation 0.010. What is the probability that the diameter will exceed 0.795?

5. The width of a slot on a duralumin forging is normally distributed with mean 0.9000 inch and standard deviation 0.0020 inch. The specification limits are given as 0.9000 ± 0.0050 inch. What percentage of forgings will be defective?

6. The measurement of pitch diameter of thread of a fitting is normally distributed with mean 0.4008 inch and standard deviation 0.0003 inch. The specifications are given as 0.4000 ± 0.0010 inch. What is the probability of a defective occurring?

7. Diameters of bolts in mass production are normally distributed with mean 0.25 inch and standard deviation 0.01 inch. Bolt specifications call for diameters of 0.24 ± 0.02 inch. What is the probability of a defective being produced?

8. In Problem 5, what is the maximum allowable standard deviation which will permit no more than three in a thousand defectives to be produced when the slot widths are normally distributed with mean 0.9000 inch?

9. In Problem 6, what is the maximum allowable standard deviation which will permit no more than one in a thousand defectives to be produced when the pitch diameters are normally distributed and centered at the nominal dimension?

10. In Problem 4, what must be the mean value of finished diameter if the probability of exceeding 0.795 is to be 0.01?

11. In Problem 6, if the standard deviation of the pitch diameter is 0.0003 inch, what must be the centering to insure that no more than 0.5% of the items produced will exceed the upper specification limit?

12. A new item is about to be produced which must conform to the specification limits 10 ± 3 units. When production is begun, it is found that the natural tolerance limits fall between 12 and 14 (the process is normal and $\alpha = 27/10,000$). (a) What percentage of the material will fall outside the specification limits? (b) What minimum adjustment on the mean value must be made (assuming the mean can be adjusted) so that a satisfactory product will be manufactured? Neglect the fraction of material that falls below the lower specification limit.

13. If the diameters of certain shafts must be less than 1.500 inches to be usable, and if the shafts are produced by a process that gives mean diameter 1.490 inches with standard deviation 0.004 inch, what percentage of shafts must be scrapped? If in addition a shaft must be larger than 1.480 inches, what percentage will be scrapped? Assume diameters are normally distributed.

14. In the situation of Problem 4, determine the mean and the standard deviation of sample averages of nine diameters and find what percentage of averages should fall below 0.770. What percentage above 0.795?

15. In the situation of Problem 5, samples of size 5 forgings are observed hourly and their means computed. What percentage of these sample averages will lie outside the specification limits? Determine limits such that the percentage falling outside these limits is 0.27%.

16. In the situation of Problem 6, samples of size 4 pitch diameters are observed. What is the probability that a sample average will fall within the specification limits? What is the probability that an average will exceed 0.4005?

17. The diameters of rivets produced by a certain machine are normally distributed with mean 6.75 and variance 0.0016, where measurements are made in hundredths of an inch. A random sample of four rivets is taken from the output of this machine. Find (a) the probability that a rivet taken at random from the output of this machine will have a diameter smaller than 6.6716 hundredths of an inch; (b) the probability that the sample mean diameter of a random sample of four rivets from the output of this machine will exceed 6.7829.

18. The diameters of subway tokens are normally distributed with mean 1.0 inch and standard deviation 0.03 inch. The specifications required by the existing token machines are 1.0 ± 0.05 inch. (a) What percentage of new tokens will be within the specification limits? (b) If a sample of 9 tokens is taken, what is the probability that the sample mean diameter will be within the specification limits?

19. The distribution of test scores on an examination is approximately normal with mean 70 and standard deviation 7. (a) What is the probability of receiving a grade of less than 80? (b) If 49 students take the test, what is the probability that the average of their scores lies between 68 and 72? (c) What is the probability that the difference in the scores of two students (picked at random) lies between -5 and $+5$?

20. Let X_1 and X_2 be independent normally distributed random variables with $\mu_1 = 0$, $\sigma_1^2 = 4$ and $\mu_2 = 1$, $\sigma_2^2 = 9$ given as the respective means and variances. (a) Find $P\{X_2 \leq 0\}$. (b) Find $P\{X_1 + X_2 > 1\}$. (c) If random samples of size 9 and 16 are taken on X_1 and X_2, respectively, what is the distribution of the random variable $Y = \bar{X}_1 - \bar{X}_2$?

21. Denote by $\bar{X}_1$ and $\bar{X}_2$ the sample means of two independent random samples of size n. The observations going into a sample mean are normally distributed with common mean, μ, and with common variance equal to 2. Determine n so that the probability will be 0.98 that $\bar{X}_1$ and $\bar{X}_2$ differ by less than 2, i.e., find n such that

$$P\{-2 \leq \bar{X}_1 - \bar{X}_2 \leq 2\} = 0.98.$$

22. The life of electric light bulbs is known to be a normally distributed random variable with unknown mean μ and standard deviation 200 hours. The value of a lot of 1,000 bulbs is $(1,000)(1/5,000)\mu$ dollars. A random sample of n bulbs are to be sampled by a prospective buyer and $1,000(1/5,000)\bar{X}$ dollars paid to the manufacturer. How large should n be so that the probability is 0.95 that the buyer does not overpay or underpay the manufacturer by more than $20?

23. The standard deviation of the diameters of a shaft and bearing have identical values of 0.001 inch. The mean distance between diameters of the part is 0.0035. What is the probability of interference, assuming that each dimension of the mating parts is normally distributed?

24. Two resistors are assembled in series. Each is nominally rated at 10 ohms. The resistors are known to be normally distributed about the nominal rating, each having a standard deviation equal to 0.5 ohm. What is the probability that an assembly will have a combined dimension in excess of 21.5 ohms? What must the standard deviations be so that the probability of a combined dimension exceeding 21.5 ohms will be 0.01?

25. A delivery truck carries loaded cartons of items. If the weight of each carton is approximately normally distributed with mean 50 pounds and standard deviation 5 pounds, how many cartons can the truck carry so that the probability of the total load exceeding 1 ton will be only 0.01?

26. Suppose X, Y, and Z are independent normally distributed random variables, each having zero mean and unit standard deviation. Find $P\{X > Y + Z\}$.

27. A shaft is made up of six different sections laid end to end. The lengths of these component sections are known to be independent, normally distributed random variables with means and variances, respectively: (8.10, 0.05), (7.25, 0.04), (9.75, 0.06), (3.45, 0.04), (17.15, 0.10), (6.20, 0.07). If the specifications call for the assembled shaft to be 52 ± 1.5 in length, what is the probability that an assembled shaft meets specifications?

28. The measurable characteristic of an electrical device is normally distributed with mean 10 units and standard deviation 3 units. The measuring device is subject to error. The probability distribution of this device is normal with mean zero and standard deviation $\frac{1}{2}$ unit. What is the probability of a measurement of a device exceeding 13 units?

29. The smog content of air is monitored hourly. The acceptable content of a particular constituent is specified as 7.7%. If the actual content of this constituent is a normally distributed random variable with mean 7.6 and variance 0.0016, and if the measurement error is a random variable with mean 0 and variance 0.0009, find: (a) the probability that a single measurement will exceed 7.7, (b) the probability that the mean of three measurements will exceed 7.7.

30. Two mating parts A and B are approximately normally distributed as follows: $\mu_A = 3.000$ inches, $\mu_B = 3.005$ inches, $\sigma_A = 0.001$ inch, and $\sigma_B = 0.002$ inch. What is the probability of interference?

31. In Problem 30, what should μ_A be so that the probability of interference is 0.01?

32. Two mating parts A and B are approximately normally distributed as follows: $\mu_A = 3.000$ inches and $\mu_B = 3.005$ inches. If $\sigma_A = \sigma_B$, determine their common value so that the probability of a clearance less than 0.001 does not exceed 1%.

33. Past information indicates the standard deviations of the dimensions of two mating parts have identical values of 0.0012 inch. It is desired that the probability of a clearance less than 0.003 inch should be 0.02. What mean difference between diameters should be specified by the designer? Assume normal distributions and random assembly.

34. Two 8-ohm resistors are placed in series in such a manner that the final assembly should be 16 ± 0.25 ohms. (a) Assuming each resistor is normally and identically distributed, determine specification limits for the components. (b) If the natural tolerance limits, which are defined to be those limits within which all but 5% of the items fall, coincide with the specification limits, what is the probability that the average resistance of five final assemblies will exceed 16.05 ohms?

35. At one stage in the manufacture of an article, a cylindrical rod with circular cross section has to fit into a circular socket. Quality control measurements show that the distributions of rod diameter and socket diameter are normal with constants:

Rod diameter: mean 5.01 centimeters, standard deviation 0.03 centimeter.
Socket diameter: mean 5.10 centimeters, standard deviation 0.04 centimeter.

If components are selected at random for assembly, what proportion of rods will not fit?

36. (a) Suppose the life in hours of a certain electronic tube is normally distributed with mean $\mu = 160$ hours. The specification limits call for the product to lie between 120 hours and 200 hours with probability 0.95. What is the maximum allowable standard deviation that the process can have and still maintain its quality? (b) If the process is at the value of σ found in (a), what is the probability of sample averages of 2 falling outside the specification limits?

37. An emf of E volts drives a current of I amperes through a resistance of R ohms. According to Ohm's law, $E = IR$. Suppose E and R are normally distributed random variables whose natural (3 standard deviation) tolerances are given by

$$E = 60 \pm 2 \text{ volts}, \qquad R = 6 \pm 0.2 \text{ ohms}.$$

Find the natural tolerances of I under the "customary assumptions." (List these assumptions.) Start with the expression $I = E/R$.

38. Two circular holes of radii R_1 and R_2 are cut from a rectangular piece of metal of thickness T, width X, and length Y. The weight of metal remaining in the bar after the holes are cut is

$$W = cT(XY - \pi R_1^2 - \pi R_2^2).$$

Suppose all dimensions (T, X, Y, R_1, and R_2) are random variables independently and normally distributed about their nominal values and such that 99.9% of each distribution lies within known specification limits. Find a reasonable approximation to the standard deviation of W.

39. It is known that

$$\text{distance} = \text{velocity} \times \text{time}.$$

Using the Taylor Expansion around the means of velocity and time, find the expression for the approximate variance of distance in terms of the variance of velocity and the variance of time.

40. The value of the constant of gravity is to be determined from $g = 2s/t^2$ by observing a body fall. The experimenter has a choice of setting s at 4 feet with t

about $\frac{1}{2}$ second or of setting s at 16 feet with t about 1 second. Assume that the standard deviation of s is 0.002 foot and that the standard deviation of recording time is 0.04 second. If the expected value of s is equal to 4 and the expected value of $t = \frac{1}{2}$, calculate the variance of g. If the expected value of s is equal to 16 and the expected value of $t = 1$, calculate the variance of g. Which value of s is better in the sense of smaller variance of determining g?

4

OTHER
PROBABILITY DISTRIBUTIONS

4.1 Introduction

Chapter 3 was concerned with the distribution of a random variable whose density function is given by

$$f_X(z) = \frac{1}{\sqrt{2\pi}\,\sigma} e^{-(z-\mu)^2/2\sigma^2}; \quad -\infty < z < \infty,$$

i.e., the normally distributed random variable with mean μ and variance σ^2. The results on linear combinations of random variables (LCN) indicated that if $X_1, X_2, \ldots, X_n$ is a random sample of n independent normally distributed random variables, each with mean μ and variance σ^2, the random variable, the sample mean ($\bar{X} = \sum_{i=1}^{n} X_i/n$), is normally distributed with mean μ and variance σ^2/n. In other words, the exact distribution of the sample mean is one which depends on the distribution of the individual random variables comprising the sample mean, and turns out to be normally distributed when the assumptions above are fulfilled. The usefulness of the distribution of the sample mean is evidenced by the following example. The mean, μ, of a normal distribution is unknown and is to be estimated on the basis of the results of a random sample. Intuitively, the sample mean $\bar{X}$ is judged as a "good" estimate of the quantity μ. How good an estimate $\bar{X}$ is depends on the probability distribution of $\bar{X}$. A measure of the "closeness" of $\bar{X}$ to μ is given by

$$P\{a \leq \bar{X} - \mu \leq b\}$$

for any a and b; i.e., it is given by the distribution of $\bar{X}$. If the standard deviation σ is known, these probabilities can be evaluated by using the identity

$$P\{a \leq \bar{X} - \mu \leq b\} = P\left\{\frac{a\sqrt{n}}{\sigma} \leq \frac{\sqrt{n}\,(\bar{X} - \mu)}{\sigma} \leq \frac{b\sqrt{n}}{\sigma}\right\}$$

$$= P\left\{\frac{a\sqrt{n}}{\sigma} \leq N \leq \frac{b\sqrt{n}}{\sigma}\right\},$$

where N is a normally distributed random variable with zero mean and unit variance. This last expression can be evaluated for any value of a, b, σ, and n by referring to Appendix Table 1.

The distribution of a function of random variables is not necessarily normal. For example, let $X_1, X_2, \ldots, X_n$ be a random sample of size n (the X's can have any common distribution). The random variable, the sample variance S^2, is defined to be

$$S^2 = \frac{\sum_{i=1}^{n}(X_i - \bar{X})^2}{n - 1}.$$

The square root of S^2 is called the sample standard deviation and is denoted by S. Neither the sample variance nor the sample standard deviation has a normal distribution; this comment is appropriate whether or not the common distribution of the underlying X's is normal. Most of the remaining sections of this chapter will be devoted to determining the distribution of random variables which are generated from (functions of) other random variables.

4.2 The Chi-Square Distribution

4.2.1 THE CHI-SQUARE RANDOM VARIABLE

This section is concerned with the distribution of a random variable which consists of sums of squares of other random variables. In particular, an experiment is performed. Let $X_1, X_2, \ldots, X_\nu$ be random variables that are assumed to be independent and normally distributed, each with zero mean and unit variance. The $X_1, X_2, \ldots, X_\nu$ can be thought of as a random sample resulting from ν trials (or ν tests) where each X_i denotes the measurable quantity associated with the ith trial (or ith test). Let χ^2 equal the sum of squares of these ν random variables, i.e., $\chi^2 = X_1^2 + X_2^2 + \ldots + X_\nu^2$. χ^2 also must be a random variable because it is a function of random variables (see Sec. 2.4). The range of this random variable is clearly the interval between zero and plus infinity since χ^2 is a sum of squares. All that remains is to determine its distribution. This distribution is not arbitrary since the distribution of the X's was specified (as normal with zero mean and unit variance). The density function of the chi-square random variable can then be derived (by the technique given in Sec. 2.12 or by other techniques which are not included in this text), and is given by

$$f_{\chi^2}(z) = \begin{cases} \dfrac{1}{2^{\nu/2}\Gamma\left(\dfrac{\nu}{2}\right)}(z)^{(\nu-2)/2}e^{-z/2}, & \text{for } z > 0 \\ 0, & \text{otherwise,} \end{cases}$$

where Γ represents the gamma function.[1]

A continuous random variable whose density function can be expressed as $f_{\chi^2}(z)$ *is said to be a chi-square random variable with* ν *degrees of freedom, or alternatively, is said to have a chi-square distribution with* ν *degrees of freedom.* The quantity ν is the number of independent random variables whose squares are being added and is designated as the degrees of freedom. ν may also be thought of as a parameter associated with the distribution.

Random variables having an approximate chi-square distribution may arise naturally in practice or may be generated in an exact fashion in the manner previously described. An example of the former is the distribution of the lives of vacuum tubes which often approximates the chi-square distribution or a distribution related to the chi-square. An example of the latter is as follows. A hole is to be drilled in a rectangular table top. The desired location of the hole is specified by giving its distance from two perpendicular edges of the top. The actual location of the hole is obtained by measuring the desired distances from the proper edges, using a steel tape. Measurement errors account for any difference between the desired location and the actual location of the hole. The distribution of measurement errors, then, is of practical importance. If the differences between the actual distance from an edge and its desired position are independent normally distributed random variables, each having zero mean and unit standard deviation, the square of the distance of the actual measured point from its desired position has a chi-square distribution with 2 degrees of freedom.

If $X_1, X_2, \ldots, X_\nu$ are independent normally distributed random variables with means $\mu_1, \mu_2, \ldots, \mu_\nu$, and variances $\sigma_1^2, \sigma_2^2, \ldots, \sigma_\nu^2$, respectively, the sum of squares of the X's does *not* have a chi-square distribution. This resulting random variable fails to fulfill the condition which requires all the

---

[1] By definition,

$$\Gamma\left(\frac{\nu}{2}\right) = \left(\frac{\nu}{2} - 1\right)!$$

$$= \begin{cases} \left(\dfrac{\nu}{2} - 1\right)\left(\dfrac{\nu}{2} - 2\right)\cdots 3\cdot 2\cdot 1, & \text{for } \nu \text{ even and } \nu > 2 \\ \left(\dfrac{\nu}{2} - 1\right)\left(\dfrac{\nu}{2} - 2\right)\cdots \dfrac{3}{2}\cdot \dfrac{1}{2}\sqrt{\pi}, & \text{for } \nu \text{ odd and } \nu > 2 \end{cases}$$

and, further,

$$\Gamma(1) = 1 \quad \text{and} \quad \Gamma\left(\frac{1}{2}\right) = \sqrt{\pi}.$$

X's to have zero mean and unit variance. However, the quantities

$$N_i = \frac{X_i - \mu_i}{\sigma_i}, \quad i = 1, 2, \ldots, \nu$$

do have a normal distribution each with zero mean and unit variance. Hence,

$$\chi^2 = \sum_{i=1}^{\nu} N_i^2 = \sum_{i=1}^{\nu} \left(\frac{X_i - \mu_i}{\sigma_i} \right)^2$$

has a chi-square distribution with ν degrees of freedom.

An example of such a random variable is as follows. An unmanned space vehicle is directed at a target location on a nearby planet. If the lateral and forward distances of the point of impact from the target location are independent normally distributed random variables each with zero mean (this implies that positive and negative miss distances in each direction are equally likely) and equal variances, the squared distance between the landing point and the target value, divided by the variance, has a chi-square distribution with 2 degrees of freedom. This navigational problem is easily extended to three dimensions.

Summarizing, then, there are many physical phenomena that have a probability distribution which is approximated by the chi-square distribution without any apparent justification. These situations are usually discovered by experience and by analyzing frequency histograms. Other random variables have chi-square distributions because they are known to satisfy the conditions of the definition, i.e., *if $X_1, X_2, \ldots, X_\nu$ are a random sample of ν independent normally distributed random variables, each with zero mean and unit variance, the random variable*

$$\chi^2 = X_1^2 + X_2^2 + \ldots + X_\nu^2$$

has a chi-square distribution with ν degrees of freedom.

In the former case, the degrees of freedom may be considered as a parameter (constant) of the probability distribution whereas in the latter case the degrees of freedom are the number of independent random variables whose squares are being added.

A sketch of the density function of the chi-square random variable is given in Fig. 4.1. It should be noted that this distribution is skewed rather than symmetric as is the normal. Furthermore, the density function has mass only between 0 and ∞. This is evident because the random variable is a sum of squares which must necessarily be non-negative. The expected value of the chi-square random variable is given by

$$E(\chi^2) = \nu$$

and the variance of the chi-square random variable is given by

$$\sigma_{\chi^2}^2 = 2\nu.$$

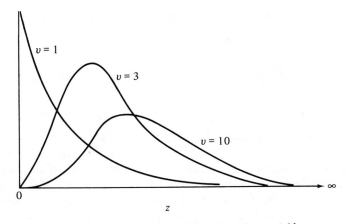

Fig. 4.1. *Density function of the χ^2 random variable.*

In other words, these moments are expressed in terms of the degrees of freedom only, the mean equaling the degrees of freedom and the variance equaling twice the degrees of freedom.

The probability that χ^2 is greater than a constant $\chi^2_{\alpha;\nu}$ is represented by

$$P\{\chi^2 \geqq \chi^2_{\alpha;\nu}\} = \int_{\chi^2_{\alpha;\nu}}^{\infty} f(z)\,dz$$

$$= \int_{\chi^2_{\alpha;\nu}}^{\infty} \frac{1}{2^{\nu/2}\Gamma(\nu/2)}(z)^{(\nu-2)/2}e^{-z/2}dz = \alpha.$$

This function has been tabulated, and values of the percentage point $\chi^2_{\alpha;\nu}$ can be found in Appendix Table 2. For example, if $\nu = 5$,

$$P\{\chi^2 \geqq \chi^2_{0.10;5}\} = P\{\chi^2 \geqq 9.236\} = 0.10.$$

4.2.2. THE ADDITION THEOREM FOR THE CHI-SQUARE DISTRIBUTION

The chi-square distribution is one of the few probability distributions that has the reproductive property; i.e., if two random variables, each having a chi-square distribution, are independent and are added together, the resulting random variable also has a chi-square distribution. In this property, it is similar to the normal distribution which also has this reproductive feature. This result is stated precisely in the following theorem on the addition of chi-square random variables.

If χ_1^2 and χ_2^2 are independent chi-square random variables with ν_1 and ν_2 degrees of freedom, respectively, the sum of these random variables,

$$\chi^2 = \chi_1^2 + \chi_2^2,$$

also has a chi-square distribution with

$$\nu = \nu_1 + \nu_2 \text{ degrees of freedom.}$$

This theorem is an immediate consequence of the definitions. The distribution of χ_1^2 is equivalent to the distribution of the sum of squares of ν_1 independent normally distributed random variables, each with zero mean and unit variance. Hence, the random variable χ_1^2 can be written as

$$\chi_1^2 = X_1^2 + X_2^2 + \ldots + X_{\nu_1}^2$$

where the X's are independent normally distributed random variables each with zero mean and unit variance.

Similarly, χ_2^2 can be written as

$$\chi_2^2 = Y_1^2 + Y_2^2 + \ldots + Y_{\nu_2}^2$$

where the Y's are independent normally distributed random variables each with zero mean and unit variance. Furthermore, the X's and the Y's are independent since χ_1^2 and χ_2^2 are assumed independent.

The random variable $\chi^2 = \chi_1^2 + \chi_2^2$ can then be written as

$$\chi^2 = X_1^2 + X_2^2 + \cdots + X_{\nu_1}^2 + Y_1^2 + Y_2^2 + \cdots + Y_{\nu_2}^2.$$

It is evident that χ^2 is just the sum of squares of $\nu_1 + \nu_2 = \nu$ independent normally distributed random variables each having zero mean and unit variance. Hence, it follows, from the definition, that the random variable χ^2 must have a chi-square distribution with ν degrees of freedom.

Since the reproductive property of the chi-square distribution holds for the sum of two independent random variables, it must hold for the sum of any finite number of independent chi-square random variables.

4.2.3 THE DISTRIBUTION OF THE SAMPLE VARIANCE, S^2

The distribution of the sample mean was discussed previously. This section discusses the distribution of the sample variance, S^2, or more correctly, the distribution of a quantity related to this random variable.

Since
$$S^2 = \frac{\sum_{i=1}^{n}(X_i - \bar{X})^2}{n - 1}$$

is called the sample variance, it is natural to expect this random variable to be used as an estimate of the variance, σ^2, of a normal distribution when σ^2 is unknown. The estimation process can be considered as follows: the variance of a normal distribution is unknown, a random sample of n observations is drawn, the outcome of the random variable S^2 is observed, and this value is used as the estimate of σ^2. How good an estimate S^2 is depends on the distribution of S^2. A measure of the closeness of S^2 to σ^2 is given by

$$P\{a \leq S^2/\sigma^2 \leq b\}$$

for any a and b; i.e., it is given by the distribution of S^2.[1] It will be shown that the distribution of S^2/σ^2 is related to the chi-square distribution.

The particular random variable which has a chi-square distribution is $(n-1)S^2/\sigma^2$. More precisely:

If $X_1, X_2, \ldots, X_n$ are independent normally distributed random variables, each having mean μ and variance σ^2,[2] i.e., a random sample, the random variable $(n-1)S^2/\sigma^2$ has a chi-square distribution with $n-1$ degrees of freedom.

This may appear to be startling since $(n-1)S^2/\sigma^2$ is not directly the sum of squares of $(n-1)$ independent normally distributed random variables, each with zero mean and unit variance. However, this result implies that, by a suitable change of variables, the random variable

$$\frac{(n-1)S^2}{\sigma^2} = \frac{\sum_{i=1}^{n}(X_i - \bar{X})^2}{\sigma^2}$$

can be expressed as

$$\sum_{i=1}^{n-1} N_i^2$$

where the N's are independent normally distributed random variables, each with zero mean and unit variance. This result is not easily proved and will be omitted. A heuristic argument follows. Consider the random variable

$$\frac{(n-1)S^2}{\sigma^2} = \frac{(X_1 - \bar{X})^2 + \cdots + (X_n - \bar{X})^2}{\sigma^2} = \frac{\sum_{i=1}^{n}(X_i - \bar{X})^2}{\sigma^2}.$$

Note that if $\bar{X}$ were replaced by μ, forming

$$\frac{\sum_{i=1}^{n}(X_i - \mu)^2}{\sigma^2},$$

this expression would have a chi-square distribution with n degrees of freedom since the $(X_i - \mu)/\sigma$ are independent normally distributed, each with zero mean and unit variance. The function $(n-1)S^2/\sigma^2$ can be broken up as follows:

---

[1] S^2/σ^2 was used as a measure of closeness instead of $S^2 - \sigma^2$ because the distribution of S^2/σ^2 is easily obtained, whereas the distribution of $S^2 - \sigma^2$ is rather difficult to obtain.

[2] The following structure is appropriate. An experiment is performed. The $X_1, X_2, \ldots, X_n$ can be thought of as a random sample resulting from n trials where each X_i denotes the measurable quantity associated with the ith trial. This interpretation can be used throughout the ensuing sections.

$$\frac{(n-1)S^2}{\sigma^2} = \frac{\sum_{i=1}^{n}(X_i - \bar{X})^2}{\sigma^2} = \frac{\sum_{i=1}^{n}[(X_i - \mu) - (\bar{X} - \mu)]^2}{\sigma^2}$$

$$= \frac{\sum_{i=1}^{n}(X_i - \mu)^2}{\sigma^2} - \frac{n(\bar{X} - \mu)^2}{\sigma^2},$$

or

$$\frac{\sum_{i=1}^{n}(X_i - \mu)^2}{\sigma^2} = \frac{(n-1)S^2}{\sigma^2} + \left(\frac{\bar{X} - \mu}{\sigma/\sqrt{n}}\right)^2.$$

But

$$\frac{\sum_{i=1}^{n}(X_i - \mu)^2}{\sigma^2}$$

has a chi-square distribution with n degrees of freedom. Also, $[(\bar{X} - \mu)/(\sigma/\sqrt{n})]^2$ has a chi-square distribution with 1 degree of freedom since $(\bar{X} - \mu)/(\sigma/\sqrt{n})$ is a normally distributed random variable with zero mean and unit variance. Hence, it is not surprising that $(n-1)S^2/\sigma^2$ has a chi-square distribution with $n-1$ degrees of freedom.

As an example of the use of this probability distribution, suppose a random sample of 10 observations is to be drawn where $X_1, X_2, \ldots, X_{10}$ are independent normally distributed random variables, each with mean μ and variance σ^2. From Appendix Table 2,

$$P\left\{\frac{9S^2}{\sigma^2} \leq 16.919\right\} = 0.95$$

since $9S^2/\sigma^2$ has a chi-square distribution with 9 degrees of freedom. It will be seen in later chapters that the probability distribution of $(n-1)S^2/\sigma^2$ can be used to get interval estimates of the variance when it is unknown. For example, the probability statement

$$P\left\{\frac{9S^2}{\sigma^2} \leq 16.919\right\} = 0.95$$

can be rewritten as

$$P\left\{\sigma^2 \geq \frac{9S^2}{16.919}\right\} = 0.95,$$

which is really a statement about the variance. Furthermore, this probability statement can be used to determine whether the variance is equal to a specified value σ_0^2. For example, if the variance is equal to σ_0^2, then $9S^2/\sigma_0^2$ should usually be less than 16.919.

One further result about the random variable S^2 will be stated without proof:

If $X_1, X_2, \ldots, X_n$ are independent normally distributed random variables with the same mean and variance, the random variables $\bar{X}$ and S^2 are independently distributed.

This implies that the distribution of S^2 is unaffected by announcing the outcome of the random variable $\bar{X}$ even though both $\bar{X}$ and S^2 are computed from the same random sample.

4.3 The *t* Distribution

4.3.1 THE *t* RANDOM VARIABLE

Another continuous random variable of importance is the random variable that has a *t* distribution. This random variable can be generated as follows. An experiment is performed and a sample space is obtained. Suppose there exist two independent random variables, denoted by X and χ^2, defined over this sample space such that X is normally distributed with zero mean and unit variance and χ^2 has a chi-square distribution with v degrees of freedom. Let the random variable t^1 be a function of X and χ^2 and be defined by

$$t = X \sqrt{v} / \sqrt{\chi^2}.$$

For example, an experiment may consist of placing a transistor on test, recording its initial diameter together with its time to failure. If it is reasonable to assume that the diameter D of the transistor is normally distributed with mean μ and variance σ^2, and the time to failure Y is exponentially distributed with parameter θ, then it follows that (1) $X = (D - \mu)/\sigma$ is normal with zero mean and unit variance, and (2) $\chi^2 = 2Y/\theta$ has a chi-square distribution with two degrees of freedom.[2] Furthermore, it may be reasonable to assume that the original diameter of the transistor does not affect the life of the transistor, so that X and χ^2 can be considered to be independent random variables. Then the random variable t is given by

$$t = X \sqrt{2} / \sqrt{\chi^2} = \frac{D - \mu}{\sigma} \bigg/ \sqrt{\frac{Y}{\theta}}.$$

Another example consists of measuring the Rockwell hardness of n specimens, each manufactured and tested under very controlled conditions, i.e., a random sample chosen. Denote by $Y_1, Y_2, \ldots, Y_n$ the Rockwell hardness of the first specimen, second specimen, $\ldots$, nth specimen, respectively. If it is reasonable to assume that the Y_i are normally distributed with

---

[1] For the first time, a lower case letter, *t*, is used to denote a random variable rather than a capital letter. This is done to conform with the customary notation for this random variable.

[2] This is easily seen by examining the CDF of $\chi^2 = 2Y/\theta$, i.e., $P\{2Y/\theta \leq b\} = P\{Y \leq b\theta/2\}$ which equals $1 - e^{-b/2}$ (since Y has an exponential distribution with parameter θ). This is just the CDF of the chi-square random variable with 2 degrees freedom.

mean μ and variance σ^2, then it follows that (1) $X = \sqrt{n}\,(\bar{Y} - \mu)/\sigma$ is normal with zero mean and unit variance and (2)

$$\chi^2 = \frac{\sum\limits_{i=1}^{n} (Y_i - \bar{Y})^2}{\sigma^2}$$

has a chi-square distribution with $(n - 1)$ degrees of freedom (see Sec. 4.2.3). Furthermore, it follows (see Sec. 4.2.3) that X and χ^2 are independent random variables. Then the random variable t is given by

$$t = X\sqrt{n-1}/\sqrt{\chi^2} = \sqrt{n}\,\frac{(\bar{Y} - \mu)}{\sigma}\sqrt{n-1}\Bigg/\sqrt{\frac{\sum\limits_{i=1}^{n}(Y_i - \bar{Y})^2}{\sigma^2}}$$

$$= \sqrt{n}\,(\bar{Y} - \mu)\Bigg/\sqrt{\frac{\sum\limits_{i=1}^{n}(Y_i - \bar{Y})^2}{n - 1}}.$$

The range of the random variable t is given by the interval minus infinity to plus infinity since the values of X lie in this interval and the values of χ^2 are non-negative. All that remains is to determine the distribution of t. This distribution is not arbitrary since the distributions of X and χ^2 are specified. The density function for the t random variable can then be derived (by the technique given in Sec. 2.12 or by other techniques which are not included in this text), and is given by

$$f_t(z) = \frac{1}{\sqrt{\pi v}}\frac{\Gamma\left(\dfrac{v + 1}{2}\right)}{\Gamma\left(\dfrac{v}{2}\right)}\left(1 + \frac{z^2}{v}\right)^{-(v+1)/2}, \qquad -\infty < z < \infty.$$

A continuous random variable whose density function can be expressed as $f_t(z)$ is said to be a t random variable with v degrees of freedom, or alternatively, is said to have a t distribution with v degrees of freedom. The degrees of freedom are those associated with the chi-square component.

Physical phenomena whose distributions can be approximated by a t distribution can arise in nature, although this approximation is rarely made. The probability density is difficult to work with and simpler approximations are usually used. The most important applications of the t distributions arise from studying random variables which are generated in the manner previously described; i.e.,

If X is a normally distributed random variable with zero mean and unit variance, and χ^2 is a random variable which is independent of X and has a chi-square distribution with v degrees of freedom, the random variable

$$t = X\sqrt{v}/\sqrt{\chi^2}$$

has a t distribution with v degrees of freedom.

The usefulness of such random variables in making decisions about the mean of a normal distribution when the variance is unknown will be described in Sec. 4.3.2.

A sketch of the density function of the t random variable is given in Fig. 4.2.

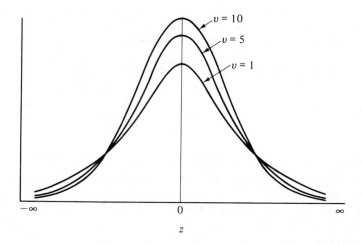

Fig. 4.2. *Distribution function of the t random variable.*

Like the normal distribution with zero mean, the t distribution is symmetric about zero. In fact, as the degrees of freedom, v, tend toward infinity, the t distribution tends toward the standardized normal distribution. The expected value of the t random variable is given by $E(t) = 0$ for $v > 1$, and the variance of the t random variable is given by $\sigma_t^2 = v/(v - 2)$ for $v > 2$.

The probability that t is greater than a constant $t_{\alpha;v}$ is represented by

$$P\{t \geq t_{\alpha;v}\} = \int_{t_{\alpha;v}}^{\infty} f_t(z)\, dz = \alpha.$$

This function has been tabulated and the values of the percentage point $t_{\alpha;v}$ can be found in Appendix Table 3.

For example, if $v = 5$,

$$P\{t \geq t_{0.05;5}\} = P\{t \geq 2.015\} = 0.05;$$

if $v = 29$,

$$P\{t \geq t_{0.05;29}\} = P\{t \geq 1.699\} = 0.05.$$

For $v \geq 30$, $t_{\alpha;v}$ is close to K_α, the percentage point of the normal distribution, except for extreme values of α. Since the t distribution is symmetric about zero, values of $t_{\alpha;v}$ for $\alpha > 0.50$ can be obtained from the relation $t_{\alpha;v} = -t_{1-\alpha;v}$.

4.3.2 THE DISTRIBUTION OF $(\bar{X} - \mu)\sqrt{n}/S$.

In Chapter 3, the distribution of the random variable $(\bar{X} - \mu)\sqrt{n}/\sigma$ was shown to be normal with zero mean and unit variance provided the X's which comprise the sample mean are independent and normally distributed, each with mean μ and variance σ^2. This random variable is useful for making decisions about μ when σ is known. For example, the probability statement

$$P\left\{-1.96 \leq \frac{(\bar{X} - \mu)\sqrt{n}}{\sigma} \leq 1.96\right\} = 0.95$$

can be rearranged to read

$$P\left\{\frac{-1.96\sigma}{\sqrt{n}} \leq \bar{X} - \mu \leq \frac{1.96\sigma}{\sqrt{n}}\right\} = 0.95.$$

Thus, the difference between the estimate, $\bar{X}$, and the quantity to be estimated, μ, lies between plus and minus $1.96\sigma/\sqrt{n}$ with probability 0.95. The statement has usefulness provided σ is known, usually from past data. However, frequently there is no record of past data so that σ is unknown. In this situation it is natural to replace σ by an estimate, S. Unfortunately, the resulting random variable

$$\frac{(\bar{X} - \mu)\sqrt{n}}{S}$$

is no longer normally distributed with zero mean and unit variance. The purpose of this section is to determine the distribution of this random variable.

If $X_1, X_2, \ldots X_n$, are independent normally distributed random variables, each having mean μ and variance σ^2, the random variable $(\bar{X} - \mu)\sqrt{n}/\sigma$ is normally distributed with zero mean and unit variance, and the random variable $(n - 1)S^2/\sigma^2$ has a chi-square distribution with $n - 1$ degrees of freedom (see Sec. 4.2.3). Furthermore, $\bar{X}$ and S^2 are independent random variables (see Sec. 4.2.3). In accordance with the definition of a t random variable given in Sec. 4.3.1,

$$\frac{\dfrac{(\bar{X} - \mu)\sqrt{n}}{\sigma}\sqrt{n-1}}{\sqrt{\dfrac{(n-1)S^2}{\sigma^2}}} = \frac{(\bar{X} - \mu)\sqrt{n}}{S}$$

has a t distribution with $n - 1$ degrees of freedom.[1]

This random variable can be used to make decisions about μ when the standard deviation is unknown. For example, if a random sample of size

---

[1] This result was previously obtained in Sec. 4.3.1 in the context of the Rockwell hardness example.

10 is used, the probability statement

$$P\left\{-t_{0.025;9} \leq \frac{(\bar{X} - \mu)\sqrt{10}}{S} \leq t_{0.025;9}\right\}$$

$$= P\left\{-2.262 \leq \frac{(\bar{X} - \mu)\sqrt{10}}{S} \leq 2.262\right\} = 0.95$$

can be rearranged to read

$$P\left\{\frac{-2.262S}{\sqrt{10}} \leq \bar{X} - \mu \leq \frac{2.262S}{\sqrt{10}}\right\} = 0.95.$$

The 95% limits on $\bar{X} - \mu$ are now $\pm 2.262S/\sqrt{10}$ instead of $\pm 1.960\sigma/\sqrt{10}$, which were the limits when σ is known.

The distribution of $(\bar{X} - \mu)\sqrt{n}/S$ can also be used to make a decision about whether μ is equal to a specified value μ_0 when σ is unknown. If $\mu = \mu_0$, one would expect the random variable $(\bar{X} - \mu_0)\sqrt{n}/S$ to lie in the interval

$$(-t_{\alpha/2;n-1}, t_{\alpha/2;n-1})$$

with probability $1 - \alpha$. Thus, for $1 - \alpha$ large, a value of this random variable falling outside this interval is an indication that μ is not equal to μ_0.

4.3.3 THE DISTRIBUTION OF THE DIFFERENCE BETWEEN TWO SAMPLE MEANS

Industrial experimentation is often concerned with comparing two treatments. A random sample of n_x items is drawn, using the standard treatment, and a random sample of n_y items is drawn, using the new treatment. The treatments are judged according to the magnitude of the value the random variable $\bar{X} - \bar{Y}$ takes on. In order to make reasonable decisions about the effectiveness of these treatments, it is necessary to examine the distribution of $\bar{X} - \bar{Y}$.

Let $X_1, X_2, \ldots X_{n_x}$ be a random sample of n_x independent normally distributed random variables, each having mean μ_x and variance σ^2; let $Y_1, Y_2, \ldots, Y_{n_y}$ be a random sample of n_y independent normally distributed random variables, each having mean μ_y and variance σ^2 (same as for the X's); and also let all the X's and Y's be independent. The distribution of the random variable

$$\bar{X} = \sum_{i=1}^{n_x} X_i/n_x$$

is then normal with mean μ_x and variance σ^2/n_x; the distribution of the random variable

$$\bar{Y} = \sum_{i=1}^{n_y} Y_i/n_y$$

is also normal with mean μ_y and variance σ^2/n_y, and is independent of $\bar{X}$; the distribution of the random variable

$$\frac{(n_x - 1)S_x^2}{\sigma^2} = \frac{\sum_{i=1}^{n_x} (X_i - \bar{X})^2}{\sigma^2}$$

is chi-square with $n_x - 1$ degrees of freedom and is independent of $\bar{X}$ and $\bar{Y}$; and, finally, the distribution of the random variable

$$\frac{(n_y - 1)S_y^2}{\sigma^2} = \frac{\sum_{i=1}^{n_y} (Y_i - \bar{Y})^2}{\sigma^2}$$

is also chi-square with $n_y - 1$ degrees of freedom and is independent of $\bar{X}$, $\bar{Y}$, and S_x^2.

Using the results on linear combinations of normally distributed variables (LCN), it is clear that $\bar{X} - \bar{Y}$ is normally distributed with mean $\mu_x - \mu_y$ and variance $\sigma^2/n_x + \sigma^2/n_y$. Hence, the random variable

$$\frac{\bar{X} - \bar{Y} - (\mu_x - \mu_y)}{\sigma\sqrt{\left(\dfrac{1}{n_x} + \dfrac{1}{n_y}\right)}}$$

is normally distributed with zero mean and unit variance. If the standard deviation σ is known, the problem is resolved. However, if σ is unknown, as in most practical situations, some further steps are necessary.

Using the reproductive property of the chi-square distribution, the random variable

$$\frac{(n_x - 1)S_x^2}{\sigma^2} + \frac{(n_y - 1)S_y^2}{\sigma^2}$$

has a chi-square distribution with $n_x + n_y - 2$ degrees of freedom, and furthermore is distributed independently of $\bar{X} - \bar{Y}$. Hence, in accordance with the definition of a t random variable given in Sec. 4.3.1, the random variable

$$\frac{\dfrac{\bar{X} - \bar{Y} - (\mu_x - \mu_y)}{\sigma\sqrt{\dfrac{1}{n_x} + \dfrac{1}{n_y}}}\sqrt{n_x + n_y - 2}}{\sqrt{[(n_x - 1)S_x^2 + (n_y - 1)S_y^2]/\sigma^2}}$$

$$= \frac{(\bar{X} - \bar{Y}) - (\mu_x - \mu_y)}{\sqrt{\dfrac{(n_x - 1)S_x^2 + (n_y - 1)S_y^2}{n_x + n_y - 2}}\sqrt{\dfrac{1}{n_x} + \dfrac{1}{n_y}}}$$

has a t distribution with $n_x + n_y - 2$ degrees of freedom. Note that this random variable does not contain σ.

4.4 The F Distribution

4.4.1 THE F RANDOM VARIABLE

The last continuous random variable to be studied is the random variable that has an F distribution. This random variable can be generated as follows. An experiment is performed and a sample space is obtained. Suppose there exists a random variable defined over this sample space which has a chi-square distribution with v_1 degrees of freedom. Denote this random variable by χ_1^2. Suppose further that there exists another random variable which is independent of χ_1^2 and has a chi-square distribution with v_2 degrees of freedom. Denote this random variable by χ_2^2. Let the random variable F be a function of χ_1^2 and χ_2^2 and be defined by

$$F = \frac{\chi_1^2/v_1}{\chi_2^2/v_2}.$$

The range of the random variable F is given by the interval zero to plus infinity since the values of χ_1^2 and χ_2^2 are all non-negative. All that remains is to determine the distribution of F. This distribution is not arbitrary since the distributions of χ_1^2 and χ_2^2 are specified. The density function for the F random variable can then be derived (by the technique given in Sec. 2.12 or by other techniques which are not included in this text), and is given by

$$f_F(z) = \begin{cases} \dfrac{\Gamma\left(\dfrac{v_1 + v_2}{2}\right) v_1^{v_1/2} v_2^{v_2/2}}{\Gamma\left(\dfrac{v_1}{2}\right)\Gamma\left(\dfrac{v_2}{2}\right)} \dfrac{(z)^{(v_1/2)-1}}{(v_2 + v_1 z)^{(v_1+v_2)/2}}, & \text{for } z > 0 \\ 0, & \text{otherwise.} \end{cases}$$

A continuous random variable whose density function can be expressed as $f_F(z)$ is said to be an F random variable with v_1 and v_2 degrees of freedom, or alternatively, is said to have an F distribution with v_1 and v_2 degrees of freedom.

Physical phenomena whose probability distribution can be approximated by an F distribution can arise in nature. In particular, the distribution of the tangent of an angle is often approximated by a distribution related to the F. However, the most important applications of this distribution arise from studying random variables which are generated in the manner previously described; i.e.,

If χ_1^2 and χ_2^2 are independent random variables which have chi-square distributions with v_1 and v_2 degrees of freedom, respectively, the random variable

$$F = \frac{\chi_1^2/v_1}{\chi_2^2/v_2}$$

has an F distribution with v_1 and v_2 degrees of freedom.

The usefulness of such random variables in making decisions about the ratio of two sample variances will be described in Sec. 4.4.2.

A sketch of the density function of the F random variable is given in Fig. 4.3. Like the chi-square distribution, this distribution is skewed, with probability mass distributed to the right of the origin only. The expected value of the F random variable is given by $E(F) = v_2/(v_2 - 2)$ for $v_2 > 2$, and the variance of the F random variable is given by

$$\sigma_F^2 = \frac{v_2^2(2v_2 + 2v_1 - 4)}{v_1(v_2 - 2)^2(v_2 - 4)} \quad \text{for } v_2 > 4.$$

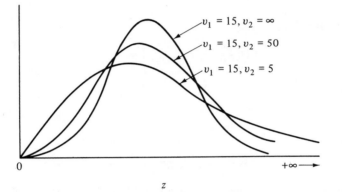

Fig. 4.3. *Density of the F random variable.*

The probability that F is greater than a constant is represented by

$$P\{F > F_{\alpha; v_1, v_2}\} = \int_{F_{\alpha; v_1, v_2}}^{\infty} f_F(z)\, dz = \alpha.$$

Since the F distribution depends on the two parameters v_1 and v_2, a three-way table is needed to tabulate the value of F corresponding to different probabilities and values of v_1 and v_2. Appendix Table 4 is such a table, where percentage points are given for values of $\alpha \leq 50\%$, i.e., the shaded area in Fig. 4.4.

If the percentage points for values of $\alpha > 50\%$ are desired, they may be obtained by using the following equality:

$$P\left\{\frac{\chi_1^2/v_1}{\chi_2^2/v_2} \geq F_{\alpha; v_1, v_2}\right\} = P\left\{\frac{\chi_2^2/v_2}{\chi_1^2/v_1} < \frac{1}{F_{\alpha; v_1, v_2}}\right\} = \alpha,$$

or

$$P\left\{\frac{\chi_2^2/v_2}{\chi_1^2/v_1} \geq \frac{1}{F_{\alpha; v_1, v_2}}\right\} = 1 - \alpha.$$

However, $(\chi_2^2/v_2)/(\chi_1^2/v_1)$ also has an F distribution but with v_2 and v_1

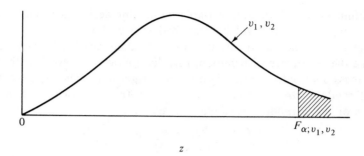

Fig. 4.4. *The percentage points for the F distribution.*

degrees of freedom so that

$$P\left\{\frac{\chi_2^2/\nu_2}{\chi_1^2/\nu_1} \geqq F_{1-\alpha;\,\nu_2,\,\nu_1}\right\} = 1 - \alpha.$$

Hence,

$$F_{1-\alpha;\,\nu_2,\,\nu_1} = \frac{1}{F_{\alpha;\,\nu_1,\,\nu_2}}$$

or

$$F_{\alpha;\,\nu_1,\,\nu_2} = \frac{1}{F_{1-\alpha;\,\nu_2,\,\nu_1}}.$$

For example, the percentage point for $\alpha = 0.95$, $\nu_1 = 6$, and $\nu_2 = 8$, $F_{0.95;\,6,\,8}$, can be obtained as follows:

$$F_{0.95;\,6,\,8} = \frac{1}{F_{0.05;\,8,\,6}} = \frac{1}{4.15} = 0.241.$$

4.4.2 THE DISTRIBUTION OF THE RATIO OF TWO SAMPLE VARIANCES

A problem arising in industrial experimentation is to compare the variability of two processes. A random sample is drawn of n_x items, using process X, and a random sample is drawn of n_y items, using process Y. The sample variances for both processes are compared, i.e., S_x^2/S_y^2. If this ratio is close to unity, the variabilities are judged to be nearly equivalent; otherwise, not. To make good decisions, and to quantify the statement "close to unity," it is necessary to examine the distribution of S_x^2/S_y^2.

Let $X_1, X_2, \ldots, X_{n_x}$ be a random sample of n_x independent normally distributed random variables, each having mean μ_x and variance σ_x^2. Let $Y_1, Y_2, \ldots, Y_{n_y}$ be a random sample of n_y independent normally distributed random variables each having mean μ_y and variance σ_y^2; and also let all the X's and Y's be independent. The distribution of

$$\frac{(n_x - 1)S_x^2}{\sigma_x^2} = \frac{\sum\limits_{i=1}^{n_x} (X_i - \bar{X})^2}{\sigma_x^2}$$

is then chi-square with $n_x - 1$ degrees of freedom (see Sec. 4.2.3). Similarly, the distribution of

$$\frac{(n_y - 1)S_y^2}{\sigma_y^2} = \frac{\sum_{i=1}^{n_y} (Y_i - \bar{Y})^2}{\sigma_y^2}$$

is also chi-square with $n_y - 1$ degrees of freedom. Furthermore, these two chi-square random variables are independent since the X's and Y's are independent. Hence, in accordance with the definition of the F random variable given in Sec. 4.4.1, the random variable

$$\frac{\dfrac{(n_x - 1)S_x^2}{\sigma_x^2}\bigg/(n_x - 1)}{\dfrac{(n_y - 1)S_y^2}{\sigma_y^2}\bigg/(n_y - 1)} = \frac{S_x^2/\sigma_x^2}{S_y^2/\sigma_y^2}$$

has an F distribution with $n_x - 1$ and $n_y - 1$ degrees of freedom.

4.5 The Binomial Distribution

4.5.1 THE BINOMIAL RANDOM VARIABLE

A discrete random variable of importance is the random variable that has a binomial distribution. An experiment is performed and a sample space is generated. *A random variable X defined over this sample space whose probability distribution can be expressed as*

$$P_X(k) = P\{X = k\} = \frac{n!}{k!(n-k)!}p^k(1-p)^{n-k}, \text{ for } k = 0, 1, \ldots, n,$$

where $0 < p < 1$ is a constant and n is a positive integer, is said to be a binomial random variable, or alternatively, is said to have a binomial distribution.[1] Note that this distribution is a function of the two parameters p and n.

Random variables having a binomial distribution may arise naturally in practice as follows: Consider the case of $n = 1$. The random variable X then takes on the values 0 and 1 with probabilities $1 - p$ and p, respectively, i.e.,

$$P_X(0) = P\{X = 0\} = 1 - p$$
$$P_X(1) = P\{X = 1\} = p.$$

Such a random variable is called a Bernoulli random variable. Examples of Bernoulli random variables are (1) the upturned face of a tossed coin, with the numbers 1 and 0 being recorded in place of heads and tails, respec-

---

[1] The term $n!/k!(n-k)!$ is often called the binomial coefficient and is denoted by the symbol $\binom{n}{k}$. Also, for more concise expression, $1 - p$ may be denoted by q. Hence, $P_X(k)$ can be expressed as $P_X(k) = \binom{n}{k}p^k q^{n-k}$.

tively, and where p denotes the probability of the upturned face being a one; (2) the quality of an item, with the numbers 1 and 0 being recorded in place of defective and non-defective, respectively, and where p denotes the probability of an item being defective; and (3) in general, for any dichotomous random variable taking on the values 1 and 0, where 1 denotes a "success" and 0 denotes a "failure," and where p is the probability of success. Whereas these examples are illustrative of Bernoulli random variables, they can be extended to become examples of more general binomial random variables by looking at random samples of Bernoulli random variables. In particular, if $X_1, X_2, \ldots, X_n$ are independent Bernoulli random variables, each having parameter p, then it can be proven that

$$X = X_1 + X_2 + \cdots + X_n$$

has a binomial distribution with parameters p and n. Thus, for the quality control example, if n items are inspected, then the number of defective items, $X = X_1 + X_2 + \cdots + X_n$, among the n is a binomially distributed random variable with parameters p and n, provided that the "quality" of each item is an independent Bernoulli random variable with parameter p.[1] If $n = 18$ and $p = 0.10$, the probability of obtaining k defectives, $k = 0, 1, 2, \ldots,$ 18, is:

k	$P_X(k)$
0	0.150
1	0.300
2	0.284
3	0.168
4	0.070
5	0.022
6	0.005
7	0.001
8	0.000
.	.
.	.
.	.
18	0.000

The probability distribution function of this binomial distribution is plotted in Fig. 4.5.

Without reference to a specific application, a typical characterization for a binomial random variable follows: An experiment is performed, consisting of n trials. Let $X_1, X_2, \ldots, X_n$ be independent dichotomous random variables denoting the outcome of the first trial, second trial, $\ldots$, nth trial, respectively.

---

[1] If p is interpreted as the actual fraction of defective items in a lot of size N, then the "quality" of each item is not an independent Bernoulli random variable with parameter p, and the number of defective items among the n looked at does not have a binomial distribution (it will have a hypergeometric distribution). However, from a practical point of view, if N is large compared to n, the quality of each item is approximately an independent Bernoulli random variable with parameter p.

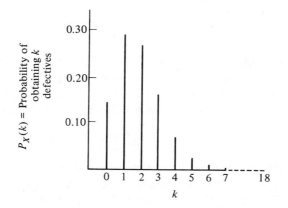

Fig. 4.5. *Probability distribution for the binomial distribution with* $n = 18$ *and* $p = 0.10$.

Let X_i take on the value 1 with probability p if the ith trial results in a success, and 0 with probability $1 - p$ if the ith trial results in a failure. Then the random variable $X = X_1 + X_2 + \cdots + X_n$ denotes the total number of "successes" in n trials and has a binomial distribution with parameters p and n.

The mean of the binomial distribution is easily obtained, i.e.,

$$E(X) = \sum_{k=0}^{n} k \frac{n!}{k!(n-k)!} p^k (1-p)^{n-k}$$

$$= np \sum_{k=1}^{n} \frac{(n-1)!}{(k-1)!(n-k)!} p^{k-1} (1-p)^{n-k}$$

$$= np.$$

The variance is obtained in a similar manner and is found equal to $np(1 - p)$.

An alternative method for obtaining these first two moments uses the property that a binomial random variable can be considered as a sum of n independent identically distributed Bernoulli random variables, each having parameter p. If X_i, $i = 1, 2, \ldots, n$, is such a Bernoulli random variable, then

$$E(X_i) = (0)(1 - p) + (1)(p) = p,$$

and

$$\mathrm{Var}(X_i) = E(X_i - p)^2 = (0 - p)^2 (1 - p) + (1 - p)^2 (p) = p(1 - p).$$

Since $X = X_1 + X_2 + \ldots + X_n$ is a binomial random variable, it follows from LCN (see Sec. 3.4.) that

$$E(X) = E\left(\sum_{i=1}^{n} X_i\right) = \sum_{i=1}^{n} E(X_i) = np$$

and

$$\text{Var}(X) = \text{Var}\left(\sum_{i=1}^{n} X_i\right) = \sum_{i=1}^{n} \text{Var}(X_i) = np(1 - p).$$

A random variable of practical importance related to the binomial is the ratio X/n. In the quality control example this random variable can be interpreted as the fraction defective in the n units sampled. The probability distribution of X/n can be obtained from the binomial as follows:

$$P\left\{\frac{X}{n} \le b\right\} = P\{X \le bn\} = \sum_{k=0}^{[bn]} \frac{n!}{k!(n-k)!} p^k(1 - p)^{n-k},$$

where $[bn]$ is the largest integer less than or equal to bn.

The moments of the random variables X/n are easily obtained, using the results of Sec. 2.11. The expected value of X/n is given by

$$E\left(\frac{X}{n}\right) = \frac{1}{n} E(X) = p,$$

and the variance of X/n is given by

$$\text{Var}\left(\frac{X}{n}\right) = \frac{1}{n^2} \text{Var}(X) = \frac{p(1 - p)}{n}.$$

*4.5.2 Tables of the Binomial Probability Distribution

The binomial distribution has been tabulated for $n = 1(1)49$ and $p = 0.01(0.01)0.50$ in *Tables of the Binomial Probability Distribution*, National Bureau of Standards, Applied Mathematics Series, 6, Washington, 1950. Harry Roming, in a book entitled *50–100 Binomial Tables*, John Wiley & Sons, Inc., New York, 1953, covers the range of n values, $n = 50(5)100$, and the range of p values, $p = 0.01(0.01)0.99$. The most extensive tabulation of the binomial distribution appears in tables of the *Cumulative Binomial Probability Distribution*, Harvard University Press, Cambridge, 1955. This book covers the parameters

$$n = 1(1)50(2)100(10)200(20)500(50)1000$$

and

$$p = 0.01(0.01)0.50,$$

$$\tfrac{1}{16}, \tfrac{1}{12}, \tfrac{1}{8}, \tfrac{1}{6}, \tfrac{3}{16}, \tfrac{5}{16}, \tfrac{1}{3}, \tfrac{3}{8}, \tfrac{5}{12}, \tfrac{7}{16}.$$

Other values may be obtained by use of the identity

$$P\{X \le b\} = \sum_{k=0}^{b} \binom{n}{k} p^k(1 - p)^{n-k} = 1 - P\left\{F < \frac{(n - b)}{(b + 1)} \frac{p}{(1 - p)}\right\}$$

where F is a random variable having the F distribution with $v_1 = 2(b + 1)$ and $v_2 = 2(n - b)$ degrees of freedom, though the usefulness of the identity is restricted by the tables available. For values of the probability distribution outside of this range, the computations may become tedious. Consequently, some form of approximation to the distribution is useful. Three types of approximations will be discussed: the normal approximation, the arc sine transformation, and the Poisson approximation.

*4.5.3 The Normal Approximation to the Binomial

It can be shown that the binomial distribution for a given value of p and increasing values of n converges to the normal distribution. This can be obtained by noting that a binomial random variable can be considered as the sum of n independent Bernoulli random variables, and then invoking the central limit theorem. Consequently, for large values of n, the binomial distribution can be approximated by the normal distribution with mean np and variance npq (where $q = 1 - p$), i.e.,

$$P\{X = k\} = \binom{n}{k} p^k q^{n-k} \simeq \frac{1}{\sqrt{2\pi npq}} e^{-(k-np)^2/2npq}.$$

In most applications, the cumulative sum is required rather than the single terms, i.e.,

$$P\{a \leq X \leq b\} = \sum_{k=a}^{b} \binom{n}{k} p^k q^{n-k}.$$

This expression can be approximated by

$$P\{a \leq X \leq b\} = \sum_{k=a}^{b} \binom{n}{k} p^k q^{n-k} \simeq \int_a^b \frac{1}{\sqrt{2\pi npq}} e^{-(z-np)^2/2npq} \, dz.$$

However, because of the change from a discontinuous to a continuous distribution, a better approximation is given by

$$P\{a \leq X \leq b\} = \sum_{k=a}^{b} \binom{n}{k} p^k q^{n-k} \simeq \int_{a-1/2}^{b+1/2} \frac{1}{\sqrt{2\pi npq}} e^{-(z-np)^2/2npq} \, dz$$

$$= \int_{(a-1/2-np)/\sqrt{npq}}^{(b+1/2-np)/\sqrt{npq}} \frac{1}{\sqrt{2\pi}} e^{-v^2/2} \, dv.$$

No exhaustive study has been made of the accuracy of the normal approximation. However, the approximation is known to be poor for $p < 1/(n + 1)$ or $p > n/(n + 1)$ and outside the interval plus or minus 3 times the standard deviation about the mean, i.e., outside the interval

$$[np - 3\sqrt{npq}, \qquad np + 3\sqrt{npq}].$$

The approximation is known to be good for p close to $\frac{1}{2}$, and Hald[1] indicates that the approximation is good when the range of application is limited by the inequality $npq > 9$.

A similar normal approximation holds for the random variable X/n, the fraction of "successes" in n trials. This fraction X/n is also approximately

---

[1] A. Hald, *Statistical Theory with Engineering Applications*, John Wiley & Sons, Inc., New York, 1952.

normally distributed with mean p and variance pq/n. Hence, making the correction for continuity,

$$P\left\{a \leq \frac{X}{n} \leq b\right\} \simeq \int_{a-1/2n}^{b+1/2n} \frac{\sqrt{n}}{\sqrt{2\pi pq}} e^{-n(z-p)^2/2pq} \, dz$$

$$= \int_{\sqrt{n}(a-1/2n-p)/\sqrt{pq}}^{\sqrt{n}(b+1/2n-p)/\sqrt{pq}} \frac{1}{\sqrt{2\pi}} e^{-v^2/2} \, dv.$$

*4.5.4 THE ARC SINE TRANSFORMATION

It is clear that both the mean and the variance of the binomial distribution depend on p. In most practical problems, p is never known, and there is some advantage in finding a transformation such that at least the variance of the transformed variable is independent of p. Let

$$H = \frac{X}{n} \text{ be the fraction of "successes" in } n \text{ trials}$$

and

$$Y = 2 \text{ arc sin } \sqrt{H}.$$

If H is approximately normally distributed with mean p and variance pq/n, then $Y = 2$ arc sin $\sqrt{H}$ is also approximately normally distributed with mean 2 arc sin $\sqrt{p}$ and variance $1/n$. Hence,

$$P\{a \leq H \leq b\} \simeq \int_{2 \text{ arc sin } \sqrt{a}}^{2 \text{ arc sin } \sqrt{b}} \frac{\sqrt{n}}{\sqrt{2\pi}} e^{-n(z - 2 \text{ arc sin } \sqrt{p})^2/2} \, dz.$$

Again, because of the change from a discontinuous to a continuous distribution, a better approximation is given by

$$P\{a \leq H \leq b\} \simeq \int_{2 \text{ arc sin } \sqrt{a-1/2n}}^{2 \text{ arc sin } \sqrt{b+1/2n}} \frac{\sqrt{n}}{\sqrt{2\pi}} e^{-n(z - 2 \text{ arc sin } \sqrt{p})^2/2} \, dz$$

$$= \int_{\sqrt{n}(2 \text{ arc sin } \sqrt{a-1/2n} - 2 \text{ arc sin } \sqrt{p})}^{\sqrt{n}(2 \text{ arc sin } \sqrt{b+1/2n} - 2 \text{ arc sin } \sqrt{p})} \frac{1}{\sqrt{2\pi}} e^{-v^2/2} \, dv.$$

There has been no systematic study of this approximation formula, but some scattered results indicate that it is as good as the normal approximation.

The approximate probability distribution of X, the number of "successes" in n trials, is obtained from the relation

$$P\{a \leq X \leq b\} = P\left\{\frac{a}{n} \leq H \leq \frac{b}{n}\right\}.$$

*4.5.5 THE POISSON APPROXIMATION TO THE BINOMIAL

For fixed p and increasing n, the binomial distribution converges to the normal distribution. If, as n increases, p approaches zero in such a manner that $np = \lambda$ becomes and remains constant, the binomial distribution converges to the Poisson distribution. The probability distribution of the Poisson random variable was introduced in Sec. 2.7 and is given by

$$P\{X = k\} = \frac{e^{-\lambda}\lambda^k}{(k)!} \qquad \text{for } k = 0, 1, 2, \ldots .$$

The mean and variance of the Poisson random variable were shown (Sec. 2.10) to be equal to λ.

The convergence of the binomial distribution to the Poisson distribution is shown as follows: for n sufficiently large, p approaches zero in such a manner that $np = \lambda$ becomes and remains constant. Hence,

$$\binom{n}{k}p^k q^{n-k} = \frac{n(n-1)\cdots(n-k+1)}{k!} \frac{\lambda^k}{n^k}\left(1 - \frac{\lambda}{n}\right)^{n-k}$$

$$= \left\{\left(1 - \frac{1}{n}\right)\left(1 - \frac{2}{n}\right)\cdots\left(1 - \frac{k-1}{n}\right)\left(1 - \frac{\lambda}{n}\right)^{-k}\right\}\frac{\lambda^k}{k!}\left(1 - \frac{\lambda}{n}\right)^n.$$

As $n \to \infty$, the terms in braces approach 1, and $(1 - (\lambda/n))^n$ approaches $e^{-\lambda}$. Hence,

$$\binom{n}{k}p^k q^{n-k} \to \frac{e^{-\lambda}\lambda^k}{(k)!}.$$

This approximation to the binomial is good when p is small and n is large. It is generally considered justifiable to use the Poisson approximation when $p < 0.1$.

PROBLEMS

1. Referring to Sec. 4.1, find the probability that the sample average shear strength of test spot welds differs from the mean of the distribution by less than 1 pound, i.e., $P\{-1.0 \leq \bar{X} - \mu \leq 1.0\}$ when the individual shear strengths of the spot welds are normally distributed with mean μ and standard deviation 10 pounds, and a sample of sixteen is observed.

2. In Problem 1, how many observations are required so that

$$P\{-1.0 \leq \bar{X} - \mu \leq 1.0\} = 0.90?$$

3. Referring to Sec. 4.1, write down the expression for the sample size in terms of σ, $K_{\alpha/2}$, and b so that

$$P\{-b \leq \bar{X} - \mu \leq b\} = 1 - \alpha.$$

4. Verify the 1% point of the chi-square distribution for $\nu = 2$ by direct integration of the density function.

5. Show that $E(\chi^2) = \nu$ and $\sigma_{\chi^2}^2 = 2\nu$.

6. The sample variance S^2 is called an "estimate of σ^2," where

$$S^2 = \frac{\Sigma(X_i - \bar{X})^2}{n - 1}.$$

Intuitively a desirable property for the sample variance to possess is for $E(S^2) = \sigma^2$.

Prove that $E(S^2) = \sigma^2$, assuming that the X_i are independent normally distributed random variables with mean μ and variance σ^2. Prove the result without assuming normality (but assuming independence and identical distribution).

7. If χ^2 has a chi-square distribution with 9 degrees of freedom, find $P\{\chi^2 \leq 8\}$, $P\{\chi^2 \geq 6\}$, and $P\{4 \leq \chi^2 \leq 9\}$.

8. If χ^2 has a chi-square distribution with 19 degrees of freedom, find $P\{\chi^2 \leq 20\}$, $P\{\chi^2 \geq 15\}$, and $P\{16 \leq \chi^2 \leq 21\}$.

9. In the example on drilling a hole in a rectangular table top given in Sec. 4.2.1, what is the probability that the distance between the measured point and desired location is less than 2.44 units?

10. In the space vehicle problem given in Sec. 4.2.1, what is the probability that the distance between the point of impact and the target exceeds 2.63 feet if $\sigma^2 = 5$?

11. X and Y are horizontal and vertical components of the deviations of a shot from the center of a target. X and Y are independently normally distributed with zero means and standard deviations 8 and 10 inches, respectively. Find the equation of an ellipse such that a shot will fall inside the ellipse with probability 0.95.

12. If χ_1^2 has a chi-square distribution with 7 degrees of freedom and χ_2^2 has an independent chi-square distribution with 3 degrees of freedom, find the probability that $\chi_1^2 + \chi_2^2$ exceeds 12.

13. If χ_1^2 has a chi-square distribution with 4 degrees of freedom and χ_2^2 has an independent chi-square distribution with 5 degrees of freedom, find the probability that $\chi_1^2 + \chi_2^2$ exceeds 19.

14. Find $P\{0.618 \leq S^2/\sigma^2 \leq 1.60\}$ when S^2 is based on a random sample of 11 observations on a normally distributed random variable with unknown mean μ and unknown variance σ^2.

15. Referring to Problem 14, how many observations are required to insure that $P\{0.618 \leq S^2/\sigma^2 \leq 1.60\} \geq 0.95$?

16. The "on" temperature of a thermostatically controlled switch is normally distributed with unknown mean and variance. A random sample is to be drawn and the sample variance computed. How many observations are required to insure that $P\{S^2/\sigma^2 \leq 1.83\} \geq 0.99$?

17. Suppose a random sample of 10 observations of the response to drug X is obtained. Each response is normally distributed and has mean 7 and variance 2. Another random sample of 10 observations of the response to drug Y is also obtained, this random sample being independent of the other. The response to drug Y is normally distributed with mean 9 and variance 2. Find $E(S_x^2 + S_y^2)$.

18. The sample standard deivation S_1 of the resistance of a certain electrical part is to be computed from a sample of seven parts. The sample standard deviation of another component S_2 which is added to the first, making the final assembly, is also computed from a sample of seven parts. It is known that each component is independent and normally distributed with common unknown variance σ^2. An estimate of the variance of the final assembly will be given by $S_1^2 + S_2^2$. What is the probability that $(S_1^2 + S_2^2)/\sigma^2$ will be less than 3?

19. In Problem 18, how many observations on each component will be required to insure that $P\{(S_1^2 + S_2^2)/\sigma^2 \leq 3\} = 0.99$?

20. Show that $\sigma_{R^2}^2 = 2\sigma^4/n$, where $R^2 = \sum_{i=1}^{n} (X_i^2)/n$ and X is normally distributed with zero mean and variance σ^2. *Hint*: Use the relation that $\sigma_{\chi^2}^2 = 2\nu$.

21. Find $E(S^2)$ and $\sigma_{S^2}^2$, where $S^2 = \sum (X_i - \bar{X})^2/(n - 1)$ and X is normally distributed with mean μ and variance σ^2. Hint: Use the relations $E(\chi^2) = \nu$ and $\sigma_{\chi^2}^2 = 2\nu$.

22. Suppose a random sample of size n is drawn where $X_1, X_2, \ldots, X_n$ have normal distributions with mean zero and variance σ^2. (a) Find $E(\sum X_i^2)$. (b) If the assumption of normality is not valid but the X's still have zero mean and variance σ^2, find $E(\sum X_i^2)$.

23. Prove that the density function of the t variate approaches the density function of the standardized normal random variable as the number of degrees of freedom ν becomes infinite. Assume that the constant approaches $1/\sqrt{2\pi}$.

24. Verify the 5% point of the t distribution for $\nu = 1$ by direct integration of the density function.

25. Show that $E(t) = 0$ for $\nu > 1$ and $\sigma_t^2 = \nu/(\nu - 2)$ for $\nu > 2$.

26. If t has a t-distribution with 9 degrees of freedom, find

$$P\{t \geq \tfrac{4}{3}\}, \qquad P\{-\tfrac{8}{3} < t\},$$

and
$$P\{-3.25 \leq t \leq 3.25\}.$$

27. If t has a t-distribution with 25 degrees of freedom, find

$$P\{t \geq \tfrac{10}{9}\}, \qquad P\{-\tfrac{20}{9} \leq t\},$$

and
$$P\{-2.787 \leq t \leq 2.787\}.$$

28. Find $P\{-1.383 \leq [(\bar{X} - \mu)\sqrt{10}]/S\}$ where $\bar{X}$ and S are based on 10 observations.

29. Find $P\{-1.383 \leq [(\bar{X} - \mu)\sqrt{10}]/S\}$ where $\bar{X}$ is based on 10 observations and S on 20 different observations.

30. Show that $t_{\alpha/2;\,\nu}^2 = F_{\alpha;1,\nu}$; i.e., show $P\{t^2 \geq t_{\alpha/2;\,\nu}^2\} = P\{F \geq F_{\alpha;1,\nu}\}$ where $t_{\alpha/2;\,\nu}^2 = F_{\alpha;1,\nu}$. Hint: Use the definitions of the t and F random variables.

31. Show that $E(F) = \nu_2/(\nu_2 - 2)$ for $\nu_2 > 2$, and

$$\sigma_F^2 = \frac{\nu_2^2(2\nu_2 + 2\nu_1 - 4)}{\nu_1(\nu_2 - 2)^2(\nu_2 - 4)} \qquad \text{for } \nu_2 > 4.$$

32. Find the value of $F_{0.05;7,9}$ such that $P\{F \geq F_{0.05;7,9}\} = 0.05$ where F is a random variable having an F distribution with 7 and 9 degrees of freedom.

33. If F is a random variable having an F distribution with 3 and 6 degrees of freedom, find $P\{F \geq 4.76\}$, $P\{F \geq 6.6\}$, and $P\{F \leq 5\}$.

34. Find the value of $F_{0.95;9,7}$, $F_{0.05;7,9}$, and $F_{0.90;4,12}$.

35. Two children throw darts (independently) at a target. Let (X_1, Y_1) denote the point of impact of the first child's dart, and let (X_2, Y_2) denote the point of impact of the second child's dart. Assume X_1 and Y_1 are independent normally distributed random variables with mean 0 and variance σ_1^2. Assume X_2 and Y_2 are independent normally distributed random variables with mean 0 and variance σ_2^2. Let $D_1 = $ distance from point of impact of first child's dart to the center of the target, and similarly for D_2. (a) What must be the ratio σ_1/σ_2 so that $D_1 < D_2$ with probability 0.90? (b) If $\sigma_1 = \sigma_2$, what is the probability that $D_1 < D_2$? Justify your answer.

36. The variability of two treatments is to be compared. A random sample of

10 is to be obtained from each treatment and the sample variance computed. Find

$$P\left\{\frac{S_1^2/\sigma_1^2}{S_2^2/\sigma_2^2} \geq 3\right\} = P\left\{\frac{\sigma_2^2}{\sigma_1^2} \geq 3\frac{S_2^2}{S_1^2}\right\}.$$

37. If $X_1, \cdots, X_{11}$ are independent normally distributed random variables, each with mean μ_x and variance σ^2, and $Y_1, \cdots, Y_{21}$ are independent normally distributed random variables, each with mean μ_y and variance σ^2, find

$$P\left\{\sum_{i=1}^{11} (X_i - \bar{X})^2 \geq 1.685 \sum_{j=1}^{21} (Y_j - \bar{Y})^2\right\}.$$

38. If $Z = \sum_{i=1}^{n} (X_i - \bar{X})^2 / \sum_{i=1}^{n} (Y_i - \bar{Y})^2$, where X_i is distributed normally mean μ_x and variance σ_x^2 and Y_i is distributed normally and independent of X_i with mean μ_y and variance σ_y^2, find $E(Z)$.

39. If $X_1, X_2, \cdots, X_n$ are independent normally distributed random variables with mean μ_x and standard deviation σ_x, and $Y_1, Y_2, \cdots, Y_m$ are also independent normally distributed random variables with mean μ_y and standard deviation σ_y (X's and Y's are also independent of each other), state whether the following random variables are normally distributed (N), t distributed (t), chi-square distributed (χ^2), F distributed (F), or none of the above (0):

(a) $\dfrac{(\bar{X} - \mu_x)\sqrt{n}}{S_x}$

(b) $\dfrac{(X_1 - \mu_x)}{S_x}$

(c) $\dfrac{X_1 - \mu_x}{\sigma_x}$

(d) $\dfrac{(n - 1)S_x^2}{\sigma_x^2} + \dfrac{\sum_{i=1}^{m} (Y_i - \bar{Y})^2}{\sigma_y^2}$

(e) S_x^2/S_y^2

(f) $\bar{X}/\bar{Y}$

(g) $\sum_{i=1}^{n} X_i^2 / \sigma_x^2$

(h) $\dfrac{\bar{X} - \bar{Y}}{\sqrt{(\sigma_x^2/n) + (\sigma_y^2/m)}}$

(i) $\dfrac{\bar{X}\sqrt{n}}{S_x}$

(j) $X_1 - Y_1$.

40. Let $X_i, i = 1, \cdots, 7$, and $Y_j, j = 1, \cdots, 9$, be independent random samples from normal distributions both with zero means and variances of $\frac{1}{4}$ and $\frac{1}{3}$, respectively. (a) Find

$$P\left\{4 \sum_{i=1}^{7} X_i^2 > 6 \sum_{j=1}^{9} Y_j^2\right\}.$$

(b) Determine the value of the constant a such that

$$P\{a\bar{X} > S_y\} = 0.05,$$

where $\bar{X}$ is the sample mean of the $X_i, i = 1, \cdots, 7$, and S_y the sample standard deviation of the $Y_j, j = 1, \cdots, 9$.

41. Suppose X_1, X_2 are independent normal random variables with means μ_1 and μ_2 and variances σ_1^2 and σ_2^2, respectively, and

$$Y = a_1 X_1^2 + a_2 X_2^2 + a_3 X_1 + a_4 X_2 + a_5.$$

(a) Find values of the a_i's such that Y will have a chi-squared distribution with 2 degrees of freedom. (b) If $\sigma_1^2 = \sigma_2^2$, find $P\{X_1 - \mu_1 > \sqrt{(X_2 - \mu_2)^2}\}$.

42. If the probability of an item being defective is 0.2, what is the probability of obtaining exactly 0, 1, 2, and 3 defectives in a sample of size 3? What is the probability of obtaining less than 2 defectives?

43. If the probability of an item being defective is 0.1, use the Poisson approximation to obtain the probability that a sample of size 50 will contain at least three defectives; exactly zero defectives.

44. Solve Problem 43, using the normal approximation.

45. In manufacturing plants it is often useful to estimate the amount of time an employee is gainfully occupied. In "work sampling" the fraction of time that an employee works (p) is often estimated by taking a random sample and finding the ratio of the number of times he is working to the total number of tries, i.e., $\hat{p} = D/n$. The estimate $\hat{p}$ is then related to the binomial random variable. Assuming the normal approximation, find the expression for the number of observations necessary such that the estimated value ($\hat{p}$) differs from the true value (p) by no more than ± 0.03 with probability greater than 0.95. In probability terms, it is desired to find an n such that

$$P\{-0.03 \leq \hat{p} - p \leq 0.03\} > 0.95.$$

If the problem has been solved correctly, the solution is a function of p. Two cases should be considered. If it is known that p is around 0.70, find the value of n. If p is completely unknown, find the value of n which is a maximum for any value of p between 0 and 1.

46. In the recent presidential polls, it was desired to predict the winner before the election. Simplify the problem by assuming that there are only two candidates, and that a candidate is elected by a majority of the popular vote. A random sample of size n is to be taken by the pollsters prior to the election, and it is necessary to determine the required sample size. The fraction of people in the sample who vote for candidate 1, V_1/n, is to be used as the estimate of p, the fraction of the population voting for candidate 1. Assuming that the binomial model is applicable and using the normal approximation to the binomial, find n such that

$$P\left\{-0.01 \leq \frac{V_1}{n} - p \leq 0.01\right\} \geq 0.90.$$

Two cases should be considered: (1) it is known that p is approximately 0.50, and (2) p is completely unknown so that it is necessary to choose the maximum n for any value of p between 0 and 1.

47. Solve Problem 43, using the arc sine transformation.

48. Solve Problem 45, using the arc sine transformation.

49. Solve Problem 46, using the arc sine transformation.

5

DECISION MAKING

5.1 Introduction

It has already been pointed out that statistics is a science which deals with making decisions based on observed data in the face of uncertainty. For example, an experimenter may wish to perform an experiment to determine whether (1) a new method of sealing vacuum tubes will increase their life, (2) a new alloy will have an increased breaking strength, (3) a new source of raw material has resulted in a change in the variability of output, (4) a special treatment of concrete will have an effect on breaking strength, (5) the average Rockwell hardness of a certain type of material is a fixed value on the B scale, etc. Each of these examples calls for making a decision based on inferences drawn from a sample. The "uncertainty" is reflected in the fact that the quantity of interest is unknown and the same experiment performed under the same conditions a second time will usually give different sample results. Consequently, if the experiment is performed only once, the decision will be a function of which sample is obtained.

There are other situations which call for decision making even in the absence of experimental data. Based on the mathematical abstraction of a physical phenomenon being studied, decisions may be required based on an analysis of alternative actions and the "consequences" of these actions. Section 5.2 will consider a framework for the analysis of such problems when experimental data is unavailable, Sec. 5.3 will then extend this framework to incorporating experimental data into the decision making process, and Sec. 5.4 will then consider in detail some special cases of decision making based on observed data.

5.2 Decision Making without Experimentation

In describing a general framework for analyzing a decision problem it is necessary to assume that the decision maker must make a choice among various alternative courses of action, and the consequences of these actions are known and dependent on the "state of the real world" (often called the "state of nature"). To illustrate these ideas, consider a variation of the Rockwell hardness example presented in Chapter 2. Suppose that a company manufactures an expensive product containing a steel shaft of special composition. The steel shafts are purchased from Mill A at a unit price of 100 dollars, and the company performs several costly operations on these shafts, such as milling, grinding, etc. Because the final Rockwell hardness is an important characteristic, each unit is tested for Rockwell hardness after all these operations are completed, testing being done at this time because the work performed on the shaft changes its initial hardness characteristics. The company has found that this final Rockwell hardness is a normally distributed random variable with mean equal to 72 on the B scale and a standard deviation of 2. If the final Rockwell hardness of a specimen falls below 68 or above 76, it is considered to be too soft or too hard, respectively, and it must be given further treatment (shot peening or annealing, respectively) at an additional unit cost of 150 dollars in order to bring its hardness into the proper range. Although this cost is relatively high compared to the original cost, it is relatively small compared to the cost of performing the additional operations on the initial shaft.

The manufacturer is offered an opportunity to buy 500 unbranded steel shafts of the type required for his product at a unit cost of 70 dollars from a competitor who is discontinuing the manufacture of this product. The fact that the shafts are unbranded implies that the manufacturer is unaware of the identity of the mill that produced the shafts, although all 500 shafts are produced by the same mill. He is also offered the option to purchase only 250 of these shafts at a unit cost of 75 dollars. Note that either alternative results in substantial savings over the 100-dollar unit price of shafts purchased directly from Mill A. However, it is common knowledge that there are only two mills that produce this special type of steel, Mill A (the current supplier) and Mill B. The company had previous dealings with Mill B and had found that the final Rockwell hardness of shafts from their stock is also normally distributed with a standard deviation of 2, but with a mean equal to 75. Mill B sells their stock at a unit price of 70 dollars. Since an order for 500 shafts is due, the manufacturer is faced with taking an action.

5.2.1 ACTION SPACE

In general, define the action space as the set of all possible actions available to the decision maker and denote this set by $\underline{A}$. Denote by a the elements of the action space, i.e., the individual actions. In the example, there are five possible actions:

a_1:   Purchase the 500 unbranded shafts being offered,
a_2:   Purchase 500 shafts direct from Mill A,

a_3:   Purchase 250 unbranded shafts and 250 shafts from Mill A,

a_4:   Purchase 250 unbranded shafts and 250 shafts from Mill B,

a_5:   Purchase 500 shafts direct from Mill B.

Presumably the manufacturer has analyzed the relative merits of Mill A versus Mill B previously and prefers Mill A. This would preclude actions a_4 and a_5, but he will take this opportunity to reexamine his decision. Thus, $\mathbf{A}$ is the set $\{a_1, a_2, a_3, a_4, a_5\}$. This implies that these five actions exhaust all the possible actions open to the manufacturer. In general, the action space can consist of a finite or infinite number of elements, depending on the problem to be considered. Other finite numbers of actions can be obtained for the Rockwell hardness example by changing the number of possible alternatives that the manufacturer can have, e.g., buy 100 of the unbranded shafts. An infinite number of actions can be obtained by modifying the example so that the manufacturer will agree to purchase all the unbranded shafts. However, he must choose a price (action) that he is willing to pay; this price may or may not be accepted by the supplier.[1]

5.2.2 STATES OF NATURE

The consequences of an action taken may depend on the "state of nature" (or, alternatively, the "state of the real world") as well as on the particular action taken. A description of these states is given by an exhaustive (mutually exclusive) enumeration of the indices which form a classification of the abstracted mathematical model of the physical phenomenon under consideration. Usually the "states of nature" are characterized by specifying a family of probability distributions; with each state represented by a possible value of *the* parameter (occasionally vector valued) of this family.[2] Thus, in the Rockwell hardness example the states of nature are characterized by the class of normal distributions with unknown mean and known standard deviation equal to 2; and the means $\mu = 72$ and $\mu = 75$ then identify the two possible states ($\mu = 72$ if the unbranded shafts come from Mill A whereas $\mu = 75$ if they come from Mill B). In general, the set of all possible states of nature, i.e., the state space, will be denoted by Θ and the individual states will be denoted by θ. Using the Rockwell hardness example again, the state space Θ contains two elements θ_1 and θ_2, where θ_1 corresponds to the mean of a

---

[1] Actually this is a finite action problem because units of money are discrete. However, if such units were on a continuous scale, this would be an infinite action problem. This phenomenon is similar to problems encountered with all measuring instruments.

[2] A parameter of a probability distribution is a constant associated with the distribution. A parameter may be a moment such as the mean of an exponential distribution or just a number such as the degrees of freedom for a chi-square distribution. For a normal distribution with *unknown* mean μ and standard deviation σ, (μ, σ) may be viewed as a vector valued parameter.

normal distribution equal to 72 ($\mu = 72$) and θ_2 corresponds to the mean of a normal distribution equal to 75 ($\mu = 75$).

The number of "states of nature" may be finite or infinite. Any finite number k of states can be obtained for this Rockwell hardness example by changing the number of possible mills from 2 to k and specifying the mean associated with each mill. An infinite number of possible states of nature can be obtained by modifying the example to indicate that the unbranded shafts are produced by an unknown mill with no previous history so that the mean Rockwell hardness can be almost any number.

5.2.3 LOSS FUNCTION

For each combination of action and state of nature, the consequence of the action that the decision maker has taken must be measured. Therefore, the existence of a loss function $l(a, \theta)$ shall be postulated which measures the "loss" incurred if action a is taken when nature is in state θ. This "loss" is usually expressed in monetary units, but the framework is applicable for other reasonable utility measures. The term loss function is used to represent the consequences even though these consequences may be measured in terms of cost, gain, profit, etc. Such measures as gain or profit may be viewed as negative losses.

Thus, for every combination of a and θ there exists a number $l(a, \theta)$. Sometimes it is reasonable to assume that these numbers are available directly, but more usually the loss incurred depends on the outcome of a random variable which, of course, depends on the true state of nature. In this case $l(a, \theta)$ is interpreted as the expected value of the loss incurred if action a is taken when nature is in state θ.

This is actually the situation for the Rockwell hardness scenario, where the losses reflect the purchase cost plus the expected cost of rework, but a detailed discussion of how the numbers $l(a, \theta)$ are obtained will be postponed until Sec. 5.2.6. For the present, it will be assumed that the loss function is known and is as presented in Table 5.1.

Table 5.1. Loss Function for Rockwell Hardness Example

Actions Purchase	States of Nature for Unbranded Shafts	
	$\theta_1: \mu = 72$	$\theta_2: \mu = 75$
a_1: 500 unbranded	38,420	58,153
a_2: 500 Mill A	53,420	53,420
a_3: 250 unbranded 250 Mill A	47,170	57,036
a_4: 250 unbranded 250 Mill B	49,536	59,403
a_5: 500 Mill B	58,153	58,153

Note that the loss for actions a_2 and a_5 are independent of the state of the unbranded shafts. This occurs because the shafts are purchased from Mill A and Mill B, respectively, and, hence, are not functions of the unpurchased unbranded shafts. It should be emphasized that the losses incurred are to be considered as given numbers associated with an action and a state of nature and are not part of the decision making mechanism.

5.2.4 CRITERIA FOR CHOOSING AMONG ACTIONS

If the best action is defined to be that action which minimizes the loss, it is possible to select the best action provided that the state of nature is known. In the Rockwell hardness example, if it is known that the mean Rockwell hardness of the unbranded shafts is 72 (i.e., they were actually produced in Mill A), the best action is a_1 and the 500 unbranded shafts should be purchased. Similarly, if it is known that the mean Rockwell hardness of the unbranded shafts is 75 (i.e., they were actually produced in Mill B), the best action is a_2 and 500 shafts should be purchased from Mill A. However, for most real world problems, and with the scenario presented for the Rockwell hardness example, the true state of nature is not known so that choosing the best action is not obvious. However, there are some rational principles that should be followed. For example, actions a_4 and a_5 will not be chosen by a rational decision maker since the losses incurred are always greater than the losses incurred using actions a_3 and a_2, respectively, for each state of nature. In the context of the example, the purchaser should never buy shafts knowingly from Mill B since he is always better off by buying from Mill A. In general, *an action a* is said to be dominated by an action a if*

$$l(a, \theta) \leq l(a^*, \theta)$$

for all possible states θ. Thus, actions a_4 and a_5 are dominated by actions a_3 and a_2, respectively. A rational decision maker will not use dominated actions. Therefore, Table 5.1 can be replaced by Table 5.2.

After eliminating the dominated actions, there still exists a need for a criterion to choose among the remaining actions. Alternative approaches that seem to have practical importance are the minimax principle and criteria that utilize prior knowledge about the possible states of nature.

**Table 5.2. Loss Function for Rockwell Hardness Example—
Dominated Actions Eliminated**

Actions Purchase	States of Nature for Unbranded Shafts	
	$\theta_1: \mu = 72$	$\theta_2: \mu = 75$
a_1: 500 unbranded	38,420	58,153
a_2: 500 Mill A	53,420	53,420
a_3: 250 unbranded 250 Mill A	47,170	57,036

The minimax principle reflects a very conservative point of view. For each action, the decision maker finds the maximum loss that can be incurred as a function of the possible states of nature. He then chooses that action which minimizes the maximum loss. In the Rockwell hardness example,

$$\max_{\mu} l(a_1, \mu) = 58,153,$$

$$\max_{\mu} l(a_2, \mu) = 53,420,$$

and
$$\max_{\mu} l(a_3, \mu) = 57,036.$$

Thus, the minimum of these values is 53,420, so that the decision maker will select action a_2 and purchase the 500 shafts from Mill A if he believes in this principle. As an aside this framework can be viewed as a problem in game theory, where nature becomes the decision maker's opponent, and all the ideas of game theory can be used to aid in obtaining a solution.

It is evident that the minimax principle is very conservative, particularly since nature is not really a conscious cpponent. Furthermore, the adoption of this principle implies that the decision maker is unwilling to describe his degree of belief about the unknown state of nature in terms of a probability distribution. If the decision maker is willing to postulate a (subjective) probability distribution over the unknown states of nature which expresses his psychological attitude, then this can be incorporated into his decision making mechanism for choosing among alternative actions.

Those who advocate utilizing previous knowledge about the possible states of nature generally state that the decision maker has accumulated experience about the possible states of nature which is expressable in terms of a probability distribution. Even though this experience is subjective in that it represents the personal beliefs of the decision maker, and different decision makers may make different assessments, they suggest that it still should be used for decision making purposes. For those situations where the decision maker feels completely ignorant about the possible states of nature it is often argued that all possible states should be considered to be equally likely so that this becomes a special probability distribution. Section 5.2.5 considers the implications, in detail, of adapting the point of view that the decision maker can describe his degree of belief about the unknown state of nature in terms of a (subjective) probability distribution.

5.2.5 BAYES PRINCIPLE

If the decision maker is willing to postulate a probability distribution over the possible states of nature, *the state of nature can be treated as a random variable; the (subjective) probability distribution, which describes his degree of belief about the possible states of nature, is called a prior probability distribution.* By utilizing the information given by the prior distribution and the loss function, the decision maker is able to establish a framework for choosing between alternative actions.

Returning to the Rockwell hardness example, recall that Mill A and Mill B are the only two mills that produce the special type of steel required by the manufacturer. Furthermore, by utilizing published information he is able to conclude that two-thirds of the production of this special steel comes from Mill A and the remaining one third from Mill B. This information will be used to establish the prior distribution. For the unbranded shafts, the probability that they were produced in Mill A (thereby having a mean Rockwell hardness of 72) is two-thirds and the probability that they were produced in Mill B (thereby having mean Rockwell hardness of 75) is one-third, i.e.,

$$P\{\mu = 72\} = \tfrac{2}{3}$$
$$P\{\mu = 75\} = \tfrac{1}{3}.[1]$$

In general, if the state of nature, θ, is a discrete random variable, denote its prior probability distribution by $P\{\theta = k\} = \pi_\theta(k)$. If the state of nature, θ, is a continuous random variable, its prior density function will be denoted by $\pi_\theta(z)$.[2]

Since the state of nature, θ, is now considered to be a random variable, $l(a, \theta)$ is also a random variable. Hence, a reasonable criterion for measuring the consequences of an action is to compute the expected loss, $E[l(a, \theta)]$, where the expectation is taken with respect to the prior probability distribution defined over the possible states of nature, i.e.,

$$l(a) = E[l(a, \theta)] = \begin{cases} \sum\limits_{\text{all } k} l(a, k)\,\pi_\theta(k), & \text{if } \theta \text{ is discrete} \\ \int_{-\infty}^{\infty} l(a, z)\,\pi_\theta(z)\,dz, & \text{if } \theta \text{ is continuous.} \end{cases}$$

For the Rockwell hardness example

$$l(a_1) = E[l(a_1, \theta)] = l(a_1, 72)\,P\{\mu = 72\} + l(a_1, 75)\,P\{\mu = 75\}$$
$$= (38{,}420)(\tfrac{2}{3}) + (58{,}153)(\tfrac{1}{3}) = 44{,}998,$$

$$l(a_2) = E[l(a_2, \theta)] = l(a_2, 72)\,P\{\mu = 72\} + l(a_2, 75)\,P\{\mu = 75\}$$
$$= (53{,}420)(\tfrac{2}{3}) + (53{,}420)(\tfrac{1}{3}) = 53{,}420,$$

$$l(a_3) = E[l(a_3, \theta)] = l(a_3, 72)\,P\{\mu = 72\} + l(a_3, 75)\,P\{\mu = 75\}$$
$$= (47{,}170)(\tfrac{2}{3}) + (57{,}036)(\tfrac{1}{3}) = 50{,}459.$$

---

[1] Note that for this example specifying the state of nature θ is equivalent to specifying the mean Rockwell hardness μ.

[2] Note that for the first time the same symbol is being used for both a probability distribution and a density function. They may be distinguished by their arguments, i.e., k and z, respectively.

The Bayes principle calls for the decision maker to choose the action that minimizes the expected loss; this action is known as the Bayes procedure or the Bayes rule. The Bayes procedure for the Rockwell hardness example is to choose action a_1 which calls for purchasing the 500 unbranded shafts, with an expected cost of 44,998 dollars.

It is rather interesting to characterize the Bayes procedure as a function of the prior distribution when the prior distribution is allowed to vary. Since this distribution is completely determined by specifying $P\{\mu = 72\}$ (since $P\{\mu = 75\} = 1 - P\{\mu = 72\}$), the expected loss can be plotted as a function of this probability, and is shown in Fig. 5.1. The heavy lines represent the Bayes procedures since they constitute the smallest expected losses. If $P\{\mu = 72\}$ is less than 0.24, the best action to take is a_2, otherwise a_1 is the preferred action. Note that a prior distribution does not exist for which action a_3 is optimal.

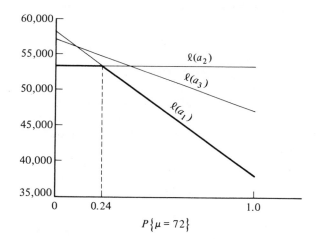

Fig. 5.1. *Plot of expected loss for the Rockwell hardness example.*

5.2.6 EVALUATION OF THE LOSS FUNCTION FOR THE ROCKWELL HARDNESS EXAMPLE

In Sec. 5.2.3, the loss function for the Rockwell hardness example was introduced and presented in Table 5.1. It was also pointed out that sometimes it is reasonable to assume that losses are available directly, but more usually the loss incurred depends on the outcome of a random variable which, of course, depends on the true state of nature. In this case, $l(a, \theta)$ is interpreted as the expected value of the loss incurred if action a is taken when nature is in state θ. In particular, in the Rockwell hardness problem, if the 500 unbranded units are purchased (action a_1 taken), the scenario indicates that the cost of producing a completed product depends on whether or not the Rockwell hardness (R.h.) of the final shaft falls in the interval 68 to 76, i.e.,

$$\text{Unit cost} = \begin{cases} \text{Unit purchase price} + \text{unit rework cost} \\ 70 + 150\,Y,^1 \end{cases}$$

where Y is a (Bernoulli) random variable which takes on the values

$$1, \text{ if R.h.} < 68 \text{ or R.h.} > 76$$

$$0, \text{ if } 68 \leq \text{R.h.} \leq 76.$$

Furthermore, the probability distribution of this random variable is given by

$$P\{Y = 1\} = P\{\text{R.h.} < 68\} + P\{\text{R.h.} > 76\} = p$$

$$= \int_{-\infty}^{68} \frac{1}{\sqrt{2\pi}(2)} e^{-(z-\mu)^2/2(4)}\, dz + \int_{76}^{\infty} \frac{1}{\sqrt{2\pi}(2)} e^{-(z-\mu)^2/2(4)}\, dz,$$

and

$$P\{Y = 0\} = P\{68 \leq \text{R.h.} \leq 76\} = 1 - p = 1 - P\{Y = 1\}$$

$$= \int_{68}^{76} \frac{1}{\sqrt{2\pi}(2)} e^{-(z-\mu)^2/2(4)}\, dz.$$

Note that p is a function of μ, and μ is either 72 or 75. The symbol p can be interpreted as the fraction of completed shafts that require rework. With the aid of Appendix Table I and from Fig. 5.2, this fraction can be evaluated and found equal to

$$p = \begin{cases} 0.0456, \text{ if } \mu = 72 \\ 0.3087, \text{ if } \mu = 75. \end{cases}$$

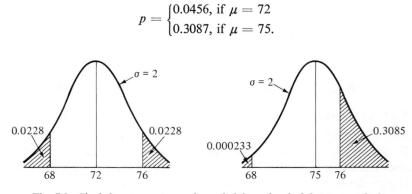

Fig. 5.2. *Shaded areas represent the probability of a shaft being reworked.*

---

[1] The costs associated with performing the additional operations on the shafts are omitted since they are common to all possible actions and, hence, have no bearing on the choice among alternatives.

Thus, it is evident that the unit cost is itself a random variable and its expectation is a function of the state of nature, i.e., action a_1 is taken,

$$\text{Expected unit cost} = 70 + 150\ E(Y)$$
$$= 70 + 150\ [(1)(p) + (0)(1 - p)]$$
$$= 70 + 150p$$
$$= \begin{cases} 76.840, \text{ if } \mu = 72 \\ 116.305, \text{ if } \mu = 75. \end{cases}$$

Since action a_1 calls for purchasing 500 unbranded shafts, the loss function for action a_1 is given by

$$l(a_1, 72) = (500)(76.840) = 38{,}420, \text{ and}$$
$$l(a_1, 75) = (500)(116.305) = 58{,}153.$$

As previously mentioned, the scenario indicates that the cost of producing a completed product is a random variable and, hence, the expected loss must be computed. Equivalently, the scenario could have stated directly that the loss function for action a_1 was given by

$$l(a_1, \mu) = (500)(70 + 150p),$$

where p is to be interpreted as the fraction of shafts that must be reworked (those shafts whose final Rockwell hardness falls outside the interval 68 to 76) as a function of the average Rockwell hardness μ. A simpler loss function could have been obtained if the scenario stated (somewhat arbitrarily) that the loss incurred is 38,500 dollars when action a_1 is taken and $\mu = 72$; and is 58,000 dollars when $\mu = 75$.[1] In these last two cases the losses are not random variables and, hence, can be used directly. A more complicated loss function could have been obtained if the scenario called for the loss for action a_1 to increase as the Rockwell hardness falls further outside the interval 68 to 76, e.g.,

$$\text{Unit cost} = \begin{cases} 70, & \text{if } 68 \leqq \text{R.h.} \leqq 76 \\ 70 + 150\ (72 - \text{R.h.})^2, & \text{if R.h.} < 68 \\ 70 + 150\ (\text{R.h.} - 72)^2, & \text{if R.h.} > 76. \end{cases}$$

For this case, the unit cost is again a random variable so that its expectation must be computed.

Returning to the Rockwell hardness example as originally stated, a table of the loss function can be constructed whose entries correspond to the expected loss incurred for each combination of action and state of nature. This is shown in Table 5.1. The entries for action a_1 have already been derived, and the remaining entries in Table 5.1 (corresponding to actions $a_2, a_3, a_4,$ and a_5) are easily calculated. For actions a_3 and a_4, the expected

---

[1] This is actually what occurred by asserting that Table 5.1 represented the losses.

unit cost for purchasing *unbranded* shafts is obtained in a manner similar to that used for action a_1, and is given by

$$75 + 150E(Y) = 75 + 150p = \begin{cases} 81.840, \text{ if } \mu = 72 \\ 121.305, \text{ if } \mu = 75. \end{cases}$$

For action a_3, the expected unit cost for purchasing shafts from Mill A is given by

$$100 + 150E(Y) = 100 + 150p = 106.840,$$

since p is always 0.0456 when shafts are produced in Mill A. Hence, if action a_3 is taken,

$$l(a_3, 72) = 250(81.840) + 250(106.840) = 47,170,$$

and

$$l(a_3, 75) = 250(121.305) + 250(106.840) = 57,036.$$

Note that the component of the loss function due to the 250 shafts purchased from Mill A is always 250(106.840), even when the *unbranded* shafts are produced in Mill B, i.e., it is independent of the state of the unbranded shafts.

For action a_4, the expected unit cost for purchasing shafts from Mill B is given by

$$70 + 150E(Y) = 70 + 150p = 116.305,$$

since p is always 0.3087 when shafts are produced in Mill B. Hence, if action a_4 is taken,

$$l(a_4, 72) = 250(81.840) + 250(116.305) = 49,536,$$

and $\quad l(a_4, 75) = 250(121.305) + 250(116.305) = 59,403.$

Again, note that the component of the loss function due to the 250 shafts purchased from Mill B is always 250(116.305) even when the unbranded shafts are produced in Mill A, i.e., it is independent of the state of the unbranded shafts.

For action a_2, the expected unit cost for purchasing shafts from Mill A is given by

$$100 + 150E(Y) = 100 + 150p = 106.840,$$

since p is always 0.0456 when shafts are produced in Mill A. Therefore,

$$l(a_2, 72) = l(a_2, 75) = 500(106.840) = 53,420,$$

and is independent of the state of the unbranded shafts.

Finally, for action a_5, the expected unit cost for purchasing shafts from Mill B is given by

$$70 + 150E(Y) = 70 + 150p = 116.305,$$

since p is always 0.3087 when shafts are produced in Mill B. The loss function is then given by

$$l(a_5, 72) = l(a_5, 75) = 500(116.305) = 58,153,$$

and is also independent of the state of the unbranded shafts.

5.2.7 FURTHER EXAMPLES

In the first example, the states of nature will be characterized by specifying a family of discrete probability distributions, and the set of possible actions will contain a countably infinite number of points. As an illustration, suppose a store advertises weekly "specials" in the Sunday newspapers which are available at a reduced price by mail if postmarked (with accompanying check) by Monday, or otherwise may be purchased at the store on Wednesday. The attractiveness of these advertised specials, from the viewpoint of potential customers, depends on numerous factors, e.g., the type of merchandise offered and the discount price, but may be categorized as "good," "average," and "poor." If the attractiveness is "good," the number of customers who appear at the store on Wednesday may be represented by a Poisson random variable with parameter equal to 24. If the attractiveness is "average," the number of customers who appear at the store on Wednesday may be represented by a Poisson random variable with parameter equal to 20. Finally, if the attractiveness of the advertisement is "poor," the number of customers who appear at the store on Wednesday may be represented by a Poisson random variable with parameter equal to 16. The attractiveness of the advertisement on the mail order business can be assumed to be similar to that on the store purchases so that the "good," "average," and "poor" classification is again appropriate. Since the level of sales is comparable for the two, the mail order response, when viewed as a function of this classification, can be considered to have the same probability distribution as the store purchase response. The problem confronting the manager of the store is to determine the number of clerks that he should have available on Wednesday to service the customers. At present the mail order aspect of the business is not causing any trouble.

There are three possible states of nature, θ_1, θ_2, and θ_3, which represent the parameters of a Poisson distribution, $\lambda = 24, \lambda = 20$, and $\lambda = 16$, respectively. The store manager must select an action which corresponds to choosing the number of clerks available on Wednesday. Hence, action a_i will be denoted by the choosing of an integer $i, i = 1, 2, \ldots$. An appropriate loss function which balances the cost of clerk idleness with the cost of customer waiting time can be chosen from queueing theory and is presented in Table 5.3. From this table it is easily seen that action a_4, making

Table 5.3. Table of the Loss Function (expressed in dollars
per day) for Store Example

Action	States of Nature		
	$\theta_1: \lambda = 24$	$\theta_2: \lambda = 20$	$\theta_3: \lambda = 16$
a_2: 2 servers	203	95	33
a_3: 3 servers	64	51	45
a_4: 4 servers	64	62	61

4 clerks available, is a dominated action. Furthermore, after completing
the calculations required for obtaining entries in Table 5.3,[1] it becomes
evident that actions which call for making 5 or more clerks available are
also dominated actions since there will be virtually no customer-waiting and
the loss is essentially due only to the clerks' idle time (their salary). Similarly,
action a_1, making a single clerk available, will also be a dominated action
because of excessive customer-waiting (the system becomes unstable).

The store manager is experienced with the effects of this form of advertising and has concluded (subjectively) that the attractiveness of the advertisement placed in last Sunday's paper (from the viewpoint of potential
customers) has a one-in-ten chance of being "good," a five-in-ten chance
of being "average," and a four-in-ten chance of being "poor." Hence, the
prior probability distribution can be interpreted as

$$P\{\lambda = 24\} = 0.1, P\{\lambda = 20\} = 0.5, \text{ and } P\{\lambda = 16\} = 0.4.$$

The expected loss associated with action i (with the expectation taken with

---

[1] As a reference, see F. S. Hillier and G. J. Lieberman, *Introduction to Operations
Research*, Holden-Day, San Francisco, 1967. The expected loss per day, $l(i, \lambda)$, is given by

$$l(i, \lambda) = (i - \lambda/\mu)k + cL$$

where

μ is the average number of customers serviced on Wednesday (taken equal to 16 for
this example),

c is the cost of a customer waiting in line to be served (taken equal to $100 per day
per waiting customer),

k is the cost of an idle clerk (taken equal to $20 per day per clerk),

$[i - (\lambda/\mu)]$ represents the average number of idle clerks per day,

L is the expected number of customers waiting in line to be served. The expression for
L is given by

$$L = \frac{P_0(\lambda/\mu)^i(\lambda/i\mu)}{i![1 - (\lambda/i\mu)]^2},$$

where

$$P_0 = \frac{1}{\sum_{j=0}^{i-1} \frac{(\lambda/\mu)^j}{j!} + \frac{(\lambda/\mu)^i}{i!}\left(\frac{1}{1 - (\lambda/i\mu)}\right)}.$$

respect to the prior probability distribution) is then given by

$$l(i) = E[l(i, \lambda)] = 0.1l(i, 24) + 0.5l(i, 20) + 0.4l(i, 16).$$

Using the losses given in Table 5.3, the expected losess are calculated to be

$$l(2) = 0.1(203) + 0.5(95) + 0.4(33) = 81.0$$
$$l(3) = 0.1(64) \quad + 0.5(51) + 0.4(45) = 49.9.$$

Hence, the Bayes procedure is to have 3 clerks available on Wednesday.

The second example to be considered in this section is the variation of the Rockwell hardness example alluded to in Sec. 5.2, where there are now an infinite number of possible states of nature. It will be assumed that mills other than Mill A and Mill B produce this special steel and that the un-branded shafts being offered for sale are produced by an unknown mill. However, it is the judgment of the decision maker that the average Rock-well hardness of this type of special steel should usually be within 72 ± 3. He agrees to interpret this statement as implying that the prior density func-tion of the average Rockwell hardness of the special steel can be assumed to be normally distributed with mean 72 and standard deviation 1.5,[1] i.e.,

$$\pi_\theta(z) = \pi_\mu(z) = \frac{1}{\sqrt{2\pi}\, 1.5}\, e^{-(z-72)^2/2(1.5)^2}, \qquad -\infty \leqq z \leqq \infty.$$

The purchaser is interested only in "estimating" the average Rockwell hardness of the mill that produced the unbranded shafts, i.e., he wants to choose a number which will be used, for decision making purposes, as an estimate of the average Rockwell hardness. Thus, the "actions" are equated to these estimates so that the set of possible actions consists of all numbers a such that $-\infty \leqq a \leqq \infty$. The loss incurred will be judged in terms of the magnitude of the difference between the action a taken and the actual value of the average Rockwell hardness μ. In particular, the "seriousness" of the size of this difference increases *quickly* as the estimate falls further away from the actual average Rockwell hardness so that the loss function,

$$l(a, \mu) = c(\mu - a)^2,$$

appears to be reasonable, where c is a positive constant. The expected loss becomes

$$l(a) = E[l(a, \mu)] = c \int_{-\infty}^{\infty} (z - a)^2 \frac{1}{\sqrt{2\pi}\, 1.5}\, e^{-(z-72)^2/2(1.5)^2}\, dz.$$

The Bayes procedure is to choose the value of a that minimizes $l(a)$. Note that $l(a)$ is just c times the second moment of the random variable μ about

---

[1] The basis for this assumption is that the decision maker interprets the interval state-ment as $P\{-3 \leqq \mu - 72 \leqq 3\} = 0.95$. Hence, $1.96\,\sigma = 3$, and $\sigma \sim 3/2 = 1.5$.

a point a. Using the parallel axis theorem discussed in Sec. 2.10, with μ now treated as a random variable, $l(a)$ can be expressed as

$$l(a) = c[\text{Variance of } \mu + (a - E(\mu))^2].$$

Hence, $l(a)$ is clearly minimized by choosing a equal to $E(\mu)$. Thus, the Bayes procedure calls for choosing the mean of the prior distribution (72) as the estimate of the average Rockwell hardness of the mill that produced the specimens. It is evident that this result holds for any prior distribution.

5.3 Decision Making with Experimentation

Throughout the examples discussed in Sec. 5.2 no mention was made about the availability of experimental data. If such information is made available, possibly at a cost, it is reasonable to assume that it should be utilized for decision making purposes. Presumably, this additional information should enable the decision maker to take "better actions." In the original Rockwell hardness example presented in Sec. 5.2, assume that the prospective purchaser of the 500 unbranded shafts is allowed to draw a random sample of size 4, perform the necessary operations, test them for their final Rockwell hardness, and then take an appropriate action.[1] Recall that taking a random sample of size 4 implies that a sequence, X_1, X_2, X_3, and X_4, of independent identically distributed random variables is generated. In this example, each of the X's is an independent and identically normally distributed random variable with standard deviation equal to 2 and with mean equal to 72 or 75, depending on whether the shafts were produced in Mill A or Mill B. Surely the outcome of the random sample should provide the purchaser with "additional information" about whether the shafts were produced in Mill A or Mill B. The utilization of such experimentation will be discussed in the ensuing sections.

5.3.1 DECISION PROCEDURES

When experimental data is to be made available in the form of a random sample $X_1, X_2, \ldots, X_n$, the decision maker (now also called an experimenter) is interested in choosing a decision procedure which is dependent on the outcome of the experiment. *A decision procedure (or decision rule) is a rule that specifies* (1) *the amount (sample size) and form of experimentation, and* (2) *the action that is to be taken for each possible outcome of the random sample obtained from experimentation.* This decision procedure is to be chosen before the experiment is performed. In order to quantify this definition, denote by X some chosen random variable which describes the "information" contained in the random sample of size n. The random variable X can

---

[1] If the 500 unbranded shafts are not purchased, an agreement is reached whereby the manufacturer is able to keep the 4 tested units at no gain or loss.

represent such random variables as the sample mean, the sample variance, the sample median, the largest observation in the sample, the random variable X_1, the vector random variable $(X_1, X_2, \ldots, X_n)$, and, in general, any function of the random sample. In order to relate the action a that is to be taken with each possible outcome of the random variable X, define a function d such that

$$a = d[x]$$

for all possible values x that the random variable X can take on, i.e., if the random variable X takes on the value x, the decision maker takes action $a = d[x]$. Therefore, it is evident that the function d is a decision procedure.

Returning to the Rockwell hardness example, a particular decision procedure chosen somewhat arbitrarily can be described as follows. Draw a random sample of size 4 and observe the random variable $\bar{X}$. (Note that the sample mean $\bar{X}$ plays the role of X in the above formulation.) If

$\bar{X} < 73$, take action a_1 (purchase 500 unbranded shafts).

$73 \leq \bar{X} \leq 74$, take action a_3 (purchase 250 unbranded shafts and 250 shafts from Mill A).

$\bar{X} > 74$, take action a_2 (purchase 500 shafts from Mill A).

Thus, a decision procedure d_1 can be defined functionally as

$$d_1[\bar{x}] = \begin{cases} a_1 \text{ if } \bar{x} < 73, \\ a_3 \text{ if } 73 \leq \bar{x} \leq 74, \\ a_2 \text{ if } \bar{x} > 74. \end{cases}$$

A second decision procedure, d_2, also chosen arbitrarily, can be described as follows: Draw a random sample of size 4 and observe the outcome of the random variable $\bar{X}$. Define $d_2[\bar{x}]$ so that

$$d_2[\bar{x}] = \begin{cases} a_1 \text{ if } \bar{x} < 73 \text{ (take action } a_1), \\ a_2 \text{ if } \bar{x} \geq 73 \text{ (take action } a_2). \end{cases}$$

A third decision procedure, d_3, which uses a different random variable to characterize the information contained in the random sample, is described as follows: Draw a random sample of size 3 and observe the outcome of the random variable, the sample median $\tilde{X}$. Define $d_3[\tilde{x}]$ so that

$$d_3[\tilde{x}] = \begin{cases} a_1 \text{ if } \tilde{x} < 73, \\ a_3 \text{ if } 73 \leq \tilde{x} \leq 75, \\ a_2 \text{ if } \tilde{x} > 75. \end{cases}$$

These are but three of the many possible decision procedures, all chosen arbitrarily. However, a decision maker is interested in taking "good" actions so that criteria for choosing among rules must be developed.

5.3.2 RISK FUNCTION

Before comparing decision procedures it is necessary to find an appropriate measure of their consequences. When experimentation is permitted, and before the data are observed, the action to be taken by using a decision procedure is a function of the random variable X, i.e.,

$$a = d[X],$$

and, hence, $d[X]$ is also a random variable.[1] The particular action chosen is given by $d[x]$. When the random variable X takes on the value x, the random variable $d[X]$ takes on the value $d[x]$.

In the absence of experimental data, a loss function $l(a, \theta)$ was defined which measured the loss incurred if action a was taken when nature was in state θ. If this is extended to include experimentation, then a natural measure of the consequences of using decision procedure $d[X]$ when nature is in state θ is given by $l(d[X], \theta)$. However, since $d[X]$ is a random variable, $l(d[X], \theta)$ is also a random variable so that its expectation, $E[l(d[X], \theta)]$, is a more reasonable measure. *This expectation, $E[l(d[X], \theta)]$, is called the risk function, $R(d, \theta)$,* i.e.,

$$R(d, \theta) = E[l(d[X], \theta)],$$

where the expectation is taken with respect to the probability distribution of the random variable X when the state of nature is θ.

Whenever the set of possible actions is discrete, either finite or countably infinite, $R(d, \theta)$ can easily be evaluated. The case for a finite number, m, of actions will be presented, but the countably infinite case is a simple extension. Let $a_1, a_2, \ldots, a_m$ denote the m possible actions and let $l(a_i, \theta)$, $i = 1, 2, \ldots, m$, denote the loss incurred if action a_i is taken when the state of nature is θ. A decision procedure d can be defined functionally as follows:

$$d[x] = \begin{cases} a_1 & \text{if } x \in B_1, \\ a_2 & \text{if } x \in B_2, \\ \quad \vdots \\ a_m & \text{if } x \in B_m, \end{cases}$$

where B_i are disjoint sets, and $B_1 \cup B_2 \cup \ldots \cup B_m$ is the set of all possible values that the random variable X can take on. The risk, $R(d, \theta)$, can now

---

[1] Since $d[X]$ is a random variable, the notation should indicate that a is also a random variable by denoting it by A. However, this will not be done because it is virtually impossible to maintain a consistent notation to differentiate between a random variable and the value the random variable takes on. Although some attempt will be made throughout the remainder of the text, there will be many instances where the rule is violated. The content of the material should enable the reader to distinguish between the random variable and the value the random variable takes on.

be expressed as

$$R(d, \theta)$$
$$= l(a_1, \theta)P\{X \in B_1\} + l(a_2, \theta)P\{X \in B_2\} + \cdots + l(a_m, \theta)P\{X \in B_m\}.^1$$

The decision procedure d_1 introduced in Sec. 5.3.1 for the Rockwell hardness example will be evaluated. Recall that the random variable $\bar{X}$ was chosen to describe the information contained in the random sample of size 4. The decision procedure d_1 was defined as follows:

$$d_1[\bar{x}] = \begin{cases} a_1 \text{ if } \bar{x} < 73, \\ a_3 \text{ if } 73 \leq \bar{x} \leq 74, \\ a_2 \text{ if } \bar{x} > 74. \end{cases}$$

Furthermore, since the Rockwell hardness of the shafts are normally distributed random variables with mean μ (where μ is either 72 or 75) and standard deviation equal to 2, the random variable $\bar{X}$, the sample mean, is also normally distributed with mean μ, but with standard deviation equal to $2/\sqrt{4} = 1$. Hence, using the losses appearing in Table 5.2 and the results for evaluating the risk function when there is a finite number of actions, $R(d_1, 72)$ is given by

$$R(d_1, 72) = (38{,}420)P\{\bar{X} < 73\} + (47{,}170)P\{73 \leq \bar{X} \leq 74\}$$
$$+ (53{,}420)P\{\bar{X} > 74\}$$

where $\bar{X}$ is a normally distributed random variable with mean equal to 72 and standard deviation equal to 1. Using Appendix Table 1, this expression is easily evaluated and found to be equal to

$$R(d_1, 72) = (38{,}420)(0.8413) + (47{,}170)(0.1359) + (53{,}420)(0.0228)$$
$$= 39{,}951.$$

In a similar manner, $R(d_1, 75)$ is given by

$$R(d_1, 75) = 58{,}153 P\{\bar{X} < 73\} + 57{,}036 P\{73 \leq \bar{X} \leq 74\} + 53{,}420 P\{\bar{X} > 74\},$$

where $\bar{X}$ is a normally distributed random variable with mean equal to 75 and standard deviation equal to 1. Using Appendix Table 1 once again, $R(d_1, 75)$ is found equal to

$$R(d_1, 75) = (58{,}153)(0.0228) + 57{,}036(0.1359) + (53{,}420)(0.8413)$$
$$= 54{,}019.$$

The decision procedure d_2, introduced in Sec. 5.3.1 for the Rockwell hardness example, can also be evaluated. Recall that the random variable $\bar{X}$ was again chosen to describe the information in the random sample of size 4.

---

[1] It has been assumed implicitly that the loss function does not depend on the actual sample outcome except through the above characterization of the decision function. This assumption will hold throughout the text unless otherwise stated (see Sec. 13.2.8).

However, the decision procedure d_2 was defined as

$$d_2[\bar{x}] = \begin{cases} a_1 & \text{if } \bar{x} < 73, \\ a_2 & \text{if } \bar{x} \geq 73. \end{cases}$$

The risks are given by

$$R(d_2, 72) = (38,420)\text{P}\{\bar{X} < 73\} + (53,420)\text{P}\{\bar{X} \geq 73\} = 40,801,$$

where $\bar{X}$ is a normally distributed random variable with mean equal to 72 and standard deviation equal to 1, and

$$R(d_2, 75) = (58,153)\text{P}\{\bar{X} < 73\} + 53,420\,\text{P}\{\bar{X} \geq 73\} = 53,528,$$

where $\bar{X}$ is a normally distributed random variable with mean equal to 75 and standard deviation equal to 1.

In comparing these two decision procedures, d_1 and d_2, it is clear that d_1 is preferred to d_2 when the average Rockwell hardness is 72 since $R(d_1, 72) < R(d_2, 72)$. However, d_2 is preferred to d_1 when the average Rockwell hardness is 75 since $R(d_2, 75) < R(d_1, 75)$. Hence, a choice between these two decision procedures cannot be made by using the criterion that the risk is smallest for both possible values of the state of nature. In fact, the experimenter would be interested in finding an optimal decision procedure in the sense that it has uniformly smallest risk for every possible value of the state of nature. Unfortunately, such an optimal rule does not generally exist, and other criteria for selecting optimal decision procedures are required. One such criterion is the minimax criterion. In comparing d_1 and d_2, the maximum risk under decision rule d_1 is incurred when $\mu = 75$ ($R(d_1, 75) = 54,019$), and the maximum risk under decision rule d_2 is also incurred when $\mu = 75$ ($R(d_2, 75) = 53,528$). Since the minimum of these two risks is 53,528, the minimax decision procedure for comparing the two rules is d_2. As noted earlier, this criterion is extremely conservative. A Bayes criterion will be presented in the next section.

5.3.3 BAYES DECISION PROCEDURES

In Sec. 5.2.5, it was pointed out that if the decision maker is willing to postulate a probability distribution over the possible states of nature, the state of nature can be treated as a random variable; the (subjective) probability distribution, which describes his degree of belief, is referred to as a prior probability distribution. Just as in the no-data case, the decision maker is able to establish a framework for choosing between alternative decision rules by utilizing the prior distribution and the risk function when experimental data become available. A reasonable criterion for measuring the consequences of any given decision procedure d is to compute the expected risk, where the expectation is taken with respect to the prior probability distribution defined over the possible states of nature. Denote this expected risk for a given decision procedure d by $B(d)$. If the state of nature, θ, is considered to be a discrete random variable having probability distribution $\pi_\theta(k) =$

$P\{\theta = k\}$, then $B(d)$ is given by

$$B(d) = \sum_{\text{all } k} R(d, k)\, \pi_\theta(k) = \sum_{\text{all } k} R(d, k)\, P\{\theta = k\}.$$

If the state of nature, θ, is considered to be a continuous random variable with probability density $\pi_\theta(z)$, then $B(d)$ is given by

$$B(d) = \int_{-\infty}^{\infty} R(d, z)\, \pi_\theta(z)\, dz.$$

In this framework and utilizing the prior distribution given in Sec. 5.2.5 for the Rockwell hardness example, decision procedures d_1 and d_2 can be compared. For this example, $P\{\mu = 72\} = \frac{2}{3}$ and $P\{\mu = 75\} = \frac{1}{3}$. The risks for d_1 and d_2 were obtained in Sec. 5.3.2, i.e.,

$$R(d_1, 72) = 39{,}951, \qquad R(d_1, 75) = 54{,}019; \quad \text{and}$$

$$R(d_2, 72) = 40{,}801, \qquad R(d_2, 75) = 53{,}528.$$

Thus,

$$B(d_1) = (39{,}951)(\tfrac{2}{3}) + (54{,}019)(\tfrac{1}{3}) = 44{,}640; \quad \text{and}$$

$$B(d_2) = (40{,}801)(\tfrac{2}{3}) + (53{,}528)(\tfrac{1}{3}) = 45{,}044,$$

so that d_1 is a better decision procedure according to this criterion.

This technique is useful for evaluating the consequences of some given decision procedures, but it does not find the "optimal" decision procedure for the experimenter. This "optimal" rule can be obtained by again using the Bayes principle. *The Bayes principle calls for the decision maker to choose the decision procedure that minimizes the expected risk, $B(d)$; this decision procedure is known as the Bayes rule or the Bayes procedure.*

This is a simple criterion, but the actual determination of the procedure d is generally difficult. Instead of finding this optimal rule which requires specifying the action that is to be taken for each possible outcome of the random sample obtained from the experimentation, an *equivalent* determination of the optimal procedure will be obtained.

The rule suggested when no experimentation occurs (see Sec. 5.2.5) calls for the decision maker to compute the expected loss for each action, where this expectation is taken with respect to the prior probability distribution, and to choose the action that minimizes the expected loss. In the presence of experimental data this procedure is still intuitively attractive, but the added information should affect the experimenter's degree of belief in the prior distribution. For example, if the purchaser is able to test 4 shafts for their Rockwell hardness, and finds that the sample mean is 74.5, this "appears to be evidence" that the unbranded shafts came from Mill B.

He may no longer be willing to give two-to-one odds (probability $= \frac{2}{3}$ versus probability $= \frac{1}{3}$) that the unbranded shafts were produced in Mill A. This evidence obtained by experimentation should be incorporated into the decision making framework in some objective fashion, i.e., the prior distribution should be "updated" by including the information contained in the outcome of the experiment. An updated prior distribution, utilizing the outcome x of the random variable X, is called a posterior distribution. The posterior distribution of θ is defined to be the conditional distribution of θ given that $X = x$, and will be denoted by

$$g_{\theta \mid X=x}(k), \text{ if } \theta \text{ is a discrete random variable,}$$

and by

$$g_{\theta \mid X=x}(z), \text{ if } \theta \text{ is a continuous random variable.}$$

The usual techniques for obtaining a conditional probability are used to get the posterior distribution, and they will be discussed in detail in the next section.

Following the methods suggested in Sec. 5.2.5 for obtaining Bayes procedures, but replacing the prior distribution with the posterior distribution, calls for the decision maker to compute the *expected loss* for each action, where the expectation is now taken with respect to the *posterior probability distribution*, and to choose the action that minimizes this expected loss. In fact, it can be shown that this method, which utilizes the outcome of experimental data, does indeed lead to the Bayes decision procedure, i.e., it is equivalent to the procedure which minimizes the expected risk. Although this method does not result in the determination of the entire decision procedure, $d[x]$, for each x, it has the advantage of requiring the evaluation of $d[x]$ only at the single point corresponding to the outcome of the random variable X. Of course, the entire function $d[x]$ can be obtained in this manner by evaluating $d[x]$ for every possible outcome of the experiment, but this is cumbersome. However, if the entire function $d[x]$ is desired, there are other methods which can be used to find this function, although they are generally computationally difficult.

To summarize, then, the technique to be used for selecting the action corresponding to the Bayes decision procedure is:

1. Choose a random variable X and observe its outcome.
2. Determine the posterior distribution of θ, i.e., the conditional distribution of θ, given that $X = x$ (methods for determining this posterior distribution are given in Sec. 5.3.4).
3. Compute the expected loss $l_g(a)$[1] for each action a, where the expectation is taken with respect to the posterior distribution, i.e.,

$$l_g(a) = E[l(a, \theta)] = \begin{cases} \displaystyle\sum_{\text{all } k} l(a, k)\, g_{\theta \mid X=x}(k), & \text{if } \theta \text{ is discrete} \\[2mm] \displaystyle\int_{-\infty}^{\infty} l(a, z)\, g_{\theta \mid X=x}(z)\, dz, & \text{if } \theta \text{ is continuous.} \end{cases}$$

---

[1] Note that the symbol $l_g(a)$ is used instead of $l(a)$ to indicate that the expectation is taken with respect to the posterior distribution.

4. Select the action that minimizes $l_g(a)$.

In the Rockwell hardness example, suppose that a random sample of size 4 was taken and the random variable $\bar{X}$ takes on the value 74.5. Also assume that the posterior distribution of θ (this is obtained in Sec. 5.3.4) is given by

$$g_{\theta | \bar{X} = 74.5}(72) = P\{\theta = 72 | \bar{X} = 74.5\} = 0.0905, \quad \text{and}$$

$$g_{\theta | \bar{X} = 74.5}(75) = P\{\theta = 75 | \bar{X} = 74.5\} = 0.9095.$$

Using the entries in Table 5.2, the expected losses are given by

$$l_g(a_1) = (38{,}420)\,(0.0905) + (58{,}153)\,(0.9095) = 56{,}367$$

$$l_g(a_2) = (53{,}420)\,(0.0905) + (53{,}420)\,(0.9095) = 53{,}420$$

$$l_g(a_3) = (47{,}170)\,(0.0905) + (57{,}036)\,(0.9095) = 56{,}143.$$

Since $l_g(a_2)$ is the smallest, the Bayes decision rule calls for taking action a_2, i.e., purchasing 500 shafts directly from Mill A. In this example, the inclusion of experimentation results in a change of action. This is not surprising in view of the posterior distribution which indicates a dramatic change from the prior distribution. A little thought reveals that an outcome of the experiment of $\bar{x} = 74.5$ would be very unlikely if the average Rockwell hardness of the unbranded shafts was 72.

5.3.4 CALCULATION OF THE POSTERIOR DISTRIBUTION

Since the posterior distribution is simply the conditional distribution of θ, given the outcome of the experiment, the content of Secs. 2.12 and 2.15 is applicable. In this context (X, θ) is to be considered as a bivariate random variable. For illustrative purposes, assume that X and θ are both discrete, i.e., (X, θ) is a discrete bivariate random variable.[1] Before presenting the expression for the posterior distribution, $g_{\theta | X = x}(k)$, it is appropriate to describe the individual terms.

Recall that the state of nature θ is now being considered as a random variable. Its probability distribution is called the prior probability distribution and is given by $\pi_\theta(k)$, i.e.,

$$\pi_\theta(k) = P\{\theta = k\}.$$

The probability distribution of the random variable X requires a modified interpretation. Because θ is being treated as a random variable, the probability distribution of X is really a *conditional* probability distribution, conditioned on nature being in some state. Therefore, the usual notation for the probability distribution will be changed to reflect this "conditioning" and show its dependence on the value of the parameter θ. Hence, let the

---

[1] Although these sections are concerned with discrete bivariate random variables and continuous bivariate random variables, the concepts easily extend to "mixed" bivariate random variables, e.g., where θ is discrete and X is continuous.

expression

$$P_{X \mid \theta = k}(j) = P\{X = j \mid \theta = k\}$$

now denote the probability distribution of the random variable X (conditioned on the state of nature). For example, if X is a Poisson random variable with $\lambda = 20$, then the conditional probability distribution of X given $\lambda = 20$ is given by

$$P_{X \mid \lambda = 20}(j) = P\{X = j \mid \lambda = 20\} = \frac{e^{-20} 20^j}{j!}, \qquad \text{for } j = 0, 1, 2, \ldots .$$

The expression for the posterior distribution $g_{\theta \mid X = x}(k)$ (the conditional distribution of θ given $X = x$) can now be presented, and is given by

$$g_{\theta \mid X = x}(k) = \frac{P_{X \mid \theta = k}(x)\, \pi_\theta(k)}{\sum\limits_{\text{all } k} P_{X \mid \theta = k}(x)\, \pi_\theta(k)}, \quad^1$$

where $P_{X \mid \theta = k}(x)$ is the aforementioned ordinary probability distribution evaluated at x (the actual outcome of the experiment).

The posterior distribution for other bivariate random variables can be expressed in a similar manner. When θ is a discrete random variable and X is a continuous random variable, the posterior distribution (the conditional distribution of θ, given $X = x$) is given by

$$g_{\theta \mid X = x}(k) = \frac{f_{X \mid \theta = k}(x)\, \pi_\theta(k)}{\sum\limits_{\text{all } k} f_{X \mid \theta = k}(x)\, \pi_\theta(k)},$$

where $\pi_\theta(k)$ is the prior distribution of the random variable θ, and $f_{X \mid \theta = k}(x)$ represents the ordinary density function of the random variable X evaluated at x (the actual outcome of the experiment) but is expressed as a conditional density to reflect its dependence on the value of the parameter θ.

When (X, θ) is a continuous bivariate random variable, i.e., both X and θ are continuous, the posterior density function (the conditional density of θ given $X = x$) is given by

$$g_{\theta \mid X = x}(z) = \frac{f_{X \mid \theta = z}(x)\, \pi_\theta(z)}{\displaystyle\int_{-\infty}^{\infty} f_{X \mid \theta = z}(x)\, \pi_\theta(z)\, dz},$$

where $\pi_\theta(z)$ is the prior density function of the random variable θ, and $f_{X \mid \theta = z}(x)$ represents the ordinary density function of the random variable X evaluated at x (the actual outcome of the experiment) but expressed as a conditional density to reflect its dependence on the value of the parameter θ.

Finally, when θ is a continuous random variable and X is a discrete random variable, the posterior density function (the conditional density of θ, given

---

[1] This expression is derived in Sec. 5.3.4.1.

$X = x$) is given by

$$g_{\theta \mid X = x}(z) = \frac{P_{X \mid \theta = z}(x)\, \pi_\theta(z)}{\displaystyle\int_{-\infty}^{\infty} P_{X \mid \theta = z}(x)\, \pi_\theta(z)\, dz},$$

where $\pi_\theta(z)$ is the prior density function of the random variable θ, and $P_{X \mid \theta = z}(x)$ represents the ordinary probability distribution of the random variable X evaluated at x (the actual outcome of the experiment) but expressed as a conditional density to reflect its dependence on the value of the parameter θ.

As an illustration of the calculations required for obtaining the posterior distribution, consider the Rockwell hardness example. Recall that a random sample of size 4 was taken and the random variable $\bar{X}$ takes on the value $\bar{x} = 74.5$. Since θ is a discrete random variable and $\bar{X}$ is a continuous random variable, the second expression given is applicable, i.e.,

$$g_{\mu \mid \bar{X} = 74.5}(k) = \frac{f_{X \mid \mu = k}(74.5)\, \pi_\mu(k)}{\displaystyle\sum_{\text{all } k} f_{X \mid \mu = k}(74.5)\, \pi_\mu(k)}, \qquad \text{for } k = 72 \text{ and } 75,$$

where $\pi_\mu(k)$ is the prior distribution of the random variable μ and $f_{X \mid \mu = k}(74.5)$ is the density function of the sample mean evaluated at 74.5, i.e., it is the density function of a normally distributed random variable with mean $\mu = k$ and standard deviation $2/\sqrt{4} = 1$. Therefore, the posterior distribution of θ is given by

$$g_{\mu \mid \bar{X} = 74.5}(72) = \frac{\dfrac{2}{3}\left[\dfrac{1}{\sqrt{2\pi}} e^{-(74.5 - 72)^2/2}\right]}{\dfrac{2}{3}\left[\dfrac{1}{\sqrt{2\pi}} e^{-(74.5 - 72)^2/2}\right] + \dfrac{1}{3}\left[\dfrac{1}{\sqrt{2\pi}} e^{-(74.5 - 75)^2/2}\right]}$$

$$= \frac{\dfrac{2}{3}}{\dfrac{2}{3} + \dfrac{1}{3}[e^3]} = 0.0905,$$

$$g_{\mu \mid \bar{X} = 74.5}(75) = \frac{\dfrac{1}{3}\left[\dfrac{1}{\sqrt{2\pi}} e^{-(74.5 - 75)^2/2}\right]}{\dfrac{2}{3}\left[\dfrac{1}{\sqrt{2\pi}} e^{-(74.5 - 72)^2/2}\right] + \dfrac{1}{3}\left[\dfrac{1}{\sqrt{2\pi}} e^{-(74.5 - 75)^2/2}\right]}$$

$$= \frac{\dfrac{1}{3}}{\dfrac{2}{3} e^{-3} + \dfrac{1}{3}} = 0.9095.$$

Although the computation of the denominator in the expression for the posterior distribution is necessary for evaluating this probability, it does not play a role in the selection of alternative actions because it is common to all the risk calculations. Omitting the denominator from the posterior distribution has the effect of inflating the risk for each action by this factor, thereby leaving the selection of the Bayes decision rule unchanged.

*5.3.4.1 *Derivation of Posterior Distribution.* The posterior distribution will be derived for the case where (X, θ) is a discrete bivariate random variable. The derivation for the other cases is similar.

Denote the joint probability distribution of the discrete bivariate random variable (X, θ) by

$$P_{X\theta}(j, k) = P\{X = j, \theta = k\}.$$

This joint distribution can be expressed in terms of the conditional distribution of X, given $\theta = k$, and the marginal distribution of θ, i.e.,

$$P_{X\theta}(j, k) = P\{X = j \mid \theta = k\} P\{\theta = k\} = P_{X\mid\theta=k}(j) \pi_\theta(k).$$

Furthermore, this expression can be used to obtain the marginal distribution $P_X(j)$ of the random variable X, i.e.,

$$P_X(j) = \sum_{\text{all } k} P_{X\theta}(j, k) = \sum_{\text{all } k} P_{X\mid\theta=k}(j) \pi_\theta(k).$$

Finally, the joint distribution can also be expressed in terms of the conditional distribution of θ, given $X = j$ (the posterior distribution), and the marginal distribution of X, i.e.,

$$P_{X\theta}(j, k) = P\{\theta = k \mid X = j\} P\{X = j\} = g_{\theta\mid X=j}(k) P_X(j).$$

Equating these two alternative expressions for $P_{X\theta}(j, k)$ leads to the expression for the posterior distribution, i.e.,

$$g_{\theta\mid X=j}(k) = \frac{P_{X\mid\theta=k}(j) \pi_\theta(k)}{P_X(j)}.$$

Finally, letting $j = x$ (the outcome of the experiment) and substituting for $P_X(j)$, leads to the equation given in Section 5.3.4, i.e.

$$g_{\theta\mid X=x}(k) = \frac{P_{X\mid\theta=k}(x) \pi_\theta(k)}{\sum_{\text{all } k} P_{X\mid\theta=k}(x) \pi_\theta(k)}.$$

5.3.5 FURTHER EXAMPLES

This section extends the examples presented in Sec. 5.2.7 which the reader is urged to review. Recall that the first example dealt with advertising in a newspaper. The attractiveness of the advertisement (from the viewpoint of potential customers) was classified as good, average, and poor where these states represented parameters of Poisson distributions equal to 24, 20, and 16, respectively. The manager had to determine the number of clerks to be made available for servicing the customers on Wednesday, where the number of customers arriving at the store was considered to be a Poisson random variable with parameter dependent on the state. Recall also that it was stated that "the mail order response, when viewed as a function of this classification (good, average, and poor), can be considered to have the same probability distribution as the store purchase response." Although the mail order business is causing no problems, the store manager now realizes that the response from the mail order data is equivalent to experimentation. Furthermore, since all the mail orders arrive by Tuesday, these data can be used to aid in the decision making process. The number of mail order responses received by Tuesday is 13, i.e., the actual outcome x of the Poisson random variable is 13. The loss function is given in Table 5.3, and the prior distribution is given by

$$P\{\lambda = 24\} = 0.1, \; P\{\lambda = 20\} = 0.5, \; \text{and} \; P\{\lambda = 16\} = 0.4.$$

The posterior distribution is required. Since (X, θ) is a discrete bivariate random variable, the initial expression for the posterior distribution given in Sec. 5.3.4 is applicable, i.e.,

$$g_{\lambda \mid X = 13}(k) = \frac{P_{X \mid \lambda = k}(13) \, \pi_\lambda(k)}{\sum\limits_{\text{all } k} P_{X \mid \lambda = k}(13) \, \pi_\lambda(k)}, \qquad \text{for } k = 24, \, 20 \text{ and } 16,$$

where $\pi_\lambda(k)$ is the prior distribution of the random variable λ, and $P_{X \mid \lambda = k}(13)$ is the probability distribution of a Poisson random variable with parameter $\lambda = k$ evaluated at the point 13. Therefore, making use of Appendix Table 5 (Tables of Poisson Distribution), the posterior distribution of λ is given by[1]

$$g_{\lambda \mid X = 13}(24) = \frac{(0.1)\left[\dfrac{(24)^{13} \, e^{-24}}{13!}\right]}{(0.1)\left[\dfrac{(24)^{13} e^{-24}}{13!}\right] + (0.5)\left[\dfrac{(20)^{13} e^{-20}}{13!}\right] + (0.4)\left[\dfrac{(16)^{13} e^{-16}}{13!}\right]}$$

$$= \frac{(0.1)(0.006)}{(0.1)(0.006) + (0.5)(0.027) + (0.4)(0.082)} = 0.013.$$

$$g_{\lambda \mid X = 13}(20) = \frac{(0.5)\left[\dfrac{(20)^{13} \, e^{-20}}{13!}\right]}{(0.1)\left[\dfrac{(24)^{13} e^{-24}}{13!}\right] + (0.5)\left[\dfrac{(20)^{13} e^{-20}}{13!}\right] + (0.4)\left[\dfrac{(16)^{13} e^{-16}}{13!}\right]}$$

$$= \frac{(0.5)(0.027)}{(0.1)(0.006) + (0.5)(0.027) + (0.4)(0.082)} = 0.288.$$

$$g_{\lambda \mid X = 13}(16) = \frac{(0.4)\left[\dfrac{(16)^{13} \, e^{-16}}{13!}\right]}{(0.1)\left[\dfrac{(24)^{13} e^{-24}}{13!}\right] + (0.5)\left[\dfrac{(20)^{13} e^{-20}}{13!}\right] + (0.4)\left[\dfrac{(16)^{13} e^{-16}}{13!}\right]}$$

$$= \frac{(0.4)(0.082)}{(0.1)(0.006) + (0.5)(0.027) + (0.4)(0.082)} = 0.699.$$

---

[1] In the notation of this example, Appendix Table 5 is a table of the cumulative sums of Poisson terms, i.e., $P\{X \leq j \mid \lambda = k\}$. To find $P\{X = j \mid \lambda = k\}$ from these tables, the identity $P\{X = j \mid \lambda = k\} = P\{X \leq j \mid \lambda = k\} - P\{X \leq j - 1 \mid \lambda = k\}$ is used.

Using the entries in Table 5.3, the expected losses are given by

$$l_g(a_2) = (0.013)(203) + (0.288)(95) + (0.699)(33) = 53.1$$
$$l_g(a_3) = (0.013)(64) + (0.288)(51) + (0.699)(45) = 47.0.$$

Since $l_g(a_3)$ is the smallest, the Bayes decision procedure calls for taking action a_3, i.e., having 3 clerks available on Wednesday. This is the same action that was selected without experimentation. Although the availability of data changed the posterior distribution rather dramatically, it was not sufficient to change the action.

The final example is the variation of the Rockwell hardness example where there exists an infinite number of possible states of nature. The scenario now indicates that mills other than Mill A and Mill B produce this special steel, and the unbranded shafts being offered for sale are produced by an unknown mill. The prior distribution of θ, $\pi_\theta(z)$, is assumed to be normally distributed with mean 72 and standard deviation 1.5, i.e.,

$$\pi_\theta(z) = \pi_\mu(z) = \frac{1}{\sqrt{2\pi}(1.5)} e^{-(z-72)^2/2(1.5)^2}, \quad -\infty \leq z \leq \infty.$$

The purchaser is interested only in "estimating" the average Rockwell hardness of the mill that produced the unbranded shafts. A random sample of size 4 is taken and the random variable $\bar{X}$ is to be observed. The loss function is given by

$$l(a, \mu) = c(\mu - a)^2,$$

where c is a positive constant. The posterior distribution is required. Since (X, θ) is a continuous bivariate random variable, the third expression for the posterior distribution given in Sec. 5.3.4 is applicable, i.e.,

$$g_{\mu \mid \bar{X} = \bar{x}}(z) = \frac{f_{\bar{X} \mid \mu = z}(\bar{x}) \pi_\mu(z)}{\int_{-\infty}^{\infty} f_{\bar{X} \mid \mu = z}(\bar{x}) \pi_\mu(z) \, dz}, \quad \text{for } -\infty \leq z \leq \infty,$$

where $\pi_\mu(z)$ is the prior distribution of the random variable μ, and $f_{\bar{X} \mid \mu = z}(\bar{x})$ is the density function of a normally distributed random variable with mean z and standard deviation $2/\sqrt{4} = 1$, evaluated at the point $\bar{x}$. The posterior distribution is then given by

$$g_{\mu \mid \bar{X} = \bar{x}}(z) = \frac{\frac{1}{\sqrt{2\pi}} e^{-(\bar{x}-z)^2/2} \frac{1}{\sqrt{2\pi}(1.5)} e^{-(z-72)^2/2(1.5)^2}}{\int_{-\infty}^{\infty} \frac{1}{\sqrt{2\pi}} e^{-(\bar{x}-z)^2/2} \frac{1}{\sqrt{2\pi}(1.5)} e^{-(z-72)^2/2(1.5)^2} \, dz}.$$

By using appropriate algebraic manipulation, the posterior distribution can be expressed as

$$g_{\mu \mid X = x}(z) = \frac{1}{\sqrt{2\pi} \sqrt{\frac{9}{13}}} e^{-(z-[288/13 + 9\bar{x}/13])^2/2(9/13)},$$

that is, it is normal with mean equal to $[\frac{288}{13} + \frac{9}{13}\bar{x}]$ and standard deviation $\sqrt{\frac{9}{13}}$.[1]

The expected loss, with the expectation taken with respect to the posterior distribution, is given by

$$l_g(a) = E[l(a, \mu)] = c \int_{-\infty}^{\infty} (z - a)^2 g_{\mu \mid \bar{X}=\bar{x}}(z) \, dz.$$

The Bayes procedure calls for selecting the action which minimizes $l_g(a)$ and, as indicated in Sec. 5.2.6, $l_g(a)$ is minimized by choosing a equal to the *mean of the posterior distribution*. In this example, the Bayes procedure is to "estimate" the average Rockwell hardness by a linear function of the sample mean, i.e., the "estimate" is $(288/13) + (9\bar{x}/13)$.

5.3.6 ASSESSMENT OF THE BAYESIAN APPROACH

The Bayesian approach has played an important role in the development of the application of statistics to engineering, science, business, and many other areas. It has formulated a model structure that fits many areas and has presented parameters in a light which is often easily understood by most practitioners, namely, a loss structure usually based on costs. It has utilized prior information, and has resulted in the determination of optimum procedures for a wide class of problems. However, the Bayesian approach asks a great deal of the decision maker. It requires that he be in a position to list for each possible state of nature and each possible action, the consequences of these actions, including all the long-run eventualities associated with the action taken. Whereas this can be done for many problems in business, problems in engineering and science are often not amenable to such an assessment of the consequences, particularly where exploratory research is to be conducted or where experimental results are to be used by future scientists and for purposes not presently anticipated. Furthermore, even when a measure of consequences should be available, the necessary information required may be difficult to obtain.

There exist philosophical arguments about the use of loss as a measure of consequences. However, other criteria have been suggested and frequently the decision making structure remains unaltered. For example, under the "loss criterion" the decision maker may be penalized in terms of incurring a loss, even if he makes the "best" decision for a given state of nature. An alternative principle calls for the decision maker to incur no penalty if he

---

[1] In general, if the prior distribution of μ is normal with mean μ_0 and standard deviation σ_μ, and if the distribution of X is also normal with mean μ and standard deviation σ_X, then the posterior distribution of μ, given $X = x$, is also normal with mean $E_g(\mu)$ and standard deviation σ_g, where

$$E_g(\mu) = \frac{\mu_0[1/\sigma_\mu^2] + x[1/\sigma_X^2]}{1/\sigma_\mu^2 + 1/\sigma_X^2}, \quad \text{and} \quad \frac{1}{\sigma_g^2} = \frac{1}{\sigma_\mu^2} + \frac{1}{\sigma_X^2}.$$

takes the "best" action for a given state of nature, and to incur a penalty for selecting any other action equal to the loss that could have been avoided had he taken the best action, i.e., the penalty represents the difference between the loss incurred and the loss incurred if the best possible action were taken. This is known as the principle of regret, and the corresponding loss is known as regret. The regret function $r(a, \theta)$ is then related to the loss function and is given by

$$r(a, \theta) = l(a, \theta) - \min_{a \in A} l(a, \theta).$$

A table of the regret function for the original Rockwell hardness example is constructed from Table 5.2 and is given in Table 5.4.

Table 5.4. Regret Function for Rockwell Hardness Example

Actions Purchase	States of Nature for Unbranded Shafts	
	$\theta_1: \mu = 72$	$\theta_2: \mu = 75$
a_1: 500 unbranded	0	4,733
a_2: 500 Mill A	15,000	0
a_3: 250 unbranded 250 Mill A	8,750	3,616

Thus, it is evident that the regret function has a minimum value of zero for each state of nature. Furthermore, it is also evident that the loss function and the regret function are not generally equal, although there are important cases where it is reasonable to assume that the loss incurred is zero when the best action is taken for a given state of nature. In these situations, the loss and the regret are equivalent. This occurs in the second example of Sec. 5.2.7 (and Sec. 5.3.5). In those situations where the loss and regret functions differ, it is reasonable to question whether or not the resultant actions may differ. It is easily shown that the Bayes procedures are always the same so that the decision making structure remains unaltered.

Another concern about the Bayesian approach relates to the concept of the prior distribution. The Bayesian approach postulates the existence of such a (subjective) probability distribution, which introduces the personal feelings of the decision maker into the structure of the problem. Clearly, two decision makers or two experimenters, studying the same problem and observing the same experimental data, may take different actions because their personal prior distributions are different. Besides the philosophical arguments about "objective" versus "subjective" decision making, there are similar arguments about whether the state of nature should be considered as a random variable.

Section 5.4 will be concerned with an alternate approach for a certain class of decision problems, where the state of nature will be viewed as an unknown constant and the loss function will take on a particular form.

5.4 Significance Tests

In Sec. 5.3.1, the concept of a decision procedure was introduced. A decision procedure was defined as a rule that specifies (a) the amount (sample size) and form of experimentation and (b) the action that is to be taken for each possible outcome of the random sample obtained from experimentation. The action taken was to be selected from a class of possible actions. However, there are many important problems where only two possible actions need to be considered, and this section deals with such problems.

Consider the Rockwell hardness example, presented in Sec. 5.2, modified as follows:

1. Generalize the problem to allow for the unbranded shafts to be produced in a mill whose average Rockwell hardness is completely unknown. The original formulation where the unbranded shafts are produced in Mill A or Mill B then becomes a special case.
2. Only two actions will be permitted, namely, a_1 and a_2. Action a_1 calls for the purchase of the 500 unbranded shafts, whereas action a_2 calls for foregoing the opportunity to purchase the 500 unbranded shafts (thereby implying that the shafts will be bought directly from Mill A).
3. The loss function is not quite as well specified in terms of knowing the cost parameters. What is assumed to be known is that the purchaser is interested in buying the unbranded shafts because of the price discount if the average Rockwell hardness of the unbranded shafts is equal to 72,[1] i.e., if the shafts were produced in Mill A or in another mill having an average Rockwell hardness of 72. However, he "prefers" to purchase the shafts directly from Mill A if the average Rockwell hardness of the unbranded shafts is not equal to 72. It is evident that this is the philosophy in the original formulation of the problem where the shafts were assumed to be produced in either Mill A or Mill B ($\mu = 72$ or $\mu = 75$). If the purchaser "knew" that the unbranded shafts were produced in Mill A, he would buy them; if he "knew" that they were produced in Mill B, he would not purchase them. Thus, action a_1 is the preferred action if the average Rockwell hardness of the shafts produced by a mill is equal to 72 ($\mu = 72$), whereas a_2 is the preferred action if the average Rockwell hardness of the shafts produced by a mill is not equal to 72 ($\mu \neq 72$). Action a_1 is then equivalent to concluding that $\mu = 72$, and action a_2 is equivalent to concluding that $\mu \neq 72$.

---

[1] This terminology is somewhat loose. What is meant by the statement, here and in subsequent discussions, that "the Rockwell hardness of the unbranded shafts is equal to 72" is that the expected value of the random variable, the final Rockwell hardness of shafts of this particular composition produced by the mill in question, is equal to 72. From the interpretation of expectation, the average of the 500 unbranded shafts should be "close" to this expected value.

With the availability of experimentation, a decision procedure is sought for the solution of this problem. Several decision procedures were presented in Sec. 5.3.1. However, since only two actions are presently available, and any average Rockwell hardness is now permitted, some additional decision procedures will be given. It should be remarked that these decision procedures are chosen arbitrarily and must be subsequently evaluated.

A decision procedure, $d_{\bar{x}}$, which uses the sample mean $\bar{X}$ to characterize the information contained in the random sample, is described as follows: Draw a random sample of size 4 and observe the outcome of the random variable $\bar{X}$. Define the decision rule $d_{\bar{x}}$ as

$$d_{\bar{x}}[\bar{x}] = \begin{cases} a_1 & \text{if} \quad 70.7 \leq \bar{x} \leq 73.3, \\ a_2 & \text{if} \quad \bar{x} < 70.7 \quad \text{or} \quad \bar{x} > 73.3. \end{cases}$$

Thus, decision procedure $d_{\bar{x}}$ calls for concluding that the average Rockwell hardness of the unbranded shafts is 72 if $70.7 \leq \bar{X} \leq 73.3$. If $\bar{X}$ falls outside this interval, the decision procedure calls for concluding that the average Rockwell hardness is not 72. Recall that a decision procedure is chosen before the experiment is performed—before any observations are taken. The value of $n = 4$ and the end points of the interval $[70.7, 73.3]$ are not chosen arbitrarily, but are integrated into the procedure. The choice of these parameters will be discussed in detail later.

A second decision procedure, $d_{\tilde{x}}$, uses the sample median $\tilde{X}$ to characterize the information contained in the random sample and is described as follows: Draw a random sample of size 3 and observe the outcome of the random variable $\tilde{X}$. Define the decision procedure $d_{\tilde{x}}$ as

$$d_{\tilde{x}}[\tilde{x}] = \begin{cases} a_1 & \text{if} \quad 70 \leq \tilde{x} \leq 74, \\ a_2 & \text{if} \quad \tilde{x} < 70 \quad \text{or} \quad \tilde{x} > 74. \end{cases}$$

This decision procedure calls for concluding that the average Rockwell hardness of the unbranded shafts is 72 if $70 \leq \tilde{X} \leq 74$. If $\tilde{X}$ falls outside this interval, the decision procedure calls for concluding that the average Rockwell hardness is not 72.

A third decision procedure, d_{x_M}, uses the maximum observation, X_M, from a random sample to characterize the information contained in the random sample and is described as follows: Draw a random sample of size 4 and observe the outcome of the random variable $X_{\max}$. Define the decision procedure d_{x_M} as

$$d_{x_M}[x_M] = \begin{cases} a_1 & \text{if} \quad 70 \leq x_M \leq 74, \\ a_2 & \text{if} \quad x_M < 70 \quad \text{or} \quad x_M > 74. \end{cases}$$

This decision procedure calls for concluding that the average Rockwell hardness of the unbranded shafts is 72 if $70 \leq X_M \leq 74$. If X_M falls outside this interval, the decision procedure calls for concluding that the average Rockwell hardness is not 72.

These are but three of a large number of possible decision procedures, all of which were chosen somewhat arbitrarily. However, they all have the following characteristics in common: a random sample of prescribed size is drawn. The random variable X is chosen to characterize the information contained in the random sample. If X falls within "some specified" limits, often referred to as an "acceptance region" A, action a_1 is taken. If X falls outside A, action a_2 is taken.[1] Thus, a decision procedure has associated with it a random variable X, a sample size, and an acceptance region (which may or may not be an interval).

5.4.1 THE OPERATING CHARACTERISTIC CURVE

In the previous section, several procedures were presented for determining whether the average Rockwell hardness was 72 on the B scale. These procedures were examples drawn from an infinite number of such procedures, and the question still remains as to which one to choose. Section 5.3.2 suggests the risk function as an appropriate measure of the consequences for any given procedure. However, this requires knowledge of the "costs" associated with each possible state of nature and action taken, including all long-run eventualities associated with the action taken. In many engineering and scientific applications, particularly where exploratory research is to be conducted, an assessment of such "costs" is virtually impossible. Hence, an alternative measure of the consequences for any given procedure is desirable. The alternative measure to be recommended is called the operating characteristic curve, and is expressed in probabilistic terms. The operating characteristic (OC) curve is defined subsequently in this section and its relationship to the risk function will be discussed in the following section.

In searching for a "good" decision procedure for the Rockwell hardness example, it is desirable to choose a procedure that leads to the correct conclusion most of the time; i.e., a procedure that selects action a_1 (thereby concluding that the average Rockwell hardness of the unbranded shafts is 72) when the true average Rockwell hardness is actually 72, and selects action a_2 (thereby concluding that the average Rockwell hardness of the unbranded shafts is not 72) when in reality the true average Rockwell hardness is some number other than 72. Since decisions depend on random variables, which by their nature have probability distributions associated with them, it is rather unusual to be able to make the correct decision *always*.

There are two types of incorrect decisions that can be made. If the average Rockwell hardness is actually 72, the decision procedure of the experimenter can call for concluding that it is not 72 (selecting action a_2). This error is known as the Type I error and the probability of its occurrence is denoted by α. It is also often called the *level of significance*. If the average Rockwell

---

[1] Standard statistical texts refer to the complement of the acceptance region, the region of rejection, as the critical region.

hardness is actually some number other than 72, say, 75, the decision procedure of the experimenter can call for concluding that it is 72 (selecting action a_1). This error is known as the Type II error and the probability of its occurrence is denoted by β.

To summarize, then, a decision procedure has associated with it a random variable X, a sample size n, and an acceptance region A. The probability of the random variable falling outside the acceptance region when the true average Rockwell hardness is 72 is the magnitude of the error of Type I (denoted by α). The probability of the random variable falling inside the acceptance region when the true average Rockwell hardness is some number other than 72 is the magnitude of the error of Type II (denoted by β).

The error of Type II is not a constant, then, but depends on the true average Rockwell hardness (the actual state of nature). If μ denotes the true average Rockwell hardness, then $\beta(\mu)$ is a better notation than the symbol β for the magnitude of the Type II error. In other words, β is a function (in the mathematical sense) of μ defined for all values of $\mu \neq 72$. The magnitude of the Type II error, $\beta(\mu)$, can be calculated by finding the probability that the appropriate random variable falls within the acceptance region when the true average Rockwell hardness is $\mu \neq 72$. This is equivalent to finding the probability of concluding the average Rockwell hardness is 72 (selecting action a_1) when in reality μ is some value other than 72.

The operating characteristic (OC) curve of a decision procedure, d, is a plot of the probability of selecting action a_1 (concluding that $\mu = 72$) versus the true state of nature (the true average Rockwell hardness). A sketch of such an OC curve is given in Fig. 5.3. Note that the OC curve coincides

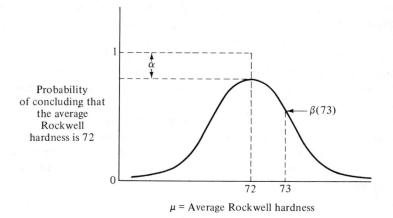

Fig. 5.3. *Operating characteristic (OC) curve.*

with $\beta(\mu)$ for all $\mu \neq 72$. When $\mu = 72$, the ordinate of the OC curve equals $(1 - \alpha)$.

The OC curve is then the measure of the consequence for a given decision procedure alluded to earlier, and procedures are chosen according to these OC curves.

Points on the OC curve for the decision procedure $d_{\bar{x}}$ are easily obtained. Action a_1 is taken whenever $70.7 \leq \bar{X} \leq 73.3$. Therefore, the probability of selecting action a_1 (concluding that the average Rockwell hardness is 72) is equivalent to

$$P\{70.7 \leq \bar{X} \leq 73.3\},$$

where $\bar{X}$ is a normally distributed random variable with mean μ and standard deviation $2/\sqrt{4} = 1$. (Recall that a random sample of size 4 was drawn and the Rockwell hardness was assumed to be a normally distributed random variable with mean equal to μ and standard deviation equal to 2.) With the aid of Appendix Table 1, this probability is easily evaluated. Several values are presented in Table 5.5. The level of significance α for this procedure is seen to be $(1 - 0.806) = 0.194$ and $\beta(73)$ is just 0.607, etc.

Table 5.5. Tabulation of OC Curves for Decision Procedure $d_{\bar{x}}$

μ	Probability of Concluding $\mu = 72$
69	0.045
70	0.242
71	0.607
72	0.806
73	0.607
74	0.242
75	0.045

5.4.2 COMPARISON OF THE OC CURVE WITH THE RISK FUNCTION

As indicated in Sec. 5.3.2, a measure of the consequences for any procedure is given by the risk function, which is just the expected loss when decision procedure d is used and the state of nature is θ, i.e.,

$$R(d, \theta) = E[l(d[X], \theta)],$$

where the expectation is taken with respect to the probability distribution of the random variable X. Because of the special structure of the problem now being studied, this expression can be simplified. Recall that two action problems are being studied where action a_1 is taken if X falls within the acceptance region A, and a_2 is taken if X falls outside A. Hence, the risk can now be expressed as

$$R(d, \theta) = l(a_1, \theta) \, P\{X \in A\} + l(a_2, \theta) \, P\{X \notin A\}.$$

Now examine the state space Θ. Partition it into two mutually exclusive sets Θ_1 and Θ_2 so that a_1 is the preferred action when the true state of nature, θ, is in Θ_1, and a_2 is the preferred action when θ is in Θ_2. If the loss function is a regret function, i.e.,

$$l(a_1, \theta) = 0 \text{ if } \theta \text{ is in } \Theta_1$$
$$l(a_2, \theta) = 0 \text{ if } \theta \text{ is in } \Theta_2,$$

the risk can then be expressed as

$$R(d, \theta) = \begin{cases} l(a_2, \theta)\, \mathrm{P}\{X \notin A\}, \text{ for } \theta \in \Theta_1 \\ l(a_1, \theta)\, \mathrm{P}\{X \in A\}, \text{ for } \theta \in \Theta_2. \end{cases}$$

If $l(a_2, \theta)$ is set equal to 1 for $\theta \in \Theta_1$ and if $l(a_1, \theta)$ is set equal to 1 for $\theta \in \Theta_2$, the risk is finally expressed as

$$R(d, \theta) = \begin{cases} \mathrm{P}\{X \notin A\}, & \text{for } \theta \in \Theta_1 \\ \mathrm{P}\{X \in A\}, & \text{for } \theta \in \Theta_2. \end{cases}$$

In the context of the present formulation of the Rockwell hardness example it is easily seen that this last expression for the risk is equivalent to the operating characteristic curve. In this example Θ_1 consists of the single point $\mu = 72$ and Θ_2 contains all other possible values of $\mu \neq 72$. Hence, the risk is given by

$$R(d, \theta) = \begin{cases} \mathrm{P}\{X \notin A\} = \alpha, & \text{for } \mu = 72 \\ \mathrm{P}\{X \in A\} = \beta(\mu), & \text{for } \mu \neq 72. \end{cases}$$

Note that the risk function coincides with the operating characteristic curve for all $\mu \neq 72$ and is just one minus the value of the OC curve at $\mu = 72$.

5.4.3 COMPARISON OF OC CURVES

Even when the OC curve of a decision procedure is used as a measure of its consequences, choosing among decision procedures requires further study. Suppose the OC curves of the procedures mentioned in Sec. 5.4 are as indicated in Fig. 5.4. Clearly, the procedure based on $\bar{X}$ is superior to the procedure based on $\tilde{X}$ except when $\mu = 72$. Thus, procedure $d_{\bar{x}}$ uniformly has a smaller error of Type II than procedure $d_{\tilde{x}}$, but has a larger error of Type I.

To make these procedures comparable, it is desirable to fix the Type I errors of both procedures and to compare their Type II errors. This can easily be done. It is apparent that the Type I error of either procedure can be made as small as desired by increasing the length of the acceptance interval. For example, in procedure $d_{\bar{x}}$, if the probability that $\bar{X}$ falls outside the interval [70.7, 73.3], when the true average Rockwell hardness is 72 (the Type I error), is α_1, and the probability that $\bar{X}$ falls outside the interval [60, 90], when the true average Rockwell hardness is 72, is α_2, then α_2 must

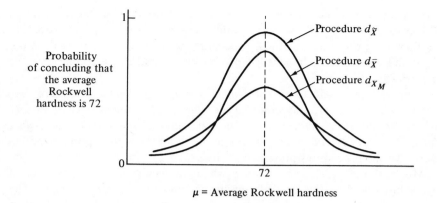

Probability of concluding that the average Rockwell hardness is 72

Procedure $d_{\tilde{X}}$

Procedure $d_{\bar{X}}$

Procedure d_{X_M}

72

μ = Average Rockwell hardness

Fig. 5.4. *Comparison of several OC curves.*

be less than α_1. A similar example can be given for procedure $d_{\tilde{x}}$. Consequently, the Type I error of either procedure can be fixed by judiciously choosing the proper acceptance interval. If this is done, the OC curves will appear as in Fig. 5.5.[1]

It is now clear that procedure $d_{\bar{x}}$ is superior to procedure $d_{\tilde{x}}$. If the true average Rockwell hardness is 72, both procedures will lead to this conclusion with the same probability. However, if the true average Rockwell hardness is *not* 72, the procedure based on $\bar{X}$ uniformly will lead to the correct

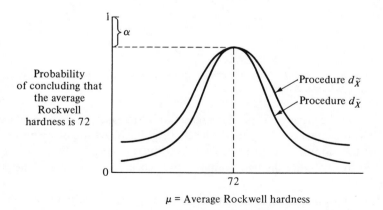

Probability of concluding that the average Rockwell hardness is 72

α

Procedure $d_{\tilde{X}}$

Procedure $d_{\bar{X}}$

72

μ = Average Rockwell hardness

Fig. 5.5. *Comparison of the OC curves of procedures $d_{\bar{X}}$ and $d_{\tilde{x}}$ for a fixed Type I error.*

---

[1] The comparable OC curve for a procedure based on the sample median, but using a sample of size 4 (with an appropriate definition for the sample median of four observations), will lie between the OC curves for the $d_{\tilde{x}}$ and $d_{\bar{x}}$ procedures.

conclusion with higher probability; i.e., the probability of saying the average Rockwell hardness is 72 when in reality it is not, is smaller for procedure $d_{\bar{x}}$ than for procedure $d_{\tilde{x}}$; thus, procedure $d_{\bar{x}}$ is better than procedure $d_{\tilde{x}}$. A similar conclusion is reached if procedures $d_{\bar{x}}$ and d_{x_M} are compared for a fixed Type I error.

A procedure is said to be *optimum* if, *for a fixed sample size, there does not exist any other procedure having the same level of significance or smaller whose OC curve lies entirely below the OC curve of the optimum procedure for all values of the abscissa.* There may be more than one optimum procedure. All the procedures presented in the ensuing chapters will be optimum.

Much of the previous discussion was concerned with a comparison of procedures holding the level of significance and the sample size fixed. The effect of varying the sample size for a given procedure, holding the level of significance fixed, can be seen by referring to Fig. 5.6.

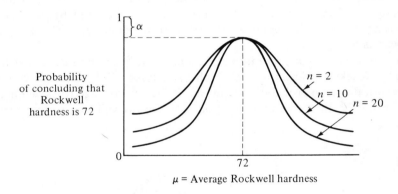

Fig. 5.6. *Effect of varying the sample size on procedure $d_{\bar{x}}$, holding the level of significance fixed.*

Note that the error of Type II decreases as n increases, as is to be expected. Similarly, the effect of changing the level of significance for a given procedure, holding the sample size fixed can be seen by referring to Fig. 5.7. This is achieved by changing the end points of the acceptance interval of the stated procedure.

It is evident that for fixed n the error of Type I can be made as small as desired. However, this is achieved at the expense of increasing the error of Type II. Similarly, for any given value of μ, the error of Type II can be made as small as desired, although at the expense of increasing the error of Type I.

Naturally, the experimenter would like to achieve an OC curve such as that in Fig. 5.8.

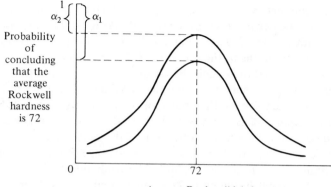

μ = Average Rockwell hardness

Fig. 5.7. *Effect of varying the level of significance on procedure d$\bar{x}$, holding the sample size fixed.*

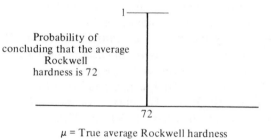

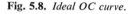

μ = True average Rockwell hardness

Fig. 5.8. *Ideal OC curve.*

If the true average Rockwell hardness is 72, the probability of coming to this conclusion is 1. If the true average Rockwell hardness is some number other than 72, the probability of concluding that it is 72 is zero. This OC curve can be achieved only by letting n become very large. This, of course, is entirely too costly, for which reason the experimenter must be satisfied with an approximation to this ideal OC curve. For example, if $\mu = 72.000003$, and the procedure led to the conclusion that $\mu = 72$, generally, this would not result in a serious error. On the other hand, saying that $\mu = 72$ when in reality it is 80 may be very serious. Consequently, the experimenter, and only the experimenter, must choose a value for the level of significance which he is willing to tolerate and a value of μ that is important from the practical point of view to detect, together with the error associated with this point. In other words, he must choose two points on the OC curve. These two points completely determine the entire OC curve. Since the OC curve depends on the sample size n and acceptance constants, this amounts to completely specifying the parameters of the procedure. Very often the experimenter must weigh the cost of taking additional observations against the advantage of decreasing the Type II error. In practice, there are two major limitations to the use of the OC curve. Very often the number of observations is fixed in advance by custom or by the limitations of testing equipment. Indeed, the statistical analysis may be of secondary importance, based on data taken for another purpose. Even in this case, a look at the OC curve is important as it gives an idea of the type of differences one is likely to detect and, hence, an indication of the sensitivity of the analysis. For example, if it is important to detect an average Rockwell hardness of 73 and a limited sample size is available, the probability of concluding that the average Rockwell hardness is 72 when in reality it is 73 may be 0.60. If this is the case, it is evident that performing the experiment is a waste of time and money unless the sample size can be increased.

In the second place, it often happens that the OC curve depends on parameters in which one is not interested. This could occur in the previous example if the standard deviation were unknown. The OC curve, as will become evident later, depends on the standard deviation as well as the true mean. However, even if only the general magnitude of the standard deviation is known, the OC curves are very useful in designing experiments. The experimenter must realize that whenever he picks a sample size he is implicitly picking an OC curve and that the more information he has available in making this decision the better. OC curves for most of the standard significance tests are presented in Chapters VI and VII.

5.4.4 Tests of Hypotheses

In many situations the experimenter is interested in testing a particular hypothesis. For example, he is interested in testing the hypothesis that the true average Rockwell hardness of the unbranded shafts is 72 on the B scale, or he is interested in testing whether a new method of sealing vacuum tubes is the same as the old method, etc. In each of these examples he has an alternative in mind. In the Rockwell hardness example, the alternative is that the average Rockwell hardness does not equal 72; i.e., it is either less than or greater than 72. This is known as a two-sided alternative and it is associated with a two-sided procedure. This was the alternative considered and is credible. A somewhat less likely alternative can arise as follows: Suppose the experimenter knows that the process followed by all mills insures that the average Rockwell hardness is at least 72. In other words, he discounts the possibility of the average Rockwell hardness falling below 72 so that the alternative (to μ being equal to 72) is that the average Rockwell hardness is greater than 72. This alternative is known as a one-sided alternative and the procedure associated with it is known as a one-sided procedure.

In the vacuum tube example an alternative to the hypothesis that the two methods of sealing are equivalent is that the new method either increases or decreases the average life of vacuum tubes. This is a two-sided alternative. Perhaps a more realistic alternative is the one-sided alternative, i.e., the new method can only increase the average life of vacuum tubes. Of course, the realism of the model depends on the particular situation, and is a problem which the experimenter faces.

A one-sided procedure can also be used in a somewhat different context. Suppose that an average Rockwell hardness below 72 will have no appreciable effect on the performance of the product and, hence, is acceptable. Now, the hypothesis to be tested is that the average Rockwell hardness is 72 or less whereas the alternative is still the one-sided alternative that the average Rockwell hardness is greater than 72.

For one-sided procedures used in either context, the alternatives mentioned dealt with the parameter being greater than a given value. In any given situation, the alternative may be that the parameter is less than the given value.

At this point it would be well to formalize the discussion about "hypotheses." For the purposes of this text a *hypothesis will be viewed as an assumption about the state of nature usually expressed as the behavior of a random variable and/or its probability distribution.* Associated with a hypothesis is an alternative which also is an assumption about the behavior of the random variable and/or the probability distribution in question. *Testing a hypothesis calls for choosing a decision procedure which leads to accepting or rejecting the hypothesis.* Rejection of the hypothesis is equivalent to accepting the alternative. Thus, in the context of a two-actions problem, accepting the hypothesis coincides with selecting action a_1 whereas rejecting the hypothesis, and thereby accepting the alternative, coincides with selecting action a_2. *The operating characteristic curve of a decision procedure is a plot of the probability of accepting the hypothesis versus the true state of nature usually represented by a parameter of the probability distribution under consideration. The error of Type I is the probability of rejecting the hypothesis when the hypothesis is true, and the error of Type II is the probability of accepting the hypothesis when it is false.* The reader will recall that these are just generalizations of the definitions given for the Rockwell hardness example.

Accepting or rejecting a hypothesis is then simply an alternative expression for selecting an action a_1 or selecting an action a_2. The important concept that the experimenter must grasp is that he is to choose a decision procedure whose OC curve reflects the "measure of consequence" or risk that he is willing to tolerate. This is best accomplished by sketching the desired OC curve and then finding the desired decision procedure (by methods to be discussed in the next chapter).

5.4.5 ONE- AND TWO-SIDED PROCEDURES

Finding an appropriate decision procedure that is formally associated with testing a particular hypothesis versus a particular alternative often causes some confusion. Problems with two-sided alternatives leave little ambiguity as to the procedure to follow. For example, in the Rockwell hardness problem a logical acceptance region is an interval if the random variable $\bar{X}$ is used to characterize the information in the random sample. Since $\bar{X}$ is an "estimate" of the average Rockwell hardness, the hypothesis that the average Rockwell hardness is 72 should be accepted if $\bar{X}$ takes on a value close to 72 (above or below). For the one-sided alternative where the possibility of decreasing the average Rockwell hardness is discounted, or in the case where an average Rockwell hardness below 72 is acceptable, the experimenter should tend to reject the hypothesis that the average Rockwell hardness equals 72 if $\bar{X}$ is too large, e.g., $\bar{X} > 72.86.$[1] A sketch of the OC curve for this procedure is found in Fig. 5.9. It is compared with the two-sided procedure having the same probability of acceptance when $\mu = 72$.

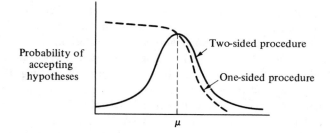

Fig. 5.9. *Comparison of the OC curves for one- and two-sided procedures.*

It is interesting to note that when the average Rockwell hardness is 72, both procedures are equivalent. When the average Rockwell hardness is greater than 72, the one-sided procedure is better. When the average Rockwell hardness is less than 72, the two-sided procedure has a better chance of concluding that it is not equal to 72. This is to be expected since it has been assumed in using the one-sided procedure that either the average Rockwell hardness *cannot* be less than 72 or, in the other situation, that accepting the hypothesis when the average Rockwell hardness is less than 72 is desirable.

For some one-sided alternative problems the definition of the level of significance needs further discussion. Recall that the one-sided alternative is appropriate in the Rockwell hardness example when there exists knowledge that μ cannot be less than 72 or when steel shafts having low Rockwell

---

[1] This particular number is chosen so that the probability of accepting the hypothesis for this procedure and procedure $d_{\bar{X}}$ is the same when $\mu = 72$.

hardness are desirable (highly impractical in the context of the example). For the case when the average Rockwell hardness cannot be below 72, presumably the OC curve is not defined for $\mu < 72$. However, for the case where an average Rockwell hardness below 72 is both permitted and desirable, the error of type I exists for all $\mu \leq 72$. The OC curve then coincides with the error of Type II for $\mu > 72$ and is given by one minus the error of Type I for $\mu \leq 72$. In this case, the *level of significance is defined to be the maximum value of the error of Type I*. Using this definition and refering to Fig. 5.9, the level of significance for the one-sided procedure is one minus the value of the OC curve at $\mu = 72$.

In one-sided alternative problems, difficulties often arise as to which procedure to use. For example, suppose the army is presently buying shoes from a manufacturer, and these shoes are known to have an average life of 12 months. A new manufacturer puts in a bid at the same price but states that the average life of his shoes is greater than 12 months. The army is going to run an experiment to ascertain whether or not this claim is valid. They will seek a decision procedure to use with their experimental data.

Using the notation of "testing a hypothesis," two types of hypotheses arise: (1) testing the hypothesis that the average life of the new company's shoes is *less than or equal to* 12 months $(H: \mu \leq 12)$ against the alternative that the average life is *greater than* 12 months $(A: \mu > 12)$, and (2) testing the hypothesis that the average life of the new company's shoes is *greater than or equal to* 12 months $(H: \mu \geq 12)$ against the alternative that the average life is *less than* 12 months $(A: \mu < 12)$.

In the first situation, i.e., $H: \mu \leq 12$; $A: \mu > 12$, accepting the hypothesis implies a retention of the old manufacturer, whereas rejecting the hypothesis implies a switch to the new company. The decision procedure associated with this case should result in an OC curve similar to the plot in Fig. 5.10. If the average life of the new company's shoes is 12 months or less, the probability of accepting the hypothesis, and thereby retaining the old manufacturer, is close to one. If the average life of the shoes is slightly greater than

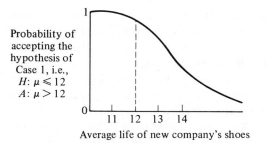

Fig. 5.10. *The OC curve of the procedure when the army is satisfied with the present manufacturer.*

12 months, the probability of retaining the old manufacturer is still quite high. If the average life of the shoes is substantially greater than 12 months, the probability of accepting the hypothesis is small (rejecting the hypothesis is large) and a switch to the new manufacturer is likely.

The physical situation leading to this hypothesis model is exemplified when the army is very satisfied with the present manufacturer. The old company sends its product on time, they are reliable, their workmanship is good, a switch causes increased red tape and administrative difficulties, etc. In other words, the burden of proof is on the new company to show that its shoes have an average life substantially greater than 12 months. The army is desirous of a switch only if a substantial increase in the average life will result. An example of a procedure which has this property is: Accept the hypothesis that $\mu \leq 12$ months if $\bar{X} \leq 15$; otherwise, reject it.

In the second situation, i.e., $H: \mu \geq 12$; $A: \mu < 12$, accepting the hypothesis implies a switch to the new company, whereas rejecting the hypothesis implies a retention of the old manufacturer. The decision procedure associated with this case should result in an OC curve similar to the plot in Fig. 5.11.

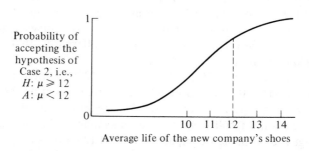

Fig. 5.11. *The OC curve of the procedure when the army is not satisfied with the present manufacturer.*

If the average life of the new company's shoes is 12 months or more, the probability of accepting the hypothesis, and thereby switching to the new manufacturer, is close to one. If the average life of the shoes is slightly less than 12 months, the probability of switching to the new manufacturer is still quite high. If the average life of the shoes is substantially less than 12 months, the probability of accepting the hypothesis is small (rejecting the hypothesis is large) and retaining the old manufacturer is likely.

The physical situation leading to this hypothesis model is exemplified when the army is unhappy with the present manufacturer, and is looking for an opportunity to change. The old company is late with shipments and generally unreliable. In other words, they are willing to change not only if the average life of the new company's shoes exceeds 12 months, but even if their average life is slightly less than 12 months. The burden of proof is on the new manufacturer to show that the average life of his shoes is not substantially less than 12 months. An example of a procedure which has this property is: Accept the hypothesis that $\mu \geq 12$ if $\bar{X} \geq 10$; otherwise, reject it.

In the two situations just described, the army indicated an inherent preference for one of the companies. A third alternative arises when the army is completely indifferent as to which company it chooses, provided the difference in the average lives of the shoes is small. If the average life of the new company's shoes is only 12 months, the army is indifferent, and is willing to tolerate an equal chance of obtaining other manufacturers. On the other hand, if there is a substantial increase over 12 months, the new company should be chosen; if there is a substantial decrease under 12 months, the old manufacturer should be retained. The hypotheses of case 1 or case 2 may be used, but with a level of significance of $\frac{1}{2}$, thereby expressing the indifference of the army when the average life of the shoes is only 12 months. The OC curve for the hypothesis of case 1, i.e., $H: \mu \leq 12$; $A: \mu > 12$, is shown in Fig. 5.12 with a level of significance equal to $\frac{1}{2}$.

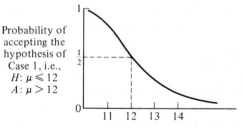

Fig. 5.12. *The OC curve of the procedure when the army is indifferent.*

Thus, if the average life of the new company's shoes is 12 months, the probability of accepting the hypothesis and thereby retaining the old manufacturer is $\frac{1}{2}$. If the average life of the shoes is substantially less than 12 months, the probability of accepting the hypothesis and thereby retaining the old manufacturer is close to one. If the average life of the shoes is substantially greater than 12 months, the probability of accepting the hypothesis and thereby retaining the old manufacturer is quite small, so that a switch will usually result. The procedure which has this property is, as one might expect: Accept the hypothesis that $\mu \leq 12$ if $\bar{X} \leq 12$.

If the OC curve of the hypothesis for case 2, i.e., $H: \mu \geq 12$; $A: \mu < 12$, were plotted, it would appear as in Fig. 5.9, but with a level of significance of $\frac{1}{2}$. The procedure which is associated with such an OC curve is: Accept the hypothesis that $\mu \geq 12$ if $\bar{X} \geq 12$. Accepting the hypothesis for this case is equivalent to switching to the new company. Thus, if $\bar{X} \geq 12$, the switch is made. This is equivalent to the procedure associated with Fig. 5.10 (case 1). In this situation if $\bar{X} \leq 12$, the hypothesis is accepted and the old manufacturer retained, or alternatively, if $\bar{X} > 12$, the switch is made.

To summarize, one-sided alternatives can be classified into two groups:

1. There exists knowledge that certain values of the unknown parameter cannot exist, e.g., the average Rockwell hardness cannot be less than 72.
2. The hypothesis is concerned with statements about values of the unknown parameter being less than or equal to a constant (or alternatively, greater than or equal to a constant), e.g., the shoe example.

In either case, it is advantageous to sketch the form of the OC curve and determine a one-sided procedure which will lead to this OC curve.

PROBLEMS

1. For the Rockwell hardness example described in Sec. 5.2 and for a unit cost given by

$$\text{Unit cost} = \begin{cases} 70, & \text{if } 68 \leq \text{R.h.} \leq 76 \\ 70 + 150(72 - \text{R.h.})^2, & \text{if} \quad \text{R.h.} < 68 \\ 70 + 150(\text{R.h.} - 72)^2, & \text{if} \quad \text{R.h.} > 76, \end{cases}$$

determine the (expected) loss function table.

2. For Problem 1, find the minimax procedure.

3. At the end of a day's production, a machine is serviced. If the servicing is done properly, the machine is in a "good" state and the quality of each unit produced can be assumed to be a Bernoulli random variable with parameter $p = 0.01$, i.e., the probability that a unit is defective is $p = 0.01$ ($P\{X = 1\} = p$). If the servicing is not done properly, the machine is in the "bad" state and the quality of each unit can be assumed to be a Bernoulli random variable with parameter $p = 0.10$. By spending an additional \$100, the actual state of the machine can be determined and then placed into the "good" state if necessary. If the machine is found to be in the "good" state already, it remains in that state. The machine produces 1,000 units each day and the cost of each defective unit produced is \$10. Let the action a_1 denote incurring the additional service cost (thereby insuring that the machine is placed into the "good" state), and let the action a_2 denote not incurring the additional service cost. The loss function is given by

a	p	
	0.01	0.10
a_1	200	200
a_2	100	1000

Find the minimax procedure.

4. A lot of 10,000 items is submitted for acceptance inspection by a manufacturer. The consumer is to select one of these two actions:

a_1 accept the lot;

a_2 accept the lot but screen the 10,000 items.

Each defective item found prior to assembly costs 20 cents to rework. In addition, if a defective item is detected after assembly, an additional cost of 70 cents is incurred. The cost of screening each item in the entire lot is 2 cents per unit. The consumer has had a long term relationship with the manufacturer and has noted that similar lots have contained fraction defectives p of 0.01, 0.05, or 0.10. Furthermore, it can be assumed that the quality of items within a given lot can be represented by a Bernoulli random variable with parameter p, i.e., the probability that an item is defective is given by p ($P\{X = 1\} = p$). (a) Find the loss table. (b) Find the minimax procedure.

5. In Problem 3, assume that there is sufficient evidence to indicate that $\frac{7}{8}$ of the time the machine is in the "good" state after servicing and $\frac{1}{8}$ of the time it is in the "bad" state. What is the Bayes procedure?

6. In Problem 4, assume that the consumer uses his "information" about the manufacturer's "process average" and expresses it as follows:

$$P\{p = 0.01\} = 0.2; \ P\{p = 0.05\} = 0.6; \text{ and } P\{p = 0.10\} = 0.2.$$

Find the Bayes procedure.

7. Find the Bayes procedure for the Rockwell hardness example given in Sec. 5.2 if the prior distribution is given by $P\{\mu = 72\} = \frac{1}{3}$ and $P\{\mu = 75\} = \frac{2}{3}$.

8. Prove that the Bayes procedure is the same if regret is used instead of loss for the case of a finite action space and a finite state space and where no experimentation is permitted.

9. Find the Bayes procedure for Problem 1 if $P\{\mu = 72\} = \frac{1}{5}$ and $P\{\mu = 75\} = \frac{4}{5}$.

10. Find the Bayes procedure for the Rockwell hardness example described in Sec. 5.2 if the prior distribution is given by $P\{\mu = 72\} = \frac{1}{3}$ and $P\{\mu = 75\} = \frac{4}{3}$.

11. For the procedure d_3 described in Sec. 5.3.1, compute the risk function. Note that for $n = 3$, $P\{\tilde{X} \leq b\} = [F(b)]^2[3 - 2F(b)]$ where $F(b)$ is the CDF of the random variable constituting an observation in the random sample, i.e., it is the CDF of a normally distributed random variable with mean equal to μ and standard deviation equal to 2.

12. For the Rockwell hardness example presented in Sec. 5.2, compute the risk function for the decision procedure d_1 (defined in Sec. 5.3.1), using a random sample of size 9.

13. For the Rockwell hardness example presented in Sec. 5.2, compute the risk function for the decision procedure d defined as

$$d[x] = \begin{cases} a_1 & \text{if} \quad x < 73 \\ a_3 & \text{if } 73 \leq x \leq 74 \\ a_2 & \text{if} \quad x > 74, \end{cases}$$

where x represents the Rockwell hardness of a single shaft tested (a random sample of size 1).

14. Compute the Bayes procedure for the Rockwell hardness example described in Sec. 5.2.5 when the sample mean of 4 observations is 73.5 and when it is 74.0.

15. Compute the Bayes procedure for the Rockwell hardness example, using the loss function given in Problem 1, observing that the sample mean from a random sample of size 4 is equal to 74.5, and using the prior distribution $P\{\mu = 72\} = \frac{2}{3}$ and $P\{\mu = 75\} = \frac{1}{3}$.

16. Compute the Bayes procedure for the Rockwell hardness example described in Sec. 5.2 when the sample mean of 4 observations is 74.5 and the prior distribution is given by $P\{\mu = 72\} = 0.95$ and $P\{\mu = 75\} = 0.05$.

17. Compute the Bayes procedure for the newspaper advertising problem given in Sec. 5.3.5 if the number of mail order responses is equal to 10.

18. Compute the Bayes procedure for the newspaper advertising problem given in Sec. 5.3.5 if the number of mail order responses is equal to 13 and the prior distribution is given by $P\{\lambda = 24\} = 0$, $P\{\lambda = 20\} = 0.10$, $P\{\lambda = 16\} = 0.90$.

19. In the context of Problems 3 and 5, suppose that immediately after servicing, but long before actual production is to begin (when it becomes too late to spend the additional \$100), two items can be produced. After observing the quality of these two items a decision can be made about whether or not to observe the state of the machine (by spending the \$100). (a) Let D denote the number of defective items. What is the (conditional) distribution of D? (b) Find the posterior distribution of p, given $D = d$. (c) For $d = 0$, 1, and 2, find the Bayes procedure.

20. In the context of Problems 4 and 6, suppose that the consumer is able to draw a random sample of 10 items before making his decision. This sampling will be done on a no-gain-no-loss basis and the manufacturer will still submit a total of 10,000 items to the consumer. (a) Let D denote the number of defective items. What is the (conditional) distribution of D? (b) Find the posterior distribution of p given that exactly one defective item is found. (c) Find the Bayes procedure.

21. Assume that the state of nature, λ, represents the parameter of a Poisson distribution, $0 < \lambda < \infty$. Let X be a Poisson random variable, i.e.,

$$P\{X = k\} = \frac{\lambda^k e^{-\lambda}}{k!}.$$

Also assume that the prior distribution for λ is exponential with parameter $\theta = 2$, i.e.,

$$f_\lambda(z) = \begin{cases} \frac{1}{2} e^{-z/2} & \text{for } z \geq 0 \\ 0 & \text{for } z < 0. \end{cases}$$

Suppose there are two possible actions given by

$$a_1: \text{say, } \lambda \leq 1$$
$$a_2: \text{say, } \lambda > 1,$$

with a loss function given as

$$l(a_1, \lambda) = \begin{cases} 0 & \text{if } \lambda \leq 1 \\ 10(\lambda - 1) & \text{if } \lambda > 1, \end{cases}$$

$$l(a_2, \lambda) = \begin{cases} 10(1 - \lambda) & \text{if } \lambda \leq 1 \\ 0 & \text{if } \lambda > 1. \end{cases}$$

Find the Bayes procedure if the value of the random variable X is observed and is equal to 0, i.e., $k = 0$. It may be remarked that this problem can be solved for any value of k and for the more general class of prior distributions called the gamma distribution, i.e.,

$$f_\lambda(z) = \begin{cases} \dfrac{1}{\Gamma(c)d^c} z^{(c-1)} e^{-z/d} & \text{for } z \geq 0 \\ 0 & \text{for } z < 0, \end{cases}$$

where c and d are both positive constants.

22. Solve Problem 21 by using the loss function

$$l(a_1, \lambda) = \begin{cases} 0 & \text{if } \lambda \leq 1 \\ 10 & \text{if } \lambda > 1, \end{cases}$$

$$l(a_2, \lambda) = \begin{cases} 10 & \text{if } \lambda \leq 1 \\ 0 & \text{if } \lambda > 1. \end{cases}$$

See remark for Problem 21.

23. Assume that the states of nature, p, represents the parameter of a Bernoulli distribution, $0 < p < 1$. A random sample of size n is obtained, and let X denote the random variable which is the number of observed successes. It has been shown that X has a binomial distribution, i.e.,

$$P\{X = k\} = \binom{n}{k} p^k (1 - p)^{n-k} \qquad \text{for } k = 0, 1, 2, \cdots, n.$$

Also assume that the prior distribution for p is uniform on the interval zero to one, i.e.,

$$f_p(z) = \begin{cases} 1 & \text{if } 0 \leq z \leq 1 \\ 0 & \text{otherwise.} \end{cases}$$

Suppose there are two possible actions given by

$$a_1 : \text{say, } p \leq p_0$$
$$a_2 : \text{say, } p > p_0,$$

with a loss function given as

$$l(a_1, p) = \begin{cases} 0 & \text{if } p \leq p_0 \\ r(p - p_0) & \text{if } p > p_0, \end{cases}$$

$$l(a_2, p) = \begin{cases} r(p_0 - p) & \text{if } p \leq p_0 \\ 0 & \text{if } p > p_0. \end{cases}$$

Find the Bayes procedure if $p_0 = 0.05$, $r = 1$, $n = 10$, and $k = 1$. Note the identity

$$\int_0^1 z^{(c-1)} (1 - z)^{(d-1)} \, dz = \frac{\Gamma(c)\Gamma(d)}{\Gamma(c + d)}, \qquad \text{for } c, d > 0.$$

It may be remarked that this problem can be solved for any values of n, k, and p_0 and for the more general class of prior distributions called the beta distribution, i.e.,

$$f_p(z) = \begin{cases} \dfrac{\Gamma(c + d)}{\Gamma(c)\Gamma(d)} z^{(c-1)} (1 - z)^{(d-1)}, & 0 \leq z \leq 1 \\ 0, & \text{elsewhere,} \end{cases}$$

where c and d are positive constants.

24. Solve Problem 23 by using the loss function

$$l(a_1, p) = \begin{cases} 0 & \text{if } p \leq p_0 \\ 1 & \text{if } p > p_0, \end{cases}$$

$$l(a_2, p) = \begin{cases} 1 & \text{if } p \leq p_0 \\ 0 & \text{if } p > p_0. \end{cases}$$

See remark for Problem 23.

25. Prove the result appearing in a footnote in Sec. 5.3.5 concerning the posterior distribution of θ when the prior distribution of θ is normal and the distribution of X is also normal.

26. For the procedure $d_{\bar{x}}$ described in Sec. 5.4, find $\beta(73.5)$.

27. For the procedure $d_{\tilde{x}}$ described in Sec. 5.4, find α and $\beta(73)$. Note that for $n = 3$,

$$P\{\tilde{X} \leq b\} = [F(b)]^2[3 - 2F(b)]$$

where $F(b)$ is the CDF of the random variable constituting an observation in the random sample, i.e, it is the CDF of a normally distributed random variable with mean equal to μ and standard deviation equal to 2.

28. Find a symmetric (about 72) acceptance region A based on the sample median of a random sample of size 3 so that $P\{\tilde{X} \in A \mid \mu = 72\} = 0.806$, i.e., the procedure has a level of significance equal to that of procedure $d_{\bar{x}}$. See remark contained in Problem 27 for the distribution of $\tilde{X}$.

29. For the procedure d_{x_M} described in Sec. 5.4, find α and $\beta(73)$. Note that for $n = 4$,

$$P\{X_M \leq b\} = [F(b)]^4,$$

where $F(b)$ is described in Problem 27.

30. Find a symmetric (about 72) acceptance region A based on the maximum of a random sample of size 4 so that

$$P\{X_M \in A \mid \mu = 72\} = 0.806,$$

i.e., the procedure has a level of significance equal to that of procedure $d_{\bar{x}}$. See remark contained in Problem 29.

31. Repeat Problem 30 with $n = 3$.

6

TESTS OF HYPOTHESES ABOUT
A SINGLE PARAMETER

6.1 Test of the Hypothesis that the Mean of a Normal Distribution Has a Specified Value when the Standard Deviation Is Known

6.1.1 CHOICE OF AN OC CURVE

In the previous chapter, the concept of an OC curve was introduced and reference was made to a problem concerning the average Rockwell hardness of unbranded steel shafts. Recall that it was assumed that there was information indicating that the distribution of Rockwell hardness can be approximated by a normal distribution with *known* standard deviation equal to 2 units on the B scale. The unbranded shafts may have been produced in any mill so that the mean of the normal distribution is unknown. If the average Rockwell hardness is equal to 72, there is a desire to purchase the unbranded shafts because of their attractive price. In the previous discussion several decision procedures were presented rather arbitrarily and without regard to the "design of the experiment," i.e., without examining the random variable used to describe the sample data, the sample size, and the acceptance region. It has been indicated that there is an operating characteristic curve associated with every decision procedure which can be used as the appropriate "measure of consequence." Hence, in designing an experiment, the experimenter should consider the operating characteristic curve. This OC curve reflects the risks that he is willing to tolerate. Instead of designating the entire OC curve, it is sufficient to choose two points on this curve: the probability of concluding the average Rockwell hardness is 72 when it really is 72; and the value of the average Rockwell hardness that it is considered important to detect, together with the risk associated with this point. For example, in determining the first point the experimenter may be seeking a procedure which will conclude that the average Rockwell hardness is 72 when it is actually 72, with probability 0.95. The number 0.95 is not chosen arbitrarily but, instead, reflects the magnitude of the Type I error ($\alpha = 1 - 0.95 = 0.05$) that the experimenter is willing to tolerate. The magnitude of the Type I error can be 0.001, or 0.01, or even 0.50, depending on the particular situation. If the decision involves determining whether or not to shut down a steel process, the error of Type I should be very small. Closing a steel plant is extremely costly and an error in this decision may be quite important.

At the other end of the spectrum is the case of choosing between two manufacturers who offer similar products at similar prices. If there is no difference in quality of the products, it is not terribly important which manufacturer is chosen and, hence, a large Type I error can be tolerated. In any event, the magnitude of the Type I error is a choice the experimenter must make, and reflects the risk that he is willing and able to tolerate.

The second point on the OC curve that the experimenter must choose is the value of the average Rockwell hardness that it is considered important to detect, together with the risk associated with this point. Obviously, deciding that an average Rockwell hardness of 72.000007 is different from 72 is not very important, whereas a similar decision about an average Rockwell hardness of 75 may be quite important. It was pointed out in the previous chapter that a correct decision is rarely made all the time, and consequently, some risk (β) must be associated with concluding that the average Rockwell hardness is 72 when it is actually 75. This risk may be 0.01, 0.1, 0.15, 0.50, etc., depending on the seriousness of the decision, and is different for different problems. Again, this point on the OC curve, both ordinate and abscissa, is chosen by the experimenter and reflects his risks. This choice is not up to the statistician. Once two points on the OC curve have been found, the proper procedure can be determined.

Formalizing these concepts, the problem of concern is to test the hypothesis that the mean of a normal distribution with known standard deviation is equal to a fixed constant, i.e., $\mu = \mu_0$.[1] Two points on the OC curve are chosen. These points are denoted by $(\mu_0, 1 - \alpha)$ and (μ_1, β) and are shown in Fig. 6.1.

6.1.2 TABLES AND CHARTS FOR DETERMINING DECISION RULES

An experiment is to be performed and a random sample of size n drawn. The resulting sequence of random variables $X_1, X_2, \ldots, X_n$ can be thought of as each representing the measurable quantity associated with a trial or test. It is assumed that the X_i are independent normally distributed random

---

[1] If the alternative is one-sided, the hypothesis may also be $\mu \leq \mu_0$ or $\mu \geq \mu_0$. The notation $\mu = \mu_0$ will be interpreted to include these cases.

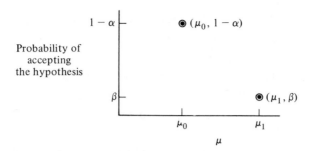

Fig. 6.1. *Two points used for determining an OC curve.*

variables, each with unknown mean μ and *known variance* σ^2. The *optimum* procedure for testing the hypothesis that the mean of the normal distribution has a specified value, i.e., $\mu = \mu_0$, is based on the random variable U, where

$$U = \frac{(\bar{X} - \mu_0)\sqrt{n}}{\sigma}.$$

A random variable used in this context is called a *test statistic*. If the value taken on by the test statistic as a result of experimentation falls into the acceptance region, the hypothesis that μ equals μ_0 is accepted; otherwise, it is rejected. This section is devoted to finding the acceptance region and the sample size for the test statistic U, using the tables and charts which are provided. One- and two-sided procedures are considered.

6.1.2.1 *Tables and Charts for Two-Sided Procedures.* Two-sided procedures are analyzed first. The rules presented assume that the risk function, which is expressed by the OC curve, is symmetric about μ_0. This implies that for any positive constant ϵ the risk associated with the value $\mu = \mu_0 + \epsilon$ is the same as the risk associated with $\mu = \mu_0 - \epsilon$, or an error incurred when $\mu = \mu_0 + \epsilon$ is as serious as an error incurred when $\mu = \mu_0 - \epsilon$. This restriction is relaxed later.

The acceptance region for the two-sided procedure is given by the interval $[-K_{\alpha/2}, K_{\alpha/2}]$, where α is the level of significance (Type I error) and $K_{\alpha/2}$ is the 100 $\alpha/2$ percentage point (normal deviate corresponding to $\alpha/2$) of the normal distribution. Thus, the hypothesis that the mean of a normal distribution equals μ_0 is accepted if

$$-K_{\alpha/2} \leq U \leq K_{\alpha/2}$$

and rejected if U lies outside this interval. For example, in testing the hypothesis that the average Rockwell hardness of the unbranded steel shafts is 72 on the B scale, assume that the point (72,0.95) on the OC curve is specified. The Type I error is then $\alpha = 0.05$ and $K_{\alpha/2} = K_{0.025} = 1.96$. Hence, the hypothesis that the average Rockwell hardness is 72 is accepted if

$$-1.96 \leq U \leq 1.96$$

and rejected if U lies outside this interval.

It is easily verified that this procedure yields the point $(\mu_0, 1 - \alpha)$ on the OC curve. Whenever $\mu = \mu_0$, the test statistic

$$U = \frac{(\bar{X} - \mu_0)\sqrt{n}}{\sigma}$$

is normally distributed with zero mean and unit variance. Hence, if $\mu = \mu_0$,

$$P\{-K_{\alpha/2} \leq U \leq K_{\alpha/2}\} = 1 - \alpha.$$

The required sample size is obtained by referring to a set of operating characteristic curves which are provided for this purpose. An OC curve for testing the hypothesis that the mean of a normal distribution is μ_0 against a two-sided alternative is usually a plot of the probability of acceptance against the true mean μ, and is symmetric about μ_0. Such a plot is shown in Fig. 6.2 for a fixed sample size n. In order for these curves to be useful in all

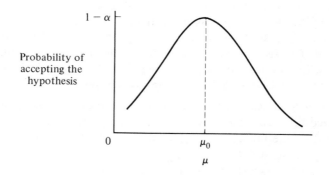

Fig. 6.2. *The OC curve for testing the hypothesis that the mean of a normal distribution with known standard deviation is μ_0 against a two-sided alternative.*

problems concerning the mean of a normal distribution, regardless of the values of μ_0 and σ, the abscissa scale is changed from μ to d, where

$$d = \frac{|\mu - \mu_0|}{\sigma}.^{[1]}$$

The OC curve of Fig. 6.2 is now replotted against d and is shown in Fig. 6.3 for the same fixed sample size n. The point $d = 0$ corresponds to the point $\mu = \mu_0$.

---

[1] The symbol $|\mu - \mu_0|$ is read as the absolute value of the difference between μ and μ_0.

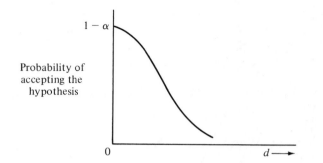

Fig. 6.3. *Standardized OC curve for testing the hypothesis that the mean of a normal distribution with known standard deviation is μ_0 against a two-sided alternative.*

A set of such OC curves for different values of n appears in Figs. 6.4 and 6.5. Figure 6.4 presents the OC curves of the two-sided normal test for a level of significance (Type I error) equal to 0.05. Figure 6.5 presents the OC curves of the two-sided normal test for a level of significance equal to 0.01.

If the level of significance of the procedure is either 0.05 or 0.01, the required sample size n is obtained by entering the appropriate figure (Fig. 6.4 if $\alpha = 0.05$ and Fig. 6.5 if $\alpha = 0.01$) with the point $(|\mu_1 - \mu_0|/\sigma, \beta)$ and

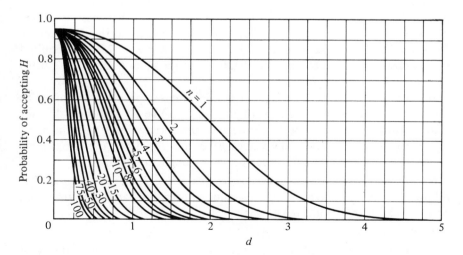

Fig. 6.4. *The OC curves for different values of n for the two-sided normal test for a level of significance $\alpha = 0.05$.*

(Reproduced by permission from "Operating Characteristics for the Common Statistical Tests of Significance" by Charles L. Ferris, Frank E. Grubbs, Chalmers L. Weaver, *Annals of Mathematical Statistics*, June, 1946.)

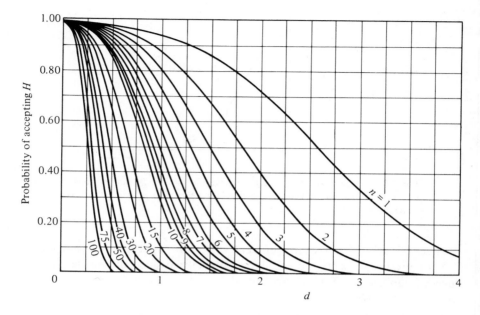

Fig. 6.5. *The OC curves for different values of n for the two-sided normal test for a level of significance* $\alpha = 0.01$.

reading out the value of the sample size associated with the OC curve that passes through this point. Returning to the Rockwell hardness example, if it is stated that the probability of accepting the hypothesis that the average Rockwell hardness is 72 should not exceed 0.10 when $\mu = 75$,[1] the required sample size is obtained by entering Fig. 6.4 with the point $(|75 - 72|/2, 0.10)$ and reading out a sample size of $n = 5$.

If a non-symmetric risk function is desired, the sample size is obtained as follows: Let the risk associated with $\mu = (\mu_0 + \epsilon)$ be β^+ and the risk associated with $\mu = (\mu_0 - \epsilon)$ be β^-. Enter the appropriate figure with the point $d = \epsilon/\sigma$ and the smaller of the values β^+ or β^-. Read out the sample size associated with the OC curve that passes through this point.

6.1.2.2 *Summary for Two-Sided Procedures Using Tables and Charts.* The following steps are taken, using a two-sided procedure for testing the mean of a normal distribution—σ known:

1. Two points on the OC curve are chosen: the probability of accepting the hypothesis when $\mu = \mu_0$, i.e., $(\mu_0, 1 - \alpha)$, and the probability of accepting the hypothesis when $\mu = \mu_1$, i.e., (μ_1, β) where μ_1 is the value of the mean considered important to detect. On the OC curve

---

[1] This implies that the Type II error at $\mu = 75$ equals 0.10 since the OC curve is a decreasing function of μ.

with the d scale, these points are translated to $(0, 1 - \alpha)$ and $(|\mu_1 - \mu_0|/\sigma, \beta)$.

2. Find the required sample size n by entering the appropriate figure (Fig. 6.4 if $\alpha = 0.05$ and Fig. 6.5 if $\alpha = 0.01$) with the point $(|\mu_1 - \mu_0|/\sigma, \beta)$ and reading out the value of the sample size associated with the OC curve that passes through this point.
3. Determine the acceptance region which is the interval $[-K_{\alpha/2}, K_{\alpha/2}]$.
4. Draw a random sample of n items.
5. Compute the value of the test statistic $U = (\bar{X} - \mu_0)\sqrt{n}/\sigma$.
6. If $-K_{\alpha/2} \leq U \leq K_{\alpha/2}$, i.e., $|U| \leq K_{\alpha/2}$, accept the hypothesis that $\mu = \mu_0$. If U lies outside this interval, reject the hypothesis and conclude that the mean is not μ_0.

6.1.2.3 *Tables and Charts for One-Sided Procedures.* If the procedure is one-sided and rejection is desired when the true mean exceeds μ_0 (this being the alternative to the hypothesis that $\mu = \mu_0$ or $\mu \leq \mu_0$), the acceptance region is given by the interval $[-\infty, K_\alpha]$, where α is the level of significance and K_α is the 100 α percentage point (normal deviate corresponding to α) of the normal distribution. Thus, the hypothesis that $\mu = \mu_0$ (or $\mu \leq \mu_0$) is accepted if $U \leq K_\alpha$ and rejected if U lies outside this interval. This acceptance procedure leads to the point $(\mu_0, 1 - \alpha)$ on the OC curve. Whenever $\mu = \mu_0$, the test statistic

$$U = \frac{(\bar{X} - \mu_0)\sqrt{n}}{\sigma}$$

is normally distributed with zero mean and unit variance. Hence, if $\mu = \mu_0$,

$$P\{U \leq K_\alpha\} = 1 - \alpha.$$

The OC curves necessary for determining the sample size n are given in Figs. 6.6 and 6.7 for the one-sided procedure where rejection is desired when the true mean exceeds μ_0. Figure 6.6 presents the OC curves for a level of significance equal to 0.05, and Fig. 6.7 presents the OC curves for a level of significance equal to 0.01. As in the two-sided case, for these curves to be useful in all such problems concerning the mean of a normal distribution, regardless of the values of μ_0 and σ, the abscissa scale is again changed from μ to d, where d is now defined as

$$d = \frac{\mu - \mu_0}{\sigma}.$$

If the level of significance of the procedure is either 0.05 or 0.01, the required sample size n is obtained by entering the appropriate figure (Fig. 6.6 if $\alpha = 0.05$ and Fig. 6.7 if $\alpha = 0.01$) with the point $((\mu_1 - \mu_0)/\sigma, \beta)$ and reading out the value of the sample size associated with the OC curve that passes through this point.

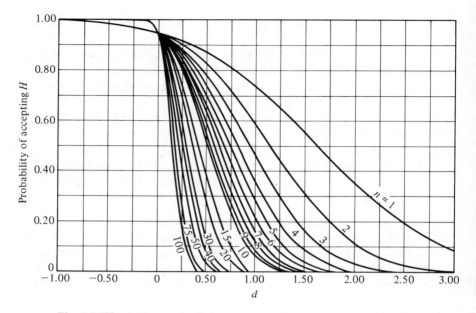

Fig. 6.6. *The OC curves for different values of n for the one-sided normal test for a level of significance α = 0.05.*

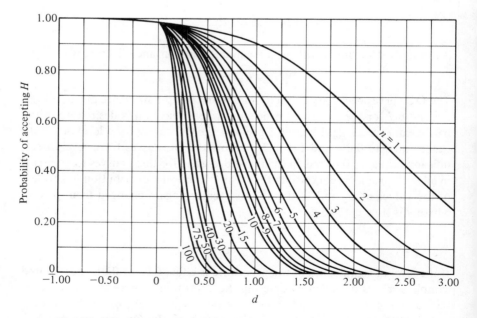

Fig. 6.7. *The OC curves for different values of n for the one-sided normal test for a level of significance α = 0.01.*

If the procedure is one-sided and rejection is desired when the true mean is less than μ_0 (this being the alternative to the hypothesis that $\mu = \mu_0$ or $\mu \geq \mu_0$), the acceptance region is given by the interval $[-K_\alpha, \infty]$. Thus, the hypothesis that $\mu = \mu_0$ (or $\mu \geq \mu_0$) is accepted if $U \geq -K_\alpha$ and rejected if U lies outside this interval. Clearly, this procedure yields the point $(\mu_0, 1 - \alpha)$ on the OC curve. Whenever $\mu = \mu_0$, the test statistic

$$U = \frac{(\bar{X} - \mu_0)\sqrt{n}}{\sigma}$$

is normally distributed with zero mean and unit variance. Hence, if $\mu = \mu_0$,

$$P\{U \geq -K_\alpha\} = 1 - \alpha.$$

The OC curves necessary for determining the sample size for the one-sided procedure where rejection is desired when the true mean is less than μ_0 are also given in Figs. 6.6 and 6.7. However, the abscissa d is now defined as

$$d = \frac{\mu_0 - \mu}{\sigma}.$$

If the level of significance of the procedure is either 0.05 or 0.01, the required sample size n is obtained by entering the appropriate figure (Fig. 6.6 if $\alpha = 0.05$ and Fig. 6.7 if $\alpha = 0.01$) with the point $((\mu_0 - \mu_1)/\sigma, \beta)$ and reading out the value of the sample size associated with the OC curve that passes through this point.

6.1.2.4 *Summary for One-Sided Procedures Using Tables and Charts.* The following steps are taken, using a one-sided procedure for testing the mean of a normal distribution:

1. Two points on the OC curve are chosen: the probability of accepting the hypothesis when $\mu = \mu_0$, i.e., $(\mu_0, 1 - \alpha)$; and the probability of accepting the hypothesis when $\mu = \mu_1$, i.e., (μ_1, β) where μ_1 is the value of the mean considered important to detect. If rejection is desired when $\mu > \mu_0$, then these points are translated to $(0, 1 - \alpha)$ and $((\mu_1 - \mu_0)/\sigma, \beta)$ on the OC curve with the d scale. If rejection is desired when $\mu < \mu_0$, these points are translated to $(0, 1 - \alpha)$ and $((\mu_0 - \mu_1)/\sigma, \beta)$ on the OC curve with the d scale.
2. Find the required sample size n by entering the appropriate figure (Fig. 6.6 if $\alpha = 0.05$ and Fig. 6.7 if $\alpha = 0.01$). If rejection is desired when $\mu > \mu_0$, the figure is entered with the point $((\mu_1 - \mu_0)/\sigma, \beta)$ and the value of the sample size associated with the OC curve that passes through this point is read out. If rejection is desired when $\mu < \mu_0$, the figure is entered with the point $((\mu_0 - \mu_1)/\sigma, \beta)$ and the value of the sample size associated with the OC curve that passes through this point is read out.
3. Determine the acceptance region which is the interval $[-\infty, K_\alpha]$ for the alternative $\mu > \mu_0$ and $[-K_\alpha, \infty]$ for the alternative $\mu < \mu_0$.

4. Draw a random sample of n items.
5. Compute the value of the test statistic $U = [(\bar{X} - \mu_0)\sqrt{n}]/\sigma$.
6. If the alternative is $\mu > \mu_0$, accept the hypothesis that $\mu = \mu_0$ (or $\mu \leq \mu_0$) whenever $U \leq K_\alpha$; otherwise, reject the hypothesis and conclude that $\mu > \mu_0$. If the alternative is $\mu < \mu_0$, accept the hypothesis that $\mu = \mu_0$ (or $\mu \geq \mu_0$) whenever $U \geq -K_\alpha$; otherwise, reject the hypothesis and conclude that $\mu < \mu_0$.

6.1.2.5 *Tables and Charts for Operating Characteristic Curves.* Whenever the sample size is fixed in advance, or after it is chosen, the operating characteristic curves given in Figs. 6.4 and 6.5 for two-sided procedures and Figs. 6.6 and 6.7 for one-sided procedures can be used to evaluate the risks involved in using these procedures. If the level of significance is either 0.05 or 0.01, the appropriate OC curve is entered with the sample size n and d, and the probability of accepting the hypothesis is read from the curve. The value of d chosen corresponds to the value μ for which the risk is to be evaluated.

The fact that the OC curves presented are for levels of significance equal to 0.01 and 0.05 does not imply that these are the only levels ever used. Often, other values are appropriate so that suitable methods for determining the sample size and acceptance region must be found. Such methods are described in Sec. 6.1.3. However, it can be said that $\alpha = 0.05$ and 0.01 are most commonly used in industry, although emphasizing that the choice of α is up to the experimenter, rather than the statistician. The level of significance reflects a tolerable risk of making an incorrect decision and can be chosen only by someone involved directly in experimentation.

The OC curves presented are calculated under the assumption that the test statistic U is normally distributed. This condition is fulfilled if $X_1, X_2, \ldots, X_n$ are independent and identically normally distributed random variables. However, the OC curves of the procedures presented are valid for more general forms of underlying probability distributions. Since the test statistic U is related to the sample mean $\bar{X}$, the conclusions of the central limit theorem are applicable. Hence, U is approximately normally distributed, regardless of the form of the distribution of the X's, provided n is sufficiently large.

A summary of the procedures for testing the hypothesis that the mean of a normal distribution has a specified value when the standard deviation is known is given in Table 6.1. For purposes of this summary, criteria for rejecting the hypothesis are given instead of criteria for acceptance.

6.1.3 Analytical Determination of Decision Rules

6.1.3.1 *Acceptance Regions and Sample Sizes.* Let $X_1, X_2, \ldots, X_n$ be a random sample of n independent normally distributed random variables

Table 6.1. Test of the Hypothesis That the Mean of a Normal Distribution Has a Specified Value when the Standard Deviation Is Known

Notation for Hypothesis $H: \mu = \mu_0$	*Test Statistic* $U = (\bar{X} - \mu_0)\sqrt{n}/\sigma$
Criteria for Rejection	*Method for Choosing Sample Size*
$\lvert U \rvert > K_{\alpha/2}$ if rejection is desired when the true mean departs from μ_0 in either direction.	Choose a value of $\mu_1 \neq \mu_0$ for which rejection of the hypothesis is desired with given high probability; calculate $d = \lvert \mu_1 - \mu_0 \rvert/\sigma$ and select n from the OC curves of Fig. 6.4 or 6.5.
$U > K_{\alpha}$ if rejection is desired when the true mean exceeds μ_0.	Choose a value of $\mu_1 > \mu_0$ for which rejection of the hypothesis is desired with given high probability; calculate $d = (\mu_1 - \mu_0)/\sigma$ and select n from the OC curves of Fig. 6.6. or 6.7.
$U < -K_{\alpha}$ if rejection is desired when the true mean is less than μ_0.	Choose a value of $\mu_1 < \mu_0$ for which rejection of the hypothesis is desired with given high probability; calculate $d = (\mu_0 - \mu_1)/\sigma$ and select n from the OC curves of Fig. 6.6 or 6.7.

Level of Significance	*For One-Sided Tests*	*For Two-Sided Tests*
$\alpha = 0.05$	$K_{0.05} = 1.645$	$K_{0.025} = 1.960$
$\alpha = 0.01$	$K_{0.01} = 2.326$	$K_{0.005} = 2.576$

each with unknown mean μ and known variance σ^2. The optimum procedure for testing the hypothesis that μ equals μ_0 is based on the test statistic U, where

$$U = \frac{(\bar{X} - \mu_0)\sqrt{n}}{\sigma}.$$

This section is devoted to finding acceptance regions and expressions for the sample size for both one- and two-sided procedures.

As was indicated in Sec. 6.1.2, the acceptance region for the two-sided procedure is the interval $[-K_{\alpha/2}, K_{\alpha/2}]$, where α is the level of significance and $K_{\alpha/2}$ is the $100\,\alpha/2$ percentage point of the normal distribution. The expression for the sample size n is given approximately by

$$n \cong \frac{(K_{\alpha/2} + K_{\beta})^2 \sigma^2}{(\mu_1 - \mu_0)^2},$$

where β is the probability of accepting the hypothesis when $\mu = \mu_1$, i.e., Type II error, and K_{β} is the $100\,\beta$ percentage point of the normal distribution. This approximation is good whenever

$$P\left\{ N < -K_{\alpha/2} - \sqrt{n}\frac{\lvert \mu_1 - \mu_0 \rvert}{\sigma} \right\}$$

is small compared to β, where N is a normally distributed random variable with zero mean and unit variance.

Returning to the Rockwell hardness example of the previous sections in this chapter, the following quantities are given:

$$\alpha = 0.05, \qquad \mu_0 = 72,$$
$$\beta = 0.10, \qquad \mu_1 = 75,$$
$$\sigma = 2.$$

From Appendix Table 1, $K_{\alpha/2} = K_{0.025} = 1.96$ and $K_\beta = K_{0.10} = 1.28$. Substituting into the expression for n,

$$n \cong \frac{(1.96 + 1.28)^2 2^2}{(75 - 72)^2} = 5 \text{ observations.}$$

Checking to see if the approximation is good,

$$P\left\{N < -1.96 - \frac{\sqrt{5}\,|75 - 72|}{2}\right\} = P\{N < -5.31\}$$

is negligible and, hence, $n = 5$ is the solution. This coincides with the value of n found by using the OC curves provided.

If the procedure is one-sided and rejection is desired when the true mean exceeds μ_0 (this being the alternative to the hypothesis that $\mu = \mu_0$ or $\mu \le \mu_0$), the acceptance region is the interval $[-\infty, K_\alpha]$, where α is the level of significance and K_α is the $100\,\alpha$ percentage point of the normal distribution. If the procedure is one-sided and rejection is desired when the true mean is less than μ_0 (this being the alternative to the hypothesis that $\mu = \mu_0$ or $\mu \ge \mu_0$), the acceptance region is the interval $[-K_\alpha, \infty]$. For both of these one-sided procedures the expression for the sample size n is given by

$$n = \frac{(K_\alpha + K_\beta)^2 \sigma^2}{(\mu_1 - \mu_0)^2},$$

where β is the probability of accepting the hypothesis when $\mu = \mu_1$, i.e., Type II error, and K_β is the $100\,\beta$ percentage point of the normal distribution.

It has already been verified that the acceptance regions presented for the one- and two-sided procedures yield a probability of $1 - \alpha$ of accepting the hypothesis that $\mu = \mu_0$ when it is true. The derivation of the expression for the sample size for the one-sided procedure where rejection is desired if $\mu > \mu_0$ will follow. The derivations of the sample sizes for the two-sided procedure and for the one-sided procedure where rejection is desired if $\mu < \mu_0$ are similar.

If the procedure is one-sided and rejection is desired when the true mean exceeds μ_0, the acceptance region is given by the interval $[-\infty, K_\alpha]$. Thus, the hypothesis that $\mu = \mu_0$ (or $\mu \le \mu_0$) is accepted if

$$U \le K_\alpha$$

and the probability of accepting the hypothesis is given by

$$P\left\{U = \frac{(\bar{X} - \mu_0)\sqrt{n}}{\sigma} \leq K_\alpha\right\}.$$

Two points on the operating characteristic curve are specified, i.e., $(\mu_0, 1 - \alpha)$ and (μ_1, β). The proper choice of the acceptance region returns the point $(\mu_0, 1 - \alpha)$ on the OC curve. The second point requires that the following equality hold, namely

$$P\left\{U = \frac{(\bar{X} - \mu_0)\sqrt{n}}{\sigma} \leq K_\alpha\right\} = \beta \qquad \text{when } \mu = \mu_1.$$

This is equivalent to stating that the probability of accepting the hypothesis that $\mu = \mu_0$ is β when the mean of the X's is actually equal to μ_1. Whenever the mean of the X's is μ_1, the expected value of $\bar{X}$ is also μ_1 so that the expected value of U is given by

$$E(U) = \frac{(\mu_1 - \mu_0)\sqrt{n}}{\sigma}.$$

Hence, when $\mu = \mu_1$, U is normally distributed with mean equal to $(\mu_1 - \mu_0)\sqrt{n}/\sigma$ and unit variance. It then follows that

$$P\{U \leq K_\alpha\} = P\left\{N \leq K_\alpha - \frac{(\mu_1 - \mu_0)\sqrt{n}}{\sigma}\right\} = \beta,$$

where N is a normally distributed random variable with zero mean and unit variance. This probability statement is illustrated in Fig. 6.8. Because of the symmetry of the normal distribution the shaded area of Fig. 6.8 is equivalent to

$$P\left\{N \geq \left[\frac{(\mu_1 - \mu_0)\sqrt{n}}{\sigma}\right] - K_\alpha\right\}$$

so that

$$P\left\{N \geq \left[\frac{(\mu_1 - \mu_0)\sqrt{n}}{\sigma}\right] - K_\alpha\right\} = \beta.$$

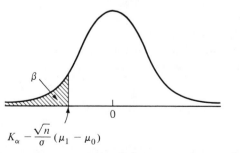

Fig. 6.8. *The shaded area indicates* $P\left\{N \leq K_\alpha - \frac{(\mu_1 - \mu_0)\sqrt{n}}{\sigma}\right\}.$

Hence, $[(\mu_1 - \mu_0)\sqrt{n}/\sigma] - K_\alpha = K_\beta$, where K_β is the $100\,\beta$ percentage point of the normal distribution. Solving this last expression for n leads to

$$n = \frac{(K_\alpha + K_\beta)^2\sigma^2}{(\mu_1 - \mu_0)^2},$$

which is the result mentioned earlier.

6.1.3.2 *The Operating Characteristic Curve.* If the sample size and acceptance region are known, the entire OC curve is easily determined. Let the true mean of the X's be μ so that the expected value of U is

$$E(U) = \frac{(\mu - \mu_0)\sqrt{n}}{\sigma}.$$

The random variable U is then normal with mean $(\mu - \mu_0)\sqrt{n}/\sigma$ and unit variance. For the one-sided procedure where rejection is desired when the true mean exceeds μ_0, the expression for the OC curve (probability of accepting the hypothesis) is given by $P\{U \leq K_\alpha\}$ which is equivalent to

$$P\left\{N \leq K_\alpha - \frac{(\mu - \mu_0)\sqrt{n}}{\sigma}\right\}$$

where N is a normally distributed random variable with zero mean and unit variance. This last expression can be evaluated for any value of μ with the aid of Appendix Table 1. It is interesting to note that for fixed n and K_α this expression for the OC curve is just a function of $d = (\mu - \mu_0)/\sigma$, which is the abscissa of the OC curves plotted in Figs. 6.6 and 6.7.

If the procedure is one-sided and rejection is desired when the true mean is less than μ_0, the OC curve is given by the expression

$$P\{U \geq -K_\alpha\} = P\left\{N \geq -K_\alpha - \frac{(\mu - \mu_0)\sqrt{n}}{\sigma}\right\}.$$

Finally, if the procedure is two-sided, the OC curve is given by the expression

$$P\{-K_{\alpha/2} \leq U \leq K_{\alpha/2}\}$$
$$= P\left\{-K_{\alpha/2} - \frac{(\mu - \mu_0)\sqrt{n}}{\sigma} \leq N \leq K_{\alpha/2} - \frac{(\mu - \mu_0)\sqrt{n}}{\sigma}\right\}.$$

As an example of the use of the curve, we return to the Rockwell hardness example and the two-sided procedure associated with it. The following data are available or have been obtained:

$$\alpha = 0.05, \qquad \sigma = 2,$$
$$K_{\alpha/2} = K_{0.025} = 1.96, \qquad n = 5,$$
$$\mu_0 = 72.$$

If the actual mean of the normal distribution is $\mu = 73$, the probability of accepting the hypothesis is given by

$$P\left\{-1.96 - \frac{(73 - 72)\sqrt{5}}{2} \leq N \leq 1.96 - \frac{(73 - 72)\sqrt{5}}{2}\right\}$$
$$= P\{-3.08 \leq N \leq 0.84\} = 0.80.$$

6.1.4 EXAMPLE

A manufacturer produces a special alloy steel with an average tensile strength of 25,800 psi. A change in the composition of the alloy is said to increase the breaking strength. The standard deviation of the tensile strength is known to be 300 psi and the change in composition is not expected to change this value. If there is no change in the average tensile strength, the manufacturer wants to reach this conclusion with probability 0.99 ($\alpha = 0.01$). If the average tensile strength is increased by as much as 250 psi, the manufacturer is willing to take only a 10% risk in not detecting it ($\beta = 0.10$).

The hypothesis to be tested is that the average tensile strength is unaffected by the change in composition in the material against the alternative that the change in composition increases the average tensile strength. The possibility of the change causing a decrease is discounted. Thus, the decision procedure is to accept the hypothesis if $U \leq K_\alpha = K_{0.01} = 2.326$ since the level of significance is given as 0.01. The required sample size is found from the OC curve in Fig. 6.7. From the information above, $d = (\mu_1 - \mu_0)/\sigma = 250/300 = 0.833$. Entering Fig. 6.7 with $d = 0.833$ and a probability of acceptance equal to 0.10, the required sample size is seen to be about 19 observations.

The experiment is performed and the sample mean $\bar{X}$ of the 19 observations is calculated and found to be 26,100 psi. It then follows that

$$U = \frac{(26,100 - 25,800)\sqrt{19}}{300} = 4.36,$$

which is greater than $K_{0.01} = 2.326$, so the hypothesis is rejected. Thus, it is concluded that the change in composition increases the tensile strength.

It is interesting to comment on this example when the maximum number of observations available is only 7; e.g., suppose funds are provided for no more than 7 specimens. In this case, Fig. 6.7 reveals that if the increase in tensile strength is as much as 250 psi, this procedure detects such a difference only 50% of the time. Hence, it may be advisable to wait until additional funds become available. If the experiment is performed with only 7 observations, and the procedure results in acceptance of the hypothesis, this does not necessarily imply that an increase in tensile strength has not been effected. Rather, it implies that the increase is not sufficiently large to detect it with a sample of size 7; another way of putting it is that there is insufficient information available to conclude that an increase has occurred.

Also note that too many observations can also cause serious problems by detecting differences that are statistically significant but practically insignificant. For example, suppose that increases in the tensile strength of 100 psi or less are of no importance from a practical point of view. Yet, if a random sample of 100 is drawn, Fig. 6.7 indicates that the decision procedure will detect an increase of 100 psi approximately 84% of the time, thereby leading to a conclusion which is statistically correct (the true average tensile strength is not 25,800 psi), but which has no practical significance.

6.2 Test of the Hypothesis That the Mean of a Normal Distribution Has a Specified Value When the Standard Deviation Is Unknown

6.2.1 CHOICE OF AN OC CURVE

Section 6.1 dealt with the problem of a test for the mean of a normal distribution having a specified value, assuming that the standard deviation is known. Such tests are often quite useful. In repetitive experiments, considerable information about variability may be accumulated and assumptions of known standard deviation validated. Such a test may be applied quickly; often it is possible to tell at a glance whether an observed mean departs by more than two or three times its standard deviation from a hypothesized value. However, both the Type I error and the Type II error depend rather sensitively on the assumed value of σ. In the example of Sec. 6.1.4, the standard deviation of tensile strength was assumed to be 300 psi and a test procedure was derived which had a Type I error (α) equal to 0.01, and a Type II error (β) equal to 0.10. If this assumed standard deviation were off by 10% and the true value were, say, 270, then the Type I error (α) would be 0.0049 instead of 0.01 and the Type II error (β) would be 0.074 instead of 0.10.

A common occurrence in experimental work is that all the information about the variability is contained in the data in hand. In a material testing laboratory, a great deal of prior experimental data may be available for hardness experiments of comparable alloys, but in much physical and chemical experimentation, decisions are made on the basis of a single experiment which contains all the relevant information. It is of great importance to provide tests of hypotheses about means which do not depend on a known standard deviation.

In testing the hypothesis that the mean of a normal distribution is equal to μ_0 when the standard deviation is unknown, the t statistic is used. This test statistic is a function of the sample mean, the sample standard deviation, and the sample size. The abscissae of operating characteristic curves for procedures based upon this t statistic depend on the quantity $(\mu - \mu_0)/\sigma$.

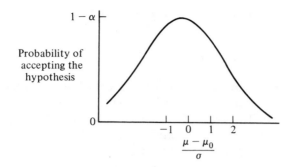

Fig. 6.9. *The OC curve of the two-sided (t) test.*

An example of such an OC curve is shown in Fig. 6.9 for a two-sided test. In order to determine the acceptance region and sample size, two points on the OC curve must be chosen. These points are denoted by $(0, 1 - \alpha)$ and $((\mu_1 - \mu_0)/\sigma, \beta)$. The first point corresponds to choosing the probability of accepting the hypothesis when it is true (when $\mu = \mu_0$), and the second point corresponds to choosing the probability of accepting the hypothesis when $(\mu_1 - \mu_0)/\sigma$ is the value of the ratio of the difference in means to the true standard deviation considered important to detect. The dependence of this second point on σ may appear to be incongruous at first since the t statistic is used specifically in situations when σ is unknown. However, a little thought will reveal that this is quite reasonable. If the experimental error (σ) is large, small differences in the mean cannot be considered important, whereas if the experimental error is small, these differences tend to become magnified. Hence, it is not the absolute values of the deviations from μ_0 which are important, but rather the relative deviations compared to σ which must be considered. For the purpose of choosing an OC curve a *rough* notion of σ is usually adequate. Before experimentation, such a notion can be obtained from past data or from some prior knowledge. If the OC curve is examined after experimentation, the sample standard deviation can be used as a "rough" estimate. It is emphasized that a rough notion is *not* equivalent to a complete knowledge of the standard deviation.

6.2.2 TABLES AND CHARTS FOR CARRYING OUT t TESTS

An experiment is to be performed and a random sample of size n drawn. Each of the resulting sequence of random variables, $X_1, X_2, \ldots, X_n$, can be thought of as representing the measurable quantity associated with a trial or test. It is assumed that the X_i are independent normally distributed random variables, each with unknown mean μ and unknown variance σ^2. The *optimum* procedure for testing the hypothesis that the mean of the normal distribution has a specified value, i.e., $\mu = \mu_0$, is based on the test statistic t, where

$$t = \frac{(\bar{X} - \mu_0)\sqrt{n}}{S}.$$

If the value taken on by the test statistic as a result of experimentation falls into the acceptance region, the hypothesis that μ equals μ_0 is accepted; otherwise, it is rejected. This section is devoted to finding the acceptance region and the sample size for the test statistic t, using the tables and charts which are provided. One- and two-sided procedures are considered.

6.2.2.1 *Tables and Charts for Two-Sided Procedures.* The two-sided procedure is analyzed first. Symmetry of the OC curve about zero will be assumed. This restriction can be relaxed by using the same methods as were presented in the section on the U test in this connection.

The acceptance region for the two-sided procedure for any sample size, n, is given by the interval $[-t_{\alpha/2;n-1}, t_{\alpha/2;n-1}]$, where α is the level of significance (Type I error) and $t_{\alpha/2;n-1}$ is the 100 $\alpha/2$ percentage point of the t distribution with $n - 1$ degrees of freedom. Thus, the hypothesis that the mean of a normal distribution equals μ_0 is accepted if $-t_{\alpha/2;n-1} \leq t \leq t_{\alpha/2;n-1}$ and rejected if t lies outside this interval. It is easily verified that this procedure yields a probability of $1 - \alpha$ of accepting the hypothesis that $\mu = \mu_0$ when it is true. When $\mu = \mu_0$, the test statistic

$$t = \frac{(\bar{X} - \mu_0)\sqrt{n}}{S}$$

has a t distribution with $n - 1$ degrees of freedom. Hence, if $\mu = \mu_0$,

$$P\{-t_{\alpha/2;n-1} \leq t \leq t_{\alpha/2;n-1}\} = 1 - \alpha.$$

The required sample size is obtained by referring to a set of operating characteristic curves which are provided for this purpose. Figure 6.10 presents the OC curves of the two-sided t test for a level of significance equal to 0.05. Figure 6.11 presents the OC curves of the two-sided t test for a level of significance equal to 0.01. The abscissa scale for these curves is d, where

$$d = \frac{|\mu - \mu_0|}{\sigma}.$$

If the level of significance of the procedure is either 0.05 or 0.01, the required sample size n is obtained by entering the appropriate figure (Fig. 6.10 if $\alpha = 0.05$ and Fig. 6.11 if $\alpha = 0.01$) with the point $(|\mu_1 - \mu_0|/\sigma, \beta)$ and reading out the value of the sample size associated with the OC curve that passes through this point.

6.2.2.2 *Summary for Two-Sided Procedures Using Tables and Charts.* The following steps are taken, using a two-sided procedure for testing the mean

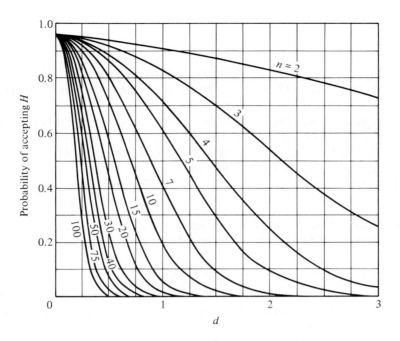

Fig. 6.10. *The OC curves for different values of n for the two-sided t test for a level of significance* $\alpha = 0.05$.
(Reproduced by permission from "Operating Characteristics for the Common Statistical Tests of Significance" by Charles L. Ferris, Frank E. Grubbs, Chalmers L. Weaver, *Annals of Mathematical Statistics, June,* 1946.)

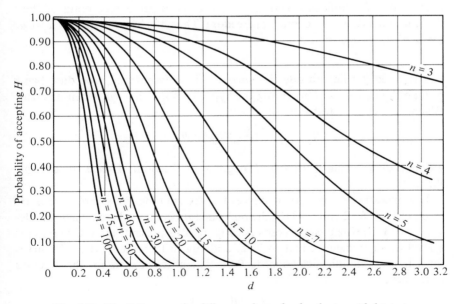

Fig. 6.11. *The OC curves for different values of n for the two-sided t test for a level of significance* $\alpha = 0.01$.

of a normal distribution—σ unknown:

1. Two points on the OC curve are chosen and are translated to the d scales of Figs. 6.10 and 6.11. On this scale, these points are denoted by $(0, 1 - \alpha)$ and $(|\mu_1 - \mu_0|/\sigma, \beta)$.
2. Find the sample size n by entering the appropriate figure (Fig. 6.10 if $\alpha = 0.05$ and Fig. 6.11 if $\alpha = 0.01$) with the point $(|\mu_1 - \mu_0|/\sigma, \beta)$ and then reading out the value of the sample size associated with the OC curve that passes through this point.
3. Determine the acceptance region which is the interval

$$[-t_{\alpha/2;n-1}, t_{\alpha/2;n-1}].$$

4. Draw a random sample of n items.
5. Compute the value of the test statistic

$$t = \frac{(\bar{X} - \mu_0)\sqrt{n}}{S}.$$

6. If $-t_{\alpha/2;n-1} \leq t \leq t_{\alpha/2;n-1}$, i.e., $|t| \leq t_{\alpha/2;n-1}$, accept the hypothesis that $\mu = \mu_0$. If t lies outside this interval, reject the hypothesis and conclude that the mean is not μ_0.

6.2.2.3 *Tables and Charts for One-Sided Procedures.* If the procedure is one-sided and rejection is desired when the true mean exceeds μ_0 (this being the alternative to the hypothesis that $\mu = \mu_0$ or $\mu \leq \mu_0$), the acceptance region for any sample size, n, is given by the interval $[-\infty, t_{\alpha;n-1}]$ where α is the level of significance and $t_{\alpha;n-1}$ is the 100 α percentage point of the t distribution with $n - 1$ degrees of freedom. Thus, the hypothesis that $\mu = \mu_0$ (or $\mu \leq \mu_0$) is accepted if

$$t \leq t_{\alpha;n-1}$$

and rejected if t lies outside this interval. This acceptance procedure has a probability of $1 - \alpha$ of accepting the hypothesis that $\mu = \mu_0$ when it is true. When $\mu = \mu_0$, the test statistic

$$t = \frac{(\bar{X} - \mu_0)\sqrt{n}}{S}$$

has a t distribution with $n - 1$ degrees of freedom. Hence, if $\mu = \mu_0$, $P\{t \leq t_{\alpha;n-1}\} = 1 - \alpha$.

The operating characteristic curves for determining the sample size n are given in Figs. 6.12 and 6.13 for the one-sided procedure where rejection is desired when the true mean exceeds μ_0. Figure 6.12 presents the OC curves for a level of significance equal to 0.05, and Fig. 6.13 presents the OC curves for a level of significance equal to 0.01. The abscissa scale for these curves is d, where

$$d = \frac{(\mu - \mu_0)}{\sigma}.$$

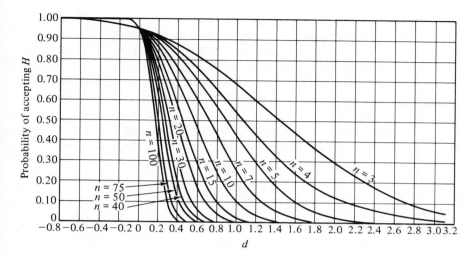

Fig. 6.12. *The OC curves for different values of n for the one-sided t test for a level of significance* α = 0.05.

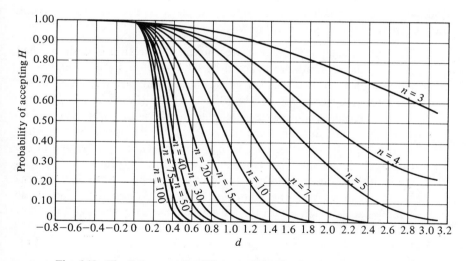

Fig. 6.13. *The OC curves for different values of n for the one-sided t test for a level of significance* α = 0.01.

If the level of significance of the procedure is either 0.05 or 0.01, the required sample size n is obtained by entering the appropriate figure (Fig. 6.12 if α = 0.05 and Fig. 6.13 if α = 0.01) with the point $((\mu_1 - \mu_0)/\sigma, \beta)$ and reading out the value of the sample size associated with the OC curve that passes through this point.

If the procedure is one-sided and rejection is desired when the true mean is less than μ_0 (this being the alternative to the hypothesis that $\mu = \mu_0$ or $\mu \geq \mu_0$), the acceptance region for any sample size n is given by the interval $[-t_{\alpha;n-1}, \infty]$. Thus, the hypothesis that $\mu = \mu_0$ (or $\mu \geq \mu_0$) is accepted if $t \geq -t_{\alpha;n-1}$ and rejected if t lies outside this interval. This acceptance procedure also has a probability of $1 - \alpha$ of accepting the hypothesis that $\mu = \mu_0$ when it is true.

The required sample size is again obtained by referring to Figs. 6.12 and 6.13. To use these curves for this one-sided procedure, the abscissa scale d is now defined as

$$d = \frac{\mu_0 - \mu}{\sigma}.$$

If the level of significance of the procedure is either 0.05 or 0.01, the required sample size n is obtained by entering the appropriate figure (Fig. 6.12 if $\alpha = 0.05$ and Fig. 6.13 if $\alpha = 0.01$) with the point $((\mu_0 - \mu_1)/\sigma, \beta)$ and reading out the value of the sample size associated with the OC curve that passes through this point.

6.2.2.4 *Summary for One-Sided Procedures Using Tables and Charts.* The following steps are taken, using a one-sided procedure for testing the mean of a normal distribution—σ unknown:

1. Two points on the OC curve are chosen. If rejection is desired when $\mu > \mu_0$, these points are denoted by $(0, 1 - \alpha)$ and $((\mu_1 - \mu_0)/\sigma, \beta)$. If rejection is desired when $\mu < \mu_0$, these points are translated to the d scale and are then denoted by $(0, 1 - \alpha)$ and $((\mu_0 - \mu_1)/\sigma, \beta)$.
2. Find the required sample size n by entering the appropriate figure (Fig. 6.12 if $\alpha = 0.05$ and Fig. 6.13 if $\alpha = 0.01$). If rejection is desired when $\mu > \mu_0$, the figure is entered with the point $((\mu_1 - \mu_0)/\sigma, \beta)$ and the value of the sample size associated with the OC curve that passes through this point is read out. If rejection is desired when $\mu < \mu_0$, the figure is entered with the point $((\mu_0 - \mu_1)/\sigma, \beta)$ and the value of the sample size associated with the OC curve that passes through this point is read out.
3. Determine the acceptance region which is the interval $[-\infty, t_{\alpha;n-1}]$ for the alternative $\mu > \mu_0$ and $[-t_{\alpha;n-1}, \infty]$ for the alternative $\mu < \mu_0$.
4. Draw a random sample of n items.
5. Compute the value of the test statistic $t = (\bar{X} - \mu_0)\sqrt{n}/S$.
6. If the alternative is $\mu > \mu_0$, accept the hypothesis that $\mu = \mu_0$ (or $\mu \leq \mu_0$) whenever $t \leq t_{\alpha;n-1}$; otherwise, reject the hypothesis and conclude that $\mu > \mu_0$. If the alternative is $\mu < \mu_0$, accept the hypothesis that $\mu = \mu_0$ (or $\mu \geq \mu_0$) whenever $t \geq -t_{\alpha;n-1}$; otherwise, reject the hypothesis and conclude that $\mu < \mu_0$.

6.2.2.5 *Tables and Charts for Operating Characteristic Curves.* Whenever the sample size is fixed in advance, or after it is chosen, the operating characteristic curves given in Figs. 6.10 and 6.11 for two-sided procedures and Figs. 6.12 and 6.13 for one-sided procedures can be used to evaluate the risks involved in using these procedures. If the level of significance is either 0.05 or 0.01, the appropriate OC curve is entered with the sample size n and d, and the probability of accepting the hypothesis is read from the curve. The value of d chosen is related to the value $(\mu - \mu_0)/\sigma$ for which the risk is to be evaluated.

If other levels of significance are desired, Appendix Table 3, which is a table of the percentage points of the t distribution, can be used to determine the acceptance region for a fixed value of n. However, analytical procedures which are beyond the scope of this text must be used to obtain the OC curve.

Although the OC curves are calculated under the assumption that the random variables $X_1, X_2, \ldots, X_n$ are independent normally (or approximately normally) distributed random variables, they are also valid for more general forms of underlying probability distributions, if the sample size is sufficiently large.

A summary of the procedures for testing the hypothesis that the mean of a normal distribution has a specified value when the standard deviation is

Table 6.2. Test of the Hypothesis That the Mean of a Normal Distribution Has a Specified Value When the Standard Deviation Is Unknown

Notation for Hypothesis $H: \mu = \mu_0$	*Test Statistic* $t = (\bar{X} - \mu_0)\sqrt{n}/S$
Criteria for Rejection	*Method for Choosing Sample Size*
$|t| > t_{\alpha/2;n-1}$ if rejection is desired when the true mean departs from μ_0 in either direction.	Determine a value of $d = |\mu_1 - \mu_0|/\sigma$ for which rejection of the hypothesis is desired with given high probability and enter Fig. 6.10 or 6.11 to find the necessary sample size.
$t > t_{\alpha;n-1}$ if rejection is desired when the true mean exceeds μ_0.	Determine a value of $d = (\mu_1 - \mu_0)/\sigma$ for which rejection of the hypothesis is desired with given high probability and enter Fig. 6.12 or 6.13 to find the necessary sample size.
$t < -t_{\alpha;n-1}$ if rejection is desired when the true mean is less than μ_0.	Determine a value of $d = (\mu_0 - \mu_1)/\sigma$ for which rejection of the hypothesis is desired with given high probability and enter Fig. 6.12 or 6.13 to find the necessary sample size.

Values of $t_{\alpha;\nu}$, the 100 α percentage points of the t distribution for ν degrees of freedom, are given in Appendix Table 3.

unknown is given in Table 6.2. For purposes of this summary, criteria for rejecting the hypothesis are given instead of criteria for acceptance.

6.2.3 EXAMPLES OF t TESTS

(a) In the manufacture of a food product, the label states that the box contains 10 lbs. The boxes are filled by machine and it is of interest to determine whether or not the machine is set properly. Previous experience has indicated that the standard deviation is approximately 0.05 lb., although this is not known precisely. The company feels that the present setting is unsatisfactory if the machine fills the boxes so that the mean weight differs from 10 lbs. by more than 0.1 lb. If this be the case, the probability of detecting a difference of this magnitude must not be less than 0.95 (i.e., $\beta = 0.05$). It is decided to tolerate a level of significance of 5% (i.e., $\alpha = 0.05$). Thus, from Fig. 6.10 and $d = 0.1/0.05 = 2$, a sample size of 6 is sufficient to insure rejection with probability 0.95 or greater if the mean weight differs from 10 lbs. by more than 0.1 lb. The data are: 9.99 lbs., 9.99 lbs., 10.00 lbs., 10.11 lbs., 10.09 lbs., 9.95 lbs.

$$\bar{X} = 10.021667.[1]$$

$$S^2 = \frac{\sum (X_i - \bar{X})^2}{n-1} = \frac{\sum X_i^2 - n\bar{X}^2}{n-1} = \frac{602.6229 - 602.6029}{5} = 0.0040;$$

$$S = 0.0632.$$

The hypothesis that the machine setting is 10 lbs. is accepted if $-t_{0.025;\,5} \leq t \leq t_{0.025;\,5}$. The value of the test statistic t is

$$t = \frac{(\bar{X} - 10)\sqrt{n}}{S} = \frac{(10.0217 - 10)\sqrt{6}}{0.0632} = 0.841.$$

From Appendix Table 3, $t_{\alpha/2;\,n-1} = t_{0.025;\,5} = 2.571$. Thus, the value of the test statistic falls within the acceptance region so that it is unnecessary to change the machine setting.

(b) According to schedule, a given operation is supposed to be performed in 6.4 minutes. A study is performed to determine whether a particular worker conforms to the standard; i.e., whether deviations from standard can be regarded as random fluctuations or whether they indicate that the performance achieved deviates systematically, either higher or lower. If the departure is as much as 0.5 minute, a change in the standard should be made with a probability of at least 0.95 (i.e., $\beta = 0.05$). Experience indicates that the standard deviation of such operations should be about 0.4 minute. A level of significance of 1% ($\alpha = 0.01$) will be satisfactory. Thus, Fig. 6.11 is entered

---

[1] It is important to compute $\bar{X}$ to more places than appears necessary for the significance of the final result. The reason becomes clear upon looking at the calculation of S^2 where a subtraction operation is involved.

with $d = 0.5/0.4 = 1.25$; a sample size of $n = 15$ is found which will guarantee 95% rejection if departures of more than 0.5 from the mean occur. A sample of 15 times of operation of a worker is taken and the data in minutes are found to be 6.10, 6.65, 7.00, 6.25, 6.35, 6.85, 7.10, 7.35, 7.05, 7.50, 6.90, 6.70, 7.20, 7.15, and 6.95. The sample statistics are found to be as follows:

$$\bar{X} = 6.87333, \quad S^2 = 0.16067, \quad \text{and} \quad S = 0.4008.$$

The value of the test statistic t is

$$t = \frac{(\bar{X} - 6.4)\sqrt{15}}{S} = \frac{(6.87333 - 6.4)\sqrt{15}}{0.4008} = 4.574.$$

This value is compared to $t_{\alpha/2;n-1} = t_{0.005;14} = 2.977$. Thus, $t = 4.574 > t_{\alpha/2;n-1} = 2.977$ and it is concluded that the fluctuation in time cannot be regarded as random and that a change must be made in the standard.

6.3 Test of the Hypothesis That the Standard Deviation of a Normal Distribution Has a Specified Value

6.3.1 CHOICE OF AN OC CURVE

The earlier sections of this chapter dealt with the problem of tests of the mean of a normal distribution. Another common problem is to test the hypothesis that the standard deviation of the normal distribution is equal to σ_0.[1] The test procedure is based upon the chi-square statistic. This test statistic is a function of the sample variance S^2, or more particularly, of the ratio of S^2 to σ_0^2. The abscissae of operating characteristic curves for procedures based on this chi-square statistic depend on the quantity $\lambda = \sigma/\sigma_0$. An example of such an OC curve is shown in Fig. 6.14 for a two-sided test.

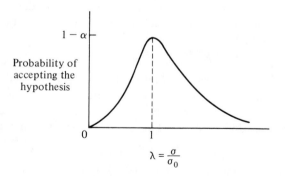

Fig. 6.14. *The OC curves of the two-sided chi-square test.*

---

[1] This is equivalent to testing the hypothesis that the variance of the normal distribution equals σ_0^2.

To determine the acceptance region and sample size, two points on the OC curve must be chosen. These points are denoted by $(1, 1 - \alpha)$ and $(\sigma_1/\sigma_0, \beta)$. The first point corresponds to choosing the probability of accepting the hypothesis when it is true (when $\sigma = \sigma_0$), and the second point corresponds to choosing the probability of accepting the hypothesis when $\sigma = \sigma_1$ is the value of the true standard deviation considered important to detect.

6.3.2 CHARTS AND TABLES TO DESIGN TESTS OF DISPERSION

An experiment is to be performed and a random sample of size n drawn. Each of the resulting sequence of random variables $X_1, X_2, \ldots, X_n$ can be thought of as representing the measurable quantity associated with a trial or test. It is assumed that the X_i are independent normally distributed random variables, each with unknown mean μ and unknown variance σ^2. The *optimum* procedure for testing the hypothesis that the standard deviation of the normal distribution has a specified value, i.e., $\sigma = \sigma_0$, is based on the test statistic χ^2, where

$$\chi^2 = \frac{(n-1)S^2}{\sigma_0^2}.$$

If the value taken on by the test statistic as a result of experimentation falls into the acceptance region, the hypothesis that $\sigma = \sigma_0$ is accepted; otherwise, it is rejected. This section is devoted to finding the acceptance region and the sample size for the test statistic χ^2, using the tables and charts which are provided. One- and two-sided procedures will be considered.

6.3.2.1 *Tables and Charts for Two-Sided Procedures.* The acceptance region for the two-sided procedure for any sample size n is given by the interval $[\chi_{1-\alpha/2;n-1}^2, \chi_{\alpha/2;n-1}^2]$, where α is the level of significance (Type I error) and $\chi_{1-\alpha/2;n-1}^2$ and $\chi_{\alpha/2;n-1}^2$ are, respectively, the $100(1 - \alpha/2)$ and $100\,\alpha/2$ percentage points of the chi-square distribution with $n - 1$ degrees of freedom. Thus, the hypothesis that the standard deviation of a normal distribution equals σ_0 is accepted if

$$\chi_{1-\alpha/2;n-1}^2 \leqq \chi^2 \leqq \chi_{\alpha/2;n-1}^2$$

and rejected if χ^2 lies outside this interval. It is easily verified that this procedure yields a probability of $1 - \alpha$ of accepting the hypothesis that $\sigma = \sigma_0$ when it is true. When $\sigma = \sigma_0$, the test statistic $\chi^2 = (n-1)S^2/\sigma_0^2$ has a chi-square distribution with $n - 1$ degrees of freedom. Hence, if $\sigma = \sigma_0$,

$$P\{\chi_{1-\alpha/2;n-1}^2 \leqq \chi^2 \leqq \chi_{\alpha/2;n-1}^2\} = 1 - \alpha.$$

The required sample size is obtained by referring to a set of operating characteristic curves which are provided for this purpose. Figure 6.15 presents the OC curves of the two-sided chi-square test for a level of significance equal to 0.05. Figure 6.16 presents the OC curves of the two-sided chi-square test for a level of significance equal to 0.01. If the level of significance of the procedure is either 0.05 or 0.01, the required sample size n is obtained by entering the appropriate figure (Fig. 6.15 if $\alpha = 0.05$ and Fig. 6.16 if $\alpha = 0.01$) with the point $(\sigma_1/\sigma_0, \beta)$ and reading out the value of the sample size associated with the OC curve that passes through this point.

6.3.2.2 *Summary for Two-Sided Procedures Using Tables and Charts.* The following steps are taken, using a two-sided procedure for testing the standard

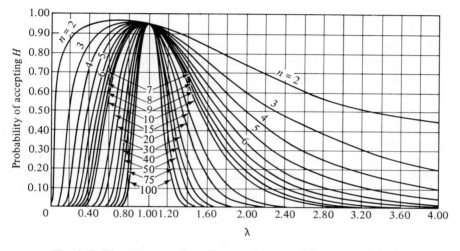

Fig. 6.15. *The OC curves for different values of n for the two-sided chi-square test for a level of significance $\alpha = 0.05$.*

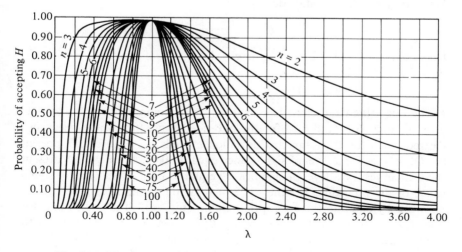

Fig. 6.16. *The OC curves for different values of n for the two-sided chi-square test for a level of significance $\alpha = 0.01$.*

deviation of a normal distribution:

1. Two points on the OC curve are chosen. These points are denoted by $(1, 1 - \alpha)$ and $(\sigma_1/\sigma_0, \beta)$.
2. Find the required sample size n by entering the appropriate figure (Fig. 6.15 if $\alpha = 0.05$ and Fig. 6.16 if $\alpha = 0.01$) with the point $(\sigma_1/\sigma_0, \beta)$ and reading out the value of the sample size associated with the OC curve that passes through this point.
3. Determine the acceptance region which is the interval $[\chi^2_{1-\alpha/2;n-1}, \chi^2_{\alpha/2;n-1}]$.
4. Draw a random sample of n items.
5. Compute the value of the test statistic $\chi^2 = (n - 1)S^2/\sigma_0^2$.
6. If $\chi^2_{1-\alpha/2;n-1} \leq \chi^2 \leq \chi^2_{\alpha/2;n-1}$, accept the hypothesis that $\sigma = \sigma_0$. If χ^2 lies outside this interval, reject the hypothesis and conclude that the standard deviation is not σ_0.

6.3.2.3 *Tables and Charts for One-Sided Procedures.* If the procedure is one-sided and rejection is desired when the true standard deviation exceeds σ_0 (this being the alternative to the hypothesis that $\sigma = \sigma_0$ or $\sigma \leq \sigma_0$), the acceptance region for any sample size, n, is given by the interval $[-\infty, \chi^2_{\alpha;n-1}]$, where α is the level of significance and $\chi^2_{\alpha;n-1}$ is the 100 α percentage point of the chi-square distribution with $n - 1$ degrees of freedom. Thus, the hypothesis that $\sigma = \sigma_0$ (or $\sigma \leq \sigma_0$) is accepted if

$$\chi^2 \leq \chi^2_{\alpha;n-1}$$

and rejected if χ^2 lies outside this interval. This acceptance procedure has a probability of $1 - \alpha$ of accepting the hypothesis that $\sigma = \sigma_0$ when it is true. When $\sigma = \sigma_0$, the test statistic

$$\chi^2 = \frac{(n - 1)S^2}{\sigma_0^2}$$

has a chi-square distribution with $n - 1$ degrees of freedom. Hence, if $\sigma = \sigma_0$,

$$P\{\chi^2 \leq \chi^2_{\alpha;n-1}\} = 1 - \alpha.$$

The operating characteristic curves for determining the sample size n are given in Figs. 6.17 and 6.18 for the one-sided procedure where rejection is desired when the true standard deviation exceeds σ_0. Figure 6.17 presents the OC curves for a level of significance equal to 0.05, and Fig. 6.18 presents the OC curves for a level of significance equal to 0.01. If the level of significance of the procedure is either 0.05 or 0.01, the required sample size n is obtained by entering the appropriate figure (Fig. 6.17 if $\alpha = 0.05$ and Fig. 6.18 if $\alpha = 0.01$) with the point $(\sigma_1/\sigma_0, \beta)$ and reading out the value of the sample size associated with the OC curve that passes through this point.

If the procedure is one-sided and rejection is desired when the true standard deviation is less than σ_0 (this being the alternative to the hypothesis that

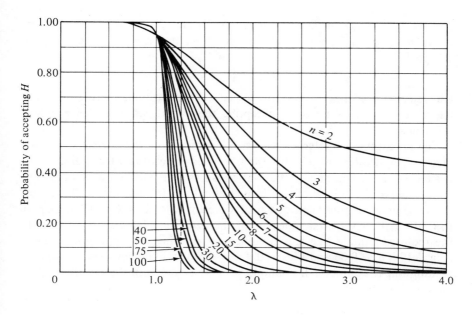

Fig. 6.17. *The OC curves for different values of n for the one-sided (upper tail) chi-square test for a level of significance* $\alpha = 0.05$.
(Reproduced by permission from "Operating Characteristics for the Common Statistical Tests of Significance" by Charles L. Ferris, Frank E. Grubbs, Chalmers L. Weaver, *Annals of Mathematical Statistics*, June, 1946.)

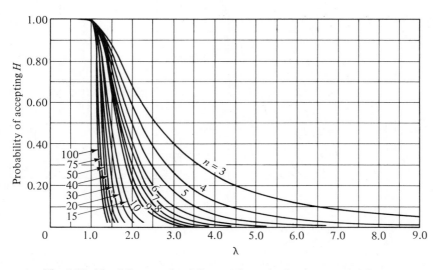

Fig. 6.18. *The OC curves for different values of n for the one-sided (upper tail) chi-square test for a level of significance* $\alpha = 0.01$.

$\sigma = \sigma_0$ or $\sigma \geq \sigma_0$), the acceptance region for any sample size, n, is given by the interval $[\chi^2_{1-\alpha;n-1}, \infty]$. Thus, the hypothesis that $\sigma = \sigma_0$ (or $\sigma \geq \sigma_0$) is accepted if

$$\chi^2 \geq \chi^2_{1-\alpha;n-1}$$

and rejected if χ^2 lies outside this interval. This acceptance procedure also has a probability of $1 - \alpha$ of accepting the hypothesis that $\sigma = \sigma_0$ when it is true.

The operating characteristic curves for determining the sample size n are given in Figs. 6.19 and 6.20 for the one-sided procedure where rejection is desired when the true standard deviation is less than σ_0. Figure 6.19 presents the OC curves for a level of significance equal to 0.05, and Fig. 6.20 presents the OC curves of this one-sided procedure for a level of significance equal to 0.01. If the level of significance of the procedure is either 0.05 or 0.01, the required sample size n is obtained by entering the appropriate figure (Fig. 6.19 if $\alpha = 0.05$ and Fig. 6.20 if $\alpha = 0.01$) with the point $(\sigma_1/\sigma_0, \beta)$ and reading out the value of the sample size associated with the OC curve that passes through this point.

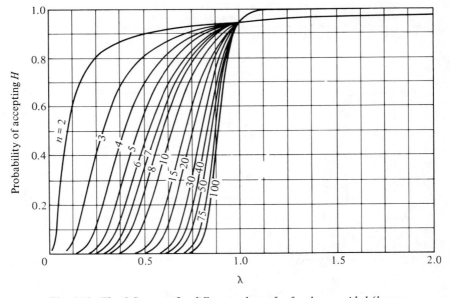

Fig. 6.19. *The OC curves for different values of n for the one-sided (lower tail) chi-square test for a level of significance $\alpha = 0.05$.*
(Reproduced by permission from "Operating Characteristics for the Common Statistical Tests of Significance" by Charles L. Ferris, Frank E. Grubbs, Chalmers L. Weaver, *Annals of Mathematical Statistics*, June, 1946.)

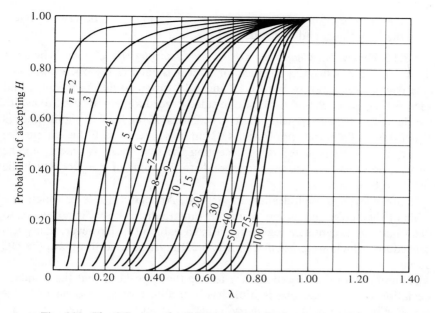

Fig. 6.20. *The OC curves for different values of n for the one-sided (lower tail) chi-square test for a level of significance* $\alpha = 0.01$.

6.3.2.4 *Summary for One-Sided Procedures Using Tables and Charts.* The following steps are taken, using a one-sided procedure for testing the standard deviation of a normal distribution:

1. Two points on the OC curve are chosen. These points are denoted by $(1, 1 - \alpha)$ and $(\sigma_1/\sigma_0, \beta)$.
2. Find the required sample size n by entering the appropriate figure. If rejection is desired when $\sigma > \sigma_0$, Fig. 6.17 (if $\alpha = 0.05$) or Fig. 6.18 (if $\alpha = 0.01$) is entered with the point $(\sigma_1/\sigma_0, \beta)$ and the value of the sample size associated with the OC curve that passes through this point is read out. If rejection is desired when $\sigma < \sigma_0$, Fig. 6.19 (if $\alpha = 0.05$) or Fig. 6.20 (if $\alpha = 0.01$) is entered with the point $(\sigma_1/\sigma_0, \beta)$ and the value of the sample size associated with the OC curve that passes through this point is read out.
3. Determine the acceptance region which is the interval $[-\infty, \chi^2_{\alpha;n-1}]$ for the alternative $\sigma > \sigma_0$ and $[\chi^2_{1-\alpha;n-1}, \infty]$ for the alternative $\sigma < \sigma_0$.
4. Draw a random sample of n items.
5. Compute the value of the test statistic $\chi^2 = (n-1)S^2/\sigma_0^2$.
6. If the alternative is $\sigma > \sigma_0$, accept the hypothesis that $\sigma = \sigma_0$ (or $\sigma \leq \sigma_0$) whenever $\chi^2 \leq \chi^2_{\alpha;n-1}$; otherwise, reject the hypothesis and conclude that $\sigma > \sigma_0$. If the alternative is $\sigma < \sigma_0$, accept the hypothesis

that $\sigma = \sigma_0$ (or $\sigma \geqq \sigma_0$) whenever $\chi^2 \geqq \chi^2_{1-\alpha;n-1}$; otherwise, reject the hypothesis and conclude that $\sigma < \sigma_0$.

6.3.2.5 *Tables and Charts for Operating Characteristic Curves.* Whenever the sample size is fixed in advance, or after it is chosen, the operating characteristic curves given in Figs. 6.15 and 6.16 for two-sided procedures and Figs. 6.17, 6.18, 6.19, and 6.20 for one-sided procedures can be used to evaluate the risks involved in using these procedures. If the level of significance is either 0.05 or 0.01, the appropriate OC curve is entered with the sample size n and λ, and the probability of accepting the hypothesis is read from the curve. The value of λ chosen corresponds to the value of σ/σ_0 for which the risk is to be evaluated.

If other levels of significance are desired, Appendix Table 2, which is a table of the percentage points of the chi-square distribution, can be used to determine the acceptance region for a fixed value of n. However, analytical procedures which are presented in Sec. 6.3.3 must be used to obtain the OC curve.

A summary of the procedures for testing the hypothesis that the standard deviation of a normal distribution has a specified value is given in Table 6.3. For purposes of this summary, criteria for rejecting the hypothesis are given instead of criteria for acceptance.

Table 6.3. Test of a Hypothesis That the Standard Deviation of a Normal Distribution Has a Specified Value

Notation for Hypothesis H: $\sigma = \sigma_0$	*Test Statistic* $\chi^2 = (n-1)S^2/\sigma_0^2$
Criteria for Rejection	*Method for Choosing Sample Size*
$\chi^2 > \chi^2_{\alpha/2;n-1}$ or $\chi^2 < \chi^2_{1-\alpha/2;n-1}$ if rejection is desired when the true standard deviation differs from σ_0 in either direction.	Choose a value of the relative error, $\lambda = \sigma_1/\sigma_0$, for which rejection of the hypothesis is desired with given high probability; enter Fig. 6.15 or 6.16 to find the required sample size.
$\chi^2 > \chi^2_{\alpha;n-1}$ if rejection is desired when the true standard deviation exceeds σ_0.	Choose a value $\sigma_1 > \sigma_0$ for which rejection of the hypothesis is desired with given high probability; calculate $\lambda = \sigma_1/\sigma_0$ and enter Fig. 6.17 or 6.18 to find the required sample size.
$\chi^2 < \chi^2_{1-\alpha;n-1}$ if rejection is desired when the true standard deviation is less than σ_0.	Choose a value $\sigma_1 < \sigma_0$ for which rejection of the hypothesis is desired with given high probability; calculate $\lambda = \sigma_1/\sigma_0$ and enter Fig. 6.19 or 6.20 to find the required sample size.

Values of $\chi^2_{\alpha;\nu}$, the $100\,\alpha$ percentage points of the chi-square distribution for ν degrees of freedom, are given in Appendix Table 2.

6.3.3 Analytical Treatment for Chi-Square Tests

In Sec. 4.2.3, it was pointed out that the quantity $(n - 1)S^2/\sigma^2$ has a chi-square distribution with $n - 1$ degrees of freedom, provided σ^2 is the variance of the X's which constitute S^2. If testing σ_0^2 against alternatives of the sort $\sigma^2 > \sigma_0^2$ is of interest, the decision procedure is to reject when $(n - 1)S^2/\sigma_0^2$ is large. If a test is desired with a Type I error equal to α, the decision procedure calls for accepting the hypothesis whenever

$$\frac{(n - 1)S^2}{\sigma_0^2} \leqq \chi^2_{\alpha;n-1}$$

or equivalently,

$$S^2 \leqq \frac{\chi^2_{\alpha;n-1}\sigma_0^2}{n - 1}.$$

It is clear that when $\sigma = \sigma_0$,

$$P\left\{\frac{(n - 1)S^2}{\sigma_0^2} \leqq \chi^2_{\alpha;n-1}\right\} = 1 - \alpha.$$

The OC curve of the test can be computed easily from a table of the chi-square distribution. Suppose that the true value of the variance is σ^2. Then $(n - 1)S^2/\sigma^2$ has a chi-square distribution with $n - 1$ degrees of freedom, and the probability of accepting the hypothesis is given by

$$P\left\{\frac{(n - 1)S^2}{\sigma_0^2} \leqq \chi^2_{\alpha;n-1}\right\} = P\left\{S^2 \leqq \frac{\chi^2_{\alpha;n-1}\sigma_0^2}{n - 1}\right\}$$

$$= P\left\{\frac{(n - 1)S^2}{\sigma^2} \leqq \chi^2_{\alpha;n-1}\frac{\sigma_0^2}{\sigma^2}\right\}.$$

Denoting the ratio σ/σ_0 by λ, the expression for the OC curve is then given by $P\{\chi^2 \leqq \chi^2_{\alpha;n-1}/\lambda^2\}$, where χ^2 denotes a random variable having a chi-square distribution with $n - 1$ degrees of freedom. Note that aside from the Type I error and number of observations, the OC curve depends only on λ. If $\lambda = 1$, i.e., the hypothesis $\sigma = \sigma_0$ is true, then

$$P\left\{\chi^2 \leqq \frac{\chi^2_{\alpha;n-1}}{\lambda^2}\right\} = P\{\chi^2 \leqq \chi^2_{\alpha;n-1}\} = 1 - \alpha.$$

The final expression for the OC curve permits easy solution of the sample size problem. Suppose that a particular value of σ_1 is given for which acceptance is desired with probability β. If $\lambda_1 = \sigma_1/\sigma_0$, this condition implies that

$$P\left\{\frac{(n - 1)S^2}{\sigma_0^2} \leqq \chi^2_{\alpha;n-1}\right\} = P\left\{\chi^2 \leqq \frac{\chi^2_{\alpha;n-1}}{\lambda_1^2}\right\} = \beta.$$

It follows from this last expression that

$$\frac{\chi^2_{\alpha;n-1}}{\lambda_1^2} = \chi^2_{1-\beta;n-1} \quad \text{or} \quad \lambda_1^2 = \frac{\chi^2_{\alpha;n-1}}{\chi^2_{1-\beta;n-1}}.$$

The sample size for this problem can be determined by going to a table of the percentage points of the chi-square distribution, Appendix Table 2, and finding the degrees of freedom for which the ratio of the 100 αth percentage point to the 100 $(1 - \beta)$th percentage point is equal to λ_1^2. For example, if $\alpha = 0.01$, $\beta = 0.05$, and $\lambda_1^2 = 4.2$, Appendix Table 2 indicates that the 15 degrees of freedom have values of $\chi_{0.01; 15}^2 = 30.578$ and $\chi_{0.95; 15}^2 = 7.261$ such that $\lambda_1^2 = 30.578/7.261 = 4.2$. Hence, the required sample size is 16. Similar analytic treatment can be provided when lower-tailed or two-tailed tests are of interest.

If rejection is desired when $\sigma < \sigma_0$, the expression for the OC curve (probability of accepting the hypothesis) is given by

$$P\left\{\chi^2 \geq \frac{\chi_{1-\alpha; n-1}^2}{\lambda^2}\right\},$$

where χ^2 denotes a random variable having a chi-square distribution with $n - 1$ degrees of freedom and $\lambda = \sigma/\sigma_0$. The expression required for determining the sample size is given by

$$\lambda_1^2 = \frac{\chi_{1-\alpha; n-1}^2}{\chi_{\beta; n-1}^2}.$$

If the two-sided chi-square test is used, the expression for the OC curve (probability of accepting the hypothesis) is given by

$$P\left\{\frac{\chi_{1-\alpha/2; n-1}^2}{\lambda^2} \leq \chi^2 \leq \frac{\chi_{\alpha/2; n-1}^2}{\lambda^2}\right\},$$

where χ^2 denotes a random variable having a chi-square distribution with $n - 1$ degrees of freedom and $\lambda = \sigma/\sigma_0$. The expression required for determining the sample size is given approximately by

$$\lambda_1^2 = \frac{\chi_{\alpha/2; n-1}^2}{\chi_{1-\beta; n-1}^2}$$

when $P\{\chi^2 \leq \chi_{1-\alpha/2; n-1}^2/\lambda_1^2\}$ is small compared to β, or by

$$\lambda_1^2 = \frac{\chi_{1-\alpha/2; n-1}^2}{\chi_{\beta; n-1}^2}$$

when $P\{\chi^2 \geq \chi_{\alpha/2; n-1}^2/\lambda_1^2\}$ is small compared to β.

6.3.4 EXAMPLE

The standard deviation of a dimension of a standard product is $\sigma = 0.1225$ inch. This standard product is somewhat obsolete and a new product is under consideration. It will be adopted if the variability of this dimension is not substantially larger than that of the standard. If the standard deviation, σ, of the new product is as large as 0.2450 inch, the probability of adopting this product should not exceed 1 % ($\beta = 0.01$). The test is to be made at the 1 % level of significance. The hypothesis to be tested, then, is that $\sigma \leq \sigma_0 = 0.1225$ against the one-sided alternative that $\sigma > \sigma_0$. From Fig. 6.18 it appears that a sample of approximately 25 will give a probability of acceptance of 0.01 for $\lambda = 0.2450/0.1225 = 2$. A sample of 25 items is drawn and S^2 is found to be 0.0384. Hence, the value of the test statistic, $\chi^2 = (n-1)S^2/\sigma_0^2$, is 61.4, which exceeds $\chi^2_{0.01;24} = 42.980$, so that the hypothesis is rejected. The new product is not adopted.

Suppose the company allotted a total of only 7 observations for the experiment and kept the 1% level of significance. The hypothesis that $\sigma \leq 0.1225$ is accepted if $\chi^2 = (n-1)S^2/\sigma_0^2 \leq \chi^2_{0.01;6} = 16.812$; otherwise it is rejected. Suppose further that S^2 computed from the seven observations was equal to the value of S^2 found for the 25 observations, namely, 0.0384. The value of the test statistic is now 15.4, which is less than $\chi^2_{0.01;6} = 16.812$ so that the hypothesis is accepted. The new product is adopted. A reference to Fig. 6.18 reveals that for $n = 7$ and $\lambda = 2$, the probability of accepting the hypothesis is approximately 35%. In other words, if the standard deviation of the new product is as large as 0.2450, only about 2 out of 3 times will the experimental procedure result in its detection. The hypothesis will be accepted 1 out of 3 times. Thus, even if the hypothesis is accepted, as in this example when $n = 7$, one is unable to feel secure that the standard deviation is not as large as 0.2450.

PROBLEMS

In the following problems it may be assumed that a random sample, $X_1, X_2, \ldots, X_n$, is drawn, and each X_i is approximately normally distributed, or alternatively, the test statistic formed from these random variables has the appropriate distribution to validate the OC curves.

1. The average warpwise breaking strength of a certain type of cloth is required to be not less than $\mu = 180$ pounds per square inch. The standard deviation based on past experience is 4 psi. A shipment of a lot of this cloth is received from a supplier and specimens are withdrawn from three pieces. These are tested with the following results:

> first piece: 182 psi;
> second piece: 172 psi;
> third piece: 177 psi.

Using a 5% level of significance, should the lot be accepted? What is the probability of accepting a lot which has a lot average warpwise breaking strength of $\mu = 170$ psi?

2. A manufacturer of synthetic rubber claims that the average hardness of his rubber is 64.3 degrees Shore. It is felt that this claim may be an overestimate or an underestimate, so an experiment is to be performed. Past experience indicates that the standard deviation of the random variable, the hardness of rubber, is 2 degrees Shore. If the true average (μ) is 64.3 degrees, the probability should be 0.95 of reaching this conclusion. Furthermore, if the average strengths differ by as much as ± 1.5 degrees, the procedure should result in the conclusion that μ does not equal 64.3 degrees with probability greater than 0.80. (a) What is the necessary sample size for this experiment? (b) If $\bar{X} = 65$ degrees and n is that value found in (a), should one conclude that the hardness is 64.3? (c) Compute exactly the probability of concluding that the average hardness of rubber is 64.3 degrees Shore when the true average hardness is 65 degrees Shore, using the sample size found in (a). Do not use the OC curves in the text, but compute the exact probability. You can check your answer with the value read off the curve.

3. In a chemical process it is very important that a certain solution to be used as a reactant has a pH of 8.30. A method for determining pH is available which for solutions of this type is known to give measurements with a known standard deviation of $\sigma = 0.02$. If the pH is really 8.30, it is desired to design the test so that this conclusion will be reached with probability equal to 0.95. On the other hand, if the pH differs from 8.30 by 0.03 (in either direction), it is desired that the probability of picking up such a difference should exceed 0.98. (a) What is the decision procedure that should be used? (b) What is the required sample size? (c) If $\bar{X} = 8.31$, what is your conclusion? (d) If the actual pH is 8.32, what is the probability of concluding that the pH is not 8.30, using the above procedure? (Calculate this probability directly, but you may check your result by using the OC curves in the text.)

4. Primers are used for initiating charges of explosives. The density of the primers affects this initiating power. A procedure was required to determine whether a batch was acceptable. The standard deviation was known from past experience to be 0.03 and the average density in normal operation was 1.54.

To ensure that good batches were nearly always accepted and bad batches nearly always rejected, it was decided that the following requirements should be satisfied: (1) If the mean density was at the normal value of 1.54, there should be a 95% chance of acceptance. (2) If the mean density was as low as 1.50, there should be a 90% chance of rejection. (a) Determine the necessary sample size and state the required decision rule that has the risks given above. (b) If the mean density was 1.52, what is the probability of accepting the lot? Do not use the figure in the text to get the solution to this part of the problem. Calculate it directly. (c) If $\bar{X} = 1.53$, would you accept the lot?

5. A shoe manufacturer claims that he can supply, at the same price now being paid, shoes that will give longer wear than the shoes now being used by the United States Army. Army records show that, for such troops, the average life of shoes is 12 months, with a standard deviation of two months. The Army has had poor relations with the present supplier and will welcome a change. To test the manufacturer's claim, the Army issues 64 pairs of the new shoes at random among troops. The (sample) average time it takes for 64 pairs to wear out is $12\frac{1}{2}$ months. Should the Army accept the manufacturer's claim at the 5% level of significance? If the true average life is $11\frac{3}{8}$ months, what is the probability of accepting the hypothesis?

6. A certain type of rocket has been kept in storage for two years. At the time of its acceptance, it was determined that the average range of these rockets was 2,500 yards with a standard deviation of 150 yards. It is necessary to make a decision regarding the disposition of the lot. If the average range has diminished as much as 150 yards, such a change should be detected with probability 0.90.

How many observations are required to run a test at the 5% level of significance? If the sample mean $\bar{X}$ is 2,401 yards, should the rockets be discarded?

7. The average shear strength of test spot welds is 400 psi and the standard deviation is 20 psi. An additive is being considered which is said to increase the average shear strength. An experiment is to be run to ascertain the properties of the new material. If the average shear strength is increased as much as 25 psi, such a change should be detected with probability 0.98. If there is no change, this should be determined with probability 0.95. (a) How many observations are required? (b) If the sample mean is 460 psi, should one conclude that the average shear strength has been increased?

8. In Problem 1, how many observations are necessary so that if the average lot warpwise breaking strength is 170 psi, there will be a probability of at least 0.99 of detecting it? If $\bar{X}$ were 177 psi and the sample size is as found above, would the lot be accepted?

9. In Problem 2, suppose that a sample of size 10 is used. If $\mu = 65.8$ degrees, what is the probability of accepting the hypothesis that the hardness is 64.3? If the true average hardness is 65 degrees, what is the probability of accepting the hypothesis? Use the analytical method to get your result.

10. In Problem 5, how many pairs of shoes are required so that if the average life, μ, is actually $11\frac{1}{2}$ months, it will be discovered with probability 0.95, i.e., the hypothesis that $\mu \geq 12$ months will be rejected? If such a sample leads to $\bar{X} = 11\frac{1}{4}$ months, would you accept the hypothesis?

11. Several complaints relative to the quality of standard No. 5 screened coke have been received by a coke plant foreman. It is suggested that the porosity factor, $\bar{X}$, measured in pounds weight differences in dry and soaked coke, may be at fault. The shipments were purchased on the assertion of an average porosity factor of 2 pounds. From the producer's point of view, both large and small average porosity factors are undesirable. It is known from past experience that X has a standard deviation of $\frac{1}{2}$ pound. Test the hypothesis that the shipments actually did conform with the claim above at the 5% level. Does the choice of sample size afford good protection against accepting the hypothesis when in fact the true average is as great as 2.10 pounds or as small as 1.90 pounds? The data are as follows:

2.16	2.17	2.34	1.98	1.97	1.89
2.19	2.23	2.15	2.47	2.31	1.94
2.31	1.86	2.25	2.14	2.15	2.16
2.30	2.48	2.11	2.15	2.24	2.14
2.25	1.90	2.04	2.09	2.08	2.25

12. In Problem 11, suppose it is desired that the assertion be made, with probability not exceeding 0.20, that the average porosity factor is 2 pounds when the true average is as great as 2.10 pounds. How large a sample size is required if the test is to be run at the 5% level of significance? Solve graphically and analytically.

13. The owner of a large number of taverns buys 30-gallon kegs of beer in large lots (hundreds of kegs at a time). Since the beer is sold in 25-cent glasses, he is interested in whether the average number of gallons in the kegs is adequate. Assume that the process which fills the kegs is such that the number of gallons in a keg is a random variable with standard deviation equal to $\frac{2}{3}$ gallon. If the average size is as small as 29 gallons, the owner would like to reject a lot with probability exceeding 0.9. If the average size is as stipulated (30 gallons), he would like to accept a lot with probability 0.99. (a) How many kegs should he be prepared to sample? (b) If $\bar{X}$ is 28.5 gallons, should the lot be accepted or rejected?

14. Foam rubber is produced in sheets by an existing process. The softness of the foam rubber (as measured on a softness scale) is a random variable which can be assumed to have mean 73 and known standard deviation σ equal to 2. A chemical engineer claims to have developed a formula superior (in the sense that the average softness is increased) to the one currently used. The plant manager wants to change formulas if the new process increases the average softness by the amount of one standard deviation, (σ), and wants to detect a change of this magnitude with probability 0.90. If, however, the mean is unaffected, he wants to maintain the present process with probability 0.95. Specify: (a) The hypothesis and alternative to be tested. (b) Test statistic and acceptance region. (c) Sample size necessary to achieve the given Type I and II error probabilities. (d) What is the probability of accepting the hypothesis if the mean has shifted 1.2 standard deviations? (Use analytic method.)

15. A manufacturer of light bulbs has developed a new production process which he hopes will increase the mean efficiency (in lumens per watt) of his product from the present mean of 9.5. The results of an experiment conducted on 10 bulbs are given below. Should the manufacturer believe that the efficiency has been increased? Use a 5% level of significance. What is the probability of saying that there was no increase if the efficiency were 11? (Use S as the estimate of σ for OC curve purposes.)

9.278	12.045
9.971	13.024
10.250	9.871
11.461	11.578
11.515	10.851

$$\sum X^2 = 1218.451898$$

16. A supplier of cloth claims that his product has an average warpwise breaking strength of 90 pounds or better. The company for which you work is anxious to buy this cloth, but decides first to test his claim at the 5% level of significance. If the test leads to acceptance of the claim, it will buy the material. Values of average warpwise strength slightly less than 90 pounds are not considered important to guard against, although values substantially less are critical. From past experience with this kind of material, the standard deviation of the warpwise breaking strength is known to be "approximately" 12 pounds. (σ is not known.) (a) The experiment calls for a sample of 25 pieces of cloth. Determine the test procedure. (b) What is the risk under (a) of advising the company to buy the yarn, when its average warpwise breaking strength is actually only 86 pounds? (c) If the cloth has actually an average warpwise breaking strength of 90 pounds, what is the risk that you might reject the claim? (d) You carry out the test as designed in (a) and find that the average strength for the sample of cloth tested is 85 pounds. Would you, or would you not, advise the company to buy the yarn? $S = 12.5$.

17. Suppose it is desired to test the hypothesis that the melting point of an alloy is 1,200 degrees centigrade. If the melting point differs by more than 20 degrees, it is desired to change the composition. Assume that $\alpha = 0.01$ and $\beta = 0.10$, and that σ is "approximately" 15 degrees. How many determinations should be made? What will the rejection rule be?

18. In the manufacture of cylindrical rods with a circular cross section which fit into a circular socket, the lathe is set so that the mean diameter is 5.01 centimeters. It was necessary to test the accuracy of the machine setting. A sample of size 10 was taken with the results as follows:

$$
\begin{array}{cc}
5.036 & 5.031 \\
5.085 & 5.064 \\
4.991 & 4.942 \\
4.935 & 5.051 \\
4.999 & 5.011
\end{array}
$$

$$\sum X^2 = 251.473971$$

(a) Test the hypothesis that the setting is correct at the 5% level of significance. (b) If the setting is off by 0.05 centimeter, what is the approximate probability of detecting such a shift? [Use value of S found in (a) as estimate of σ.] (c) How many observations are required to insure, (with probability 0.95), that a setting off by 0.05 centimeter will be detected?

19. It is desired to test whether a special treatment on concrete has an effect (in either direction) on the average strength of this concrete which is known to be 3,000 psi. A small experiment is to be made from a given batch of raw materials using the new treatment. The procedure is to be such that if the new concrete has the same tensile strength as the old, the probability will be 0.99 of reaching this conclusion. Moreover, if the average strength differs by as much as 250 psi, the procedure should result in the conclusion that they differ with probability greater than 0.75. A very rough estimate of the standard deviation is 200 psi. (a) What is the necessary sample size? (b) Suppose that

$$\sum_{i=1}^{n} (X_i - \bar{X})^2 = 40,103, \qquad \bar{X} = 3,393,$$

and n is that value found in (a), should one conclude the special treatment has no effect on the strength?

20. In Problem 15, how many observations are required so that if the mean efficiency is increased by as much as 1 lumen per watt, it will be detected with probability 0.90? A rough estimate of σ is $\frac{1}{2}$ lumen per watt. If $\bar{X} = 11$, $S = 0.503$, and the number of observations are as found above, would the hypothesis be accepted?

21. A certain plastic company supplies plastic sheets for industrial use. A new type plastic has been produced and the company would like to claim that the average stress resistance of this new product is at least 30.0 where stress resistance, X, is measured in pounds per square inch necessary to crack the sheet. The following random sample was drawn off the production line. Based on this sample would you say, at the 5% level, that the company should make its claim?

30.1	32.7	22.5	27.5
27.7	29.8	28.9	31.4
31.2	24.3	26.4	22.8
29.1	33.4	32.5	21.7

$$\sum X^2 = 12984.54$$

22. A manufacturer of gauges claims that the standard deviation in the use of his gauge does not exceed 0.0003. Unknown to an analyst who is using this gauge, you have him measure the same items nine times during the course of his regular testing duties and find that the sample standard deviation of the nine measurements is 0.0006. (a) In view of these results, would you be justified in rejecting the manufacturer's claim? Use a 1% level of significance. (b) What is the probability of rejecting the hypothesis that $\sigma = 0.0003$ when the true standard deviation is 0.0005? Get both the analytical and graphical solutions.

23. You wish to test whether the standard deviation of measurements made by a certain thermometer is 0.010 degree against the alternative that it is greater than 0.010 degree. A "tolerance standard deviation" of 0.005 degree is given and you wish to have a 0.90 probability of rejection when this tolerance is exceeded. How many measurements would you make and what would your acceptance criteria be if you used a 1% level of significance? Would you accept the hypothesis if $S = 0.012$?

24. Metal panels of standard thickness were coated with a known thickness of film. The coated panels were then stored and the strength of the resin film was assessed by a machine which screwed a steel ball into the panel, thereby stretching the film on the reverse side until the film ruptured. When rupture occurred, electric contact was completed through the panel and a relay operated to stop the machine. The depth of penetration of the ball was a measure of the film strength. The reproducibility of the original method, as measured by the standard deviation, was known *accurately* to be 4 units from a large number of trials but was not considered satisfactory, and various tests were conducted with a view to improvement. For these tests, panels were prepared under standard conditions. In one typical experiment an attempt was made to improve reproducibility by specially preparing the surface of the panels prior to coating them with the resin. The level of significance chosen was $\alpha = 0.05$. It was decided that a suitable test would be one which would give a probability of 0.80 of detecting a reduction in variance to half its normal value. (a) What is the test procedure that should be used? (b) What is the required sample size? (c) If $S^2 = 12.2$, what is your conclusion?

25. A particular method of analysis had been in use for a certain substance, and a proposal was made to replace it by another method which was considered to have certain advantages. A question to be decided by experimentation was whether the new method was as precise as the old one. From experience it was known that the standard deviation of the old method was 0.04 unit, and it was decided that the new method would have sufficient advantages in rapidity and economy of materials to justify its adoption, provided that it was not markedly less precise than the old. The test was to be designed so that if the standard deviation of the new method was as great as 0.10 unit the risk would not be more than 0.05 of failing to discover such a change when the test of significance was made at the $\alpha = 0.01$ level. (a) Determine the required number of observations. (b) If the true value of the standard deviation, σ, was 0.07, what is the probability of accepting the hypothesis? Solve analytically. (c) If $S = 0.06$, would you accept the hypothesis?

26. A manufacturer of gauges claims that the standard deviation in the use of his gauge does not exceed $\sigma_0 = 0.0003$. Unknown to an analyst who is using this gauge, you have him measure the same items a number of times during the course of his regular testing duties. You do not wish to run a risk of more than 5% of contradicting the manufacturer's claims when they are true. However, if the actual standard deviation σ_1 is such that $\sigma_1/\sigma_0 = 2$, you wish to conclude that the manufacturer's claims are false with probability greater than 0.9. (a) What is the required sample size? (b) If $(n - 1)S^2/(0.0003)^2$ is 25.3, would you accept or reject the manufacturer's claims?

27. A producer claims that the lengths of the parts he manufactures have a standard deviation not exceeding 0.05 inch. You measure a sample of nine parts and find that their standard deviation is 0.10 inch. Could you reasonably reject the producer's claim at the 5% level of significance? Discuss the OC curve for your test.

28. Given a sample of size 20 from a normal distribution with $\sum (X - \bar{X})^2 = 0.225$, test the hypothesis that $\sigma = 0.1$ with error of the first kind equal to 0.01. If $\sigma = 0.2$, what is the probability that you will accept the hypothesis that $\sigma = 0.1$?

29. A pharmaceutical house produces a certain drug item whose weight has an inherent standard deviation of 5 milligrams. The company's research team has proposed a new method of producing the drug. However, this entails some costs, and will be adopted only if there is strong evidence the standard deviation of the weight of the items is less than 5 milligrams. In fact, if the new standard deviation is as small as 3 milligrams, the switch should be made with probability exceeding 0.8. How many observations are required if $\alpha = 0.05$? The company runs the experiment, using $n = 10$. The data in grams are given below. With these data would the new method be adopted? Was the choice of n good, fair or poor?

$$\begin{array}{cc} 5.728 & 5.731 \\ 5.722 & 5.719 \\ 5.727 & 5.724 \\ 5.718 & 5.726 \\ 5.723 & 5.722 \end{array}$$

$$\sum (X - \bar{X})^2 = 148$$

30. The Space Agency is developing a new rocket to place a satellite in orbit, and is interested in determining whether it is precise with respect to the short or long distance (e.g., only the X coordinate of distance). The mean of the distribution can be easily adjusted so that the primary problem deals with the variance of the distribution. It is desired that, when aimed properly, at least 99% of the rockets fall within 2 miles of the aimed-at value. (a) What is the maximum allowable standard deviation? (b) If the true standard deviation of the distance is twice the value found in (a), the rocket should be returned to the contractor for further development. This is to be determined by experimentation and the Space Agency is willing to assume a risk of no more than 10%. On the other hand, if the rocket is satisfactory, it should be ascertained with probability no less than 0.99. What is the required sample size? (c) If $S^2 = 2.25$, what should be the disposition of the rocket? (d) If σ is $\frac{3}{2}$ the value found in (a), determine analytically the probability of accepting the rocket.

31. In Problem 14, test the hypothesis that the standard deviation of the new formula is 2 against the alternative that the new formula results in an increased standard deviation at the 1% level of significance. Assume that an increase from $\sigma = 2$ to $\sigma = 4$ must be detected with probability 0.80. (a) Find the test statistic and acceptance region. (b) Determine the sample size necessary to achieve the given Type I and II error probabilities. (c) If $S^2 = 11$, what decision is called for?

32. Derive the expression for the sample size of the two-sided procedure for testing the hypothesis that $\mu = \mu_0$ when the standard deviation is known.

33. Derive the expression for the sample size of the two-sided procedure for testing the hypothesis that $\sigma = \sigma_0$.

34. Suppose that the random variable X represents the life of bulbs and has the following density function

$$f_X(z) = \begin{cases} \dfrac{1}{\theta} e^{-z/\theta}, & 0 \leq z \leq \infty \\ 0, & \text{otherwise.} \end{cases}$$

(a) To test the hypothesis that $\theta = 1{,}000$ against the alternative that $\theta > 1000$, the following test is proposed: Obtain one observation from X, say, X_1, and reject if $X_1 > 1{,}000$. Obtain an expression for the OC curve. (b) What is the level of significance of the test described in (a)?

7

TESTS OF HYPOTHESES

ABOUT TWO PARAMETERS

7.1 Test of the Hypothesis that the Means of Two Normal Distributions Are Equal when Both Standard Deviations Are Known

7.1.1 CHOICE OF AN OC CURVE

The tests of significance discussed so far are concerned with determining whether a parameter of a probability distribution is equal to a specified value. Another problem which often arises in experimental work is to determine whether two probability distributions have some parameters which are the same, without specifying the common values of these parameters. The simplest example of such tests is testing for the equality of two means.

Suppose that an experiment is to be performed to determine whether surface finish has an effect on the endurance limit of steel; in particular, several unpolished and polished smooth-turned specimens are to be compared. The data will consist of the experimentally determined endurance limits (reverse bending) from both kinds of specimens. The statistical problem is to decide whether the expected values of the observations from each set are equal or are different, i.e., to determine whether the average endurance limit of the polished specimens differs from the average endurance limit of the unpolished specimens. This is equivalent to finding whether surface finish has a real effect.

This section is concerned with testing the hypothesis that means of two normal distributions, μ_x and μ_y, are equal when both standard deviations, σ_x and σ_y, are known. The statistic used for testing the hypothesis that $\mu_x = \mu_y$ (or $\mu_x - \mu_y = 0$) is a function of the difference between the sample means. The abscissae of operating characteristic curves for procedures based on this test statistic naturally depend on the difference in the true means,

225

i.e., $\mu_x - \mu_y$. To determine the acceptance region and sample size, two points on the OC curve must be chosen. These points are denoted by $(0, 1 - \alpha)$ and $(\mu_x - \mu_y, \beta)$. The first point corresponds to choosing the probability of accepting the hypothesis when it is true (when $\mu_x = \mu_y$), and the second point corresponds to choosing the probability of accepting the hypothesis when $\mu_x - \mu_y$ is the value of the difference in means considered important to detect.

7.1.2 TABLES AND CHARTS FOR DETERMINING DECISION RULES

An experiment is to be performed and random samples of size n_x and size n_y are to be drawn. Each of the resulting sequence of random variables, $X_1, X_2, \ldots, X_{n_x}$, can be thought of as representing the measurable quantity associated with a trial or test of the "x characteristic," whereas each of the random variables, $Y_1, Y_2, \ldots, Y_{n_y}$, can be thought of as representing the measurable quantity associated with a trial or test of the "y characteristic." It is assumed that the X_i are independent normally distributed random variables, each with unknown mean μ_x and known standard deviation σ_x, and that the Y_i are independent normally distributed random variables, each with unknown μ_y and known standard deviation σ_y. Furthermore, let all the X's and Y's be independent. The *optimum* procedure for testing the hypothesis that $\mu_x = \mu_y$ is based on the random variable U, where

$$U = \frac{\bar{X} - \bar{Y}}{\sqrt{\sigma_x^2/n_x + \sigma_y^2/n_y}}.$$

If the value taken on by the test statistic as a result of experimentation falls into the acceptance region, the hypothesis that $\mu_x = \mu_y$ is accepted; otherwise, it is rejected. This section is devoted to finding the acceptance region and sample size for the test statistic U, using the tables and charts which are provided. One- and two-sided procedures are considered.

7.1.2.1 *Tables and Charts for Two-Sided Procedures.* Two-sided procedures are analyzed first. Symmetry of the OC curve about zero is assumed. This restriction can be relaxed by using the same methods as were presented in Sec. 6.1.2.1 in connection with the single-parameter problem. The acceptance region for the two-sided procedure is given by the interval $[-K_{\alpha/2}, K_{\alpha/2}]$ where α is the level of significance and $K_{\alpha/2}$ is the $100\,\alpha/2$ percentage point of the normal distribution. Thus, the hypothesis that $\mu_x = \mu_y$ is accepted if

$$-K_{\alpha/2} \leqq U \leqq K_{\alpha/2},$$

and rejected if U lies outside this interval. It is easily verified that this procedure yields the point $(0, 1 - \alpha)$ on the OC curve. Whenever $\mu_x = \mu_y$, the test statistic

$$U = \frac{\bar{X} - \bar{Y}}{\sqrt{\sigma_x^2/n_x + \sigma_y^2/n_y}}$$

is normally distributed with zero mean and unit variance. Hence, if the hypothesis is true, i.e., $\mu_x = \mu_y$,

$$P\{-K_{\alpha/2} \leq U \leq K_{\alpha/2}\} = 1 - \alpha.$$

The required sample size is obtained by referring to the set of operating characteristic curves used in Chapter VI, namely, Figs. 6.4 and 6.5. In order to use these curves the abscissa scale, d, is now defined as

$$d = \frac{|\mu_x - \mu_y|}{\sqrt{\sigma_x^2 + \sigma_y^2}},$$

and it is necessary to choose $n_x = n_y = n$. If the level of significance of the procedure is either 0.05 or 0.01, the required sample size n is obtained by entering the appropriate figure (Fig. 6.4 if $\alpha = 0.05$ and Fig. 6.5 if $\alpha = 0.01$) with the point $(|\mu_x - \mu_y|/\sqrt{\sigma_x^2 + \sigma_y^2}, \beta)$ and reading out the value of the sample size associated with the OC curve that passes through this point.

Occasionally, the cost of obtaining observations from one distribution may be substantially greater than from the other as, for example, in comparing an experimental item with a production item or in comparing a use test with a specification test. In such situations it may be undesirable to choose n_x equal to n_y. A proper choice of n_x and n_y, having the property that the resultant OC curve passes through the two prescribed points, can be found by a method of trial and error. One guesses at values of n_x and n_y. An equivalent value of n is computed from the expression

$$n = \frac{\sigma_x^2 + \sigma_y^2}{\sigma_x^2/n_x + \sigma_y^2/n_y}.$$

The OC curve of Fig. 6.4 or 6.5 for

$$n = \frac{\sigma_x^2 + \sigma_y^2}{\sigma_x^2/n_x + \sigma_y^2/n_y}$$

is entered with $|\mu_x - \mu_y|/\sqrt{\sigma_x^2 + \sigma_y^2}$ and the probability of accepting the hypothesis is read from the curve. If this probability is β, the chosen values of n_x and n_y are satisfactory. If this probability is not β, changes in n_x and or n_y are made and the procedure is repeated.

An alternative procedure is to enter the OC curves of Fig. 6.4 or 6.5 at the point

$$\left(\frac{|\mu_x - \mu_y|}{\sqrt{\sigma_x^2 + \sigma_y^2}}, \beta \right)$$

and read out the value of the "equivalent" sample size, n, associated with the OC curve that passes through this point. The appropriate values of n_x and n_y must then satisfy

$$n = \frac{\sigma_x^2 + \sigma_y^2}{\sigma_x^2/n_x + \sigma_y^2/n_y}.$$

In fact, if the unit costs of taking observations on X and Y are c_x and c_y, respectively, and if the fixed equivalent sample size is equal to n, then the (optimal) values of n_x and n_y that minimize the total cost of sampling are given by

$$n_x = \frac{(\sigma_x^2 + \sigma_x\sigma_y\sqrt{c_y/c_x})n}{\sigma_x^2 + \sigma_y^2},$$

and

$$n_y = \frac{(\sigma_y^2 + \sigma_x\sigma_y\sqrt{c_x/c_y})n}{\sigma_x^2 + \sigma_y^2}.$$

These results are obtained by treating n_x and n_y as continuous variables so that suitable modifications must be made due to the discreteness of these variables.

7.1.2.2 *Summary for Two-Sided Procedures using Tables and Charts.* The following steps are taken, using a two-sided procedure for testing the hypothesis that $\mu_x = \mu_y$ with both σ_x and σ_y known.

1. Two points on the OC curve are chosen: the probability of accepting the hypothesis when it is true, i.e., $(0, 1 - \alpha)$; and the probability of accepting the hypothesis when $\mu_x - \mu_y$ is the value of the difference in means considered important to detect, i.e., $(\mu_x - \mu_y, \beta)$. On the OC curve with the d scale, these points are translated to $(0, 1 - \alpha)$ and

$$\left(\frac{|\mu_x - \mu_y|}{\sqrt{\sigma_x^2 + \sigma_y^2}}, \beta \right).$$

2. If $n_x = n_y = n$, the required sample size n is found by entering the appropriate figure (Fig. 6.4 if $\alpha = 0.05$ and Fig. 6.5 if $\alpha = 0.01$) with the point

$$\left(\frac{|\mu_x - \mu_y|}{\sqrt{\sigma_x^2 + \sigma_y^2}}, \beta \right)$$

and reading out the value of the sample size associated with the OC curve that passes through this point. If $n_x \neq n_y$, the appropriate values of n_x and n_y are found by trial and error, using an equivalent value of n given by

$$n = \frac{\sigma_x^2 + \sigma_y^2}{\sigma_x^2/n_x + \sigma_y^2/n_y}.$$

3. Determine the acceptance region which is the interval $[-K_{\alpha/2}, K_{\alpha/2}]$.
4. Draw a random sample of n_x and n_y items.
5. Compute the value of the test statistic

$$U = \frac{\bar{X} - \bar{Y}}{\sqrt{\sigma_x^2/n_x + \sigma_y^2/n_y}}.$$

6. If $-K_{\alpha/2} \leq U \leq K_{\alpha/2}$, i.e., $|U| \leq K_{\alpha/2}$, accept the hypothesis that $\mu_x = \mu_y$. If U lies outside this interval, reject the hypothesis and conclude that $\mu_x \neq \mu_y$.

7.1.2.3 *Summary for One-Sided Procedures using Tables and Charts.* The use of the tables and charts for one-sided procedures for testing the equality of two means is similar to their use for the two-sided procedures. Hence, only a summary is given.

The following steps are taken, using a one-sided procedure for testing the hypothesis that $\mu_x = \mu_y$ with both σ_x and σ_y known.

1. Two points on the OC curve are chosen: the probability of accepting the hypothesis when it is true, i.e., $(0, 1 - \alpha)$; and the probability of accepting the hypothesis when $\mu_x - \mu_y$ is the value of the difference in means considered important to detect, i.e., $(\mu_x - \mu_y, \beta)$. If rejection is desired when $\mu_x > \mu_y$, these points are translated to $(0, 1 - \alpha)$ and

$$\left(\frac{\mu_x - \mu_y}{\sqrt{\sigma_x^2 + \sigma_y^2}}, \beta \right)$$

on the OC curve with the d scale. If rejection is desired when $\mu_x < \mu_y$, these points are translated to $(0, 1 - \alpha)$ and

$$\left(\frac{\mu_y - \mu_x}{\sqrt{\sigma_x^2 + \sigma_y^2}}, \beta \right)$$

on the OC curve with the d scale.

2. If $n_x = n_y = n$, the required sample size is found by entering the appropriate figure (Fig. 6.6 if $\alpha = 0.05$ and Fig. 6.7 if $\alpha = 0.01$). If rejection is desired when $\mu_x > \mu_y$, the figure is entered with the point

$$\left(\frac{\mu_x - \mu_y}{\sqrt{\sigma_x^2 + \sigma_y^2}}, \beta \right)$$

and the value of the sample size associated with the OC curve that passes through this point is read out. If rejection is desired when $\mu_x < \mu_y$, the figure is entered with the point

$$\left(\frac{\mu_y - \mu_x}{\sqrt{\sigma_x^2 + \sigma_y^2}}, \beta \right)$$

and the value of the sample size associated with the OC curve that passes through this point is read out. If $n_x \neq n_y$, the appropriate values of n_x and n_y are found by trial and error, using an equivalent value of n given by

$$n = \frac{\sigma_x^2 + \sigma_y^2}{\sigma_x^2/n_x + \sigma_y^2/n_y}$$

as described in Sec. 7.1.2.1.

3. Determine the acceptance region which is the interval $[-\infty, K_\alpha]$ for the alternative $\mu_x > \mu_y$ and $[-K_\alpha, \infty]$ for the alternative $\mu_x < \mu_y$.
4. Draw a random sample of n_x and n_y items.

5. Compute the value of the test statistic

$$U = \frac{\bar{X} - \bar{Y}}{\sqrt{\sigma_x^2/n_x + \sigma_y^2/n_y}}.$$

6. If the alternative is $\mu_x > \mu_y$, accept the hypothesis that $\mu_x = \mu_y$ (or $\mu_x \leq \mu_y$) whenever $U \leq K_\alpha$; otherwise, reject the hypothesis and conclude that $\mu_x > \mu_y$. If the alternative is $\mu_x < \mu_y$, accept the hypothesis that $\mu_x = \mu_y$ (or $\mu_x \geq \mu_y$) whenever $U \geq -K_\alpha$; otherwise, reject the hypothesis and conclude that $\mu_x < \mu_y$.

7.1.2.4 *Tables and Charts for Operating Characteristic Curves.* If $n_x = n_y = n$ and this sample size, n, is fixed in advance or after it is chosen, the operating characteristic curves given in Figs. 6.4 and 6.5 for two-sided procedures and Figs. 6.6 and 6.7 for one-sided procedures can be used to evaluate the risks involved in using this procedure. If the level of significance is either 0.05 or 0.01, the appropriate OC curve is entered with the sample size, n, and d, and the probability of accepting the hypothesis is read from the curve. The value of d chosen corresponds to the value of $\mu_x - \mu_y$ for which the risk is to be evaluated.

If $n_x \neq n_y$ and these values are fixed in advance or after they are chosen, the OC curve can still be used to evaluate the risks as follows. The OC curve corresponding to an equivalent value of

$$n = \frac{\sigma_x^2 + \sigma_y^2}{\sigma_x^2/n_x + \sigma_y^2/n_y}$$

is entered with d calculated for the value of $\mu_x - \mu_y$ for which the risk is to be evaluated; the desired risk (probability of accepting the hypothesis) is read from the curve.

If values of α other than 0.05 or 0.01 are desired, Appendix Table 1 (Table of the Percentage Points of the Normal Distribution) can be used to determine the acceptance region. However, analytical procedures which are given in Sec. 7.1.3 must be used to obtain the OC curve.

The OC curves presented are calculated under the assumption that the test statistic U is normally distributed. This assumption is satisfied if the conditions stated at the beginning of Sec. 7.1.2 are fulfilled. It is also satisfied for more general forms of underlying probability distributions if the sample size is sufficiently large because the conclusions of the central limit theorem are applicable (U is a function of sample means).

A summary of the procedures for testing the hypothesis that $\mu_x = \mu_y$ when both σ_x and σ_y are known is given in Table 7.1.

7.1.3 ANALYTICAL DETERMINATION OF DECISION RULES

7.1.3.1 *Acceptance Regions and Sample Sizes.* Let $X_1, X_2, \ldots, X_{n_x}$ be a set of n_x independent random variables, each with unknown mean μ_x

Table 7.1. Test of the Hypothesis that the Means of Two Normal Distributions Are Equal when Both Standard Deviations Are Known

Notation for the Hypothesis $H: \mu_x = \mu_y$	Test Statistic $U = (\bar{X} - \bar{Y})/\sqrt{\sigma_x^2/n_x + \sigma_y^2/n_y}$
Criteria for Rejection	*Method for Choosing Sample Size, Assuming $n_x = n_y = n$*
$|U| > K_{\alpha/2}$ if rejection is desired when the true difference in means is either positive or negative.	Choose a value of $\mu_x - \mu_y$ for which rejection of the hypothesis is desired with given high probability; calculate[1] $$d = \frac{|\mu_x - \mu_y|}{\sqrt{\sigma_x^2 + \sigma_y^2}}$$ and select n from the OC curves of Fig. 6.4 or 6.5.
$U > K_\alpha$ if rejection is desired when $\mu_x > \mu_y$.	Choose a value of $\mu_x - \mu_y > 0$ for which rejection of the hypothesis is desired with given high probability; calculate[1] $$d = \frac{\mu_x - \mu_y}{\sqrt{\sigma_x^2 + \sigma_y^2}}$$ and select n from the OC curves of Fig. 6.6 or 6.7.
$U < -K_\alpha$ if rejection is desired when $\mu_x < \mu_y$.	Choose a value of $\mu_x - \mu_y < 0$ for which rejection of the hypothesis is desired with given high probability; calculate[1] $$d = \frac{\mu_y - \mu_x}{\sqrt{\sigma_x^2 + \sigma_y^2}}$$ and select n from the OC curves of Fig. 6.6 or 6.7.

Values of K_α, the 100 α percentage points of the normal distribution, are given in Appendix Table 1.

[1] To find the required sample size for the OC curves given, it is necessary to choose $n = n_x = n_y$. However, if n_x and n_y are fixed in advance, $n_x \neq n_y$, the resulting protection, using the above rule, can be obtained by entering the curves using d and the equivalent sample size $n = (\sigma_x^2 + \sigma_y^2)/[\sigma_x^2/n_x + \sigma_y^2/n_y]$.

and known standard deviation σ_x. Let $Y_1, Y_2, \ldots, Y_{n_y}$ be a set of n_y independent random variables, each with unknown mean μ_y and known standard deviation σ_y. Furthermore, let all the X's and Y's be independent. The optimum procedure for testing the hypothesis that $\mu_x = \mu_y$ is based on the test statistic U, where

$$U = \frac{\bar{X} - \bar{Y}}{\sqrt{\sigma_x^2/n_x + \sigma_y^2/n_y}}.$$

This section is devoted to finding acceptance regions and expressions for the sample size for both one- and two-sided procedures.

Under the above-mentioned assumptions (or from the central limit theorem) $\bar{X}$ is normally distributed with mean μ_x and variance σ_x^2/n_x, and $\bar{Y}$ is normally distributed with mean μ_y and variance σ_y^2/n_y. Moreover, $\bar{X}$ and $\bar{Y}$ are independent. It follows from the result on linear combinations of normally distributed random variables (LCN) that the difference of two independent normal variables is normal with mean equal to the difference of the means, and variance equal to the sum of the variances. Therefore, the random variable $\bar{X} - \bar{Y}$ is normally distributed with mean $\mu_x - \mu_y$ and variance equal to $\sigma_x^2/n_x + \sigma_y^2/n_y$, and the random variable

$$\frac{\bar{X} - \bar{Y} - (\mu_x - \mu_y)}{\sqrt{\sigma_x^2/n_x + \sigma_y^2/n_y}}$$

has a normal distribution with zero mean and unit variance. Thus, if the hypothesis is true, i.e., $\mu_x = \mu_y$,

$$U = \frac{\bar{X} - \bar{Y}}{\sqrt{\sigma_x^2/n_x + \sigma_y^2/n_y}}$$

has a normal distribution with zero mean and unit variance, and probabilities associated with various values of U can be obtained directly from tables of the percentage points of the normal distribution (Appendix Table 1). Hence, all the following acceptance regions, given by (1) $[-K_{\alpha/2}, K_{\alpha/2}]$ for the two-sided procedure, (2) $[-\infty, K_\alpha]$ for the one-sided procedure where rejection is desired when $\mu_x > \mu_y$, and (3) $[-K_\alpha, \infty]$ for the one-sided procedure where rejection is desired when $\mu_x < \mu_y$, have a probability of accepting the hypothesis when it is true (level of significance) equal to $1 - \alpha$, where K_α is the 100 α percentage point of the normal distribution.

For the two-sided procedure, the expression for the sample size $n = n_x = n_y$ is given approximately by

$$n \simeq \frac{(K_{\alpha/2} + K_\beta)^2(\sigma_x^2 + \sigma_y^2)}{(\mu_x - \mu_y)^2}$$

where the Type II error β is the probability of accepting the hypothesis when the true difference in means equals $\mu_x - \mu_y$, and K_β is the 100 β percentage point of the normal distribution. This approximation is good whenever

$$P\left\{N < -K_{\alpha/2} - \frac{|\mu_x - \mu_y|\sqrt{n}}{\sqrt{\sigma_x^2 + \sigma_y^2}}\right\}$$

is small compared to β, where N is a normally distributed random variable with zero mean and unit variance.

For one-sided procedures, the expression for the sample size $n = n_x = n_y$ is given by

$$n = \frac{(K_\alpha + K_\beta)^2(\sigma_x^2 + \sigma_y^2)}{(\mu_x - \mu_y)^2}.$$

The derivation of the expressions for the sample size is similar to that given in Sec. 6.1.3.1 and, hence, is omitted.

7.1.3.2 *The Operating Characteristic Curve.* If n_x and n_y and the acceptance region are known, the entire OC curve is easily determined. Let the true difference in means of the X's and Y's be given by $\mu_x - \mu_y$. Thus, the expected value of U is

$$E(U) = \frac{\mu_x - \mu_y}{\sqrt{\sigma_x^2/n_x + \sigma_y^2/n_y}} \cdot$$

U is then normally distributed with mean

$$\frac{\mu_x - \mu_y}{\sqrt{\sigma_x^2/n_x + \sigma_y^2/n_y}}$$

and unit variance. For the one-sided procedure where rejection is desired when $\mu_x > \mu_y$, the expression for the OC curve (probability of accepting the hypothesis) is given by

$$P\{U \leq K_\alpha\}$$

which is equivalent to

$$P\left\{N \leq K_\alpha - \frac{(\mu_x - \mu_y)\sqrt{n}}{\sqrt{\sigma_x^2 + \sigma_y^2}}\right\} \qquad \text{where } n_x = n_y = n.$$

This last expression can be evaluated for any value of $\mu_x - \mu_y$ with the aid of Appendix Table 1. It is interesting to note that for fixed n and K_α this expression for the OC curve is just a function of

$$d = \frac{\mu_x - \mu_y}{\sqrt{\sigma_x^2 + \sigma_y^2}}$$

and d is the abscissa of the OC curves plotted in Figs. 6.6 and 6.7.

For the single-parameter problem discussed in Chapter 6, the OC curve for testing the hypothesis that $\mu = \mu_0$ was given by

$$P\left\{N \leq K_\alpha - \frac{(\mu - \mu_0)\sqrt{n}}{\sigma}\right\} \quad \text{or} \quad P\{N \leq K_\alpha - d\sqrt{n}\}$$

where d was defined as $(\mu - \mu_0)/\sigma$.

For testing the hypothesis that $\mu_x = \mu_y$, the OC curve described above is given by

$$P\left\{N \leq K_\alpha - \frac{(\mu_x - \mu_y)\sqrt{n}}{\sqrt{\sigma_x^2 + \sigma_y^2}}\right\} \quad \text{or} \quad P\{N \leq K_\alpha - d\sqrt{n}\}$$

where d is presently defined as

$$\frac{\mu_x - \mu_y}{\sqrt{\sigma_x^2 + \sigma_y^2}} \cdot$$

It is now evident why the same OC curves given in Figs. 6.6 and 6.7 can be used for both problems.

If the procedure is one-sided and rejection is desired when $\mu_x < \mu_y$, the OC curve is given by the expression

$$P(U \geq -K_\alpha) = P\left\{N \geq -K_\alpha - \frac{(\mu_x - \mu_y)\sqrt{n}}{\sqrt{\sigma_x^2 + \sigma_y^2}}\right\}$$

where $n_x = n_y = n$.

Finally, if the procedure is two-sided, the OC curve is given by the expression

$$P\{-K_{\alpha/2} \leq U \leq K_{\alpha/2}\}$$
$$= P\left\{-K_{\alpha/2} - \frac{(\mu_x - \mu_y)\sqrt{n}}{\sqrt{\sigma_x^2 + \sigma_y^2}} \leq N \leq K_{\alpha/2} - \frac{(\mu_x - \mu_y)\sqrt{n}}{\sqrt{\sigma_x^2 + \sigma_y^2}}\right\}$$

where $n_x = n_y = n$.

7.1.4 EXAMPLE

In Sec. 7.1.1 an experiment to determine whether surface finish has an effect on the endurance limit of steel was discussed. There exists a theory that polishing changes the average endurance limit (reverse bending). An experiment is performed on 0.4% carbon steel, using both unpolished and polished smooth-turned specimens. The finish on the smooth-turned polished specimens was obtained by polishing with No. 0 and No. 00 emery cloth.

If the average endurance limit of polished specimens is denoted by μ_x and the average endurance limit of unpolished specimens by μ_y, the hypothesis to be tested is that $\mu_x = \mu_y$ against the two-sided alternative that $\mu_x \neq \mu_y$. If polishing fails to have an effect, assume that this should be noted with probability 0.95 ($\alpha = 0.05$). Moreover, it is important, from a practical point of view, to detect a change, if it is as much as $\pm 6,800$ psi, with probability at least equal to 0.9. Furthermore, polishing should not have any effect on the standard deviation of the endurance limit, which is known from the performance of numerous endurance limit experiments to be 4,000 psi. From Fig. 6.4 (5% level of significance) with $d = 6,800/\sqrt{(4,000)^2 + (4,000)^2} = 1.20$, it is found that about eight observations on each group of specimens are required to have a probability of 0.9 of detecting a difference as large as 6,800 psi. The data are as follows:

Endurance Limit for Polished 0.4% Carbon Steel (X)	Endurance Limit for Unpolished 0.4% Carbon Steel (Y)
86,500	82,600
91,900	82,400
89,400	81,700
84,000	79,500
89,900	79,400
78,700	69,800
87,500	79,900
83,100	83,400
$\bar{X} = 86,375$	$\bar{Y} = 79,838$

$$U = \frac{6537}{\sqrt{(4,000)^2/8 + (4,000)^2/8}} = 3.27; \quad K_{0.025} = 1.96$$

The acceptance region is given by the interval $[-1.96, 1.96]$. Since $U = 3.27$ exceeds 1.96, the hypothesis is rejected and it is concluded that the average endurance limit of polished 0.4% carbon steel specimens differs from the average endurance limit of unpolished specimens.

7.2 Test of the Hypothesis that the Means of Two Normal Distributions Are Equal, Assuming that the Standard Deviations Are Unknown but Equal

7.2.1 CHOICE OF AN OC CURVE

This is the same problem as the one discussed in Sec. 7.1 except that the standard deviations are now assumed to be unknown but equal, i.e., $\sigma_x = \sigma_y = \sigma$, where σ is unknown.[1] In testing the hypothesis that $\mu_x = \mu_y$ in this situation, a t statistic is used. This test statistic is a function of the difference in the sample means, the sample standard deviation, and the sample size. The abscissae of operating characteristic curves for procedures based on this t statistic depend on the quantity $(\mu_x - \mu_y)/\sigma$. As in the t test for a single parameter, it is not the absolute value of the difference in means which is important for OC curve purposes, but rather the relative difference, relative to the inherent variability, σ. In order to determine the acceptance region and sample size, two points on the OC curve must be chosen. These points will be denoted by $(0, 1 - \alpha)$ and $((\mu_x - \mu_y)/\sigma, \beta)$. The first point corresponds to choosing the probability of accepting the hypothesis when it is true (when $\mu_x = \mu_y$), and the second point corresponds to choosing the probability of accepting the hypothesis when $(\mu_x - \mu_y)/\sigma$ is the ratio of the difference in means to the standard deviation considered important to detect. For the purpose of choosing an OC curve a *rough* notion of σ is usually adequate. As indicated previously, such a notion can be obtained from past data or some prior knowledge before experimentation; or if the OC curve is examined after experimentation, the sample estimate of the standard deviation can be used as a "rough" estimate.

7.2.2 TABLES AND CHARTS FOR CARRYING OUT TWO-SAMPLE t TESTS

An experiment is to be performed and random samples of size n_x and size n_y are to be drawn. Each of the resulting sequence of random variables, $X_1, X_2, \ldots, X_{n_x}$, can be thought of as representing the measurable quantity associated with a trial or test of the "x characteristic" whereas each of the random variables, $Y_1, Y_2, \ldots, Y_{n_y}$, can be thought of as representing the measurable quantity associated with a test or trial of the "y characteristic." It is assumed that the X_i and Y_i are independent normally distributed random variables. The mean of each of the X's is μ_x (unknown) and the mean of each of the Y's is μ_y (unknown); the standard deviation of the X's is σ_x and the standard deviation of the Y's is σ_y, where σ_x and σ_y are both unknown but equal to a common value, σ. The *optimum* procedure for testing the

---

[1] A procedure for testing the hypothesis that $\sigma_x = \sigma_y$ is given in Sec. 7.6.

hypothesis that $\mu_x = \mu_y$ is based on the random variable t, where

$$t = \frac{\bar{X} - \bar{Y}}{\sqrt{\dfrac{1}{n_x} + \dfrac{1}{n_y}} \sqrt{\dfrac{\sum\limits_{i=1}^{n_x} (X_i - \bar{X})^2 + \sum\limits_{i=1}^{n_y} (Y_i - \bar{Y})^2}{n_x + n_y - 2}}}.$$

If the value the test statistic takes on as a result of experimentation falls into the acceptance region, the hypothesis that $\mu_x = \mu_y$ is accepted; otherwise, it is rejected. This section is devoted to finding the acceptance region and sample size for the test statistic t, using the tables and charts which are provided. One- and two-sided procedures are considered.

7.2.2.1 *Tables and Charts for Two-Sided Procedures.* Two-sided procedures are analyzed first. Symmetry of the OC curve about zero is assumed. This restriction can be relaxed by using the same methods that were presented in Sec. 6.1.2.1. The acceptance region for the two-sided procedure for any sample sizes, n_x and n_y, is given by the interval $[-t_{\alpha/2;\,n_x+n_y-2},\, t_{\alpha/2;\,n_x+n_y-2}]$, where α is the level of significance and $t_{\alpha/2;\,n_x+n_y-2}$ is the $100\,\alpha/2$ percentage point of the t distribution with $n_x + n_y - 2$ degrees of freedom. Thus, the hypothesis that $\mu_x = \mu_y$ is accepted if

$$-t_{\alpha/2;\,n_x+n_y-2} \leqq t \leqq t_{\alpha/2;\,n_x+n_y-2}$$

and rejected if t lies outside this interval. It is easily verified that this procedure yields a probability of $1 - \alpha$ of accepting the hypothesis that $\mu_x = \mu_y$ when it is true. It was shown in Sec. 4.3.3 that, when $\mu_x = \mu_y$, the test statistic

$$t = \frac{\bar{X} - \bar{Y}}{\sqrt{\dfrac{1}{n_x} + \dfrac{1}{n_y}} \sqrt{\dfrac{\sum\limits_{i=1}^{n_x} (X_i - \bar{X})^2 + \sum\limits_{i=1}^{n_y} (Y_i - \bar{Y})^2}{n_x + n_y - 2}}}$$

has a t distribution with $n_x + n_y - 2$ degrees of freedom. Hence, if $\mu_x = \mu_y$,

$$P\{-t_{\alpha/2;\,n_x+n_y-2} \leqq t \leqq t_{\alpha/2;\,n_x+n_y-2}\} = 1 - \alpha.$$

The required sample size is obtained by referring to the set of operating characteristic curves used in Chapter 6, namely, Figs. 6.10 and 6.11. To use these curves, the abscissa scale, d, is now defined as

$$d = \frac{|\mu_x - \mu_y|}{2\sigma}$$

and it is necessary to choose $n_x = n_y = n$. If the level of significance of the procedure is either 0.05 or 0.01, the required sample size n is obtained by entering the appropriate figure (Fig. 6.10 if $\alpha = 0.05$ and Fig. 6.11 if $\alpha = 0.01$) with the point $(|\mu_x - \mu_y|/2\sigma,\, \beta)$ and reading out the value associated with the OC curve that passes through this point. Denote this quantity by n'. The required sample size n is given by the expression

$$n = \frac{n' + 1}{2}.$$

7.2.2.2 *Summary for Two-Sided Procedures using Tables and Charts.* The following steps are taken, using a two-sided procedure for testing the hypothesis that $\mu_x = \mu_y$ with both σ_x and σ_y unknown but equal to σ.

1. Two points on the OC curve are chosen, namely, points $(0, 1 - \alpha)$ and $((\mu_x - \mu_y)/\sigma, \beta)$. These points are translated to the d scale and are denoted by $(0, 1 - \alpha)$ and $(|\mu_x - \mu_y|/2\sigma, \beta)$ on this scale.
2. If $n_x = n_y = n$, the required sample size n is found by entering the appropriate figure (Fig. 6.10 if $\alpha = 0.05$ and Fig. 6.11 if $\alpha = 0.01$) with the point $(|\mu_x - \mu_y|/2\sigma, \beta)$ and reading out the value associated with the OC curve that passes through this point. Denote the quantity by n'. The required sample size, n, is given by the expression $n = (n' + 1)/2$.
3. Determine the acceptance region which is the interval

$$[-t_{\alpha/2; n_x + n_y - 2}, t_{\alpha/2; n_x + n_y - 2}].$$

4. Draw a random sample of n_x and n_y items.
5. Compute the value of the test statistic

$$t = \frac{\bar{X} - \bar{Y}}{\sqrt{\frac{1}{n_x} + \frac{1}{n_y}} \sqrt{\frac{\sum\limits_{i=1}^{n_x} (X_i - \bar{X})^2 + \sum\limits_{i=1}^{n_y} (Y_i - \bar{Y})^2}{n_x + n_y - 2}}}.$$

6. If $t_{\alpha/2; n_x + n_y - 2} \leq t \leq t_{\alpha/2; n_x + n_y - 2}$, i.e., $|t| \leq t_{\alpha/2; n_x + n_y - 2}$, accept the hypothesis that $\mu_x = \mu_y$. If t lies outside this interval, reject the hypothesis and conclude that $\mu_x \neq \mu_y$.

7.2.2.3 *Summary for One-Sided Procedures using Tables and Charts.* The use of the tables and charts for one-sided procedures for testing the equality of two means is similar to their use for the two-sided procedures. Hence, only a summary is given.

The following steps are taken, using a one-sided procedure for testing the hypothesis that $\mu_x = \mu_y$ with both σ_x and σ_y unknown but equal to σ.

1. Two points on the OC curve are chosen, namely points $(0, 1 - \alpha)$ and $(\mu_x - \mu_y, \beta)$. If rejection is desired when $\mu_x > \mu_y$, these points are translated to $(0, 1 - \alpha)$ and $((\mu_x - \mu_y)/2\sigma, \beta)$ on the OC curve with the d scale. If rejection is desired when $\mu_x < \mu_y$, then these points are translated to $(0, 1 - \alpha)$ and $((\mu_y - \mu_x)/2\sigma, \beta)$ on the OC curve with the d scale.

2. If $n_x = n_y = n$, the required sample size n is found by entering the appropriate figure (Figs. 6.12 and 6.13). If rejection is desired when $\mu_x > \mu_y$, the figure is entered with the point $((\mu_x - \mu_y)/2\sigma, \beta)$ and the value associated with the OC curve that passes through this point is read out. If rejection is desired when $\mu_x < \mu_y$, the figure is entered with the point $((\mu_y - \mu_x)/2\sigma, \beta)$ and the value associated with the OC curve that passes through this point is read out. Denote these quantities by n'. The required sample size, n, is obtained from the expression $n = (n' + 1)/2$.

3. Determine the acceptance region which is that interval $[-\infty, t_{\alpha; n_x + n_y - 2}]$ for the alternative $\mu_x > \mu_y$ and $[-t_{\alpha; n_x + n_y - 2}, \infty]$ for the alternative $\mu_x < \mu_y$.

4. Draw a random sample of n_x and n_y items.

5. Compute the value of the test statistic

$$t = \frac{\bar{X} - \bar{Y}}{\sqrt{\dfrac{1}{n_x} + \dfrac{1}{n_y}} \sqrt{\dfrac{\displaystyle\sum_{i=1}^{n_x} (X_i - \bar{X})^2 + \sum_{i=1}^{n_y} (Y_i - \bar{Y})^2}{n_x + n_y - 2}}}.$$

6. If the alternative is $\mu_x > \mu_y$, accept the hypothesis that $\mu_x = \mu_y$ (or $\mu_x \leq \mu_y$) whenever $t \leq t_{\alpha; n_x + n_y - 2}$; otherwise, reject the hypothesis and conclude that $\mu_x > \mu_y$. If the alternative is $\mu_x < \mu_y$, accept the hypothesis that $\mu_x = \mu_y$ (or $\mu_x \geq \mu_y$) whenever $t \geq -t_{\alpha; n_x + n_y - 2}$; otherwise, reject the hypothesis and conclude that $\mu_x < \mu_y$.

7.2.2.4 *Tables and Charts for Operating Characteristic Curves.* If $n_x = n_y = n$ and this sample size, n, is fixed in advance or after it is chosen, the operating characteristic curves given in Figs. 6.10 and 6.11 for two-sided procedures and Figs. 6.12 and 6.13 for one-sided procedures can be used to evaluate risks involved in using these procedures. If the level of significance is either 0.05 or 0.01, the appropriate OC curve associated with the value $n' = 2n - 1$ is entered with d, and the probability of accepting the hypothesis is read from the curve. The value of d chosen is related to the value $(\mu_x - \mu_y)/\sigma$ for which the risk is to be evaluated.

If other levels of significance are desired, Appendix Table 3, which is a table of the percentage points of the t distribution, can be used to determine the acceptance region for a fixed value of n. However, analytical procedures, which are beyond the scope of this text, must be used to obtain the OC curve.

Although the OC curves are calculated under the assumption that the X's and Y's are independent normally (or approximately normally) distributed random variables, they are also valid for more general forms of underlying probability distributions if the sample sizes are sufficiently large.

A summary of the procedures for testing the hypothesis that $\mu_x = \mu_y$ when both σ_x and σ_y are unknown but equal is given in Table 7.2.

Table 7.2. Test of the Hypothesis that the Means of Two Normal Distributions Are Equal, Assuming that the Standard Deviations Are Unknown but Equal

Notation for the Hypothesis $H: \mu_x = \mu_y$	Test Statistic
	$$t = \dfrac{\bar{X} - \bar{Y}}{\sqrt{\dfrac{1}{n_x} + \dfrac{1}{n_y}} \sqrt{\dfrac{\sum_{i=1}^{n_x}(X_i - \bar{X})^2 + \sum_{i=1}^{n_y}(Y_i - \bar{Y})^2}{n_x + n_y - 2}}}$$

Criteria for Rejection	Method for Choosing Sample Size, Assuming $n_x = n_y = n$
$\lvert t \rvert > t_{\alpha/2; n_x + n_y - 2}$ if rejection is desired when μ_x is not equal to μ_y.	Choose a value of $(\mu_x - \mu_y)/\sigma$ for which rejection of the hypothesis is desired with given high probability. Calculate $d = \lvert \mu_x - \mu_y \rvert / 2\sigma$ and enter Fig. 6.10 or 6.11 to find n'. The required sample size is $(n' + 1)/2$.
$t > t_{\alpha; n_x + n_y - 2}$ if rejection is desired when $\mu_x > \mu_y$.	Choose a value of $(\mu_x - \mu_y)/\sigma$ for which rejection of the hypothesis is desired with given high probability. Calculate $d = (\mu_x - \mu_y)/2\sigma$ and enter Fig. 6.12 or 6.13 to find n'. The required sample size is $(n' + 1)/2$.
$t < -t_{\alpha; n_x + n_y - 2}$ if rejection is desired when $\mu_x < \mu_y$.	Choose a value of $(\mu_x - \mu_y)/\sigma$ for which rejection of the hypothesis is desired with given high probability. Calculate $d = (\mu_y - \mu_x)/2\sigma$ and enter Fig. 6.12 or 6.13 to find n'. The required sample size is $(n' + 1)/2$.

Values of $t_{\alpha; \nu}$, the 100α percentage points of the t distribution with ν degrees of freedom, are given in Appendix Table 3.

7.2.3 EXAMPLE

A manufacturer of electric irons produces these items in two plants. Both plants have the same suppliers of small parts. A saving can be made by purchasing thermostats for plant B from a local supplier. A single lot was purchased from the local supplier and it was desired to test whether or not these new thermostats were as accurate as the old. The thermostats were to be tested on the irons on the 550°F setting, and the actual temperatures were to be read to the nearest 0.1° with a thermocouple. The level of significance (α) chosen is 5%. With the old supplier, very few complaints were received, and the manufacturer feels that a switch is undesirable if the average temperature changes by more than 10.5°, with the risk of making an incorrect decision not exceeding 0.10. The order of magnitude of the standard deviation is roughly 10° for the old supplier, and there is no reason to suspect that it will be different for the new supplier. For $d = 10.5/(2 \times 10) = 0.525$, and from Fig. 6.10, $n' = 45$, corresponding to a probability of 0.9

of detecting a change of 10.5° or more. Hence, $n = 23$. The data are:

New Supplier X (Degrees F)		Old Supplier Y (Degrees F)	
530.3	559.1	559.7	564.6
559.3	555.0	534.7	554.5
549.4	538.6	554.8	553.0
544.0	551.1	545.0	538.4
551.7	565.4	544.6	548.3
566.3	554.9	538.0	552.9
549.9	550.0	550.7	535.1
556.9	554.9	563.1	555.0
536.7	554.7	551.1	544.8
558.8	536.1	553.8	558.4
538.8	569.1	538.8	548.7
543.3			560.3

$$\sum_{i=1}^{n_x} X_i = 12{,}674.3 \qquad \sum_{i=1}^{n_y} Y_i = 12{,}648.3$$

$$\bar{X} = 551.056522 \qquad \bar{Y} = 549.926086$$

$$\sum_{i=1}^{n_x} X_i^2 = 6{,}986{,}586.07 \qquad \sum_{i=1}^{n_y} Y_i^2 = 6{,}957{,}333.23$$

$$\sum_{i=1}^{n_x} (X_i - \bar{X})^2 = \sum_{i=1}^{n_x} X_i^2 - n\bar{X}^2 = 2{,}330.391$$

$$\sum_{i=1}^{n_y} (Y_i - \bar{Y})^2 = \sum_{i=1}^{n_y} Y_i^2 - n\bar{Y}^2 = 1{,}703.130$$

$$\frac{\sum_{i=1}^{n_x} (X_i - \bar{X})^2 + \sum_{i=1}^{n_y} (Y_i - \bar{Y})^2}{n_x + n_y - 2} = \frac{4{,}033.521}{44} = 91.670932.$$

The hypothesis that $\mu_x = \mu_y$ is accepted if $|t| \leq t_{\alpha/2; n_x + n_y - 2}$;

$$|t| = \frac{551.056522 - 549.926086}{\sqrt{2/23}\sqrt{91.670932}}$$

$$= 0.400 \leq t_{\alpha/2; n_x + n_y - 2} = t_{0.025; 44} = 2.02.$$

Therefore, the hypothesis that there is no difference in the mean temperatures of the two suppliers is accepted at the 5% level of significance, and a switch is made.

7.3 Test of the Hypothesis that the Means of Two Normal Distributions Are Equal, Assuming that the Standard Deviations Are Unknown and Not Necessarily Equal

7.3.1 TEST PROCEDURE

The assumption that σ_x equals σ_y (both unknown) is often unwarranted. Unfortunately, an exact procedure based on a t statistic is unavailable to cover this situation. However, a procedure exists, based on a test statistic

t', which has the property that, when $\mu_x = \mu_y$, t' has an approximate t distribution. The statistic t' is given by

$$t' = \frac{\bar{X} - \bar{Y}}{\sqrt{S_x^2/n_x + S_y^2/n_y}}$$

and the associated degrees of freedom are

$$\nu = \frac{(S_x^2/n_x + S_y^2/n_y)^2}{\dfrac{(S_x^2/n_x)^2}{(n_x + 1)} + \dfrac{(S_y^2/n_y)^2}{(n_y + 1)}} - 2.$$

The probability distribution of t' has not been determined when μ_x does not equal μ_y. Hence, only one point on the OC curve can be guaranteed, namely, the probability of accepting the hypothesis that $\mu_x = \mu_y$ when it is true. A summary of this procedure is given in Table 7.3.

Table 7.3. Test of the Hypothesis that the Means of Two Normal Distributions Are Equal when the Standard Deviations Are Unknown and Not Necessarily Equal

Notation for the Hypothesis $H: \mu_x = \mu_y$	Test Statistic
	$t' = \dfrac{\bar{X} - \bar{Y}}{\sqrt{S_x^2/n_x + S_y^2/n_y}}$
Criteria for Rejection	
$\|t'\| > t_{\alpha/2;\,\nu}$ if rejection is desired when μ_x is not equal to μ_y.	*Formula for Obtaining the Degrees of Freedom ν*
$t' > t_{\alpha;\,\nu}$ if rejection is desired when $\mu_x > \mu_y$.	$\nu = \dfrac{(S_x^2/n_x + S_y^2/n_y)^2}{(S_x^2/n_x)^2/(n_x + 1) + (S_y^2/n_y)^2/(n_y + 1)} - 2$
$t' < -t_{\alpha;\,\nu}$ if rejection is desired when $\mu_x < \mu_y$.	

Values of $t_{\alpha;\,\nu}$, the 100α percentage point of the t distribution with ν degrees of freedom, are given in Appendix Table 3.

7.3.2 EXAMPLE

A manufacturer of automobile crankshafts was troubled with the problem of bend in the final shaft. Bend may be caused by the length and weight of the shaft, by improper setup of machine tools, by the heat treating process, or by some combination of these causes. It is suspected that nitriding the shaft, i.e., the process that hardens the surface of the shaft by a heat and nitrous oxide reaction with the steel, is the main cause for bend. Twenty-five shafts were measured before nitriding at the front main center journal by a dial indicator gauge accurate to the 0.0001 inch. Similarly, another 25 shafts were measured at the same spot after the nitriding operation. It cannot be assumed that the variability in the bend is the same before and after nitriding. The hypothesis to be tested (at the 5% level of significance) is that the mean value of the bend is the same before and after nitriding, where rejection is desired if the average bend after nitriding is larger. Hence, the acceptance region is $[-t_{\alpha; n_x+n_y-2}, \infty]$. Denoting by X the values before nitriding and by Y the values after nitriding, the following results were obtained:

$$\sum_{i=1}^{25} X_i = 203 \times 10^{-4} \qquad\qquad \sum_{i=1}^{25} Y_i = 453 \times 10^{-4}$$

$$\bar{X} = 0.0008120 \qquad\qquad\qquad \bar{Y} = 0.001812$$

$$\sum_{i=1}^{25} X_i^2 = 2{,}933 \times 10^{-8} \qquad\qquad \sum_{i=1}^{25} Y_i^2 = 14{,}362 \times 10^{-8}$$

$$S_x^2 = \frac{2{,}933 \times 10^{-8} - 25(0.0008120)^2}{24} \qquad S_y^2 = \frac{14{,}362 \times 10^{-8} - 25(0.001812)^2}{24}$$

$$= 53.5 \times 10^{-8} \qquad\qquad\qquad = 256 \times 10^{-8}$$

$$\frac{S_x^2}{n_x} = 2.14 \times 10^{-8} \qquad\qquad\qquad \frac{S_y^2}{n_y} = 10.24 \times 10^{-8}$$

$$\nu = \frac{[(2.14)10^{-8} + (10.24)10^{-8}]^2}{(2.14)^2(10^{-16})/26 + (10.24)^2(10^{-16})/26} - 2 = 34$$

$$t' = \frac{0.0008120 - 0.001812}{\sqrt{(2.14)10^{-8} + (10.24)10^{-8}}} = -2.84$$

$$t_{0.05; 34} = 1.69.$$

Since $t' < -1.69$, the hypothesis that the means are the same before and after nitriding is rejected at the 5% significance level, and it is concluded that the average bend is larger after nitriding.

7.4 Test for Equality of Means when the Observations Are Paired

7.4.1 TEST PROCEDURE

An important special case arises when the observations are taken in pairs, each pair being taken under the same experimental conditions, with the conditions varying from pair to pair. Consider an experiment comparing the corrosion of pipe with two kinds of coating. Many factors besides the coating may effect the corrosion rate: type of soil, length of time of burial, angle of burial, etc. If one takes a number of specimens of pipe with each kind of coating and buries them in various kinds of soil for various lengths of time, an analysis by the method of the last section might be misleading. If the experiment were not arranged carefully, one type of coating might be associated with a particular type of soil or one type of coating might be left in the ground longer than the others, and the apparent difference in coating might, in fact, be due to some other factor. There are two ways to guard against such bias, namely, control and randomization. Control is achieved by taking observations in pairs, each pair to consist of specimens of each of the two coatings and thus one pair of specimens might be buried for five years in clay, another eight years in sand, etc. It is practically impossible to control all variables; such things as size of pipe, orientation, etc., may be of some importance but are not major factors. These factors should be randomized, i.e., assigned to the test specimen by a random device such as coin tossing or random numbers. If this is done, the differences between the specimens will be due to the difference in coating with the effect of randomized factors contributing to increasing the experimental error.

The test procedure involves the difference between each pair of observations. Let $(X_1, Y_1), (X_2, Y_2), \ldots, (X_n, Y_n)$ be a set of n paired random variables which denote the measurable quantities associated with the $1, 2, \ldots, n$ trials, respectively. Form the difference between each pair of observations, i.e., $D_1 = X_1 - Y_1, D_2 = X_2 - Y_2, \ldots, D_n = X_n - Y_n$. It is assumed that these differences are independent normally distributed random variables, each with common unknown mean μ_D and common unknown standard deviation σ_D.[1] In the corrosion example μ_D is the difference in effect of the two kinds of coating. It is likely to be the same from pair to pair even if the actual means of the X's and Y's differ from pair to pair due to changing experimental conditions. The standard deviation, σ_D, although unknown, depends on σ_x^2 and σ_y^2 [and actually equals $(\sigma_x^2 + \sigma_y^2)^{1/2}$ if X and Y are independent] with σ_x not necessarily equal to σ_y. Hence, the standard deviation of observations with one type of coating may differ from the standard deviation of the observations with the other type of coating, and both may be unknown, but these values must be constant from pair to pair.

With these assumptions it is evident that testing for the equality of means is equivalent to testing the hypothesis that $\mu_D = 0$ when the standard deviation, σ_D, is unknown. Hence, the results of Sec. 6.2, testing the hypothesis that the mean of a normal distribution has a specified value when the standard deviation is unknown, are applicable. The test statistic t of Sec. 6.2 becomes, on replacing the X's by D's,

$$t = \frac{\bar{D}\sqrt{n}}{S_D}$$

where

$$\bar{D} = \frac{\sum_{i=1}^{n} D_i}{n}$$

and

$$S_D = \sqrt{\frac{\sum_{i=1}^{n} (D_i - \bar{D})^2}{n - 1}}.$$

---

[1] A sufficient condition for this assumption to hold true is that all X_i and Y_i be independent normally distributed random variables; for the ith pair, the expected value of X_i is μ_i and the expected value of Y_i is $\mu_i + \mu_D$, i.e., the expected values differ by a constant independent of i; the standard deviation of X_i is σ_x and the standard deviation of Y_i is σ_y, with σ_x not necessarily equal to σ_y but σ_x and σ_y constant for all i.

Thus, for the case of paired observations the reader is referred to Sec. 6.2, using the D's as the random variables of interest.

It is important to note that the assumptions necessary for the two-sample t test given in Sec. 7.2 automatically satisfy the requirements for the use of paired observations when the observations are paired at random; i.e., an X and Y are chosen at random and paired. Since all the X_i have the same distribution and all the Y_i have the same distribution, the difference in means between any X and Y is the same regardless of which X and Y are chosen. The effect of this random pairing is seen by noting that the degrees of freedom used for the two-sample t test as suggested in Sec. 7.2 is $n_x + n_y - 2$ or $2(n - 1)$ if $n_x = n_y = n$; whereas the degrees of freedom for the randomly paired observations is just $n - 1$. Since the OC curve of the procedure is essentially related to the degrees of freedom, it is evident that unnecessary random pairing results in a poorer OC curve. However, if the problem described in this section, which required pairing, is solved by using the two-sample t test, the results are meaningless. They are meaningless because the denominator of the two-sample t-test statistic,

$$\sqrt{\frac{2}{n}} \sqrt{\frac{\sum (X_i - \bar{X})^2 + \sum (Y_i - \bar{Y})^2}{2n - 2}},$$

is not an estimate of the standard deviation of $\bar{X} - \bar{Y}$ when effects of extraneous factors exist. If the two-sample t test is applied when paired observations are called for, it is evident that as the magnitude of these effects increases, the sum of squares above becomes larger, and the more difficult it is to judge the distribution means significant even when they do differ. On the other hand,

$$\sqrt{\frac{\sum (D_i - \bar{D})^2}{n(n - 1)}}$$

is always an estimate of the standard deviation of $\bar{D}$ because pairing "controls" the effects of the extraneous factors so that the test statistic based on the D's is appropriate. Hence, if pairing is required for the problem, the paired observations are necessary for analysis of the data; if pairing is not required, it should not be used.

7.4.2 EXAMPLE[1]

In Sec. 7.4.1, an experiment comparing the corrosion of pipe with two kinds of coating was discussed. Pairs of specimens are to be inspected for the amount of corrosion. One specimen of each type of coating is to be included in each pair; each pair is to be buried in the same soil, at the same depth,

---

[1] This example is based upon an example taken from H. A. Freeman, *Industrial Statistics*, John Wiley & Sons, Inc., New York, 1942, p. 8.

in a similar position, and for the same length of time. It is important, from a practical point of view, to detect a difference in the average depth of maximum pits for each coating of 0.011 inch with probability greater than or equal to 0.95. The test is to be run at a 5% level of significance. From past experience, the standard deviation of depth of maximum pit on similar coatings is known to have a very approximate value of 0.008 inch. Hence, σ_D is approximately equal to 0.0113 inch, i.e.,

$$\sigma_D = \sqrt{\sigma_x^2 + \sigma_y^2} = \sqrt{(0.008)^2 + (0.008)^2} = 0.0113.$$

From Figure 6.10 with $d = |0.011 - 0|/0.0113 = 0.97$, it is seen that about fifteen pairs of observations (fifteen differences) are required to have a probability of 0.95 of detecting a difference as large as 0.011 inch. The results of the sample are as follows:

Depth of Maximum Pits (Expressed in Thousandths of an Inch)

Coating A	Coating B	Difference
73	51	+22
43	41	+ 2
47	43	+ 4
53	41	+12
58	47	+11
47	32	+15
52	24	+28
38	43	− 5
61	53	+ 8
56	52	+ 4
56	57	− 1
34	44	− 10
55	57	− 2
65	40	+25
75	68	+ 7

It then follows that

$$\bar{D} = \frac{\sum_{i=1}^{n} D_i}{n} = \frac{120}{15} = 8$$

and the estimate of the variance,

$$S_D = \sqrt{\frac{\sum_{i=1}^{n} (D_i - \bar{D})^2}{n - 1}}$$
$$= \sqrt{121.571}.$$

Thus, the test statistic t is equal to

$$t = \frac{8\sqrt{15}}{\sqrt{121.571}} = 2.810.$$

This value is compared with $t_{\alpha/2;n-1}$, for a 5% level of significance, i.e., compared with $t_{0.025;14} = 2.145$. Thus,

$$2.810 > t_{0.025;14} = 2.145$$

and it is concluded that there is a significant difference in corrosion rates between the two types of coating.

7.5 Non-Parametric Tests

The procedures discussed in previous sections of this chapter are, strictly speaking, valid only when the original observations are assumed to come from a normal distribution, and are only approximately valid when the data arise from non-normal distributions which are reasonably well behaved. Unfortunately, it is not possible to give a precise delineation of situations in which they are valid; such judgments must be made by experimenters on the basis of their experience. Recently, more and more procedures have been coming into use which do not depend on the assumption of normality; such tests are called non-parametric tests, and usually assume only continuity of the probability distribution. These procedures are also useful in that they are easily applied.

7.5.1 THE SIGN TEST

Consider the problem of the corrosion experiment in Sec. 7.4. If the type of coating had no effect on corrosion, half of the differences would be expected to be positive and half to be negative. If there is a preponderance of plus or minus signs, one would suspect a difference due to coating. In order to quantify this statement, it is necessary to determine the distribution of the number of plus signs and minus signs in a sample of size n. Under suitable conditions (to be discussed subsequently) this number is a random variable having a binomial distribution with parameters n and $p = \frac{1}{2}$ when coating has no effect on corrosion, so that a statistical test of significance can easily be constructed.

This test of significance is called the sign test and can be carried out formally as follows: Examine the difference $D_i = X_i - Y_i$ for each paired observation. Record a plus $(+)$ or minus $(-)$ for each pair according to whether D_i is positive or negative, respectively. Denote by R the number of times the *less frequent sign* occurs. If the value that the random variable R takes on is less than or equal to a "critical value," the hypothesis that X and Y have the same distribution is rejected,[1] e.g., in the corrosion experiment this would imply the existence of a difference due to coating. The critical values of R for the 10%, 5%, 2% and 1% levels of significance are

---

[1] Actually, if the sign test is to be used to test this hypothesis, it must also be assumed that the underlying distributions of each pair of observations may differ only in their means.

given in Table 7.4. Strictly speaking, the hypothesis to be tested is that each difference between the paired observations has a probability distribution (which need not be the same for all differences) with median equal to zero, i.e., $P\{X_i - Y_i > 0\} = \frac{1}{2}$ for all i. If the underlying distributions of the individual observations are symmetric, the sign test can be used to test the hypothesis that $\mu_{x_i} = \mu_{y_i}$ or equivalently, the mean of the difference, μ_D, equals zero. If it is assumed that the underlying distributions of each pair of observations may differ only in their means, the sign test can also be used to test the hypothesis that $\mu_{x_i} = \mu_{y_i}$, this being equivalent to testing the hypothesis that the probability distributions of each pair are the same. Thus, if R, the number of times the less frequent sign occurs, is less than or equal

Table 7.4. Critical Values for R (the Number of Times that the Less Frequent Sign Occurs) for the Sign Test [1]

n \ α	10%	5%	2%	1%	n \ α	10%	5%	2%	1%
5	0	–	–	–	25	7	7	6	5
6	0	0	–	–	26	8	7	6	6
7	0	0	0	–	27	8	7	7	6
8	1	0	0	0	28	9	8	7	6
9	1	1	0	0	29	9	8	7	7
10	1	1	0	0	30	10	9	8	7
11	2	1	1	0	31	10	9	8	7
12	2	2	1	1	32	10	9	8	8
13	3	2	1	1	33	11	10	9	8
14	3	2	2	1	34	11	10	9	9
15	3	3	2	2	35	12	11	10	9
16	4	3	2	2	36	12	11	10	9
17	4	4	3	2	37	13	12	10	10
18	5	4	3	3	38	13	12	11	10
19	5	4	4	3	39	13	12	11	11
20	5	5	4	3	40	14	13	12	11
21	6	5	4	4	41	14	13	12	11
22	6	5	5	4	42	15	14	13	12
23	7	6	5	4	43	15	14	13	12
24	7	6	5	5	44	16	15	13	13
					45	16	15	14	13
					46	16	15	14	13
					47	17	16	15	14
					48	17	16	15	14
					49	18	17	15	15

For $n \geq 50$, R is approximately normally distributed with mean $n/2$ and variance $n/4$.

[1] This table was prepared from entries appearing in *Tables of the Binomial Probability Distribution*, National Bureau of Standards Applied Mathematics Series, 6, 1950. U. S. Government Printing Office.

to the critical value given in the table, the hypothesis that $P\{X_i - Y_i > 0\}$ $= \frac{1}{2}$, or $\mu_{x_i} = \mu_{y_i}$ when applicable, is rejected, and it is concluded that there are treatment effects. Table 7.4 is easily constructed because it is based on the binomial distribution with parameters equal to n and $p = \frac{1}{2}$. Since the sign test is essentially based on the number of times a plus (or minus) appears in n trials, it is equivalent to observing the number of successes in n independent Bernoulli trials. When the hypothesis is true, this statistic has a binomial distribution with parameters n and $p = \frac{1}{2}$. Because of the discreteness of the binomial distribution, the critical values of R given in Table 7.4 do not correspond exactly to the stated levels of significance. The test, in fact, is stricter than the level indicated.

It is interesting to note the results of applying the sign test to the data presented in Sec. 7.4.2. The number of negative differences (the less frequent sign) is equal to 4; i.e., $R = 4$. From Table 7.4, it is seen that, for $n = 15$, the critical value at the 5% level is 3 and it is concluded, using the sign test, that there is no difference in the coating. This apparent contradiction to the results of the paired t test for equality of means is due to the fact that the sign test is not as sensitive a test, since it is the sign and not the magnitude of the difference which is taken into account.

This raises the interesting question about how good a "test" is the sign test compared to competitors. There is no answer to this question without specifying the underlying distributions, and even if the underlying distributions are normal, so that the t test is applicable, suitable simple measures are not readily available. A somewhat useful measure called the Pitman asymptotic efficiency of one decision procedure relative to another can be interpreted loosely as the limit of the ratio of the sample sizes required to achieve the same OC curve as the sample sizes get large. For the sign test relative to the t test (when applicable) the relative asymptotic efficiency for shift in mean alternatives is equal to $2/\pi$, which is surprisingly good, considering that the sign test does not make use of the magnitude of the differences; only the signs.

The sign test has the important advantage of being simple to perform and is useful even when data are qualitative in nature. Moreover, this test does not require X and Y to be independent random variables; it requires only that $P\{X_i - Y_i > 0\}$ be constant for each i.

Because of the useful properties that the sign test possesses, practitioners are often interested in applying this test when large quantities of data are available. It is therefore useful to indicate that Table 7.4 can be extended for $n \geq 50$ by noting that R is asymptotically normally distributed with mean $n/2$ and standard deviation $\sqrt{n}/2$ when the hypothesis is true (see Sec. 4.5.3 and the relevant remarks concerning corrections for continuity).

Although ties in the data, i.e., $D_i = 0$, should not appear because the random variables are assumed to be continuous, the discreteness of measuring instruments often causes such an occurrence. There are several methods for dealing with these ties, although it appears that the best way to treat this phenomenon is to disregard these ties, thereby applying the test to the remaining data. Another acceptable alternative is to count a tie as half a plus and half a minus.

Table 7.4 can also be used for one-sided procedures. Let R denote either the number of plus signs or the number of minus signs, depending on which of the two is expected to appear least frequently according to the alternative to the hypothesis. In other words, if the number of minus signs is expected to be less than the number of plus signs, let R represent the number of minus signs. If R is less than or equal to the critical value in the table, the hypothesis is rejected at a level of significance equal to half the value ($\frac{1}{2}\%$, 1%, $2\frac{1}{2}\%$, or 5%) given in the table heading.

7.5.2 THE WILCOXON SIGNED RANK TEST

The sign test in the previous section is simple to apply but takes no account of the magnitude of the differences whatever. A better non-parametric test can be based on the signed rank of the differences; i.e., differences are first ranked without regard to sign and then these ranks are given the corresponding sign of the difference. The hypothesis to be tested in this problem is the same as for the sign test, i.e., $P\{X_i - Y_i > 0\} = \frac{1}{2}$ or $\mu_{x_i} = \mu_{y_i}$ for all i, provided the required assumptions that the distributions are symmetric or differ only in their means are fulfilled.

For the corrosion data, this ranking would be as follows:

Difference	Rank	Signed Rank
22	13	13
2	$2\frac{1}{2}$	$2\frac{1}{2}$
4	$4\frac{1}{2}$	$4\frac{1}{2}$
12	11	11
11	10	10
15	12	12
28	15	15
-5	6	-6
8	8	8
4	$4\frac{1}{2}$	$4\frac{1}{2}$
-1	1	-1
-10	9	-9
-2	$2\frac{1}{2}$	$-2\frac{1}{2}$
25	14	14
7	7	7

Table 7.5. Significance Points for the Absolute Value of the Smaller Sum of Signed Ranks (T) Obtained from Paired Observations[1]

n \ α	10%	5%	2%	1%
4				
5	0			
6	2	0		
7	3	2	0	
8	5	3	1	0
9	8	5	3	1
10	10	8	5	3
11	13	10	7	5
12	17	13	9	7
13	21	17	12	9
14	25	21	15	12
15	30	25	19	15
16	35	29	23	19
17	41	34	27	23
18	47	40	32	27
19	53	46	37	32
20	60	52	43	37
21	67	58	49	42
22	75	65	55	48
23	83	73	62	54
24	91	81	69	61
25	100	89	76	68
26	110	98	84	75
27	119	107	92	83
28	130	116	101	91
29	140	126	110	100
30	151	137	120	109
31	163	147	130	118
32	175	159	140	128
33	187	170	151	138
34	200	182	162	148
35	213	195	173	159
36	227	208	185	171
37	241	221	198	182
38	256	235	211	194
39	271	249	224	207
40	286	264	238	220
41	302	279	252	233
42	319	294	266	247
43	336	310	281	261
44	353	327	296	276
45	371	343	312	291
46	389	361	328	307
47	407	378	345	322
48	426	396	362	339
49	446	415	379	355
50	466	434	397	373

For $n > 50$, T is approximately normally distributed with mean $n(n + 1)/4$ and variance $n(n + 1)(2n + 1)/24$. A continuity correction of $-\frac{1}{2}$ can be added to the mean, if desired.

[1] Abridged and reproduced by permission from "Extended Tables of the Wilcoxon Matched Pair Signed Rank Statistic" by Robert L. McCornack, *Journal of the American Statistical Association*, Vol. 60, Sept., 1965.

The smallest observation is given rank 1 and the ties are assigned average ranks. Affix to each rank the sign of the difference.

If there is no difference between coatings, one would expect the sum of the positive ranks approximately to equal numerically the sum of the negative ranks. Since

$$1 + 2 + 3 + \cdots + n = \frac{n(n+1)}{2},$$

the absolute value of the sum of either the positive or negative signed ranks should be near $n(n+1)/4$ or about 60 for these data. Actually, the sum of negative ranks is $18\frac{1}{2}$ and the sum of positive ranks is $101\frac{1}{2}$. To see whether such a difference could arise by chance or whether it indicates a significant difference, the absolute value of the smaller sum of ranks (T) is compared to a value in Table 7.5. If T is less than or equal to the critical value given in the table, the hypothesis is rejected. In this case, $T = 18\frac{1}{2}$ is less than 25, which is the critical value corresponding to a 5% level of significance for $n = 15$, so that the hypothesis is rejected at the 5% level of significance, and it is concluded that there is a difference due to coating.

Table 7.5 can also be used for one-sided procedures. Let T denote either the sum of the positive ranks or the absolute value of the sum of the negative ranks, depending on which of the two is expected to be smaller according to the alternative to the hypothesis. In other words, if the sum of the positive ranks is expected to be less than the absolute value of the negative ranks, let T represent the sum of the positive ranks. If T is less than or equal to the critical value in the table, the hypothesis is rejected at a level of significance equal to half the value $(5\%, 2\frac{1}{2}\%, 1\%, \frac{1}{2}\%)$ given in the table heading.

The signed rank test possesses almost all the advantages of the sign test, and appears to utilize the data more efficiently. This is reflected in a value of approximately 0.95 for the asymptotic efficiency of the signed rank test relative to the t test (when applicable) for "shift in mean" alternatives. Therefore, this test is often appealing to many practitioners. Table 7.5 can be extended for large n by noting that T is asymptotically normally distributed with mean $n(n+1)/4$ and variance $n(n+1)(2n+1)/24$ when the hypothesis is true.

7.5.3 WILCOXON TEST FOR TWO INDEPENDENT SAMPLES

The sign test and the signed rank test are applicable to paired samples. When observations are not paired and the two probability distributions are assumed to differ only in their means, a useful procedure for testing the hypothesis that the means are the same is: First arrange all observations from both samples in order of magnitude and rank them. For the endurance limit example of Sec. 7.1.4, the data would appear as follows:

Sample	Endurance Limit	Rank
2	69,800	1
1	78,700	2
2	79,400	3
2	79,500	4
2	79,900	5
2	81,700	6
2	82,400	7
2	82,600	8
1	83,100	9
2	83,400	10
1	84,000	11
1	86,500	12
1	87,500	13
1	89,400	14
1	89,900	15
1	91,900	16

If there is no difference between samples, one would expect them to inter-mingle in a regular way and, if they are both of the same size, we would expect the sum of ranks to be about the same for both. In this case the ranks in sample 2 are 1, 3, 4, 5, 6, 7, 8, and 10, which total 44; the ranks in sample 1 total 92.

To see whether the difference is significant, Table 7.6 is entered with

$$R_1 = \text{sum of ranks of smaller sample}$$

and
$$R'_1 = n_1(n_1 + n_2 + 1) - R_1$$

and rejection is called for if either is less than or equal to the tabled critical value, R_1^{++}, where n_1 is the size of sample 1 and n_2 the size of sample 2. If the samples are unequal in size, sample 1 is to be the smaller; thus $n_1 \leq n_2$. In this case $R_1 = 44$ and $R'_1 = 92$. The 5% point for $n_1 = 8$ and $n_2 = 8$ is 49. Hence, the departure is significant at the 5% level (since $R_1 < 49$).

In the case of ties, average ranks are assigned. Table 7.6 can also be used for one-sided procedures. If under the alternative to the hypothesis, R_1 should be large, reject if R'_1 is less than or equal to the critical value given in the table. If under the alternative, R_1 should be small, reject if R_1 is less than or equal to the critical value. The level of significance equals half the values ($2\frac{1}{2}\%$ and $\frac{1}{2}\%$) given in the table heading.

The Wilcoxon test for two independent samples has an asymptotic efficiency relative to the two-sample t test (when applicable) for "shift in mean" alternatives of approximately 0.95. Therefore, this test is also appealing to practitioners. Table 7.6 can be extended for large n_1 and n_2 by noting that R_1 is approximately normally distributed with mean $\frac{1}{2}n_1(n_1 + n_2 + 1)$ and variance $\frac{1}{12}n_1 n_2(n_1 + n_2 + 1)$ when the distributions of X and Y are the same.

Table 7.6. 5% Critical Points of Rank Sums (R_1^+)[1]

n_2 \ $n_1 \to$	2	3	4	5	6	7	8	9	10	11	12	13	14	15
4			10											
5		6	11	17										
6		7	12	18	26									
7		7	13	20	27	36								
8	3	8	14	21	29	38	49							
9	3	8	15	22	31	40	51	63						
10	3	9	15	23	32	42	53	65	78					
11	4	9	16	24	34	44	55	68	81	96				
12	4	10	17	26	35	46	58	71	85	99	115			
13	4	10	18	27	37	48	60	73	88	103	119	137		
14	4	11	19	28	38	50	63	76	91	106	123	141	160	
15	4	11	20	29	40	52	65	79	94	110	127	145	164	185
16	4	12	21	31	42	54	67	82	97	114	131	150	169	
17	5	12	21	32	43	56	70	84	100	117	135	154		
18	5	13	22	33	45	58	72	87	103	121	139			
19	5	13	23	34	46	60	74	90	107	124				
20	5	14	24	35	48	62	77	93	110					
21	6	14	25	37	50	64	79	95						
22	6	15	26	38	51	66	82							
23	6	15	27	39	53	68								
24	6	16	28	40	55									
25	6	16	28	42										
26	7	17	29											
27	7	17												
28	7													

For large n_1 and n_2, R_1 is approximately normally distributed with mean $\frac{1}{2}n_1(n_1 + n_2 + 1)$ and variance $\frac{1}{12}n_1 n_2(n_1 + n_2 + 1)$. A continuity correction of $-\frac{1}{2}$ can be added to the mean, if desired.

[1] Reprinted by permission from Colin White, "The Use of Ranks in a Test of Significance for Comparing Two Treatments," *Biometrics*, 1952, Vol. 8, p. 37.

Table 7.6 (continued). 1% Critical Points of Rank Sums (R_1^{++})

n_2 \ $n_1 \rightarrow$	2	3	4	5	6	7	8	9	10	11	12	13	14	15
5				15										
6			10	16	23									
7			10	17	24	32								
8			11	17	25	34	43							
9		6	11	18	26	35	45	56						
10		6	12	19	27	37	47	58	71					
11		6	12	20	28	38	49	61	74	87				
12		7	13	21	30	40	51	63	76	90	106			
13		7	14	22	31	41	53	65	79	93	109	125		
14		7	14	22	32	43	54	67	81	96	112	129	147	
15		8	15	23	33	44	56	70	84	99	115	133	151	171
16		8	15	24	34	46	58	72	86	102	119	137	155	
17		8	16	25	36	47	60	74	89	105	122	140		
18		8	16	26	37	49	62	76	92	108	125			
19	3	9	17	27	38	50	64	78	94	111				
20	3	9	18	28	39	52	66	81	97					
21	3	9	18	29	40	53	68	83						
22	3	10	19	29	42	55	70							
23	3	10	19	30	43	57								
24	3	10	20	31	44									
25	3	11	20	32										
26	3	11	21											
27	4	11												
28	4													

7.6 Test of the Hypothesis that the Standard Deviations of Two Normal Distributions Are Equal

7.6.1 CHOICE OF AN OC CURVE

It frequently is of interest to test the hypothesis that the standard deviations of two normal distributions are equal.[1] This test may be preliminary to a two-sample t test or may arise in circumstances in which the mean can be set but the variability cannot, e.g., in machine settings. The test procedure is based on an F statistic. This test statistic is a function of the ratio of sample variances. The abscissae of operating characteristic curves for procedures

---

[1] This is equivalent to testing the hypothesis that the variances of two normal distributions are equal.

based on this F statistic depend on the quantity $\lambda = \sigma_x/\sigma_y$, the true ratio of the standard deviations.

To determine the acceptance region and sample size, two points on the OC curve must be chosen. These points will be denoted by $(1, 1 - \alpha)$ and $((\sigma_x/\sigma_y), \beta)$. The first point corresponds to choosing the probability of accepting the hypothesis when it is true (when $\sigma_x = \sigma_y$), and the second point corresponds to choosing the probability of accepting the hypothesis when σ_x/σ_y is the ratio of the true standard deviations considered important to detect.

7.6.2 CHARTS AND TABLES FOR CARRYING OUT F TESTS

An experiment is to be performed and random samples of size n_x and size n_y are to be drawn. Each of the resulting sequence of random variables, $X_1, X_2, \ldots, X_{n_x}$, can be thought of as representing the measurable quantity associated with a trial or test of the "x characteristic," whereas each of the random variables $Y_1, Y_2, \ldots, Y_{n_y}$ can be thought of as representing the measurable quantity associated with a trial or test of the "y characteristic." It is assumed that the X_i are independent normally distributed random variables, each with unknown mean μ_x and unknown standard deviation σ_x, and that the Y_i are independent normally distributed random variables, each with unknown mean μ_y and unknown standard deviation σ_y. Furthermore, let all the X's and Y's be independent. The optimum procedure for testing the hypothesis that $\sigma_x = \sigma_y$ is based on the random variable

$$F = \frac{S_x^2}{S_y^2}$$

where
$$S_x^2 = \frac{\sum_{i=1}^{n_x} (X_i - \bar{X})^2}{n_x - 1} \quad \text{and} \quad S_y^2 = \frac{\sum_{i=1}^{n_y} (Y_i - \bar{Y})^2}{n_y - 1} .$$

If the value taken on by the test statistic as a result of experimentation falls into the acceptance region, the hypothesis that $\sigma_x = \sigma_y$ is accepted; otherwise, it is rejected. This section is devoted to finding the acceptance region and the sample size for the test statistic F, using the tables and charts which are provided. One- and two-sided procedures are considered.

7.6.2.1 *Tables and Charts for Two-Sided Procedures.* The acceptance region for the two-sided procedure for any sample sizes, n_x and n_y, is given by the interval $[F_{1-\alpha/2; n_x-1, n_y-1}, F_{\alpha/2; n_x-1, n_y-1}]$, which is equivalent (see Sec. 4.4.1) to the interval $[1/F_{\alpha/2; n_y-1, n_x-1}, F_{\alpha/2; n_x-1, n_y-1}]$, where α is the level of significance and $F_{\alpha/2; v_1, v_2}$ is the $100\,\alpha/2$ percentage point of the F distribution with v_1 and v_2 degrees of freedom.[1] Thus, the hypothesis that $\sigma_x = \sigma_y$ is accepted if

$$1/F_{\alpha/2; n_y-1, n_x-1} \leqq F \leqq F_{\alpha/2; n_x-1, n_y-1}$$

---

[1] If the notation X is assigned to the data so that S_x^2 denotes the larger sample variance, the acceptance region is also given by the interval $[-\infty, F_{\alpha/2; n_x-1, n_y-1}]$.

and rejected if F lies outside this interval. It is easily verified that this procedure yields a probability of $1 - \alpha$ of accepting the hypothesis that $\sigma_x = \sigma_y$ when it is true. In Sec. 4.4.2, it was shown that whenever $\sigma_x = \sigma_y$, the test statistic $F = S_x^2/S_y^2$ has an F distribution with $n_x - 1$ and $n_y - 1$ degrees of freedom. Hence, if $\sigma_x = \sigma_y$,

$$P\{1/F_{\alpha/2;\,n_y-1,\,n_x-1} \leq F \leq F_{\alpha/2;\,n_x-1,\,n_y-1}\} = 1 - \alpha.$$

The required sample size is obtained by referring to a set of operating characteristic curves which are provided for this purpose. Figure 7.1 presents the OC curves of the two-sided F test for a level of significance equal to 0.05. Figure 7.2 presents OC curves of the two-sided F test for a level of significance equal to 0.01. To use these curves it is necessary to choose $n_x = n_y = n$. If the level of significance of the procedure is either 0.05 or 0.01, the required sample size n is obtained by entering the appropriate figure (Fig. 7.1 if $\alpha = 0.05$ and Fig. 7.2 if $\alpha = 0.01$) with the point $((\sigma_x/\sigma_y), \beta)$ and reading out the value of the sample size associated with the OC curve that passes through this point.

7.6.2.2 *Summary for Two-Sided Procedures using Tables and Charts.* The following steps are taken, using a two-sided procedure for testing the hypothesis that $\sigma_x = \sigma_y$:

1. Two points on the OC curve are chosen. These points are denoted by $(1, 1 - \alpha)$ and $((\sigma_x/\sigma_y), \beta)$.
2. If $n_x = n_y = n$, the required sample size, n, is found by entering the appropriate figure (Fig. 7.1 if $\alpha = 0.05$ and Fig. 7.2 if $\alpha = 0.01$) with

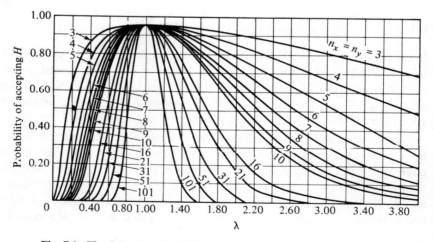

Fig. 7.1. *The OC curves for different values of n for the two-sided F test for a level of significance* $\alpha = 0.05$.

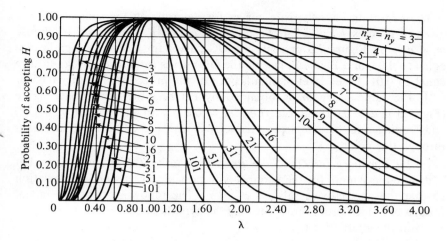

Fig. 7.2. *The OC curves for different values of n for the two-sided F test for a level of significance* α = 0.01.

the point $((\sigma_x/\sigma_y), \beta)$ and reading out the value of the sample size associated with the OC curve that passes through this point.

3. Determine the acceptance region which is the interval

$$[1/F_{\alpha/2;n_y-1,n_x-1}, F_{\alpha/2;n_x-1,n_y-1}].$$

4. Draw a random sample of n_x and n_y items.

5. Compute the value of the test statistic, $F = S_x^2/S_y^2$.

6. If $1/F_{\alpha/2;n_y-1,n_x-1} \leq F \leq F_{\alpha/2;n_x-1,n_y-1}$, accept the hypothesis and conclude that $\sigma_x = \sigma_y$.

7.6.2.3 *Tables and Charts for One-Sided Procedures.* The notation X and Y is arbitrary in a one-sided test. Let X denote the symbol for the variable with possible larger standard deviation so that the alternative is always $\sigma_x > \sigma_y$. The acceptance region with any sample sizes, n_x and n_y, for testing the hypothesis that $\sigma_x = \sigma_y$ (or $\sigma_x \leq \sigma_y$) is given by the interval $[-\infty, F_{\alpha;n_x-1,n_y-1}]$, where α is the level of significance and $F_{\alpha;n_x-1,n_y-1}$ is the 100 α percentage point of the F distribution with $n_x - 1$ and $n_y - 1$ degrees of freedom. Thus, the hypothesis that $\sigma_x = \sigma_y$ (or $\sigma_x \leq \sigma_y$) is accepted if

$$F \leq F_{\alpha;n_x-1,n_y-1}$$

and rejected if F lies outside this interval. This acceptance procedure has a probability of $1 - \alpha$ of accepting the hypothesis that $\sigma_x = \sigma_y$ when it is true. Whenever $\sigma_x = \sigma_y$, the test statistic $F = S_x^2/S_y^2$ has an F distribution with $n_x - 1$ and $n_y - 1$ degrees of freedom. Hence, if $\sigma_x = \sigma_y$,

$$P(F \leq F_{\alpha;n_x-1,n_y-1}) = 1 - \alpha.$$

The operating characteristic curves for determining the sample size are given in Figures 7.3 and 7.4. To use these curves it is necessary to choose $n_x = n_y = n$. Figure 7.3 presents the OC curves for a level of significance equal to 0.05, and Fig. 7.4 presents the OC curves for a level of significance equal to 0.01. If the level of significance of the procedure is either 0.05 or 0.01, the sample size, n, is obtained by entering the appropriate figure (Fig. 7.3 if $\alpha = 0.05$ and Fig. 7.4 if $\alpha = 0.01$) with the point $((\sigma_x/\sigma_y), \beta)$ and reading out the sample size associated with the OC curve that passes through this point.

7.6.2.4 *Summary for One-Sided Procedures using Tables and Charts.* Let X denote the symbol for the variable with possible larger standard deviation so that the alternative is always $\sigma_x > \sigma_y$. The following steps are taken, using a one-sided procedure for testing the hypothesis that $\sigma_x = \sigma_y$ (or $\sigma_x \leq \sigma_y$) with the alternative that $\sigma_x > \sigma_y$:

1. Two points on the OC curve are chosen. These points are denoted by $(1, 1 - \alpha)$ and $((\sigma_x/\sigma_y), \beta)$.
2. If $n_x = n_y = n$, the required sample size, n, is found by entering the appropriate figure (Fig. 7.3 if $\alpha = 0.05$ and Fig. 7.4 if $\alpha = 0.01$) with the point $((\sigma_x/\sigma_y), \beta)$ and reading out the value of the sample size associated with the OC curve that passes through this point.
3. Determine the acceptance region which is the interval

$$[-\infty, F_{\alpha; n_x - 1, n_y - 1}].$$

4. Draw a random sample of n_x and n_y items.
5. Compute the value of the test statistic, $F = S_x^2/S_y^2$.
6. If $F \leq F_{\alpha; n_x - 1, n_y - 1}$, accept the hypothesis that $\sigma_x = \sigma_y$ (or $\sigma_x \leq \sigma_y$). If F lies outside this interval, reject the hypothesis and conclude that $\sigma_x > \sigma_y$.

7.6.2.5 *Tables and Charts for Operating Characteristic Curves.* If $n_x = n_y = n$ and this sample size is fixed in advance or after it is chosen, the operating characteristic curves given in Figs. 7.1 and 7.2 for two-sided procedures, and Figs. 7.3 and 7.4 for one-sided procedures can be used to evaluate the risks involved in using these procedures. If the level of significance is either 0.05 or 0.01, the OC curve associated with the sample size n is entered with λ, and the probability of accepting the hypothesis is read from the curve. The value of λ chosen corresponds to the value of σ_x/σ_y for which the risk is to be evaluated.

The F statistic for testing the hypothesis that $\sigma_x = \sigma_y$ is relatively insensitive. If we want to detect a λ of 2 with probability 0.10 (and $\alpha = 0.05$), Fig. 7.1 reveals that about 26 observations are necessary; a λ of 1.6 requires 50 observations. To detect a 10 or 20% change in σ requires hundreds of observations.

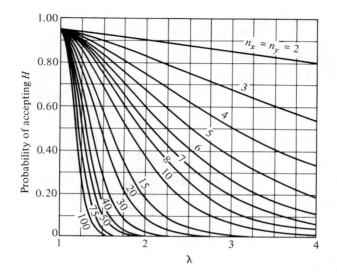

Fig. 7.3. *The OC curves for different values of n for the one-sided F test for a level of significance* $\alpha = 0.05$.

(Reproduced by permission from "Operating Characteristics for the Common Statistical Tests of Significance" by Charles L. Ferris, Frank E. Grubbs, Chalmers L. Weaver, *Annals of Mathematical Statistics*, June, 1946.)

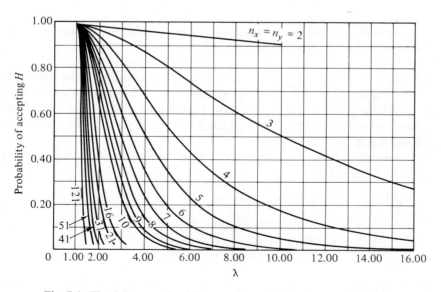

Fig. 7.4. *The OC curves for different values of n for the one-sided F test for a level of significance* $\alpha = 0.01$.

**Table 7.7. Test of the Hypothesis that the Standard Deviations of
Two Normal Distributions Are Equal**

Notation for the Hypothesis	*Test Statistic*
$H: \sigma_x^2 = \sigma_y^2$ or $\sigma_x = \sigma_y$	$F = \dfrac{\sum\limits_{i=1}^{n_x} (X_i - \bar{X})^2/(n_x - 1)}{\sum\limits_{i=1}^{n_y} (Y_i - \bar{Y})^2/(n_y - 1)} = \dfrac{S_x^2}{S_y^2}$

Criteria for Rejection	*Method for Choosing Sample Sizes*
$F < 1/F_{\alpha/2;\, n_y-1,\, n_x-1}$ or $F > F_{\alpha/2;\, n_x-1,\, n_y-1}$ if rejection is desired when σ_x is not equal to σ_y.	The OC curve depends on $\lambda = \sigma_x/\sigma_y$ and is given in Figs. 7.1 and 7.2 for the case $n_x = n_y = n$. Choose a value of λ for which rejection of the hypothesis is desired with given high probability. Enter Fig. 7.1 or 7.2 to find the required sample size.
The notation X and Y is arbitrary in a one-sided test; let X be the symbol for the variable with possible larger variance. $F > F_{\alpha;\, n_x-1,\, n_y-1}$ if rejection is desired when $\sigma_x > \sigma_y$.	The OC curve depends on $\lambda = \sigma_x/\sigma_y$ and is given in Figs. 7.3 and 7.4 for the case $n_x = n_y = n$. OC curves for $n_x \neq n_y$ are given in Figs. 10.10—10.17 for $n_x - 1 = \nu_1 = 1, 2, \ldots, 8$ and various values of $n_y - 1 = \nu_2$. Choose a value of λ for which rejection of the hypothesis is desired with given high probability. Enter the appropriate figure to find the required sample size.

Values of $F_{\alpha;\, \nu_1,\, \nu_2}$, the $100\,\alpha$ percentage points of the F distribution with ν_1 and ν_2 degrees of freedom, are given in Appendix Table 4.

If levels of significance other than 0.05 or 0.01 are desired, Appendix Table 4, which is a table of the percentage points of the F distribution, can be used to determine the acceptance region for fixed values of n_x and n_y. However, analytical methods which are presented in Sec. 7.6.3 must be used to obtain the OC curve.

A summary of the procedures for testing the hypothesis that $\sigma_x = \sigma_y$ is given in Table 7.7.

*7.6.3 ANALYTICAL TREATMENT FOR TESTS

In Sec. 4.4.2, it was pointed out that the quantity

$$\frac{S_x^2/\sigma_x^2}{S_y^2/\sigma_y^2} = \frac{S_x^2}{S_y^2} \cdot \frac{\sigma_y^2}{\sigma_x^2}$$

has an F distribution with $n_x - 1$ and $n_y - 1$ degrees of freedom. If one is interested in testing the hypothesis that $\sigma_x = \sigma_y$ against the alternative that $\sigma_x > \sigma_y$, the test statistic S_x^2/S_y^2 has an F distribution with $n_x - 1$ and $n_y - 1$

degrees of freedom whenever $\sigma_x = \sigma_y$. If a test with a Type I error equal to α is desired, the rule of acceptance is to accept the hypothesis if

$$\frac{S_x^2}{S_y^2} \leq F_{\alpha;\,n_x-1,\,n_y-1}.$$

It is clear that whenever $\sigma_x = \sigma_y$,

$$P\left\{\frac{S_x^2}{S_y^2} \leq F_{\alpha;\,n_x-1,\,n_y-1}\right\} = 1 - \alpha.$$

The OC curve of the test can be computed easily from a table of the F distribution. Suppose that the true value of the ratio of the standard deviations is $\lambda = \sigma_x/\sigma_y$. Then,

$$\frac{S_x^2}{S_y^2} \cdot \frac{\sigma_y^2}{\sigma_x^2} = \frac{S_x^2}{S_y^2} \cdot \frac{1}{\lambda^2}$$

has an F distribution with $n_x - 1$ and $n_y - 1$ degrees of freedom; and the probability of accepting the hypothesis is given by

$$P\left\{\frac{S_x^2}{S_y^2} \leq F_{\alpha;\,n_x-1,\,n_y-1}\right\} = P\left\{\frac{S_x^2}{S_y^2} \cdot \frac{1}{\lambda^2} \leq \frac{F_{\alpha;\,n_x-1,\,n_y-1}}{\lambda^2}\right\}$$

$$= P\left\{F \leq \frac{F_{\alpha;\,n_x-1,\,n_y-1}}{\lambda^2}\right\}$$

where F now denotes any random variable having an F distribution with $n_x - 1$ and $n_y - 1$ degrees of freedom.[1] The OC curve is given by

$$P\left\{F \leq \frac{F_{\alpha;\,n_x-1,\,n_y-1}}{\lambda^2}\right\}.$$

The expression in this form permits easy solution of the sample size problem. Suppose that a particular value of λ, say, λ_1, is given, for which acceptance is desired only with probability β. This condition implies that

$$P\left\{\frac{S_x^2}{S_y^2} \leq F_{\alpha;\,n_x-1,\,n_y-1}\right\} = P\left\{F \leq \frac{F_{\alpha;\,n_x-1,\,n_y-1}}{\lambda_1^2}\right\} = \beta,$$

where F is a random variable having an F distribution and β is the Type II error associated with λ_1. It follows from this last expression that

$$\frac{F_{\alpha;\,n_x-1,\,n_y-1}}{\lambda_1^2} = F_{1-\beta;\,n_x-1,\,n_y-1} = \frac{1}{F_{\beta;\,n_y-1,\,n_x-1}}$$

or

$$\lambda_1^2 = (F_{\alpha;\,n_x-1,\,n_y-1})(F_{\beta;\,n_y-1,\,n_x-1}).$$

If $n_x = n_y = n$, this reduces to

$$\lambda_1^2 = (F_{\alpha;\,n-1,\,n-1})(F_{\beta;\,n-1,\,n-1}).$$

---

[1] In the immediately preceding sections, F denoted the quantity S_x^2/S_y^2, which has an F distribution only when $\sigma_x = \sigma_y$. In the remainder of this section (Sec. 7.6.3), F denotes a random variable which has an F distribution with $n_x - 1$ and $n_y - 1$ degrees of freedom, and must not be confused with the quantity S_x^2/S_y^2.

The sample size for this problem can be determined by going to the table of the F distribution, appendix Table 4, and finding the degrees of freedom for which the product of the 100 α percentage point and the 100 β percentage point is equal to λ_1^2. The sample size n is one more than the degrees of freedom. For example, if $\alpha = 0.05$, $\beta = 0.10$, and $\lambda_1^2 = 3.0$, Appendix Table 4 indicates that for 30 degrees of freedom, $F_{0.05;30,30} = 1.84$ and $F_{0.10;30,30} = 1.61$ and $\lambda_1^2 = (1.84)(1.61) = 3.0$. Hence, the required sample size $n_x = n_y = n = 31$. Similar analytic treatment can be provided for two-sided procedures.

For a two-sided procedure, the expression for the OC curve (probability of accepting the hypothesis) is given by

$$P\left\{\frac{1}{(F_{\alpha/2;n_y-1,n_x-1})\lambda^2} \leq F \leq \frac{F_{\alpha/2;n_x-1,n_y-1}}{\lambda^2}\right\},$$

where F denotes a random variable having an F distribution and $\lambda = \sigma_x/\sigma_y$. When $n_x = n_y = n$, the expression required for determining the sample size is given approximately by

$$\lambda_1^2 \cong (F_{\alpha/2;n-1,n-1})(F_{\beta;n-1,n-1})$$

when $P\{F \leq 1/(F_{\alpha/2;n-1,n-1})\lambda_1^2\}$ is small compared to β, and by

$$\lambda_1^2 \cong 1/[(F_{\alpha/2;n-1,n-1})(F_{\beta;n-1,n-1})]$$

when $P\{F \geq (F_{\alpha/2;n-1,n-1}/\lambda_1^2)\}$ is small compared to β.

7.6.4 EXAMPLE

The standard deviation of a particular dimension of a metal component is small enough so that it is satisfactory in subsequent assembly; a new supplier of metal plate is under consideration and will be preferred if the standard deviation of his product is not larger, because the cost of his product is lower than that of the present supplier. Hence, the hypothesis to be tested is that the standard deviation of the new supplier is less than or equal to the standard deviation of the old supplier, and the alternative is that the standard deviation of the new supplier exceeds that of the old supplier. A one-sided procedure is called for, and the symbol X is reserved for the new supplier. Without consulting an OC curve, it is decided to base the decision on a sample of 100 items from each supplier since data on dimensions are relatively easy to obtain and it is known that small numbers of observations give relatively little protection against erroneous decisions. The test is to be run at the 5% level of significance. The following are computed:

New Supplier: $\quad S_x^2 = 0.00041$

Current Supplier: $\quad S_y^2 = 0.00057$

$$F = \frac{0.00041}{0.00057} = 0.72.$$

Table 7.8. Upper 1 Percentage Points of the Ratio of the Largest to the Sum of k Independent Estimates of Variance, Each of Which is Based on n Observations[1]

k \ n	2	3	4	5	6	7	8	9	10	11	17	37	145	∞
2	0.9999	0.9950	0.9794	0.9586	0.9373	0.9172	0.8988	0.8823	0.8674	0.8539	0.7949	0.7067	0.6062	0.5000
3	0.9933	0.9423	0.8831	0.8335	0.7933	0.7606	0.7335	0.7107	0.6912	0.6743	0.6059	0.5153	0.4230	0.3333
4	0.9676	0.8643	0.7814	0.7212	0.6761	0.6410	0.6129	0.5897	0.5702	0.5536	0.4884	0.4057	0.3251	0.2500
5	0.9279	0.7885	0.6957	0.6329	0.5875	0.5531	0.5259	0.5037	0.4854	0.4697	0.4094	0.3351	0.2644	0.2000
6	0.8828	0.7218	0.6258	0.5635	0.5195	0.4866	0.4608	0.4401	0.4229	0.4084	0.3529	0.2858	0.2229	0.1667
7	0.8376	0.6644	0.5685	0.5080	0.4659	0.4347	0.4105	0.3911	0.3751	0.3616	0.3105	0.2494	0.1929	0.1429
8	0.7945	0.6152	0.5209	0.4627	0.4226	0.3932	0.3704	0.3522	0.3373	0.3248	0.2779	0.2214	0.1700	0.1250
9	0.7544	0.5727	0.4810	0.4251	0.3870	0.3592	0.3378	0.3207	0.3067	0.2950	0.2514	0.1992	0.1521	0.1111
10	0.7175	0.5358	0.4469	0.3934	0.3572	0.3308	0.3106	0.2945	0.2813	0.2704	0.2297	0.1811	0.1376	0.1000
12	0.6528	0.4751	0.3919	0.3428	0.3099	0.2861	0.2680	0.2535	0.2419	0.2320	0.1961	0.1535	0.1157	0.0833
15	0.5747	0.4069	0.3317	0.2882	0.2593	0.2386	0.2228	0.2104	0.2002	0.1918	0.1612	0.1251	0.0934	0.0667
20	0.4799	0.3297	0.2654	0.2288	0.2048	0.1877	0.1748	0.1646	0.1567	0.1501	0.1248	0.0960	0.0709	0.0500
24	0.4247	0.2871	0.2295	0.1970	0.1759	0.1608	0.1495	0.1406	0.1338	0.1283	0.1060	0.0810	0.0595	0.0417
30	0.3632	0.2412	0.1913	0.1635	0.1454	0.1327	0.1232	0.1157	0.1100	0.1054	0.0867	0.0658	0.0480	0.0333
40	0.2940	0.1915	0.1508	0.1281	0.1135	0.1033	0.0957	0.0898	0.0853	0.0816	0.0668	0.0503	0.0363	0.0250
60	0.2151	0.1371	0.1069	0.0902	0.0796	0.0722	0.0668	0.0625	0.0594	0.0567	0.0461	0.0344	0.0245	0.0167
120	0.1225	0.0759	0.0585	0.0489	0.0429	0.0387	0.0357	0.0334	0.0316	0.0302	0.0242	0.0178	0.0125	0.0083
∞	0	0	0	0	0	0	0	0	0	0	0	0	0	0

[1] Reproduced from C. Eisenhart, M.W. Hastay, W. A. Wallis, *Techniques of Statistical Analysis*, Chapter 15, McGraw-Hill Book Company, Inc., New York, 1947.

Table 7.8. (continued). Upper 5 Percentage Points of the Ratio of the Largest to the Sum of k Independent Estimates of Variance, Each of Which is Based on n Observations

k \ n	2	3	4	5	6	7	8	9	10	11	17	37	145	∞
2	0.9985	0.9750	0.9392	0.9057	0.8772	0.8534	0.8332	0.8159	0.8010	0.7880	0.7341	0.6602	0.5813	0.5000
3	0.9669	0.8709	0.7977	0.7457	0.7071	0.6771	0.6530	0.6333	0.6167	0.6025	0.5466	0.4748	0.4031	0.3333
4	0.9065	0.7679	0.6841	0.6287	0.5895	0.5598	0.5365	0.5175	0.5017	0.4884	0.4366	0.3720	0.3093	0.2500
5	0.8412	0.6838	0.5981	0.5441	0.5065	0.4783	0.4564	0.4387	0.4241	0.4118	0.3645	0.3066	0.2513	0.2000
6	0.7808	0.6161	0.5321	0.4803	0.4447	0.4184	0.3980	0.3817	0.3682	0.3568	0.3135	0.2612	0.2119	0.1667
7	0.7271	0.5612	0.4800	0.4307	0.3974	0.3726	0.3535	0.3384	0.3259	0.3154	0.2756	0.2278	0.1833	0.1429
8	0.6798	0.5157	0.4377	0.3910	0.3595	0.3362	0.3185	0.3043	0.2926	0.2829	0.2462	0.2022	0.1616	0.1250
9	0.6385	0.4775	0.4027	0.3584	0.3286	0.3067	0.2901	0.2768	0.2659	0.2568	0.2226	0.1820	0.1446	0.1111
10	6.6020	0.4450	0.3733	0.3311	0.3029	0.2823	0.2666	0.2541	0.2439	0.2353	0.2032	0.1655	0.1308	0.1000
12	0.5410	0.3924	0.3264	0.2880	0.2624	0.2439	0.2299	0.2187	0.2098	0.2020	0.1737	0.1403	0.1100	0.0833
15	0.4709	0.3346	0.2758	0.2419	0.2195	0.2034	0.1911	0.1815	0.1736	0.1671	0.1429	0.1144	0.0889	0.0667
20	0.3894	0.2705	0.2205	0.1921	0.1735	0.1602	0.1501	0.1422	0.1357	0.1303	0.1108	0.0879	0.0675	0.0500
24	0.3434	0.2354	0.1907	0.1656	0.1493	0.1374	0.1286	0.1216	0.1160	0.1113	0.0942	0.0743	0.0567	0.0417
30	0.2929	0.1980	0.1593	0.1377	0.1237	0.1137	0.1061	0.1002	0.0958	0.0921	0.0771	0.0604	0.0457	0.0333
40	0.2370	0.1576	0.1259	0.1082	0.0968	0.0887	0.0827	0.0780	0.0745	0.0713	0.0595	0.0462	0.0347	0.0250
60	0.1737	0.1131	0.0895	0.0765	0.0682	0.0623	0.0583	0.0552	0.0520	0.0497	0.0411	0.0316	0.0234	0.0167
120	0.0998	0.0632	0.0495	0.0419	0.0371	0.0337	0.0312	0.0292	0.0279	0.0266	0.0218	0.0165	0.0120	0.0083
∞	0	0	0	0	0	0	0	0	0	0	0	0	0	0

Since the value of $F = 0.72$ is less than 1, it does not exeed $F_{0.05;99,99}$, and the hypothesis of equality of standard deviations is accepted.

It is interesting to examine the implications of this procedure with respect to the OC curve. From Fig. 7.3, it is seen that if $\lambda = 1.3$, the probability of accepting the hypothesis is approximately 0.10. Hence, since the experimental data led to acceptance of the hypothesis, it is safe to assume (with a chance of error smaller than 10%) $\lambda = \sigma_x/\sigma_y$ does not exceed 1.3. If ratios of standard deviations less than 1.3 are relatively unimportant and the above-mentioned risks are satisfactory, a sample of 100 observations from each supplier is sufficient.

7.7 Cochran's Test for the Homogeneity of Variances

It is often necessary to decide whether several variances are equal, i.e., whether $\sigma_1^2 = \sigma_2^2 = \cdots = \sigma_k^2$. The F test can be used for $k = 2$, but a different procedure is required for larger values of k. Such a procedure is given by Cochran's test. Let $S_1^2, S_2^2, \ldots, S_k^2$ be independent estimates of $\sigma_1^2, \sigma_2^2, \ldots, \sigma_k^2$, respectively, each based on n independent normally distributed random variables; and let

$$G = \frac{\text{largest } S^2}{S_1^2 + S_2^2 + \cdots + S_k^2}$$

be the ratio of the largest S^2 to their total. The hypothesis that $\sigma_1^2 = \sigma_2^2 = \cdots = \sigma_k^2$ is accepted if

$$G \leq g_\alpha$$

where g_α is given in Table 7.8 for levels of significance, α, equal to 0.05 and 0.01. This table is entered with n, the number of observations within each group, and k, the number of variances being considered.

PROBLEMS

In the following problems it may be assumed (unless otherwise stated) that independent random samples are drawn, and each random variable is approximately normally distributed, or alternatively, the test statistic formed from these random variables has the appropriate distribution to validate the OC curves.

1. Two machines, X and Y, are used for filling bottles and are supposed to fill with a net volume of 5.65 liters. The filling process for machine X has a standard deviation of 0.015 liter and the filling process for machine Y has a standard deviation of 0.016 liter. A question of whether or not the two machines are doing the same job has arisen. The control engineer claims that because of the relative agreement of standard deviations and the results of other measures, the two are both filling on the average with the same amount, whether or not this amount is the desired 5.65 liters. A random sample is taken from each machine. Compute the sample means and state whether or not you think the control engineer is right. Then test at the 5% level and compare your decisions. Assuming equal sample size, how many observations would be required to offer protection of $\beta = 0.05$ when the true difference is 0.03?

	5.63	5.61	5.65	5.65	5.62
X:	5.61	5.63	5.68	5.62	5.60
	5.68	5.64	5.66	5.61	5.68
	5.61	5.62	5.61	5.65	5.63

	5.68	5.65	5.59	5.64	5.66
	5.61	5.62	5.64	5.63	5.61
Y:	5.64	5.66	5.60	5.65	5.63
	5.67	5.64	5.60	5.60	5.65
	5.60	5.65	5.60	5.63	5.60

2. An experiment is performed to test the difference in effectiveness of two methods of cultivating wheat. Ten patches of ground are treated with shallow plowing and fifteen with deep plowing. The sample mean yield per acre of the first group is 44.3 bushels and the sample mean for the second group is 44.7. Assume that the standard deviation of the shallow planting is 0.6 bushel and for the deep is 0.8. Test for equality of treatment at the 1% level. Sketch the OC curve of this test.

3. A new type of ammunition was submitted for test. Twenty rounds of standard ammunition and ten rounds of test ammunition were fired in random sequence. The sample average barrel pressure of the standard was 41,900 psi and the sample average pressure for the test was 44,200 psi. The standard deviation for both the standard and test was known to be 2,050 psi. (a) Test the hypothesis that the average barrel pressures are the same at the 1% level. (b) What difference in the expected values of the barrel pressures will be detected with probability 0.90 using the procedure in (a)?

4. A manufacturer of wooden ladders claims that the side pieces for the ladders have greater shear (parallel to the grain of the wood) if the raw lumber is air dried rather than kiln dried. However, his operating expenses could be appreciably reduced by using kiln-dried lumber. Therefore, the manufacturer has decided to conduct a test at the 5% level to determine whether or not the air-dried lumber has greater shear than the kiln-dried. He is anxious to switch to kiln-dried lumber if the experimental results justify this action. However, he feels that he must continue to use air-dried raw material if the shear of the kiln-dried material is as much as 100 psi less than that of the air-dried, and he desires that the risk of making an incorrect switch in this situation not exceed 5%. From past experience with air-dried lumber he is willing to assume that the standard deviation of its shear is 45 psi. He believes that the standard deviation for the shear of kiln-dried material is the same. (a) Determine the appropriate test for the manufacturer to use. (b) Determine the sample size required. (c) Assuming the following data result from the test in (a), using the sample size found in (b), what decision would the manufacturer make?

$$\text{Air-dried } (psi) \qquad \text{Kiln-dried } (psi)$$
$$\bar{X} = 1,170 \qquad\qquad \bar{Y} = 1,105$$

5. In a certain experimental laboratory a method, X_1, for producing gasoline from a crude oil is being investigated. Before completing experimentation, a new method, X_2, is proposed. All other things being equal, it was decided to abandon X_1 in favor of X_2 only if the average yield of the latter was substantially greater. There has been insufficient time to ascertain the true standard deviations of the yield of both processes, although there appears to be no reason why they cannot be assumed equal. Cost considerations impose size limits on the size of samples that can be obtained. If a 5% level of significance is allowed, what would be your recommendation, based on the following random samples? The numbers represent percent yield of crude oil.

X_1: 23.2, 26.6, 24.4, 23.5, 22.6, 25.7, 25.5; $\sum (X_1 - \bar{X}_1)^2 = 13.16$
X_2: 25.7, 27.7, 26.2, 27.9, 25.0, 21.4, 26.1; $\sum (X_2 - \bar{X}_2)^2 = 28.23$

Using the test procedure above, what values of $(\mu_{x_2} - \mu_{x_1})/\sigma$ will be detected with probability 0.80?

6. Two identical production lines, X and Y, produce transistors. It is desired to test the hypothesis that line X produces more units per day than line Y at the 1% level. If the average daily production of X exceeds the average daily production of Y by as much as 250 units, it is desired to detect this fact with probability 0.90. (a) How many observations should be taken? (Assume $\sigma_x = \sigma_y$ but are unknown. Use the approximate value of $\sigma = 110$ to find n.) (b) Suppose the data obtained are:

Line X	Line Y
$\bar{X} = 2,800$	$\bar{Y} = 2,680$
$\sum (X_i - \bar{X})^2 = 103,600$	$\sum (Y_i - \bar{Y})^2 = 99,800$

Perform the test and indicate decision.

7. The vice-president for production of a transistor manufacturing company which operates two plants reviews the production of the plants and notes that Plant A appears to produce consistently more per day than Plant B, even though both are designed to produce the same number of transistors in a given period of time. In a test to determine whether or not Plant A, on the average, produces more per day than Plant B, the vice-president specifies a level of significance of 0.01. If the average production of Plant A exceeds that of Plant B by 250 units, the vice-president indicates that the test should detect this difference with probability 0.90. From past experience it is known that the standard deviation of the quantity of daily production in each plant is equal and its approximate magnitude is 120 units. (a) How many observations from each plant are required to insure the risks above? (b) If the sample size is as found in (a) and the data are:

Plant A	Plant B
X = daily production	Y = daily production
$\bar{X} = 2,830$	$\bar{Y} = 2,680$
$\sum (X_i - \bar{X})^2 = 103,600$	$\sum (Y_i - \bar{Y})^2 = 99,800$

would you conclude that Plant A produces more than Plant B?

8. A new cure has been developed for Portland cement. Tests are run to determine if the new cure has an effect (positive or negative) on the strength. A single batch has been produced and subjected to both standard and experimental cures. The compressive strengths (psi) are given as:

Standard Cure X	Experimental Cure Y
4,125	4,250
4,225	3,950
4,350	3,900
3,575	4,075
3,875	4,550
3,825	4,450
3,975	4,150
3,800	4,550
3,775	3,700
3,850	4,250

$\sum X_i^2 = 155{,}740{,}625$ $\sum Y_i^2 = 175{,}663{,}125$

(a) Test for effect on strength of change in cure at the 5% level of significance. (b) Assuming that a common value of σ is approximately 280, and using the procedure in (a), find the sample size necessary to detect a change of 360 psi with probability 0.95. (Assume $\alpha = 0.05$.)

9. It is argued that the resistance of wire A is greater than the resistance of wire B. You make tests on each wire with the following results:

Wire A	Wire B
0.140 ohm	0.135 ohm
0.138	0.143
0.145	0.136
0.142	0.142
0.144	0.138
0.137	0.140

Assuming equal variances, what conclusions do you draw? Justify your answer. (*Remark*: It may help to choose a convenient computing origin.)

10. You wish to determine whether the shear strength of yarn A is different from that of yarn B. The standard deviation in shear strength is known to be somewhere near 10 pounds for both types of yarn. You are willing to run a maximum risk of 0.01 of saying that the yarns are different when they are actually the same. On the other hand, you do not wish to run a risk of more than 0.10 of saying that they are the same when actually the average strengths differ by as much as 15 pounds. (a) Determine the sample size and state the test procedure you would use in your test. (b) If $\bar{X}_A - \bar{X}_B = 6.3$ and

$$\sqrt{\frac{2}{n}} \sqrt{\frac{\sum (X_A - \bar{X}_A)^2 + \sum (X_B - \bar{X}_B)^2}{2n - 2}} = 3.1,$$

would you conclude that the average strengths are equal?

11. A laboratory test was devised for estimating the ease of filtering a certain product on the plant scale. The test consisted of measuring the time taken to filter a given volume of material under standard conditions. Six samples of plant magma were taken during spells when filtering was judged to be in State B and State A. The variance, although unknown, is assumed to be the same for both cases.

Filtration on Plant

State A	State B
8	9
10	10
12	10
13	4
9	7
14	8

(a) It is required to test at the 5% significance level whether the difference is significant. (b) Assuming that the calculated sample variance is a good estimate of the true variance, what is the probability of accepting the hypothesis when the true means differ by 4?

12. Suppose you are in a laboratory supervising a number of experimental scientists. One young man reveals that he has made a great discovery. By placing element A in steel he can increase its tensile strength. You are somewhat skeptical so the scientist presents the data to you and explains that he has made a t test (assuming equal variances) on the data, which verifies his conclusion. After looking at the results you ask him if the basic assumptions regarding the t test are satisfied. He looks amazed and says he thought that this test can always be used. You mention that this is not true and proceed to point out all the assumptions. In fact, you show him that one assumption is definitely not satisfied. Perform the t test (at the 5% level of significance), mention the assumptions, and prove that at least one assumption is not valid, given these data:

Tensile Strength

Steel with A	Steel without A
4	4
14	6
6	4
12	6
8	5
4	5
4	5
12	

If the t' test is used, would the results be changed?

13. The following are 16 independent determinations of the melting point of a compound, eight made by Analyst I and eight made by Analyst II. Data are in degrees centigrade.

Analyst I	Analyst II
164.5	163.5
169.7	162.8
169.2	163.0
169.5	163.2
161.8	160.7
168.7	161.5
169.5	160.9
163.9	162.0

Would you conclude from these data that there was a tendency for one analyst to get higher results than the other? Test at the 1% level. (Assume that the standard deviation of each analyst is the same.) (*Remark:* Subtract 160 from each observation.)

14. In Problem 13, how many observations by each analyst would be required to have a probability of 0.90 of detecting an analyst bias of 5 degrees centigrade? (Assume the common value of σ is approximately 3.20 degrees centigrade.)

15. Test the hypothesis of equality of means in Problem 13, assuming normality but without assuming equal variances.

16. An engineer in the design section of an aircraft manufacturing plant has presented theoretical evidence that painting the exterior of a particular combat airplane reduces its cruising speed. He convinces the chief design engineer that the next nine aircraft off the assembly line should be test flown to determine cruising speed prior to paint, then painted, and finally test flown to ascertain cruising speed after they are painted. The following data are obtained.

	Cruising Speed (knots)	
Aircraft	*Not Painted*	*Painted*
1	426	416
2	418	403
3	424	420
4	438	431
5	440	432
6	421	404
7	412	398
8	409	405
9	427	422

Design a test at the 5% level and complete the computation necessary to evaluate the design engineer's evidence.

17. In Problem 16, how many observations would be required to make the test procedure indicate rejection with probability 0.90 if the true cruising speed drops by 12.5 knots? Assume that the standard deviation of differences (σ_D) is about 5 knots.

18. Five men are subjected to a slight change in diet and are weighed before and after a lapse of three months. Weight in pounds:

Man	*Before*	*After*
1	162	166
2	192	196
3	138	136
4	182	190
5	159	160

(a) Test the hypothesis that there is no gain in weight at the 1% level of significance.
(b) How many men would be needed to have a probability of 0.90 of rejecting the hypothesis if the true mean weight changes by 3 pounds? (Assume σ_D is approximately 3 for this calculation.)

19. A new type of mold for concrete has been developed. Suppose that the new mold has certain advantages over the standard mold, such as faster hardening, etc., but some people have expressed doubts as to the final strength of the finished product. It is economically feasible to take only three observations on each mold, and the data are:

Compressive Strength (psi)

Batch	Standard Mold	New Mold
1	4,680	4,020
2	4,650	3,940
3	4,530	3,980

(a) Determine (at the 5% level of significance) if there is a loss of compressive strength due to the use of the new mold. (b) Assuming σ_D is approximately equal to 90, how many observations would be necessary to detect a loss of 200 psi compressive strength at least 90% of the time? (c) Discuss the OC curve for your test.

20. Two analysts took repeated readings on the hardness of city water. Determine whether one analyst has a tendency to read higher than the other. Use both normal and non-parametric methods.

Coded Measures of Hardness

Analyst X	Analyst Y
0.42	0.82
0.62	0.61
0.37	0.89
0.40	0.51
0.44	0.33
0.58	0.48
0.48	0.23
0.53	0.25
	0.67
	0.88

$$\sum (X - \bar{X})^2 = 0.0558 \qquad \sum (Y - \bar{Y})^2 = 0.55981$$

21. Ten pairs of duplicate spectrochemical determinations for nickel are presented below. The readings in the first column were taken with one type of measuring instrument and those in the second column were taken with another type.

Sample	Duplicates	
1	1.94	2.00
2	1.99	2.09
3	1.98	1.95
4	2.07	2.03
5	2.03	2.08
6	1.96	1.98
7	1.95	2.03
8	1.96	2.03
9	1.92	2.01
10	2.00	2.12

(a) Determine at the 5% level of significance whether the different equipment leads to different results. Use the signed rank test. (b) Determine at the 0.2% (approximately) level of significance whether the different equipment leads to different results. Use the sign test.

22. In research on rocket propellants aimed at reducing the delay time between the application of the firing current and explosion, it was thought that the substitution of a grade T propellant for the normal grade C should have a favorable effect. The approximate standard deviation of the normal grade C (as well as the grade T) is 0.03. An experiment was planned in which n shots would be made with the C grade and n shots with the T grade. (a) A reduction in the difference of means of 0.06 second would make the necessary change in manufacture worthwhile. It was important, however, not to recommend grade T if no improvement really resulted and α was therefore set at 0.01. The risk β of failing to detect a reduction of 0.06 second was set at 0.10. How many observations are needed to give the risks above? (b) If $\bar{C} = 0.261$ second, $\bar{T} = 0.250$ second, $S_C^2 = 0.0128$, and $S_T^2 = 0.0132$, would you adopt grade T? (c) If all the T values were smaller than the C values, except for one observation which was the maximum of all the observations, test the hypothesis of equality of means, using a non-parametric test.

23. Two treatments are to be compared at the 1% level of significance. The data are definitely not "normal" and are

Treatment A	Treatment B
16.3	
14.7	18.7
12.3	17.5
13.5	17.9
16.0	18.3
17.1	18.0
17.3	16.9
15.2	

Determine whether the treatments differ.

24. Perform a non-parametric test, using the data of Problem 16. Use the 2.5% level.

25. The following are the burning times in seconds of floating smoke pots of two different types:

Type X		Type Y	
481	572	526	537
506	561	511	582
527	501	556	601
661	487	542	558
500	524	491	578
$\sum X^2 = 2{,}856{,}698$		$\sum Y^2 = 3{,}015{,}520$	

(a) Would you conclude that one type of smoke pot tends to burn longer than the other, assuming the standard deviations are the same (use a 5% level of significance)? (b) Suppose that the standard deviations could not be assumed to be equal. Would your conclusion be changed? (c) Suppose that the necessary normality assumption could not be made, and analyze the data, using non-parametric methods.

26. An experiment was run to see if the amount of metal removed was the same for two different temperatures of the pickling bath. The data are below; each observation represents the thickness of metal lost, expressed in 0.001 inch:

Amount of Metal Removed

Temperature:	90°	120°
	2.3	2.2
	2.7	2.4
	2.9	2.0
	2.7	1.9
	2.6	2.1
	2.4	2.0

(a) Test the hypothesis of no temperature effect at the 5% level, assuming normality and common variance. (b) Perform a non-parametric test at the 5% level of significance.

27. The following table gives percent loss in tensile strength, following immersion in a corrosive solution, of paired samples of an alloy—one specimen subject to stress and one not:

Test Number	Unstressed	Stressed
1	6.4	9.2
2	4.6	7.9
3	4.6	7.3
4	6.4	8.0
5	3.2	5.7
6	5.2	7.6
7	6.5	5.7
8	4.9	4.1
9	4.3	8.1
10	6.5	5.6
11	3.7	6.9
12	4.6	6.0

What conclusions would you draw from the data about the effect of stressing on loss in tensile strength? (Use a non-parametric test at the 5% level of significance.)

28. Suppose an experiment is to be performed to determine whether a treatment has an effect (either positive or negative) on the wear resistance of a certain material. The experimenter labels two specimens from the first test-piece $1H$ and $1T$, those from the second test-piece $2H$ and $2T$, and so on for nine pairs of test-pieces. He then tosses a coin nine times. If the result of the first toss is a head, he sets aside specimen $1H$ for treatment; if a tail, he sets aside specimen $1T$. The result of the second toss decides the fate of specimens $2H$ and $2T$, and so on. The treatment is then applied to the selected nine specimens, and the abrasion resistance of treated and untreated specimens is assessed by the wear test machine. The nine differences, abrasion resistance of treated specimen minus abrasion resistance of untreated specimen, were:

$$2.6, \quad 3.1, \quad -0.2, \quad 1.7, \quad 0.6, \quad 1.2, \quad 2.2, \quad 1.1, \quad 0.1$$

(a) Using the sign test, determine if treatment has an effect on the average wear resistance at the 5% (approximately) level of significance. (b) Using the signed rank test, determine whether treatment has an effect on the average wear resistance at the 5% (approximately) level of significance.

29. It is desired to know what type of filter should be used over the screen of a cathode ray oscilloscope in order to have a radar operator easily pick out targets on the presentation. A test to accomplish this has been set up. A noise is first applied to the scope in order to make it difficult to pick out a target. A second signal, representing the target, is put into the scope, and its intensity is increased from zero until detected by the observer. The intensity setting at which the observer first notices the target signal is then recorded. This experiment is repeated again, with a different observer, on the assumption that all people do not see in exactly the same manner. After a set of readings has been taken with one type of filter on the scope, another set of readings, using the same observers, is taken with a different type of filter. The numerical value of each reading listed in the table of data is proportional to the target intensity at the time the operator first detects the target.

Observer	Filter No. 1	Filter No. 2
1	90	88
2	87	90
3	93	97
4	96	87
5	94	90
6	88	96
7	90	94
8	84	90
9	101	100
10	96	93
11	90	95
12	82	86
13	93	89
14	90	92
15	96	98
16	87	95
17	99	102
18	101	105
19	79	85
20	98	97
21	81	88

(a) Assuming normality, test the hypothesis that the filters are the same ($\alpha = 0.05$).
(b) Use a non-parametric procedure to test the hypothesis that the filters are the same ($\alpha = 0.05$).

30. A small motor is being manufactured for a special purpose. The important characteristic of the motor in this application is its starting torque. Two testers are used to evaluate the starting torque of the motors. It is desired to know whether the two testers are obtaining equivalent results or if, because of slight differences in their testing procedure, one of them is obtaining consistently higher results than the other. The results are:

Motor	A	B
1	0.41	0.38
2	0.45	0.40
3	0.36	0.32
4	0.49	0.50
5	0.39	0.31
6	0.54	0.52
7	0.38	0.32
8	0.43	0.36

(a) Using the sign test, determine whether the testers differ at the 5% level of significance. (b) Using the signed rank test, determine whether the testers differ at the 5% level of significance.

31. Nine pairs of "identical" specimens are subjected to two types of stress. Their performance is measured and the following results obtained:

Type 1:	92	86	93	91	93	90	86	89	91	88
Type 2:	88	85	82	90	81	93	87	92	86	85

Test the hypothesis that there is no difference in performance for the two types of stresses at the 5% level by both normal and non-parametric methods.

32. It is desired to test whether a special treatment of concrete has an effect on the average strength of the concrete. A small experiment is to be made from a given batch of raw materials. Samples are to be divided into two equal groups at random, with group II receiving the special treatment. The procedure is to be such that if there is no difference in the average strengths of the two groups, the probability will be 0.95 of reaching this conclusion. Furthermore, if the average strengths differ by as much as 12 kg/cm², the procedure should result in the conclusion that they differ with probability greater than 0.75. It is felt that the treatment will not have any effect on the variability and consequently the variability for each group can be assumed to be the same. An estimate of the standard deviation is 3 kg/cm². (a) What is the necessary sample size for each group? (b) Suppose that

$$\sum_{i=1}^{n} (X_{1i} - \bar{X}_1)^2 = 402 \qquad \bar{X}_1 = 295$$

$$\sum_{i=1}^{n} (X_{2i} - \bar{X}_2)^2 = 54 \qquad \bar{X}_2 = 315$$

and n is that value found in (a). Should we conclude that the special treatment has no effect on the strength? (c) Do the data above indicate that $\sigma_1 = \sigma_2$?

33. A new type of core, A, has been suggested for electrical resistors on the grounds that it improves a certain characteristic. A decrease in this characteristic represents an improvement. It is desired to determine whether or not core A is an improvement on a standard type core, B. Nine resistors from each type of core are selected. One resistor with core A and one resistor with core B are tested at wattage level 1 and the characteristic is measured on both resistors at this wattage. Also, an A-core resistor and a B-core resistor are tested at wattage level 2, another pair at wattage level 3, another at level 4, ..., and the remaining pair at wattage level 9. Let X denote the random variable associated with the characteristic for an A-core resistor, and Y the random variable for the B-core resistor. Let X_i, $i = 1, 2, \ldots, 9$, be the reading of the characteristic at wattage level i for the A-core resistor, and similarly for Y_i, $i = 1, 2, \ldots, 9$. Suppose

$$
\begin{array}{llll}
X_1 = 1 & Y_1 = 2 & X_6 = 2 & Y_6 = 3 \\
X_2 = 2 & Y_2 = 3 & X_7 = 3 & Y_7 = 1 \\
X_3 = 3 & Y_3 = 1 & X_8 = 1 & Y_8 = 2 \\
X_4 = 2 & Y_4 = 4 & X_9 = 1 & Y_9 = 2. \\
X_5 = 2 & Y_5 = 3 & &
\end{array}
$$

(a) Assuming only that the distributions of X and Y are symmetric and continuous, test the hypothesis at the 2.5% level of significance (use two different tests):

$$ H: \mu_x \geq \mu_y $$
$$ A: \mu_x < \mu_y. $$

(b) Assuming that X and Y are normally distributed with variance equal to 2, test the hypothesis

$$ H: \mu_x \geq \mu_y $$
$$ A: \mu_x < \mu_y $$

at the 2.5% level. (Use the data given.)

34. The faculty of a school wishes to compare a new method of teaching with the standard method. To control as much as possible the effects due to variation in ability among students, n sets of identical twins are chosen from among the students. One twin from each set is trained by the new method, the other twin by the standard method. At the end of a year, an examination is given.

Let X_i denote the test score of the twin in the ith set who has been trained by using the new method, and let Y_i denote the test score of the other twin in the ith set. Assume that the test scores from different sets of twins are independent. Assume that $X_1, \ldots, X_n$ and $Y_1, \ldots, Y_n$ are normally distributed random variables and that $E(X_i) - E(Y_i) = \mu$ for $i = 1, \ldots, n$. Assume that each difference $X_i - Y_i$ has standard deviation σ, and that σ is thought to be near 10. (a) It is wished to determine if there is any difference in effectiveness between the two methods. If there is no difference, this conclusion should be reached with probability 0.95. If there is a difference in effectiveness of 10, this conclusion should be reached with probability 0.95 also. Describe the test procedure and sample size which should be used. (b) If $\bar{X} = 90$, $\bar{Y} = 95$, and if $\sum_{i=1}^{n} (X_i - Y_i)^2 = 2,400$, what conclusion is reached? (c) Suppose all but 6 of the Y_i are greater than the corresponding X_i. Use a non-parametric procedure to test the hypothesis at the 5% level of significance.

35. A sprinter at a well-known university runs 10 races during each of the 1969 and 1970 seasons. Denote the 1969 observations by $X_1, X_2, \ldots, X_{10}$ and the 1970 observations by $Y_1, Y_2, \ldots, Y_{10}$. Between the two track seasons the sprinter contracts a rare tropical disease and loses his luxurious head of hair. School officials contend that due to decreased wind resistance, he can run significantly faster in 1970 than in 1969, i.e., they contend $\mu_Y > \mu_X$.

1969	1970
9.6	9.7
9.9	9.7
10.1	9.9
9.8	10.0
9.8	9.9
9.9	10.0
10.1	9.8
9.9	9.8
10.0	9.7
9.9	9.5

$$\bar{X} = 9.9 \quad \sum X_i^2 = 980.30 \qquad \bar{Y} = 9.8 \quad \sum Y_i^2 = 960.62$$

(a) Test the hypothesis that $\sigma_X^2 = \sigma_Y^2$ at the 5% level of significance. (b) Keeping in mind the conclusions of part (a), test the hypothesis that $\mu_X = \mu_Y$ against the alternative $\mu_Y > \mu_X$ at the 5% level of significance.

36. In Problem 4, it was decided to test for equality of variances. Verify that this hypothesis is tenable at the 1% level if $S_x^2 = 960$ and $S_y^2 = 1330$.

37. In Problem 22, verify the assumption of equality of variances. What must the ratio of σ_C / σ_T be before there is a probability of 0.8 of detecting a difference in the standard deviations?

38. You wish to determine whether operator B is turning out a product that is more variable (measured in terms of standard deviation) than the product of operator A. You are willing to run a maximum risk of 0.05 of saying that B's product is more variable than A's when actually it is not. On the other hand, you do not wish to run a risk of more than 0.30 of saying that the variability of B's product is equal to or less than that of A's when it is actually one-and-one-half times greater than A's. Determine the sample sizes and acceptance limit you would use in your test.

39. The amount of surface wax on each side of waxed paper bags is a random variable. There is reason to believe that there is greater variation in the amount on the inner side of the paper than on the outside. A sample of 75 observations of the amount of wax on each side of these bags is obtained and the following data recorded:

Wax in Pounds per Unit Area of Sample

Outside Surface	Inside Surface
$\bar{X} = 0.948$	$\bar{Y} = 0.652$
$\sum X_i^2 = 91$	$\sum Y_i^2 = 84$

Conduct a test to determine whether or not the variability of the amount of wax on the inner surface is greater than the amount on the outer surface ($\alpha = 0.01$).

40. In testing the equality of two variances, determine the sample sizes and acceptance region to use when you want a maximum risk of 0.05 of saying one variance is greater than the other when they are equal, and 0.01 of saying they are the same when one is four times the other.

41. In testing the equality of two variances, determine the sample sizes and acceptance region to use when you want a maximum risk of 0.05 of saying one variance is greater than the other when they are equal, and 0.10 of saying they are the same when one is 0.36 of the other.

42. The following results are calculated for two samples, each of whose observations are normally distributed:

$$x: \quad n_x = 8, \quad \Sigma\, X = 12, \quad \Sigma\, X^2 = 48$$

$$y: \quad n_y = 11, \quad \Sigma\, Y = 22, \quad \Sigma\, Y^2 = 80$$

Test the hypothesis $\sigma_x = \sigma_y$ against the alternative $\sigma_x > \sigma_y$, at the 1% level of significance.

43. Let X and Y be normally distributed random variables. Suppose a sample of 25 observations is taken from each, and a test is conducted of the hypothesis that $\sigma_x = \sigma_y$ at the 1% level. Give the rejection rule and sketch the OC curve of the test for the three cases: (a) Rejection is desired only when $\sigma_x > \sigma_y$. (b) Rejection is desired only when $\sigma_x < \sigma_y$. (c) Rejection is desired whenever $\sigma_x \neq \sigma_y$.

44. Suppose that 10 trials are made for each of three treatments with the following sample variances:

$$6.4 \quad 1.2 \quad 1.8.$$

Test the hypothesis that the variances are equal at the 1% level.

45. The variability of six machines is to be tested. Five observations are taken on each machine, and the sample variances are calculated. The results are:

$$16.2, \quad 13.8, \quad 14.6, \quad 11.8, \quad 16.5, \quad 16.0.$$

Do all the machines have the same variance (use $\alpha = 0.05$)?

46. Derive the expression for the sample size of the one-sided procedure for testing the hypothesis that $\mu_x = \mu_y$ when both standard deviations are known.

47. Derive the expression for the sample size of the two-sided procedure for testing the hypothesis that $\mu_x = \mu_y$ when both standard deviations are known.

48. Derive the expression for the sample size of the two-sided procedure for testing the hypothesis that $\sigma_x = \sigma_y$.

8

ESTIMATION

8.1 Introduction

Industrial experimentation is often performed to estimate some unknown values. These values are usually parameters (constants) of a probability distribution or functions of these parameters. For example, consider the problem presented in Sec. 6.2.3. A packaging machine is preset to fill containers with a fixed amount of food. Because there is an inherent variation in the amount which is placed in each container, the manufacturer is often interested in estimating the average weight. High average weights are costly, whereas low average weights are often illegal. Parameters other than the mean may also be of interest. In the filling problem, a determination of the variability in the machine setting is important information. This chapter is devoted to the problem of estimation and considers point estimates, confidence interval estimates, and tolerance limit estimates.

8.2 Point Estimation

The term "estimate" has been used in previous chapters in a somewhat intuitive manner. For example, in Chapter 6 the sample standard deviation was used as an "estimate" of σ for the purposes of using the OC curves for the t test. In this context the value that the random variable $\sqrt{S^2}$ takes on is a point estimate of σ. In general, a point estimate of a parameter θ is simply a unique choice for the value of the parameter. A point estimate rarely coincides with the unknown quantity to be estimated; it usually suffices for the estimate to be "close" to the unknown quantity. The general problem of point estimation may be formulated as follows: There exists a random variable X with an associated cumulative distribution function $F_X(b, \theta)$[1]

---

[1] The symbol $F_X(b, \theta)$ is now used to denote the CDF to highlight the fact that the parameter θ is unknown and to be estimated. Although most of the discussion will be concerned with single parameters and univariate random variables, much of the material presented will be applicable to distributions with multiple parameters and for multivariate random variables.

which is completely determined except for knowledge of the parameter of θ. A random sample, $X_1, X_2, \ldots, X_n$, is to be drawn and a function $\hat{\theta} = \hat{\theta}(X_1, X_2, \ldots, X_n)$ of this sample is to be chosen. The value that $\hat{\theta}$ takes on will be used as the choice of the unknown parameter θ. *The random variable $\hat{\theta}$ is called an estimator of θ and the value that $\hat{\theta}$ takes on is called the (point) estimate of θ.* It is evident that there exist many estimators for a parameter θ. For example, suppose that the random variable X is normally distributed with unknown mean μ and known standard deviation equal to 2, and μ is to be estimated from a random sample of size n. The sample mean, the sample median, the smallest observation, the largest observation, the second observation, one-half the sample mean plus the sample median, etc., can be considered as estimators of μ. How, then, is one to choose among these estimators? The obvious answer is to pick the estimator which is "generally closest" to the true value μ. It is important to note that an estimator is a random variable, and thereby cannot be judged on the basis of its performance in a particular case. Instead, it should be judged on its "long run" performance. Hence, to choose among estimators, some "measure of consequence" must be described.

8.2.1 COMPARISON OF ESTIMATORS

Recall that in Chapter 5 a general framework for decision making was introduced. In particular, a version of the Rockwell hardness example was concerned with the problem of estimation, i.e., the situation where the unbranded shafts can be produced in a mill whose average Rockwell hardness is completely unknown. This corresponds to an infinite action and infinite state of nature situation. What, if any, is the relationship between this formulation introduced in Chapter 5 and the formulation of the problem of point estimation introduced in Sec. 8.2? They are, of course, identical. An estimator is just a decision procedure, i.e., $\hat{\theta} = d[X_1, X_2, \ldots, X_n]$, which selects an action, that action being the estimate of θ for every possible outcome of the random sample. For example, the Rockwell hardness of the unbranded shafts as described above can be assumed to be normally distributed with unknown mean μ and with standard deviation equal to 2. A random sample of size 10 is to be drawn and μ is to be estimated. As mentioned earlier, there are many possible estimators, with the following being typical examples: Let $\hat{\theta}_1$ be the sample mean, $\hat{\theta}_2$ be the second observation, and $\hat{\theta}_3$ be the largest observation. Before choosing among them, a loss function must be presented.

Clearly, a loss function can be chosen only by the decision maker or experimenter. However, a "sensible" loss function is given by the quadratic function

$$l(\hat{\theta}, \theta) = c(\theta)(\hat{\theta} - \theta)^2,$$

where $c(\theta)$ is a positive function of θ. Although this loss function has been chosen rather arbitrarily, it does possess some attractive features. In particular, the loss increases rapidly as $\hat{\theta}$ departs from θ in either direction. In addition, a quadratic function of this form can often be used to approximate many well-behaved functions in the neighborhood of $\hat{\theta} = \theta$. Finally, this loss function is also a regret function.

Since the loss function $l(\hat{\theta}, \theta)$ is a random variable, the risk function, $R(\hat{\theta}, \theta)$, should be used as the "measure of consequence." The risk function is given by

$$R(\hat{\theta}, \theta) = E[l(\hat{\theta}, \theta)],$$

where the expectation is taken with respect to the distribution of the estimator $\hat{\theta}$. For the quadratic loss function, the risk function becomes

$$R(\hat{\theta}, \theta) = c(\theta)E(\hat{\theta} - \theta)^2.$$

Note that the risk is just $c(\theta)$ times the mean squared error of the estimator $\hat{\theta}$. In practice, $c(\theta)$ may be difficult to evaluate. Fortunately, however, in comparing the risks associated with two estimators, $c(\theta)$ does not play a role, e.g., the ratio of the risks is equal to the ratio of their mean squared errors. Hence, throughout the remainder of this chapter, $c(\theta)$ is taken equal to one so that the risk function for an estimator is just its mean squared error, i.e.,

$$R(\hat{\theta}, \theta) = E(\hat{\theta} - \theta)^2.$$

Returning to the Rockwell hardness example where there were three estimators suggested ($\hat{\theta}_1$ is the sample mean, $\hat{\theta}_2$ is the second observation, and $\hat{\theta}_3$ is the largest of the 10 observations), their density functions are shown in Fig. 8.1.[1] Although it appears evident that $\hat{\theta}_1$ is the "best" estimator, this is generally not obvious without evaluating the expected mean squared error for each estimator. Furthermore, in any given situation, the estimate obtained using $\hat{\theta}_2$ (or $\hat{\theta}_3$) may be closer to μ than the estimate obtained using $\hat{\theta}_1$. However, in a long series of trials, the estimates obtained from the estimator

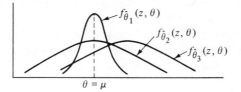

Fig. 8.1. *Densities of three estimators of the mean μ.*

---

[1] The symbol $f_{\hat{\theta}}(z, \theta)$ is now used to denote the density function of the random variable $\hat{\theta}$ to highlight the fact that the density function is dependent on the unknown parameter θ.

$\hat{\theta}_1$ will generally be closer to μ than the estimates obtained from the estimators $\hat{\theta}_2$ or $\hat{\theta}_3$ (in the sense of having smaller mean squared error). This statement can be made precise by saying that $\hat{\theta}_1$ has the smallest mean squared error, i.e., $E(\hat{\theta}_1 - \mu)^2$ is the minimum of $E(\hat{\theta}_1 - \mu)^2$, $E(\hat{\theta}_2 - \mu)^2$, and $E(\hat{\theta}_3 - \mu)^2$, where

$$E(\hat{\theta}_i - \mu)^2 = \int_{-\infty}^{\infty} (z - \mu)^2 f_{\hat{\theta}_i}(z, \theta) \, dz$$

is the mean squared error of $\hat{\theta}_i$ and $f_{\hat{\theta}_i}(z, \theta)$ is the density function of the estimator $\hat{\theta}_i$, $i = 1, 2, 3$.

The mean squared error can be interpreted as follows: Suppose a game is to be played by three players. There exists a normal probability distribution with mean equal to μ, and standard deviation equal to 2. The value of μ is known only to a referee. A sample of size 10 is drawn and player 1 uses $\hat{\theta}_1$ (the sample mean) to obtain his estimate of μ; player 2 uses $\hat{\theta}_2$ (the second observation) to obtain his estimate of μ; and player 3 uses $\hat{\theta}_3$ (the largest observation) to obtain his estimate of μ. Each player pays the other two the square of the difference between his estimate and μ (computed by the referee), multiplied by one dollar. If this game is played over and over again, the mean squared error corresponds to the average amount of money each player pays to his two opponents. Therefore, the player with the smallest mean squared error wins the most money; in this case the largest winner is player 1.

The comparison between two estimators can be made quantitative by comparing their relative efficiency. If estimator $\hat{\theta}_1$ has mean squared error $E(\hat{\theta}_1 - \theta)^2$ and if a second estimator $\hat{\theta}_2$ has mean squared error $E(\hat{\theta}_2 - \theta)^2$, the efficiency of $\hat{\theta}_2$ relative to $\hat{\theta}_1$ is defined to be $E(\hat{\theta}_1 - \theta)^2/E(\hat{\theta}_2 - \theta)^2$. If the efficiency of $\hat{\theta}_2$ relative to $\hat{\theta}_1$ is less than one, then $\hat{\theta}_1$ may reasonably be regarded as a better estimator of θ than $\hat{\theta}_2$. Note that if $E(\hat{\theta}) = \theta$, the mean squared error of θ, $E(\hat{\theta} - \theta)^2$, coincides with the variance of $\hat{\theta}$. Thus, in the Rockwell hardness example, the mean squared error of $\hat{\theta}_1$, the sample mean, coincides with the variance of the sample mean and is equal to $\sigma^2/n = 4/10$. The mean squared error of $\hat{\theta}_2$, the second observation, coincides with the variance of the second observation and is equal to 4 (which is the variance of any observation). Therefore, the efficiency of $\hat{\theta}_2$ relative to $\hat{\theta}_1$ is 1/10 and $\hat{\theta}_1$ is a better estimator of θ than $\hat{\theta}_2$. It can be shown that $E(\hat{\theta}_3) = 1.539\sigma + \mu = 3.078 + \mu$ (where $\hat{\theta}_3$ is the largest observation) and that $E(\hat{\theta}_3 - \mu)^2 = 2.712\sigma^2 = 10.848$. (The variance of $\hat{\theta}_3$ and the mean squared error of $\hat{\theta}_3$ are not equal.) Thus, the efficiency of $\hat{\theta}_3$ relative to $\hat{\theta}_1$ is

$$\frac{4/10}{10.848} < 1$$

so that $\hat{\theta}_1$ can be regarded as a better estimator of θ than $\hat{\theta}_3$.

An estimator $\hat{\theta}_0$ will be called optimal if its mean squared error does not exceed the mean squared error of any other estimator $\hat{\theta}$ for all values of θ. The sample mean, $\bar{X}$, is the optimal estimator of the mean μ of the normal distribution. Unfortunately, there exist many problems for which optimal estimators do not exist. Hence, it is desirable to examine properties other than risk for estimators, and some are considered in the next three sections.

8.2.2 UNBIASED ESTIMATORS

Suppose that an experiment is repeated several times in which a parameter θ is to be estimated. For each experiment a random sample is drawn, and $\hat{\theta}$ is chosen as the estimator of θ. Thus, each experiment results in an estimate of θ. A property that may be desirable is that the arithmetic mean of these estimates should tend to θ as the number of times the experiment is repeated becomes large. Using the customary notion of relating the arithmetic mean to its expectation, this property can be stated precisely by requiring that the estimator $\hat{\theta}$ satisfy the condition that

$$E(\hat{\theta}) = \theta.$$

The estimator $\hat{\theta}$ is said to be an unbiased estimator of θ if its expected value equals θ, i.e., if

$$E(\hat{\theta}) = \theta,$$

for all θ. If an estimator is not unbiased, it is called a biased estimator and

$$\theta - E(\hat{\theta})$$

is known as the bias. In the Rockwell hardness example, the sample mean and the second observation are then unbiased estimators of μ, whereas the largest observation is a biased estimator of μ (recall that $E(\hat{\theta}_3) = 3.078 + \mu$).

The property of unbiasedness is related to the property of mean squared error in that the mean squared error of $\hat{\theta}$ can be expressed in terms of the variance of $\hat{\theta}$ and the bias, i.e.,

$$E(\hat{\theta} - \theta)^2 = E[\hat{\theta} - E(\hat{\theta})]^2 + [\theta - E(\hat{\theta})]^2.$$

Hence, for an unbiased estimator, its mean squared error coincides with its variance. In the Rockwell hardness example, the unbiased estimator $\hat{\theta}_1$ (the sample mean) also happens to be the optimal estimator in the sense of having minimum mean squared error, but this should not be taken as a general principle. If $\hat{\theta}_1$ and $\hat{\theta}_2$ have densities as shown in Fig. 8.2, it is evident that the biased estimator, $\hat{\theta}_2$, is a better estimator of μ than the unbiased estimator, $\hat{\theta}_1$. Clearly, $\hat{\theta}_2$ has smaller mean squared error than $\hat{\theta}_1$.

Occasionally, unbiased estimators which do not have smallest mean squared error are used in practice because they possess the unbiasedness property. For example, in estimating the variance, σ^2, of a normally distrib-

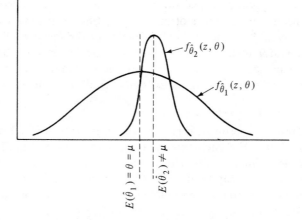

Fig. 8.2. *Distribution of two estimators of μ.*

uted random variable with unknown mean and variance, the unbiased estimator

$$S^2 = \frac{\sum_{i=1}^{n}(X_i - \bar{X})^2}{n - 1}$$

is used even though the biased estimator

$$S'^2 = \frac{\sum_{i=1}^{n}(X_i - \bar{X})^2}{n + 1}$$

has uniformly smaller mean squared error.

8.2.3 CONSISTENT ESTIMATORS

The property of unbiasedness was interpreted in terms of repeated experiments. The property of consistency is concerned with the behavior of an estimator in a single experiment when the sample size is allowed to become very large. *The estimator $\hat{\theta}(n)$ is said to be a consistent estimator if it approaches (in a probability sense) the parameter to be estimated as the sample size n gets large.* The symbol $\hat{\theta}(n)$ is used to indicate that the estimator may be a function of the size of the random sample. For example, if the estimator is the sample mean, then $\hat{\theta}(1) = \bar{X}_1 = X_1$ if $n = 1$; $\hat{\theta}(2) = \bar{X}_2 = (X_1 + X_2)/2$ if $n = 2$; $\hat{\theta}(3) = \bar{X}_3 = (X_1 + X_2 + X_3)/3$ if $n = 3$, etc. In the Rockwell hardness example, it is evident that $\hat{\theta}_1$ (the sample mean) is consistent whereas $\hat{\theta}_2$ (the second observation) is not consistent. It also can be shown that $\hat{\theta}_3$ (the largest observation) is not a consistent estimator.

In practice, estimators that are consistent usually have risks (in the sense of mean squared error) that tend to zero as the sample size gets large. However, it is always true that estimators whose risks tend to zero as the sample size gets large are consistent estimators. Furthermore, if they are biased estimators, their biases tend to zero.

8.2.4 EFFICIENT UNBIASED ESTIMATORS

If the estimators of a parameter θ are constrained to belong to the class of unbiased estimators, the risk function becomes the variance of the estimator. Hence, the optimal estimator among this class is the estimator, $\hat{\theta}_0$, whose variance does not exceed the variance of any other unbiased estimator for all values of θ, if such an estimator exists. The estimator, $\hat{\theta}_0$, is generally called the minimum variance unbiased estimator of θ and, furthermore, is the "efficient" estimator among this class.

In searching for such an estimator, a lower bound on the variance of all unbiased estimators would be useful. Such a lower bound is given by an inequality known as the Cramér-Rao Inequality. This bound can be described as follows: Denote by $\hat{\theta}$ an unbiased estimator of the parameter θ obtained from a random sample of size n. Let $f_X(z, \theta)$ denote the common density function of the continuous random variable X. Assuming that the density function satisfies certain conditions (which is the case in most practical situations), the inequality on the variance of an unbiased estimator is given by

$$\text{Variance } \hat{\theta} \geq 1/nE\left\{\left[\frac{d}{d\theta} \ln f_X(X, \theta)\right]^2\right\}.^1$$

If X is a discrete random variable, the density function, $f_X(X, \theta)$, is replaced by the probability distribution, $P_X(X, \theta)$.

Unfortunately, an estimator that actually achieves the Cramér-Rao lower bound does not always exist. However, in many cases an estimator, $\hat{\theta}_0$, can be exhibited that achieves this lower bound. In these situations, $\hat{\theta}_0$ must be the optimal unbiased estimator.

Returning to the Rockwell hardness example, the Cramér-Rao lower bound is sought for the mean μ of a normal distribution with known variance $\sigma^2 = 4$ (the calculations will be performed by using any known value σ^2).

$$\ln f_X(X, \theta) = \ln\left(\frac{1}{\sqrt{2\pi}\,\sigma}e^{-(X-\mu)^2/2\sigma^2}\right) = -\ln\sqrt{2\pi}\,\sigma - (X-\mu)^2/2\sigma^2.$$

Taking the derivative of $\ln f_X(X, \theta)$ with respect to μ leads to

$$\frac{d}{d\mu}[-\ln\sqrt{2\pi}\,\sigma - (X-\mu)^2/2\sigma^2] = [X-\mu]/\sigma^2.$$

The expectation of the square of this last expression is given by

$$\frac{E[X-\mu]^2}{\sigma^4} = \frac{\sigma^2}{\sigma^4} = \frac{1}{\sigma^2}.$$

---

[1] Note that $f_X(X, \theta)$ is the density function of X, evaluated at $z = X$ and, hence, is a random variable. Also, the natural logarithm is used.

Therefore,

$$1/nE\left\{\left[\frac{d}{d\theta}\ln f_X(X,\theta)\right]^2\right\} = \frac{1}{n/\sigma^2} = \frac{\sigma^2}{n}.$$

For the Rockwell hardness example, $\sigma^2/n = 4/10 = 0.4$. Recall that in general the variance of the sample mean is equal to σ^2/n, so that the sample mean achieves the Cramér-Rao lower bound. Therefore, $\hat{\theta}_1$ (the sample mean) is the minimum variance unbiased estimator of the mean of the normal distribution with known variance.

8.2.5 Estimation by the Method of Maximum Likelihood

One of the most useful techniques for deriving point estimators is the method of maximum likelihood. This is best illustrated in terms of the following example: A production process is said to be "in control" if the random variable X, which denotes the quality of each item produced, can be considered to be an independent Bernoulli random variable with parameter p, i.e.,

$$P\{\text{item is defective}\} = P\{X = 1\} = p$$

and

$$P\{\text{item is non-defective}\} = P\{X = 0\} = 1 - p,$$

where p is generally called the process average. An alternative expression for the probability distribution of this Bernoulli random variable is given by

$$P\{X = x\} = p^x(1 - p)^{1-x}, \quad \text{for} \quad x = 0, 1.$$

Note that this last expression yields the former expression for the probability distribution, but is a more concise representation. Suppose that p is to be estimated from a random sample of size 5 drawn from this process, i.e., five items are to be inspected and classified as defective or non-defective. Furthermore, assume that the outcomes of the random variables X_1, X_2, X_3, X_4, X_5, are given by 0, 1, 0, 0, 0, respectively. Since a random sample was drawn, the probability of obtaining these particular outcomes is given by

$$\begin{aligned}
P\{X_1 &= 0, X_2 = 1, X_3 = 0, X_4 = 0, X_5 = 0\} \\
&= P\{X_1 = 0\}P\{X_2 = 1\}P\{X_3 = 0\}P\{X_4 = 0\}P\{X_5 = 0\} \\
&= [p^0(1 - p)^1][p^1(1 - p)^0][p^0(1 - p)^1][p^0(1 - p)^1][p^0(1 - p)^1] \\
&= p(1 - p)^4.
\end{aligned}$$

This probability can be viewed as a function of p. The value of p that maximizes the probability of obtaining this observed sample (maximizes $p(1 - p)^4$) is then found. This value will be shown subsequently to be $1/5$, and is the maximum likelihood estimate of p.

This procedure can be easily formalized. A parameter θ is to be estimated from a random sample of size n. Let $P_X(k, \theta)$ denote the common probability distribution of the discrete random variable X. Denote by $x_1, x_2, \cdots, x_n$ the outcomes of the random variables $X_1, X_2, \cdots, X_n$, respectively. Define the likelihood function, L, by

$$L = P\{X_1 = x_1, X_2 = x_2, \cdots, X_n = x_n\}$$
$$= P\{X_1 = x_1\}P\{X_2 = x_2\} \cdots P\{X_n = x_n\},$$

where $P\{X_i = x_i\}$ is the probability distribution, with parameter θ, of the random variable X_i evaluated at the sample outcome x_i, i.e., $P_{X_i}(x_i, \theta)$. The likelihood function then represents the probability of obtaining the observed sample and is a function of θ. Note that the likelihood function is also the joint probability distribution of the random sample evaluated at the observed outcomes. For the "process control" example,

$$L = [p^{x_1}(1 - p)^{1-x_1}][p^{x_2}(1 - p)^{1-x_2}] \cdots [p^{x_n}(1 - p)^{1-x_n}] = p^{\Sigma x_i}(1 - p)^{n-\Sigma x_i}.$$

Note that L reduces to $p(1 - p)^4$ when $n = 5$ and $x_1 = 0$, $x_2 = 1$, $x_3 = 0$, $x_4 = 0$, and $x_5 = 0$.

The maximum likelihood principle calls for determining the value of θ that maximizes the likelihood viewed solely as a function of θ (the outcomes, $x_1, x_2, \ldots, x_n$, are considered to be known constants). This value is then the maximum likelihood estimate of θ and, furthermore, is the value of θ that maximizes the probability of obtaining the observed sample.

It is often simpler to maximize the logarithm of the likelihood function rather than the likelihood function itself. This is permissible since the logarithm is monotone, increasing so that the maxima of both functions occur at the same value of θ. The symbol $\mathscr{L}$ will denote the logarithm of the likelihood function. For the "process control" example, it is simpler to work with $\mathscr{L}$, i.e.,

$$\mathscr{L} = \log[p^{\Sigma x_i}(1 - p)^{n-\Sigma x_i}] = \sum x_i \log p + (n - \sum x_i) \log(1 - p).$$

In order to maximize $\mathscr{L}$, it is necessary to find the derivative of $\mathscr{L}$ and set it equal to zero, i.e.,

$$\frac{d\mathscr{L}}{dp} = \frac{\sum x_i}{p} - \frac{(n - \sum x_i)}{1 - p} = 0.$$

If $\hat{p}$ denotes the solution of this equation, then $\hat{p}$ is given by

$$\hat{p} = \sum x_i/n = \bar{x}.$$

The maximum likelihood estimate of p is then the average number of defective items found in the sample (since $\sum x_i$ represents the total number of defective items found in the sample), and the maximum likelihood estimator of p is simply

$$\hat{p} = \sum X_i/n.$$

For continuous random variables, the formal procedure for finding maximum likelihood estimates is identical to that for discrete random variables with the density function replacing the probability distribution in the likelihood function. Let $X_1, X_2, \cdots, X_n$ denote a random sample of continuous random variables and let $x_1, x_2, \cdots, x_n$ denote their sample outcomes. The likelihood function is now defined by

$$L = f_{X_1}(x_1, \theta) f_{X_2}(x_2, \theta) \cdots f_{X_n}(x_n, \theta),$$

where $f_{X_i}(x_i, \theta)$ represents the density function, with parameter θ, of the continuous random variable X_i evaluated at the sample outcome x_i. Again note that the likelihood function is just the joint density of the random sample evaluated at the observed outcomes. The interpretation of the likelihood function causes some difficulty. Clearly the probability of obtaining any observed sample must be zero. However, the likelihood can be viewed as a measure of the frequency with which the random variables will tend to assume values in small intervals near the observed outcomes. The maximum likelihood principle calls for determining the value of θ that maximizes the likelihood viewed solely as a function of θ. This value is then the maximum likelihood estimate of θ.

Returning to the Rockwell hardness example, the maximum likelihood estimate is easily obtained. Recall that X_i is normally distributed with mean μ and known variance $\sigma^2 = 4$. Thus, the likelihood function is given by

$$L = \left[\frac{1}{\sqrt{2\pi}(2)}e^{-(x_1-\mu)^2/(2)(4)}\right]\left[\frac{1}{\sqrt{2\pi}(2)}e^{-(x_2-\mu)^2/(2)(4)}\right]\cdots$$
$$\cdots\left[\frac{1}{\sqrt{2\pi}(2)}e^{-(x_n-\mu)^2/(2)(4)}\right] = \left(\frac{1}{\sqrt{2\pi}(2)}\right)^n e^{-\sum\limits_{i=1}^{n}(x_i-\mu)^2/(2)(4)}.$$

Again, it is simpler to work with $\mathscr{L}$, the logarithm of the likelihood, i.e.

$$\mathscr{L} = -n\ln\sqrt{2\pi}(2) - \sum_{i=1}^{n}(x_i - \mu)^2/(2)(4).$$

In order to maximize $\mathscr{L}$, it is necessary to find the derivative of $\mathscr{L}$ and set it equal to zero, i.e.,

$$\frac{d\mathscr{L}}{d\mu} = \sum_{i=1}^{n}(x_i - \mu)/4 = 0.$$

If $\hat{\mu}$ denotes the solution of this equation, then $\hat{\mu}$ is given by

$$\hat{\mu} = \sum x_i/n = \bar{x},$$

so that the sample mean is the maximum likelihood estimate of μ and $\bar{X}$ is the maximum likelihood estimator. Some remarks are in order:

1. The same estimate would have been obtained for any value of the variance σ^2, so long as it was known.
2. The maximum likelihood estimator for the mean of a normal distribution with known variance is the minimum variance unbiased estimator of μ. However, maximum likelihood estimators are not generally unbiased.

To summarize, maximum likelihood estimates are obtained from the likelihood function L. The likelihood, L, or $\mathscr{L}$, the logarithm of the likelihood, is then viewed as a function of θ and the maximum likelihood estimate is the value of θ which maximizes L or $\mathscr{L}$. This is usually achieved by finding the solution to either of the equations

$$\frac{dL}{d\theta} = 0, \quad \text{or} \quad \frac{d\mathscr{L}}{d\theta} = 0,$$

whichever is more convenient to solve. The solution must be checked to insure that maxima are actually obtained, however. (This should have been done in the previous examples.)

Maximum likelihood estimators can be found for distributions containing two or more parameters. This will be illustrated for the case of two parameters, θ_1 and θ_2. The likelihood function is obtained in exactly the same manner as for the one-parameter situation, but is now a function of the two parameters, θ_1 and θ_2. The maximum likelihood estimates of θ_1 and θ_2 are those values that maximize L. This is often obtained by simultaneously solving the equations

$$\frac{\partial L}{\partial \theta_1} = \frac{\partial L}{\partial \theta_2} = 0.$$

As an example, consider the Rockwell hardness case where the variance is now assumed to be unknown, i.e., the Rockwell hardness of the unbranded shafts are normally distributed with unknown mean μ and unknown variance σ^2. The maximum likelihood estimates of μ and σ^2 are to be obtained from a random sample of size n. The likelihood function is given by

$$L = \left[\frac{1}{\sqrt{2\pi}\sigma}e^{-(x_1-\mu)^2/2\sigma^2}\right]\left[\frac{1}{\sqrt{2\pi}\sigma}e^{-(x_2-\mu)^2/2\sigma^2}\right] \cdots \left[\frac{1}{\sqrt{2\pi}\sigma}e^{-(x_n-\mu)^2/2\sigma^2}\right]$$

$$= \left(\frac{1}{2\pi\sigma^2}\right)^{n/2} e^{-\sum_{i=1}^{n}(x_i-\mu)^2/2\sigma^2}.$$

Once again, it is simpler to work with $\mathscr{L}$, the logarithm of the likelihood, i.e.,

$$\mathscr{L} = -\frac{n}{2}\ln(2\pi) - \frac{n}{2}\ln(\sigma^2) - \sum_{i=1}^{n}\frac{(x_i - \mu)^2}{2\sigma^2}.$$

Taking the partial derivatives and setting them equal to zero leads to

$$\frac{\partial \mathscr{L}}{\partial \mu} = \sum_{i=1}^{n}\frac{(x_i - \mu)}{\sigma^2} = 0,$$

$$\frac{\partial \mathscr{L}}{\partial \sigma^2} = -\frac{n}{2\sigma^2} + \frac{\sum_{i=1}^{n}(x_i - \mu)^2}{2\sigma^4} = 0.$$

These equations are then solved simultaneously, and the maximizing values of μ and σ^2 are given by $\hat{\mu}$ and $\hat{\sigma}^2$, where

$$\hat{\mu} = \sum x_i/n = \bar{x}, \quad \text{and} \quad \hat{\sigma}^2 = \frac{\sum(x_i - \bar{x})^2}{n}.$$

These are checked to confirm that they are, indeed, the maximizing solution. Thus, the maximum likelihood estimators for the mean μ and variance σ^2 of a normally distributed random variable are given by

$$\hat{\mu} = \bar{X}, \quad \text{and} \quad \hat{\sigma}^2 = \frac{\sum (X_i - \bar{X})^2}{n}.$$

Note that the maximum likelihood estimator of the variance is neither unbiased nor optimal. However, for large n, this estimator does tend to have these properties. In fact, maximum likelihood estimators, in general, have desirable asymptotic properties. Maximum likelihood estimators are consistent and are asymptotically efficient. An estimator, $\hat{\theta}$, is said to be asymptotically efficient if $\sqrt{n}\,\hat{\theta}$ has a variance which approaches the Cramér-Rao lower bound for large n, i.e.,

$$\lim_{n \to \infty} \text{Variance} (\sqrt{n}\,\hat{\theta}) = 1/E\left\{\left[\frac{d}{d\theta} \ln f_X(X, \theta)\right]^2\right\}.$$

Furthermore, maximum likelihood estimators tend to be asymptotically normally distributed in the sense that for large n the random variable $\sqrt{n}\,(\hat{\theta} - \theta)$ approaches the distribution of a normally distributed random variable with zero mean and the aforementioned variance.

The method of maximum likelihood is frequently used to obtain estimators. The required calculations are often simple and the resultant estimators possess the previously described desirable properties. One additional property that enhances the ease of obtaining maximum likelihood estimators is the invariance property, which can be described as follows: If $\hat{\theta}$ is the maximum likelihood estimator of θ, and $T(\theta)$ is a function of θ possessing a single inverse, then $T(\hat{\theta})$ is the maximum likelihood estimator of $T(\theta)$. The usefulness of this invariance property is easily illustrated by obtaining the maximum likelihood estimator of the standard deviation σ of a normally distributed random variable with unknown mean and variance. Since σ is the positive square root of the variance (a function with a single inverse), the maximum likelihood estimator of the standard deviation is simply

$$\hat{\sigma} = \sqrt{\frac{\sum (X_i - \bar{X})^2}{n}}.$$

The alternative method for obtaining this solution requires finding the partial derivatives of L with respect to μ and σ, setting them equal to zero, and then solving them simultaneously.

8.2.6 ESTIMATION BY THE METHOD OF MOMENTS

The method of moments is a rather simple technique for obtaining estimators. Unfortunately, it often leads to estimators that do not have desirable properties. Before describing the method some definitions are required. Let X be a random variable having a CDF $F_X(b, \theta_1, \theta_2, \cdots, \theta_p)$ having p unknown parameters. Suppose a random sample of size n is drawn from this distribution. The kth sample moment M_k, taken about the origin, is defined to be

$$M_k = \frac{\sum\limits_{i=1}^{n} X_i^k}{n}.$$

The kth moment, about the origin, of the distribution has already been defined (Sec. 2.10) and is denoted by $E(X^k)$. These moments will generally be functions of the p unknown parameters. The method of moments calls for equating the first p moments of the distribution to the corresponding sample moments, i.e.,

$$M_1 = E(X)$$
$$M_2 = E(X^2)$$
$$\vdots$$
$$M_p = E(X^p),$$

and solving these equations simultaneously for the p unknown parameters. The solution is denoted by $\hat{\theta}_1, \hat{\theta}_2, \ldots, \hat{\theta}_p$, and are the estimators obtained by the method of moments.

As an example, consider the Rockwell hardness problem when the Rockwell hardness is assumed to be normally distributed with unknown mean μ and unknown variance σ^2. Estimators of μ and σ^2, determined by the method of moments, are desired. For a normally distributed random variable it is known that

$$E(X) = \mu, \quad \text{and} \quad E(X^2) = \sigma^2 + \mu^2.$$

Equating the first two sample moments with the first two moments of the distribution leads to

$$\bar{X} = \mu$$

$$\frac{\sum\limits_{i=1}^{n} X_i^2}{n} = \sigma^2 + \mu^2.$$

Solving these equations simultaneously results in the method-of-moments estimators, i.e.,

$$\hat{\mu} = \bar{X}$$

$$\hat{\sigma}^2 = \frac{\sum X_i^2}{n} - \bar{X}^2 = \frac{\sum (X_i - \bar{X})^2}{n}.$$

Note that these estimators coincide with the maximum likelihood estimators for this particular example. This is not true, in general. However, estimators obtained by the method of moments are consistent and are approximately normally distributed for large sample sizes. However, they may not be asymptotically efficient.

8.2.7 ESTIMATION BY THE METHOD OF BAYES

When there exists prior information concerning the states of nature, this information may be utilized to obtain an estimator of θ. This was exactly the second example treated in Secs. 5.2.7 and 5.3.5. Recall that this problem dealt with the situation where the unbranded shafts were produced in a mill whose average Rockwell hardness was completely unknown. However, the decision maker was prepared to treat this parameter as a random variable. He was willing to assume that he had knowledge about the behavior of the average Rockwell hardness over the possible mills, i.e., he was ready to postulate the existence of a prior (subjective) probability distribution. It was shown in this example that with a risk function proportional to the mean squared error, the Bayes decision procedure, or the Bayes estimator of a parameter θ in the present context, is just the expected value of the parameter θ, where the expectation is taken with respect to the posterior distribution of θ, given the outcome of the random sample, i.e.,

$$\hat{\theta} = E(\theta) = \int_{-\infty}^{\infty} z g_{\theta|X=x}(z) \, dz,$$

where $g_{\theta|X=x}(z)$ is this posterior density function of θ. Therefore, if a Bayes estimator is desired, the posterior distribution of the parameter θ, given the outcome of the random sample, must be obtained. The Bayes estimator is then simply the expectation of θ taken with respect to this posterior distribution. In the context of the Rockwell hardness example, the prior distribution of μ was assumed to be normally distributed with mean equal to 72 and standard deviation equal to 1.5. The distribution of the Rockwell hardness of the unbranded shafts was also assumed to be normally distributed with unknown mean μ and standard deviation equal to 2. The posterior distribution of θ, given the value of the sample mean obtained from a random sample of size 4, was then shown to be normal with mean equal to $\frac{288}{13} + \frac{9}{13}\bar{x}$ and variance equal to $\frac{9}{13}$. Therefore, the Bayes estimate is just the mean of this posterior distribution, i.e., $\frac{288}{13} + \frac{9}{13}\bar{x}$, and the Bayes estimator is given by

$$\hat{\theta} = \frac{288}{13} + \frac{9}{13}\bar{X}.$$

Note that the Bayes estimator based on a sample of size 4 does not coincide with the maximum likelihood estimator, although they will tend to agree as the sample size gets large. This characteristic is valid with almost any prior distribution. Furthermore, Bayes estimators are generally consistent and are approximately normally distributed for large sample sizes.

There exists an interesting relationship between Bayes estimates and maximum likelihood estimates for finite sample sizes. If the prior distribution is uniform, the maximum likelihood estimate of θ coincides with the value of θ that maximizes the posterior distribution of θ given the outcome of the random sample.

Table 8.1. Bayesian Estimates Given the Outcome of a Random Sample of Size n from the Sampling Distribution with Mean Squared Error Loss Function

Sampling Distribution of Interest	Parameter to be Estimated[1]	Outcome of Observed Random Variable	Prior Distribution of Parameter	Posterior Distribution of Parameter	Bayesian Estimate of Parameter
$\frac{1}{\sqrt{2\pi}\sigma}e^{-(z-\mu)^2/2\sigma^2}$ Normal (σ known; $-\infty \leqq \mu \leqq \infty$)	μ	$\bar{x}$	$\frac{1}{\sqrt{2\pi}D}e^{-(\mu-C)^2/2D^2}$ Normal (C and $D>0$ are known)	$\frac{\sqrt{\sigma^2+nD^2}}{\sqrt{2\pi}D\sigma}e^{-(\mu-M)^2(\sigma^2+nD^2)/2D^2\sigma^2}$ where $M=\dfrac{C/D^2+n\bar{x}/\sigma^2}{1/D^2+n/\sigma^2}$ Normal	$\dfrac{C/D^2+n\bar{x}/\sigma^2}{1/D^2+n/\sigma^2}$
$\frac{1}{\sqrt{2\pi}\sigma}e^{-(z-\mu)^2/2\sigma^2}$ Normal (σ known; $-\infty \leqq \mu \leqq \infty$)	μ	$\bar{x}$	Uniform over interval $(-\infty, \infty)$	$\frac{n}{\sqrt{2\pi}\sigma}e^{-n(\mu-\bar{x})^2/2\sigma^2}$ Normal	$\bar{x}$
$\eta e^{-\eta z}$ Exponential ($z>0; \eta>0$)	η	$\sum x_i$	$\frac{1}{\Gamma(C)D^C}\eta^{(C-1)}e^{-\eta/D}$ Gamma ($C>0$ and $D>0$ are known)	$\dfrac{\eta^{(C+n-1)}e^{-(D\sum x_i+1)\eta/D}}{\Gamma(C+1)[D/(D\sum x_i+1)]^{C+n}}$ Gamma	$\dfrac{(C+n)D}{(D\sum x_i+1)}$
$p^k(1-p)^{1-k}$ Bernoulli ($k=0,1; 0\leqq p\leqq 1$)	p	$\sum x_i$	$\frac{\Gamma(C+D)}{\Gamma(C)\Gamma(D)}p^{(C-1)}(1-p)^{D-1}$ Beta ($C>0$ and $D>0$ are known)	$\dfrac{\Gamma(C+D+n)}{\Gamma(C+\sum x_i)\Gamma(D+n-\sum x_i)}p^{(C+\sum x_i-1)}(1-p)^{(D+n-\sum x_i-1)}$ Beta	$\dfrac{C+\sum x_i}{C+D+n}$
$\frac{\lambda^k e^{-\lambda}}{k!}$ Poisson ($k=0,1,2,\cdots; \lambda>0$)	λ	$\sum x_i$	$\frac{1}{\Gamma(C)D^C}\lambda^{(C-1)}e^{-\lambda/D}$ Gamma ($C>0$ and $D>0$ are known)	$\dfrac{\lambda^{(C+\sum x_i-1)}e^{-(nD+1)\lambda/D}}{\Gamma(C+\sum x_i)[D/(nD+1)]^{(C+\sum x_i)}}$ Gamma	$\dfrac{(\sum x_i+C)D}{(nD+1)}$

[1] The maximum likelihood estimator for each of these parameters is the sample mean.

293

Some typical Bayesian estimates are presented in Table 8.1, using a loss function of mean squared error.

8.3 Confidence Interval Estimation

A point estimate is often inadequate as an estimate of a parameter since it rarely coincides with the parameter. An alternative type of estimate is an interval estimate of the form $[L, U]$, where L is the lower bound and U is the upper bound. The interval estimators considered will be functions of the sample, i.e., they are random variables having the property that, prior to experimentation, probability statements about the intervals including the parameter can be made. Therefore, if the parameter to be estimated is denoted by θ, an interval estimator of θ is $[L, U]$, where there is a given probability, $1 - \alpha$, that $L \leq \theta \leq U$. L and U will be called $100(1 - \alpha)\%$ confidence limits for the given parameter and the interval between them a $100(1 - \alpha)\%$ confidence interval. The fraction $(1 - \alpha)$, used in this context, is often called the confidence coefficient. The meaning of these statements can be best described by an example such as the problem of determining the average setting μ of a machine for filling packages. Suppose that a 95% confidence interval, $[L, U]$ is to be computed (using the techniques described in the following sections). This implies that, in the long run, 95% of the limits so computed can be expected to include the true average machine setting μ. If in each instance it was asserted that μ lies within the limits computed, correct statements would be expected 95 times in 100 and erroneous statements expected 5 times in 100; i.e., the statement "the true average machine setting lies within the range so computed" has a probability of 0.95 of being correct. But there would be no operational meaning in the following statement made in any one instance: "the probability is 0.95 that the true average setting falls within the limits computed in this case" since μ either does or does not fall within the limits. It should also be emphasized that even in repeated sampling the confidence interval will vary in position and width (if the variability of machine weighings is unknown) from sample to sample. This phenomenon is illustrated in Fig. 8.3, where confidence intervals computed from 13 separate trials are shown.

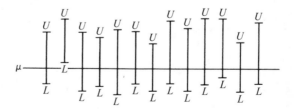

Fig. 8.3. *Confidence interval estimates of the true average setting of a weighing machine.*

If 95% confidence intervals are obtained, it is expected that 95 times in 100 these intervals will include μ; i.e., if it is stated that μ lies in the interval $[L, U]$, such statements will be correct, on the average, 95 times in 100. However, in practice an experiment is generally performed only *once*. A random sample is drawn, and a confidence interval $[L, U]$ is obtained, based on the outcome of the experiment. A typical announcement made by the experimenter is: "The parameter μ lies in the computed interval $[L, U]$; this statement is made with confidence (coefficient) 0.95." The interpretation of the experimenter's announcement is as described previously.

8.3.1 Confidence Interval for the Mean of a Normal Distribution when the Standard Deviation Is Known

Let X be a normally distributed random variable with mean μ and standard deviation σ (known). Consider a random sample of size n. Let $\bar{X}$ be the sample mean. The $100(1 - \alpha)\%$ confidence interval for μ is given by $\bar{X} \pm K_{\alpha/2}\sigma/\sqrt{n}$, where $K_{\alpha/2}$ is the $100\alpha/2$ percentage point of the normal distribution. This result is obtained by considering the distribution of $(\bar{X} - \mu)\sqrt{n}/\sigma$, which is normally distributed with mean 0 and standard deviation 1. Hence,

$$P\left\{-K_{\alpha/2} \leqq \frac{(\bar{X} - \mu)\sqrt{n}}{\sigma} \leqq K_{\alpha/2}\right\} = 1 - \alpha.$$

This probability statement is equivalent to

$$P\left\{-K_{\alpha/2}\frac{\sigma}{\sqrt{n}} \leqq \bar{X} - \mu \leqq K_{\alpha/2}\frac{\sigma}{\sqrt{n}}\right\} = 1 - \alpha$$

or

$$P\left\{\bar{X} - K_{\alpha/2}\frac{\sigma}{\sqrt{n}} \leqq \mu \leqq \bar{X} + K_{\alpha/2}\frac{\sigma}{\sqrt{n}}\right\} = 1 - \alpha.$$

Thus,

$$L = \bar{X} - K_{\alpha/2}\frac{\sigma}{\sqrt{n}} \quad \text{and} \quad U = \bar{X} + K_{\alpha/2}\frac{\sigma}{\sqrt{n}}$$

and the $100(1 - \alpha)\%$ confidence interval is given by

$$\bar{X} \pm K_{\alpha/2}\frac{\sigma}{\sqrt{n}}.$$

The length of the confidence interval is given by

$$2K_{\alpha/2}\frac{\sigma}{\sqrt{n}}.$$

Therefore, the sample size can be chosen appropriately so as to make the confidence interval be of prescribed length.

Upper one-sided confidence intervals are obtained by letting $L = -\infty$ and replacing $K_{\alpha/2}$ by K_{α}. A $100(1 - \alpha)\%$ upper confidence interval is given by

$\bar{X} + K_\alpha \sigma / \sqrt{n}$. This result follows from the probability statement

$$P\left\{\frac{(\bar{X} - \mu)\sqrt{n}}{\sigma} \geqq -K_\alpha\right\} = 1 - \alpha.$$

Similarly, a $100(1 - \alpha)\%$ lower confidence interval is given by

$$\bar{X} - K_\alpha \frac{\sigma}{\sqrt{n}}.$$

It should be noted that there exists a relationship between confidence intervals and significance tests. Suppose that one is interested in testing the hypothesis that $\mu = \mu_0$ against the alternative that $\mu \neq \mu_0$ at the α level of significance. The procedure given in Chapter 6 is equivalent to the following: accept the hypothesis that $\mu = \mu_0$ if the confidence interval $\bar{X} \pm K_{\alpha/2} \sigma / \sqrt{n}$ includes μ_0; otherwise, reject it. Similar relationships exist for all the decision procedures presented in Chapters 6 and 7, although no specific mention will be made of them in the ensuing sections.

Example. In the example of Sec. 6.1.4, the introduction of a new constituent was supposed to increase the average tensile strength of a material. An experiment was run using a sample of size $n = 19$; $\bar{X}$ was found to be 26,100 psi. Tensile strength was assumed to be normally distributed with mean μ and standard deviation $\sigma = 300$ psi. A 95% lower one-sided confidence interval for μ is given by

$$L = \bar{X} - 1.645 \frac{\sigma}{\sqrt{n}}$$

where 1.645 is the 5% point of the normal distribution. Hence,

$$L = 26,100 - (1.645)\frac{(300)}{\sqrt{19}} = 25,986.8.$$

8.3.2 CONFIDENCE INTERVAL FOR THE MEAN OF A NORMAL DISTRIBUTION WHEN THE STANDARD DEVIATION IS UNKNOWN

Let X be a normally distributed random variable with mean μ and standard deviation σ (unknown). Consider a random sample of size n. Let $\bar{X}$ be the sample mean and S the sample standard deviation. The $100(1 - \alpha)\%$ confidence interval for μ is given by

$$\bar{X} \pm t_{\alpha/2; n-1} \frac{S}{\sqrt{n}}$$

where $t_{\alpha/2; n-1}$ is the $100\alpha/2$ percentage point of the t distribution with $n - 1$ degrees of freedom. The result is obtained by considering the distribution of $(\bar{X} - \mu)\sqrt{n}/S$ which has been shown in Chapter 4 to have a t distribution with $n - 1$ degrees of freedom.

Hence,

$$P\left\{-t_{\alpha/2;\,n-1} \leq \frac{(\bar{X} - \mu)\sqrt{n}}{S} \leq t_{\alpha/2;\,n-1}\right\} = 1 - \alpha.$$

This probability statement is equivalent to

$$P\left\{-t_{\alpha/2;\,n-1}\frac{S}{\sqrt{n}} \leq \bar{X} - \mu \leq t_{\alpha/2;\,n-1}\frac{S}{\sqrt{n}}\right\} = 1 - \alpha$$

or

$$P\left\{\bar{X} - t_{\alpha/2;\,n-1}\frac{S}{\sqrt{n}} \leq \mu \leq \bar{X} + t_{\alpha/2;\,n-1}\frac{S}{\sqrt{n}}\right\} = 1 - \alpha.$$

Thus,

$$L = \bar{X} - t_{\alpha/2;\,n-1}\frac{S}{\sqrt{n}} \quad \text{and} \quad U = \bar{X} + t_{\alpha/2;\,n-1}\frac{S}{\sqrt{n}}$$

and the $100(1 - \alpha)\%$ confidence interval is given by

$$\bar{X} \pm t_{\alpha/2;\,n-1}\frac{S}{\sqrt{n}}.$$

A $100(1 - \alpha)\%$ upper one-sided confidence interval is given by $\bar{X} + t_{\alpha;\,n-1}S/\sqrt{n}$, whereas a $100(1 - \alpha)\%$ lower one-sided confidence interval is given by $\bar{X} - t_{\alpha;\,n-1}S/\sqrt{n}$.

Example. In the example of Sec. 6.2.3, a machine weighing problem was encountered. The experimental results led to $\bar{X} = 10.022$ and $S = 0.0632$. Assume that the weighings are normally distributed. Then, with these values of $\bar{X}$ and S a 95% confidence interval for μ is given by

$$\bar{X} \pm 2.571\frac{S}{\sqrt{n}}$$

where 2.571 is the $2\frac{1}{2}\%$ point of the t distribution with 5 degrees of freedom. Hence, the confidence interval becomes

$$\left[10.022 - (2.571)\frac{(0.0632)}{\sqrt{6}}, \quad 10.022 + (2.571)\frac{(0.0632)}{\sqrt{6}}\right]$$
$$= [9.956, 10.088].$$

8.3.3 Confidence Interval for the Standard Deviation of a Normal Distribution

Let X be a normally distributed random variable with mean μ (unknown) and standard deviation σ (unknown). Consider a random sample of size n. Let $\bar{X}$ be the sample mean and S, where

$$S = \sqrt{\sum_{i=1}^{n}(X_i - \bar{X})^2/(n - 1)},$$

be the sample standard deviation. The $100(1 - \alpha)\%$ confidence interval for σ is given by

$$\left[S\sqrt{\frac{n-1}{\chi^2_{\alpha/2;\,n-1}}}, \quad S\sqrt{\frac{n-1}{\chi^2_{1-\alpha/2;\,n-1}}} \right]$$

where $\chi^2_{\alpha/2;\,n-1}$ and $\chi^2_{1-\alpha/2;\,n-1}$ are, respectively, the $100\alpha/2$ and $100(1 - \alpha/2)$ percentage points of the chi-square distribution with $n - 1$ degrees of freedom.

This result is obtained by considering the distribution of $(n - 1)S^2/\sigma^2$ which has been shown in Chapter 4 to have a chi-square distribution with $n - 1$ degrees of freedom. Hence,

$$P\left\{\chi^2_{1-\alpha/2;\,n-1} \leqq \frac{(n-1)S^2}{\sigma^2} \leqq \chi^2_{\alpha/2;\,n-1}\right\} = 1 - \alpha.$$

This probability statement is equivalent to

$$P\left\{\frac{1}{\chi^2_{\alpha/2;\,n-1}} \leqq \frac{\sigma^2}{S^2(n-1)} \leqq \frac{1}{\chi^2_{1-\alpha/2;\,n-1}}\right\} = 1 - \alpha$$

or

$$P\left\{S\sqrt{\frac{n-1}{\chi^2_{\alpha/2;\,n-1}}} \leqq \sigma \leqq S\sqrt{\frac{n-1}{\chi^2_{1-\alpha/2;\,n-1}}}\right\} = 1 - \alpha.$$

A $100(1 - \alpha)\%$ upper one-sided confidence interval is given by

$$U = S\sqrt{\frac{n-1}{\chi^2_{1-\alpha;\,n-1}}}$$

and a $100(1 - \alpha)\%$ lower one-sided confidence interval is given by

$$L = S\sqrt{\frac{n-1}{\chi^2_{\alpha;\,n-1}}}.$$

Example. In the example of Sec. 6.3.4, a problem on the variability of a new product was encountered. A sample of 25 was taken and S^2 was found to be 0.0384. Assuming a normal distribution of the measurements, a 99% confidence interval for σ is given by

$$\left[S\sqrt{\frac{n-1}{\chi^2_{\alpha/2;\,n-1}}}, \quad S\sqrt{\frac{n-1}{\chi^2_{1-\alpha/2;\,n-1}}} \right]$$

$$= \left[0.1960\sqrt{\frac{24}{45.558}}, \quad 0.1960\sqrt{\frac{24}{9.886}} \right]$$

$$= [0.1423, 0.3060],$$

8.3.4 CONFIDENCE INTERVAL FOR THE DIFFERENCE BETWEEN THE MEANS OF TWO NORMAL DISTRIBUTIONS WHEN THE STANDARD DEVIATIONS ARE BOTH KNOWN

Let X be a normally distributed random variable with mean μ_x and standard deviation σ_x (known), and Y be an independent (of X) normally distributed random variable with mean μ_y and standard deviation σ_y (known). Let $\bar{X}$ denote the sample mean of a random sample of n_x observations on X, and $\bar{Y}$ denote the sample mean of a random sample of n_y observations on Y. The $100(1 - \alpha)\%$ confidence interval for $\mu_x - \mu_y$ is given by

$$\bar{X} - \bar{Y} \pm K_{\alpha/2}\sqrt{(\sigma_x^2/n_x) + (\sigma_y^2/n_y)}$$

where $K_{\alpha/2}$ is the $100\alpha/2$ percentage point of the normal distribution. This result is obtained by considering the distribution of

$$\frac{\bar{X} - \bar{Y} - (\mu_x - \mu_y)}{\sqrt{(\sigma_x^2/n_x) + (\sigma_y^2/n_y)}},$$

which has previously been shown to have a normal distribution with mean 0 and standard deviation 1. Hence,

$$P\left\{-K_{\alpha/2} \leqq \frac{\bar{X} - \bar{Y} - (\mu_x - \mu_y)}{\sqrt{(\sigma_x^2/n_x) + (\sigma_y^2/n_y)}} \leqq K_{\alpha/2}\right\} = 1 - \alpha.$$

This is clearly equivalent to

$$P\{\bar{X} - \bar{Y} - K_{\alpha/2}\sqrt{(\sigma_x^2/n_x) + (\sigma_y^2/n_y)} \leqq \mu_x - \mu_y$$
$$\leqq \bar{X} - \bar{Y} + K_{\alpha/2}\sqrt{(\sigma_x^2/n_x) + (\sigma_y^2/n_y)}\} = 1 - \alpha$$

so that the $100(1 - \alpha)\%$ confidence interval is given by

$$\bar{X} - \bar{Y} \pm K_{\alpha/2}\sqrt{(\sigma_x^2/n_x) + (\sigma_y^2/n_y)}.$$

A $100(1 - \alpha)\%$ upper one-sided confidence interval is given by

$$\bar{X} - \bar{Y} + K_\alpha\sqrt{(\sigma_x^2/n_x) + (\sigma_y^2/n_y)}$$

whereas a $100(1 - \alpha)\%$ lower one-sided confidence interval is given by

$$\bar{X} - \bar{Y} - K_\alpha\sqrt{(\sigma_x^2/n_x) + (\sigma_y^2/n_y)}.$$

Example. In the example of Sec. 7.1.4, the effect of polishing on the endurance limit of steel was studied. It was felt that polishing could have only a positive effect on the average endurance limit. Eight specimens of polished steel were measured and their average was found to be 86,375 psi. Similarly, 8 specimens of unpolished steel were measured and their average was found to be 79,838 psi. All the measurements are assumed to be normally distributed with standard deviation 4,000 psi (both the polished and unpolished specimens). A 90% lower one-sided confidence limit for the difference in means is given by

$$L = \bar{X} - \bar{Y} - 1.28\sqrt{(\sigma_x^2/n_x) + (\sigma_y^2/n_y)}$$

$$= (86,375 - 79,838) - 1.28\sqrt{\frac{(4,000)^2}{8} + \frac{(4,000)^2}{8}}$$

$$= 6,537 - 1.28(2,000)$$

$$= 3977.$$

8.3.5 CONFIDENCE INTERVAL FOR THE DIFFERENCE BETWEEN THE MEANS OF TWO NORMAL DISTRIBUTIONS WHERE THE STANDARD DEVIATIONS ARE BOTH UNKNOWN BUT EQUAL

Let X be a normally distributed random variable with mean μ_x and standard deviation σ (unknown) and Y be an independent (of X) normally distributed random variable with mean μ_y and standard deviation σ (unknown). Let $\bar{X}$ be the sample mean of a random sample of n_x observations on X, and $\bar{Y}$ be the sample mean of a random sample of n_y observations on Y. The $100(1 - \alpha)\%$ confidence interval for $\mu_x - \mu_y$ is given by

$$\bar{X} - \bar{Y} \pm t_{\alpha/2;\, n_x+n_y-2}\sqrt{\frac{1}{n_x} + \frac{1}{n_y}}\,\sqrt{\frac{\sum (X_i - \bar{X})^2 + \sum (Y_i - \bar{Y})^2}{n_x + n_y - 2}}$$

where $t_{\alpha/2;\, n_x+n_y-2}$ is the $100\alpha/2$ percentage point of the t distribution with $n_x + n_y - 2$ degrees of freedom.

This result is obtained by considering the distribution of

$$\frac{\bar{X} - \bar{Y} - (\mu_x - \mu_y)}{\sqrt{\dfrac{1}{n_x} + \dfrac{1}{n_y}}\,\sqrt{\dfrac{\sum (X_i - \bar{X})^2 + \sum (Y_i - \bar{Y})^2}{n_x + n_y - 2}}}$$

which has been shown in Chapter 4 to have a t distribution with $n_x + n_y - 2$ degrees of freedom. Hence,

$$\mathrm{P}\left\{-t_{\alpha/2;\, n_x+n_y-2} \leqq \frac{\bar{X} - \bar{Y} - (\mu_x - \mu_y)}{\sqrt{\dfrac{1}{n_x} + \dfrac{1}{n_y}}\,\sqrt{\dfrac{\sum (X_i - \bar{X})^2 + \sum (Y_i - \bar{Y})^2}{n_x + n_y - 2}}}\right.$$

$$\left. \leqq t_{\alpha/2;\, n_x+n_y-2}\right\} = 1 - \alpha.$$

This is clearly equivalent to

$$\mathrm{P}\left\{\bar{X} - \bar{Y} - t_{\alpha/2;\, n_x+n_y-2}\sqrt{\frac{1}{n_x} + \frac{1}{n_y}}\,\sqrt{\frac{\sum (X_i - \bar{X})^2 + \sum (Y_i - \bar{Y})^2}{n_x + n_y - 2}}\right.$$

$$\leqq \mu_x - \mu_y$$

$$\left. \leqq \bar{X} - \bar{Y} + t_{\alpha/2;\, n_x+n_y-2}\sqrt{\frac{1}{n_x} + \frac{1}{n_y}}\,\sqrt{\frac{\sum (X_i - \bar{X})^2 + \sum (Y_i - \bar{Y})^2}{n_x + n_y - 2}}\right\}$$

$$= 1 - \alpha,$$

so that the $100(1 - \alpha)\%$ confidence interval is given by

$$\bar{X} - \bar{Y} \pm t_{\alpha/2;\, n_x+n_y-2}\sqrt{\frac{1}{n_x} + \frac{1}{n_y}}\,\sqrt{\frac{\sum (X_i - \bar{X})^2 + \sum (Y_i - \bar{Y})^2}{n_x + n_y - 2}}.$$

A $100(1 - \alpha)\%$ upper one-sided confidence interval is given by

$$\bar{X} - \bar{Y} + t_{\alpha;\, n_x+n_y-2}\sqrt{\frac{1}{n_x} + \frac{1}{n_y}}\,\sqrt{\frac{\sum (X_i - \bar{X})^2 + \sum (Y_i - \bar{Y})^2}{n_x + n_y - 2}};$$

a $100(1 - \alpha)\%$ lower one-sided confidence interval is given by

$$\bar{X} - \bar{Y} - t_{\alpha;\, n_x+n_y-2}\sqrt{\frac{1}{n_x} + \frac{1}{n_y}}\, \sqrt{\frac{\sum(X_i - \bar{X})^2 + \sum(Y_i - \bar{Y})^2}{n_x + n_y - 2}}.$$

Example. In the example of Sec. 7.2.3, a comparison of two manufacturers' thermostats was made. Twenty-three thermostats from the new supplier were measured and their average computed to be 551.491. Similarly, 23 thermostats from the old supplier were measured and their average computed to be 549.926. It is assumed that the thermostat readings of each manufacturer are independent normally distributed random variables with common unknown standard deviation. The quantity

$$\frac{\sum(X_i - \bar{X})^2 + \sum(Y_i - \bar{Y})^2}{n_x + n_y - 2}$$

was found to be 87.683. A 90% confidence interval for the difference in means is given by

$$551.491 - 549.926 \pm t_{0.05;\, 44}\sqrt{\frac{1}{23} + \frac{1}{23}}\sqrt{87.683}$$

$$= 1.565 \pm 1.680(0.2949)(9.3639)$$

$$= [-3.074, 6.204].$$

8.3.6 CONFIDENCE INTERVAL FOR THE RATIO OF STANDARD DEVIATIONS OF TWO NORMAL DISTRIBUTIONS

Let X be a normally distributed random variable with mean μ_x and standard deviation σ_x, and Y be an independent (of X) normally distributed random variable with mean μ_y and standard deviation σ_y. Let S_x^2 be the sample variance based on a random sample of n_x observations on X, and let S_y^2 be the sample variance based on a random sample of n_y observations on Y. The $100(1 - \alpha)\%$ confidence interval for σ_x/σ_y is given by

$$\left[\frac{S_x}{S_y}\sqrt{\frac{1}{F_{\alpha/2;\, n_x-1, n_y-1}}}, \quad \frac{S_x}{S_y}\sqrt{F_{\alpha/2;\, n_y-1, n_x-1}}\right]$$

where $F_{\alpha/2;\, n_x-1, n_y-1}$ is the $100\alpha/2$ percentage point of the F distribution with $n_x - 1$ and $n_y - 1$ degrees of freedom and $F_{\alpha/2;\, n_y-1, n_x-1}$ is the $100\alpha/2$ percentage point of the F distribution with $n_y - 1$ and $n_x - 1$ degrees of freedom.

This result is obtained by considering the distribution of

$$\frac{S_y^2/S_x^2}{\sigma_y^2/\sigma_x^2}$$

which has been shown in Chapter 4 to have an F distribution with $n_y - 1$ and $n_x - 1$ degrees of freedom. Hence,

$$P\left\{F_{1-\alpha/2;\, n_y-1, n_x-1} \leqq \frac{S_y^2/S_x^2}{\sigma_y^2/\sigma_x^2} \leqq F_{\alpha/2;\, n_y-1, n_x-1}\right\} = 1 - \alpha.$$

This is clearly equivalent to

$$P\left\{\frac{S_x}{S_y}\sqrt{F_{1-\alpha/2;\,n_y-1,\,n_x-1}} \leq \frac{\sigma_x}{\sigma_y} \leq \frac{S_x}{S_y}\sqrt{F_{\alpha/2;\,n_y-1,\,n_x-1}}\right\} = 1 - \alpha.$$

It has also been shown that

$$F_{1-\alpha/2;\,n_y-1,\,n_x-1} = 1/F_{\alpha/2;\,n_x-1,\,n_y-1}.$$

Hence, the $100(1-\alpha)\%$ confidence interval is given by

$$\left[\frac{S_x}{S_y}\sqrt{\frac{1}{F_{\alpha/2;\,n_x-1,\,n_y-1}}},\ \frac{S_x}{S_y}\sqrt{F_{\alpha/2;\,n_y-1,\,n_x-1}}\right].$$

A $100(1-\alpha)\%$ upper one-sided confidence interval is given by

$$\frac{S_x}{S_y}\sqrt{F_{\alpha;\,n_y-1,\,n_x-1}}$$

whereas a $100(1-\alpha)\%$ lower one-sided confidence interval is given by

$$\frac{S_x}{S_y}\sqrt{\frac{1}{F_{\alpha;\,n_x-1,\,n_y-1}}}.$$

Example. In the example of Sec. 7.6.4, a comparison of the variabilities of two suppliers was discussed. The new supplier had a sample variance of $S_x^2 = 0.00041$, based on 100 observations, and the current supplier a sample variance of $S_y^2 = 0.00057$, also based on 100 observations. A 90% lower one-sided confidence limit for σ_x/σ_y is given by

$$\frac{S_x}{S_y}\sqrt{\frac{1}{F_{0.10;\,99,\,99}}} = \frac{\sqrt{0.00041}}{\sqrt{0.00057}}\sqrt{\frac{1}{1.29}}$$
$$= 0.747.$$

8.3.7 A Table of Point Estimates and Interval Estimates

The point estimates and confidence interval estimates presented are summarized in Table 8.2.

8.3.8 Approximate Confidence Intervals

An additional bonus of using the method of maximum likelihood to obtain point estimators is that their asymptotic properties may be exploited for confidence interval estimation. If $\hat{\theta}$ is the maximum likelihood estimator of θ, then, as noted in Sec. 8.2.5, $\hat{\theta}$ is approximately normally distributed with mean equal to θ and variance equal to the Cramér-Rao lower bound (denoted by σ_b^2) when the sample size is large. Using this normal theory, a $100(1-\alpha)\%$ confidence interval for θ is given by

$$\hat{\theta} \pm K_{\alpha/2}\sigma_b.$$

Unfortunately, σ_b is often a function of θ. However, the asymptotic theory is still valid if θ is replaced by $\hat{\theta}$ in the expression for σ_b. As an example, consider the "process control" problem introduced in Sec. 8.2.5. Based on a

Table 8.2 Summary of Estimates by Confidence Intervals and by Single Points

Parameters	Notation	Qualifying Conditions	Formula	Tables To Find Percentage Points	Point Estimates
Mean of a normal distribution	μ	Known σ	$\bar{X} - K_{\alpha/2}\dfrac{\sigma}{\sqrt{n}} \leqq \mu \leqq \bar{X} + K_{\alpha/2}\dfrac{\sigma}{\sqrt{n}}$	Normal table, Table 1	$\bar{X}$
Standard deviation of a normal distribution	σ		$S\sqrt{\dfrac{n-1}{\chi^2_{\alpha/2;\,n-1}}} \leqq \sigma \leqq S\sqrt{\dfrac{n-1}{\chi^2_{1-\alpha/2;\,n-1}}}$	Chi-square table, Table 2	S
Mean of a normal distribution	μ	Unknown σ	$\bar{X} - t_{\alpha/2;\,n-1}\dfrac{S}{\sqrt{n}} \leqq \mu \leqq \bar{X} + t_{\alpha/2;\,n-1}\dfrac{S}{\sqrt{n}}$	t table, Table 3	$\bar{X}$
Difference between the means of two normal distributions	$\mu_x - \mu_y$	$X_1,\cdots,X_{n_x}$ are observations in first sample; $Y_1,\cdots,Y_{n_y}$ are observations in second sample; σ_x and σ_y are known	$\bar{X} - \bar{Y} - K_{\alpha/2}\sqrt{\dfrac{\sigma_x^2}{n_x} + \dfrac{\sigma_y^2}{n_y}} \leqq \mu_x - \mu_y \leqq \bar{X} - \bar{Y} + K_{\alpha/2}\sqrt{\dfrac{\sigma_x^2}{n_x} + \dfrac{\sigma_y^2}{n_y}}$		$\bar{X} - \bar{Y}$
Difference between the means of two normal distributions	$\mu_x - \mu_y$	$\sigma_x = \sigma_y = \sigma$ Unknown σ	$(\bar{X} - \bar{Y}) - t_{\alpha/2;\,n_x+n_y-2}\sqrt{\dfrac{1}{n_x} + \dfrac{1}{n_y}}\sqrt{\dfrac{\sum (X_i - \bar{X})^2 + \sum (Y_i - \bar{Y})^2}{n_x + n_y - 2}}$ $\leqq \mu_x - \mu_y \leqq (\bar{X} - \bar{Y}) + t_{\alpha/2;\,n_x+n_y-2}\sqrt{\dfrac{1}{n_x} + \dfrac{1}{n_y}}\sqrt{\dfrac{\sum (X_i - \bar{X})^2 + \sum (Y_i - \bar{Y})^2}{n_x + n_y - 2}}$		$\bar{X} - \bar{Y}$
Ratio of standard deviations of two normal distributions	$\dfrac{\sigma_x}{\sigma_y}$	S_x is the first sample standard deviation; S_y is the second sample standard deviation	$\dfrac{S_x}{S_y}\sqrt{\dfrac{1}{F_{\alpha/2;\,n_x-1,\,n_y-1}}} \leqq \dfrac{\sigma_x}{\sigma_y} \leqq \dfrac{S_x}{S_y}\sqrt{F_{\alpha/2;\,n_y-1,\,n_x-1}}$	F table, Table 4	$\dfrac{S_x}{S_y}$

303

random sample of size n(large), the maximum likelihood estimator for p, the parameter of the Bernoulli random variable, was shown to be given by

$$\hat{p} = \bar{X} = \text{fraction of defectives found in the sample of size } n.$$

It is easily verified that the Cramér-Rao lower bound is given by

$$\sigma_b^2 = \frac{p(1 - p)}{n}.$$

An approximate $100(1 - \alpha)\%$ confidence interval for p is then given by

$$\hat{p} \pm K_{\alpha/2}\sqrt{\frac{\hat{p}(1 - \hat{p})}{n}},$$

where $\hat{p}$ is substituted for p in the expression for the standard deviation.

8.3.9 SIMULTANEOUS CONFIDENCE INTERVALS.

In each of the aforementioned problems, a confidence coefficient $(1 - \alpha)$ was chosen and a statement made about a parameter based on the outcome of an experiment. However, in many experimental situations, more than one parameter is of interest and several statements are required. Three typical situations are:

1. Two new treatments $(X$ and $Y)$ are to be compared with a standard (Z). The performances of these treatments and the standard are measured by their means, μ_x, μ_y, and μ_z, respectively. Confidence interval estimates for $(\mu_z - \mu_x)$ and $(\mu_z - \mu_y)$ are desired so that the probability is $(1 - \alpha)$ that both intervals simultaneously lead to correct statements. A correct statement is one in which the confidence interval *includes* the appropriate difference in means.
2. Three new treatments are to be compared. The performances of these treatments are measured by their means μ_x, μ_y, and μ_z, respectively. Confidence interval estimates for $(\mu_x - \mu_y)$, $(\mu_x - \mu_z)$, and $(\mu_y - \mu_z)$ are desired so that the probability is $(1 - \alpha)$ that all three intervals simultaneously lead to correct statements.
3. The mean, μ, and variance, σ^2, of a normally distributed random variable are both of interest. Confidence interval estimates for μ and σ^2 are desired so that the probability is $(1 - \alpha)$ that both intervals simultaneously lead to correct statements.

These three problems would be simple to solve if each confidence interval estimate statement were independent of the others. For example, consider the case of situation (1). If a confidence interval estimate for $\mu_z - \mu_x$ were available that was independent of a confidence interval estimate for $\mu_z - \mu_y$, then $100\sqrt{(1 - \alpha)}\%$ intervals could be obtained for each so that the probability is $(1 - \alpha)$ that both intervals lead to correct statements. Unfortunately, in most situations, independent confidence intervals are *not* available. However, in these situations a very useful inequality, called the Bonferroni inequality, can be applied. The Bonferroni inequality provides a bound on the probability that all the intervals lead to correct statements. In all three examples individual confidence statements are available. In (1) and (2) it will be assumed that the usual conditions required for the two-sample t test to be applicable are satisfied, i.e., if X, Y, and Z denote the responses to the three treatments [the third treatment being the standard in (1)], then X, Y and Z are assumed to be approximately normally distributed with unknown means μ_x, μ_y, and μ_z, respectively, and common unknown variance σ^2. Using the confidence intervals based on the two-sample t test, the *individual* $100(1 - \alpha)\%$ confidence interval for $(\mu_z - \mu_x)$ in example (1) is given by

$$\bar{Z} - \bar{X} \pm t_{\alpha/2; \, n_z + n_x - 2} \sqrt{\frac{1}{n_z} + \frac{1}{n_x}} \sqrt{\frac{\sum (Z_i - \bar{Z})^2 + \sum (X_i - \bar{X})^2}{n_z + n_x - 2}}, \, {}^{1}$$

and the $100(1 - \alpha)\%$ confidence interval for $\mu_z - \mu_y$ in example (1) is given by

$$\bar{Z} - \bar{Y} \pm t_{\alpha/2; \, n_z - n_y - 2} \sqrt{\frac{1}{n_z} + \frac{1}{n_y}} \sqrt{\frac{\sum (Z_i - \bar{Z})^2 + \sum (Y_i - \bar{Y})^2}{n_z + n_y - 2}}.$$

Since both intervals depend on the Z's, it is evident that they are not independent. What then can be said about the probability of both statements being correct? The Bonferroni inequality states that the probability is *at least* $(1 - 2\alpha)$ that both intervals lead to correct statements. Thus, if an overall probability of at least 0.95 is desired, individual 97.5% confidence intervals should be used.

In general, the Bonferroni inequality can be presented as follows. Suppose that a total of k confidence intervals are desired, and the probability that the ith statement is incorrect is given by α_i. The Bonferroni inequality states that

$$P\{\text{all } k \text{ statements are correct}\} \geq 1 - (\alpha_1 + \alpha_2 + \cdots + \alpha_k). \, {}^{2}$$

---

[1] Generally, a shorter confidence interval can be obtained by utilizing all the data for estimating σ^2, i.e.,

$$\frac{\sum (Z_i - \bar{Z})^2 + \sum (X_i - \bar{X})^2}{n_z + n_x - 2}$$

is replaced by

$$\frac{\sum (Z_i - \bar{Z})^2 + \sum (X_i - \bar{X})^2 + \sum (Y_i - \bar{Y})^2}{n_z + n_x + n_y - 3}$$

and $t_{\alpha/2; \, n_z + n_x - 2}$ by $t_{\alpha/2; \, n_z + n_x + n_y - 3}$. However, this is not usually important for reasonable sample sizes.

[2] The proof is simple and given for $k = 2$. For this case, $1 - $ P{both statements are correct} = P{statement 1 is incorrect or statement 2 is incorrect or both are incorrect} $\leq$ P{statement 1 is incorrect} + P{statement 2 is incorrect} = $\alpha_1 + \alpha_2$. For $k > 2$, the proof is similar.

If the right-hand side is fixed in advance by the experimenter and set equal to $(1 - \alpha)$, the α_i can be chosen as desired provided that

$$\sum_{i=1}^{k} \alpha_i = \alpha.$$

However, they are usually chosen to be equal, i.e., $\alpha_i = \alpha/k$.

The Bonferroni inequality can be applied to cases (2) and (3). With the appropriate normality assumptions case (2) is similar to case (1) except that three confidence intervals are required, i.e., a confidence interval estimate of $\mu_x - \mu_y$ must also be included. The form of the confidence interval is exactly the same as for case (1) except that $100(1 - \alpha/3)\%$ confidence intervals are needed for the individual statements. For example, the appropriate confidence interval for $\mu_x - \mu_y$ is given by

$$\bar{X} - \bar{Y} \pm t_{\alpha/6;\, n_x + n_y - 2} \sqrt{\frac{1}{n_x} + \frac{1}{n_y}} \sqrt{\frac{\sum (X_i - \bar{X})^2 + \sum (Y_i - \bar{Y})^2}{n_x + n_y - 2}},$$

if the overall probability is to be at least $(1 - \alpha)$ that all three intervals lead to correct statements. If an experiment requires comparing r means, there are

$$k = \binom{r}{2} = \frac{r!}{2!(r-2)!}$$

pairwise comparisons so that $100(1 - \alpha/k)\%$ confidence intervals are needed for the individual statements.

Applying the Bonferroni inequality to case (3) leads to individual $100(1 - \alpha/2)\%$ confidence intervals for μ and σ^2 as follows:

$$\bar{X} \pm t_{\alpha/4;\, n-1} \frac{S}{\sqrt{n}}, \qquad \text{for } \mu$$

$$\left[\frac{(n-1)S^2}{\chi^2_{\alpha/4;\, n-1}}, \frac{(n-1)S^2}{\chi^2_{1-\alpha/4;\, n-1}} \right], \qquad \text{for } \sigma^2.$$

The overall probability is at least $(1 - \alpha)$ that both intervals lead to correct statements.

As noted previously, confidence statements can be used as tests of hypotheses. For case (1), the hypothesis to be tested at the α level of significance is

H: $\mu_z = \mu_x$ and $\mu_z = \mu_y$

A: At least one of the above is false.

The hypothesis is accepted if both the individual confidence intervals include zero (since $\mu_z = \mu_x$ and $\mu_z = \mu_y$ are equivalent to $\mu_z - \mu_x = 0$ and $\mu_z - \mu_y = 0$).

For case (2), the hypothesis to be tested at the α level of significance is

H: $\mu_x = \mu_y = \mu_z$

A: At least one mean differs.

The hypothesis is accepted if all three individual confidence intervals include zero.

For case (3), the hypothesis to be tested at the α level of significance is

$$H: \quad \mu = \mu_0 \text{ and } \sigma^2 = \sigma_0^2$$
$$A: \quad \text{At least one of the above is false.}$$

The hypothesis is accepted if the individual confidence interval for μ includes μ_0 and if the individual confidence interval for σ^2 includes σ_0^2.

There are alternative solutions to the three situations presented, many of which lead to *exact* probability statements rather than inequalities, e.g., an exact solution to case (2) is given in Chapter 10. Although the Bonferroni inequality leads to a lower bound on the probability of making correct statements, it is surprisingly good, particularly when α_i is small (near 0.01) and k is not too large (in the vicinity of 6). In these situations the confidence intervals obtained by using Bonferroni inequalities compare favorably, and are often superior, to exact solutions. Furthermore, the Bonferroni technique is particularly useful since it is applicable to a broad class of problems.

Example. Consider the example of Sec. 8.3.5 with a third producer of thermostats added. A comparison among all three, stated in terms of confidence intervals, is desired. This is case (2) described previously. For manufacturers X and Y, the data given in Sec. 8.3.5 are applicable. The data for manufacturer Z are:

$$n_z = 23$$
$$\bar{Z} = 556.260$$
$$\sum (Z_i - \bar{Z})^2 = 2017.360.$$

Suppose that confidence interval estimates for the differences in means are required so that the probability is at least 0.90 that all three intervals simultaneously result in correct statements. Applying the Bonferroni inequality leads to individual $100(1 - 0.10/3)\% = 96.67\%$ confidence intervals as follows:

$$\bar{X} - \bar{Y} \pm t_{0.0167; 44} \sqrt{\frac{1}{n_x} + \frac{1}{n_y}} \sqrt{\frac{\sum (X_i - \bar{X})^2 + \sum (Y_i - \bar{Y})^2}{n_x + n_y - 2}}$$

$$= 551.491 - 549.926 \pm 2.240 \sqrt{\frac{1}{23} + \frac{1}{23}} \sqrt{\frac{2330.391 + 1703.130}{44}}$$

$$= [-4.76, 7.88], \text{ for } \mu_x - \mu_y,$$

$$\bar{X} - \bar{Z} \pm t_{0.0167; 44} \sqrt{\frac{1}{n_x} + \frac{1}{n_z}} \sqrt{\frac{\sum (X_i - \bar{X})^2 + \sum (Z_i - \bar{Z})^2}{n_x + n_z - 2}}$$

$$= 551.491 - 556.260 \pm 2.240 \sqrt{\frac{1}{23} + \frac{1}{23}} \sqrt{\frac{2330.391 + 2017.360}{44}}$$

$$= [-11.33, 1.79], \text{ for } \mu_x - \mu_z, \text{ and}$$

$$\bar{Y} - \bar{Z} \pm t_{0.0167;\,44}\sqrt{\frac{1}{n_y} + \frac{1}{n_z}}\,\sqrt{\frac{\sum (Y_i - \bar{Y})^2 + \sum (Z_i - \bar{Z})^2}{n_y + n_z - 2}}$$

$$= 549.926 - 556.260 \pm 2.240\sqrt{\frac{1}{23} + \frac{1}{23}}\,\sqrt{\frac{1703.130 + 2017.360}{44}}$$

$$= [-12.40,\ -0.26],\ \text{for}\ \mu_y - \mu_z.$$

Note that the confidence interval for $\mu_x - \mu_y$ is now wider than that obtained in Sec. 8.3.5. This is to be expected since a 96.67% confidence interval for this difference is obtained instead of a 90% interval. It may also be pointed out that it is possible to state that μ_y differs from μ_z (μ_y is smaller) since zero is not included in the confidence interval.

8.3.10 BAYESIAN INTERVALS

An alternative to the confidence interval approach for finding interval estimates is to obtain a Bayesian interval. Recall that using a Bayesian approach assumes the existence of a prior distribution for the parameter θ. A random sample of size n is drawn, and the posterior distribution of θ given the outcome of the random sample is computed. This posterior distribution can be used for obtaining probability statements about the parameter θ; in particular, limits L and U can be found such that

$$P\{L \leqq \theta \leqq U\} = 1 - \alpha.$$

The interval $[L, U]$ is called a $100(1 - \alpha)\%$ Bayesian interval for θ corresponding to the given prior distribution.

As an example, return to the Rockwell hardness problem. The unbranded shafts are assumed to be normally distributed with unknown mean μ and known standard deviation equal to 2. The prior distribution of μ is also assumed to be normally distributed with mean equal to 72 and standard deviation equal to 1.5. A random sample of size 4 was drawn. The posterior

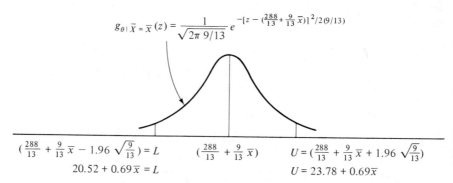

Fig. 8.4. *The 95% Bayesian interval for θ corresponding to a normal prior distribution.*

distribution of μ, given $\bar{x}$, was obtained and found to be normally distributed with mean equal to $[(288/13) + (9/13)\bar{x}]$ and variance equal to $9/13$. The posterior distribution is shown in Fig. 8.4, and the limits $L = 20.52 + 0.69\bar{x}$ and $U = 23.78 + 0.69\bar{x}$ form the 95 % Bayesian interval for μ corresponding to the normal prior distribution.

Note that this 95 % Bayesian interval differs from the 95 % confidence interval for the mean of a normal distribution with known variance $\sigma^2 = 4$ (given by $L = \bar{x} - 1.96$ and $U = \bar{x} + 1.96$). However, these two intervals will coincide if the prior distribution of μ is uniform. Unfortunately, this is the only prior distribution for which this phenomenon occurs.

Some posterior distributions are presented in Table 8.1 which can be used to obtain Bayesian interval estimates.

8.4 Statistical Tolerance Limits

The quality of a manufactured product is often specified by setting a range, the bounds of which are called tolerance limits. These limits have the property that a certain percentage of the product may be expected to fall within these limits. In Chapter 3, a study of tolerance limits was made under the assumption that the quality is normally distributed with *known* mean, μ, and *known* standard deviation, σ. Tolerance limits were defined as those limits within which $100(1 - \alpha)\%$ of the product falls. These limits were formed by adding to and subtracting from μ the quantity $K_{\alpha/2}\sigma$. Unfortunately, in practice, the values μ and σ are rarely known and frequently even assumptions about the nature of the distribution are unwarranted. However, data in the form of the outcome of a random sample may be available. Hence, some other statement is desirable about limits within which much of the distribution is expected to fall, based on these data. Two solutions will be presented, one where normality can be assumed and one which is distribution free.

8.4.1 STATISTICAL TOLERANCE LIMITS BASED ON THE NORMAL DISTRIBUTION

Suppose a random variable X is normally distributed with unknown mean μ and unknown standard deviation σ, and a random sample of size n is observed. A natural way to proceed is to replace the tolerance limits $\mu \pm K_{\alpha/2}\sigma$ defined in Chapter 3 by substituting the estimates $\bar{X}$ and S for the unknown parameters. Whereas previously it could be stated that 95 % of the distribution lies between $\mu \pm 1.96\sigma$, this statement cannot be extended to the interval $\bar{X} \pm 1.96S$. Since $\bar{X}$ and S are random variables, the limits $\bar{X} \pm 1.96S$ depend on the particular outcome of the sample. Different samples will lead to different limits. The closeness of these limits to $\mu \pm 1.96\sigma$ depends on the closeness of the estimates to their true value.

It is evident, then, that the fraction of the distribution included between $\bar{X} \pm K_{\alpha/2}S$ will not always be $1 - \alpha$. However, it is possible to determine a constant K such that, in a large series of samples from a normal distribution, a fixed proportion γ of the intervals $\bar{X} \pm KS$ will include $100(1 - \alpha)\%$ or more of the distribution. Thus, *statistical tolerance limits for a normal distribution are given by* $[L = \bar{X} - KS, U = \bar{X} + KS]$ *and have the property that the probability is equal to a preassigned value γ that the interval includes at least a specified proportion $1 - \alpha$ of the distribution.* Note that in most practical situations γ is usually a large number close to 1. Statistical tolerance limits should not be confused with a confidence interval for the mean of the normal distribution. Confidence limits for the mean of a normal distribution are such that in a given fraction, say, 0.95, of the samples from which they are computed, the interval bounded by the limits will include the true mean of the distribution.

Values of K for the parameters $n = 2(1)50$; $\gamma = 0.75, 0.90, 0.95, 0.99$; and $\alpha = 0.25, 0.10, 0.05, 0.01, 0.001$ are given in Table 8.3.

Example. In the manufacture of electron tubes to be used in stable amplifiers, it is desired to know, with probability 0.99, limits within which 90% of the future tube transconductances lie. The required tests are made on 9 tubes and the transconductances, observed in micro ohms, are 4,330; 4,287; 4,450; 4,295; 4,340; 4,407; 4,295; 4,388; and 4,356. From Table 8.3, the value of K corresponding to $n = 9$, $\gamma = 0.99$ and $\alpha = 0.10$ is 3.822. For this example, $\bar{X}$ is found to be 4,349.8 and S is 56.18. Thus, the tolerance limits are given by $4,349.8 \pm (3.822)(56.18) = [4,135.1, 4,564.5]$.

8.4.2 ONE-SIDED STATISTICAL TOLERANCE LIMITS BASED ON THE NORMAL DISTRIBUTION

Instead of specifying two limits, it is sometimes appropriate to specify a single limit, $\bar{X} - KS$, such that a fixed proportion of the distribution will be greater than this limit, or to specify a single limit, $\bar{X} + KS$, such that a fixed proportion of the distribution will be smaller than this limit. Such single limits are called one-sided tolerance limits.

Values of K for the parameters $n = 3(1)25$ and $30(5)50$; $\gamma = 0.75, 0.90, 0.95, 0.99$; and $\alpha = 0.25, 0.10, 0.05, 0.01, 0.001$ are given in Table 8.4.

Example. A manufacturer of light bulbs would like to specify a single lower limit above which he can be assured, with probability of 0.95, that 99% of his production will lie. A sample of 30 bulbs is taken and the sample mean and sample standard deviation are found to be 987.2 and 5.963, respectively. A value of $K = 3.064$ corresponding to $n = 30$, $\gamma = 0.95$, and $\alpha = 0.01$ is obtained from Table 8.4. The required lower tolerance limit is given by $\bar{X} - KS = 987.2 - (3.064)(5.963) = 968.9$.

8.4.3 DISTRIBUTION-FREE TOLERANCE LIMITS

There exist tolerance limits which are independent of the form of the underlying distribution. However, tolerance limits based on the normal distribution are substantially shorter for a fixed sample size than those based on no assumption about the underlying distribution. However, if the assumption of normality is violated, the limits obtained in the manner described above may be irrelevant. Distribution-free tolerance limits have limited industrial use since they usually require an enormous sample size for reasonable probability statements. The distribution-free tolerance limits given in this section are based on the smallest observation and the largest observation in a sample of size n.

For two-sided tolerance limits, the number of observations required so that the probability is γ that at least $100(1 - \alpha)\%$ of the distribution will lie between the smallest and largest observations of the sample is given approximately by

$$n \cong \left(\frac{2 - \alpha}{\alpha}\right)\left(\frac{\chi^2_{1-\gamma;4}}{4}\right) + \frac{1}{2}$$

where $\chi^2_{1-\gamma;4}$ is the $100(1 - \gamma)$ percentage point of the chi-square distribution with 4 degrees of freedom. To find distribution-free tolerance limits which contain at least 90% of the distribution with probability 0.99, i.e., $\alpha = 0.10$, $\gamma = 0.99$, and $\chi^2_{0.01;4} = 13.277$, a sample of size

$$n \cong (1.9/0.1)(13.277/4) + \tfrac{1}{2} = 64$$

is required. Thus, if a sample of size 64 is taken, the probability is 0.99 that at least 90% of the distribution will lie between the smallest and largest observations of the sample. This is to be compared with the example of Sec. 8.4.1 which required a sample of size 9 to be assured with probability 0.99 that at least 90% of the distribution will lie between $\bar{X} \pm 3.822S$ if the distribution is normal.

For one-sided tolerance limits, the number of observations required so that the probability is γ that at least $100(1 - \alpha)\%$ of the distribution will exceed the smallest observation (or be less than the largest observation) of the sample is given by

$$n = \frac{\log(1 - \gamma)}{\log(1 - \alpha)}.^{[1]}$$

In order to find the lower distribution-free tolerance limit which contains at least 99% of the distribution with probability 0.95, i.e., $\alpha = 0.01$ and $\gamma = 0.95$, a sample of size

$$n = \frac{\log(0.05)}{\log(0.99)} = 300$$

is required. Thus, if a sample of size 300 is taken, the probability is 0.95 that at least 99% of the distribution will exceed the smallest observation of

---

[1] The logarithm to any base may be used.

Table 8.3 Tolerance Factors for Normal Distributions[1]

Factors K such that the Probability is γ that at least a Proportion $1 - \alpha$ of the Distribution Will Be Included between $\bar{X} \pm KS$, where $\bar{X}$ and S Are Estimators of the Mean and the Standard Deviation Computed from a Random Sample of Size n.

n	γ = 0.75					γ = 0.90				
α	0.25	0.10	0.05	0.01	0.001	0.25	0.10	0.05	0.01	0.001
2	4.498	6.301	7.414	9.531	11.920	11.407	15.978	18.800	24.167	30.227
3	2.501	3.538	4.187	5.431	6.844	4.132	5.847	6.919	8.974	11.309
4	2.035	2.892	3.431	4.471	5.657	2.932	4.166	4.943	6.440	8.149
5	1.825	2.599	3.088	4.033	5.117	2.454	3.494	4.152	5.423	6.879
6	1.704	2.429	2.889	3.779	4.802	2.196	3.131	3.723	4.870	6.188
7	1.624	2.318	2.757	3.611	4.593	2.034	2.902	3.452	4.521	5.750
8	1.568	2.238	2.663	3.491	4.444	1.921	2.743	3.264	4.278	5.446
9	1.525	2.178	2.593	3.400	4.330	1.839	2.626	3.125	4.098	5.220
10	1.492	2.131	2.537	3.328	4.241	1.775	2.535	3.018	3.959	5.046
11	1.465	2.093	2.493	3.271	4.169	1.724	2.463	2.933	3.849	4.906
12	1.443	2.062	2.456	3.223	4.110	1.683	2.404	2.863	3.758	4.792
13	1.425	2.036	2.424	3.183	4.059	1.648	2.355	2.805	3.682	4.697
14	1.409	2.013	2.398	3.148	4.016	1.619	2.314	2.756	3.618	4.615
15	1.395	1.994	2.375	3.118	3.979	1.594	2.278	2.713	3.562	4.545
16	1.383	1.977	2.355	3.092	3.946	1.572	2.246	2.676	3.514	4.484
17	1.372	1.962	2.337	3.069	3.917	1.552	2.219	2.643	3.471	4.430
18	1.363	1.948	2.321	3.048	3.891	1.535	2.194	2.614	3.433	4.382
19	1.355	1.936	2.307	3.030	3.867	1.520	2.172	2.588	3.399	4.339
20	1.347	1.925	2.294	3.013	3.846	1.506	2.152	2.564	3.368	4.300
21	1.340	1.915	2.282	2.998	3.827	1.493	2.135	2.543	3.340	4.264
22	1.334	1.906	2.271	2.984	3.809	1.482	2.118	2.524	3.315	4.232
23	1.328	1.898	2.261	2.971	3.793	1.471	2.103	2.506	3.292	4.203
24	1.322	1.891	2.252	2.959	3.778	1.462	2.089	2.489	3.270	4.176
25	1.317	1.883	2.244	2.948	3.764	1.453	2.077	2.474	3.251	4.151
26	1.313	1.877	2.236	2.938	3.751	1.444	2.065	2.460	3.232	4.127
27	1.309	1.871	2.229	2.929	3.740	1.437	2.054	2.447	3.215	4.106
28	1.305	1.865	2.222	2.920	3.728	1.430	2.044	2.435	3.199	4.085
29	1.301	1.860	2.216	2.911	3.718	1.423	2.034	2.424	3.184	4.066
30	1.297	1.855	2.210	2.904	3.708	1.417	2.025	2.413	3.170	4.049
31	1.294	1.850	2.204	2.896	3.699	1.411	2.017	2.403	3.157	4.032
32	1.291	1.846	2.199	2.890	3.690	1.405	2.009	2.393	3.145	4.016
33	1.288	1.842	2.194	2.883	3.682	1.400	2.001	2.385	3.133	4.001
34	1.285	1.838	2.189	2.877	3.674	1.395	1.994	2.376	3.122	3.987
35	1.283	1.834	2.185	2.871	3.667	1.390	1.988	2.368	3.112	3.974
36	1.280	1.830	2.181	2.866	3.660	1.386	1.981	2.361	3.102	3.961
37	1.278	1.827	2.177	2.860	3.653	1.381	1.975	2.353	3.092	3.949
38	1.275	1.824	2.173	2.855	3.647	1.377	1.969	2.346	3.083	3.938
39	1.273	1.821	2.169	2.850	3.641	1.374	1.964	2.340	3.075	3.927
40	1.271	1.818	2.166	2.846	3.635	1.370	1.959	2.334	3.066	3.917
41	1.269	1.815	2.162	2.841	3.629	1.366	1.954	2.328	3.059	3.907
42	1.267	1.812	2.159	2.837	3.624	1.363	1.949	2.322	3.051	3.897
43	1.266	1.810	2.156	2.833	3.619	1.360	1.944	2.316	3.044	3.888
44	1.264	1.807	2.153	2.829	3.614	1.357	1.940	2.311	3.037	3.879
45	1.262	1.805	2.150	2.826	3.609	1.354	1.935	2.306	3.030	3.871
46	1.261	1.802	2.148	2.822	3.605	1.351	1.931	2.301	3.024	3.863
47	1.259	1.800	2.145	2.819	3.600	1.348	1.927	2.297	3.018	3.855
48	1.258	1.798	2.143	2.815	3.596	1.345	1.924	2.292	3.012	3.847
49	1.256	1.796	2.140	2.812	3.592	1.343	1.920	2.288	3.006	3.840
50	1.255	1.794	2.138	2.809	3.588	1.340	1.916	2.284	3.001	3.833

[1] Reproduced from C. Eisenhart, M. W. Hastay, and W. A. Wallis, *Techniques of Statistical Analysis*, Chapter 2, McGraw-Hill Book Company, Inc., New York, 1947.

Table 8.3 (continued). Tolerance Factors for Normal Distributions

n	$\gamma = 0.95$					$\gamma = 0.99$				
α	0.25	0.10	0.05	0.01	0.001	0.25	0.10	0.05	0.01	0.001
2	22.858	32.019	37.674	48.430	60.573	114.363	160.193	188.491	242.300	303.054
3	5.922	8.380	9.916	12.861	16.208	13.378	18.930	22.401	29.055	36.616
4	3.779	5.369	6.370	8.299	10.502	6.614	9.398	11.150	14.527	18.383
5	3.002	4.275	5.079	6.634	8.415	4.643	6.612	7.855	10.260	13.015
6	2.604	3.712	4.414	5.775	7.337	3.743	5.337	6.345	8.301	10.548
7	2.361	3.369	4.007	5.248	6.676	3.233	4.613	5.488	7.187	9.142
8	2.197	3.136	3.732	4.891	6.226	2.905	4.147	4.936	6.468	8.234
9	2.078	2.967	3.532	4.631	5.899	2.677	3.822	4.550	5.966	7.600
10	1.987	2.839	3.379	4.433	5.649	2.508	3.582	4.265	5.594	7.129
11	1.916	2.737	3.259	4.277	5.452	2.378	3.397	4.045	5.308	6.766
12	1.858	2.655	3.162	4.150	5.291	2.274	3.250	3.870	5.079	6.477
13	1.810	2.587	3.081	4.044	5.158	2.190	3.130	3.727	4.893	6.240
14	1.770	2.529	3.012	3.955	5.045	2.120	3.029	3.608	4.737	6.043
15	1.735	2.480	2.954	3.878	4.949	2.060	2.945	3.507	4.605	5.876
16	1.705	2.437	2.903	3.812	4.865	2.009	2.872	3.421	4.492	5.732
17	1.679	2.400	2.858	3.754	4.791	1.965	2.808	3.345	4.393	5.607
18	1.655	2.366	2.819	3.702	4.725	1.926	2.753	3.279	4.307	5.497
19	1.635	2.337	2.784	3.656	4.667	1.891	2.703	3.221	4.230	5.399
20	1.616	2.310	2.752	3.615	4.614	1.860	2.659	3.168	4.161	5.312
21	1.599	2.286	2.723	3.577	4.567	1.833	2.620	3.121	4.100	5.234
22	1.584	2.264	2.697	3.543	4.523	1.808	2.584	3.078	4.044	5.163
23	1.570	2.244	2.673	3.512	4.484	1.785	2.551	3.040	3.993	5.098
24	1.557	2.225	2.651	3.483	4.447	1.764	2.522	3.004	3.947	5.039
25	1.545	2.208	2.631	3.457	4.413	1.745	2.494	2.972	3.904	4.985
26	1.534	2.193	2.612	3.432	4.382	1.727	2.469	2.941	3.865	4.935
27	1.523	2.178	2.595	3.409	4.353	1.711	2.446	2.914	3.828	4.888
28	1.514	2.164	2.579	3.388	4.326	1.695	2.424	2.888	3.794	4.845
29	1.505	2.152	2.554	3.368	4.301	1.681	2.404	2.864	3.763	4.805
30	1.497	2.140	2.549	3.350	4.278	1.668	2.385	2.841	3.733	4.768
31	1.489	2.129	2.536	3.332	4.256	1.656	2.367	2.820	3.706	4.732
32	1.481	2.118	2.524	3.316	4.235	1.644	2.351	2.801	3.680	4.699
33	1.475	2.108	2.512	3.300	4.215	1.633	2.335	2.782	3.655	4.668
34	1.468	2.099	2.501	3.286	4.197	1.623	2.320	2.764	3.632	4.639
35	1.462	2.090	2.490	3.272	4.179	1.613	2.306	2.748	3.611	4.611
36	1.455	2.081	2.479	3.258	4.161	1.604	2.293	2.732	3.590	4.585
37	1.450	2.073	2.470	3.246	4.146	1.595	2.281	2.717	3.571	4.560
38	1.446	2.068	2.464	3.237	4.134	1.587	2.269	2.703	3.552	4.537
39	1.441	2.060	2.455	3.226	4.120	1.579	2.257	2.690	3.534	4.514
40	1.435	2.052	2.445	3.213	4.104	1.571	2.247	2.677	3.518	4.493
41	1.430	2.045	2.437	3.202	4.090	1.564	2.236	2.665	3.502	4.472
42	1.426	2.039	2.429	3.192	4.077	1.557	2.227	2.653	3.486	4.453
43	1.422	2.033	2.422	3.183	4.065	1.551	2.217	2.642	3.472	4.434
44	1.418	2.027	2.415	3.173	4.053	1.545	2.208	2.631	3.458	4.416
45	1.414	2.021	2.408	3.165	4.042	1.539	2.200	2.621	3.444	4.399
46	1.410	2.016	2.402	3.156	4.031	1.533	2.192	2.611	3.431	4.383
47	1.406	2.011	2.396	3.148	4.021	1.527	2.184	2.602	3.419	4.367
48	1.403	2.006	2.390	3.140	4.011	1.522	2.176	2.593	3.407	4.352
49	1.399	2.001	2.384	3.133	4.002	1.517	2.169	2.584	3.396	4.337
50	1.396	1.969	2.379	3.126	3.993	1.512	2.162	2.576	3.385	4.323

313

Table 8.4 One-Sided Tolerance Factors for Normal Distributions

Factors K such that the Probability Is γ that at least a Proportion 1 − α of the Distribution Will Be Less than $\bar{X} + KS$ (or Greater than $\bar{X} − KS$), where $\bar{X}$ and S Are Estimators of the Mean and the Standard Deviation Computed from a Random Sample of Size n.

n	$\gamma = 0.75$					$\gamma = 0.90$				
α	0.25	0.10	0.05	0.01	0.001	0.25	0.10	0.05	0.01	0.001
3	1.464	2.501	3.152	4.396	5.805	2.602	4.258	5.310	7.340	9.651
4	1.256	2.134	2.680	3.726	4.910	1.972	3.187	3.957	5.437	7.128
5	1.152	1.961	2.463	3.421	4.507	1.698	2.742	3.400	4.666	6.112
6	1.087	1.860	2.336	3.243	4.273	1.540	2.494	3.091	4.242	5.556
7	1.043	1.791	2.250	3.126	4.118	1.435	2.333	2.894	3.972	5.201
8	1.010	1.740	2.190	3.042	4.008	1.360	2.219	2.755	3.783	4.955
9	0.984	1.702	2.141	2.977	3.924	1.302	2.133	2.649	3.641	4.772
10	0.964	1.671	2.103	2.927	3.858	1.257	2.065	2.568	3.532	4.629
11	0.947	1.646	2.073	2.885	3.804	1.219	2.012	2.503	3.444	4.515
12	0.933	1.624	2.048	2.851	3.760	1.188	1.966	2.448	3.371	4.420
13	0.919	1.606	2.026	2.822	3.722	1.162	1.928	2.403	3.310	4.341
14	0.909	1.591	2.007	2.796	3.690	1.139	1.895	2.363	3.257	4.274
15	0.899	1.577	1.991	2.776	3.661	1.119	1.866	2.329	3.212	4.215
16	0.891	1.566	1.977	2.756	3.637	1.101	1.842	2.299	3.172	4.164
17	0.883	1.554	1.964	2.739	3.615	1.085	1.820	2.272	3.136	4.118
18	0.876	1.544	1.951	2.723	3.595	1.071	1.800	2.249	3.106	4.078
19	0.870	1.536	1.942	2.710	3.577	1.058	1.781	2.228	3.078	4.041
20	0.865	1.528	1.933	2.697	3.561	1.046	1.765	2.208	3.052	4.009
21	0.859	1.520	1.923	2.686	3.545	1.035	1.750	2.190	3.028	3.979
22	0.854	1.514	1.916	2.675	3.532	1.025	1.736	2.174	3.007	3.952
23	0.849	1.508	1.907	2.665	3.520	1.016	1.724	2.159	2.987	3.927
24	0.845	1.502	1.901	2.656	3.509	1.007	1.712	2.145	2.969	3.904
25	0.842	1.496	1.895	2.647	3.497	0.999	1.702	2.132	2.952	3.882
30	0.825	1.475	1.869	2.613	3.454	0.966	1.657	2.080	2.884	3.794
35	0.812	1.458	1.849	2.588	3.421	0.942	1.623	2.041	2.833	3.730
40	0.803	1.445	1.834	2.568	3.395	0.923	1.598	2.010	2.793	3.679
45	0.795	1.435	1.821	2.552	3.375	0.908	1.577	1.986	2.762	3.638
50	0.788	1.426	1.8 11	2.538	3.358	0.894	1.560	1.965	2.735	3.604

n	γ = 0.95					γ = 0.99				
α	0.25	0.10	0.05	0.01	0.001	0.25	0.10	0.05	0.01	0.001
3	3.804	6.158	7.655	10.552	13.857					
4	2.619	4.163	5.145	7.042	9.215					
5	2.149	3.407	4.202	5.741	7.501					
6	1.895	3.006	3.707	5.062	6.612	2.849	4.408	5.409	7.334	9.540
7	1.732	2.755	3.399	4.641	6.061	2.490	3.856	4.730	6.411	8.348
8	1.617	2.582	3.188	4.353	5.686	2.252	3.496	4.287	5.811	7.566
9	1.532	2.454	3.031	4.143	5.414	2.085	3.242	3.971	5.389	7.014
10	1.465	2.355	2.911	3.981	5.203	1.954	3.048	3.739	5.075	6.603
11	1.411	2.275	2.815	3.852	5.036	1.854	2.897	3.557	4.828	6.284
12	1.366	2.210	2.736	3.747	4.900	1.771	2.773	3.410	4.633	6.032
13	1.329	2.155	2.670	3.659	4.787	1.702	2.677	3.290	4.472	5.826
14	1.296	2.108	2.614	3.585	4.690	1.645	2.592	3.189	4.336	5.651
15	1.268	2.068	2.566	3.520	4.607	1.596	2.521	3.102	4.224	5.507
16	1.242	2.032	2.523	3.463	4.534	1.553	2.458	3.028	4.124	5.374
17	1.220	2.001	2.486	3.415	4.471	1.514	2.405	2.962	4.038	5.268
18	1.200	1.974	2.453	3.370	4.415	1.481	2.357	2.906	3.961	5.167
19	1.183	1.949	2.423	3.331	4.364	1.450	2.315	2.855	3.893	5.078
20	1.167	1.926	2.396	3.295	4.319	1.424	2.275	2.807	3.832	5.003
21	1.152	1.905	2.371	3.262	4.276	1.397	2.241	2.768	3.776	4.932
22	1.138	1.887	2.350	3.233	4.238	1.376	2.208	2.729	3.727	4.866
23	1.126	1.869	2.329	3.206	4.204	1.355	2.179	2.693	3.680	4.806
24	1.114	1.853	2.309	3.181	4.171	1.336	2.154	2.663	3.638	4.755
25	1.103	1.838	2.292	3.158	4.143	1.319	2.129	2.632	3.601	4.706
30	1.059	1.778	2.220	3.064	4.022	1.249	2.029	2.516	3.446	4.508
35	1.025	1.732	2.166	2.994	3.934	1.195	1.957	2.431	3.334	4.364
40	0.999	1.697	2.126	2.941	3.866	1.154	1.902	2.365	3.250	4.255
45	0.978	1.669	2.092	2.897	3.811	1.122	1.857	2.313	3.181	4.168
50	0.961	1.646	2.065	2.863	3.766	1.096	1.821	2.296	3.124	4.096

the sample. This is to be compared with the example of Sec. 8.4.2 which required a sample of size 30 to be assured with probability 0.95 that at least 99% of the distribution will exceed $\bar{X} - 3.064S$ if the distribution is normal.

PROBLEMS

In the following problems it may be assumed (unless otherwise stated) that a random sample is drawn and each random variable is approximately normally distributed. When two random samples are required, it may be assumed that the second random sample is independent of the first and each random variable is also approximately normally distributed.

1. Suppose that T_1 and T_2 are estimators of θ with $E(T_1) = \theta$, $E(T_2) = \theta/2$, $\text{Var}(T_1) = 6$, and $\text{Var}(T_2) = 2$. Which is a better estimator of θ? Why?

2. In sampling from a binomial distribution, show that the proportion of events observed in a sample is an unbiased estimator of the true probability that an event will occur. What is the variance of the observed proportion?

3. Suppose that T_1 and T_2 are estimators of θ. $E(T_1) = \theta$ and $E(T_2) > \theta$. The variance of $T_1 = 4$, the variance of $T_2 = 2$, and $E(T_2 - \theta)^2 = 7$. Which is a "better" estimator of θ? Why?

4. Suppose that T is an estimator of θ based on a random sample of size n. If $E(X_i) = \theta$ and $T = \sum a_i X_i$, what constraint must a_i satisfy in order that T be an unbiased estimator of θ?

5. Prove that $S'^2 = \sum_{i=1}^{n} (X_i - \bar{X})^2/n$ is a biased estimator of σ^2.

6. The number of defective rivets, D, on an airplane wing can be assumed to have a Poisson distribution with parameter λ, i.e.,

$$P_D(k, \lambda) = P\{D = k\} = \frac{\lambda^k e^{-\lambda}}{k!}, \qquad \text{for } k = 0, 1, 2, \ldots.$$

A random sample of n wings is observed. (a) Find $\hat{\lambda}$, the maximum likelihood estimator of λ. (b) Is this estimator unbiased? (c) Find the Cramér-Rao lower bound on the variance of unbiased estimators. Note that $E(D) = \lambda$ and $E(D)^2 = \lambda^2 + \lambda$. (d) Is $\hat{\lambda}$ the minimum variance unbiased estimator of λ? Justify your answer. (e) Find the method-of-moments estimator of λ.

7. The life of vacuum tubes follows an exponential distribution; i.e., the density function is given by

$$f_X(z, \theta) = \begin{cases} \dfrac{1}{\theta} e^{-z/\theta} & \text{for } z \geq 0, \\ 0 & \text{otherwise.} \end{cases}$$

A random sample of size n is taken. (a) Find an estimator of θ, using the method of moments. (b) Find the maximum likelihood estimator, $\hat{\theta}$, of θ. (c) Find the maximum likelihood estimator of $\eta = 1/\theta$. (d) The reliability for a vacuum tube at time τ_0 is defined to be

$$R(\tau_0) = P\{X > \tau_0\} = e^{-\tau_0/\theta}.$$

Find the maximum likelihood estimator $\hat{R}(\tau_0)$ of $R(\tau_0)$.

8. In Problem 7, find the Cramér-Rao lower bound on the variance of unbiased estimators. Does the maximum likelihood estimator coincide with the minimum variance unbiased estimator of θ?

9. The tensile strength of rubber is known to be normally distributed with unknown mean μ and known standard deviation equal to 80 psi. The lower specification limit L for this material is 2840 psi. The fraction defective, p, is then defined as the fraction of items falling below this lower limit. A random sample of size 25 is to be drawn. Find the expression for the maximum likelihood estimator of p. (*Hint*: First find the maximum likelihood estimator of μ.)

10. In Sec. 8.2.5, the maximum likelihood estimator, $\hat{p}$, for the parameter p of a Bernoulli random variable was obtained based upon a random sample of size n. It is also known that the sum, T, of n independent identically Bernoulli distributed random variables with parameter p has a binomial distribution with parameters n and p, i.e.,

$$P\{T = k\} = \binom{n}{k}p^k(1 - p)^{n-k}, \text{ for } k = 0, 1, \ldots, n.$$

Find the maximum likelihood estimator of p based on a single observation on the random variable T. Does it coincide with $\hat{p}$?

11. The time to failure, T, of transistors is assumed to have a Weibull distribution, i.e., the density function is given by

$$f_T(z) = \begin{cases} \dfrac{\beta}{\eta}z^{\beta-1}e^{-z^\beta/\eta} & \text{for } z \geq 0, \\ 0 & \text{otherwise,} \end{cases}$$

where $\beta, \eta > 0$. A random sample of size n is observed. If β is a known constant, (a) Find the maximum likelihood estimator of η. (b) The reliability of a transistor at time τ_0 is defined to be

$$R(\tau_0) = P\{T > \tau_0\} = e^{-\tau_0^\beta/\eta}.$$

Find the maximum likelihood estimator, $\hat{R}(\tau_0)$, of $R(\tau_0)$.

12. In the "process control" example introduced in Sec. 8.2.5, the maximum likelihood estimator was obtained of the parameter p of a Bernoulli random variable based on a random sample of size n. (a) Find the Cramér-Rao lower bound on the variance of all unbiased estimators. (b) Does the maximum likelihood estimator coincide with the minimum variance unbiased estimator of p? (c) Find the method-of-moments estimator of p.

13. In Problem 6, find the Bayes estimator of λ, using mean squared error as a loss function and assuming a uniform prior (over the interval $[-\infty, \infty]$) density function. (*Hint*: In Table 8.1, this prior distribution is a special case of the gamma with $c = 1$ and D approaching infinity.)

14. In Problem 6, find the Bayes estimator of λ, using mean squared error as a loss function and assuming that the prior distribution of λ is given by $e^{-\lambda}$ for $\lambda \geq 0$. (*Hint*: Use the results given in Table 8.1.)

15. In Problem 7, find the Bayes estimator of $\eta = 1/\theta$, using mean squared error as a loss function and assuming a uniform prior (over the interval $[-\infty, \infty]$) density function. (*Hint*: In Table 8.1, this prior distribution is a special case of the gamma with $c = 1$ and D approaching infinity.)

16. In Problem 7, find the Bayes estimator of $\eta = 1/\theta$, using mean squared error as a loss function and assuming that the prior distribution of η is given by $e^{-\eta}$ for $\eta \geq 0$. (*Hint*: Use the results given in Table 8.1.)

17. In the "process control" example introduced in Sec. 8.2.5, the maximum likelihood estimator was obtained of the parameter p of a Bernoulli random variable based on a random sample of size n. Find the Bayes estimator of p, using mean squared error as a loss function and assuming that the prior distribution of p is uniform over the interval $[0, 1]$. Find the Bayes estimator for p, assuming that the prior distribution of p is given by $6p(1 - p)$ over the interval $[0, 1]$. (*Hint*: Use the results given in Table 8.1.)

18. A 99% lower one-sided confidence statement is to be made about the average life of shoes. How many observations are necessary in order that the lower limit will not differ from the true average life by more than one-half month with probability 0.95? Assume that the standard deviation of shoe life is equal to 3/2 months.

19. The distribution of observations from carload lots of a given chemical can be shown to be normally distributed about the true average density of a given lot with standard deviation 0.005 g/cc. To obtain an estimate based on the mean of n samples which is within 0.002 g/cc of the true average density for the lot in 95% of the cases, what is the required sample size?

20. The warpwise breaking strength of a certain type of cloth was measured on four specimens from a lot with the following results (in psi): 182, 172, 176, 178. The standard deviation based on past experience is 5 psi. Find a 99% confidence interval for the lot average warpwise breaking strength.

21. The density of 27 explosive primers was determined and the sample average density was 1.53; the standard deviation is known to be 0.03. Find a 95% lower confidence interval for the density.

22. The sample average range of a random sample of 100 stored rockets was 2,208 yards. The standard deviation is known to be 40 yards. Find a 99% confidence interval for the true average range.

23. In Problem 22, how many observations would be required to make the length of the confidence interval 20 yards? 5 yards? 1 yard?

24. The following data represent porosity measurements on a sample from a shipment of coke. Find a 95% confidence interval for the true mean. Assume $\sigma = \frac{1}{4}$.

2.16	2.07	2.34	1.97	1.97	1.90
2.19	2.23	2.15	2.47	2.31	1.94
2.31	1.86	2.25	2.14	2.15	2.16
2.30	2.48	2.11	2.15	2.24	2.04
2.21	1.91	2.01	2.09	2.07	2.25

25. Work Problem 24 without assuming $\sigma = \frac{1}{4}$. (Note that $S = 0.153$.)

26. Work Problem 22, assuming that 40 yards is an estimate of σ based on the sample. How much does your answer change?

27. Work Problem 20 without assuming $\sigma = 5$.

28. Work Problem 21, assuming that $S = 0.04$ based on the sample.

29. Find a 99% confidence interval for the diameter of the rivet based on these data:

6.68	6.66	6.62	6.72
6.76	6.67	6.70	6.72
6.78	6.66	6.76	6.72
6.76	6.70	6.76	6.76
6.74	6.74	6.81	6.66
6.64	6.79	6.72	6.82
6.81	6.77	6.60	6.72
6.74	6.70	6.64	6.78
6.70	6.70	6.75	6.79

$\bar{X} = 6.7236$ $\sum X^2 = 1627.5595$

30. If $S = 0.04$ based on a sample of 27 primers, find a 90% confidence interval for σ.

31. Find a 95% confidence interval for σ based on the data of Problem 29.

32. The finished diameter on armored electric cable is normally distributed. A sample of 20 yields a mean of 0.790 and a sample standard deviation 0.010. Find 95% confidence intervals for both μ and σ (individually).

33. For a sample of 15, $\sum (X_i - \bar{X})^2 = 17.8$. Find upper, lower, and two-sided 95% confidence intervals for σ.

34. Given $S = 4$, find a 95% confidence interval for σ if S is based on 10 degrees of freedom; also 25.

35. An experiment is performed to test the difference in effectiveness of two methods of cultivating wheat. Ten patches of ground are treated with shallow plowing and fifteen with deep plowing. The mean yield per acre of the first group is 40.8 bushels and the mean for the second group is 44.7. Assume that the standard deviation of the shallow planting is 0.6 bushel and for the deep it is 0.8. Find a 90% confidence interval for the difference of yield.

36. Twenty rounds of standard ammunition and ten rounds of test ammunition were fired in random sequence. The average barrel pressure of the standard was 41,900 psi and the average pressure for the test was 44,200 psi. The standard deviations for both the standard and test were known to be 2,250 psi. Find upper, lower, and two-sided 95% confidence intervals for the change in pressure.

37. In the manufacture of wooden ladders, the side pieces for the ladders have greater shear (parallel to the grain of the wood) if the raw lumber is air dried rather than kiln dried. The standard deviation of both can be taken to be 25 psi. A sample of 25 observations of each type was taken resulting in the following means: air-dried (psi) 1,170 and kiln-dried (psi) 1,105. Find a 95% confidence interval for the increase.

38. In research on rocket propellants aimed at reducing the delay time between the application of the firing current and explosion, tests were run on a grade T propellant and a grade C. The standard deviation of both grades can be assumed to be 0.04. If $\bar{C} = 0.261$ second and $\bar{T} = 0.250$ second for 14 observations on each, find a 95% confidence interval for the change in times.

39. Two analysts took repeated readings on the hardness of city water. Find a 95% confidence interval for the analyst difference, assuming unknown but equal variance.

Coded Measures of Hardness

Analyst A X	Analyst B Y
0.46	0.82
0.62	0.61
0.37	0.89
0.40	0.51
0.44	0.33
0.58	0.48
0.48	0.23
0.53	0.25
	0.67
	0.88

$$\bar{X} = 0.485 \qquad \bar{Y} = 0.567$$
$$\sum (X_i - \bar{X})^2 = 0.0524 \qquad \sum (Y_i - \bar{Y})^2 = 0.5598$$

40. The following are 16 independent determinations of the melting point of a compound, eight made by analyst I and eight made by analyst II. Data are in degrees centigrade.

Analyst I (X)	Analyst II (Y)
164.4	163.5
169.7	162.8
169.2	163.0
169.5	163.2
161.8	160.7
168.7	161.5
169.5	160.9
163.9	162.1

$$\bar{X} = 167.0875 \qquad \bar{Y} = 162.2125$$
$$\sum (X_i - \bar{X})^2 = 70.81 \qquad \sum (Y_i - \bar{Y})^2 = 8.09$$

Find a 90% confidence interval for the mean difference between analysts, assuming unknown but equal variances.

41. The following are the burning times in seconds of floating smoke pots of two different types:

Type I (X)		Type II (Y)	
481	572	526	537
506	561	511	582
527	501	556	605
661	487	542	558
501	524	491	578

$$\bar{X} = 532.1 \qquad \bar{Y} = 548.6$$
$$\sum (X_i - \bar{X})^2 = 26394.9 \qquad \sum (Y_i - \bar{Y})^2 = 10724.4$$

Find a 95% confidence interval for the mean difference in burning times, assuming unknown but equal variances.

42. Work Problem 38, if unknown but equal variances are assumed and $S_C^2 = 0.00125$ and $S_T^2 = 0.00132$ are obtained from the sample.

43. Let X and Y be normally distributed with means μ_x and μ_y, variances σ_x^2 and σ_y^2. Let $\bar{X}$, $\bar{Y}$, S_x^2 and S_y^2 be the usual estimators based on n_x and n_y observations. How would you find a confidence interval for $\mu_x - \mu_y$ without assuming $\sigma_x = \sigma_y$?

44. In a study of reflex reactions, the strength of patellar reflex measured in degrees of arc for 15 men under two conditions yielded the following results:

Tensed	Relaxed
31	35
19	15
22	19
26	29
36	34
30	26
29	19
36	37
34	24
33	27
19	14
19	19
26	30
15	7
18	13

Find a 95% confidence interval for the change in reflex. *Hint*: Use paired t test.

45. The following table gives percent loss in tensile strength, following immersion in a corrosive solution, of paired samples of an alloy. One is subjected to stress and one is not.

Test number	Unstressed	Stressed
1	6.4	9.2
2	4.6	7.9
3	4.6	7.3
4	6.4	8.0
5	3.2	6.7
6	5.2	7.6
7	6.5	5.7
8	4.9	4.1
9	4.3	8.1
10	5.6	6.5
11	3.7	6.9
12	4.6	6.0

Find a 95% confidence interval for the expected difference in tensile strength.

46. In Problem 42, find a 95% confidence interval for σ_T/σ_C.

47. The following results are calculated for two samples from normal distributions:

$$X: \quad n_x = 8, \quad \sum X = 12, \quad \sum X^2 = 46$$
$$Y: \quad n_y = 11, \quad \sum Y = 22, \quad \sum Y^2 = 80$$

Find a 95% confidence interval for σ_x/σ_y.

48. Suppose that 15 trials are made on each of two treatments with the sample standard deviation ratio $S_x/S_y = 3.5$. Find upper, lower, and two-sided 90% confidence intervals for σ_x/σ_y.

49. Let $S_x^2 = 20$ and $S_y^2 = 6$. Find a 99% confidence interval for σ_x/σ_y, assuming that the sample variances are based on 25 degrees of freedom. Also 100 and 500.

50. In Problem 6, find the asymptotic distribution of $\sqrt{n}(\hat{\lambda} - \lambda)$. Find an approximate (based on this asymptotic distribution) 95% two-sided confidence interval for λ.

51. Referring to Problems 7 and 8, find the asymptotic distribution of $\sqrt{n}(\hat{\theta} - \theta)$. Find an approximate (based on the asymptotic distribution) 95% two-sided confidence interval for θ.

52. In Problem 11, find the asymptotic distribution of the maximum likelihood estimator of η and find an approximate 99% confidence interval for this parameter.

53. In Sec. 8.3.8, an approximate confidence interval for the parameter p of a Bernoulli random variable was presented, i.e.,

$$\hat{p} \pm K_{\alpha/2}\sqrt{\frac{\hat{p}(1 - \hat{p})}{n}}.$$

This was obtained from the expression

$$P\left\{-K_{\alpha/2} \leq \frac{\hat{p} - p}{\sqrt{p(1 - p)/n}} \leq K_{\alpha/2}\right\} = 1 - \alpha,$$

where $\hat{p}$ was substituted for p in the denominator. An alternative expression can be obtained by solving the expression in brackets directly for p. Show that this leads to

$$\frac{1}{(1 + K_{\alpha/2}^2/n)}\left\{\hat{p} + \frac{K_{\alpha/2}^2}{2n} \pm K_{\alpha/2}\left[\frac{\hat{p}(1 - \hat{p})}{n} + \frac{K_{\alpha/2}^2}{4n^2}\right]^{1/2}\right\}.$$

54. In Problem 39, a third analyst took ten repeated readings on the hardness of city water. The results for the third analyst are $\bar{Z} = 0.541$ and $\sum (Z_i - \bar{Z})^2 = 0.5124$. Find 95% simultaneous confidence limits for all pairwise differences between analysts (assume the variances are unknown but equal).

55. In Problem 40, two additional analysts each took eight independent determinations of the melting point of a compound. The results for the third analyst are $\bar{Z} = 163.0410$ and $\sum (Z_i - \bar{Z})^2 = 52.70$. The results for the fourth analyst are $\bar{W} = 154.0263$ and $\sum (W_i - \bar{W})^2 = 33.16$. Find 90% simultaneous confidence limits for all pairwise differences between analysts (assume the variances are unknown but equal).

56. In Problem 41, a third type of pot is included. The data are $\bar{Z} = 501.3$ and $\sum (Z_i - \bar{Z})^2 = 15214.7$ (based on 10 observations). Find 95% simultaneous confidence limits for all pairwise differences in pots (assume the variances are unknown but equal).

57. Suppose that 15 trials are made on each of three treatments with the sample variances given by $S_x^2 = 12$, $S_y^2 = 30$, and $S_z^2 = 39$. Find 85% simultaneous confidence limits for all pairwise ratios of the variances.

58. Suppose that 20 trials are made on each of four treatments with the sample variance given by $S_x^2 = 25$, $S_y^2 = 40$, $S_z^2 = 60$, and $S_w^2 = 85$. Find 94% simultaneous confidence limits for all pairwise ratios of the variances.

59. Referring to Problem 32, find a 95% simultaneous confidence interval for μ and σ.

60. In Problem 17, find the expression for a 95% Bayesian interval for p, assuming that the prior distribution is uniform over the interval [0, 1]. Do not evaluate. (*Hint*: Use the results given in Table 8.1.)

61. In Problem 13, find the expression for a 90% Bayesian interval for λ. Do not evaluate. (*Hint:* Use the results given in Table 8.1.)

62. In Problem 14, find the expression for a 99% Bayesian interval for λ. Do not evaluate. (*Hint:* Use the results given in Table 8.1.)

63. In Problem 15, find the expression for a 95% Bayesian interval for η. Do not evaluate. (*Hint*: Use the results given in Table 8.1.)

64. In Problem 16, find the expression for a 90% Bayesian interval for η. Do not evaluate. (*Hint*: Use the results given in Table 8.1.)

65. In the manufacture of cylindrical rods with a circular cross section which fits into a circular socket, it is desired to find a tolerance interval for the diameter. A sample of size 10 was taken with these results:

$$
\begin{array}{cc}
5.036 & 5.031 \\
5.085 & 5.064 \\
4.991 & 4.942 \\
4.935 & 5.051 \\
4.999 & 5.011
\end{array}
$$

$$\bar{X} = 5.0145 \qquad \sum (X_i - \bar{X})^2 = 0.0218685$$

Find a 90% tolerance interval for diameters with $\gamma = 0.95$.

66. The following observations were taken on the net volume in liters of bottles:

$$
\begin{array}{ccccc}
5.68 & 5.65 & 5.59 & 5.64 & 5.66 \\
5.61 & 5.62 & 5.64 & 5.63 & 5.61 \\
5.64 & 5.66 & 5.60 & 5.65 & 5.63 \\
5.67 & 5.64 & 5.60 & 5.60 & 5.65 \\
5.60 & 5.65 & 5.60 & 5.63 & 5.60
\end{array}
$$

$$\bar{X} = 5.63 \qquad \sum (X_i - \bar{X})^2 = 0.19605$$

Find a volume interval which will include at least 99% of the bottles with $\gamma = 0.95$.

67. A sample of 40 observations was taken on time of ignition of a rocket propellant with these results: $\bar{T} = 0.250$ second and $S_T^2 = 0.0132$. Find an interval which will contain at least 90% of the times with $\gamma = 0.99$.

68. In measuring the percent shrinkage on drying, 40 plastic clay test specimens produced these results:

19.3	20.5	17.9	17.3
15.8	16.9	17.1	19.5
20.7	18.5	22.5	19.1
18.4	18.7	18.8	17.5
14.9	12.3	19.4	16.8
17.3	19.5	17.4	16.3
21.3	23.4	18.5	19.0
16.1	18.8	17.5	18.2
18.6	18.3	16.5	17.4
20.5	16.9	17.5	18.2

$$\bar{X} = 18.2275 \qquad \sum (X_i - \bar{X})^2 = 154.16$$

Find an interval which will contain at least 75% of the product with probability 0.99.

69. In the manufacture of forgings used as terminal blocks at the end of an airplane wing span, the critical dimension was the lower limit of the slot width. In particular, it was desired that a single lower limit be specified above which it can be assured with probability 0.95 that at least 99% of the slot widths will lie. A sample of 30 forgings was taken and the sample mean and sample standard deviations were found to be 0.8750 and 0.0025 inch, respectively. Find the desired tolerance limit. How many observations would be required to find a non-parametric tolerance interval with these properties?

70. With the data in Problem 66, find a lower tolerance limit which at least 99% of the volume will exceed with probability 0.95. How many observations would be required for a non-parametric tolerance interval with these properties?

71. With the data of Problem 68, find an upper tolerance limit such that at least 75% of the product will lie below it with probability 0.99.

72. Suppose the largest observation in a sample of 300 observations is 21.7 and the smallest is 18.2. If one is interested in an interval which covers at least 99% of the product, what probability can be associated with the interval 18.2 to 21.7?

73. In Sec. 8.4.2 on one-sided tolerance factors for normal distributions with unknown mean and variance, factors K were presented as a function of n such that

$$P\left\{ \int_{\bar{X}-KS}^{\infty} \frac{1}{\sqrt{2\pi}\sigma} e^{-(z-\mu)^2/2\sigma^2}\, dz \geq 1 - \alpha \right\} = \gamma,$$

that is, the probability is γ that at least a proportion $1 - \alpha$ of the distribution will be greater than $\bar{X} - KS$. Suppose σ is known. Derive a factor K' such that when $n = 16$, the probability will be 0.95 that at least a proportion 9/10 of the distribution will be greater than $\bar{X} - K'\sigma$, i.e., find K' such that

$$P\left\{ \int_{\bar{X}-K'\sigma}^{\infty} \frac{1}{\sqrt{2\pi}\sigma} e^{-(z-\mu)^2/2\sigma^2}\, dz \geq 0.9 \right\} = 0.95.$$

9

FITTING STRAIGHT LINES

9.1 Introduction

Engineering problems often require the presentation of data showing the observed relationship between two variables. This chapter is devoted to a consideration of some of the problems that arise in expressing such a relationship.

The analysis of pairs of data often requires that they be plotted. If the relationship between the measurements is very nearly exact, a plot by eye may be sufficient. For example, if experimental results showing the relationship between proportional limit and tensile strength of dental alloys are as shown in Fig. 9.1, a statistical analysis is probably unnecessary.

Similarly, if experimental data calibrating a new method of determining calcium in the presence of large amounts of magnesium are as shown in Fig. 9.2, the analysis is very straightforward.

Different people observing these data would probably draw the same graph, and such plots would be representative of the true relationship between the variables. On the other hand, there are instances where the relationship between two variables is less certain, and plotting by eye results in graphs which depend on the judgment of the people who are involved. Different people observing the *same* data will produce significantly different plots. This can be seen by graphing the data obtained from an experiment measuring certain mechanical properties of cast and wrought gold dental alloys.[1] Several relationships were found between the mechanical properties; in particular, twenty-five pairs of observations were obtained on proportional

---

[1] S. H. Bush, "A Statistical Analysis of the Mechanical Properties of Cast and Wrought Gold Dental Alloys," *American Society for Testing Materials Bulletin*, Oct., 1952, No. 185, pp. 46–50.

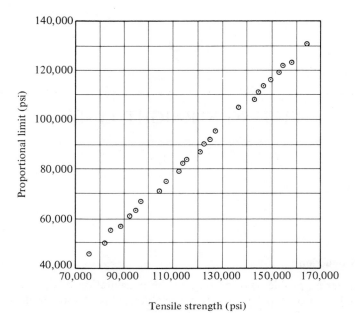

Fig. 9.1. *Relationship between proportional limit and tensile strength.*

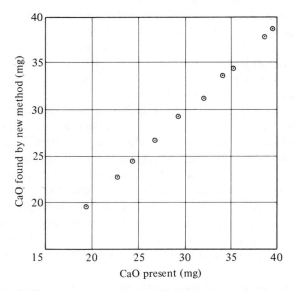

Fig. 9.2. *Gravimetric determination of calcium in the presence of magnesium.*

limit and tensile strength. The data are given in Table 9.1 and are shown plotted in Fig. 9.3.

As a further example, suppose the data obtained for calibration of a new method of determining calcium in the presence of large amounts of magnesium are as shown in Table 9.2 and as plotted in Fig. 9.4. Ten different samples containing known amounts of calcium oxide were analyzed by the new method.

In these last two examples, it is evident that plotting by eye is inadequate and that a more systematic method of determining the relationship between two variables is needed. Such a method should not only result in determining a relationship but should also describe how the observed data can be used to predict results about future observations. The following sections deal with these problems.

Table 9.1. **Mechanical Properties of Dental Gold Alloys**[1]

Tensile Strength, psi	Proportional Limit, psi
114,800	72,400
163,800	129,300
167,800	129,600
129,200	92,500
142,200	107,800
128,500	94,800
115,200	89,300
135,700	107,700
86,700	55,000
115,800	87,000
108,200	66,900
90,700	51,700
75,200	49,500
111,500	69,200
130,700	101,400
152,800	128,200
135,700	100,700
140,500	97,800
112,700	84,900
107,300	67,000
130,800	95,300
81,200	49,700
126,000	79,800
100,800	65,700
87,500	58,000

[1] Data are from S. H. Bush, "A Statistical Analysis of the Mechanical Properties of Cast and Wrought Gold Dental Alloys," *American Society for Testing Materials Bulletin*, Oct., 1952, No. 185, pp. 46–50.

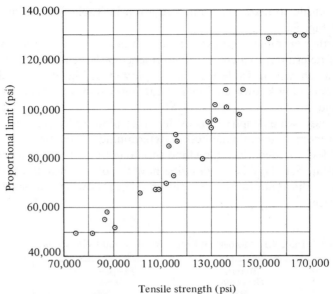

Tensile strength (psi)

Fig. 9.3. *Relationship between proportional limit and tensile strength.*
(Data from S. H. Bush, *ASTM Bulletin,* October, 1952)

Table 9.2. Gravimetric Determination of Calcium in the Presence of Magnesium

CaO Present, mg	CaO Found by New Method, mg
20.0	19.8
22.5	22.8
25.0	24.5
28.5	27.3
31.0	31.0
33.5	35.0
35.5	35.1
37.0	37.1
38.0	38.5
40.0	39.0

9.2 Types of Linear Relationships

There are two different ways in which pairs of measurements can occur: namely, when there is an underlying physical relationship and when there is a degree of association.

In the first case, a functional relationship between two variables is assumed.

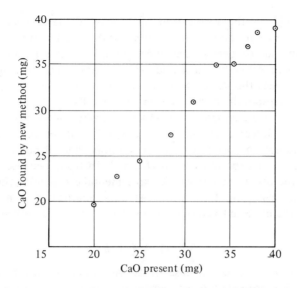

Fig. 9.4. *Gravimetric determination of calcium in the presence of magnesium.*

If an experiment is performed, n pairs of observations will be obtained. In principle, these observations should conform exactly to the assumed functional relationship, e.g., if the relationship is linear, then all points should lie on a straight line. However, in reality, the observations will deviate from the assumed functional relationship due to measurement errors or other uncontrollable experimental errors. It is often possible to assume that one of the variables can be recorded without error, and it is designated as x. The remaining variable is designated as Y and is considered to be a random variable. Naturally, the error in x is never zero, practically speaking, but it is sufficient that the error in x be small compared to the variability in Y. As an example, consider an experiment which measures the effect of time of aging on the strength of cement. The time corresponds to the x variate, the values of which are predetermined in the experiment. For a given value of time, the yield (corresponding to the Y variate) is the random variable. Most problems which involve time usually fall into the framework above. Time can frequently be measured to a sufficient degree of accuracy so that it can be assumed to be known without error.

Other examples of a functional relationship are problems which involve calibration against a known standard. A series of observations may be taken by a laboratory on material whose contents are known accurately by design. The random variable Y can be considered as the laboratory measurement, whereas the true composition may be regarded as the x variate. The example of calibrating a new method of determining CaO falls into this category.

A final example is the verification of a physical "law" such as Young's modulus. Young's modulus states that stress divided by strain is a constant. This constant can be determined experimentally. Since strain represents a distance, it can often be measured without error and corresponds to the x variate, whereas the stress can be assumed to be subject to "experimental error" and corresponds to the random variable Y.

In each of these examples, the underlying physical relationship may be assumed to be linear if the proper scales are used for the measurements. Such an assumption does not necessarily imply that all points will lie on a straight line if an experiment is performed. Instead, because of measurement error and other uncontrollable experimental errors in Y, the underlying physical relationship implies that a reasonable mathematical model for this case is that the expected value of the random variable Y, given x, is linear, i.e.,

$$E(Y|x) = A + Bx.$$

The degree of association case deals with random variables X and Y, each of which represents a measurement on a different characteristic. These two random variables are dependent, in a probability sense, but there is no unique value of one associated with a particular value of the other. Yet, they are related in some quantitative manner in that there exists a predictable range of values of one corresponding to a particular value of the other. For example, consider the characteristics, abrasion loss and hardness, for some metal specimens. One would not expect a direct functional relationship to exist between abrasion loss and hardness. Instead, for a specimen possessing a particular hardness, a spectrum of possible abrasion losses is anticipated. This is due primarily to the existence of the inherent variability among the metal specimens although experimental error may play some role.

Interest is centered on determining the relationship between two dependent variables for any of a number of reasons. Measurements on one variable, say, X, may be relatively inexpensive compared with measurements of Y, thereby resulting in a monetary saving if Y can be predicted from a knowledge that $X = x$. For example, determining abrasion loss is difficult, whereas measuring hardness by means of a Rockwell hardness machine is relatively simple. The example of proportional limit and tensile strength also falls into this category. In some situations measuring a performance characteristic is tantamount to destroying the unit, so that the recording of a "related" characteristic may be useful. For example, measuring the life of a transistor Y is equivalent to destructive testing, whereas determining the amount of impurities present, X, may be achieved without destroying the transistor. If these two random variables are "associated," then it may be possible to predict Y, given that $X = x$.

Since X and Y are both random variables in this model, they are interchangeable. However, the convention will be adopted that the variable Y will be designated as the characteristic to be predicted from the knowledge that $X = x$.

In this degree of association model it has been asserted that the two variables are related in a quantitative manner in that there exists a predictable range of values of one corresponding to a particular value of the other. A reasonable mathematical model for this "association" between X and Y is to assume the functional form of the expected value of the random variable Y given that the random variable X equals x, i.e., $E(Y|x)$. If this functional form is linear (as will be assumed in most of the chapter), then

$$E(Y|x) = A + Bx.$$

In the abrasion loss and hardness example, assuming the above form implies that expected abrasion loss for all those specimens having hardness equal to x is $A + Bx$.

It should be noted that both the underlying physical relationship and degree of association models lead to the same representation given by

$$E(Y|x) = A + Bx,$$

that is, interest is centered on determining the average value of the random variable Y as a linear function of x. Therefore, the analysis of these two models is almost identical. Hence, once the experimenter recognizes the distinction in models, the formal mechanics may be carried out without regard to the situation that exists. It is worth noting that $E(Y|x)$ is often called the regression of Y on x, and when

$$E(Y|x) = A + Bx$$

the regression is called linear regression.

9.3 Least Squares Estimators of the Slope and Intercept

9.3.1 Formulation of the Problem and Results

Both models presented in the previous section have the property that the expected value of the random variable Y can be expressed as a linear function of a known variate x, i.e.,

$$E(Y|x) = A + Bx.$$

The values of A and B are usually unknown, and are to be estimated using experimental data. The estimators of A and B are denoted by a and b, respectively.[1] Thus, the estimated relationship will be of the form

$$\tilde{Y} = a + bx.$$

Returning to the properties of gold dental alloys, the underlying (unknown) relationship between proportional limit and tensile strength may be expressed as:

Expected value of proportional limit given tensile strength

$$= A + B \times \text{(tensile strength)}.$$

A and B are estimated from the experimental data, and the estimated relationship is:

Estimated proportional limit given tensile strength

$$= a + b \times \text{(tensile strength)}.$$

In the calibration of CaO, the underlying (unknown) relationship between the amount of CaO present and the amount determined by the new method may be expressed as:

Expected value of CaO determined by the new method, given CaO

$$\text{present} = A + B \times \text{(CaO present)}.$$

The data will yield an estimated relation of the form:

Estimated value of CaO determined by the new method given CaO

$$\text{present} = a + b \times \text{(CaO present)}.$$

The proportional limit versus tensile strength example illustrates the degree of association model, whereas the new CaO method versus the known quantity illustrates the underlying physical relation model.

The usual method of estimating the intercept A and the slope B is the method of least squares. The values of a and b are determined so that the sum of the squares of the deviations of Y about the fitted line $\tilde{Y} = a + bx$ is a minumum; i.e., so that $(Y_1 - \tilde{Y}_1)^2 + (Y_2 - \tilde{Y}_2)^2 + \cdots + (Y_n - \tilde{Y}_n)^2$ is a minimum. If $(x_1, Y_1), (x_2, Y_2), \ldots, (x_n, Y_n)$ are the n pairs of observatons to be obtained experimentally,

$$b = \frac{(x_1 - \bar{x})(Y_1 - \bar{Y}) + (x_2 - \bar{x})(Y_2 - \bar{Y}) + \cdots + (x_n - \bar{x})(Y_n - \bar{Y})}{(x_1 - \bar{x})^2 + (x_2 - \bar{x})^2 + \cdots + (x_n - \bar{x})^2}$$

$$= \frac{\sum_{i=1}^{n} (x_i - \bar{x})(Y_i - \bar{Y})}{\sum_{i=1}^{n} (x_i - \bar{x})^2}$$

---

[1] Again the notation that distinguishes between random variables and the values that random variables take on breaks down. Although a and b are estimators, and will subsequently be shown to be functions of the random variables Y_i, they will still be denoted by lower-case letters. Similarly, A and B are fixed unknown constants, yet are designated by capital letters.

and
$$a = \bar{Y} - b\bar{x}$$

are the least squares estimators of B and A, respectively,

where
$$\bar{Y} = \frac{Y_1 + Y_2 + \cdots + Y_n}{n} = \frac{\sum\limits_{i=1}^{n} Y_i}{n}$$

and
$$\bar{x} = \frac{x_1 + x_2 + \cdots + x_n}{n} = \frac{\sum\limits_{i=1}^{n} x_i}{n}.$$

A computational procedure for calculating a and b is given in Sec. 9.12.

The method of least squares for determining estimators of the slope and intercept is intuitively appealing, but nevertheless requires further justification. Least squares estimators possess several desirable properties which will be described.

The estimators a and b can also be expressed as linear combinations of the Y's (the coefficients depending only on the x's which are assumed to be fixed constants). It is then simple to show that a and b are unbiased estimators of A and B, respectively, i.e., $E(a) = A$ and $E(b) = B$. Moreover, if the Y's are independent (in a probability sense) and the variance of Y is constant, say, σ^2, for all x (for a discussion of this point, see Sec. 9.11), the variance of the estimated line $\tilde{Y} = a + bx$ will be less than the variance of any other line whose parameters are expressed as linear combinations of the Y's and which are unbiased estimators of A and B. The least squares estimator is, therefore, also called the best linear unbiased estimator of the line $E(Y|x) = A + Bx$. This is often referred to as the Gauss-Markov property.

The previously discussed measures made no mention of the underlying distribution of Y. If the additional assumption is imposed that this random variable is normally distributed, i.e., Y_i is a normally distributed random variable with mean $E(Y_i|x_i) = A + Bx_i$ and variance σ^2 (independent of x_i), then the maximum likelihood estimators of the slope and intercept coincide with the least squares estimators.

Finally, the method of least squares is easily generalized to handle any polynomial regression function in one or more variables.

*9.3.2 THEORY

The least squares estimators are easily obtained, using elementary calculus. Let

$$Q = \sum_{i=1}^{n} (Y_i - A - Bx_i)^2.$$

The principle of least squares requires the estimators of A and B to minimize Q. Thus, a and b are the values of A and B which make

$$\frac{\partial Q}{\partial A} = 0$$

and $$\frac{\partial Q}{\partial B} = 0.$$

These partial derivatives are

$$\frac{\partial Q}{\partial A} = -2 \sum_{i=1}^{n} (Y_i - A - Bx_i);$$

$$\frac{\partial Q}{\partial B} = -2 \sum_{i=1}^{n} x_i(Y_i - A - Bx_i).$$

By setting the partial derivatives equal to zero, performing the indicated summations, and replacing the values of A and B by their estimates a and b, the equations

$$n\bar{Y} - na - bn\bar{x} = 0$$

and $$\sum x_i Y_i - na\bar{x} - b \sum x_i^2 = 0$$

are obtained.

The first equation leads to

$$a = \bar{Y} - b\bar{x},$$

and by substituting this value into the second equation, the desired result is obtained, i.e.,

$$b = \frac{\sum_{i=1}^{n} x_i Y_i - n\bar{x}\bar{Y}}{\sum_{i=1}^{n} x_i^2 - n\bar{x}^2} = \frac{\sum_{i=1}^{n} (x_i - \bar{x})(Y_i - \bar{Y})}{\sum_{i=1}^{n} (x_i - \bar{x})^2}.$$

It is easily verified that these solutions yield a maximum.

The expressions for b and a as linear combinations of the Y's are obtained as follows: The slope estimator, b, can be written as

$$b = \frac{\sum_{i=1}^{n} (x_i - \bar{x})(Y_i - \bar{Y})}{\sum_{i=1}^{n} (x_i - \bar{x})^2} = \frac{\sum_{i=1}^{n} (x_i - \bar{x})Y_i}{\sum_{i=1}^{n} (x_i - \bar{x})^2} - \frac{\bar{Y} \sum_{i=1}^{n} (x_i - \bar{x})}{\sum_{i=1}^{n} (x_i - \bar{x})^2}$$

$$= \frac{\sum_{i=1}^{n} (x_i - \bar{x})Y_i}{\sum_{i=1}^{n} (x_i - \bar{x})^2} \quad \text{since } \bar{Y} \sum_{i=1}^{n} (x_i - \bar{x}) = 0.$$

Let $$\frac{x_i - \bar{x}}{\sum_{i=1}^{n} (x_i - \bar{x})^2} = c_i;$$

the c_i are known constants because the x_i are assumed to be known constants and not random variables. Hence,

$$b = \sum_{i=1}^{n} c_i Y_i$$

is a linear combination of the Y's. Similarly, a can be written as

$$a = \bar{Y} - b\bar{x} = \frac{\sum_{i=1}^{n} Y_i}{n} - \left[\sum_{i=1}^{n} c_i Y_i\right]\bar{x} = \sum_{i=1}^{n}\left(\frac{1}{n} - c_i\bar{x}\right) Y_i.$$

Let

$$\frac{1}{n} - c_i\bar{x} = d_i;$$

the d_i are then fixed constants also. Hence,

$$a = \sum_{i=1}^{n} d_i Y_i$$

is a linear combination of the Y's.

The property of unbiasedness of b and a can be verified as follows:

$$b = \sum_{i=1}^{n} c_i Y_i$$

so that

$$E(b) = \sum_{i=1}^{n} c_i E(Y_i) = \sum_{i=1}^{n} c_i(A + Bx_i) = A\sum_{i=1}^{n} c_i + B\sum_{i=1}^{n} c_i x_i$$

$$= 0 + \frac{B\sum_{i=1}^{n}(x_i - \bar{x})x_i}{\sum_{i=1}^{n}(x_i - \bar{x})^2} = \frac{B\sum_{i=1}^{n}(x_i - \bar{x})(x_i - \bar{x})}{\sum_{i=1}^{n}(x_i - \bar{x})^2} = B;$$

similarly,

$$a = \bar{Y} - b\bar{x}$$

and

$$E(a) = E(\bar{Y} - b\bar{x}) = E(\bar{Y}) - E(b\bar{x}) = A + B\bar{x} - B\bar{x} = A.$$

The maximum likelihood estimators are easily obtained, provided that the Y_i are assumed to be independent normally distributed random variables each with mean $E(Y_i|x_i) = A + Bx_i$ and variance σ^2 (independent of i). The likelihood function is given by

$$L = \prod_{i=1}^{n}\left\{\frac{1}{\sqrt{2\pi}\,\sigma} e^{-(Y_i - (A + Bx_i))^2/2\sigma^2}\right\} = \frac{1}{(\sqrt{2\pi}\,\sigma)^n} e^{-\Sigma(Y_i - (A + Bx_i))^2/2\sigma^2}.$$

Denoting the logarithm of the likelihood by $\mathscr{L}$, then

$$\mathscr{L} = -n\ln\sqrt{2\pi}\,\sigma - \Sigma(Y_i - (A + Bx_i))^2/2\sigma^2.$$

The equations

$$\frac{\partial\mathscr{L}}{\partial A} = \frac{\partial\mathscr{L}}{\partial B} = 0$$

are equivalent to the equations obtained by using the method of least squares.

9.4 Confidence Interval Estimators of the Slope and Intercept

9.4.1 FORMULATION OF THE PROBLEM AND RESULTS

The method of least squares for obtaining estimators of the slope and intercept does not depend on any assumptions about the distribution of Y. Furthermore, it was pointed out in Sec. 9.3.1 that, under very weak conditions, these estimators are the best linear unbiased point estimators of the slope and intercept. It was also indicated that the least squares estimators coincide with the maximum likelihood estimators, provided that Y is assumed to be normally distributed. Such an assumption is also required if the estimated line is to be used for predictive purposes. Therefore, for the remainder of this chapter the following assumptions will prevail:

(a) n pairs of independent random variables $(x_1, Y_1), (x_2, Y_2), \ldots,$ (x_n, Y_n) are to be observed (the x_i are to be viewed as fixed constants).

(b) Each Y_i is normally distributed with unknown mean $E(Y_i \mid x_i) = A + Bx_i$ and unknown variance σ^2 (independent of the value of x_i).

Since the estimator, b, of the slope and the estimator, a, of the intercept can be represented as linear combinations of the Y_i, and since the Y_i are now assumed to be normally distributed, the random variables b and a have normal distributions with means B and A, respectively. The variance of b is given by

$$\sigma_b^2 = \frac{\sigma^2}{\sum\limits_{i=1}^{n} (x_i - \bar{x})^2}$$

and the variance of a is given by

$$\sigma_a^2 = \sigma^2 \left[\frac{1}{n} + \frac{\bar{x}^2}{\sum\limits_{i=1}^{n} (x_i - \bar{x})^2} \right].$$

Therefore, in addition to having b and a as point estimators of B and A, respectively, confidence interval estimators can also be found. If σ^2 is known, b and its variance, σ_b^2, can be used to form a confidence interval for B, and a and its variance, σ_a^2, can be used to form a confidence interval for A; just as the sample mean $\bar{X}$ and its variance $\sigma_{\bar{x}}^2$ are used to form a confidence interval for the mean μ of a normal distribution when the variance σ^2 is known. However, since σ^2 is usually unknown, confidence limits must be found which do not depend on σ^2. Such a confidence interval for B with confidence coefficient $1 - \alpha$ is given by

$$b \pm t_{\alpha/2; n-2} \frac{S_{y \mid x}}{\sqrt{\sum\limits_{i=1}^{n} (x_i - \bar{x})^2}}$$

and for A is given by

$$a \pm t_{\alpha/2;n-2} S_{y/x} \sqrt{\frac{1}{n} + \frac{\bar{x}^2}{\sum\limits_{i=1}^{n} (x_i - \bar{x})^2}}$$

where $t_{\alpha/2;n-2}$ is the $100\,\alpha/2$ percentage point of Student's t distribution with $n-2$ degrees of freedom (see Appendix Table 3). $S_{y|x}$ is an estimate of the variability about the line and is given by

$$S_{y|x} = \sqrt{\frac{\sum\limits_{i=1}^{n} (Y_i - \tilde{Y}_i)^2}{n-2}}.$$

A computational form for $S_{y|x}$ is given in Sec. 9.12. It is evident that $\sum_{i=1}^{n} (Y_i - \tilde{Y}_i)^2$ is just the sum of squares of the deviations about the fitted line.

The confidence interval for B is a minimum whenever $\sum_{i=1}^{n} (x_i - \bar{x})^2$ is a maximum. Since the x's are assumed to be fixed, the choice of values of x's is often available. The value of $\sum_{i=1}^{n} (x_i - \bar{x})^2$ is maximized when the x observations are divided equally at the two extreme points of the range of the x's of interest. However, this should be done only when there is strong *a priori* knowledge that the relationship is linear since such a division of the points precludes picking up any non-linear effects. A method of testing for non-linearity is given in Sec. 9.10.

*9.4.2 THEORY

The distributions of b and a are easily obtained. It has already been shown that the estimators of the slope and intercept are linear combinations of independent normally distributed random variables, i.e.,

$$b = \sum_{i=1}^{n} c_i Y_i \quad \text{and} \quad a = \sum_{i=1}^{n} d_i Y_i$$

where c_i and d_i are constants and the Y's are independent normally distributed random variables. It then follows from the results on linear combinations of independent normally distributed random variables that b and a are also normally distributed. The expected value of b is B and the expected value of a is A. The variance of b (a linear combination of independent Y's, each of which has variance σ^2) is

$$\sigma_b^2 = \sum_{i=1}^{n} c_i^2 \sigma^2 = \frac{\sigma^2}{\sum\limits_{i=1}^{n} (x_i - \bar{x})^2},$$

and the variance of a is

$$\sigma_a^2 = \sum_{i=1}^{n} d_i^2 \sigma^2 = \sigma^2 \sum_{i=1}^{n} \left(\frac{1}{n} - c_i \bar{x} \right)^2 = \sigma^2 \sum_{i=1}^{n} \left[\frac{1}{n^2} - \frac{2c_i \bar{x}}{n} + c_i^2 \bar{x}^2 \right]$$

$$= \sigma^2 \left[\frac{1}{n} - 0 + \frac{\bar{x}^2}{\sum\limits_{i=1}^{n} (x_i - \bar{x})^2} \right] = \sigma^2 \left[\frac{1}{n} + \frac{\bar{x}^2}{\sum\limits_{i=1}^{n} (x_i - \bar{x})^2} \right].$$

It can be shown (beyond the scope of this text) that

$$\frac{(n-2)S_{y|x}^2}{\sigma^2}$$

has a chi-square distribution with $n-2$ degrees of freedom and is distributed independently of a and b. Hence, the expression for the confidence limits is easily derived from the basic definitions of the t distribution; i.e., the ratio of the two independent random variables, the numerator normal with mean zero and standard deviation one and the denominator the square root of a chi-square random variable divided by its degrees of freedom. Thus,

$$\left[\frac{b-B}{\sigma\big/\sqrt{\sum_{i=1}^{n}(x_i-\bar{x})^2}}\right]\bigg/\sqrt{\frac{(n-2)S_{y|x}^2}{\sigma^2(n-2)}} = \frac{b-B}{S_{y|x}\big/\sqrt{\sum_{i=1}^{n}(x_i-\bar{x})^2}}$$

has a t distribution with $n-2$ degrees of freedom.

Hence,

$$P\left\{-t_{\alpha/2;n-2} \le \frac{b-B}{S_{y|x}\big/\sqrt{\sum_{i=1}^{n}(x_i-\bar{x})^2}} \le t_{\alpha/2;n-2}\right\} = 1-\alpha$$

and

$$P\left\{b - t_{\alpha/2;n-2}\frac{S_{y|x}}{\sqrt{\sum_{i=1}^{n}(x_i-\bar{x})^2}} \le B \le b + t_{\alpha/2;n-2}\frac{S_{y|x}}{\sqrt{\sum_{i=1}^{n}(x_i-\bar{x})^2}}\right\} = 1-\alpha$$

or

$$b \pm t_{\alpha/2;n-2}\frac{S_{y|x}}{\sqrt{\sum_{i=1}^{n}(x_i-\bar{x})^2}}$$

is the confidence interval for B with confidence coefficient $1-\alpha$.
The confidence interval for A is found in a similar manner.

9.5 Point Estimators and Confidence Interval Estimators of the Average Value of Y for a Given x

9.5.1 FORMULATION OF THE PROBLEM AND RESULTS

Very often estimators of linear relationships are desired in order to get point estimators or interval estimators of the average value of Y corresponding to a given x. For example, in the calcium oxide calibration, the experimenters, in presenting their results professionally, may be interested in estimating by a confidence interval, or by a point, the expected value of CaO present as determined by their new method for a fixed amount of CaO present. Another example deals with the relationship between average tensile strength of cement as a function of curing time, t; i.e., tensile strength is related to time by the function $Ce^{-B/t}$, or, in linear form, the expected value of ln (strength) $= \ln(C) - B/t$, where $\ln(C)$ corresponds to A, $1/t$ corresponds to x, and B and C are constants. A cement manufacturer is interested in either a point estimator or an interval estimator of the expected tensile strength of his cement after a particular period of time. A point estimator for the expected value of Y for a given value of x, say, x^*, is given by

$$\tilde{Y}^* = a + bx^*.$$

A confidence interval for the average value of Y for a given value of x, say, x^*, with confidence coefficient $1 - \alpha$ is given by

$$a + bx^* \pm t_{\alpha/2;\,n-2}S_{y|x}\sqrt{\frac{1}{n} + \frac{(x^* - \bar{x})^2}{\sum\limits_{i=1}^{n}(x_i - \bar{x})^2}}$$

where values of $t_{\alpha/2;\,n-2}$, the $100\,\alpha/2$ percentage point of the t distribution with $n - 2$ degrees of freedom, can be found in Appendix Table 3. This interval becomes increasingly large as x^* becomes more distant from the value of x used in the experiment. This is intuitively sound since estimates are expected to become poorer when one extrapolates away from the range used in the original experiment. The interval is narrowest whenever $x^* = \bar{x}$.

*9.5.2 THEORY

The expression for the confidence interval of the average value of Y for a given x, say, x^*, is obtained from the distribution of the random variable $\tilde{Y}^*$.

$$\tilde{Y}^* = a + bx^* = \sum_{i=1}^{n} d_i Y_i + \left(\sum_{i=1}^{n} c_i Y_i\right)x^*$$

$$= \sum_{i=1}^{n}(d_i + c_i x^*)Y_i.$$

Thus, $\tilde{Y}^*$ is a linear combination of independent normally distributed random variables and, therefore, is normally distributed with mean

$$E(\tilde{Y}^*) = E(a + bx^*) = A + Bx^*.$$

The variance of $\tilde{Y}^*$ (a linear combination of independent Y's each of which has variance σ^2) is

$$\sigma_{\tilde{y}^*}^2 = \sigma^2 \sum_{i=1}^{n}(d_i + c_i x^*)^2$$

$$= \sigma^2 \sum_{i=1}^{n}[d_i^2 + 2c_i d_i x^* + (c_i x^*)^2]$$

$$= \sigma^2\left[\frac{1}{n} + \frac{\bar{x}^2}{\sum\limits_{i=1}^{n}(x_i - \bar{x})^2} - \frac{2x^*\bar{x}}{\sum\limits_{i=1}^{n}(x_i - \bar{x})^2} + \frac{x^{*2}}{\sum\limits_{i=1}^{n}(x_i - \bar{x})^2}\right]$$

$$= \sigma^2\left[\frac{1}{n} + \frac{(x^* - \bar{x})^2}{\sum\limits_{i=1}^{n}(x_i - \bar{x})^2}\right].$$

Hence, $(\tilde{Y}^* - A - Bx^*)/\sigma_{y^*}$ is normally distributed with mean 0 and variance 1; $[(n - 2)S_{y|x}^2/\sigma^2]$ is independently distributed as chi-square with $n - 2$ degrees of freedom; and

$$\frac{\tilde{Y}^* - A - Bx^*}{S_{y|x}\sqrt{\dfrac{1}{n} + \dfrac{(x^* - \bar{x})^2}{\sum\limits_{i=1}^{n}(x_i - \bar{x})^2}}}$$

has a t distribution with $n - 2$ degrees of freedom. It follows that

$$P\left\{-t_{\alpha/2;n-2} \leqq \frac{\tilde{Y}^* - A - Bx^*}{S_{y|y}\sqrt{\dfrac{1}{n} + \dfrac{(x^* - \bar{x})^2}{\sum\limits_{i=1}^{n}(x_i - \bar{x})^2}}} \leqq t_{\alpha/2;n-2}\right\} = 1 - \alpha$$

and the confidence interval for $A + Bx^*$ with confidence coefficient $1 - \alpha$ is given by

$$\tilde{Y}^* \pm t_{\alpha/2;n-2}S_{y|x}\sqrt{\frac{1}{n} + \frac{(x^* - \bar{x})^2}{\sum\limits_{i=1}^{n}(x_i - \bar{x})^2}}.$$

9.6 Point Estimators and Interval Estimators of the Independent Variable x Associated with an Observation on the Dependent Variable Y

A point estimate or an interval estimate of the independent variable associated with an observation on the dependent variable Y is often appropriate when there is an underlying physical relation. For example, in the determination of calcium oxide, the experimenter may wish to estimate the true amount of CaO present, given a determination of this amount using the new method. A point estimate of x for a given observation on Y, say, Y', is given by

$$x' = \frac{Y' - a}{b}.$$

A confidence interval estimate of x for a given observation on Y, say, Y', with confidence coefficient $1 - \alpha$ is given by

$$\bar{x} + \frac{b(Y' - \bar{Y})}{c} \pm t_{\alpha/2;n-2}\frac{S_{y|x}}{c}\sqrt{\left(1 + \frac{1}{n}\right)c + \frac{(Y' - \bar{Y})^2}{\sum\limits_{i=1}^{n}(x_i - \bar{x})^2}}$$

where

$$c = b^2 - t_{\alpha/2;n-2}^2\frac{S_{y|x}^2}{\sum\limits_{i=1}^{n}(x_i - \bar{x})^2};$$

values of $t_{\alpha/2;n-2}$ can be found in Appendix Table 3. An approximate confi-

Table 9.3 Summary of Point Estimators, Confidence Interval Estimators and Prediction Interval Estimators

Parameter	Symbol for Estimator	Computation Formula for Estimator	$100(1-\alpha)$ Percent Confidence Interval
B	b	$\dfrac{\sum_{i=1}^{n}(x_i-\bar{x})(Y_i-\bar{Y})}{\sum_{i=1}^{n}(x_i-\bar{x})^2}$	$b \pm t_{\alpha/2;n-2}\dfrac{S_{y\mid x}}{\sqrt{\sum_{i=1}^{n}(x_i-\bar{x})^2}}$
A	a	$\bar{Y}-b\bar{x}$	$a \pm t_{\alpha/2;n-2}S_{y\mid x}\sqrt{\dfrac{1}{n}+\dfrac{\bar{x}^2}{\sum_{i=1}^{n}(x_i-\bar{x})^2}}$
Average value of Y for a given value of $x=x^*$	Y^*	$a+bx^*$	$a+bx^* \pm t_{\alpha/2;n-2}S_{y\mid x}\sqrt{\dfrac{1}{n}+\dfrac{(x^*-\bar{x})^2}{\sum_{i=1}^{n}(x_i-\bar{x})^2}}$
Value of x corresponding to an observed value of Y' for the case where there is an underlying physical relationship	x'	$\dfrac{Y'-a}{b}$	$\dfrac{Y'-a}{b} \pm \dfrac{t_{\alpha/2;n-2}S_{y\mid x}}{b}\sqrt{1+\dfrac{1}{n}+\dfrac{\left(\dfrac{Y'-a}{b}-\bar{x}\right)^2}{\sum_{i=1}^{n}(x_i-\bar{x})^2}}$ §

Random Variable	Symbol for Predictor	Computation Formula for Predictor	$100(1-\alpha)$ Percent Prediction Interval
Future observation on Y, corresponding to $x=x^0$	Y^0	$a+bx^0$	$a+bx^0 \pm t_{\alpha/2;n-2}S_{y\mid x}\sqrt{1+\dfrac{1}{n}+\dfrac{(x^0-\bar{x})^2}{\sum_{i=1}^{n}(x_i-\bar{x})^2}}$

§ Approximate form.

341

dence interval is given by the expression

$$\frac{Y' - a}{b} \pm \frac{t_{\alpha/2;\, n-2} S_{y|x}}{b} \sqrt{1 + \frac{1}{n} + \frac{\left(\dfrac{Y' - a}{b} - \bar{x}\right)^2}{\sum\limits_{i=1}^{n} (x_i - \bar{x})^2}}.$$

This approximation is good when $t_{\alpha/2;\, n-2}^2 S_{y|x}^2/b \sum_{i=1}^{n} (x_i - \bar{x})^2$ is small. The proofs for the results above are somewhat complicated, and hence will be omitted.

A summary of point estimates and confidence interval estimates is given in Table 9.3.

9.7 Prediction Interval for a Future Observation on the Dependent Variable

9.7.1 FORMULATION OF THE PROBLEM AND RESULTS

Section 9.5 dealt with determining a confidence interval for the mean value of Y corresponding to a given x, say, x^*. The average value of Y is a fixed but unknown constant, and a confidence interval is appropriate. It is often the case, however, that a confidence statement about the mean value is unimportant, whereas a probability statement about a future observation is relevant. For example, in the discussion about the relationship between tensile strength of cement and curing time, it was pointed out that the cement manufacturer is interested in the average tensile strength of his cement after a particular period of time, i.e., a confidence statement, whereas a builder is interested in the tensile strength of his particular batch of cement to determine whether it will carry the required load. The builder will find a confidence statement inadequate. He requires assurance that after a specified period of time, say, 28 days, the probability is $1 - \alpha$ that the tensile strength of his batch of cement will lie in a specified interval. In the CaO determination, the experimenter may be interested in making a probability statement about an interval containing a future determination by the new method for a fixed concentration of CaO rather than making a confidence statement about the average value of CaO determined by the new method for this given concentration.

Finally, in the determination of the mechanical properties of gold dental alloys, the particular dentist using the alloy may be interested in the average value of proportional limit for an alloy having a fixed tensile strength, whereas the patient is interested in the proportional limit of the gold in his mouth for a given tensile strength; i.e., the patient is interested in a prediction interval in which the proportional limit of the gold alloy in his mouth will lie.

The following statement can be made: The probability is $1 - \alpha$ that a future observation Y^0 corresponding to x^0 will lie in the interval

$$a + bx^0 \pm t_{\alpha/2;n-2} S_{y|x} \sqrt{1 + \frac{1}{n} + \frac{(x^0 - \bar{x})^2}{\sum\limits_{i=1}^{n} (x_i - \bar{x})^2}}$$

where values of $t_{\alpha/2;n-2}$ can be found in Appendix Table 3. Again, this interval is narrowest when $x^0 = \bar{x}$, and increases as x^0 becomes more distant from $\bar{x}$.

If more than one prediction interval is desired, they can be obtained by means of a Bonferroni inequality. In particular, if k predictions about future values of Y are desired based on the same fitted line, possibly corresponding to different x's, each can be obtained from the aforementioned prediction intervals by replacing α with α/k (the value of t now becomes $t_{\alpha/2k;n-2}$). Furthermore, the probability is at least $(1 - \alpha)$ that all k future observations will lie in their predicted intervals.

If an unspecified number of prediction intervals based on the same line is desired, the above technique is not applicable. Instead, an alternative technique, using simultaneous tolerance intervals, provides a solution. Predictions using simultaneous tolerance intervals can be obtained as follows: A prediction interval is constructed for the first future observation Y_1^0 corresponding to x_1^0. Another prediction interval is then constructed for the second future observation Y_2^0 corresponding to x_2^0. This is continued for as many intervals as the experimenter desires, with each interval constructed in a manner to be described below.

If, based on the same fitted line, the experimenter asserts that the proportion of future observations falling within the given prediction intervals (for any x^0) is at least $(1 - \alpha)$, and similar statements are made repeatedly based on different fitted lines, then for $100P\%$ of the different lines the statements will be correct.

The prediction intervals are obtained from the expression

$$a + bx^0 \pm c^{**} S_{y|x} \sqrt{\frac{1}{n} + \frac{(x^0 - \bar{x})^2}{\sum (x_i - \bar{x})^2}},$$

where c^{**} is given in Table 9.4.

*9.7.2 THEORY

The results for a prediction interval for a single future observation will be derived. The expression for the prediction interval is obtained from the distribution of

$$Y^0 - \tilde{Y}^0.$$

The random variable Y^0, being a (future) observation, is normally distributed with mean $A + Bx^0$ and variance σ^2. The random variable $\tilde{Y}^0$, the

Table 9.4. Values of c^{**}

n	$\alpha = 0.50$	$\alpha = 0.25$	$\alpha = 0.10$	$\alpha = 0.05$	$\alpha = 0.01$	$\alpha = 0.001$
			$P = 0.90$			
4	7.471	10.160	13.069	14.953	18.663	23.003
6	5.380	7.453	9.698	11.150	14.014	17.363
8	5.037	7.082	9.292	10.722	13.543	16.837
10	4.983	7.093	9.366	10.836	13.733	17.118
12	5.023	7.221	9.586	11.112	14.121	17.634
14	5.101	7.394	9.857	11.447	14.577	18.232
16	5.197	7.586	10.150	11.803	15.057	18.856
18	5.300	7.786	10.449	12.165	15.542	19.484
20	5.408	7.987	10.747	12.526	16.023	20.104
			$P = 0.95$			
4	10.756	14.597	18.754	21.445	26.760	32.982
6	6.652	9.166	11.899	13.669	17.167	21.266
8	5.933	8.281	10.831	12.484	15.750	19.568
10	5.728	8.080	10.632	12.286	15.553	19.369
12	5.684	8.093	10.701	12.391	15.724	19.619
14	5.711	8.194	10.880	12.617	16.045	20.050
16	5.771	8.337	11.107	12.898	16.431	20.559
18	5.848	8.499	11.357	13.204	16.845	21.097
20	5.937	8.672	11.619	13.521	17.272	21.652
			$P = 0.99$			
4	24.466	33.019	42.398	48.620	60.500	74.642
6	10.444	14.285	18.483	21.215	26.606	32.920
8	8.290	11.453	14.918	17.166	21.652	26.860
10	7.567	10.539	13.796	15.911	20.097	24.997
12	7.258	10.182	13.383	15.479	19.579	24.403
14	7.127	10.063	13.267	15.355	19.485	24.316
16	7.079	10.055	13.306	15.410	19.582	24.467
18	7.074	10.111	13.404	15.552	19.794	24.746
20	7.108	10.198	13.566	15.745	20.065	25.122

Reprinted by permission from "Simultaneous Tolerance Intervals in Regression," by G. J. Lieberman and R. G. Miller, *Biometrika*, **50**, 1 and 2, 164 (1963).

point on the fitted line corresponding to $x = x^0$, has been shown to be normally distributed with mean $A + Bx^0$ and variance

$$\sigma^2 \left[\frac{1}{n} + \frac{(x^0 - \bar{x})^2}{\sum\limits_{i=1}^{n} (x_i - \bar{x})^2} \right] \quad \text{(see Sec. 9.5.2).}$$

Furthermore, Y^0 and $\tilde{Y}^0$ are independent in a probability sense since $\tilde{Y}^0$ is derived from the first n observations and Y^0 corresponds to a future observation. Hence, $Y^0 - \tilde{Y}^0$ is normally distributed with mean zero and variance

$$\sigma^2 \left[1 + \frac{1}{n} + \frac{(x^0 - \bar{x})^2}{\sum\limits_{i=1}^{n} (x_i - \bar{x})^2} \right].$$

Since $(n-2)S_{y|x}^2/\sigma^2$ has a chi-square distribution with $n-2$ degrees of freedom and is distributed independently of $Y^0 - \tilde{Y}^0$,

$$\frac{Y^0 - \tilde{Y}^0}{S_{y|x}\sqrt{1 + \dfrac{1}{n} + \dfrac{(x^0 - \bar{x})^2}{\displaystyle\sum_{i=1}^{n}(x_i - \bar{x})^2}}}$$

has a t distribution with $n-2$ degrees of freedom. It then follows that

$$P\left\{-t_{\alpha/2;\,n-2} \leqq \frac{Y^0 - \tilde{Y}^0}{S_{y|x}\sqrt{1 + \dfrac{1}{n} + \dfrac{(x^0 - \bar{x})^2}{\displaystyle\sum_{i=1}^{n}(x_i - \bar{x})^2}}} \leqq t_{\alpha/2;\,n-2}\right\} = 1 - \alpha$$

or

$$P\left\{a + bx^0 - t_{\alpha/2;\,n-2}S_{y|x}\sqrt{1 + \dfrac{1}{n} + \dfrac{(x^0 - \bar{x})^2}{\displaystyle\sum_{i=1}^{n}(x_i - \bar{x})^2}} \leq Y^0 \leq a + bx^0 \right.$$

$$\left. + t_{\alpha/2;\,n-2}S_{y|x}\sqrt{1 + \dfrac{1}{n} + \dfrac{(x^0 - \bar{x})^2}{\displaystyle\sum_{i=1}^{n}(x_i - \bar{x})^2}}\right\} = 1 - \alpha.$$

9.8 Tests of Hypotheses about the Slope and Intercept

It is sometimes useful to determine whether the slope or intercept is equal to some hypothesized value. For example, in the CaO determination, the experimenters may wish to determine whether their new method of determination was "perfect," i.e., $A = 0$, or $B = 1$, or both $A = 0$ and $B = 1$ simultaneously. The fact that experimental results do or do not lead to $a = 0$ and/or $b = 1$ may be attributable to experimental error. Hence, some objective procedures are needed.

The hypothesis that $A = A_0$ is rejected whenever

$$\left| \frac{a - A_0}{S_{y|x}\sqrt{\dfrac{1}{n} + \dfrac{\bar{x}^2}{\displaystyle\sum_{i=1}^{n}(x_i - \bar{x})^2}}} \right| > t_{\alpha/2;\,n-2}.$$

The hypothesis that $B = B_0$ is rejected whenever

$$\left| \frac{b - B_0}{S_{y|x}\sqrt{\dfrac{1}{\displaystyle\sum_{i=1}^{n}(x_i - \bar{x})^2}}} \right| > t_{\alpha/2;\,n-2}.$$

Values of $t_{\alpha/2;\,n-2}$, the $100\,\alpha/2$ percentage point of the t distribution with $n-2$ degrees of freedom, can be found in Appendix Table 3.

The hypothesis that $A = A_0$ and $B = B_0$ simultaneously is rejected whenever

$$F = \left[n(a - A_0)^2 + 2n\bar{x}(a - A_0)(b - B_0) + \sum_{i=1}^{n} x_i^2(b - B_0)^2 \right] \bigg/ 2S_{y|x}^2 > F_{\alpha;2,n-2}$$

where $F_{\alpha;2,n-2}$ is the $100\,\alpha$ percentage point of the F distribution with 2 and $n - 2$ degrees of freedom and can be found in Appendix Table 4.

In each of the above three tests the probability is $(1 - \alpha)$ that the decision rule will lead to acceptance when the hypothesis is true: i.e., the probability of rejection when the hypothesis is true (the Type I error) is α.

Note that a Bonferroni inequality can also be used to test the hypothesis that $A = A_0$ and $B = B_0$ simultaneously (a and b are dependent random variables). This requires that each of the two individual hypotheses, $A = A_0$ and $B = B_0$, be tested at the $\alpha/2$ level of significance. The probability is then at least $(1 - \alpha)$ that the hypothesis that $A = A_0$ and $B = B_0$ simultaneously will be accepted when it is, in fact, true. This test can also be used to get simultaneous confidence intervals for A and B with confidence coefficient at least $(1 - \alpha)$.[1]

The results about testing the hypotheses that $A = A_0$ or $B = B_0$ follow from Sec. 9.4.2 where it was shown that

$$\frac{a - A_0}{S_{y|x}\sqrt{\dfrac{1}{n} + \dfrac{\bar{x}^2}{\sum\limits_{i=1}^{n}(x_i - \bar{x})^2}}}$$

and

$$\frac{b - B_0}{S_{y|x}\sqrt{\dfrac{1}{\sum\limits_{i=1}^{n}(x_i - \bar{x})^2}}}$$

have the t distribution with $n - 2$ degrees of freedom.

The OC curves for these procedures are summarized in Table 9.5.

9.9 Estimation of the Slope B when A Is Known To Be Zero

Situations frequently arise in which it is possible to assume beforehand that the intercept A is zero. The expected value of Y is given by

$$E(Y \mid x) = Bx.$$

---

[1] To get an exact joint simultaneous confidence interval estimate of A and B with confidence coefficient $1 - \alpha$, the sample values, a, b, and $s_{y|x}^2$, are substituted into the expression for F. The confidence region is defined by $F \leq F_{\alpha;2,n-2}$. To determine whether a given pair (A_0, B_0) is included in the confidence region, one substitutes these given values of A_0 and B_0 into F; then, if $F \leq F_{\alpha;2,n-2}$, the pair (A_0, B_0) is included in the confidence region.

Table 9.5. Significance Tests for Coefficients of Straight Lines

Hypothesis	Test Statistic	Criteria for Rejection	OC Curve	Notation
$A = A_0$	$$t = \frac{a - A_0}{S_{y\mid x}\sqrt{\dfrac{1}{n} + \dfrac{\bar{x}^2}{\sum (x_i - \bar{x})^2}}}$$	$\lvert t \rvert > t_{\alpha/2;\, n-2}$	Compute $$d = \frac{\lvert A_0 - A_1 \rvert}{\sigma\sqrt{\left(\dfrac{1}{n} + \dfrac{\bar{x}^2}{\sum (x_i - \bar{x})^2}\right)(n-1)}}$$ and refer to Fig. 6.10 or 6.11, using the curve for $n - 1$.	Here σ is the standard deviation of Y about the mean $A + Bx$, $S_{y\mid x}^2$ is the estimate of σ^2, and is given by $$S_{y\mid x}^2 = \frac{1}{n-2}\sum[Y_i - (a + bx_i)]^2$$ $$= \frac{1}{n-2}\left\{\sum(Y_i - \bar{Y})^2 - \frac{[\sum(x_i - \bar{x})(Y_i - \bar{Y})]^2}{\sum (x_i - \bar{x})^2}\right\};$$ A_1 is the value of the intercept for which the probability of acceptance is desired.
$B = B_0$	$$t = \frac{b - B_0}{S_{y\mid x}\sqrt{\dfrac{1}{\sum (x_i - \bar{x})^2}}}$$	$\lvert t \rvert > t_{\alpha/2;\, n-2}$	Compute $$d = \frac{\lvert B_0 - B_1 \rvert}{\sigma\sqrt{\dfrac{n-1}{\sum (x_i - \bar{x})^2}}}$$ and refer to Fig. 6.10 or 6.11, using the curve for $n - 1$.	Here B_1 is the value of the slope for which the probability of acceptance is desired.

347

The least squares estimate for B is found by setting the derivative of $\sum_{i=1}^{n}(Y_i - Bx_i)^2$ with respect to B equal to zero. The value of B satisfying the equation is given by

$$b = \frac{\sum_{i=1}^{n} x_i Y_i}{\sum_{i=1}^{n} x_i^2}.$$

This is similar to the previous result except that the values of x and Y are used in the summation rather than the deviations about their mean. For this case, it is very clear that b is normally distributed with mean B and variance $\sigma^2/\sum_{i=1}^{n} x_i^2$ (by noting that the x's are assumed to be fixed constants).

The estimated line is given by $\tilde{Y} = bx$. A confidence interval for B with confidence coefficient $1 - \alpha$ is given by

$$b \pm \frac{t_{\alpha/2;\,n-1} S_{y|x}}{\sqrt{\sum_{i=1}^{n} x_i^2}}$$

where the value of $t_{\alpha/2;\,n-1}$, the $100\,\alpha/2$ percentage point of the t distribution with $n - 1$ degrees of freedom, can be found in Appendix Table 3. The random variable $S_{y|x}$ is an estimate of the variability about the line and is given by

$$S_{y|x} = \sqrt{\frac{\sum_{i=1}^{n}(Y_i - \tilde{Y}_i)^2}{n-1}} = \sqrt{\frac{\sum_{i=1}^{n} Y_i^2 - \left[\left(\sum_{i=1}^{n} x_i Y_i\right)^2 \Big/ \sum_{i=1}^{n} x_i^2\right]}{n-1}}.$$

A point estimator of the expected value of Y for a given value of x, say, x^*, is given by $\tilde{Y}^* = bx^*$. A confidence interval estimate for the expected value of Y for a given value of x, say, x^*, with confidence coefficient $1 - \alpha$, is given by

$$bx^* \pm t_{\alpha/2;\,n-1} S_{y|x} \sqrt{\frac{x^{*2}}{\sum_{i=1}^{n} x_i^2}}$$

where the values of $t_{\alpha/2;\,n-1}$ and $S_{y|x}$ can be found as described above.

A point estimator of the value of x corresponding to a given observation on Y, say, Y', is given by $x' = Y'/b$. A confidence interval for the value of x corresponding to a given observation on Y, say, Y', with confidence coefficient $1 - \alpha$, is given approximately by

$$\frac{Y'}{b} \pm \frac{t_{\alpha/2;\,n-1} S_{y|x}}{b} \sqrt{1 + \frac{(Y'/b)^2}{\sum_{i=1}^{n} x_i^2}}$$

where $t_{\alpha/2;\,n-1}$ and $S_{y|x}$ are as described above.

A prediction interval, having the property that the probability is $1 - \alpha$

that a future observation Y^0 corresponding to x^0 will lie in the interval, is

given by
$$bx^0 \pm t_{\alpha/2;\,n-1} S_{y|x} \sqrt{1 + \frac{(x^0)^2}{\sum\limits_{i=1}^{n} x_i^2}}.$$

Finally, the hypothesis that $B = B_0$ is rejected whenever

$$\left| \frac{b - B_0}{\sqrt{S_{y|x}^2 \Big/ \sum\limits_{i=1}^{n} x_i^2}} \right| > t_{\alpha/2;\,n-1}.$$

The probability of rejection when the hypothesis is true, i.e., the Type I error, is α. The derivations of the mathematical results of the statements above are similar to those presented for the case where the intercept is not assumed to be zero.

9.10 Ascertaining Linearity

In all the previous sections, it was assumed that the curve to be estimated was linear in x. If the experiment is performed so that there are k values of Y for each of the x's, a test for linearity can be made.[1] Let $\bar{Y}_i$ be the mean of the k observations on the Y_i, and $Y_{i\nu}$ be the νth of the k observations of Y corresponding to x_i. Thus, if there are four Y's corresponding to x_3, these random variables are designated by Y_{31}, Y_{32}, Y_{33}, Y_{34}, and $\bar{Y}_3$ is the average of these four observations.

The least squares estimate of the line can be obtained in the usual fashion, using the nk individual observations, or the equivalent result may be obtained from the relations

$$b = \frac{\sum\limits_{i=1}^{n} (x_i - \bar{x})(\bar{Y}_i - \bar{\bar{Y}})}{\sum\limits_{i=1}^{n} (x_i - \bar{x})^2}$$

and
$$a = \bar{\bar{Y}} - b\bar{x}$$

where
$$\bar{\bar{Y}} = \frac{\sum\limits_{i=1}^{n} \bar{Y}_i}{n} = \frac{\sum\limits_{i=1}^{n} \sum\limits_{\nu=1}^{k} Y_{i\nu}}{nk}.$$

The hypothesis of linearity is rejected if

$$F = \frac{k \sum\limits_{i=1}^{n} (\bar{Y}_i - \tilde{Y}_i)^2 \Big/ (n - 2)}{\sum\limits_{i=1}^{n} \sum\limits_{\nu=1}^{k} (Y_{i\nu} - \bar{Y}_i)^2 \Big/ n(k - 1)} > F_{\alpha;\,n-2,\,n(k-1)}$$

---

[1] Actually, all that is necessary is that there be more than one value of Y for some x.

where values of $F_{\alpha; n-2, n(k-1)}$, the 100 α percentage point of the F distribution with $n - 2$ and $n(k - 1)$ degrees of freedom are given in Appendix Table 4.

This test is reasonable in that it compares the variability about the fitted line (numerator) with the inherent variability of the Y's which is independent of the form of the relationship (denominator). If linearity exists, these variabilities should compare favorably.

Furthermore, if linearity exists, these two variances can be combined to give the estimator

$$S_{y|x}^2 = \frac{k \sum_{i=1}^{n} (\bar{Y}_i - \tilde{Y}_i)^2 + \sum_{i=1}^{n} \sum_{v=1}^{k} (Y_{iv} - \bar{Y}_i)^2}{nk - 2} = \frac{\sum_{i=1}^{n} \sum_{v=1}^{k} (Y_{iv} - \tilde{Y}_i)^2}{nk - 2}.$$

Note that this is the same result for $S_{y|x}^2$ that would have been obtained if the least squares relationship had been estimated from the nk individual observations.

The numerator when multiplied by $(n - 2)/\sigma^2$ has a chi-square distribution with $(n - 2)$ degrees of freedom. The denominator is just a weighted average of the *estimators* of σ^2 associated with each x_i (the same value of σ^2 is assumed associated with each x_i) and, when multiplied by $n(k - 1)/\sigma^2$, the denominator has a chi-square distribution with $n(k - 1)$ degrees of freedom. Furthermore, the numerator and denominator are independent (in a probability sense). Therefore, when these quantities are divided by their respective degrees of freedom and their ratio formed, the resultant expression is just F, thereby enabling one to conclude that this quantity has an F distribution with $n - 2$ and $n(k - 1)$ degrees of freedom.

The test depends on having more than one Y for some x. If there is just one Y for each x, a supplementary investigation should be carried out by studying the quantities $(Y_i - \tilde{Y}_i)/S_{y|x}$. These values are approximately normally distributed with mean 0 and variance 1 for large n provided the relationship is linear with constant variability.

Finally, in the degree of association model, the underlying joint distribution of the two random variables may provide a sufficient condition for linearity. Two variables X and Y are said to have a bivariate normal distribution if their joint density can be expressed as

$$f_{XY}(u, v) = \frac{1}{2\pi\sigma_x\sigma_y\sqrt{1 - \rho^2}} e^{-\left\{\frac{(u-\mu_x)^2}{\sigma_x^2} - 2\rho\frac{(u-\mu_x)(v-\mu_y)}{\sigma_x\sigma_y} + \frac{(v-\mu_y)^2}{\sigma_y^2}\right\}/2(1-\rho^2)},$$

where μ_x and σ_x^2 are the mean and variance of X; μ_y and σ_y^2 are the mean and variance of Y; and ρ is the correlation coefficient (a constant which is discussed in detail in Sec. 9.14). It is easily verified that the conditional distribution of Y, given $X = x$, is normal with mean

$$E(Y|x) = \mu_y + \rho\frac{\sigma_y}{\sigma_x}(x - \mu_x) = A + Bx$$

and variance $\sigma_y^2(1 - \rho^2) = \sigma^2$, which is exactly the linear model under consideration. Hence, a sufficient condition for $E(Y|x)$ to be linear is that X and Y have a bivariate normal distribution.

9.11 Transformation to a Straight Line

If there is prior knowledge that the relationship is non-linear, or if this has been determined from a test similar to the one in the previous section, it is often possible to transform the data to a linear form. For example, such a transformation was discussed previously. The relationship between tensile strength of cement and time was given by

$$Y = Ce^{-B/t}.$$

Taking logarithms of both sides, one obtains a linear relation of the form

$$\text{average value of } \ln(Y) = \ln(C) - B/t,$$

where $\ln(C)$ corresponds to A and $1/t$ to x. Generally, then, it is often possible by simple transformations of the variables to represent the relationship as a straight line in the transformed variates. The transformations are usually chosen on the basis of a graphical analysis of the observations, using all the prior knowledge available about the theoretical underlying relation.

In some instances, Y may have a linear relationship with some function of x, and no transformation is required. For example, the average value of Y may be written as $B \cos z$ where $\cos z$ corresponds to x and $A = 0$. In any case, the assumptions underlying the use of the previously described techniques must be satisfied for Y or the function of Y, whichever is the appropriate variate.

It sometimes occurs that although the underlying relationship is linear, the variability is not constant for all x. Variance stabilizing transformations are often applied in this case. If the variability in Y tends to increase linearly with x, plotting the square roots (or higher roots) of the data often stabilizes the variability. If the variability increases at a higher rate, a logarithm transformation is sometimes appropriate. If the variability tends to increase and then decrease, an inverse sine transformation often is fruitful. It must be emphasized that a constant variance is not the only condition sought, and precautions are still necessary when using the techniques described previously with the transformed variables. Fortunately, however, the transformation of scale to meet the condition of constant variability often has the effect of improving the closeness of the distribution to normality; a correlation of variability of Y with x in the original scale often supplies excessive skewness which tends to be eliminated after the transformation. But the validity of any assumption of normality should be watched, for while moderate departures from normality are known not to be serious, any large departures are likely to affect the validity of the probability statements made.

9.12 Work Sheets for Fitting Straight Lines

A worksheet for fitting straight lines is given in Table 9.6. This worksheet has been arranged for ease of computation when using a desk calculator. Completion of this table will result in obtaining all the quantities described in the previous sections.

9.13 Illustrative Examples

Example 1. The data on proportional limit and tensile strength of gold dental alloys given in Table 9.1, Sec. 9.1, will be analyzed. The following quantities will be obtained:

1. least squares estimates of A and B;
2. 95% confidence interval estimates of A and B;
3. a point estimate of the average value of the proportional limit (Y^*) corresponding to a tensile strength of 129,000 psi (x^*);
4. a 90% confidence interval estimate of the average value of the proportional limit corresponding to a tensile strength of 129,000 psi (x^*);
5. a 95% prediction interval for the proportional limit corresponding to a tensile strength of 129,000 psi (x^0);
6. a test of the hypothesis that $B = 1.5$ such that the probability of accepting the hypothesis when it is true is 0.95.

From the data given on the worksheet for Example 1, the following are obtained:

1. The least squares estimates of A and B are computed directly on the worksheet; the values of $a = -30,793.792$ and $b = 0.96982743$ are obtained.
2. To compute the confidence interval estimates of A and B, the value of $t_{\alpha/2;\, n-2} = t_{0.025;\, 23} = 2.0687$ is obtained from Appendix Table 3. Substitute into the formulas given in Sec. 9.4; the confidence interval for A is

$$-30,793.792 \pm (2.0687)(6,242.8268)$$
$$= -30,793.792 \pm 12,914.536$$
$$= [-43,708.328, \quad -17,879.256].$$

Similarly, the confidence interval estimate for B is

$$0.96982743 \pm (2.0687)(0.051148740)$$
$$= 0.96982743 \pm 0.10581140$$
$$= [0.864, \quad 1.076].$$

3. A point estimate of the average value of the proportional limit corresponding to a tensile strength of $x^* = 129,000$ is given by

$$Y^* = a + bx^* = -30,793.792 + (0.96982743)(129,000) = 94,313.946.$$

4. To compute the confidence interval estimate of the average value of the proportional limit associated with a given tensile strength of $x^* = 129,000$, the value of $t_{\alpha/2;\,n-2} = t_{0.05;\,23} = 1.7139$ is obtained from Appendix Table 3. One substitutes into the formula given in Sec. 9.5; the required confidence interval is given by

$$-30,793.792 + (0.96982743)(129,000) \pm (1.7139)(1,321.5532)$$
$$= 94,313.946 \pm 2,265.0100$$
$$= [92,048.936, \quad 96,578.956].$$

5. To compute the prediction interval for the proportional limit corresponding to a tensile strength of $x^0 = 129,000$, a value of $t_{\alpha/2;\,n-2} = t_{0.025;\,23} = 2.0687$ is obtained from Appendix Table 3. Then substituting into the formula given in Sec. 9.7, one obtains for the prediction interval

$$-30,793.792 - (0.96982743)(129,000) \pm (2.0687)(6,300.2909)$$
$$= 94,313.946 \pm 13,033.412$$
$$= [81,280.534, \quad 107,347.358].$$

6. The hypothesis that $B = 1.5$ is rejected at the 95% level if

$$\left| \frac{b - 1.5}{S_{y|x}\sqrt{\dfrac{1}{\sum (x_i - \bar{x})^2}}} \right| = \left| \frac{0.96982743 - 1.5}{0.051148740} \right| = |-10.3653|$$

exceeds $t_{\alpha/2;\,n-2} = t_{0.025;\,23} = 2.0687$. Thus,

$$|-10.3653| = 10.3653 > 2.0687$$

and therefore the hypothesis that $B = 1.5$ is rejected.

Example 2. The data on calibrating the new method of determining CaO given in Table 9.2 will be analyzed. The following quantities will be obtained:

1. least squares estimates of A and B;
2. 90% confidence interval estimates of A and B;
3. a point estimate of the average value of the amount of CaO present obtained by the new method (Y^*) when there is actually 30 mg present (x^*);
4. a 99% confidence interval estimate of the average value of the amount of CaO present obtained by the new method when there is actually 30 mg present (x^*);
5. a 95% confidence interval estimate of the true amount of CaO present when the new method gives a quantity equal to 29.6 mg (Y');
6. a 95% prediction interval for the amount of CaO present determined by the new method when there is actually 30 mg present (x^0);
7. a test of the hypothesis that $B = 1$ such that the probability of accepting the hypothesis when it is true is 0.90;

Table 9.6. Worksheet for the Fitting of Straight Lines

GENERAL DATA

x represents _____ :

$\sum x_i = $ _____ ; $\bar{x} = \sum x_i/n = $ _____ ; $(\sum x_i)\bar{x} = $ _____ ;

$\sum x_i^2 = $ _____ ;

$\sum x_i^2 - (\sum x_i)\bar{x} = \sum (x_i - \bar{x})^2 = $ _____ .

Y represents _____ :

$\sum Y_i = $ _____ ; $\bar{Y} = \sum Y_i/n = $ _____ ; $(\sum Y_i)\bar{Y} = $ _____ ;

$\sum Y_i^2 = $ _____ ; $\sum Y_i^2 - (\sum Y_i)\bar{Y} = \sum (Y_i - \bar{Y})^2 = $ _____ .

$n = $ _____ ; $1/n = $ _____ ; $(\sum x_i)(\sum Y_i)/n = \bar{x}\sum Y_i = $ _____ ;

$\sum x_i Y_i = $ _____ ;

$\sum x_i Y_i - (\sum x_i)(\sum Y_i)/n = \sum (x_i - \bar{x})(Y_i - \bar{Y}) = $ _____ ;

$[\sum (x_i - \bar{x})(Y_i - \bar{Y})]^2 = $ _____ ;

$\dfrac{[\sum (x_i - \bar{x})(Y_i - \bar{Y})]^2}{\sum (x_i - \bar{x})^2} = $ _____ .

DATA FOR EQUATION OF LINE:

$b = \dfrac{\sum (x_i - \bar{x})(Y_i - \bar{Y})}{\sum (x_i - \bar{x})^2} = $ _____ ; $a = \bar{Y} - b\bar{x} = $ _____ ;

equation of line $= \bar{Y} = a + bx = $ _____ $+$ _____ x.

ESTIMATION OF STANDARD DEVIATION:

$(n - 2)S_{y|x}^2 = \sum (Y_i - \bar{Y})^2 - \dfrac{[\sum (x_i - \bar{x})(Y_i - \bar{Y})]^2}{\sum (x_i - \bar{x})^2} = $ _____ ;

$S_{y|x}^2 = $ _____ ; $S_{y|x} = $ _____ .

DATA FOR SIGNIFICANCE TESTS AND FOR CONFIDENCE INTERVALS FOR A AND B:

$\dfrac{1}{\sum (x_i - \bar{x})^2} = $ _____ ; $\sqrt{\dfrac{1}{\sum (x_i - \bar{x})^2}} = $ _____ ;

$S_{y|x}\sqrt{\dfrac{1}{\sum (x_i - \bar{x})^2}} = $ _____ ; $\bar{x}^2 = $ _____ ;

$\dfrac{\bar{x}^2}{\sum (x_i - \bar{x})^2} = $ _____ ; $\dfrac{1}{n} + \dfrac{\bar{x}^2}{\sum (x_i - \bar{x})^2} = $ _____ ;

$\sqrt{\dfrac{1}{n} + \dfrac{\bar{x}^2}{\sum (x_i - \bar{x})^2}} = $ _____ ;

$S_{y|x}\sqrt{\dfrac{1}{n} + \dfrac{\bar{x}^2}{\sum (x_i - \bar{x})^2}} = $ _____ .

Table 9.6 (continued). Worksheet for the Fitting of Straight Lines

DATA FOR CONFIDENCE INTERVALS FOR AVERAGE VALUE OF Y CORRESPONDING TO $x = x^*$:

$x^* = $ _____; $(x^* - \bar{x})^2 = $ _____;

$\dfrac{(x^* - \bar{x})^2}{\sum (x_i - \bar{x})^2} = $ _____; $\dfrac{1}{n} + \dfrac{(x^* - \bar{x})^2}{\sum (x_i - \bar{x})^2} = $ _____;

$\sqrt{\dfrac{1}{n} + \dfrac{(x^* - \bar{x})^2}{\sum (x_i - \bar{x})^2}} = $ _____;

$S_{y|x} \sqrt{\dfrac{1}{n} + \dfrac{(x^* - \bar{x})^2}{\sum (x_i - \bar{x})^2}} = $ _____.

DATA FOR CONFIDENCE INTERVALS FOR x CORRESPONDING TO AN OBSERVED VALUE $Y = Y'$ FOR THE CASE WHEN THERE IS AN UNDERLYING PHYSICAL RELATIONSHIP:

$Y' = $ _____; $x' = \dfrac{Y' - a}{b} = $ _____; $\left(\dfrac{Y' - a}{b} - \bar{x}\right)^2 = $ _____;

$\dfrac{\left(\dfrac{Y' - a}{b} - \bar{x}\right)^2}{\sum (x_i - \bar{x})^2} = $ _____;

$1 + \dfrac{1}{n} + \dfrac{\left(\dfrac{Y' - a}{b} - \bar{x}\right)^2}{\sum (x_i - \bar{x})^2} = $ _____;

$\sqrt{1 + \dfrac{1}{n} + \dfrac{\left(\dfrac{Y' - a}{b} - \bar{x}\right)^2}{\sum (x_i - \bar{x})^2}} = $ _____;

$S_{y|x} \sqrt{1 + \dfrac{1}{n} + \dfrac{\left(\dfrac{Y' - a}{b} - \bar{x}\right)^2}{\sum (x_i - \bar{x})^2}} = $ _____.

DATA FOR PREDICTION INTERVALS FOR A FUTURE VALUE OF Y CORRESPONDING TO $x = x^0$:

$x^0 = $ _____; $(x^0 - \bar{x})^2 = $ _____; $\dfrac{(x^0 - \bar{x})^2}{\sum (x_i - \bar{x})^2} = $ _____;

$1 + \dfrac{1}{n} + \dfrac{(x^0 - \bar{x})^2}{\sum (x_i - \bar{x})^2} = $ _____;

$\sqrt{1 + \dfrac{1}{n} + \dfrac{(x^0 - \bar{x})^2}{\sum (x_i - \bar{x})^2}} = $ _____;

$S_{y|x} \sqrt{1 + \dfrac{1}{n} + \dfrac{(x^0 - \bar{x})^2}{\sum (x_i - \bar{x})^2}} = $ _____.

Worksheet for the Fitting of Straight Lines for Example 1

GENERAL DATA:

x represents tensile strength:

$\sum x_i = 2{,}991{,}300;\quad \bar{x} = 119{,}652;\quad (\sum x_i)\bar{x} = 357{,}915{,}027{,}600;$

$\sum x_i^2 = 372{,}419{,}750{,}000;$

$\sum x_i^2 - (\sum x_i)\bar{x} = \sum (x_i - \bar{x})\bar{x} = 14{,}504{,}722{,}400.$

Y represents proportional limit:

$\sum Y_i = 2{,}131{,}200;\quad \bar{Y} = 85{,}248;\quad (\sum Y_i)\bar{Y} = 181{,}680{,}537{,}600;$

$\sum Y_i^2 = 196{,}195{,}960{,}000;$

$\sum Y_i^2 - (\sum Y_i)\bar{Y} = \sum (Y_i - \bar{Y})^2 = 14{,}515{,}422{,}400.$

$n = 25;\quad 1/n = 0.04;\quad (\sum x_i)(\sum Y_i)/n = \bar{x} \sum Y_i = 255{,}002{,}342{,}400;$

$\sum x_i Y_i = 269{,}069{,}420{,}000;$

$\sum x_i Y_i - (\sum x_i)(\sum Y_i)/n = \sum (x_i - \bar{x})(Y_i - \bar{Y}) = 14{,}067{,}077{,}600;$

$[\sum (x_i - \bar{x})(Y_i - \bar{Y})]^2 = (1{,}978{,}826{,}722)10^{11};$

$\dfrac{[\sum (x_i - \bar{x})(Y_i - \bar{Y})]^2}{\sum (x_i - \bar{x})^2} = 13{,}642{,}637{,}670.$

DATA FOR EQUATION OF LINE:

$$b = \frac{\sum (x_i - \bar{x})(Y_i - \bar{Y})}{\sum (x_i - \bar{x})^2} = 0.96982743;\quad a = \bar{Y} - b\bar{x} = -30{,}793.792;$$

equation of line $= \hat{Y} = a + bx = -30{,}793.792 + 0.96982743x.$

ESTIMATION OF STANDARD DEVIATION:

$$(n - 2)S_{y|x}^2 = \sum (Y_i - \bar{Y})^2 - \frac{[\sum (x_i - \bar{x})(Y_i - \bar{Y})]^2}{\sum (x_i - \bar{x})^2}$$
$$= 872{,}784{,}730;\quad S_{y|x}^2 = 6{,}160.1268.$$

$S_{y|x}^2 = 37{,}947{,}162.17;\quad S_{y|x} = 6{,}160.1268.$

DATA FOR SIGNIFICANCE TESTS AND FOR CONFIDENCE INTERVALS FOR A AND B:

$$\frac{1}{\sum (x_i - \bar{x})^2} = (0.68943064)10^{-10};$$

$$\sqrt{\frac{1}{\sum (x_i - \bar{x})^2}} = (0.83031960)10^{-5};$$

$$S_{y|x}\sqrt{\frac{1}{\sum (x_i - \bar{x})^2}} = 0.051148740;\quad \bar{x}^2 = 14{,}316{,}601{,}104;$$

$$\frac{\bar{x}^2}{\sum (x_i - \bar{x})^2} = 0.98703034;\quad \frac{1}{n} + \frac{\bar{x}^2}{\sum (x_i - \bar{x})^2} = 1.02703034;$$

$$\sqrt{\frac{1}{n} + \frac{\bar{x}^2}{\sum (x_i - \bar{x})^2}} = 1.01342505;$$

$$S_{y|x}\sqrt{\frac{1}{n} + \frac{\bar{x}^2}{\sum (x_i - \bar{x})^2}} = 6{,}242.8268.$$

Worksheet for the Fitting of Straight Lines for Example 1 (continued)

DATA FOR CONFIDENCE INTERVALS FOR AVERAGE VALUE OF Y CORRESPONDING TO $x = x^*$:

$x^* = 129{,}000$; $(x^* - \bar{x})^2 = 87{,}385{,}104$;

$$\frac{(x^* - \bar{x})^2}{\sum (x_i - \bar{x})^2} = 0.0060245968; \quad \frac{1}{n} + \frac{(x^* - \bar{x})^2}{\sum (x_i - \bar{x})^2} = 0.046024597;$$

$$\sqrt{\frac{1}{n} + \frac{(x^* - \bar{x})^2}{\sum (x_i - \bar{x})^2}} = 0.21453344;$$

$$S_{y|x}\sqrt{\frac{1}{n} + \frac{(x^* - \bar{x})^2}{\sum (x_i - \bar{x})^2}} = 1{,}321.5532.$$

DATA FOR PREDICTION INTERVALS FOR A FUTURE VALUE OF Y CORRESPONDING TO $x = x^0$:

$x^0 = 129{,}000$; $(x^0 - \bar{x})^2 = 87{,}385{,}104$;

$$\frac{(x^0 - \bar{x})^2}{\sum (x_i - \bar{x})^2} = 0.0060245968;$$

$$1 + \frac{1}{n} + \frac{(x^0 - \bar{x})^2}{\sum (x_i - \bar{x})^2} = 1.04602460;$$

$$\sqrt{1 + \frac{1}{n} + \frac{(x^0 - \bar{x})^2}{\sum (x_i - \bar{x})^2}} = 1.02275344;$$

$$S_{y|x}\sqrt{1 + \frac{1}{n} + \frac{(x^0 - \bar{x})^2}{\sum (x_i - \bar{x})^2}} = 6{,}300.2909.$$

DATA FOR CONFIDENCE INTERVALS FOR x CORRESPONDING TO AN OBSERVED VALUE OF $Y = Y'$ FOR THE CASE WHEN THERE IS AN UNDERLYING PHYSICAL RELATIONSHIP:

$$Y' = \underline{\hspace{1cm}}; \quad x' = \frac{Y' - a}{b} = \underline{\hspace{1cm}}; \quad \left(\frac{Y' - a}{b} - \bar{x}\right)^2 = \underline{\hspace{1cm}};$$

$$\frac{\left(\dfrac{Y' - a}{b} - \bar{x}\right)^2}{\sum (x_i - \bar{x})^2} = \underline{\hspace{1cm}};$$

$$1 + \frac{1}{n} + \frac{\left(\dfrac{Y' - a}{b} - \bar{x}\right)^2}{\sum (x_i - \bar{x})^2} = \underline{\hspace{1cm}};$$

$$\sqrt{1 + \frac{1}{n} + \frac{\left(\dfrac{Y' - a}{b} - \bar{x}\right)^2}{\sum (x_i - \bar{x})^2}} = \underline{\hspace{1cm}};$$

$$S_{y|x}\sqrt{1 + \frac{1}{n} + \frac{\left(\dfrac{Y' - a}{b} - \bar{x}\right)^2}{\sum (x_i - \bar{x})^2}} = \underline{\hspace{1cm}}.$$

8. a joint test of the hypothesis that $A = 0$ and $B = 1$ such that the probability of accepting the hypothesis when it is true is 0.90.

From the data given on the worksheet for Example 2, the following are obtained:

1. The least squares estimates of A and B are computed directly on the worksheet; the values of $a = -0.29277822$ and $b = 1.0065202$ are obtained.

2. To compute the confidence interval estimates of A and B, the value $t_{\alpha/2; n-2} = t_{0.05; 8} = 1.8595$ is obtained from Appendix Table 3. Substitute into the formulas given in Sec. 9.4; the confidence interval for A is

$$-0.29277822 \pm 1.8595(1.2611638)$$
$$= -0.29277822 \pm 2.3451341$$
$$= [-2.6379123, \quad 2.0523559].$$

Similarly, the confidence interval estimate for B is

$$1.0065202 \pm (1.8595)(0.03968357)$$
$$= 1.0065202 \pm 0.073791613$$
$$= [0.93272859, \quad 1.0803118].$$

3. A point estimate of the average value of the amount of CaO present obtained by the new method when there are actually $x^* = 30$ mg present is given by

$$\tilde{Y}^* = -0.29277822 + (1.0065202)(30) = 29.902828.$$

4. To compute the confidence interval estimate of the average value of the amount of CaO present obtained by the new method when there are actually $x^* = 30$ mg present, the value of $t_{\alpha/2; n-2} = t_{0.005; 8} = 3.3554$ is obtained from Appendix Table 3. One substitutes into the formula given in Sec. 9.5; the required confidence interval is given by

$$-0.29277822 + (1.0065202)(30) \pm (3.3554)(0.26323107)$$
$$= 29.902828 \pm 0.88324553$$
$$= [29.019582, \quad 30.786073].$$

5. To compute the confidence interval estimate of the true amount of CaO present when the new method gives a quantity $Y' = 29.6$ mg, the value of $t_{\alpha/2; n-2} = t_{0.025; 8} = 2.3060$ is obtained from Appendix Table 3. One substitutes into the formula given in Sec. 9.6; the required confidence interval is

$$\frac{29.6 - (-0.29277822)}{1.0065202} \pm \frac{2.3060}{1.0065202}(0.86274368)$$
$$= 29.699134 \pm 1.9765991$$
$$= [27.722535, \quad 31.675733].$$

6. To compute the prediction interval for the amount of CaO present determined by the new method corresponding to an actual value $x^0 = 30$ mg present, a value of $t_{\alpha/2; n-2} = t_{0.025; 8} = 2.3060$ is obtained from Appendix Table 3. Then substituting into the formula given in Sec. 9.7, one obtains for the prediction interval

$$-0.29277822 + (1.0065202)(30) \pm (2.3060)(0.86205668)$$
$$= 29.902828 \pm 1.9879027$$
$$= [27.914925, \quad 31.890731].$$

7. To test the hypothesis that $B = 1$, substitute into the expression given in Sec. 9.8;

$$\left| \frac{b - 1}{S_{y|x} \frac{1}{\sqrt{\sum (x_i - \bar{x})^2}}} \right| = \left| \frac{0.0065202}{0.039683578} \right| = |0.164305| = 0.164305.$$

This number is compared with the value of $t_{\alpha/2; n-2} = t_{0.05; 8} = 1.8595$ which is obtained from Appendix Table 3. Thus,

$$0.164305 < 1.8595$$

and the hypothesis that $B = 1$ is not rejected at the 90% level.

8. To test the joint hypothesis, $A = 0$ and $B = 1$, substitute into the expression for F given in Sec. 9.8. For this example, one obtains

$$F = [10(-0.29277822)^2 + 20(31.1)(-0.29277822)(0.0065202)$$
$$+ 10,100(0.0065202)^2]/1.3477022$$
$$= 0.0991913/1.3477022$$
$$= 0.0736003.$$

This value of F is compared with a value of $F_{\alpha; 2, n-2} = F_{0.10; 2, 8} = 3.11$ obtained from Appendix Table 4. Thus,

$$F = 0.0736003 < 3.11$$

and the joint hypothesis, $A = 0$ and $B = 1$, is not rejected.

Example 3. As a final example, assume that the data in Table 9.2 are augmented so that, corresponding to a given quantity of CaO present (x_i), there are two values of CaO determined by the new method (Y_{i1} and Y_{i2}). The data are given below. Using the expressions for a and b given in Sec. 9.10, i.e.,

$$a = \bar{\bar{Y}} - b\bar{x}$$

and

$$b = \frac{\sum_{i=1}^{n} (x_i - \bar{x})(\bar{Y}_i - \bar{\bar{Y}})}{\sum_{i=1}^{n} (x_i - \bar{x})^2},$$

Worksheet for the Fitting of Straight Lines for Example 2

GENERAL DATA:

x represents CaO present:

$\sum x_i = 311; \quad \bar{x} = \sum x_i/n = 31.1; \quad (\sum x_i)\bar{x} = 9{,}672.1;$

$\sum x_i^2 = 10{,}100;$

$\sum x_i^2 - (\sum x_i)\bar{x} = \sum (x_i - \bar{x})^2 = 427.9.$

Y represents CaO found by new method:

$\sum Y_i = 310.1; \quad \bar{Y} = \sum Y_i/n = 31.01; \quad (\sum Y_i)\bar{Y} = 9{,}616.201;$

$\sum Y_i^2 = 10{,}055.09; \quad \sum Y_i^2 - (\sum Y_i)\bar{Y} = \sum (Y_i - \bar{Y})^2 = 438.889.$

$n = 10; \quad 1/n = 0.1; \quad (\sum x_i)(\sum Y_i)/n = \bar{x}\sum Y_i = 9{,}644.11;$

$\sum x_i Y_i = 10{,}074.8;$

$\sum x_i Y_i - (\sum x_i)(\sum Y_i)/n = \sum (x_i - \bar{x})(Y_i - \bar{Y}) = 430.69;$

$[\sum (x_i - \bar{x})(Y_i - \bar{Y})]^2 = 185{,}493.88;$

$\dfrac{[\sum (x_i - \bar{x})(Y_i - \bar{Y})]^2}{\sum (x_i - \bar{x})^2} = 433.49820.$

DATA FOR EQUATION OF LINE:

$b = \dfrac{\sum (x_i - \bar{x})(Y_i - \bar{Y})}{\sum (x_i - \bar{x})^2} = 1.0065202; \quad a = \bar{Y} - b\bar{x} = -0.29277822;$

equation of line $= \hat{Y} = a + bx = -0.29277822 + 1.0065202x.$

ESTIMATION OF STANDARD DEVIATION:

$(n - 2)S_{y|x}^2 = \sum (Y_i - \bar{Y})^2 - \dfrac{[\sum (x_i - \bar{x})(Y_i - \bar{Y})]^2}{\sum (x_i - \bar{x})^2} = 5.3908086;$

$S_{y|x}^2 = 0.67385108; \quad S_{y|x} = 0.82088433.$

DATA FOR SIGNIFICANCE TESTS AND FOR CONFIDENCE INTERVALS FOR A AND B:

$\dfrac{1}{\sum (x_i - \bar{x})^2} = 0.0023369946; \quad \sqrt{\dfrac{1}{\sum (x_i - \bar{x})^2}} = 0.048342472;$

$S_{y|x}\sqrt{\dfrac{1}{\sum (x_i - \bar{x})^2}} = 0.039683578; \quad \dfrac{1}{n} + \dfrac{\bar{x}^2}{\sum (x_i - \bar{x})^2} = 2.3603646;$

$\dfrac{\bar{x}^2}{\sum (x_i - \bar{x})^2} = 2.2603646; \quad \dfrac{1}{n} + \dfrac{\bar{x}^2}{\sum (x_i - \bar{x})^2} = 2.3603646; \quad \bar{x}^2 = 967.21;$

$\sqrt{\dfrac{1}{n} + \dfrac{\bar{x}^2}{\sum (x_i - \bar{x})^2}} = 1.5363478;$

$S_{y|x}\sqrt{\dfrac{1}{n} + \dfrac{\bar{x}^2}{\sum (x_i - \bar{x})^2}} = 1.2611638.$

Worksheet for the Fitting of Straight Lines for Example 2 (continued)

DATA FOR CONFIDENCE INTERVALS FOR AVERAGE VALUE OF Y CORRESPONDING TO $x = x^*$:

$x^* = 30$; $(x^* - \bar{x})^2 = 1.21$;

$\dfrac{(x^* - \bar{x})^2}{\sum (x_i - \bar{x})^2} = 0.0028277635$; $\dfrac{1}{n} + \dfrac{(x^* - \bar{x})^2}{\sum (x_i - \bar{x})^2} = 0.10282776$;

$\sqrt{\dfrac{1}{n} + \dfrac{(x^* - \bar{x})^2}{\sum (x_i - \bar{x})^2}} = 0.32066768$;

$S_{y|x}\sqrt{\dfrac{1}{n} + \dfrac{(x^* - \bar{x})^2}{\sum (x_i - \bar{x})^2}} = 0.26323107.$

DATA FOR PREDICTION INTERVALS FOR A FUTURE VALUE OF Y CORRESPONDING TO $x = x^0$:

$x^0 = 30$; $(x^0 - \bar{x})^2 = 1.21$; $\dfrac{(x^0 - \bar{x})^2}{\sum (x_i - \bar{x})^2} = 0.0028277635$;

$1 + \dfrac{1}{n} + \dfrac{(x^0 - \bar{x})^2}{\sum (x_i - \bar{x})^2} = 1.1028278$;

$\sqrt{1 + \dfrac{1}{n} + \dfrac{(x^0 - \bar{x})^2}{\sum (x_i - \bar{x})^2}} = 1.0501561$;

$S_{y|x}\sqrt{1 + \dfrac{1}{n} + \dfrac{(x^0 - \bar{x})^2}{\sum (x_i - \bar{x})^2}} = 0.86205668.$

DATA FOR CONFIDENCE INTERVALS FOR x CORRESPONDING TO AN OBSERVED VALUE OF $Y = Y'$ FOR THE CASE WHEN THERE IS AN UNDERLYING PHYSICAL RELATIONSHIP:

$Y' = 29.6$; $x' = \dfrac{Y' - a}{b} = 29.699134$;

$\left(\dfrac{Y' - a}{b} - \bar{x}\right)^2 = 1.9624255$;

$\dfrac{\left(\dfrac{Y' - a}{b} - \bar{x}\right)^2}{\sum (x_i - \bar{x})^2} = 0.0045861778$;

$1 + \dfrac{1}{n} + \dfrac{\left(\dfrac{Y' - a}{b} - \bar{x}\right)^2}{\sum (x_i - \bar{x})^2} = 1.1045862$;

$\sqrt{1 + \dfrac{1}{n} + \dfrac{\left(\dfrac{Y' - a}{b} - \bar{x}\right)^2}{\sum (x_i - \bar{x})^2}} = 1.0509930$;

$S_{y|x}\sqrt{1 + \dfrac{1}{n} + \dfrac{\left(\dfrac{Y' - a}{b} - \bar{x}\right)^2}{\sum (x_i - \bar{x})^2}} = 0.86274368.$

the least squares line is given by

$$\tilde{Y} = -0.67925285 + 1.0146062x.$$

A test for linearity is called for.

x_i	Y_{i1}	Y_{i2}	$\bar{Y}_i$	$\tilde{Y}_i$	$(\bar{Y}_i - \tilde{Y}_i)^2$	$(Y_{i1} - \bar{Y}_i)^2$	$(Y_{i2} - \bar{Y}_i)^2$	$x_i \bar{Y}_i$
20.0	19.8	19.6	19.70	19.61	0.0081	0.0100	0.0100	394.000
22.5	22.8	22.1	22.45	22.15	0.0900	0.1225	0.1225	505.125
25.0	24.5	24.3	24.40	24.69	0.0841	0.0100	0.0100	610.000
28.5	27.3	28.4	27.85	28.24	0.1521	0.3025	0.3025	793.725
31.0	31.0	30.0	30.50	30.77	0.0729	0.2500	0.2500	945.500
33.5	35.0	33.0	34.00	33.31	0.4761	1.0000	1.0000	1139.000
35.5	35.1	35.0	35.05	35.34	0.0841	0.0025	0.0025	1244.275
37.0	37.1	36.8	36.95	36.86	0.0081	0.0225	0.0225	1367.150
38.0	38.5	38.0	38.25	37.88	0.1369	0.0625	0.0625	1453.500
40.0	39.0	40.2	39.60	39.90	0.0900	0.3600	0.3600	1584.000
					1.2024	2.1425	2.1425	10036.275

It is desired to test at the 5% level. Appendix Table 4 is entered and the value of $F_{\alpha; n-2, n(k-1)} = F_{0.05; 8, 10} = 3.071$ is obtained. Then substituting into the formula given in Sec. 9.10, one obtains

$$\frac{2(1.2024)/8}{4.2850/10} = 0.7015 < 3.071.$$

Thus, the hypothesis of linearity is accepted.

9.14 Correlation

In engineering problems, interest is sometimes centered in determining the distribution of two related variables and the degree of association between them rather than in estimating one variable from another. When X and Y have a bivariate normal distribution (see Sec. 9.10), the measure of the degree of association is the correlation ρ.[1] The correlation can be positive or negative. When it is positive, one variable tends to increase as the other increases; when it is negative, one variable tends to decrease as the other increases. ρ lies between the limits -1 and $+1$. A high absolute value of ρ indicates a high degree of association whereas a small absolute value indicates a small degree of association. When the absolute value of ρ is 1, the relationship is perfect. When $\rho = 0$, the variables are independent.

An estimator of ρ is given by the sample correlation coefficient r, which

---

[1] The covariance between two random variables is defined to be $E(X - \mu_x)(Y - \mu_y)$ and is denoted by σ_{xy}. For the bivariate normal distribution the correlation ρ is equivalent to $\sigma_{xy}/\sigma_x \sigma_y$.

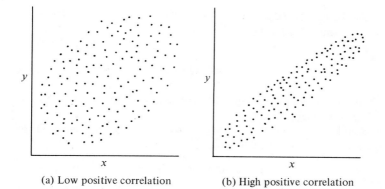

(a) Low positive correlation   (b) High positive correlation

Fig. 9.5. *Data leading to low and high positive correlation coefficients.*

is defined as

$$r = \frac{\sum_{i=1}^{n} (x_i - \bar{x})(Y_i - \bar{Y})}{\sqrt{\sum_{i=1}^{n} (x_i - \bar{x})^2 \sum_{i=1}^{n} (Y_i - \bar{Y})^2}}.$$

Data which will lead to a sample correlation coefficient having low positive correlation are indicated in Fig. 9.5(a) whereas data which will lead to a sample correlation coefficient having high positive correlation are indicated in Fig. 9.5(b).

A test of the hypothesis that $\rho = 0$ is given by rejecting when

$$|t| = \left| \frac{r}{\sqrt{1 - r^2}} \right| \sqrt{n - 2} \geq t_{\alpha/2;\,n-2}$$

where values of $t_{\alpha/2;\,n-2}$ are given in Appendix Table 3. It must be emphasized that the discussion above pertains to the correlation coefficient from a bivariate normal distribution. The correlation coefficient is of less value when the bivariate distribution is not normal, but it can still be used as an approximate measure of association.

The sample correlation coefficient can be derived from the slope of the fitted least squares line, e.g.,

$$r = b\sqrt{\frac{\sum (x_i - \bar{x})^2}{\sum (Y_i - \bar{Y})^2}}.$$

Consequently, it is clear that the sample correlation coefficient does not contain any additional information. In fact, a significance test for $\rho = 0$ is equivalent to testing whether $B = 0$, i.e., whether or not a relationship exists. These relationships indicate that correlation and fitting lines are mathematically equivalent, although these techniques are used for different types of problems. In engineering applications, the correlation coefficient does not play a very important role.

PROBLEMS

In each of the problems, assume that the assumptions stated in Sec. 9.4.1 are satisfied, unless otherwise noted.

1. A chemical reagent (x) is used to obtain a precipitate of a particular substance (Y) in a given solution. The data are:

Y	x	Y	x
8.4	7.2	8.4	6.0
5.4	4.8	9.5	6.7
6.3	5.2	10.4	7.0
6.8	4.9	12.7	8.0
8.0	5.4	10.3	7.3
11.1	6.4	7.0	4.6
12.3	6.8	5.1	4.2
13.3	8.0		

$$\sum (x_i - \bar{x})^2 = 21.8533 \qquad \sum (Y_i - \bar{Y})^2 = 97.4 \qquad \sum (x_i - \bar{x})(Y_i - \bar{Y}) = 41.52$$

(a) Estimate the line by the method of least squares. (b) If $x = 7.1$, estimate the expected amount of precipitate. (c) Determine a 95% confidence limit for the expected amount of precipitate corresponding to $x = 7.1$. (d) Find a prediction interval such that the probability is 95% that the value of precipitate corresponding to 7.1 units of reagent will lie in it. (e) Find 95% simultaneous prediction intervals for the values of precipitate corresponding to 6.9 and 7.1 units of reagent. (f) Test the hypothesis, at the 1% level of significance, that $B = 2$. (g) If the actual slope is 3.0, what is the probability of accepting the hypothesis in (f)?

2. The following results on the tensile strength of specimens of cold drawn copper have been recorded in a laboratory.

Y	x
Tensile Strength (psi)	Brinell Hardness Number
38,871	104.2
40,407	106.1
39,887	105.6
40,821	106.3
33,701	101.7
39,481	104.4
33,003	102.0
36,999	103.8
37,632	104.0
33,213	101.5
33,911	101.9
29,861	100.6
39,451	104.9
40,647	106.2
35,131	103.1

$$\sum (x_i - \bar{x})^2 = 49.5574 \qquad \sum (Y_i - \bar{Y})^2 = 168,845,370 \qquad \sum (x_i - \bar{x})(Y_i - \bar{Y}) = 89,159.72$$

(a) Estimate the relationship by the method of least squares. (b) Test the hypothesis that $A = -200,000$ at the 5% level of significance. (c) Test the hypothesis that $B = 2,000$ at the 5% level of significance. (d) Get 95% confidence intervals for B; for A; the mean value of tensile strength corresponding to a Brinell hardness of 105.0; and σ. (e) Find an interval such that the probability is 0.90 that the value of tensile strength for a Brinell hardness of 105.0 will lie in this interval. (f) Using simultaneous tolerance intervals with $(1 - \alpha) = 0.90$ and $P = 0.95$, compute a prediction interval for tensile strength corresponding to a Brinell hardness of 105.0.

3. It has been noted that the number of missing rivets (x) found at aircraft final inspection is associated with the number of errors of alignment (Y) also observed at final inspection. A study was undertaken to help estimate the average number of alignment errors from the number of missing rivets. The data are:

x	Y	x	Y
13	7	9	4
15	7	17	9
17	8	15	8
28	15	13	7
13	8	19	10
14	7	16	9
7	3	25	14
33	18	13	8
20	10	25	13
25	12	15	8

$$\sum (x_i - \bar{x})^2 = 824.8 \qquad \sum (Y_i - \bar{Y})^2 = 249.75 \qquad \sum (x_i - \bar{x})(Y_i - \bar{Y}) = 446$$

(a) Estimate the relationship by the method of least squares. (b) Estimate the expected number of alignment errors when an aircraft has 14 missing rivets by a point estimate and a 90% confidence interval estimate. (c) Determine a 95% prediction interval for the number of alignment errors when an aircraft has 12 missing rivets.

4. Determining shear strength of spot welds is difficult, whereas measuring weld diameter of spot welds is relatively simple. It would be advantageous if shear strength could be predicted from a measurement of weld diameter. The data are:

(Y) Shear Strength (psi)	(x) Weld Diameter $(0.0001\ in.)$
370	400
780	800
1,210	1,250
1,560	1,600
1,980	2,000
2,450	2,500
3,070	3,100
3,550	3,600
3,940	4,000
3,950	4,000
Totals: 22,860	23,250

The following can be computed from the data:

$$\sum (x_i - \bar{x})^2 = 15,686,250$$
$$\sum (Y_i - \bar{Y})^2 = 15,461,440$$
$$\sum (x_i - \bar{x})(Y_i - \bar{Y}) = 15,573,000$$

(a) Determine the least squares line. (b) Test the hypothesis that the slope (B) equals 1, using a level of significance equal to 0.01. (c) If the true slope equals 1.01, what is the probability of accepting the hypothesis in (b)? (d) Estimate the expected value of shear strength when the weld diameter is 0.2300. (e) Find a prediction interval such that the probability is 95% that the value of shear strength corresponding to a weld diameter of 0.2300 inch will be in it. (f) Using simultaneous tolerance intervals with $(1 - \alpha) = 0.95$ and $P = 0.90$, compute a prediction interval for shear strength corresponding to a weld diameter of 0.2300 inch.

5. Consider the data below:

x	1	2	3	4
Y	8	20	31	29
	10	23	28	28
$\bar{Y}$	9	21.5	29.5	28.5

$$\sum_{i=1}^{4} (Y_i - \bar{\bar{Y}})^2 = 267.7, \qquad \sum_{i=1}^{4} \sum_{v=1}^{2} (Y_{iv} - \bar{Y}_i)^2 = 11.5,$$

$$\sum_{i=1}^{4} (x_i - \bar{x})^2 = 5, \qquad \sum_{i=1}^{4} (\bar{Y}_i - \tilde{Y}_i)^2 = 46.6,$$

$$\sum_{i=1}^{4} (x_i - \bar{x})(\bar{Y}_i - \bar{\bar{Y}}) = 33.25, \qquad \bar{x} = 2.5,$$

$$\bar{\bar{Y}} = \frac{\sum_{i=1}^{4} \bar{Y}_i}{4} = 22.125,$$

x represents the temperature in degrees centigrade (coded) of a certain part of a process, and the two values of Y repeat determinations on the yield at that temperature (coded). We are interested in the effect of temperatures on the yield. (a) Determine the least squares line. (b) Check for linearity at the 5% level of significance. (c) Test the hypothesis that $B = 6$, using a level of significance of 0.05. (d) If the true $B = 8$, what is the probability of accepting the hypothesis in (c)? (e) Find a prediction interval such that the probability is 99% that the average value of two determinations corresponding to a temperature of 2.75 degrees centigrade will lie in it.

6. A screw manufacturer is interested in giving out data to his customers on the relation between nominal and actual lengths. The following results (in inches) were observed:

Nominal x	Actual Y		
$\frac{1}{4}$	0.252	0.262	0.245
$\frac{1}{2}$	0.496	0.512	0.490
$\frac{3}{4}$	0.743	0.744	0.751
1	0.976	1.010	1.004
$1\frac{1}{4}$	1.265	1.254	1.252
$1\frac{1}{2}$	1.498	1.518	1.504
$1\frac{3}{4}$	1.738	1.759	1.750
2	2.005	1.992	1.999

If the manufacturing process were perfect, it would be expected that $A = 0$ and $B = 1$. Note that, without using regression, one could make a confidence statement about the average length of "one-inch" screws on the basis of the three numbers in the table; but a much more precise statement can be made by using all the data, together with the fact that linear regression is applicable. (a) Estimate the above relation. (b) For nominal one-inch screws, find a confidence interval for the average actual value of screw length. (c) For a one-inch screw, find an interval such that the actual value will lie in this interval with probability 0.95. (d) Check for linearity at the 1% level of significance.

7. The corrosion of a certain metallic substance has been studied in dry oxygen at 500 degrees centigrade. In this experiment the gain in weight after various periods of exposure was used as a measure of the amount of oxygen which had reacted with the sample. The data are given as:

x (hours)	$Y = $ weight gain (coded)
1	1
2	2
3	6

(a) Determine the least squares line. (b) If $x = 2$, estimate the expected value of weight gain by a 95% confidence interval. (c) If $x = 2$, estimate the weight gain by a 95% prediction interval. (d) Find 95% simultaneous prediction intervals for the weight gains corresponding to 2 and 4 hours.

8. The determination of abrasion loss is difficult, whereas measuring hardness by means of a Rockwell hardness machine is relatively simple. It would therefore be advantageous if abrasion loss could be predicted from a measurement of hardness. A test was run and data were obtained from 30 observations. Let hardness be denoted by x and abrasion loss by Y. The data are summarized as follows:

$$\bar{x} = 70.27 \qquad \bar{Y} = 175.4$$
$$\sum (x_i - \bar{x})^2 = 4,300, \qquad \sum (Y_i - \bar{Y})^2 = 225,011$$
$$\sum (x_i - \bar{x})(Y_i - \bar{Y}) = -22,946$$
$$S^2_{y|x} = 3,663.$$

The least squares line is

$$\tilde{Y} = 550.4 - 5.336x.$$

(a) Test the hypothesis that $B = -5$, using a level of significance of 0.01. (b) If the slope equals -7, what is the probability of accepting the hypothesis in (a)? (c) Estimate the expected value of abrasion loss when the hardness is 72 by a 95% confidence interval. (d) Find a prediction interval such that the probability is 95% that the value of abrasion loss corresponding to a hardness of 72 will lie in it.

9. The production of carbon steel (x) and alloy steel (Y) in tons for eight quarters during 1971–1972 for a certain steel plant is given as:

x:	1,536	2,004	3,569	4,118	1,621	2,241	3,321	4,432
Y:	211	561	341	915	248	589	602	834

$$\sum (x_i - \bar{x})^2 = 9,172,763.5$$
$$\sum (Y_i - \bar{Y})^2 = 466,807.875$$
$$\sum (x_i - \bar{x})(Y_i - \bar{Y}) = 1,570,395.75$$

(a) Determine the relation between Y and x by the method of least squares. (b) Predict with probability 0.95 the number of tons of alloy that would be produced, given that 3,100 tons of carbon steel were produced. (c) Estimate by a 95% confidence interval the expected value of the number of tons of alloy steel, given that 3,100 tons of carbon steel were produced.

10. A gauge is to be calibrated, using several dead weights. Denote by x the standard and by Y the actual gauge reading. A sample of ten measurements is taken and the following are computed from the data:

$$\bar{x} = 230.5, \qquad \bar{Y} = 226.1$$
$$\sum (x_i - \bar{x})^2 = 1,560.63, \qquad \sum (Y_i - \bar{Y})^2 = 1,539.15$$
$$\sum (x_i - \bar{x})(Y_i - \bar{Y}) = 1,532.41.$$

(a) A completely accurate gauge would be expected to yield a line with slope 1. Test the hypothesis that the slope is 1 at the 1% level of significance. (b) If the true slope is 1.2, what is the probability of accepting the hypothesis in (a)? (c) Estimate by a 95% confidence interval the expected gauge reading when the amount of pressure is 250. (d) Find a prediction interval such that the probability is 95% that the gauge reading will lie in it when the amount of pressure is 250. (e) Estimate the expected amount of pressure by a 95% confidence interval when the gauge reads 25 psi.

11. A test was performed to determine the relationship between the chemical content of a particular constituent (Y) in solution (phosphorus) and the crystallization temperature (x). The data are as follows:

Degrees Centigrade	Grams per Liter
x	Y
−1.7	1.1
−0.4	2.3
0.2	3.2
1.1	4.3
2.3	5.4
3.1	6.6
4.2	7.8
5.3	8.8

$$\sum (x_i - \bar{x})^2 = 40.07875 \qquad \sum (Y_i - \bar{Y})^2 = 51.19875$$
$$\sum (x_i - \bar{x})(Y_i - \bar{Y}) = 45.24125 \qquad S_{y|x} = 0.1472$$

(a) Estimate the relationship by the method of least squares. (b) Test the hypothesis that $B = 0$ at the 5% level of significance. (c) Determine a joint 95% confidence interval for A and for B. (d) Estimate by a point estimate and by a 95% confidence interval the expected value of Y, given $x = 5$. (e) Estimate by a 95% prediction interval the amount of phosphorus that will be found in a future trial if the crystallization temperature is 6 degrees centigrade. (f) Using simultaneous tolerance intervals with $(1 - \alpha) = 0.95$ and $P = 0.95$, compute a prediction interval for the amount of phosphorus corresponding to a temperature of 6 degrees.

12. The elongation of a piece of boiler plate (Y) is related to the amount of force applied in tons per square inch (x). The data are:

x:	1.33	2.68	3.57	4.46	5.35	6.24	7.14	8.93	9.82	10.70
Y:	27	50	67	83	101	117	134	150	188	206

$$\sum (x_i - \bar{x})^2 = 88.14996 \qquad \sum (Y_i - \bar{Y})^2 = 30,620.1 \qquad \sum (x_i - \bar{x})(Y_i - \bar{Y}) = 1,633.624$$

(a) Estimate the least squares line. (b) Test the hypothesis that $B = 20$ at the 1% level. (c) If the true slope is 22, what is the probability of accepting the hypothesis in (b)? (d) If the elastic limit is reached when $x = 14$, predict the expected elongation for this limit by a 95% confidence limit. (e) If the elongation is 170, estimate the amount of force applied by a 95% confidence limit.

13. Under a certain set of conditions, five temperatures were read on a pyrometer and compared with true temperatures. The data are:

(Y) Pyrometer readings, °C:	950	975	1,000	1,025	1,050
(x) True temperatures, °C:	1,082	1,118	1,156	1,193	1,231

$$\sum (Y_i - \bar{Y})^2 = 6{,}250 \qquad \sum (x_i - \bar{x})^2 = 13{,}914 \qquad \sum (x_i - \bar{x})(Y_i - \bar{Y}) = 9{,}325$$

(a) Estimate the relationship by the method of least squares. (b) Estimate the true temperature for a pyrometer reading of 1,200 degrees centigrade by a 95% confidence limit. (c) Test the hypothesis that $B = 1$ at the 5% level. (d) If the true slope is 0.75, what is the probability of accepting the hypothesis in (c)?

14. A piece of metal $\frac{1}{2}$ inch in diameter and 4 inches in length is subjected to a test to determine how much strain is set up by specified values of twisting movement. The data are:

x-Twisting movement (lb/in.):	100	300	500	700	1,000	1,200	1,360
Y-Strain (torsion units):	111	329	551	769	1,103	1,322	1,497

$$\sum (x_i - \bar{x})^2 = 1{,}325{,}942.32 \qquad \sum (Y_i - \bar{Y})^2 = 1{,}608{,}665.43$$
$$\sum (x_i - \bar{x})(Y_i - \bar{Y}) = 1{,}460{,}474.32$$

(a) Estimate the relationship by the method of least squares. (b) Determine a 99% confidence interval for the expected value of strain if the elastic limit of 1,400 lb/in. has been reached. (c) Test the hypothesis that the increase in Y is 1.1 times the increase in x (i.e., $B = 1.1$) at the 5% level of significance. (d) Estimate by a 95% prediction interval the strain if the elastic limit of 1,400 lb/in. has been reached.

15. The gauging laboratory of a manufacturing firm obtained the following data by measuring a group of standard gauge blocks with a micrometer used by the inspection department.

Standard x (inches)	Micrometer Reading Y (inches)
0.5000	0.501
0.5500	0.552
0.6000	0.597
0.6500	0.646
0.7000	0.703
0.7500	0.754
0.8000	0.798
0.8500	0.853
0.9000	0.902
1.0000	1.006

$$\sum (x_i - \bar{x})^2 = 0.231 \qquad \sum (Y_i - \bar{Y})^2 = 0.2356736 \qquad \sum (x_i - \bar{x})(Y_i - \bar{Y}) = 0.23329$$

(a) Compute estimates of A and B. (b) Plot the data above. Plot the line derived in (a). (c) Determine the confidence interval for B, using a 95% confidence coefficient. (d) Test the hypothesis that $B = 1$ ($\alpha = 0.01$). (e) Determine the confidence interval for the expected value of Y, given that $x = x^* = 0.8000$. Use $\alpha = 0.05$. (f) Estimate the true value x by a 95% confidence interval when the micrometer reading is 0.80.

16. In the construction of mutual characteristic curves for the Type 6J5 triode, the plate voltage is held constant at 100 volts. The grid voltage is varied and the plate current observed. (The grid voltage can be very accurately determined.) The following data are obtained:

Grid Voltage (x) (volts)	Plate Current (Y) (milliamperes)
−4.00	1.32
−3.50	1.73
−3.00	2.71
−2.50	3.42
−2.00	4.59
−1.50	5.41
−1.00	6.33
−0.50	9.29
0.00	12.00

$$\sum (x_i - \bar{x})^2 = 15 \qquad \sum (Y_i - \bar{Y})^2 = 101.125$$
$$\sum (x_i - \bar{x})(Y_i - \bar{Y}) = 37.315 \qquad S_{y|x} = 1.08791$$

(a) Plot the given raw data. (b) Fit a straight line to the data above. (c) Determine a confidence interval for the slope. Use $\alpha = 0.01$. (d) Estimate the plate current by a 95% prediction interval when the grid voltage is −2.25 volts. (e) Find 95% simultaneous prediction intervals for the plate current when the grid voltages are −2.20, −2.25, and −2.50.

17. The following data concern case hardening a machined part by induction hardening at 1,750 degrees Fahrenheit.

(x) Time Heated (seconds)	(Y) Depth of Hardening (inches)
5	0.12
10	0.23
15	0.29
20	0.37
25	0.46
30	0.61

$$\sum (x_i - \bar{x})^2 = 437.5 \qquad \sum (Y_i - \bar{Y})^2 = 0.15093334 \qquad \sum (x_i - \bar{x})(Y_i - \bar{Y}) = 8.05$$

(a) Fit a straight line to the data. (b) Find a 95% confidence interval for the slope. (c) Test the hypothesis that $B = 0.02$ for $\alpha = 0.05$. (d) What is the probability of accepting the hypothesis in (c) when the true slope is 0.03? (e) Find a 95% prediction interval for the depth of hardening after 20 seconds. (f) Using simultaneous tolerance intervals with $(1 - \alpha) = 0.95$ and $P = 0.9$, compute a prediction interval for the depth after 20 seconds.

18. The chief production engineer of a manufacturing firm desires to ascertain whether or not there is a linear relation between the number of products assembled by the women who staff his assembly line and the length of time they work without being allowed an interruption. His time study engineer by random sampling obtains the following data:

Time Worked Without Interruption (hours)	Pieces Completed
0.5	56
	61
	54
1.0	123
	101
	110
1.5	176
	160
	151
2.0	200
	215
	207
2.5	270
	243
	251
3.0	345
	370
	353
3.5	401
	426
	418
4.0	432
	461
	428

(a) Using the data above, test for linearity ($\alpha = 0.05$). (b) If linearity is found in (a), estimate the equation of the line. If linearity is not found in (a), how would you determine whether a quadratic relation exists?

19. The following are data given on results of a study on accelerated solidification of ingots.

Depth of Solidification	Time (minutes)
4.8	2.865
6.2	3.700
7.5	4.295
8.5	4.860
9.7	5.450
11.8	6.145

Let Y represent the depth of solidification, and let x represent the square root of time of cooling. Suppose that the model that seems likely is $E(Y|x) = A + Bx$. (a) Derive point estimates of A and B by the method of least squares. (b) Construct a 95% confidence interval for B. How is this related to a test of the hypothesis that $B = 10$? (c) Estimate by a point estimate and by a 95% confidence interval the expected depth of solidification after 100 minutes. (d) An ingot is cooled for 87 minutes. Estimate the depth of solidification by a 95% confidence interval.

20. The following data represent the results of a calibration of a Heath Model U5A V.T.W.M. Voltmeter against a standard, which is a Sensitive Research Voltmeter. Three readings on the Heath Model were taken for each standard voltage.

Standard	Heath		
x	Y		
20	19.5	20.3	20.5
30	29.5	30	30.5
40	38.8	39.9	39.8
50	48.5	49	49.5
60	58.5	59	58.9
70	68	68.5	68.4
80	78.5	79	78.6
90	87.5	87.9	88
100	97	97	97

(a) Estimate the least squares line. (b) Test for linearity at the 5% level of significance. (c) Test the hypothesis that $B = 1$ at a 1% level of significance. (d) If $B = 1.1$, what is the probability of accepting the hypothesis in (c)? (e) If $\bar{Y}$ based upon a set of Heath meter readings is 58.7, estimate the true reading by a 95% confidence interval. (f) Test, at the 5% level of significance, the hypothesis that the variance of Y for each x is constant.

21. The following data represent the results of a calibration test on a Crosby gauge.

Pressure Gauge Calibration Data

	Standard	Run 1 (spin-down)	Run 2 (spin-up)	Run 3 (spin-level)
1.	10	8	6	7
2.	15	14	12	13
3.	20	18	16	17
4.	30	28	26	28
5.	40	40	37	37
6.	45	45	42	43
7.	55	55	51	53
8.	65	65	61	63
9.	70	70	66	69
10.	80	80	76	78
11.	90	90	87	89
12.	95	96	92	94
13.	105	106	101	104
14.	115	116	112	113
15.	120	122	117	119
16.	130	132	129	131
17.	140	143	138	140
18.	145	147	144	145
19.	150	150	148	149

$$\sum (x_i - \bar{x})^2 = 38,000 \qquad \sum \sum (Y_{iv} - \bar{Y}_i)^2 = 122.6667$$
$$\sum (x_i - \bar{x})(\bar{Y}_i - \bar{Y}) = 38,758.3333 \qquad \sum (\bar{Y}_i - \bar{Y}_i)^2 = 4.656$$

(a) Estimate the least squares line. (b) Test for linearity at the 1% level of significance. (c) Test the hypothesis that $B = 1$ at the 5% level of significance. (d) If the average Crosby gauge reading is 85, derive a 95% confidence interval for the reading on the standard.

22. It has been determined that the relation between stress (S) and the number of cycles to failure (N) for a particular type alloy is given by

$$S = \frac{A}{N^m}$$

where A and m are unknown constants. An experiment is run, yielding these data:

Stress (thousand psi)	N (million cycles to failure)
55.0	0.223
50.5	0.925
43.5	6.75
42.5	18.1
42.0	29.1
41.0	50.5
35.7	126
34.5	215
33.0	445
32.0	420

$$\sum (x_i - \bar{x})^2 = 0.2902015 \qquad \sum (Y_i - \bar{Y})^2 = 58.928746$$
$$\sum (x_i - \bar{x})(Y_i - \bar{Y}) = -4.0679194$$

(a) Letting $x = \ln S$ and $Y = \ln N$, determine the relationship between Y and x. (b) Estimate m by a 99% confidence interval. (c) For a stress of 40,000 psi, estimate by a 95% prediction interval the number of cycles to failure.

23. The following data represent the results of an experiment of small creep tests of a precision cast high temperature alloy. The experiment was run at 725 degrees centigrade.

Stress (S), tons/in.^{2}	Fracture-time (T), hours
20	43
18	141
16	385
14	2099

The underlying relationship is of the form

$$\ln T = A + BS.$$

(a) Estimate the relationship above by the method of least squares. (b) For a stress of 16 tons/in.^{2}, estimate the fracture time by a point estimate. (c) For a stress of 16 tons/in.^{2}, estimate the fracture time by a 95% prediction interval.

24. The following data represent the effect of time on the loss of hydrogen from samples of steel stored at 20 degrees centigrade.

Time (T), hours	Hydrogen Content (H), (ppm)	
1	7.7	8.5
2	7.5	8.1
6	6.2	6.8
17	5.7	5.3
30	4.2	4.6

(a) Estimate the relationship $H = A + B \ln T$ by the method of least squares. (b) Test the relationship in (a) at the 10% level of significance. (c) Estimate by a 95% prediction interval the hydrogen content at 20 hours. (d) Estimate by a 95% confidence interval the expected hydrogen content at 20 hours.

25. The following are data given as results of a study of the effect of primary crystallization temperature on the phosphorus content of a solution.

Phosphorus, grains per liter (Y)	Primary Crystallization Temperature, degrees C (x)
10.9	25
9.3	20
8.2	15
7.5	12
6.2	9
5.8	6
4.2	3
3.9	0
2.8	−3
2.0	−6

$\sum (x_i - \bar{x})^2 = 908.9$ $\sum (Y_i - \bar{Y})^2 = 75.896$ $\sum (x_i - \bar{x})(Y_i - \bar{Y}) = 261.82$

(a) Plot the data and draw a freehand line representing the regression. (b) Assuming that the relationship is linear, find the least squares estimate of the regression line of phosphorus content on temperature. (c) Test the hypothesis that $B = 0$ at a 10% level of significance. (d) Set up a 90% confidence interval for A. (e) Estimate by a point estimate and by a 90% confidence interval the expected phosphorus content for a primary crystallization temperature of 7 degrees centigrade. (f) Estimate by a 95% prediction interval the amount of phosphorus that will be found on a future observation if the primary crystallization temperature is 4 degrees centigrade. (g) Using simultaneous tolerance intervals with $(1 - \alpha) = 0.95$ and $P = 0.95$, compute a prediction interval for the phosphorus if the temperature is 4 degrees.

26. Given the following observations of Y vs. x:

x	Y	
1	−1.9	$\sum x = 18$
2	−1.0	$\sum Y = 2.9$
3	−0.1	$\sum xY = 33.1$
5	2.0	$\sum x^2 = 88$
7	3.9	$\sum Y^2 = 23.83$

suppose that for each x, values of Y are normally distributed with mean $A + Bx$ and variance σ^2, and that the Y's corresponding to different x's are independent random variables. (a) Find the least squares regression line of Y on x. (b) Find a 99% prediction interval for a future observation of Y corresponding to $x = 6$. (c) Find 99% simultaneous prediction intervals for Y if $x = 4$ and 6.

27. Suppose for each x the values of Y are normally distributed with mean $A + Bx$ and variance σ^2, and the Y's corresponding to different x's are independent. Given the following data:

x	Y	x^2	Y^2	xY
0	−2.8	0	7.84	0
1	−1.7	1	2.89	−1.7
2	−0.9	4	0.81	−1.8
3	0.2	9	0.04	0.6
4	1.1	16	1.21	4.4
5	2.3	25	5.29	11.5
6	3.1	36	9.61	18.6
7	4.2	49	17.64	29.4
8	5.3	64	28.09	42.4

$\sum x_i = 36 \qquad \sum Y_i = 10.8 \qquad \sum x_i^2 = 204 \qquad \sum Y_i^2 = 73.42 \qquad \sum xY = 103.4$

(a) Find least squares regression of Y on x. (b) Find a 95% confidence interval for average Y, given $x = 5$. (c) Find a 95% prediction interval for a future value of Y, given $x = 5$.

28. If a particle is dropped at time $t = 0$, physical theory indicates that the relationship between π, the distance traveled, and t, the time elapsed, is $\pi = gt^k$ for some positive constants g and k. A transformation to linearity can be obtained by taking logarithms:

$$\log \pi = \log g + k \log t.$$

Letting $Y = \log \pi$, $A = \log g$, $x = \log t$, this relation becomes $Y = A + kx$. Due to random error in measurement, however, it can be stated that only $E(Y \mid x) = A + kx$. Assume Y is normally distributed with mean $A + kx$ and variance σ^2.

A physicist who wishes to estimate k and g performs the following experiment: At time 0 the particle is dropped. At time t, the distance π is measured. He performs this experiment 5 times, obtaining the following data:

$Y = \log \pi$	$x = \log t$
−3.95	−2.0
−2.12	−1.0
0.08	0.0
2.20	+1.0
3.87	+2.0

(a) Obtain least squares estimates for k and $\log g$. (b) Find a 95% confidence interval for $E(\log \pi \mid x)$ when $x^* = 0$. (c) Find a 95% prediction interval for π

when $x^0 = 0$. The following data has been calculated:

$$\bar{Y} = 0.016$$
$$\sum (x_i - \bar{x})^2 = 10$$
$$\sum (x_i - \bar{x})(Y_i - \bar{Y}) = 19.96$$
$$\sum (Y_i - \hat{Y}_i)^2 = 0.0858$$
$$\frac{S_{y|x}}{\sqrt{\sum (x_i - \bar{x})^2}} = 0.053, \quad S_{y|x}\sqrt{1/5} = 0.076.$$

Note: All logarithms are to the base 10.

29. Suppose for each x the values of Y are normally distributed with mean $A + Bx + Cx^2$ and variance σ^2, and the Y's corresponding to different x's are independent. Derive the least squares estimators of A, B, and C.

30. Consider a relation between Y and x given by

$$E(Y|x) = \gamma x, \quad \gamma > 0.$$

Suppose one observation on Y is taken at each of n distinct points on the $x > 0$ axis, resulting in a sample $(Y_1, x_1), (Y_2, x_2), \cdots, (Y_n, x_n)$. Suppose further that the Y's are independent and are distributed as

$$f_Y(z) = \begin{cases} \dfrac{1}{\gamma x} e^{-z/\gamma x}, & \text{for } z \geq 0 \\ 0, & \text{otherwise.} \end{cases}$$

(a) Find the least squares estimator for γ. (b) Find the maximum likelihood estimator for γ.

31. Consider the following model:

$$E(Y|x) = A + B(x - \bar{x}).$$

Suppose one observation on Y is taken at each of n distinct points on the x axis, resulting in a sample $(Y_1, x_1), (Y_2, x_2), \cdots, (Y_n, x_n)$. (a) Find the least squares estimators of A and B. (b) Assuming that the Y's are independent and normally distributed with mean $A + B(x - \bar{x})$ and variance σ^2, find the distribution of the estimators in (a). (c) Assuming independence of the estimators in part (a) (which can be proven to be true), construct a 90% prediction interval for a single future value of Y at $x = x^0$. (d) Construct a 90% simultaneous prediction interval for three future values of Y corresponding to x_1^0, x_2^0, and x_3^0.

32. For the bivariate normal distribution introduced in Sec. 9.10, (a) Find the marginal density of X. What is its form? (b) Find the conditional density of $Y|X = x$. What is its form and what are the parameters?

33. Suppose that the relation between Y and x is given by

$$E(Y|x) = Bx$$

where Y is assumed to be normally distributed with mean Bx and *known* variance σ^2. A total of n independent pairs of observations are taken. (a) Find the least squares estimator of B. (b) Find the variance of the least squares estimator of B. (c) Find a 95% two-sided confidence interval for the expected value of Y corresponding to x^*. (d) Find a 95% prediction interval for a future observation corresponding to x^0.

10

ANALYSIS OF VARIANCE

10.1 Introduction

In Chapter 7, testing whether the means of two distributions are equal was considered. The t test was used for making such a comparison. This problem is actually a special case of the more general problem of comparing several means. A possible solution to this general problem is to enumerate every possible pair, and test whether each pair differs, using the t test. The disadvantage of this procedure is clear after a little reflection. Even if the distribution means are the same, it is evident that if there are enough sample means, some will be extremely large and some extremely small. This will lead to one or more fictitious significant differences with probability much higher than α. This distortion in the significance level introduces another difficulty. Suppose that the 45 possible comparisons among ten means have all been made, and four are significant by a 0.05 level t test; the experimenter must now wonder *which* (if any) of these four significant differences are really indicative of true differences.

Although a solution was proposed in Sec. 8.3.9, using a Bonferroni inequality, the appropriate solution to the problem is the analysis of variance. The analysis of variance, however, has a much wider application than this simple generalization, and is probably the most powerful procedure in the field of experimental statistics. To carry out the analysis, it is necessary to formulate a mathematical model in terms of the unknown parameters and the associated random variables. The following sections of this chapter deal with the concepts involved in the analysis of variance.

10.2 Model for the One-Way Classification

10.2.1 Fixed Effects Model

An interlaboratory calibration check on horizontal tension testing machines is to be made. Four laboratories are involved in the test and each

**Table 10.1. Interlaboratory Calibration, Horizontal
Tension Testing Machines**

Laboratory					
A	73	73	73	75	75
B	74	74	74	74	75
C	68	69	69	69	70
D	71	71	72	72	73

laboratory makes five tests on No. 16 wire. It can be assumed that the reproducibility among repeat tests is the same for all laboratories. The results of the testing are given in Table 10.1.

Before analyzing these data, it is helpful to formulate a mathematical model. This has the advantage of requiring the experimenter to define the effects which he estimates and the interpretation that he will propose.

As a first model, consider these four laboratories as the only laboratories possessing horizontal tension testing machines which are of interest. This may arise when they are the only laboratories within a certain locale, and hence receive all the business from the neighboring industrial establishments. The results of the calibration, then, will apply only to these four laboratories and any conclusions derived will not be extended to laboratories in general.

A mathematical model describing this situation is obtained by assuming that the tension reading of a specimen consists of the sum of two components: a fixed effect due to the machine of the particular laboratory tested and a random effect which represents the variation of the particular observation from the fixed laboratory effect. The random effects are assumed to be independently normally distributed with mean 0 and variance σ^2. Thus, the third specimen from laboratory B has a tensile strength equal to

$74 =$ fixed effect due to machine in laboratory B $+$ random effect.

All the specimens from laboratory B have the same fixed effect and differ only because of the random effect.

More generally, these assumptions can be summarized as follows:

$$X_{ij} = \zeta_i + \epsilon_{ij}$$

where X_{ij} is the outcome of the jth specimen corresponding to the ith treatment (or laboratory in the example given); $i = 1, 2, \ldots, r; j = 1, 2, \ldots, c;$[1] ζ_i is the fixed effect attributed to the ith treatment; ϵ_{ij} is the random effect assumed to be independently normally distributed with mean 0 and variance σ^2.

---

[1] Note that the subscript i pertains to the rows which denote the treatments in the one-way classification, there being a total of r rows. The subscript j pertains to the columns which represent the repeat tests, there being a total of c columns (repeat tests).

A few words about the selection of the specimens of wire to be used in the experiment are in order. The tensile strength of wire depends on many factors besides the particular machine used for testing: the temperature, presence of alloys, and so forth. It is important in running experiments which calibrate machines of different laboratories to make sure that differences which are asserted to be due to machines do not arise from some other factors, for example, from the fact that all the specimens used in laboratory A were treated at a higher temperature than the others. In some cases, other factors which affect the wire may be isolated as factors in the experiment and their effect removed by the techniques of two-way, three-way, and other analysis of variance techniques discussed in the subsequent sections. If this is not done, the specimens to be used at each laboratory should be chosen at random with respect to the other factors which are not specified; this random choice will insure that the conclusions are not vitiated by a coincidence of effects. A table of random numbers is useful in assigning specimens to the specific treatments. If these precautions are not taken and the fluctuation of the data is not random, statistical methods based on the assumption of randomness should not be used.

An equivalent way of stating the model above is further to subdivide ζ_i into $\zeta. + \varphi_i$. Thus, the fixed effect ζ_i can be expressed as a component $\zeta.$ common to all laboratories and a fixed effect φ_i peculiar only to the ith laboratory. This can easily be done by defining $\zeta.$ as the average of all the ζ_i; i.e.,

$$\zeta. = \frac{\sum_{i=1}^{r} \zeta_i}{r},$$

and defining φ_i as the deviation of the ζ_i from the average; i.e.,

$$\varphi_i = \zeta_i - \zeta.$$

Thus, φ_i has the property that

$$\sum_{i=1}^{r} \varphi_i = 0.$$

X_{ij} can then be expressed as

$$X_{ij} = \zeta. + \varphi_i + \epsilon_{ij}$$

where X_{ij} is normally distributed with mean $\zeta. + \varphi_i$ and variance σ^2.

The purpose of the entire experiment is to draw some conclusions about whether the laboratory machines read the same and, if not, which differ. This is equivalent to making statements about the equality of the φ_i. If the φ_i are equal, the laboratories are the same. Furthermore, if the φ_i are equal, they all must be equal to zero, since $\sum_{i=1}^{r} \varphi_i = 0$.

This model will be called the *fixed effects* model because the φ_i are fixed effects and the conclusion derived from the experiment will be extended only to the treatments (laboratories in this example) considered.

10.2.2 RANDOM EFFECTS MODEL

In the calibration example, suppose that a large number of laboratories are to be calibrated instead of four. However, only four laboratories (not necessarily the four of the previous section) are to be included in the experiment. But inferences about all the laboratories are to be made from the results of the experiment which will be based on a random sample of four chosen from the many.

The mathematical model may still be written as

$$X_{ij} = \zeta. + \varphi_i + \epsilon_{ij},$$

but the interpretation of the φ_i is different. The φ_i are random variables each having a probability distribution which is the distribution of laboratory effects (from which a random sample of four is to be drawn). The φ_i are usually assumed to be normally distributed with zero mean and variance σ_φ^2. If the experiment were to be repeated, a new random sample would be drawn and different laboratories might be included. In the fixed effects model, repeating the experiment would entail using the same four laboratories. Thus, the φ_i are random variables prior to performing the experiment. Once a laboratory is chosen, the observations within the laboratory differ only because of the random effects ϵ_{ij} (which are assumed to be normally distributed with mean 0 and variance σ^2 and to be distributed independently of φ_i). The quantity $\zeta.$ is a constant common to all the observations. In this model, knowledge about the particular φ_i is rather useless. Inferences are to be drawn about *all* the laboratories, i.e., the distribution of φ_i which is assumed normal with mean 0 and variance σ_φ^2. If $\sigma_\varphi^2 = 0$, all the laboratories are equivalent; if σ_φ^2 is very large, there are large discrepancies between laboratories. Hence, it is important to estimate σ_φ^2 and to make inferences about its magnitude.

This model is called the *random effects* model because the φ_i are random variables, and the conclusions derived from the experiment are extended to all the treatments (laboratories in the example) from which the random sample was drawn. In this model, it is evident that the X_{ij} are normally distributed with mean $\zeta.$ and variance $\sigma^2 + \sigma_\varphi^2$.

10.2.3 FURTHER EXAMPLES OF FIXED EFFECTS AND OF THE RANDOM EFFECTS MODELS

(a) A study was made recently to determine whether the ash content of coal from a certain source varied according to the four origins within the source. The coal was delivered from each origin in separate trucks. Since each truck had a different origin, the coal was sampled from each truck (prior to loading). This one-way analysis of variance fits the framework of the fixed effects model since the four origins are presumably fixed and inferences are to be made about these origins only. Repeated experiments would require the use of the same origins.

(b) A special study was made on lots of metallic oxide. Samples were drawn at random from each of eighteen lots and the specimens were analyzed for the percent metal content by weight. Assuming these eighteen lots to be random samples from a larger number of lots, the problem fits into the random component model. Repeated experiments would use different lots.

10.2.4 COMPUTATIONAL PROCEDURE, ONE-WAY CLASSIFICATION

Before carrying out any further analysis, it is important to complete the analysis of variance table. This is done following the computational procedure described below. The results are summarized in the Analysis of Variance Table 10.2.

Computational Procedure

1. Calculate totals for each treatment:

$$R_1, R_2, \ldots, R_r.$$

2. Calculate over-all total:

$$T = R_1 + R_2 + \cdots + R_r.$$

3. Compute crude total sum of squares:

$$\sum_{i=1}^{r} \sum_{j=1}^{c} X_{ij}^2 = X_{11}^2 + X_{12}^2 + \cdots + X_{rc}^2.$$

4. Calculate crude sum of squares between treatments:

$$\frac{\sum_{i=1}^{r} R_i^2}{c} = (R_1^2 + R_2^2 + \cdots + R_r^2)/c.$$

5. Calculate correction factor due to mean $= T^2/rc$. From the quantities given above compute:

6. $$SS_3 = (4) - (5) = \sum_{i=1}^{r} \frac{R_i^2}{c} - \frac{T^2}{rc},$$

7. $$SS = (3) - (5) = \sum_{i=1}^{r} \sum_{j=1}^{c} X_{ij}^2 - \frac{T^2}{rc},$$

8. $$SS_2 = (7) - (6) = SS - SS_3.$$

By referring to Table 10.2 it can be verified that $SS = SS_2 + SS_3$. In fact, this was used in obtaining SS_2 in step 8 in the computational procedure. It should be pointed out that subtracting a constant from each observation will not change the analysis of variance table.

Table 10.2. Analysis of Variance Table, One-Way Classification

Source	Sum of Squares	Degrees of Freedom D.F.	Mean Square	Expected Mean Square for the Fixed Effects Model	Expected Mean Square for the Random Effects Model
Between treatments	$SS_3 = c\sum_{i=1}^{r}(\bar{X}_{i\cdot} - \bar{X}_{\cdot\cdot})^2$	$r - 1$	$SS_3^* = SS_3/(r-1)$	$\sigma^2 + \dfrac{c\sum_{i=1}^{r}(\varphi_i)^2}{r-1}$	$\sigma^2 + c\sigma_\varphi^2$
Within treatments	$SS_2 = \sum_{i=1}^{r}\sum_{j=1}^{c}(X_{ij} - \bar{X}_{i\cdot})^2$	$r(c-1)$	$SS_2^* = SS_2/r(c-1)$	σ^2	σ^2
Total	$SS = \sum_{i=1}^{r}\sum_{j=1}^{c}(X_{ij} - \bar{X}_{\cdot\cdot})^2$	$rc - 1$	$SS^* = SS/(rc-1)$	—	—

In this table $\bar{X}_{i\cdot} = \sum_{j=1}^{c}\dfrac{X_{ij}}{c}$; $\bar{X}_{\cdot\cdot} = \sum_{i=1}^{r}\sum_{j=1}^{c}\dfrac{X_{ij}}{rc}$

10.2.5 THE ANALYSIS OF VARIANCE PROCEDURE

10.2.5.1 *A Heuristic Justification.* Looking at the analysis of variance table, it is evident that the terms in the "Mean Square" column, SS_3^* and SS_2^*, are unbiased estimators of σ^2 when the hypothesis is true. Thus, in the absence of row effects (in both models) the ratio SS_3^*/SS_2^* should be close to one. When the hypothesis is false, the denominator still estimates σ^2, whereas the numerator tends to overestimate this value. Hence, the analysis of variance procedure involves rejection of the hypothesis when the ratio is too large compared to one. To quantify the term "large" it is necessary to obtain the distribution of the ratio SS_3^*/SS_2^*. This is usually done by proving that the terms in the "Sum of Squares" column, SS_3 and SS_2, when divided by σ^2, each have independent chi-square distributions with given degrees of freedom when the hypothesis is true. It then follows that SS_3^*/SS_2^* has an F distribution. The chi-square distributions of SS_3 and SS_2 are usually obtained from the partition theorem which is stated in Sec. 10.2.5.2. This theorem gives criteria by which "Sums of Squares" can be judged as chi-square random variables. In particular, requirements are given in terms of degrees of freedom so that a method of obtaining degrees of freedom is necessary. The partition theorem is not a requirement for the understanding of the ensuing sections, but rather a justification for the characterization of ratios of mean squares as random variables having an F distribution. Similar procedures, as outlined above, are used for the two-way classification presented in the later sections, and the partition theorem provides a rationale for the F test.

Since the "degrees of freedom" play such an important role in the analysis of variance, it is important that some further comments be made. The number of degrees of freedom for n variables may be defined as the number of variables minus the number of linear relations (constraints) between them. For example, the number of degrees of freedom associated with SS_2 is rc minus the number of linear constraints between these variables. For each treatment (row) the sum of the deviations about the row mean must be zero, i.e.,

$$\sum_{j=1}^{c}(X_{ij} - \bar{X}_{i.}) = 0 \quad \text{for} \quad i = 1, 2, \ldots, r.$$

This is equivalent to saying that we have imposed one linear restriction on the observations for each row. Since there are r rows, there are r such constraints. Hence, the number of degrees of freedom associated with SS_2 is $rc - r$ or $r(c - 1)$.

***10.2.5.2** *The Partition Theorem.* Degrees of freedom having been defined in the previous section, the following partition theorem is stated without proof.

Let $Y_1, Y_2, \ldots, Y_n$ be independent identically distributed normal random variables each with mean 0 and variance 1. Let the sum of squares of the

n variables be partitioned into a sum of k sums of squares, $T_1, T_2, \ldots, T_k$, with $v_1, v_2, \ldots, v_k$ degrees of freedom, respectively, so that

$$\sum Y_i^2 = T_1 + T_2 + \cdots + T_k.$$

The necessary and sufficient condition that $T_1, T_2, \ldots, T_k$ are independently distributed as chi-square random variables with $v_1, v_2, \ldots, v_k$ degrees of freedom, respectively, is that

$$v_1 + v_2 + \cdots + v_k = n.$$

The importance of this theorem can be seen by considering the random variables X_{ij} in the one-way analysis of variance. Again, if there are no treatment effects (no laboratory differences), each X_{ij} is normally distributed with mean μ and variance σ^2 (under either model).

Let $Y_{ij} = (X_{ij} - \mu)/\sigma$ so that when there are no treatment effects Y_{ij} is normally distributed with mean 0 and variance 1.

Thus, the following equalities are obtained:

$$\sum \sum Y_{ij}^2 = \frac{\sum \sum (X_{ij} - \mu)^2}{\sigma^2}$$

$$\sum \sum (Y_{ij} - \bar{Y}..)^2 = \frac{\sum \sum (X_{ij} - \bar{X}..)^2}{\sigma^2}$$

$$(\bar{Y}..)^2 = \frac{(\bar{X}.. - \mu)^2}{\sigma^2}$$

$$\sum (\bar{Y}_{i.} - \bar{Y}..)^2 = \frac{\sum (\bar{X}_{i.} - \bar{X}..)^2}{\sigma^2}.$$

It is also evident that

$$\frac{\sum \sum (X_{ij} - \mu)^2}{\sigma^2} = \frac{\sum \sum (X_{ij} - \bar{X}..)^2}{\sigma^2} + \frac{rc(\bar{X}.. - \mu)^2}{\sigma^2}$$

$$= \frac{SS}{\sigma^2} + \frac{rc(\bar{X}.. - \mu)^2}{\sigma^2}.$$

However, since $SS = SS_2 + SS_3$, it follows that

$$\frac{\sum \sum (X_{ij} - \mu)^2}{\sigma^2} = \frac{\sum \sum (X_{ij} - \bar{X}_{i.})^2}{\sigma^2}$$

$$+ \frac{c \sum (\bar{X}_{i.} - \bar{X}..)^2}{\sigma^2} + \frac{rc(\bar{X}.. - \mu)^2}{\sigma^2}.$$

Using the equalities about the Y's, the following identity holds:

$$\sum \sum Y_{ij}^2 = \sum \sum (Y_{ij} - \bar{Y}_{i.})^2 + c \sum (\bar{Y}_{i.} - \bar{Y}..)^2 + rc(\bar{Y}..)^2.$$

It follows from the definition of degrees of freedom that

$\sum \sum Y_{ij}^2$	has	rc	degrees of freedom;
$\sum \sum (Y_{ij} - \bar{Y}_{i.})^2$	has	$r(c-1)$	degrees of freedom;
$c \sum (\bar{Y}_{i.} - \bar{Y}..)^2$	has	$r-1$	degrees of freedom;
$n(\bar{Y}..)^2$	has	1	degree of freedom.

Hence, from the partition theorem and from

$$rc = 1 + r(c - 1) + (r - 1),$$

it follows that SS_2/σ^2 and SS_3/σ^2 are independently distributed as chi-square random variables with $r(c - 1)$ and $r - 1$ degrees of freedom, respectively.

10.2.6 ANALYSIS OF THE FIXED EFFECTS MODEL, ONE-WAY CLASSIFICATION

A major problem arising in the analysis of variance for the fixed effects model is in testing the hypothesis that the r effects are all equal. This hypothesis can be written as

$$H: \quad \varphi_1 = \varphi_2 = \varphi_3 = \cdots = \varphi_r.$$

In fact, if the hypothesis is true, each φ_i must equal zero. It is seen by referring to the analysis of variance table, Table 10.2, that the expected mean squares for both the between-treatments and within-treatments sources, $E(SS_3^*)$ and $E(SS_2^*)$, respectively, equal σ^2 if the hypothesis is true. $E(SS_2^*)$ equals σ^2 even if the hypothesis is false. However, if the hypothesis is false, $E(SS_3^*)$ exceeds σ^2 by the amount $c \sum_{i=1}^{n} \varphi_i^2/(r - 1)$. Hence, if the ratio SS_3^*/SS_2^* is computed, a value far from unity would indicate that the hypothesis is false. It can be shown (using the partition theorem) that SS_3/σ^2 and SS_2/σ^2 are independently distributed as chi-square random variables with $r - 1$ and $r(c - 1)$ degrees of freedom, respectively, when the hypothesis is true. Hence, the ratio SS_3^*/SS_2^* is distributed as an F random variable with $r - 1$ and $r(c - 1)$ degrees of freedom when the hypothesis is true. Thus, the hypothesis of equality of the φ's is rejected if

$$F = SS_3^*/SS_2^* > F_{\alpha; r-1, r(c-1)}$$

where $F_{\alpha; r-1, r(c-1)}$ is the upper α percentage point of the F distribution given in Appendix Table 4. Note that a one-sided test is used since this ratio can be too large only if the hypothesis is false.

If the hypothesis

$$H: \quad \varphi_1 = \varphi_2 = \varphi_3 = \cdots = \varphi_r$$

is rejected, it is possible to make statements about which means differ. For a long time there was no satisfactory solution to this problem. However, Tukey[1] and Scheffé[2] have presented solutions whereby it becomes possible to make contrasts of the form $\varphi_m - \varphi_n$. Tukey's procedure leads to probability statements in the form of confidence intervals for the contrasts $\varphi_m - \varphi_n$;

---

[1] J. Tukey, *Allowances for Various Types of Error Rates*, unpublished invited address presented before a joint meeting of the Institute of Mathematical Statistics and the Eastern North American Region of the Biometric Society on March 19, 1952, at Blacksburg, Va.

[2] H. Scheffé, "A Method for Judging All Contrasts in the Analysis of Variance," *Biometrika*, June, 1953, Vol. 40.

i.e., the probability is $1 - \alpha$ that the values $\varphi_m - \varphi_n$ for all such contrasts $(m = 1, 2, \ldots, r; n = 1, 2, \ldots, r)$ simultaneously satisfy

$$\bar{X}_m - \bar{X}_n - k \leqq \varphi_m - \varphi_n \leqq \bar{X}_m - \bar{X}_n + k$$

where k is a factor obtained from Table 10.3 or 10.4, depending on whether $\alpha = 0.05$ or 0.01, and from the analysis of variance table. The means are said to differ significantly when the appropriate confidence interval fails to include 0. Furthermore, if the interval includes only positive values, the difference is said to differ significantly from zero, and to be positive. Similarly, if the confidence interval includes only negative values, the difference is said to differ significantly from zero, and to be negative. If the interval includes 0, the means are said not to differ significantly. Summarizing, then, the F test is made for the homogeneity of the means. If this hypothesis is rejected, Tukey's procedure is applied to determine which of the means differ.

A possible short-cut method for determining the homogeneity of all the treatment effects is to check whether $\bar{X}_{max} - \bar{X}_{min} \pm k$ includes 0. If this contrast includes 0, every other contrast of this type includes 0, and hence all the treatment effects are called equal. If the contrast above fails to include zero, the corresponding treatment effects, and perhaps others, differ significantly. This short-cut method has a level of significance exactly equal to α. However, the data in the analysis of variance table are necessary for determining the factor k, and for other results, so that this table should be completed anyway.

The factor k can be obtained from the data in the analysis of variance table and Table 10.3 or 10.4, depending on the level of significance. Table 10.3 or 10.4 is entered with the indices, degrees of freedom and the number of treatments studied (r). The proper degrees of freedom are those corresponding to the degrees of freedom of the within treatments source in the analysis of variance table, i.e., $r(c - 1)$. The factor k^* is read from the table; k is obtained from $k = k^* \sqrt{SS_2^*/c}$.

Scheffé's procedure also leads to confidence statements and if his test is applied when the F test of homogeneity is rejected, the level of significance is preserved. On the other hand, if Tukey's test is applied when the F test of homogeneity is rejected, the significance level is increased *slightly*. However, if the only type comparison of interest has the form $\varphi_m - \varphi_n$, Tukey's procedure will result in shorter confidence intervals for $\varphi_m - \varphi_n$ than would Scheffé's procedure.[1] Consequently, in the ensuing sections the Tukey method will be used and one will proceed as if the significance level suffered a negligible increase.

---

[1] If other types of contrasts are also desired, e.g., $(\zeta_1 + \zeta_2 + \zeta_3)/3 - (\zeta_4 + \zeta_5 + \zeta_6)/3$, Scheffé's procedure should be used.

Table 10.3. Table of Factors k^* (5% Significance Level)[1]

ν \ n^a	2	3	4	5	6	7	8	9	10	11	12	13	14	15	16	17	18	19	20
1	18.0	26.7	32.8	37.2	40.5	43.1	45.4	47.3	49.1	50.6	51.9	53.2	54.3	55.4	56.3	57.2	58.0	58.8	59.6
2	6.09	8.28	9.80	10.89	11.73	12.43	13.03	13.54	13.99	14.39	14.75	15.08	15.38	15.65	15.91	16.14	16.36	16.57	16.77
3	4.50	5.88	6.83	7.51	8.04	8.47	8.85	9.18	9.46	9.72	9.95	10.16	10.35	10.52	10.69	10.84	10.98	11.12	11.24
4	3.93	5.00	5.76	6.31	6.73	7.06	7.35	7.60	7.83	8.03	8.21	8.37	8.52	8.67	8.80	8.92	9.03	9.14	9.24
5	3.64	4.60	5.22	5.67	6.03	6.33	6.58	6.80	6.99	7.17	7.32	7.47	7.60	7.72	7.83	7.93	8.03	8.12	8.21
6	3.46	4.34	4.90	5.31	5.63	5.89	6.12	6.32	6.49	6.65	6.79	6.92	7.04	7.14	7.24	7.34	7.43	7.51	7.59
7	3.34	4.16	4.68	5.06	5.36	5.61	5.82	6.00	6.16	6.30	6.43	6.55	6.66	6.76	6.85	6.94	7.02	7.09	7.17
8	3.26	4.04	4.53	4.89	5.17	5.40	5.60	5.77	5.92	6.05	6.18	6.29	6.39	6.48	6.57	6.65	6.73	6.80	6.87
9	3.20	3.95	4.42	4.76	5.02	5.24	5.43	5.60	5.74	5.87	5.98	6.09	6.19	6.28	6.36	6.44	6.51	6.58	6.65
10	3.15	3.88	4.33	4.66	4.91	5.12	5.30	5.46	5.60	5.72	5.83	5.93	6.03	6.12	6.20	6.27	6.34	6.41	6.47
11	3.11	3.82	4.26	4.58	4.82	5.03	5.20	5.35	5.49	5.61	5.71	5.81	5.90	5.98	6.06	6.14	6.20	6.27	6.33
12	3.08	3.77	4.20	4.51	4.75	4.95	5.12	5.27	5.40	5.51	5.61	5.71	5.80	5.88	5.95	6.02	6.09	6.15	6.21
13	3.06	3.73	4.15	4.46	4.69	4.88	5.05	5.19	5.32	5.43	5.53	5.63	5.71	5.79	5.86	5.93	6.00	6.06	6.11
14	3.03	3.70	4.11	4.41	4.64	4.83	4.99	5.13	5.25	5.36	5.46	5.56	5.64	5.72	5.79	5.86	5.92	5.98	6.03
15	3.01	3.67	4.08	4.37	4.59	4.78	4.94	5.08	5.20	5.31	5.40	5.49	5.57	5.65	5.72	5.79	5.85	5.91	5.96
16	3.00	3.65	4.05	4.34	4.56	4.74	4.90	5.03	5.15	5.26	5.35	5.44	5.52	5.59	5.66	5.73	5.79	5.84	5.90
17	2.98	3.62	4.02	4.31	4.52	4.70	4.86	4.99	5.11	5.21	5.31	5.39	5.47	5.55	5.61	5.68	5.74	5.79	5.84
18	2.97	3.61	4.00	4.28	4.49	4.67	4.83	4.96	5.07	5.17	5.27	5.35	5.43	5.50	5.57	5.63	5.69	5.74	5.79
19	2.96	3.59	3.98	4.26	4.47	4.64	4.79	4.92	5.04	5.14	5.23	5.32	5.39	5.46	5.53	5.59	5.65	5.70	5.75
20	2.95	3.58	3.96	4.24	4.45	4.62	4.77	4.90	5.01	5.11	5.20	5.28	5.36	5.43	5.50	5.56	5.61	5.66	5.71
24	2.92	3.53	3.90	4.17	4.37	4.54	4.68	4.81	4.92	5.01	5.10	5.18	5.25	5.32	5.38	5.44	5.50	5.55	5.59
30	2.89	3.48	3.84	4.11	4.30	4.46	4.60	4.72	4.83	4.92	5.00	5.08	5.15	5.21	5.27	5.33	5.38	5.43	5.48
40	2.86	3.44	3.79	4.04	4.23	4.39	4.52	4.63	4.74	4.82	4.90	4.98	5.05	5.11	5.17	5.22	5.27	5.32	5.36
60	2.83	3.40	3.74	3.98	4.16	4.31	4.44	4.55	4.65	4.73	4.81	4.88	4.94	5.00	5.06	5.11	5.15	5.20	5.24
120	2.80	3.36	3.69	3.92	4.10	4.24	4.36	4.47	4.56	4.64	4.71	4.78	4.84	4.90	4.95	5.00	5.04	5.09	5.13
∞	2.77	3.32	3.63	3.86	4.03	4.17	4.29	4.39	4.47	4.55	4.62	4.68	4.74	4.80	4.84	4.89	4.93	4.97	5.01

[1] This table is taken by permission from "Extended and Corrected Tables of the Upper Percentage Points of the Studentized Range," by Joyce M. May, *Biometrika*, Vol. 39, Parts 1 & 2, April 1952; and from "Corrigenda to Tables of Percentage Points of the Studentized Range," by H. O. Hartley. *Biometrika*, Vol. 40, Parts 1 & 2, June 1953.

[a] Here n is the number of effects being studied.

Table 10.4. Table of Factors k^* (1% Significance Level)[1]

ν \ n^a	2	3	4	5	6	7	8	9	10	11	12	13	14	15	16	17	18	19	20
1	90.0	134	164	186	202	216	227	237	246	253	260	266	272	277	282	286	291	295	298
2	14.0	18.9	22.3	24.7	26.6	28.2	29.5	30.7	31.7	32.6	33.4	34.2	34.8	35.5	36.0	36.5	37.0	37.5	38.0
3	8.26	10.56	12.17	13.34	14.25	15.00	15.65	16.20	16.69	17.13	17.53	17.89	18.23	18.54	18.83	19.09	19.33	19.56	19.79
4	6.51	8.08	9.17	9.97	10.58	11.10	11.55	11.93	12.26	12.56	12.84	13.09	13.32	13.53	13.73	13.92	14.09	14.25	14.40
5	5.70	6.97	7.80	8.42	8.91	9.32	9.67	9.97	10.24	10.48	10.70	10.89	11.08	11.24	11.40	11.55	11.68	11.81	11.93
6	5.24	6.32	7.03	7.56	7.97	8.31	8.61	8.87	9.10	9.30	9.49	9.65	9.81	9.95	10.08	10.21	10.32	10.43	10.54
7	4.95	5.92	6.54	7.01	7.37	7.68	7.94	8.17	8.37	8.55	8.71	8.86	9.00	9.12	9.24	9.35	9.46	9.55	9.65
8	4.74	5.63	6.20	6.63	6.96	7.24	7.47	7.68	7.86	8.03	8.18	8.31	8.44	8.55	8.66	8.76	8.86	8.95	9.03
9	4.60	5.42	5.96	6.35	6.66	6.91	7.13	7.33	7.50	7.65	7.79	7.91	8.02	8.13	8.23	8.33	8.41	8.50	8.58
10	4.48	5.26	5.77	6.14	6.43	6.67	6.88	7.06	7.22	7.36	7.49	7.60	7.71	7.81	7.91	7.99	8.08	8.15	8.23
11	4.39	5.14	5.62	5.98	6.25	6.47	6.67	6.84	6.99	7.13	7.25	7.36	7.46	7.56	7.65	7.73	7.81	7.88	7.95
12	4.32	5.04	5.50	5.84	6.10	6.32	6.51	6.67	6.81	6.94	7.06	7.17	7.26	7.36	7.44	7.52	7.60	7.67	7.73
13	4.26	4.96	5.40	5.73	5.98	6.19	6.37	6.53	6.67	6.79	6.90	7.01	7.10	7.19	7.27	7.35	7.42	7.49	7.55
14	4.21	4.89	5.32	5.64	5.88	6.08	6.26	6.41	6.54	6.66	6.77	6.87	6.96	7.05	7.13	7.20	7.27	7.34	7.40
15	4.17	4.83	5.25	5.56	5.80	5.99	6.16	6.31	6.44	6.55	6.66	6.76	6.85	6.93	7.00	7.07	7.14	7.20	7.26
16	4.13	4.78	5.19	5.49	5.72	5.91	6.08	6.22	6.35	6.46	6.56	6.66	6.74	6.82	6.90	6.97	7.03	7.09	7.15
17	4.10	4.73	5.14	5.43	5.66	5.85	6.01	6.15	6.27	6.38	6.48	6.57	6.66	6.73	6.81	6.87	6.94	7.00	7.05
18	4.07	4.70	5.09	5.38	5.60	5.79	5.95	6.08	6.20	6.31	6.41	6.50	6.58	6.65	6.73	6.79	6.85	6.91	6.97
19	4.05	4.66	5.05	5.34	5.55	5.73	5.89	6.02	6.14	6.25	6.34	6.43	6.51	6.58	6.65	6.72	6.78	6.84	6.89
20	4.02	4.63	5.02	5.30	5.51	5.69	5.84	5.97	6.09	6.19	6.28	6.37	6.45	6.52	6.59	6.66	6.71	6.77	6.82
24	3.96	4.54	4.91	5.17	5.37	5.54	5.69	5.81	5.92	6.02	6.11	6.19	6.26	6.33	6.39	6.45	6.51	6.57	6.61
30	3.89	4.45	4.80	5.05	5.24	5.40	5.53	5.65	5.76	5.85	5.93	6.01	6.08	6.14	6.20	6.26	6.31	6.36	6.41
40	3.82	4.36	4.70	4.93	5.11	5.26	5.39	5.50	5.60	5.69	5.77	5.84	5.90	5.96	6.02	6.07	6.12	6.17	6.21
60	3.76	4.28	4.60	4.82	4.99	5.13	5.25	5.36	5.45	5.53	5.60	5.67	5.73	5.78	5.83	5.88	5.93	5.98	6.01
120	3.70	4.20	4.50	4.71	4.87	5.00	5.12	5.21	5.30	5.38	5.44	5.50	5.56	5.61	5.66	5.71	5.75	5.79	5.83
∞	3.64	4.12	4.40	4.60	4.76	4.88	4.99	5.08	5.16	5.23	5.29	5.35	5.40	5.45	5.49	5.53	5.57	5.61	5.64

[1] This table is taken by permission from "Extended and Corrected Tables of the Upper Percentage Points of the Studentized Range," by Joyce M. May, *Biometrika*, Vol. 39, Parts 1 & 2, April 1952; and from "Corrigenda to Tables of Percentage Points of the Studentized Range," by H. O. Hartley. *Biometrika*, Vol. 40, Parts 1 & 2, June 1953, p. 236.

a Here n is the number of effects being studied.

10.2.7 The OC Curve of the Analysis of Variance for the Fixed Effects Model

There has been no discussion of the implications of the analysis of variance procedure. This procedure guarantees that if the treatments do not differ, $100(1 - \alpha)\%$ of the time this conclusion will be reached. On the other hand, there have been no probability statements attached to the procedure if the treatments do differ. For example, suppose that all but one of the testing machines had the same expected value. Denote the expected value of the odd laboratory by M and the other means by N. Furthermore, suppose it were possible to plot the OC curve of the analysis of variance procedure as in Fig. 10.1.

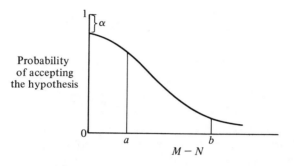

Fig. 10.1. *The OC curve for laboratory tests.*

If $M = N$ (all machines are the same), the probability of accepting the hypothesis is $(1 - \alpha)$. If $(M - N) = b$, there is a large probability of rejecting the hypothesis. However, if $(M - N) = a$, the probability of accepting the hypothesis is quite high. If differences of this magnitude are important to detect, a procedure having this OC curve is inadequate. In fact, running an experiment under this setup is rather wasteful of funds if a difference of $(M - N) = a$ is important.

Consequently, some indication of the OC curve for the analysis of variance procedure is important. A problem arises in a suitable choice of an abscissa scale. If laboratories differ, what measure of difference can be used? In the example just described, it was assumed that all but one of the means were the same. Rarely can such an assumption be made in practice. A measure for which an OC curve can be constructed is

$$\Phi = \frac{\sqrt{\sum_{i=1}^{r} \varphi_i^2 / r}}{\sigma / \sqrt{c}}.$$

Two interpretations for this measure can be given. If the laboratories do not differ, the standard deviation of an observation chosen at random is just σ. On the other hand, if the laboratories do differ, the standard deviation of an observation chosen at random is increased to

$$\sqrt{\sigma^2 + \left(\sum_{i=1}^{r} \varphi_i^2/r\right)}.$$

Consequently, if the experimenter chooses a fixed percentage, say, P, for the increase in standard deviation of an observation beyond which he wishes to reject the hypothesis of equality of the φ_i, this is equivalent to choosing

$$\frac{\sqrt{\sigma^2 + \left(\sum_{i=1}^{r} \varphi_i^2/r\right)}}{\sigma} = 1 + 0.01P$$

(where the increase in σ, P, is expressed in percent) or

$$\frac{\sqrt{\sum_{i=1}^{r} \varphi_i^2/r}}{\sigma} = \sqrt{(1 + 0.01P)^2 - 1}$$

so that

$$\Phi = \frac{\sqrt{\sum_{i=1}^{r} \varphi_i^2/r}}{\sigma/\sqrt{c}} = (\sqrt{(1 + 0.01P)^2 - 1})\sqrt{c}$$

If a probability β of rejecting the hypothesis is associated with the chosen value of P, two points on the OC curve are fixed, the other point being the error of Type I associated with $\Phi = 0$.

A second method for determining a critical value of the measure Φ is to find upper and lower bounds for Φ. This information can be obtained if the range (largest minus the smallest) of the φ_i is specified. In the laboratory example this entails choosing the largest difference between any two laboratories which are tolerable, and beyond which the hypothesis of equality of the φ_i should be rejected. Denote the largest difference by W.

$$\Phi\,(\text{min}) = \frac{W}{\sigma}\sqrt{\frac{c}{2r}}$$

$$\Phi\,(\text{max}) = \begin{cases} \dfrac{W}{2\sigma}\sqrt{c} & \text{for } r \text{ even} \\[2ex] \dfrac{W}{2\sigma}\sqrt{\dfrac{c(r^2 - 1)}{r^2}} & \text{for } r \text{ odd.} \end{cases}$$

Thus, a bound on Φ can be obtained if W is given and some estimate of σ is available. Such an estimate can often be obtained from previous data.

Operating characteristic curves for the analysis of the variance for the fixed effects model are presented in Figs. 10.2 to 10.9.[1] These curves are

---

[1] By permission, Figs. 10.2–10.9 are from "Power Function for Analysis of Variance Tests Derived from the Non-Central F Distribution," by E. S. Pearson and H. O. Hartley, *Biometrika*, Vol. 38, Part 1 and 2, June, 1951.

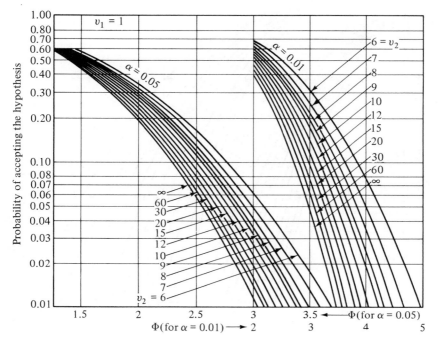

Fig. 10.2. *The OC curves for the analysis of variance with* $v_1 = 1$.

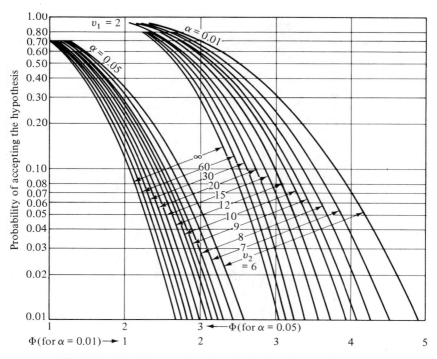

Fig. 10.3. *The OC curves for the analysis of variance with* $v_1 = 2$.

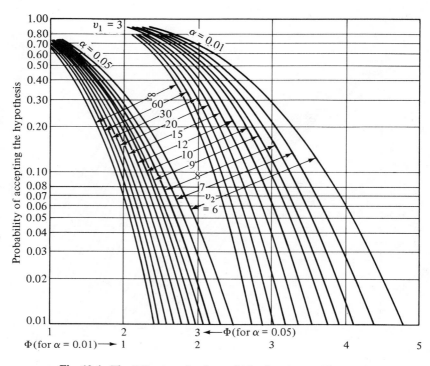

Fig. 10.4. *The OC curves for the analysis of variance with $v_1 = 3$.*

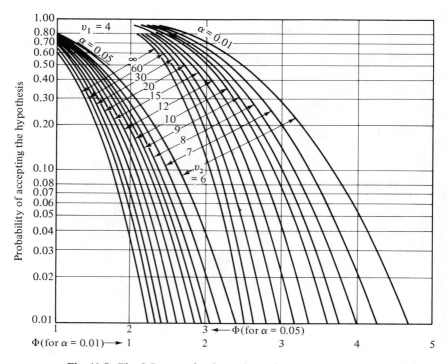

Fig. 10.5. *The OC curves for the analysis of variance with $v_1 = 4$.*

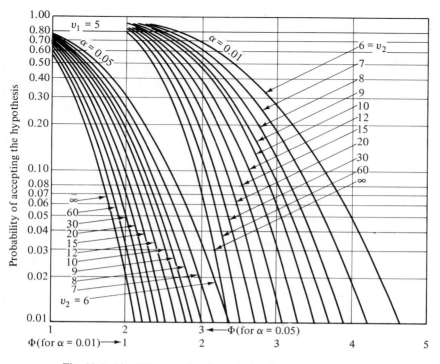

Fig. 10.6. *The OC curves for the analysis of variance with* $v_1 = 5$.

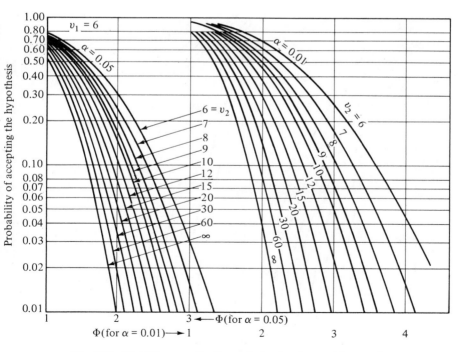

Fig. 10.7. *The OC curves for the analysis of variance with* $v_1 = 6$.

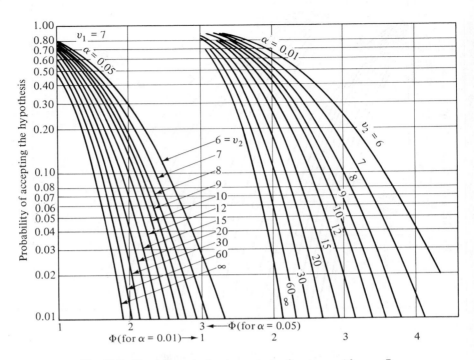

Fig. 10.8. *The OC curves for the analysis of variance with $v_1 = 7$.*

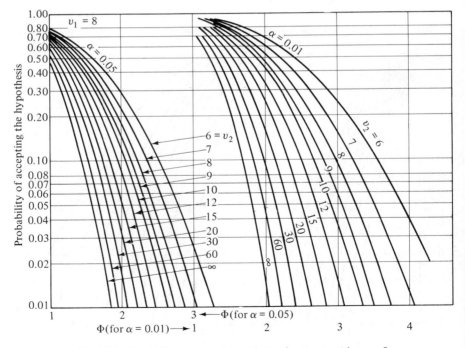

Fig. 10.9. *The OC curves for the analysis of variance with $v_1 = 8$.*

indexed according to the degrees of freedom for the treatments $v_1 = r - 1$ and degrees of freedom for the error term $v_2 = r(c - 1)$. These curves can be used directly for finding the probability of accepting the hypothesis given the number of observations used in the experiment, i.e., r and c, or can be used to design an experiment by means of trial and error. In this latter case, if r is fixed (number of treatments), the appropriate value of c can be obtained as follows if two points are specified. A value c is chosen; Φ is then computed and the OC curves are entered with v_1 and v_2 for the appropriate value of α. The probability of acceptance corresponding to Φ is read out. If this coincides with the desired value, a solution is reached; if not, the next higher value of c is chosen.

10.2.8 EXAMPLE USING THE FIXED EFFECTS MODEL

If the laboratories are treated as fixed effects, the following is an analysis of the data given in Table 10.1.

Computational Procedure

1. $R_1 = 369, \quad R_2 = 371, \quad R_3 = 345, \quad R_4 = 359.$

2. $T = \sum_{i=1}^{4} R_i = 1{,}444$

3. $\sum_{i=1}^{4} \sum_{j=1}^{5} X_{ij}^2 = 104{,}352$

4. $\dfrac{\sum_{i=1}^{4} R_i^2}{5} = 104{,}341.6$

5. $\dfrac{T^2}{20} = 104{,}256.8$

6. $SS_3 = 84.8$

7. $SS = 95.2$

8. $SS_2 = 10.4$

Thus, the value of F for testing whether the laboratories are equal is given by

$$F = SS_3^* / SS_2^* = 43.49.$$

Since $F_{0.05; 3, 16} = 3.24$ and $F = 43.49 > 3.24$, the hypothesis that $\varphi_1 = \varphi_2$

Table 10.5. Analysis of Variance Table for Laboratory Calibration

Source	Sum of Squares	Degrees of Freedom	Mean Square
Between treatments	84.8	3	28.27
Within treatments	10.4	16	0.65
Total	95.2	19	5.01

$= \varphi_3 = \varphi_4$ is rejected and it is concluded that the laboratories differ. To determine the magnitude of these differences, compute k for $r = 4$ and $r(c - 1) = 16$. Here $k^* = 4.05$, where the factor 4.05 is obtained from Table 10.3. Therefore, $k = 4.05\sqrt{0.65/5} = 1.46$ and any two means differing by this much differ significantly at $\alpha = 0.05$. The four laboratory means are: 73.8 for Laboratory A; 74.2 for Laboratory B; 69.0 for Laboratory C; and 71.8 for Laboratory D. Thus, the following differences are significant:

> Laboratory A from Laboratory C
> Laboratory A from Laboratory D
> Laboratory B from Laboratory C
> Laboratory B from Laboratory D
> Laboratory C from Laboratory D.

In obtaining the OC curve for the test, suppose the criterion about the increase in the standard deviation is to be used. If the difference in the laboratories is such that a random observation has its standard deviation increased by as much as 20%, the resultant value of Φ is

$$[\sqrt{(1.2)^2 - 1}]\sqrt{5} = \sqrt{(0.44)5} = \sqrt{2.20} = 1.48;$$

$v_1 = r - 1 = 3$ and $v_2 = r(c - 1) = (4)(4) = 16$. Hence, the probability of accepting the hypothesis at the 5% level of significance is found to be approximately 0.43 from Fig. 10.4. If, however, prior to the experiment a probability of acceptance of only 0.10 could be tolerated, a sample of 9 observations is required from each laboratory. This is attained by trial and error. If $c = 6$ is chosen first and computations similar to that above are made, the probability of acceptance is found to equal 0.30. Continuing in this manner, the desired probability of acceptance is reached with $c = 9$.

$$\Phi = [\sqrt{(1.2)^2 - 1}]\sqrt{9} = 1.99.$$

Entering Fig. 10.4 with $v_1 = 3$ and $v_2 = 4(8) = 32$, the probability of accepting the hypothesis is approximately 0.10, using the 5% level of significance.

10.2.9 ANALYSIS OF THE RANDOM EFFECTS MODEL

In the random effects model, testing the hypothesis about the homogeneity of treatments is equivalent to testing the hypothesis

$$H: \quad \sigma_\varphi^2 = 0.$$

Again referring to the analysis of variance table, Table 10.2, it is evident that both $E(SS_3^*)$ and $E(SS_2^*)$ equal σ^2 if the hypothesis is true. $E(SS_2^*)$ equals σ^2 even if the hypothesis is false, whereas $E(SS_3^*)$ exceeds σ^2 by an amount equal to $c\sigma_\varphi^2$ when the hypothesis is false. Hence, a large value of SS_3^*/SS_2^* compared to unity indicates that the hypothesis is false. Just as in the fixed effects model,

$$F = SS_3^*/SS_2^*$$

is distributed as an F random variable with $r - 1$ and $r(c - 1)$ degrees of freedom. Thus, the hypothesis that $\sigma_\varphi^2 = 0$ (implying homogeneity of all treatments) is rejected if

$$SS_3^*/SS_2^* > F_{\alpha; r-1, r(c-1)}.$$

To determine a measure of how much the treatments differ, σ_φ^2 can be estimated. The analysis of variance table indicates that SS_2^* is an unbiased estimator of σ^2 and that SS_3^* is an unbiased estimator of $\sigma^2 + c\sigma_\varphi^2$. Setting these estimators equal to the parameters that are being estimated, i.e., $SS_2^* = \sigma^2$ and $SS_3^* = \sigma^2 + c\sigma_\varphi^2$, an estimator of σ_φ^2 can be obtained from these equations. Such an estimator is given by

$$(SS_3^* - SS_2^*)/c.$$

If the value this estimator takes on is negative, the estimate is taken to be zero.

10.2.10 The OC Curve for the Random Effects Model

The analysis of variance procedure for the random effects model was presented in Sec. 10.2.9. The hypothesis that $\sigma_\varphi^2 = 0$ was accepted if the quantity

$$SS_3^*/SS_2^* \leq F_{\alpha; r-1, r(c-1)}.$$

The random variable SS_3^*/SS_2^* has an F distribution whenever $\sigma_\varphi^2 = 0$, and the probability of acceptance is $(1 - \alpha)$. In general, the probability of accepting the hypothesis that $\sigma_\varphi^2 = 0$ can be plotted as a function of

$$\lambda = \sqrt{\frac{\sigma^2 + c\sigma_\varphi^2}{\sigma^2}}.^1$$

The resulting OC curves are similar to the curves presented in Figs. 7.3 and 7.4. Unfortunately, these OC curves were presented for equal degrees of freedom and, hence, their use in the analysis of variance is limited. Further

---

[1] If the hypothesis is false, SS_3^*/SS_2^* does not have the F distribution. However, the quantity

$$\frac{SS_3^*/(\sigma^2 + c\sigma_\varphi^2)}{SS_2^*/\sigma^2}$$

does have an F distribution with $r - 1$ and $r(c - 1)$ degrees of freedom. Hence, the criterion for acceptance of

$$SS_3^*/SS_2^* \leq F_{\alpha; (r-1), r(c-1)}$$

is equivalent to

$$\frac{SS_3^*/(\sigma^2 + c\sigma_\varphi^2)}{SS_2^*/\sigma^2} \leq F_{\alpha; (r-1), r(c-1)} \frac{\sigma^2}{\sigma^2 + c\sigma_\varphi^2}.$$

The probability distribution of

$$\frac{SS_3^*/(\sigma^2 + c\sigma_\varphi^2)}{SS_2^*/\sigma^2}$$

is known and, consequently, the probability of acceptance can be plotted as a function of $\lambda = \sqrt{(\sigma^2 + c\sigma_\varphi^2)/\sigma^2}$.

OC curves are presented in Figs. 10.10 to 10.17. These are indexed for degrees of freedom $v_1 = r - 1$ and $v_2 = r(c - 1)$.

A physical interpretation of

$$\lambda = \sqrt{\frac{\sigma^2 + c\sigma_\varphi^2}{\sigma^2}} = \sqrt{1 + \frac{c\sigma_\varphi^2}{\sigma^2}}$$

can be obtained as follows: Consider the example of the laboratories where the same four laboratories are chosen from many. If the laboratories are homogeneous, the standard deviation of an observation chosen at random is just σ. On the other hand, if the laboratories are not homogeneous, the standard deviation of an observation chosen at random is increased to

$$\sqrt{\sigma^2 + \sigma_\varphi^2}.$$

Consequently, if the experimenter chooses a fixed percentage, say, P, for the increase in standard deviation of an observation beyond which he wishes to reject the hypothesis of homogeneity, it is equivalent to choosing

$$\sqrt{(\sigma^2 + \sigma_\varphi^2)/\sigma^2} = 1 + 0.01P$$

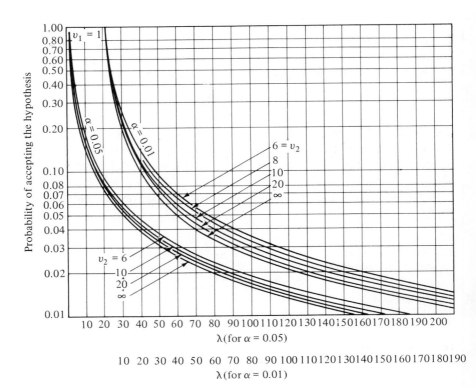

Fig. 10.10. *The OC curves for the random effects model with $v_1 = 1$.*

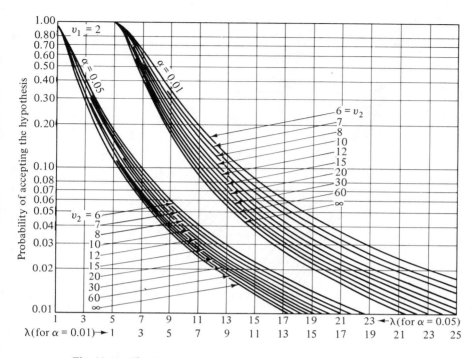

Fig. 10.11. *The OC curves for the random effects model with $v_1 = 2$.*

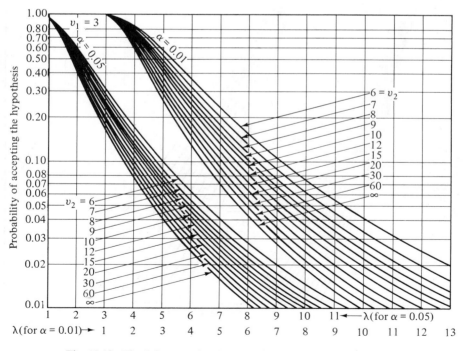

Fig. 10.12. *The OC curves for the random effects model with $v_1 = 3$.*

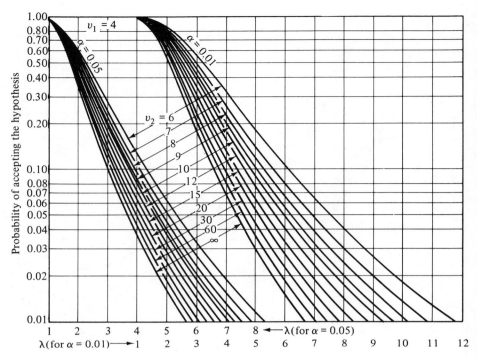

Fig. 10.13. *The OC curves for the random effects model with $v_1 = 4$.*

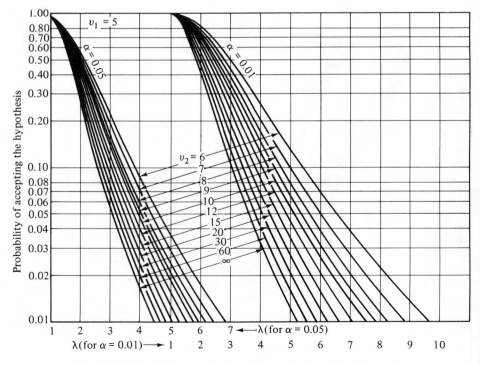

Fig. 10.14. *The OC curves for the random effects model with $v_1 = 5$.*

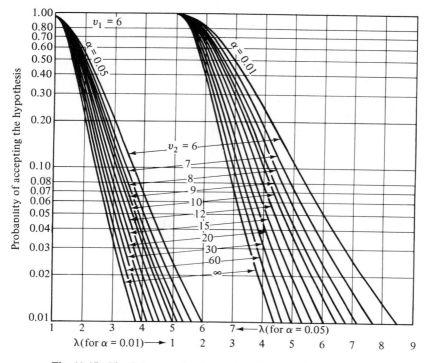

Fig. 10.15. *The OC curves for the random effects model with $v_1 = 6$.*

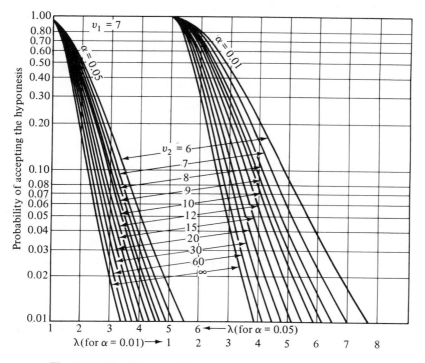

Fig. 10.16. *The OC curves for the random effects model with $v_1 = 7$.*

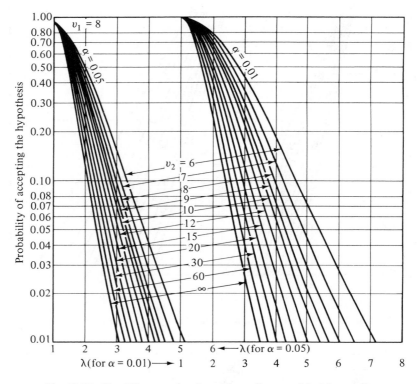

Fig. 10.17. *The OC curves for the random effects model with $v_1 = 8$.*

(where the increase in σ, P, is expressed in percent) or

$$\sigma_\varphi^2/\sigma^2 = (1 + 0.01P)^2 - 1;$$

hence, $$\lambda = \sqrt{1 + c[(1 + 0.01P)^2 - 1]}.$$

If a probability β of rejecting the hypothesis is associated with the chosen value of P, two points on the OC curve are fixed, the other point being the error of Type I associated with $\lambda = 1$.

The probabilities of acceptance for a given experimental setup can be evaluated in the manner above by choosing a P and then finding the probability of acceptance. On the other hand, if P, β, and r (number of treatments) are chosen in advance, the proper value of c can be found by trial and error in a similar manner as described for the OC curve of the fixed effects model.

10.2.11 EXAMPLE USING THE RANDOM EFFECTS MODEL

If it is assumed that there were a large number of laboratories to be calibrated and a random sample of four laboratories is chosen, the example of Sec. 10.2.8 would fit the random effects model. The formal calculations including the computational procedure, analysis of variance table, and F tests are identical with the results found in Sec. 10.2.8. The fact that $F = SS_3^*/SS_2^*$ exceeded the critical value indicates that all laboratories are not homogeneous. Determining which of the four laboratories included in the experiment differ is irrelevant, and determining which of the laboratories differ among all laboratories is impossible. However, an estimate of the variability of laboratories, σ_φ^2, can be obtained, and is found to be

$$(SS_3^* - SS_2^*)/c = 5.524.$$

When a 20% increase in the variability of the random variable occurs, the probability of accepting the hypothesis that $\sigma_\varphi^2 = 0$, using the F procedure, is 0.59 for a 5% level of significance. This value is found as follows:

$$\lambda = \sqrt{1 + 5[(1.2)^2 - 1]}$$
$$= \sqrt{1 + 2.2} = \sqrt{3.2} = 1.79.$$

From Fig. 10.12 ($v_1 = 3$), the value of 0.59 is read out corresponding to $\lambda = 1.79$.

If a probability of acceptance of only 0.10 can be tolerated for $P = 35\%$, a value of $c = 16$ is required. This value is found by trial and error.

10.2.12 RANDOMIZATION TESTS IN THE ANALYSIS OF VARIANCE

It has been emphasized in the discussion of the F test in the analysis of variance that the random effects ϵ_{ij} are assumed to be normally and independently distributed. These tests actually have a wider justification if they are considered special cases of what are called permutation tests. Suppose, for example, one is interested in testing the difference between two effects, say, percentage of carbon, and suppose that the experimenter takes eight samples (let them be named a, b, c, etc.) and chooses four with a table of random numbers to receive treatment A and the other four treatment B. The results of this experiment are:

A		B	
a	2	f	17
h	15	e	19
b	10	c	18
d	9	g	12

He could test the hypothesis that there is no difference between the treatments by calculating a t test or an F test, or consider it a one-way analysis of variance, calculating an F with 1 and 6 degrees of freedom; the value that the F random variable takes on is 6 in this problem. Suppose he is interested in a level of significance of 0.05. Even if he is unwilling to assume normality, he can obtain a test of the hypothesis as follows: Consider all possible pairs of samples that can be formed with the eight numbers *above*; there are $\binom{8}{4}$

= 70 different ways for the experiment to turn out (one for each way of selecting four of the eight numbers for treatment A). If there is no difference between treatments, each of these outcomes is equally likely. For each of the 70 pairs of samples, compute the value that the F random variable takes on. Suppose the procedure is adopted that the hypothesis of equality of treatments will be rejected if one of the samples with the four largest F values is actually obtained. The probability of rejection when the hypothesis is true is $4/70 = 0.057$, since all 70 possibilities are equally likely. If the hypothesis is false, pairs of samples having large F values are more likely to occur.

A partial list of the 70 outcomes (starting with most extreme ones) is given in Table 10.6.

Table 10.6.

A	B	F
2 9 10 12 15 17 18 19	15 17 18 19 2 9 10 12	14.84
2 9 10 15 12 17 18 19	12 17 18 19 2 9 10 15	6.00
2 9 10 17 12 15 18 19	12 15 18 19 2 9 10 17	3.61
2 9 12 15 10 17 18 19	10 17 18 19 2 9 12 15	3.61

Thus, the proposed test calls for rejection of the hypothesis if the value that the F random variable takes on is at least 6 (one of the 4 sample pairs having an F value of 6.00 or 14.84 is obtained). The significance level of the test is 0.057. Furthermore, with the particular sample obtained ($F = 6$) rejection is called for.

If the experimenter were willing to assume normality and perform the analysis of variance at a 5% level of significance, then the decision procedure would call for rejection of the hypothesis if $F > F_{0.05; 1,6} = 5.99$. Note that this procedure is identical to the proposed test, which does not require an assumption of normality.

This kind of test is called a randomization test. It has the drawback that for medium or large-sized samples the computational burden is enormous. For example, if there are 20 objects divided at random into two groups of 10, then the number of possibilities is $\binom{20}{10} = 184,756$, and listing the 5% most extreme cases would be prohibitive.

Both numerical and analytical investigations[1] have shown, however, that for medium and large samples, the distribution of F obtained through enu-

---

[1] The agreement between the two values 0.05 and 0.057 is not merely by coincidence.

meration of the possibilities is closely approximated by the tabulated F distribution. So, without assuming normality, the hypothesis of no treatment effects can be tested by making a randomization test; but instead of exactly evaluating the significance level by actual counting, one is content with an approximate evaluation as yielded by the F tables.

10.3 Two-Way Analysis of Variance, One Observation per Combination

Experiments can be conducted in such a way as to study the effects of several variables in the same experiment. For each variable a number of levels may be chosen for study. When observations are made for all possible combinations of levels the experiment is called a factorial experiment. This section is concerned with a factorial experiment where the effects of two variables at several levels (or using several treatments) are to be studied when there is one observation per combination of levels.

10.3.1 FIXED EFFECTS MODEL

In this section only fixed levels of the factors are considered; i.e., the categories or levels of a given factor chosen are the only ones of interest. Furthermore, for each combination of levels only one observation is taken. For example, Youden[1] reports the following experiment. A study was made of chemical cells used as a means of setting up a reference temperature. The sealed cells, containing about a pound of chemical, were heated just sufficiently to melt the chemical and were then carefully insulated. The temperature of the triple point was maintained very closely for a day or so. Considerable time was required for the resistance thermometers to attain thermal equilibrium when placed in wells extending into the chemical. Consequently, only one thermometer could be employed on a given melting of the cell. Differences in readings found among a number of cells could also arise if the thermometers were not in agreement. When all the thermometers in turn were tried with one cell, the work spread over several days, thereby introducing possible variations in the bridge measurements that did not occur among readings made on the same day. Four chemical cells and four thermometers were chosen for the study. Each thermometer was placed in every chemical cell, the total experiment extending over a period of four days. The possible day effect was randomized out by assigning the day in which a particular thermometer was placed in a given cell by means of a random device.[2] If the purpose of the study was to calibrate these four thermometers and these

---

[1] This example is based on an example appearing in W. J. Youden, *Statistical Methods for Chemists*, John Wiley & Sons, Inc., New York, 1951, pp. 96–98.

[2] Actually, Youden used a design called a Latin Square, which isolated the day effect. The randomization of the day effect is an acceptable alternative.

Table 10.7. Pairing of Cells and Thermometers

Chemical Cell	Thermometers I	II	III	IV	Totals	Average
1	36	38	36	30	140	35.0
2	17	18	26	17	78	19.5
3	30	39	41	34	144	36.0
4	30	45	38	33	146	36.5
Totals	113	140	141	114	508	
Average	28.25	35.0	35.25	28.5		

four chemical cells, and the results interpreted to hold for the chosen thermometers and chemical cells only, these quantities may be considered as fixed effects.

The data are given in Table 10.7. Only one reading is taken for each combination of thermometer and chemical cell. The entries are the readings converted to degrees Centigrade. Only the third and fourth decimal places are given, as the readings agree up to the last two places.

The mathematical model can be written as

$$X_{ij} = \zeta_{ij} + \epsilon_{ij}; \quad i = 1, 2, 3, 4; \quad j = 1, 2, 3, 4;$$

where X_{ij} is the temperature reading in the ith chemical cell using the jth thermometer; ζ_{ij} is the mean (of the distribution) temperature reading in the ith chemical cell, using the jth thermometer; ϵ_{ij} is the random effect which is assumed to be independently normally distributed with mean 0 and variance (unknown) σ^2.

Before proceeding to make inferences about the structure of the ζ_{ij}, it is useful to refer to Table 10.8, which is a table of expected values.

Table 10.8. Table of Expected Values for Temperature Readings

Chemical Cell	Thermometers I	II	III	IV	Average over Thermometers	Differences
1	ζ_{11}	ζ_{12}	ζ_{13}	ζ_{14}	$\zeta_{1\cdot}$	$\varphi_1 = \zeta_{1\cdot} - \zeta_{\cdot\cdot}$
2	ζ_{21}	ζ_{22}	ζ_{23}	ζ_{24}	$\zeta_{2\cdot}$	$\varphi_2 = \zeta_{2\cdot} - \zeta_{\cdot\cdot}$
3	ζ_{31}	ζ_{32}	ζ_{33}	ζ_{34}	$\zeta_{3\cdot}$	$\varphi_3 = \zeta_{3\cdot} - \zeta_{\cdot\cdot}$
4	ζ_{41}	ζ_{42}	ζ_{43}	ζ_{44}	$\zeta_{4\cdot}$	$\varphi_4 = \zeta_{4\cdot} - \zeta_{\cdot\cdot}$
Average over Cells:	$\zeta_{\cdot 1}$	$\zeta_{\cdot 2}$	$\zeta_{\cdot 3}$	$\zeta_{\cdot 4}$	$\zeta_{\cdot\cdot}$	
Differences:	$\gamma_1 = \zeta_{\cdot 1} - \zeta_{\cdot\cdot}$	$\gamma_2 = \zeta_{\cdot 2} - \zeta_{\cdot\cdot}$	$\gamma_3 = \zeta_{\cdot 3} - \zeta_{\cdot\cdot}$	$\gamma_4 = \zeta_{\cdot 4} - \zeta_{\cdot\cdot}$		

Here

$$\zeta_{i\cdot} = \frac{1}{4}\sum_{j=1}^{4}\zeta_{ij}; \quad \zeta_{\cdot j} = \frac{1}{4}\sum_{i=1}^{4}\zeta_{ij}; \quad \zeta_{\cdot\cdot} = \frac{1}{16}\sum_{i=1}^{4}\sum_{j=1}^{4}\zeta_{ij}.$$

Thus, a mean ζ_{ij} can be written

$$\zeta_{ij} = \zeta_{\cdot\cdot} + (\zeta_{i\cdot} - \zeta_{\cdot\cdot}) + (\zeta_{\cdot j} - \zeta_{\cdot\cdot}) + (\zeta_{ij} - \zeta_{i\cdot} - \zeta_{\cdot j} + \zeta_{\cdot\cdot})$$
$$= \zeta_{\cdot\cdot} + \varphi_i + \gamma_j + \eta_{ij}$$

where $\eta_{ij} = \zeta_{ij} - \zeta_{i\cdot} - \zeta_{\cdot j} + \zeta_{\cdot\cdot}$ and is known as the interaction term.

An interaction between two effects, φ_i and γ_j, is said to exist if the joint effect of the two taken simultaneously is different from the sum of their separate effects. Thus, either lye or muriatic acid may be an effective cleanser, but taken together they are ineffective, and would be said to interact. Also, suppose there are five machines together with four workmen and it is desired to test whether machines differ in the number of units produced per day. It is quite possible that the second man may work better on the third machine than any other machine because he has worked for 20 years on such a machine. The resultant increased production cannot be assumed to be a characteristic of this man or of this particular machine, but a combination of only this man with only this machine, and we would say that interaction was present.

Returning to the table of means, the following linear relationships are obtained:

$$\sum_{i=1}^{4}\varphi_i = 0; \quad \sum_{j=1}^{4}\gamma_j = 0; \quad \sum_{i=1}^{4}\eta_{ij} = 0; \quad \sum_{j=1}^{4}\eta_{ij} = 0.$$

Furthermore, instead of the original 16 means, new parameters have been introduced which will supply the results. By writing $\zeta_{ij} = \zeta_{\cdot\cdot} + \varphi_i + \gamma_j + \eta_{ij}$ the mean for the ith chemical cell and jth thermometer can be written as a constant, $\zeta_{\cdot\cdot}$, plus an effect due to the ith cell which is constant over all columns, φ_i, plus an effect due to the jth thermometer which is constant over all rows, γ_j, plus the interaction between the ith cell and jth thermometer, η_{ij}. Testing whether (1) the cells are homogeneous, (2) the thermometers are homogeneous, and (3) whether there is any interaction, is equivalent to testing:

(1) $\qquad\qquad\qquad \varphi_1 = \varphi_2 = \varphi_3 = \varphi_4 = 0;$

(2) $\qquad\qquad\qquad \gamma_1 = \gamma_2 = \gamma_3 = \gamma_4 = 0;$

(3) $\qquad\qquad\qquad \eta_{ij} = 0$

for all i and j.

More generally, these assumptions can be summarized as

$$X_{ij} = \zeta_{\cdot\cdot} + \varphi_i + \gamma_j + \eta_{ij} + \epsilon_{ij}$$

where X_{ij} is the outcome of the experiment, using the ith row effect and jth column effect, $i = 1, 2, \ldots, r$; $j = 1, 2, \ldots, c$; $\zeta..$ is a general mean; φ_i is the effect of adding the ith row treatment, $\sum_{i=1}^{r} \varphi_i = 0$; γ_j is the effect of adding the jth column treatment, $\sum_{j=1}^{c} \gamma_j = 0$; η_{ij} is the interaction of the ith row treatment with the jth column treatment, $\sum_{i=1}^{r} \eta_{ij} = \sum_{j=1}^{c} \eta_{ij} = 0$; the ϵ_{ij} are the random effects, which are independently normally distributed each with mean 0 and variance σ^2.

10.3.2 RANDOM EFFECTS MODEL

In the chemical cells and thermometer example, a more realistic assumption about the cells and thermometers is that the four used in experimentation were chosen at random from a large number of cells and thermometers. Furthermore, inferences drawn from the experiment are to extend to chemical cells and thermometers of these types in general, rather than be confined to the ones used in the experiment. In other words, the cells and thermometers are actually random variables whose values depend on the outcome of the selection process. The mathematical model may still be written as

$$X_{ij} = \zeta.. + \varphi_i + \gamma_j + \eta_{ij} + \epsilon_{ij}; \quad i = 1, 2, \ldots, r; j = 1, 2, \ldots, c;$$

but the interpretations of the φ_i, the row effect, γ_j, the column effect, and η_{ij}, the interaction effect, are different. The φ_i, γ_j, and η_{ij} are independent random variables, all having mean 0 and variance σ_φ^2, σ_γ^2, and σ_η^2, respectively. These random variables are usually assumed to be normally distributed. The outcome of these effects (values associated with the thermometers and cells chosen) depends on their probability distributions; interest is centered on certain statistical properties characterizing these distributions, and not specifically on the particular individuals which happen to be drawn. Consider, for example, the thermometer effects: Is one really interested in (1) whether the four thermometers chosen read differently, or (2) the variation of all thermometers from which the four thermometers may be imagined drawn? If the answer is (1), the thermometer effects must be considered as fixed effects. If the answer is (2), the thermometer effects must be considered as random effects. The interactions, such as chemical cells and thermometers, represent differential effects of one of the factors caused by different levels of the other. In deciding about the interactions among two factors, the interaction effect is considered as a fixed effect if the factors involved in the interaction are fixed effects; otherwise it is treated as a random effect. Of course, in the cell-thermometer example, the interaction may still be assumed zero even if considered as a random effect.

Thus, inferences about row effects, column effects, and interaction effects are equivalent to making inferences about σ_φ^2, σ_γ^2, and σ_η^2, respectively. For example, if $\sigma_\varphi^2 = 0$, all of the chemical cells are homogeneous.

10.3.3 MIXED FIXED EFFECTS AND RANDOM EFFECTS MODEL

The previous sections dealt with the two-way analysis of variance when the factors were fixed effects and random effects, respectively. A third possibility to consider is one effect fixed and one effect random. This model is denoted as the mixed effect model. In the chemical cells and thermometer problem, the cell effects can be considered as a random sample from a large number of such effects, whereas the thermometers can be considered as fixed effects. This implies that the experimental results will lead to conclusions about the distribution of chemical cells. The conclusion about thermometers will extend only to the four thermometers used. It must be emphasized that the decision as to whether a factor is at fixed or random levels is a practical one and not up to the statistician. If the experimenter considers chemical cells as a random factor and wishes to generalize the results of his experiment beyond the few he has chosen, he must take measures to assure that the cells are really chosen at random from the large number of chemical cells available.

The general mathematical model may still be written as

$$X_{ij} = \zeta.. + \varphi_i + \gamma_j + \eta_{ij} + \epsilon_{ij}; \quad i = 1, 2, \ldots, r; j = 1, 2, \ldots, c;$$

where $\zeta..$ is the over-all mean.

φ_i is the row effect and is taken to be a normally distributed random variable with mean 0 and variance σ_φ^2. This corresponds to chemical cells in the description of the problem given in this section. γ_j is the column effect and will be taken to be the fixed effect of this model. $\sum_{j=1}^c \gamma_j = 0$. This corresponds to the thermometers in the description of the problem given in this section. η_{ij} is the interaction effect and is taken to be a normally distributed random variable with mean 0 and variance σ_η^2. This is a random variable rather than a fixed effect because one of the factors (rows) is a random effect. However, it also seems reasonable to assume that $\sum_{j=1}^c \eta_{ij} = 0$ since all the η_{ij} (for fixed i) are in the sample.

Thus, inferences about row effects are equivalent to making inferences about σ_φ^2; inferences about column effects are equivalent to making inferences about γ_j for all j; and inferences about interaction effects are equivalent to making inferences about σ_η^2.

10.3.4 COMPUTATIONAL PROCEDURE, TWO-WAY CLASSIFICATION, ONE OBSERVATION PER COMBINATION

Before carrying out any further analyses, it is necessary to complete the analysis of variance table. This is done by following the computational procedure described below. The results are summarized in the analysis of variance table, Table 10.9.

Table 10.9. Analysis of Variance Table, Two-Way Classification, One Observation per Combination

Source	Sum of Squares	Degrees of Freedom	Mean Square	Expected Mean Square for the Fixed Effects Model	Expected Mean Square for the Random Effects Model	Expected Mean Square for the Mixed Effects Model
Between columns	$SS_4 = r \sum_{j=1}^{c}(\bar{X}_{.j} - \bar{X}_{..})^2$	$c-1$	$SS_4^* = SS_4/(c-1)$	$\sigma^2 + \dfrac{r}{c-1}\sum_{j=1}^{c}\gamma_j^2$	$\sigma^2 + \sigma_\eta^2 + r\sigma_\gamma^2$	$\sigma^2 + \sigma_\eta^2 + \dfrac{r}{c-1}\sum_{j=1}^{c}\gamma_j^2$
Between rows	$SS_3 = c \sum_{i=1}^{r}(\bar{X}_{i.} - \bar{X}_{..})^2$	$r-1$	$SS_3^* = SS_3/(r-1)$	$\sigma^2 + \dfrac{c}{r-1}\sum_{i=1}^{r}\varphi_i^2$	$\sigma^2 + \sigma_\eta^2 + c\sigma_\varphi^2$	$\sigma^2 + c\sigma_\varphi^2$
Residual	$SS_2 = \sum_{i=1}^{r}\sum_{j=1}^{c}(X_{ij} - \bar{X}_{i.} - \bar{X}_{.j} + \bar{X}_{..})^2$	$(r-1)(c-1)$	$SS_2^* = SS_2/(r-1)(c-1)$	$\sigma^2 + \dfrac{1}{(r-1)(c-1)}\sum_{i=1}^{r}\sum_{j=1}^{c}\eta_{ij}^2$	$\sigma^2 + \sigma_\eta^2$	$\sigma^2 + \sigma_\eta^2$
Total	$SS = \sum_{i=1}^{r}\sum_{j=1}^{c}(X_{ij} - \bar{X}_{..})^2$	$rc-1$	$SS^* = SS/(rc-1)$	—	—	—

$\bar{X}_{.j} = \sum_{i=1}^{r} X_{ij}/r; \quad \bar{X}_{i.} = \sum_{j=1}^{c} X_{ij}/c; \quad \bar{X}_{..} = \sum_{i=1}^{r}\sum_{j=1}^{c} X_{ij}/rc.$

Computational Procedure

1. Calculate row totals:
$$R_1, R_2, \ldots, R_r.$$

2. Calculate column totals:
$$C_1, C_2, \ldots, C_c.$$

3. Calculate over-all total:
$$T = R_1 + R_2 + \cdots + R_r.$$

4. Calculate crude total sum of squares:
$$\sum_{i=1}^{r} \sum_{j=1}^{c} X_{ij}^2 = X_{11}^2 + X_{12}^2 + \cdots + X_{cr}^2.$$

5. Calculate crude sum of squares between rows:
$$\sum_{i=1}^{r} \frac{R_i^2}{c} = \frac{(R_1^2 + R_2^2 + \cdots + R_r^2)}{c}.$$

6. Calculate crude sum of squares between columns:
$$\sum_{j=1}^{c} \frac{C_j^2}{r} = \frac{(C_1^2 + C_2^2 + \cdots + C_c^2)}{r}.$$

7. Calculate correction factor due to mean $= T^2/rc$.
From the quantities above compute:

8.
$$SS_4 = (6) - (7) = \sum_{j=1}^{c} \frac{C_j^2}{r} - \frac{T^2}{rc},$$

9.
$$SS_3 = (5) - (7) = \sum_{i=1}^{r} \frac{R_i^2}{c} - \frac{T^2}{rc},$$

10.
$$SS = (4) - (7) = \sum_{i=1}^{r} \sum_{j=1}^{c} X_{ij}^2 - \frac{T^2}{rc},$$

11.
$$SS_2 = (10) - (9) - (8) = SS - SS_3 - SS_4.$$

Note that $SS = SS_2 + SS_3 + SS_4$. In fact this was used in obtaining SS_2 in step 11 in the computational procedure.

10.3.5 ANALYSIS OF THE FIXED EFFECTS MODEL, TWO-WAY CLASSIFICATION, ONE OBSERVATION PER COMBINATION

To determine whether there is a row effect or a column effect it is necessary to test the hypotheses that

$$H: \quad \varphi_1 = \varphi_2 = \cdots = \varphi_r = 0$$
$$H: \quad \gamma_1 = \gamma_2 = \cdots = \gamma_c = 0,$$

respectively.

Referring to the analysis of variance table, Table 10.9, the expected mean square for the residual equals σ^2 plus some interaction terms. The expected mean square for the row effects equals σ^2 plus some row effect terms. Both SS_2^* and SS_3^* estimate σ^2 only if the interaction terms are zero and there are no row effects, respectively. Hence, to get a reasonable test on row effects in the fixed effects model with one observation per combination, it is necessary to assume that the interaction term is zero, i.e.,

$$X_{ij} = \zeta.. + \varphi_i + \gamma_j + \epsilon_{ij}.$$

A similar assumption is necessary to get a reasonable test for column effects. There is often some prior knowledge available about the physical situation which allows for this assumption.

In the thermometer and chemical cell example there is no reason to postulate an interaction between chemical cells and thermometers. If there is a ranking among the thermometers, it is certainly not an affair of the cells, and so one may confidently assume the interactions between thermometers and chemical cells to be zero, i.e., $\eta_{ij} = 0$ for all i and j.

Henceforth, the interaction will be assumed to equal zero so that $E(SS_2^*) = \sigma^2$ always. The expected mean squares for both row effects and the residual sources, $E(SS_3^*)$ and $E(SS_2^*)$, respectively, equal σ^2 if the hypothesis is true. $E(SS_2^*)$ equals σ^2 even if the hypothesis is false. However, in this event, $E(SS_3^*)$ exceeds σ^2 by the amount $c \sum_{i=1}^{r} \varphi_i^2/(r-1)$. Hence, if the ratio SS_3^*/SS_2^* is computed, a value far from unity would indicate that the hypothesis is false. It can be shown that (using the partition theorem) SS_3/σ^2 and SS_2/σ^2 are independently distributed as chi-square random variables with $(r-1)$ and $(r-1)(c-1)$ degrees of freedom, respectively, when the hypothesis is true. Hence, the ratio SS_3^*/SS_2^* is distributed as an F random variable with $r-1$ and $(r-1)(c-1)$ degrees of freedom when the hypothesis is true. Thus, the hypothesis of equality of the φ's is rejected if

$$F = SS_3^*/SS_2^* > F_{\alpha;(r-1),(r-1)(c-1)}.$$

Using an analogous argument, the hypothesis of no column effects, i.e., the hypothesis that $\gamma_1 = \gamma_2 = \cdots = \gamma_c = 0$, is rejected if

$$F = SS_4^*/SS_2^* > F_{\alpha;(c-1),(r-1)(c-1)}.$$

If the hypothesis that there are no row effects or no column effects, or both, is rejected, Tukey's method can again be applied to get confidence intervals on the difference between two row effects or two column effects. The probability is $1 - \alpha$ that the values $(\varphi_m - \varphi_n)$ for all such contrasts (for $m = 1, 2, \ldots, r; n = 1, 2, \ldots, r$) simultaneously satisfy

$$\bar{X}_{m\cdot} - \bar{X}_{n\cdot} - k \leq \varphi_m - \varphi_n \leq \bar{X}_{m\cdot} - \bar{X}_{n\cdot} + k,$$

and the probability is $1 - \alpha$ that the values $(\gamma_f - \gamma_g)$ for all such contrasts (for $f = 1, 2, \ldots, c; g = 1, 2, \ldots, c$) simultaneously satisfy

$$\bar{X}_{\cdot f} - \bar{X}_{\cdot g} - k \leq \gamma_f - \gamma_g \leq \bar{X}_{\cdot f} - \bar{X}_{\cdot g} + k.$$

Again two mean effects are judged significantly different when 0 is not included in the confidence interval.

The factor k for row effects and for column effects is obtained from the analysis of variance table and from Table 10.3 or Table 10.4, depending on the level of significance. Table 10.3 or 10.4 is entered with the indices, degrees of freedom, and the number of row effects being studied (r) if row effects are of interest. The proper degrees of freedom are those corresponding to the degrees of freedom of the residual source in the analysis of variance table, i.e., $(r-1)(c-1)$. The factor k^* is read from the table, and

$$k \text{ (for rows)} = k^*\sqrt{SS_2^*/c}.$$

Similarly, if column effects are of interest, the table is entered with indices, degrees of freedom and the number of column effects being studied (c). The factor k^* is read from the table and

$$k \text{ (for columns)} = k^*\sqrt{SS_2^*/r}.$$

Note that the expected mean square column indicates which ratios to use in the F test.

10.3.6 The OC Curve of the Analysis of Variance for the Fixed Effects Model, Two-Way Classification, One Observation per Combination

If the OC curve for testing the hypothesis about row effects is desired, the appropriate measure for which an OC curve can be constructed is

$$\Phi = \frac{\sqrt{\sum_{i=1}^{r} \varphi_i^2/r}}{\sigma/\sqrt{c}}.$$

Similarly, if the OC curve for testing the hypothesis about column effects is desired, the appropriate measure for which an OC curve can be constructed is

$$\Phi = \frac{\sqrt{\sum_{j=1}^{c} \gamma_j^2/c}}{\sigma/\sqrt{r}}.$$

The OC curves as functions of v_1, v_2 and Φ are given in Figs. 10.2 to 10.9. The appropriate values of v_1 are $(r-1)$ if row effects are being considered and $(c-1)$ if column effects are being considered. v_2 corresponds to the degrees of freedom that are associated with the residual and is given by $(r-1)(c-1)$. If the number of row effects and column effects are fixed and only one observation for each combination of levels is allowed, these OC curves can be used to study the probabilities of making the correct decisions as a function of Φ. If more than one observation per combination is allowed, the appropriate number to guarantee a fixed OC curve is given in the results of Sec. 10.4.

A physical interpretation of Φ as a measure can be given by finding upper and lower bounds for Φ as a function of the range of the φ_i [$\varphi(\max) - \varphi(\min)$ if row effects are being studied] beyond which rejection of homogeneity should be assured with a given probability. Denote this largest difference by W. Bounds on Φ are given as follows:

$$\Phi(\min) = \frac{W}{\sigma}\sqrt{\frac{c}{2r}}$$

$$\Phi(\max) = \begin{cases} \dfrac{W}{2\sigma}\sqrt{c} & \text{for } r \text{ even} \\[2ex] \dfrac{W}{2\sigma}\sqrt{\dfrac{c(r^2-1)}{r^2}} & \text{for } r \text{ odd.} \end{cases}$$

Thus, if W is specified and some estimate of σ is available, the probability of accepting the hypothesis of homogeneity of φ_i can be obtained from Figs. 10.2 to 10.9.

Similarly, if column effects are being studied,

$$\Phi(\min) = \frac{W}{\sigma}\sqrt{\frac{r}{2c}}$$

$$\Phi(\max) = \begin{cases} \dfrac{W}{2\sigma}\sqrt{r} & \text{for } c \text{ even} \\[2ex] \dfrac{W}{2\sigma}\sqrt{\dfrac{r(c^2-1)}{c^2}} & \text{for } c \text{ odd.} \end{cases}$$

10.3.7 EXAMPLE USING THE FIXED EFFECTS MODEL

If the chemical cells and thermometers are treated as fixed effects, the following is an analysis of the data given in Table 10.7. The following computations lead to the analysis of variance table.

1. $\qquad R_1, R_2, R_3, R_4 = 140, 78, 144, 146$

2. $\qquad C_1, C_2, C_3, C_4 = 113, 140, 141, 114$

3. $\qquad T = 140 + 78 + 144 + 146 = 508$

4. $\displaystyle\sum_{i=1}^{4}\sum_{j=1}^{4} X_{ij}^2 = (36)^2 + (38)^2 + \cdots + (33)^2 = 17{,}230$

5. $\displaystyle\sum_{i=1}^{4} \frac{R_i^2}{4} = \frac{(140)^2 + (78)^2 + (144)^2 + (146)^2}{4} = 16{,}934$

6. $\displaystyle\sum_{j=1}^{4} \frac{C_i^2}{4} = \frac{(113)^2 + (140)^2 + (141)^2 + (114)^2}{4} = 16{,}311.5$

7. $\qquad T^2/16 = (508)^2/16 = 16{,}129$

8. $SS_4 = 16{,}311.5 - 16{,}129 = 182.5$

9. $SS_3 = 16{,}934 - 16{,}129 = 805$

10. $SS = 17{,}230 - 16{,}129 = 1{,}101$

11. $SS_2 = 1{,}101 - 805 - 182.5 = 113.5$

Table 10.10. Analysis of Variance Table for Temperatures of Cells

Source	Sum of Squares	Degrees of Freedom	Mean Square	Test
Between thermometers	182.5	3	60.83	$F = 4.82$
Between chemical cells	805	3	268.33	$F = 21.3$
Residual	113.5	9	12.61	
Total	1,101.0	15	73.4	

Thus, the values of F for testing the hypotheses that the thermometers and cells are the same are 4.82 and 21.3, respectively. Since both of these values exceed $F_{0.05; 3, 9} = 3.86$, it can be concluded that the thermometers differ significantly and the chemical cells also differ significantly. To determine the magnitude of these differences, k is computed for thermometers and cells. k^* (for thermometers) $= 4.42$, where the factor 4.42 is obtained from Table 10.3, entering with $n = c = 4$ and $v = (r - 1)(c - 1) = 9$. Therefore, $k = 4.42\sqrt{12.61/4} = 7.86$, and any two thermometer sample means differing by this much differ significantly at $\alpha = 0.05$. However, no two of the thermometer means differ by more than 7.86, so that it is not possible to detect the unusual thermometers using this technique. Although each confidence interval includes zero, meaning that such pairs *may* be the same, it does not preclude any (or all) of the true differences to be other than zero. Thus, in this situation, which is rather rare, this technique is unable to pick any unusual thermometers even though the F test indicates that the thermometers differ.

Since there are also four cells, k (for cells) is the same as for thermometers, i.e., k (for cells) $= 7.86$ and any two cell sample means differing by this much differ significantly at $\alpha = 0.05$. From Table 10.7 it is evident that cell number 2 differs from the other three.

If a maximum difference between thermometers of 0.0007 degrees Centigrade ($W = 7$) is specified and σ is approximately equal to 3,

$$\Phi(\max) = (7/6)\sqrt{4} = 2.33,$$

$$\Phi(\min) = (7/3)\sqrt{4/8} = 1.65.$$

Hence, the probability of accepting the hypothesis that the thermometers are homogeneous when $W = 7$ lies between 0.11 and 0.41 as is found by entering Fig. 10.4 with $v_1 = 3$ and $v_2 = 9$.

It is of interest to comment on the design of the experiment. If thermometers are considered to be a "nuisance" parameter, then an alternative design is to run the experiment as a one-way analysis of variance (cells being the main effect), randomizing out the possible thermometer effect by assigning a thermometer to a cell by means of a random device (as was done for the possible day effect). This results in the inclusion of any possible thermometer effect as part of the experimental error. If the random device resulted in the assignments used in the actual experiment, then the same data would have been obtained. The subsequent analysis for the cell effects would also be the same except that the "residual" (experimental error) would contain the thermometer effects, i.e., the sum of squares for the residual becomes $(SS_2 + SS_4)$, thereby making detection of cell effects more difficult to judge because the denominator in the F test is inflated.

In particular, for this data

$$F = \frac{268.33}{(182.5 + 113.5)/12} = 10.9,$$

which is still significant ($F_{0.05; 3, 12} = 3.49$) because of the overwhelming cell differences. However, in similar situations, this one-way design could mask the possible main effects.

10.3.8 ANALYSIS OF THE RANDOM EFFECTS MODEL, TWO-WAY CLASSIFICATION, ONE OBSERVATION PER COMBINATION

In the random effects model, testing the hypothesis about the homogeneity of row effects is equivalent to testing the hypothesis

$$H: \quad \sigma_\varphi^2 = 0,$$

and testing the hypothesis about the homogeneity of column effects is equivalent to testing the hypothesis

$$H: \quad \sigma_\gamma^2 = 0.$$

Referring to the analysis of variance table, Table 10.9, the expected mean square for the residual in this model equals $\sigma^2 + \sigma_\eta^2$, i.e., the sum of the error variance and the interaction variance. The expected mean square for the row effects equals $\sigma^2 + \sigma_\eta^2 + c\sigma_\varphi^2$, i.e., the sum of the error variance, interaction variance and a constant times the row effect variance. If the hypothesis of no row effects ($\sigma_\varphi^2 = 0$) is true, the expected mean square for both row effects and the residual sources, $E(SS_3^*)$ and $E(SS_2^*)$, respectively, equals $\sigma^2 + \sigma_\eta^2$. Thus, using the ratio SS_3^*/SS_2^* leads to a reasonable procedure in testing for row effects in this model without making any assumptions about the presence or absence of interaction. This is quite different from the fixed effects model where it was necessary to postulate the absence of interaction. The random effects model, then, leads to a test for row effects whether or not interaction is present, using the ratio SS_3^*/SS_2^* as a statistic. It can be shown (using the partition theorem) that $SS_3/(\sigma^2 + \sigma_\eta^2)$ and $SS_2/(\sigma^2 + \sigma_\eta^2)$ are independently distributed as chi-square random variables with $(r - 1)$ and $(r - 1)(c - 1)$ degrees of freedom, respectively, when the hypothesis is true. Hence, the ratio SS_3^*/SS_2^* is distributed as an F random variable with $r - 1$ and $(r - 1)(c - 1)$ degrees of freedom when the hypothesis is true. Thus, the hypothesis that $\sigma_\varphi^2 = 0$ is rejected if

$$F = SS_3^*/SS_2^* > F_{\alpha; (r-1), (r-1)(c-1)}.$$

Using an analogous argument, the hypothesis of no column effect, i.e., the hypothesis that $\sigma_\gamma^2 = 0$, is rejected if

$$F = SS_4^*/SS_2^* > F_{\alpha; (c-1), (r-1)(c-1)}.$$

To determine a measure of how much the row effects differ, σ_φ^2 can be estimated. The analysis of variance table indicates that SS_2^* is an unbiased estimator of $\sigma^2 + \sigma_\eta^2$ and that SS_3^* is an unbiased estimator of $\sigma^2 + \sigma_\eta^2 + c\sigma_\varphi^2$. Setting these estimators equal to the parameters being estimated, i.e., $SS_2^* = \sigma^2 + \sigma_\eta^2$ and $SS_3^* = \sigma^2 + \sigma_\eta^2 + c\sigma_\varphi^2$, an estimator of σ_φ^2 can be obtained. Such an estimator is given by $(SS_3^* - SS_2^*)/c$. If the value this estimator takes on is negative, the estimate is taken to be zero. Similarly, an estimator of σ_γ^2 is given by $(SS_4^* - SS_2^*)/r$.

10.3.9 THE OC CURVE FOR THE RANDOM EFFECTS MODEL, TWO-WAY CLASSIFICATION

The hypothesis that $\sigma_\varphi^* = 0$ is accepted if

$$SS_3^*/SS_2^* \leqq F_{\alpha; (r-1), (r-1)(c-1)}.$$

The random variable SS_3^*/SS_2^* has an F distribution whenever $\sigma_\varphi^2 = 0$, and the probability of acceptance is $(1 - \alpha)$. In general, the probability of accepting the hypothesis that $\sigma_\varphi^2 = 0$ can be plotted as a function of

$$\lambda = \sqrt{1 + \frac{c\sigma_\varphi^2}{\sigma^2 + \sigma_\eta^2}}.\text{[1]}$$

---

[1] If the hypothesis is false,

$$\frac{SS_3^*/(\sigma^2 + \sigma_\eta^2 + c\sigma_\varphi^2)}{SS_2^*/(\sigma^2 + \sigma_\eta^2)}$$

has an F distribution with $(r - 1)$ and $(r - 1)(c - 1)$ degrees of freedom. Since the procedure of accepting the hypothesis when

$$SS_3^*/SS_2^* \leqq F_{\alpha; (r-1), (r-1)(c-1)}$$

is equivalent to accepting the hypothesis when

$$\frac{SS_3^*/(\sigma^2 + \sigma_\eta^2 + c\sigma_\varphi^2)}{SS_2^*/(\sigma^2 + \sigma_\eta^2)} \leqq F_{\alpha; r-1, (r-1)(c-1)}\left[\frac{(\sigma^2 + \sigma_\eta^2)}{\sigma^2 + \sigma_\eta^2 + c\sigma_\varphi^2}\right],$$

the probability of acceptance can be determined since the probability distribution of the left-hand side of the last inequality is known (F distribution). Hence, the OC curves of Figs. 10.10 to 10.17 are applicable when

$$\lambda = \sqrt{\frac{\sigma^2 + \sigma_\eta^2 + c\sigma_\varphi^2}{\sigma^2 + \sigma_\eta^2}} = \sqrt{1 + \frac{c\sigma_\varphi^2}{\sigma^2 + \sigma_\eta^2}}.$$

The OC curves of Figs. 10.10 to 10.17 are again applicable when λ is defined in this manner. Note that the OC curve is dependent only on the ratio $\sigma_\varphi^2/(\sigma^2 + \sigma_\eta^2)$.

Similarly, if column effects are being studied, the OC curve is a function of

$$\lambda = \sqrt{1 + \frac{r\sigma_\gamma^2}{\sigma^2 + \sigma_\eta^2}}.$$

The OC curves of Figs. 10.10 to 10.17 are entered with degrees of freedom $v_1 = r - 1$ if row effects are of interest, $v_1 = c - 1$ if column effects are of interest, and $v_2 = (r - 1)(c - 1)$.

10.3.10 Example Using the Random Effects Model

If the four chemical cells and thermometers are considered to be random samples from a large number of cells and thermometers, the example presented in Sec. 10.3.7 would fit the random effects model. The formal calculations including the computational procedure and the analysis of variance table are identical with the results of Sec. 10.3.7. The numerical values of the F test are also the same as are the conclusions drawn from the F test (adapted to the present model). However, in this model it is unnecessary to assume that there is no interaction between cells and thermometers. An estimate of σ_φ^2 is given by

$$(SS_3^* - SS_2^*)/c = 63.9$$

and an estimate of σ_γ^2 is given by

$$(SS_4^* - SS_2^*)/r = 12.1.$$

A point on the OC curve when $\sigma_\varphi^2/(\sigma^2 + \sigma_\eta^2) = 1.2$ is found as follows: The corresponding value of λ is 2.4. Figure 10.12 is entered with $v_1 = 3$, $v_2 = 9$, and $\lambda = 2.4$. The probability of accepting the hypothesis of chemical cell homogeneity when $\lambda = 2.4$ is then seen to be 0.40.

10.3.11 Analysis of the Mixed Effects Model, Two-Way Classification, One Observation per Combination

In the mixed effects model, the row effects will be assumed to be the random effects and the column effects will be assumed to be the fixed effects. Testing the hypothesis about the homogeneity of row effects is equivalent to testing the hypothesis

$$H: \quad \sigma_\varphi^2 = 0.$$

Testing the hypothesis about the homogeneity of column effects is equivalent to testing the hypothesis

$$H: \quad \gamma_1 = \gamma_2 = \cdots = \gamma_c = 0.$$

Referring to the analysis of variance table, Table 10.9, the expected mean square for the residual in this model equals $\sigma^2 + \sigma_\eta^2$, i.e., the sum of the error variance and the interaction variance. The expected mean square for the row effects (random effects) equals $\sigma^2 + c\sigma_\varphi^2$, i.e., the sum of the error variance and a constant times the row effect variance. If the hypothesis that σ_φ^2 is zero is true, $E(SS_3^*) = \sigma^2$, whereas $E(SS_2^*) = \sigma^2$ only if there is no interaction present. Hence, to test for the row effects (random effects) it is necessary to assume that $\sigma_\eta^2 = 0$. Under this assumption, the ratio SS_3^*/SS_2^* is a reasonable statistic for use in testing for row effects. It can be shown (using the partition theorem) that SS_3/σ^2 and SS_2/σ^2 are independently distributed as chi-square random variables with $(r - 1)$ and $(r - 1)(c - 1)$ degrees of freedom, respectively, when the hypothesis is true. Hence, the ratio SS_3^*/SS_2^* is distributed as an F random variable with $(r - 1)$ and $(r - 1)(c - 1)$ degrees of freedom when the hypothesis is true. Thus, the hypothesis that $\sigma_\varphi^2 = 0$ is rejected if

$$F = SS_3^*/SS_2^* > F_{\alpha;\,(r-1),\,(r-1)(c-1)}.$$

In the absence of interaction, an estimator of σ_φ^2 is available, i.e.,

$$(SS_3^* - SS_2^*)/c.$$

In testing for the column (fixed) effects, the problem is somewhat different. The expected mean square, $E(SS_4^*)$, for the column source is given by

$$\sigma^2 + \sigma_\eta^2 + \frac{r}{c-1}\sum_{j=1}^{c}\gamma_j^2.$$

As before, the expected mean square for the residual sum of squares, $E(SS_2^*)$, is given by $\sigma^2 + \sigma_\eta^2$. If there is no column effect, both expected values equal $\sigma^2 + \sigma_\eta^2$. Thus, using the ratio SS_4^*/SS_2^* leads to a reasonable procedure for testing for column (fixed) effects without making any assumption about the presence or absence of interaction. This is quite different from testing for the random effects in this model. It can be shown (by using the partition theorem) that SS_4^*/SS_2^* is distributed as an F random variable with $(c - 1)$ and $(r - 1)(c - 1)$ degrees of freedom when the hypothesis is true. Therefore, the hypothesis that $\gamma_1 = \gamma_2 = \cdots = \gamma_c = 0$ is rejected if

$$F = SS_4^*/SS_2^* > F_{\alpha;\,(c-1),\,(r-1)(c-1)}.$$

If the hypothesis that there are no column effects is rejected, Tukey's method can again be applied to get confidence intervals for the difference between two column effects. The probability is $1 - \alpha$ that the value $(\gamma_f - \gamma_g)$ for *all* such contrasts for $f = 1, 2, \ldots, c;\; g = 1, 2, \ldots, c$ simultaneously satisfy

$$\bar{X}._f - \bar{X}._g - k \leq \gamma_f - \gamma_g \leq \bar{X}._f - \bar{X}._g + k.$$

Two column effects are judged significantly different when 0 is not included in the confidence interval.

The factor k for column effects is obtained from the analysis of variance table and from Table 10.3 or Table 10.4, depending on the level of significance. Table 10.3 or Table 10.4 is entered with the indices, degrees of freedom (ν) and the number of column effects (c) being studied. The proper degrees of freedom are those corresponding to the degrees of freedom of the residual source in the analysis of variance table, i.e., $(r - 1)(c - 1)$. The factor k^* is read from the table, and

$$k = k^*\sqrt{SS_2^*/r}.$$

10.3.12 THE OC CURVE OF THE ANALYSIS OF VARIANCE FOR THE MIXED EFFECTS MODEL, TWO-WAY CLASSIFICATION, ONE OBSERVATION PER CELL

In finding the OC curve for the random effects (rows) when the absence of interaction is postulated, the procedures become identical with that given in Sec. 10.3.9. The OC curve is still a function of λ where

$$\lambda = \sqrt{\frac{\sigma^2 + c\sigma_\varphi^2}{\sigma^2}} = \sqrt{1 + \frac{c\sigma_\varphi^2}{\sigma^2}}.$$

The OC curves of Figs. 10.10 to 10.17 are entered with degrees of freedom $\nu_1 = (r - 1)$ and $\nu_2 = (r - 1)(c - 1)$.

The OC curves for testing the hypothesis about the fixed effects (columns) is a function of the measure Φ, where

$$\Phi = \sqrt{\frac{1}{c}\sum_{j=1}^{c}\gamma_j^2} \bigg/ \sqrt{\frac{\sigma^2 + \sigma_\eta^2}{r}}.$$

This is exactly the same procedure as outlined for column effects in Sec. 10.3.6 with the value σ replaced by $\sqrt{\sigma^2 + \sigma_\eta^2}$.

10.3.13 EXAMPLE USING THE MIXED EFFECTS MODEL

If the chemical cells are treated as a random sample from a large number of cells and the thermometers are treated as fixed, the chemical cell and thermometer example may be treated as a mixed effects model. The formal calculations including the computational procedure and the analysis of variance table are identical with the results of Sec. 10.3.7. The numerical values of the F test are also the same, but in testing for the random effect cells it is necessary to assume the interaction nonexistent, whereas the test for the thermometers does not depend on any assumptions about the interaction.

10.4 Two-Way Analysis of Variance, n Observations per Combination

This section is very similar to Sec. 10.3, except that it allows for more than one observation per combination.

10.4.1 DESCRIPTION OF THE VARIOUS MODELS

The following experiment was performed to determine the effect of time of aging on the strength of cement.[1] Three mixes of cement were prepared and six specimens were made from each mix. Three specimens from each mix were tested after two days and after seven days. The test specimens were 2-inch cubes that yielded under the indicated load and were in units of 10 pounds. They are shown in Table 10.11.

Table 10.11. Yield Load for Cement Specimens

	2-Days Test	7-Days Test	$\bar{x}_{i.}$
Mix 1	574 564 550	1,092 1,086 1,065	821.8
Mix 2	524 573 551	1,028 1,073 998	791.2
Mix 3	576 540 592	1,066 1,045 1,055	812.3
$\bar{x}_{.j}$	560.4	1,056.4	

There are three alternative ways to consider this problem, i.e., fixed effects model, random effects model, or mixed effects model. The mixes are actually random components, a sample of three drawn from a large number. The results of the experiment should extend to the distribution of mixes. On the other hand, the times of aging are fixed effects. The conclusions of the experiment will reveal whether the yield loads differ after 2 or 7 days, these periods being fixed. Hence, this model is actually a mixed model, although it could conceivably be analyzed as a fixed effects model (results pertain to these three mixes only) or a random effects model (the two aging periods being chosen at random). There are three observations for each mix and aging

---

[1] This example is based on an example appearing in W. J. Youden, *Statistical Methods for Chemists*, John Wiley & Sons, Inc., New York, 1951, pp. 64–65.

combination. This will enable the experimenter to test for the presence of interaction. When only one observation per combination existed, either interaction was assumed to be zero or its presence was intermixed with the error term. Its presence could not be detected analytically.

The mathematical model for this situation is:

$$X_{ijv} = \zeta.. + \varphi_i + \gamma_j + \eta_{ij} + \epsilon_{ijv}: \qquad i = 1, 2, \ldots, r,$$
$$j = 1, 2, \ldots, c,$$
$$v = 1, 2, \ldots, n,$$

where $\zeta..$ is the general mean; ϵ_{ijv} are the experimental errors which are assumed to be independently normally distributed each with mean 0 and variance σ^2.

In the fixed effects model:

φ_i is the effect of adding the ith fixed row treatment,

$$\sum_{i=1}^{r} \varphi_i = 0.$$

γ_j is the effect of adding the jth fixed column treatment,

$$\sum_{j=1}^{c} \gamma_j = 0.$$

η_{ij} is the fixed interaction of the ith row treatment with the jth column treatment,

$$\sum_{i=1}^{r} \eta_{ij} = \sum_{j=1}^{c} \eta_{ij} = 0.$$

In the random effects model:

φ_i is the effect of adding the ith row treatment and is a random variable whose probability distribution is the distribution of row treatments. Each φ_i is assumed to be normally distributed with mean 0 and variance σ_φ^2.

γ_j is the effect of adding the jth column treatment and is a random variable whose probability distribution is the distribution of column treatments. Each γ_j is assumed to be normally distributed with mean 0 and variance σ_γ^2.

η_{ij} is the interaction of the ith row treatment with the jth column treatment and is also a random variable. Each η_{ij} is assumed to be normally distributed with mean 0 and variance σ_η^2.

In the mixed effects model:

φ_i is the effect of adding the ith row treatment and is a random variable whose probability distribution is the distribution of row treatments. Each φ_i is assumed to be normally distributed with mean 0 and variance σ_φ^2.

γ_j is the effect of adding the jth fixed column treatment,

$$\sum_{j=1}^{c} \gamma_j = 0.$$

η_{ij} is the interaction of the ith random row treatment with the jth fixed column treatment and is a random variable. Each η_{ij} is assumed to be normally distributed with mean 0 and variance σ_η^2.

10.4.2 COMPUTATIONAL PROCEDURE, TWO-WAY CLASSIFICATION, n OBSERVATIONS PER COMBINATION

Before carrying out any further analysis it is necessary to complete the analysis of variance table, Table 10.12. This is done following the computational procedure described below. The results are summarized in the analysis of variance table.

Computational Procedure

1. Calculate row totals:

$$R_1, R_2, \ldots, R_r.$$

2. Calculate column totals:

$$C_1, C_2, \ldots, C_c.$$

3. Calculate within combination totals:

$$w_{11}, w_{12}, \ldots, w_{rc}.$$

4. Calculate over-all total:

$$T = R_1 + R_2 + \cdots + R_r.$$

5. Calculate crude sum of squares:

$$\sum_{i=1}^{r} \sum_{j=1}^{c} \sum_{v=1}^{n} X_{ijv}^2 = X_{111}^2 + X_{112}^2 + \cdots + X_{rcn}^2.$$

6. Calculate crude sum of squares between columns:

$$\sum_{j=1}^{c} \frac{C_j^2}{rn} = \frac{(C_1^2 + C_2^2 + \cdots + C_c^2)}{rn}.$$

7. Calculate crude sum of squares between rows:

$$\sum_{i=1}^{r} \frac{R_i^2}{cn} = \frac{(R_1^2 + R_2^2 + \cdots + R_r^2)}{cn}.$$

8. Calculate crude sum of squares between combinations:

$$\sum_{i=1}^{r} \sum_{j=1}^{c} \frac{w_{ij}^2}{n} = \frac{(w_{11}^2 + w_{12}^2 + \cdots + w_{rc}^2)}{n}.$$

9. Calculate correction factor due to mean $= T^2/rcn$.

From the quantities above compute:

10. $SS_4 = (6) - (9) = \displaystyle\sum_{j=1}^{c} \frac{C_j^2}{rn} - \frac{T^2}{rcn},$

11. $SS_3 = (7) - (9) = \displaystyle\sum_{i=1}^{r} \frac{R_i^2}{cn} - \frac{T^2}{rcn},$

12. $SS_1 = (5) - (8) = \sum_{i=1}^{r} \sum_{j=1}^{c} \sum_{v=1}^{n} X_{ijv}^2 - \sum_{i=1}^{r} \sum_{j=1}^{c} \frac{w_{ij}^2}{n}$,

13. $\quad SS = (5) - (9) = \sum_{i=1}^{r} \sum_{j=1}^{c} \sum_{v=1}^{n} X_{ijv}^2 - \frac{T^2}{rcn}$,

14. $SS_2 = (13) - (12) - (11) - (10) = SS - SS_1 - SS_3 - SS_4$.

10.4.3 ANALYSIS OF THE FIXED EFFECTS MODEL, TWO-WAY CLASSIFICATION,
n OBSERVATIONS PER COMBINATION

In the situation where there is only one observation per combination (Sec.
10.3.9), it was pointed out that to test for main effects, it was necessary to
assume the absence of interaction. In that model, there was no way to test
for the interaction. When there are n ($n > 1$) observations per combination
of level, a test for interaction does exist. Referring to the analysis of variance
table, Table 10.12, the expected mean square for the "within combinations"
source equals σ^2. The expected mean square for the "interaction" source
equals σ^2 plus some interaction effect terms. If there is no interaction present,
i.e., $\eta_{ij} = 0$ for all i and j, the expected mean square for both "interaction
effects" and "within combinations" sources, $E(SS_2^*)$ and $E(SS_1^*)$, respectively,
equals σ^2. Thus, using the ratio SS_2^*/SS_1^* leads to a reasonable procedure for
testing for interaction effects. It can be shown (using the partition theorem)
that SS_2^*/SS_1^* is distributed as an F random variable with $(r-1)(c-1)$
and $rc(n-1)$ degrees of freedom when the hypothesis of no interaction is
true. Thus, the hypothesis that $\eta_{ij} = 0$ for all i and j is rejected if

$$F = SS_2^*/SS_1^* > F_{\alpha;\,(r-1)(c-1),\,rc(n-1)}.$$

If this hypothesis is rejected, the tests for significance of row effects and
column effects are meaningless under the present formulation of the problem.
For example, suppose that there is an experiment being performed involving
men and machines and there is interaction present between workmen and
machines. Furthermore, suppose there are also machine effects present. If a
particular machine effect is positive, and interaction is present, there is no
guarantee that this machine will produce more units consistently. Yet, this
is what is implied by saying that there are machine effects. Similarly, if there
are no machine effects, but interaction is present in the form that a particular
man works best on one particular machine, it can be concluded that machines
differ when this man is the operator. Consequently, if interaction is present,
tests for row effects and column effects are irrelevant. In this situation, it is
possible to test whether the set of levels of one treatment differs for a given
level of the other treatment, e.g., to test whether machines differ when a
particular man is the operator rather than test the more general result that
machines differ regardless of who is the operator. This is accomplished by
performing a one-way analysis of variance over the fixed level (making a
one-way analysis of variance of the data for a particular operator).

Table 10.12. Analysis of Variance Table, Two-Way Classification, n Observations per Combination

Source	Sum of Squares	Degrees of Freedom	Mean Square	Expected Mean Square for the Fixed Effects Model	Expected Mean Square for the Random Effects Model	Expected Mean Square for the Mixed Effects Model
Between columns	$SS_4 = rn \sum_{j=1}^{c} (\bar{X}_{\cdot j \cdot} - \bar{X}_{\cdots})^2$	$c-1$	$SS_4^* = SS_4/(c-1)$	$\sigma^2 + \dfrac{rn}{c-1} \sum_{j=1}^{c} \gamma_j^2$	$\sigma^2 + n\sigma_\eta^2 + rn\sigma_\gamma^2$	$\sigma^2 + n\sigma_\eta^2 + \dfrac{rn}{c-1}\sum_{j=1}^{c}\gamma_j^2$
Between rows	$SS_3 = cn \sum_{i=1}^{r} (\bar{X}_{i\cdots} - \bar{X}_{\cdots})^2$	$r-1$	$SS_3^* = SS_3/(r-1)$	$\sigma^2 + \dfrac{cn}{r-1} \sum_{i=1}^{r} \varphi_i^2$	$\sigma^2 + n\sigma_\eta^2 + cn\sigma_\varphi^2$	$\sigma^2 + nc\sigma_\varphi^2$
Inter-action	$SS_2 = n\sum_{i=1}^{r}\sum_{j=1}^{c}(\bar{X}_{ij\cdot} - \bar{X}_{i\cdots} - \bar{X}_{\cdot j\cdot} + \bar{X}_{\cdots})^2$	$(r-1)(c-1)$	$SS_2^* = SS_2/(r-1)(c-1)$	$\sigma^2 + \dfrac{n}{(r-1)(c-1)}\sum_{i=1}^{r}\sum_{j=1}^{c}\eta_{ij}^2$	$\sigma^2 + n\sigma_\eta^2$	$\sigma^2 + n\sigma_\eta^2$
Within combination	$SS_1 = \sum_{i=1}^{r}\sum_{j=1}^{c}\sum_{v=1}^{n}(X_{ijv} - \bar{X}_{ij\cdot})^2$	$rc(n-1)$	$SS_1^* = SS_1/rc(n-1)$	σ^2	σ^2	σ^2
Total	$SS = \sum_{i=1}^{r}\sum_{j=1}^{c}\sum_{v=1}^{n}(X_{ijv} - \bar{X}_{\cdots})^2$	$rcn-1$	$SS^* = SS/(rcn-1)$	—	—	—

In this table $\bar{X}_{\cdot j\cdot} = \sum_{i=1}^{r}\sum_{v=1}^{n} X_{ijv}/rn$; $\bar{X}_{i\cdots} = \sum_{j=1}^{c}\sum_{v=1}^{n} X_{ijv}/cn$; $\bar{X}_{ij\cdot} = \sum_{v=1}^{n} X_{ijv}/n$; $\bar{X}_{\cdots} = \sum_{i=1}^{r}\sum_{j=1}^{c}\sum_{v=1}^{n} X_{ijv}/rcn$.

If there is more than one observation per cell and it is assumed before the experiment is performed that the interaction is zero, the interaction and within-cells sums of squares are combined into a new row, where $SS'_1 = SS_1 + SS_2$. In all ensuing formulas SS_1 and SS^*_1 should be replaced by SS'_1 and SS'^*_1, respectively.

Table 10.13. Residual with Interaction Assumed Zero

Source	Sum of Squares	Degrees of Freedom	Mean Square	Average Mean Square
Residual	$SS'_1 = SS_1 + SS_2$	$(r-1)(c-1)+rc(n-1)$ $=rcn-r-c+1$	$SS'^*_1=SS'_1/(rcn-r-c+1)$	σ^2

If the hypothesis that there is no interaction is accepted (or it is assumed to be nonexistent prior to experimentation because of physical considerations), tests for row effects and column effects can be made. Referring to the analysis of variance table, Table 10.12, the expected mean square for the "within combination" source equals σ^2. The expected mean square for the "row effects" source equals σ^2 plus some row effects terms. If there are no row effects present, i.e., $\varphi_i = 0, i = 1, 2, \ldots, r$, the expected mean square for both "row effects" and "within combination" sources, $E(SS^*_3)$ and $E(SS^*_1)$, respectively, equals σ^2. Thus, using the ratio SS^*_3/SS^*_1 leads to a reasonable procedure for testing for row effects. It can be shown (using the partition theorem) that SS^*_3/SS^*_1 is distributed as an F random variable with $(r-1)$ and $rc(n-1)$ degrees of freedom when the hypothesis of no row effects is true. Thus, the hypothesis that $\varphi_1 = \varphi_2 = \cdots = \varphi_r = 0$ is rejected if

$$F = SS^*_3/SS^*_1 > F_{\alpha;\,(r-1),\,rc(n-1)}.$$

Using an analogous argument, the hypothesis of no column effects, i.e., the hypothesis that $\gamma_1 = \gamma_2 = \cdots = \gamma_c = 0$, is rejected if

$$F = SS^*_4/SS^*_1 > F_{\alpha;\,(c-1),\,rc(n-1)}.$$

If the interactions are not significant and the row effects and/or the column effects are significant, the procedure for comparing row and/or column effects is exactly the same as that for the two-way classification, with one observation per cell, except for the following changes.

The factor k for row effects and for column effects is obtained from the analysis of variance table and from Table 10.3 or Table 10.4, depending on the level of significance. Table 10.3 or 10.4 is entered with the indices, degrees of freedom and the number of row effects being studied (r) if row effects are of interest. The proper degrees of freedom are those corresponding to the degrees of freedom of the "within cell" source in the analysis of variance table, i.e., $rc(n - 1)$. The factor k^* is read from the table and

$$k \text{ (for rows)} = k^*\sqrt{SS_1^*/nc}.$$

Similarly, if column effects are of interest, the table is entered with the indices, degrees of freedom and the number of column effects being studied (c). The factor k^* is read from the table and

$$k \text{ (for columns)} = k^*\sqrt{SS_1^*/nr}.$$

Of course, for finding confidence intervals, the calculated terms that are of interest are $\bar{X}_{i\cdot}$. and $\bar{X}_{\cdot j\cdot}$.

10.4.4 THE OC CURVE OF THE ANALYSIS OF VARIANCE FOR THE FIXED EFFECTS MODEL, TWO-WAY CLASSIFICATION, n OBSERVATIONS PER CELL

The OC curve for this model is obtained in a manner similar to that described in Sec. 10.3.6. If the OC curve for testing the hypothesis about row effects is desired, the appropriate measure for which an OC curve can be constructed is

$$\Phi = \frac{\sqrt{\sum_{i=1}^{r} \varphi_i^2/r}}{\sigma/\sqrt{nc}}.$$

Similarly, if the OC curve for testing the hypothesis about column effects is desired, the appropriate measure for which an OC curve can be constructed is

$$\Phi = \frac{\sqrt{\sum_{j=1}^{c} \gamma_j^2/c}}{\sigma/\sqrt{nr}}.$$

Finally, if the OC curve for testing the hypothesis about interaction effects is desired, the appropriate measure for which an OC curve can be constructed is

$$\Phi = \frac{1}{\sigma}\sqrt{\frac{n}{(r-1)(c-1)+1}\sum_{i=1}^{r}\sum_{j=1}^{c} \eta_{ij}^2}.$$

The OC curves as functions of v_1, v_2 and Φ are given in Figs. 10.2 to 10.9. The appropriate values of v_1 are $(r - 1)$ if row effects are being considered, $(c - 1)$ if column effects are being considered, and $(r - 1)(c - 1)$ if interaction effects are being considered; v_2 corresponds to the degrees of freedom associated with the "within combinations" source and is given by $rc(n - 1)$. For fixed r, c, and n, the OC curve of the procedure can be obtained by calculating Φ, v_1, and v_2 and entering the appropriate figure. On the other hand, if two points on the OC curve are specified, the appropriate value of n can be obtained by trial and error.

A physical interpretation of Φ as a measure can be given by finding upper and lower bounds for Φ as a function of the range of the $\varphi_i[\varphi(\max) - \varphi(\min)$ if row effects are being studied] beyond which rejection of homogeneity should be assured with a given probability. Denote this largest difference by W. Bounds on Φ are given as

$$\Phi(\min) = \frac{W}{\sigma}\sqrt{\frac{nc}{2r}}$$

$$\Phi(\max) = \begin{cases} \dfrac{W}{2\sigma}\sqrt{nc} & \text{for } r \text{ even} \\[2ex] \dfrac{W}{2\sigma}\sqrt{\dfrac{nc(r^2 - 1)}{r^2}} & \text{for } r \text{ odd.} \end{cases}$$

Similarly, if column effects are being studied,

$$\Phi(\min) = \frac{W}{\sigma}\sqrt{\frac{nr}{2c}}$$

$$\Phi(\max) = \begin{cases} \dfrac{W}{2\sigma}\sqrt{nr} & \text{for } c \text{ even} \\[2ex] \dfrac{W}{2\sigma}\sqrt{\dfrac{nr(c^2 - 1)}{c^2}} & \text{for } c \text{ odd.} \end{cases}$$

Finally, if interaction effects are being studied,

$$\Phi(\min) = \frac{W}{\sqrt{2}\,\sigma}\sqrt{\frac{n}{(r - 1)(c - 1) + 1}}$$

$$\Phi(\max) = \begin{cases} \dfrac{W}{2\sigma}\sqrt{\dfrac{nrc}{(r - 1)(c - 1) + 1}} & \text{for } rc \text{ even} \\[2ex] \dfrac{W}{2\sigma}\sqrt{\dfrac{n(r^2c^2 - 1)}{[(r - 1)(c - 1) + 1]rc}} & \text{for } rc \text{ odd.} \end{cases}$$

10.4.5 Example Using the Fixed Effects Model, Two-Way Classification, Three Observations per Combination

Returning to the experiment on the effect of time of aging in the strength of cement described in Sec. 10.4.1, if time and mixes are considered as fixed effects (making the unrealistic assumption that only the three mixes are of interest), the data can be analyzed according to this model. Interaction cannot be ignored since it is quite possible that the three mixes differ after a long period of time, without differing after a short period. This is equivalent to saying that the effect of an additional period of time is different for the three mixes, i.e., interaction is present.

The following computations lead to the analysis of variance table.

1. $R_1, R_2, R_3 = 4,931; 4,747; 4,874$

2. $C_1, C_2 = 5,044; 9,508$

3. $w_{11}, w_{21}, w_{31}, w_{12}, w_{22}, w_{32} = 1{,}688; 1{,}648; 1{,}708; 3{,}243; 3{,}099; 3{,}166$

4. $$T = 4{,}931 + 4{,}747 + 4{,}874 = 14{,}552$$

5. $\displaystyle\sum_{i=1}^{3}\sum_{j=1}^{2}\sum_{v=1}^{3} X_{ijv}^{2} = (574)^2 + (564)^2 + \cdots + (1{,}055)^2 = 12{,}882{,}026$

6. $\displaystyle\sum_{j=1}^{2} \frac{C_j^2}{9} = \frac{(5{,}044)^2 + (9{,}508)^2}{9} = 12{,}871{,}556$

7. $\displaystyle\sum_{i=1}^{3} \frac{R_i^2}{6} = \frac{(4{,}931)^2 + (4{,}747)^2 + (4{,}874)^2}{6} = 11{,}767{,}441$

8. $\displaystyle\sum_{i=1}^{3}\sum_{j=1}^{2} \frac{w_{ij}^2}{3} = \frac{(1{,}688)^2 + (1{,}648)^2 + \cdots + (3{,}166)^2}{3} = 12{,}875{,}639$

9. $\displaystyle\frac{T^2}{18} = \frac{211{,}760{,}704}{18} = 11{,}764{,}484$

10. $SS_4 = 12{,}871{,}556 - 11{,}764{,}484 = 1{,}107{,}072$

11. $SS_3 = 11{,}767{,}441 - 11{,}764{,}484 = 2{,}957$

12. $SS_1 = 12{,}882{,}026 - 12{,}875{,}639 = 6{,}387$

13. $SS = 12{,}882{,}026 - 11{,}764{,}484 = 1{,}117{,}542$

14. $SS_2 = 1{,}117{,}542 - 6{,}387 - 2{,}957 - 1{,}107{,}072 = 1{,}126$

Table 10.14. Analysis of Variance Table for Yield Loads on Cement Specimens

Source	Sum of Squares	Degrees of Freedom	Mean Square	Test
Between times	1,107,072	1	1,107,072	$F = 2{,}079.98$
Between mixes	2,957	2	1,478.5	$F = 2.78$
Interaction	1,126	2	563	$F = 1.06$
Within combinations	6,387	12	532.25	
Total	1,117,542	17	65,737.8	

The F values in the table for interaction and mixes are well below the critical 5% value for 2 and 12 degrees of freedom, i.e., $F_{0.05;2,12} = 3.89$. Hence, it is concluded that the data do not give sufficient evidence of the presence of interaction or any mix effect. On the other hand, the F test for times is significant so that it is concluded that there is an effect from the additional five days of aging. Furthermore, a comparison of column means indicates that

$$\bar{X}_{\cdot 2} - \bar{X}_{\cdot 1} - k \leqq \gamma_2 - \gamma_1 \leqq \bar{X}_{\cdot 2} - \bar{X}_{\cdot 1} + k$$

$(1{,}056.4 - 560.4) - 23.7 = 472.3$

$\leqq$ mean effect for 7 days $-$ mean effect for 2 days

$\leqq (1{,}056.4 - 560.4) + 23.7 = 519.7$

where k is obtained from Table 10.3, using as indices $c = 2$; degrees of freedom $= rc(n - 1) = 12$; $k^* = 3.08$, so that

$$k = 23.7.$$

The following is an example of the use of the OC curves. Suppose it is required to find the probability of accepting the hypothesis that there is no time effect when the difference in time effect is as large as 450 psi. Since there are only two time periods, this difference coincides with the range (largest minus smallest); 450 psi corresponds to 45 units so that $W = 45$. An estimate of σ is obtained from the data; σ is approximately equal to $\sqrt{SS_1^*}$ $= 23.2$. In this special case $(c = 2)$, $\Phi(\min)$ coincides with $\Phi(\max)$ and is equal to

$$\Phi = (45/23.2)\sqrt{(3)(3)/(2)(2)} = 2.91.$$

Entering Fig. 10.2 with $\nu_1 = 1$, $\nu_2 = 12$, and $\Phi = 2.91$, a probability equal to 0.03 of accepting the hypothesis that there are no time effects is read out.

10.4.6 ANALYSIS OF THE RANDOM EFFECTS MODEL, TWO-WAY CLASSIFICATION, n OBSERVATIONS PER COMBINATION

In this model a test for interaction, i.e., $H: \sigma_\eta^2 = 0$, is also available. Referring to the analysis of variance table, Table 10.12, the expected mean square for the "within combinations" source equals σ^2. The expected mean square for the "interaction" source equals σ^2 plus some interaction effect terms. If there is no interaction present, i.e., $\sigma_\eta^2 = 0$, the expected mean square for both "interaction effects" and "within combination" sources, $E(SS_2^*)$ and $E(SS_1^*)$, respectively, equals σ^2. Thus, using the ratio SS_2^*/SS_1^* leads to a reasonable procedure for testing for interaction effects. It can be shown (using the partition theorem) that SS_2^*/SS_1^* is distributed as an F random variable with $(r - 1)(c - 1)$ and $rc(n - 1)$ degrees of freedom when the hypothesis of no interaction is true. Thus, the hypothesis that $\sigma_\eta^2 = 0$ is rejected if

$$F = SS_2^*/SS_1^* > F_{\alpha;(r-1)(c-1),rc(n-1)}.$$

If interaction is present in this model, one can still talk about the magnitude of main effects, say, σ_φ^2. Having σ_η^2 greater than zero does not imply anything about σ_φ^2. The test for main effects, say, row effects, is different from that given in the fixed effects model. Again, referring to the analysis of variance table, Table 10.12, the expected mean square for the "interaction" source equals $\sigma^2 + n\sigma_\eta^2$. The expected mean square for the "row effect" source equals $\sigma^2 + n\sigma_\eta^2 + cn\sigma_\varphi^2$.

If there are no row effects present, i.e., $\sigma_\varphi^2 = 0$, the expected mean square for both "row effects" and "interaction effects" sources, $E(SS_3^*)$ and $E(SS_2^*)$, respectively, equals $\sigma^2 + n\sigma_\eta^2$. Thus, using the ratio SS_3^*/SS_2^* leads to a reasonable procedure for testing for row effects. This is different from the fixed effects model which uses SS_3^*/SS_1^*. It can be shown (using the partition theorem) that SS_3^*/SS_2^* is distributed as an F random variable with $(r-1)$ and $(r-1)(c-1)$ degrees of freedom when the hypothesis of no row effects is true. Thus, the hypothesis that $\sigma_\varphi^2 = 0$ is rejected if

$$F = SS_3^*/SS_2^* > F_{\alpha;\,(r-1),\,(r-1)(c-1)}.$$

Using an analogous argument, the hypothesis that $\sigma_\gamma^2 = 0$ is rejected if

$$F = SS_4^*/SS_2^* > F_{\alpha;\,(c-1),\,(r-1)(c-1)}.$$

Using the expected mean square column in the analysis of variance table, estimators of σ_η^2, σ_φ^2, and σ_γ^2 can be obtained. An estimator of σ_η^2 is given by

$$(SS_2^* - SS_1^*)/n.$$

An estimator of σ_φ^2 is given by

$$(SS_3^* - SS_2^*)/cn.$$

An estimator of σ_γ^2 is given by

$$(SS_4^* - SS_2^*)/rn.$$

If values that any of these estimators take on are negative, the estimate is taken to be zero.

10.4.7 The OC Curve of the Random Effects Model, Two-Way Classification, n Observations per Combination

The hypothesis of no interaction is accepted if

$$SS_2^*/SS_1^* \leq F_{\alpha;\,(r-1)(c-1),\,rc(n-1)}.$$

The random variable SS_2^*/SS_1^* has an F distribution whenever $\sigma_\eta^2 = 0$, and the probability of acceptance is $(1 - \alpha)$. In general, the probability of accepting the hypothesis that $\sigma_\eta^2 = 0$ can be plotted as a function of $\lambda = \sqrt{1 + n\sigma_\eta^2/\sigma^2}$.[1] The OC curves of Figs. 10.10 to 10.17 are again applicable when λ is defined in this manner. Similarly, testing for row effects involves

---

[1] The random variable

$$\frac{SS_2^*/(\sigma^2 + n\sigma_\eta^2)}{SS_1^*/\sigma^2}$$

always has an F distribution so that the probability of accepting the hypothesis, i.e.,

$$P\{SS_2^*/SS_1^* \leq F_{\alpha;\,(r-1)(c-1),\,rc(n-1)}\}$$

can always be expressed in terms of the probability that

$$\frac{SS_2^*/(\sigma^2 + n\sigma_\eta^2)}{SS_1^*/\sigma^2} \leq F_{\alpha;\,(r-1)(c-1),\,rc(n-1)}\left[\frac{\sigma^2}{\sigma^2 + n\sigma_\eta^2}\right].$$

This last expression is just a statement about an F random variable and can be expressed as a function of

$$\lambda = \sqrt{\frac{\sigma^2 + n\sigma_\eta^2}{\sigma^2}} = \sqrt{1 + \frac{n\sigma_\eta^2}{\sigma^2}}.$$

the statistic SS_3^*/SS_2^*, and the OC curve can be expressed as a function of

$$\lambda = \sqrt{1 + \frac{nc\sigma_\varphi^2}{\sigma^2 + n\sigma_\eta^2}}.$$

Finally, testing for column effects involves the statistic SS_4^*/SS_2^*, and the OC curve can be expressed as a function of

$$\lambda = \sqrt{1 + \frac{nr\sigma_\gamma^2}{\sigma^2 + n\sigma_\eta^2}}.$$

The OC curves of Figs. 10.10 to 10.17 are entered with degrees of freedom $v_1 = (r - 1)(c - 1)$ and $v_2 = rc(n - 1)$ if interaction effects are relevant, with $v_1 = (r - 1)$ and $v_2 = (r - 1)(c - 1)$ if row effects are relevant, and $v_1 = (c - 1)$ and $v_2 = (r - 1)(c - 1)$ if column effects are desired.

10.4.8 EXAMPLE USING THE RANDOM EFFECTS MODEL

If in the example of the effect of time of aging on the strength of cement, both time and mixes are considered to be random effects, this will provide an example of the model. That mixes are random is a reasonable assumption, but time being random is rather unrealistic. However, for the purpose of the example these assumptions will be made. The formal calculations including the computational procedure and the analysis of variance table are identical with the results of Sec. 10.4.5. The test for interaction is the same, but the tests for main effects are different, the latter using SS_2^* in the denominator. The results are:

mix effect $SS_3^*/SS_2^* = 2.63 < 19.0 = F_{0.05;2,2}$

time effect $SS_4^*/SS_2^* = 1{,}966.38 > 18.5 = F_{0.05;1,2}$

which leads to the same conclusions as the fixed effects model. An estimate of σ_η^2 is 10.25, an estimate of σ_φ^2 is 152.58, and an estimate of σ_γ^2 is 122,945.44. The fact that the hypothesis that $\sigma_\varphi^2 = 0$ was accepted does not necessarily imply that the hypothesis is true, and hence, an estimate of σ_φ^2 may be desirable. Accepting the hypothesis may just be an indication that there is insufficient evidence to reject it, and must be interpreted according to the OC curve.

10.4.9 ANALYSIS OF THE MIXED EFFECTS MODEL, TWO-WAY CLASSIFICATION, n OBSERVATIONS PER CELL

In the mixed effects model, the row effects will be assumed to be the random effects and the column effects will be assumed to be the fixed effects. In this model, a test for interaction is also available. The interaction, η_{ij}, is a random variable since one of the factors, row effects, is a random variable. Referring to the analysis of variance table, Table 10.12, the expected mean square for the "within combinations" source equals σ^2. The expected mean square for the "interaction" source equals σ^2 plus some interaction effect terms. If there is no interaction present, i.e., $\sigma_\eta^2 = 0$, the expected mean square for both "interaction effects" and "within combination" sources, $E(SS_2^*)$ and $E(SS_1^*)$, respectively, equals σ^2. Thus, using the ratio SS_2^*/SS_1^* leads to a reasonable procedure for testing for interaction effects. It can be shown (using the partition theorem) that SS_2^*/SS_1^* is distributed as an F random variable with $(r-1)(c-1)$ and $rc(n-1)$ degrees of freedom when the hypothesis of no interaction is true. Thus, the hypothesis that $\sigma_\eta^2 = 0$ is rejected if

$$F = SS_2^*/SS_1^* > F_{\alpha;\,(r-1)(c-1),\,rc(n-1)}.$$

Testing for main effects again requires studying the analysis of variance table, Table 10.12. For row effects (random effects), the expected mean square, $E(SS_3^*)$, equals $\sigma^2 + nc\sigma_\varphi^2$. Hence, the logical test is to use the ratio SS_3^*/SS_1^* since $E(SS_1^*) = \sigma^2$. It can be shown (using the partition theorem) that SS_3^*/SS_1^* has an F distribution with $(r-1)$ and $rc(n-1)$ degrees of freedom when the hypothesis is true. Thus, the hypothesis that $\sigma_\varphi^2 = 0$ is rejected if

$$F = SS_3^*/SS_1^* > F_{\alpha;\,(r-1),\,rc(n-1)}.$$

An estimator of σ_φ^2 is obtained from

$$(SS_3^* - SS_1^*)/nc.$$

For column effects (fixed effects) the procedure is somewhat different. The expected mean square for columns, $E(SS_4^*)$, equals

$$\sigma^2 + n\sigma_\eta^2 + \frac{rn}{c-1}\sum_{j=1}^{c}\gamma_j^2.$$

Since the expected mean square for the interaction term, $E(SS_2^*)$, equals $\sigma^2 + n\sigma_\eta^2$, the logical test to use is SS_4^*/SS_2^*. It can be shown (using the partition theorem) that the random variable SS_4^*/SS_2^* has an F distribution with $(c-1)$ and $(r-1)(c-1)$ degrees of freedom when the hypothesis of no column effect is true. The hypothesis that $\gamma_1 = \gamma_2 = \cdots = \gamma_c = 0$ is rejected if

$$F = SS_4^*/SS_2^* > F_{\alpha;\,(c-1),\,(r-1)(c-1)}.$$

Tukey's method can again be applied to get confidence intervals on the difference between two column effects. The probability is $1 - \alpha$ that the value $(\gamma_f - \gamma_g)$ for *all* such contrasts for $f = 1, 2, \ldots, c$, $g = 1, 2, \ldots, c$ simultaneously satisfies

$$\bar{X}_{\cdot f\cdot} - \bar{X}_{\cdot g\cdot} - k \leq \gamma_f - \gamma_g \leq \bar{X}_{\cdot f\cdot} - \bar{X}_{\cdot g\cdot} + k.$$

Two mean effects are judged significantly different when 0 is not included in the confidence interval.

The factor k for column effect is obtained from the analysis of variance table and from Table 10.3 or Table 10.4, depending on the level of significance. Table 10.3 or Table 10.4 is entered with the indices, degrees of freedom and the number of column effects (c) being studied. The proper degrees of freedom are those corresponding to the degrees of freedom of the "interaction" source in the analysis of variance table, i.e., $(r - 1)(c - 1)$. The factor k^* is read from the table, and

$$k = k^*\sqrt{SS_2^*/nr}.$$

10.4.10 THE OC CURVE OF THE ANALYSIS OF VARIANCE FOR THE MIXED EFFECTS MODEL, TWO-WAY CLASSIFICATION, n OBSERVATIONS PER CELL

The OC curve for the interaction effects is exactly the same as for the interaction in the random effects model, i.e.,

$$\lambda = \sqrt{\frac{\sigma^2 + n\sigma_\eta^2}{\sigma^2}} = \sqrt{1 + \frac{n\sigma_\eta^2}{\sigma^2}}.$$

The OC curves are given in Figs. 10.10 to 10.17 and are entered with degrees of freedom $v_1 = (r - 1)(c - 1)$ and $v_2 = rc(n - 1)$.

The OC curves for the random effects (rows) are found in Figs. 10.10 to 10.17, using the index

$$\lambda = \sqrt{\frac{\sigma^2 + nc\sigma_\varphi^2}{\sigma^2}} = \sqrt{1 + \frac{nc\sigma_\varphi^2}{\sigma^2}}.$$

These curves are entered with degrees of freedom $v_1 = r - 1$ and $v_2 = rc(n - 1)$.

Finally, the OC curve for testing the hypothesis about the fixed effects (columns) is a function of the measure Φ, where

$$\Phi = \frac{\sqrt{\sum_{j=1}^{c} \gamma_j^2/c}}{\sqrt{(\sigma^2 + n\sigma_\eta^2)/rn}}.$$

This is exactly the same procedure as outlined for column effects in Sec. 10.4.4 with the value σ replaced by $\sqrt{\sigma^2 + n\sigma_\eta^2}$. The degrees of freedom are $v_1 = c - 1$ and $v_2 = (r - 1)(c - 1)$.

10.4.11 EXAMPLE USING THE MIXED EFFECTS MODEL

The problem about the effect of aging on the strength of cement presented in Sec. 10.4.1 is actually a mixed effects model. The mixes are considered to be a random sample from a distribution of mixes, whereas the two times of aging are fixed effects. The calculations carried out in Sec. 10.4.3 coincide with the required calculations for this model. However, the numerical values of the F tests are different. The test for interaction is the same, but the test for the random effects involves SS_3^*/SS^*, and the test for fixed effects involves SS_4^*/SS_2^*.

Table 10.15. Table of Summary and Tests

One-Way Classification

Source	Fixed Effects		Random Effects	
	Test Statistic	Ratio of Expected Mean Squares	Test Statistic	Ratio of Expected Mean Squares
Row effects	$\dfrac{SS_3^*}{SS^*}$	$\dfrac{\sigma^2+\dfrac{c}{r-1}\sum\varphi_i^2}{\sigma^2}$	$\dfrac{SS_3^*}{SS_2^*}$	$\dfrac{\sigma^2+c\sigma_\varphi^2}{\sigma^2}$
Column effects	—	—	—	—
Interaction	—	—	—	—

Two-Way Classification, One Observation per Combination

Source	Fixed Effects		Random Effects		Mixed Effects	
	Test Statistic	Ratio of Expected Mean Squares	Test Statistic	Ratio of Expected Mean Squares	Test Statistic	Ratio of Expected Mean Squares
Row effects	$\dfrac{SS_3^*}{SS_2^*}$	$\dfrac{\sigma^2+\dfrac{c}{r-1}\sum\varphi_i^2}{\sigma^2}$	$\dfrac{SS_3^*}{SS_2^*}$	$\dfrac{\sigma^2+\sigma_\eta^2+c\sigma_\varphi^2}{\sigma^2+\sigma_\eta^2}$	$\dfrac{SS_3^*}{SS^*}$	$\dfrac{\sigma^2+c\sigma_\varphi^2}{\sigma^2}$
Column effects	$\dfrac{SS_4^*}{SS_2^*}$	$\dfrac{\sigma^2+\dfrac{r}{c-1}\sum\gamma_j^2}{\sigma^2}$	$\dfrac{SS_4^*}{SS_2^*}$	$\dfrac{\sigma^2+\sigma_\eta^2+r\sigma_\gamma^2}{\sigma^2+\sigma_\eta^2}$	$\dfrac{SS_4^*}{SS_2^*}$	$\dfrac{\sigma^2+\sigma_\eta^2+\dfrac{r}{c-1}\sum\gamma_j^2}{\sigma^2+\sigma_\eta^2}$
Interaction	Interaction assumed to be zero		No test for interaction available but no assumption required about its presence		No test for interaction available. To test for row effects interaction must be assumed zero. No such assumption required for column effects	

Two-Way Classification, n Observations per Combination

Source	Fixed Effects		Random Effects		Mixed Effects	
	Test Statistic	Ratio of Expected Mean Squares	Test Statistic	Ratio of Expected Mean Squares	Test Statistic	Ratio of Expected Mean Squares
Row effects	$\dfrac{SS_3^*}{SS_1^*}$	$\dfrac{\sigma^2+\dfrac{cn}{r-1}\sum\varphi_i^2}{\sigma^2}$	$\dfrac{SS_3^*}{SS_2^*}$	$\dfrac{\sigma^2+n\sigma_\eta^2+nc\sigma_\varphi^2}{\sigma^2+n\sigma_\eta^2}$	$\dfrac{SS_3^*}{SS_1^*}$	$\dfrac{\sigma^2+nc\sigma_\varphi^2}{\sigma^2}$
Column Effects	$\dfrac{SS_4^*}{SS_1^*}$	$\dfrac{\sigma^2+\dfrac{rn}{c-1}\sum\gamma_j^2}{\sigma^2}$	$\dfrac{SS_4^*}{SS_2^*}$	$\dfrac{\sigma^2+n\sigma_\eta^2+nr\sigma_\gamma^2}{\sigma^2+n\sigma_\eta^2}$	$\dfrac{SS_4^*}{SS_2^*}$	$\dfrac{\sigma^2+n\sigma_\eta^2+\dfrac{rn}{c-1}\sum\gamma_j^2}{\sigma^2+n\sigma_\eta^2}$
Interaction	$\dfrac{SS_2^*}{SS_1^*}$	$\dfrac{\sigma^2+\dfrac{n}{(r-1)(c-1)}\sum\sum\eta_{ij}^2}{\sigma^2}$	$\dfrac{SS_2^*}{SS_1^*}$	$\dfrac{\sigma^2+n\sigma_\eta^2}{\sigma^2}$	$\dfrac{SS_2^*}{SS_1^*}$	$\dfrac{\sigma^2+n\sigma_\eta^2}{\sigma^2}$

The results are:

$$\text{mix effect} \quad SS_3^*/SS_1^* = 2.78 < 3.89 = F_{0.05;\,2,\,12}$$
$$\text{time effect} \quad SS_4^*/SS_2^* = 1{,}966.38 > 18.5 = F_{0.05;\,1,\,2}$$

which leads to the same conclusions as before.

10.5 Summary of Models and Tests

In Table 10.15 a summary of the test statistics and the ratios of the expected mean squares is presented.

PROBLEMS

Unless otherwise specified, assume that all the necessary assumptions are satisfied for the analysis of variance technique to be applicable.

1. It is suspected that four filling machines in a plant are turning out products of non-uniform weight. An experiment is run and the data in ounces are as follows:

Machine	Net Weight				
A	12.25	12.27	12.24	12.25	12.20
B	12.18	12.25	12.26	12.22	12.19
C	12.24	12.23	12.23	12.20	12.16
D	12.20	12.17	12.19	12.18	12.16

$$SS_3 = 0.0099 \qquad SS = 0.0229$$

(a) Write out the analysis of variance table. (b) Test to see if the machines differ at the 5% level of significance. (c) Test to see if Machine A differs from Machine D. Note that this is merely testing for the difference between *two* means. (d) If there is a machine effect shown, then compute the "Tukey" contrasts. (e) It is desired that if there is a difference in machines such that a random observation has its standard deviation increased by as much as 25%, the analysis of variance procedure should lead to the rejection of the hypothesis that the machines are the same with probability exceeding 0.80. How many observations on each machine are required?

2. Suppose four different quantities of carbon are to be added to a batch of raw material in the manufacture of steel. Six specimens are taken for each of the four quantities, and it is desired to determine the effect of these percentages on the tensile strength of the resultant steel. The data are given in the following table.

Tensile Strength of Hot Rolled Steel for Different Percentages of Carbon

								$\bar{X}$
Percentage 1:	0.10%	23,050	36,000	31,100	32,650	30,900	31,400	30,850
Percentage 2:	0.20%	41,850	25,650	46,700	34,500	36,650	31,450	36,133
Percentage 3:	0.40%	47,050	43,450	43,000	38,650	41,850	35,450	41,575
Percentage 4:	0.60%	49,650	73,900	66,450	74,550	62,400	63,750	65,117

$$SS_3 = 4{,}111{,}498{,}600 \qquad SS = 4{,}979{,}619{,}063$$

(a) Does the percentage of carbon have any effect at the 1% level of significance on the tensile strength? Write out the analysis of variance table. (b) If the result of (a) is affirmative, determine which steels differ. (c) If the maximum difference in steels is 11,000 psi, what is the approximate probability of accepting the hypothesis of no differences? Use the "within treatment" mean square as the estimate of σ^2.

3. In evaluating the effect of cloud seeding in Santa Clara County, the following problem was encountered. Past data for one particular station was taken, beginning with the year 1922. It was later discovered that the rain gauges had been moved three times since that date. One possible procedure to evaluate the effect of the change in environment is to compare the successive annual precipitation amounts at one station with the corresponding average. Tabulated below are ratios of station precipitation to average precipitation for the near-by stations for each year.

Location 1		Location 2		Location 3		Location 4	
Year	Ratio	Year	Ratio	Year	Ratio	Year	Ratio
1922	1.05	1929	1.03	1936	0.98	1943	0.97
1923	1.02	1930	0.96	1937	0.91	1944	0.94
1924	0.93	1931	0.96	1938	0.92	1945	1.02
1925	1.00	1932	0.99	1939	0.86	1946	0.99
1926	1.06	1933	1.04	1940	0.97	1947	0.98
1927	1.01	1934	0.97	1941	1.00	1948	1.05
1928	0.95	1935	1.00	1942	0.90	1949	1.01

$$SS_3 = 0.020839 \qquad SS = 0.064068$$

Using the analysis of variance, determine if the change in rain gauge location had any effect on the ratio at the 5% level. Write out the analysis of variance table.

4. The tensile strength of a certain rubber seal shows the following variation with different conditions of production. Data are in pounds per square inch.

Condition of Production

A	B	C	D
3,800	4,200	3,800	3,500
4,100	4,200	3,900	3,700
4,000	4,400	3,700	3,600
3,800	4,300	3,800	3,700

$$SS_3 = 906,875 \qquad SS = 1,049,375$$

(a) Do the production conditions have any effect on the tensile strength of the rubber seal (use $\alpha = 0.05$)? Write out the analysis of variance table. (b) If the production conditions do differ, use Tukey's method to make confidence statements about which conditions differ. (c) An underlying assumption was that σ^2 was constant for all production conditions. Test this hypothesis. (d) If it is important to detect a maximum difference between production conditions of 300 psi with probability greater than or equal to 0.9, how many observations are required for each condition of production? Use the "within treatment" mean square as the estimate of σ^2.

5. Three examiners are checked on the time they take to use a certain gauge. The record of gauging times, in seconds, is as follows:

Examiners

1	2	3
12	16	13
14	16	13
12	14	15
14	14	11

(a) Fill in the analysis of variance table and determine whether the examiners differ at the 5 % level of significance. (b) If the examiners differ, which ones can be said to be different? (c) If the maximum difference in examiners is 2 seconds, what is the approximate probability of accepting the hypothesis of no differences? Use the "within treatment" mean square as the estimate of σ^2. (d) Verify the assumption that the within variability is the same for the three examiners. Use a 5 % level of significance.

6. Among the classrooms in the public schools of a given city there are 12 different lighting techniques. Each of these techniques is supposed to provide the same level of illumination. To determine whether or not the illumination they provide is uniform, the following data were compiled from a random sample of four lighting techniques. The classrooms are known to be homogeneous and hence can be discounted as a possible source of variability.

Lighting Techniques	Foot-Candles on Desk Surface				
1	31	38	38	33	31
2	31	34	27	27	29
3	34	35	39	35	30
4	37	34	27	31	26

$$SS_3 = 89.35 \qquad SS = 302.55$$

(a) Are the lighting techniques homogeneous? Use $\alpha = 0.01$. (b) Is it possible to determine which techniques differ, if any, among all 12 lighting techniques? Why? (c) Is it possible to estimate the variability among all lighting techniques? If so, estimate this variability. (d) What is the probability of accepting the hypothesis that the lighting techniques are homogeneous when $\sigma_\varphi^2 = 30$ and σ^2 is estimated from the "within treatments" mean square?

7. A small laboratory has many thermometers which are used interchangeably to make temperature measurements. To perform an experiment with all the thermometers is too costly, and a random sample of four thermometers is drawn to determine whether thermometers differ. These are placed in a cell which is kept at constant temperature. The data are as follows in degrees Centigrade and are obtained from three readings on each thermometer.

Thermometers

1	2	3	4
0.95	0.33	−2.15	1.05
1.06	−1.46	1.70	1.27
1.96	0.20	0.48	−2.05

$$SS_3 = 4.637 \qquad SS = 21.879$$

(a) Write out the analysis of variance table. (b) Are the thermometers homogeneous? ($\alpha = 0.05$) (c) Estimate the variability among the thermometers. (d) How many observations on each thermometer would be required if it is desired to detect a 40% increase in the total variability of a random observation with probability 0.9?

8. An experiment was run to check on the variability of batches of cement briquettes. Five specimens from each of nine batches were analyzed for their breaking strength. The data are:

Breaking Strength (pounds tension) of Nine Batches of Cement Briquettes

1	2	3	4	5	6	7	8	9
553	553	510	520	543	492	542	581	578
550	599	580	559	500	530	550	550	531
568	579	529	539	562	528	580	529	562
541	545	535	510	540	510	545	570	525
537	540	537	540	535	571	520	524	549

$SS_3 = 5,036.9$ $SS = 23,955.2$

(a) Write out the analysis of variance table. (b) Do the batches differ ($\alpha = 0.01$)? (c) Estimate the batch variability. (d) What is the probability of accepting the hypothesis that the batches are homogeneous when $\sigma_\varphi^2 = 500$ and σ^2 is estimated from the "within treatment" mean square?

9. A new cure has been developed for Portland cement. Its production has caused problems since the variability of the compressive strength from batch to batch appears to be excessive. An experiment has been run to study this batch variability. The data are given below.

Batch	Compressive Strength of Duplicate Measure (psi)	
1	4,150	4,250
2	4,225	3,950
3	4,025	3,900
4	3,900	4,075
5	3,875	4,550
6	3,850	4,450
7	3,975	4,150
8	3,800	4,550
9	3,775	3,700

(a) Write out the analysis of variance table. (b) Do the batches differ? Use $\alpha = 0.05$. (c) Estimate the batch variability. (d) How many observations from each batch would be required if it is desired to detect a 25% increase in the total variability of a random observation with probability 0.8?

10. The following represents the yield of a certain chemical process under varying conditions. The results are excesses over 40.0 pounds, in tenths of a pound.

Temperature	Concentration of Inert Solvent		
	A	B	C
Low	4	7	2
Medium	1	5	0
High	5	8	3

$SS_4 = 0.38889$ $SS_3 = 0.17556$ $SS = 0.56889$

Assume there is no interaction between temperature and concentration of inert solvent. (a) Does the concentration of inert solvent have a significant effect on the yield at the 1% level of significance? (b) Does the temperature have a significant effect on yield at the 1% level? (c) If there are differences in yield due to concentration, determine which differ. (d) If there are differences in yield due to temperature, determine which differ. (e) When the maximum difference in yields caused by concentration is $\frac{1}{2}$ pound, what are the bounds on the probability of accepting the hypothesis that there is no concentration effect? Use the residual mean square as the estimate of σ^2.

11. A study has been made on pyrethrum flowers to determine the content of pyrethrin, a chemical used in insecticides. Four methods of extracting the chemical are used and samples are obtained from flowers stored under three conditions, namely, fresh flowers, flowers stored for one year, and flowers stored for one year but treated. It is assumed that there is no interaction present. The data are:

Pyrethrin Content, Percent

Storage Condition	Method			
	A	B	C	D
1	1.35	1.13	1.06	0.98
2	1.40	1.23	1.26	1.22
3	1.49	1.46	1.40	1.35

$SS_4 = 0.086225$ $SS_3 = 0.17405$ $SS = 0.282425$

(a) Write out the analysis of variance table. (b) Do the methods of extraction differ? Do the storage conditions affect the content ($\alpha = 0.01$)? (c) If there are differences found in (b), determine which differ. (d) When the maximum difference in pyrethrin content due to methods is 0.20, what are the bounds on the probability of accepting the hypothesis that there is no method effect? Use the residual mean square as the estimate of σ^2.

12. The following is an example of an experiment on the testing of leather soles in a wear tester. The machine consists of four arms attached to a shaft, on each of which is placed a shoe. The shaft rotates so that the shoes brush over an abrasive surface, abrading the test specimens. The loss in weight after a given number of cycles is used as a criterion of resistance to abrasion. The position of the shoes is changed after a fixed number of cycles so that each shoe appears on each arm for a fixed number of cycles. The order is randomized. The data are given below where the entries denote the loss in weight (unit of 0.1 milligram) in a run of standard length. Four different types of leather are tested.

Leather Specimen	Arm				Average
	1	2	3	4	
A	264	260	258	241	255.75
B	208	231	216	185	210.00
C	220	263	219	225	231.75
D	217	226	215	224	220.50
Average:	227.25	245.00	227.00	218.75	229.50

$SS_4 = 1,468.5$ $SS_3 = 4,621.5$ $SS = 7,444.0$

(a) Test for specimen effects and position effects at the 5% level of significance. (b) If the specimens differ, determine which are different. (c) Give approximate bounds on the probability of accepting the hypothesis of "no specimen effects" when the largest difference in specimen means is 30. Use the residual mean square as an estimate of σ^2.

13. An experiment was run to test the effect of adding salt to the water used in quenching aluminium casts. These casts are subjected to standard heat treatments and their tensile strength measured. There are five such treatments. The data are given below in units of 1,000 psi.

Heat Treatment	Salt Water Quench	Water Quench
A	33	31
B	46	44
C	45	46
D	40	43
E	36	35

(a) Write out the analysis of variance table. *Hint:* Subtract 30 from each observation. (b) At the 5% level of significance, determine if there are any heat treatment effects and quenching effects. (c) How does this test on quenching compare with a paired t test for quenching effects? (d) If there are any heat treatment effects, determine which differ. (e) If the true difference in tensile strengths due to quenching is 2,000 psi, use the OC curve for the paired t test to find the probability of accepting the hypothesis that there are no quenching effects. Use the residual mean square as an estimate of σ^2.

14. An experiment designed to test 4 unleaded gasolines and 3 types of carburetors gave the following mileage results (coded).

		F_1	F_2	F_3	F_4	Total
	C_1	15	18	20	20	73
Carburetor	C_2	10	8	12	15	45
	C_3	17	15	20	22	74
Total		42	41	52	57	192 = Overall Total

$$\sum\sum X_{ij}^2 = 3,280$$
$$\sum \text{(Squared Row Totals)} = 12,830$$
$$\sum \text{(Squared Col. Totals)} = 9,398$$

(a) Write out the analysis of variance table. Test at the 5% level for carburetor effects and fuel effects. (b) If fuels and/or carburetors differ, use the Tukey method to determine which differ significantly at the 5% level. (c) Give approximate bounds on the probability of accepting the hypothesis "no fuel effects" if the largest difference in fuel means is 5.0.

15. In a large metals processing company which specializes in manufacturing components for sensitive mechanisms, it is noticed that there appears to be considerable variability in the finished items. Each department of this firm conducts its own inspection and has its own standard $\frac{1}{4}$-inch gauge blocks with which to check its measuring devices. It is believed that there may be too much variation in the gauge blocks and/or the various measuring devices. Therefore, four gauge blocks and four micrometers are chosen at random, and the following data obtained by the chief quality control engineer.

Gauge Blocks	Micrometer 1	2	3	4
1	0.0251	0.0253	0.0259	0.0250
2	0.0248	0.0254	0.0254	0.0249
3	0.0247	0.0249	0.0255	0.0244
4	0.0245	0.0248	0.0246	0.0252

(a) Write out the analysis of variance table. *Hint:* Subtract 0.0244 from each observation. (b) Test at the 0.05 level for homogeneity of micrometers and gauge blocks. (c) Estimate the gauge block variability and micrometer variability. (d) What value of micrometer variance will be detected with probability 0.8, using the procedure of (b)? Use the residual mean square as an estimate of σ^2. Assume there is no interaction present.

16. The plant manager has decided to run an experiment to determine whether the material received at different times has the same tensile strength. Five randomly chosen time periods are to be considered and five randomly chosen men work on the material.

Time Period	Man 1	2	3	4	5	Mean
1	7.6	11.8	17.6	8.8	17.9	12.74
2	21.4	12.9	12.4	15.0	20.6	16.46
3	16.0	9.7	7.4	18.4	16.6	13.62
4	16.0	18.3	23.6	27.4	25.2	22.10
5	23.3	30.5	25.8	24.5	26.6	26.14
Mean:	16.86	16.64	17.36	18.82	21.38	

Analysis of Variance Table

Source	S.S.	D.F.	M.S.
Between time periods..	660.34	4	165.09
Between men	77.15	4	19.29
Residual	—	—	—
Total:	1,039.39	24	

(a) Test at the 1 % level for homogeneity of time periods and homogeneity of men. (b) Estimate the time period variability and the men variability. (c) What value of time variability will be detected with probability 0.75 using the procedure of (a)? Use the residual mean square as an estimate of $\sigma^2 + \sigma_\eta^2$.

17. Receivers of a certain type have been causing trouble because of malfunctioning. A series of tests were performed to evaluate the variations of the signal generator output. These receivers had in their design two different types of vacuum tubes, each of which had an influence on the signal generator output. A single receiver was chosen for this study and five vacuum tubes were chosen randomly from each type. The signal generator output in KUV data is:

Vacuum Tube, Type B	Vacuum Tube, Type A				
	1	2	3	4	5
1	18	21	25	28	18
2	21	18	14	22	20
3	23	21	21	23	25
4	32	19	21	20	24
5	17	14	20	15	23

(a) Test for tube-type A effects at the 5% level of significance. (b) Test for tube-type B effects at the 5% level of significance. (c) Estimate the component of the variance due to tube-type A. (d) Estimate the component of the variance due to tube-type B. (e) What value of tube variability will be detected with probability 0.8, using the procedure of (a)? Use the residual mean square as an estimate of $\sigma^2 + \sigma_\eta^2$.

18. Design engineers of a radio receiver manufacturing firm desire to determine whether or not the output at a given point of a certain type of receiver differs when transistors are used instead of vacuum tubes. Seven receivers are selected at random and operated at the given point, first with vacuum tubes. Then they are reassembled, using transistors, and operated. The following table of data is obtained:

	Configuration-Output Voltage	
Receiver	Vacuum Tubes	Transistors
1	19	17
2	16	20
3	18	19
4	11	15
5	21	23
6	24	30
7	21	25

(a) Determine whether or not there is a significant difference between output voltages when vacuum tubes are used as opposed to when transistors are used. (Use $\alpha = 0.01$.) (b) If the true difference between vacuum tubes and transistors is 2 volts, what is the probability of concluding there is no difference? Use the residual mean square as an estimate of $\sigma^2 + \sigma_\eta^2$.

19. Modification of the carburetor for standard model automobile engines recommended by an engine rebuilder will allegedly increase the mileage (miles per gallon) of the automobile in which this type of engine is installed. Seven automobiles are chosen at random and are test driven over the same test run, by the same driver, under the same conditions of test (rpm, mixture setting, etc.). This table of data is obtained:

	Carburetor-Mileage in mi/gal	
Auto	Original	Modified
1	17	21
2	19	19
3	17	23
4	15	15
5	16	23
6	18	20
7	16	18

(a) Assuming no interaction, determine for $\alpha = 0.05$ whether or not there is a significant difference in mileage for the seven cars used in the test. (b) Is it possible to test for interaction, using the data above? If not, design an experiment for this situation in which a test for interaction between automobiles and the carburetors could be conducted. (c) Determine at the 5% level whether there are any carburetor effects. (d) Assuming no interaction, what value of the automobile variability will be detected with probability 0.8? Use the residual mean square as an estimate of σ^2.

20. An engineer in the design section of an aircraft manufacturing plant has presented theoretical evidence that painting the exterior of a particular combat airplane reduces its cruising speed. He convinces the chief design engineer that the next nine aircraft off the assembly line should be test flown to determine cruising speed prior to paint, then painted, and finally test flown to ascertain cruising speed after they are painted. The following data are obtained:

| | Cruising Speed (knots) | |
Aircraft	Not Painted	Painted
1	426	416
2	418	400
3	424	420
4	428	421
5	440	432
6	421	404
7	412	401
8	409	405
9	427	422

(a) Design a test at the 5% level and complete the computations necessary to evaluate the design engineer's evidence. (b) What value of the difference in cruising speeds between the painted and unpainted planes will be detected with probability 0.9? Use the residual sum of squares as an estimate of $\sigma^2 + \sigma_\eta^2$.

21. The rate of flow of fuel through three different types of nozzles was the subject of a recent experiment. Five different operators chosen from a group of 25 were used to test each nozzle. The data recorded were:

**Cubic Centimeters of Fuel through Three Nozzles
for Five Trials**

| | Nozzle Type | | |
Operator	A	B	C
1	96.5	96.5	97.1
2	97.4	96.1	96.4
3	96.0	97.9	95.6
4	97.8	96.3	95.7
5	97.2	96.8	97.3

$SS_4 = 0.7854$ $SS_3 = 0.6427$ $SS = 7.3094$

(a) Analyze the data as a two-way analysis of variance and write down the analysis of variance table. (b) Test for effects of operators and nozzle types at the 1% level of significance. (c) Estimate the variability due to operators and determine which nozzles differ, if any. (d) What is the approximate probability of accepting the hypothesis of no nozzle effects when the maximum difference between nozzles is actually 2.4 cc? Use the residual mean square as an estimate of $\sigma^2 + \sigma_\eta^2$. (e) Assuming no interaction, what value of operator variability will be detected with probability 0.75? Use the residual mean square as an estimate of σ^2.

22. It is desired to know what type of filter should be used over the screen of a cathode ray oscilloscope in order to have a radar operator easily pick out targets on the presentation. A test to accomplish this has been set up. A noise is first applied to the scope to make it difficult to pick out a target. A second signal, representing the target, is put into the scope, and its intensity is increased from zero until detected by the observer. The intensity setting at which the observer first notices the target signal is then recorded. This experiment is repeated again, with a different observer, on the assumption that all people do not see in exactly the same manner. A total of 20 observers is chosen at random. After a set of readings has been taken with one type of filter on the scope, another set of readings, using the same observers, is taken with a different type of filter. The procedure above is then carried out with a third filter. The numerical value of each reading listed in the table of data is proportional to the target intensity at the time the operator first detects the target.

Observer	Filter No. 1	Filter No. 2	Filter No. 3
1	90	88	95
2	87	90	95
3	93	97	89
4	96	87	98
5	94	90	96
6	88	96	81
7	90	90	92
8	84	90	79
9	101	100	105
10	96	93	98
11	90	95	92
12	82	86	85
13	93	89	97
14	90	92	90
15	96	98	87
16	87	95	90
17	99	102	101
18	101	105	100
19	79	85	84
20	98	97	102

(a) Test the hypothesis that the filters are the same with $\alpha = 0.05$. (b) What value of the difference in intensity between filters will be detected with probability 0.8? Use the residual sum of squares as an estimate of $\sigma^2 + \sigma_\eta^2$.

23. Three different procedures for measuring runoff in three types of terrain were employed to determine whether or not these procedures were all of equal value. The tests were carried out over a period of four years, and the results are presented below (coded).

	Procedure A		Procedure B		Procedure C	
Area D {	30	54	−68	−60	−80	18
	−26	81	4	123	30	42
Area E {	−50	26	36	−15	−83	−30
	59	−12	−6	122	−42	−55
Area F {	38	110	74	−20	−54	−9
	68	60	64	50	−42	63

$$SS_4 = 21{,}624.2223 \qquad SS_3 = 8{,}556.2223$$
$$SS_1 = 75{,}354.000 \qquad SS = 115{,}655.5556$$

(a) Write out the analysis of variance table. (b) Test for the significance of inter-action. If interaction is absent, test for procedure effects and area effects at $\alpha = 0.05$. What would be the physical meaning of interaction in this problem if any exists? (c) If there is no interaction, determine which procedures differ and which areas differ. Obtain confidence intervals for those effects which differ. (d) How many years of readings are required to detect a maximum difference between procedures equal to 100 units with probability 0.7? Obtain this result by using $\Phi(\min)$ and using the "within combinations" mean square as an estimate of σ^2.

24. The design engineer of a home automatic clothes washer is uncertain as to the length of time that he should specify for the washing cycle. However, he believes that it should be no less than 20 minutes and no greater than 30 minutes. With the aid of a home economist he designs and conducts a test in which the following data are obtained.

Soap or Detergent	Washing Time		
	20 min	25 min	30 min
1	11	12	12
	10	11	12
	12	13	13
2	10	12	14
	14	15	15
	12	15	17
3	12	10	11
	13	12	12
	11	11	11

Dirt Removed (10^{-2} lb)

Three equally sized and equally soiled loads of clothes were washed for each combination of soap or detergent and washing time in the same machine. (a) Consider the effects as fixed and analyze the above data for interaction at $\alpha = 0.05$. If no interaction is present, test for main effects. (b) If there is no interaction present, determine which soaps differ. (c) If interaction is absent, how many observations are required to detect a maximum difference between soaps equal to 0.03 lb with probability 0.8? Obtain this result by using $\Phi(\min)$ and using the "within combinations" mean square as an estimate of σ^2.

25. An experiment was run to determine the abrasive wear of several laminated bearing materials. The bearings used were either oven dried or moisture saturated, with the shaft used in the experiment made of a high carbon steel. The test was conducted on an Amsler testing machine and it can be assumed that the shaft exhibited negligible wear. The machine ran for a fixed amount of time and the depth of bearing wear in units of 10^{-3} was recorded.

Heat Treatment	Bearing Material			
	A	B	C	D
Oven dried	38	40	38	44
	40	37	41	37
Moisture saturated	32	38	32	36
	44	42	33	38

(a) Write out the analysis of variance table. (b) Test for interaction. Also test for bearing material effects and heat treatment effects if interaction is absent. If materials differ, determine which differ. Use $\alpha = 0.05$. (c) If interaction is absent, what is the approximate probability of detecting a maximum difference between bearing materials equal to 0.01? Use the "within combinations" mean square as an estimate of σ^2. If there were four observations per combination taken, what is the approximate probability?

26. An electronic manufacturer was having difficulty with a particular type of tube, where the mutual conductance was running below the bogie value. An experiment was performed to determine the effect of the exhaust variables, plate temperature, and filament lighting on the electrical characteristics of the tubes. Two levels of plate temperature and four levels of filament lighting current were used.

Transconductance

	Filament Lighting Conditions			
	L_1	L_2	L_3	L_4
Plate temperature T_1	3,774	4,710	4,176	4,540
	4,364	4,180	4,140	4,530
	4,374	4,514	4,398	3,964
Plate temperature T_2	4,216	3,828	4,122	4,484
	4,524	4,170	4,280	4,332
	4,136	4,180	4,226	4,390

$$SS_4 = 85,942 \qquad SS_3 = 25,090.6$$
$$SS_1 = 825,075 \qquad SS = 1,189,776$$

(a) Write out the analysis of variance table. (b) Test for interaction at $\alpha = 0.01$. If none is present, test for filament lighting conditions effects and plate temperature effects. If there are filament lighting effects, determine which differ. (c) If interaction is absent, how many observations per combination are required to insure detection of a difference in plate temperatures equal to 300 units with probability 0.75? Use the "within combinations" mean square as an estimate of σ^2.

27. An experiment was run to isolate and determine the magnitude of the major sources of variability contributing to differences in the breaking strength of some wool serge material. It had already been ascertained that the material is very homogeneous so that material was not felt to be a contributing factor. The remaining factors which appeared to be causing trouble were operators and machines. Four experienced operators and four Scott Testers were picked at random, and each operator was paired with each machine. The experiment was run on a one-yard section of the wool serge and the data are:

Breaking Strength, pounds

Operator	*Machines* *A*	*B*	*C*	*D*
1	110	101	108	113
	116	102	109	118
2	112	115	111	115
	111	106	109	114
3	114	107	113	117
	112	109	110	114
4	116	109	115	118
	114	110	112	116

(a) Write out the analysis of variance table. (b) Test for interaction, operator effects and machine effects at a 5% level of significance. (c) Estimate the interaction variability, operator variability and machine variability. (d) How many observations per combination are required to detect a machine variance equal to $2\sigma^2$ with probability 0.8? Use the value found in (c) as an estimate of the interaction variance. Use the "within combinations" mean square as an estimate of σ^2.

28. In an electronic components manufacturing firm there are 20 automatic machines which assemble identical transistors. The variability of the measured characteristic is a source of trouble. There are 12 identical test units with which the finished items are checked for proper operation. It is desired to determine whether or not there is homogeneity among the assembly machines and among the test units. Also it is desired to know whether or not there is interaction among the assembly machines and test units. The data below are obtained from three assembly machines and three test units, and are randomly selected.

Assembly Machine	*Test Unit* 1	2	3
1	−2.5	−3.7	−3.1
	−3.4	−2.8	−3.2
	−3.5	−3.6	−3.5
2	−3.5	−3.9	−3.3
	−2.6	−3.8	−3.4
	−3.6	−3.4	−3.5
3	−2.4	−3.5	−2.6
	−2.7	−3.2	−2.6
	−2.8	−3.5	−2.5

I_c (the instantaneous transistor collector current) in milliamperes for instantaneous emitter-to-base voltage drop equals 0.4 volt, and instantaneous transistor emitter current equals 1.0 milliampere.

(a) Test for main effects and interaction at the 0.05 level. (b) Estimate the test unit variability, the assembly machine variability, and the interaction variability. (c) What value of assembly machine variability can be detected with probability 0.75, using the procedure of (a)? Use the value found in (b) as an estimate of the interaction variability. Use the "within combinations" mean square as an estimate of σ^2. What is your answer if there were five observations per combination?

29. A small motor is being manufactured for a special purpose. The important characteristic of the motor in this application is its starting torque. Several testers are used to evaluate the starting torque of the motors. It is desired to know whether the testers are obtaining equivalent results. An experiment is performed, using three testers and eight motors, each chosen at random. Duplicate measurements are obtained. The results are as follows (coded data):

	Tester					
Motor	A		B		C	
1	9.60	9.72	11.23	10.90	9.00	10.34
2	10.81	9.90	10.98	10.55	10.81	9.65
3	9.69	10.15	10.10	10.99	9.57	8.57
4	8.43	10.12	11.01	9.68	9.03	9.05
5	8.50	9.87	9.75	8.93	8.58	9.67
6	10.92	9.23	11.14	11.69	10.31	9.76
7	9.98	9.26	9.44	9.89	9.56	11.29
8	10.31	8.95	10.04	12.72	9.80	9.21

$SS_4 = 8.4734 \qquad SS_3 = 8.0889$

$SS_1 = 15.5995 \qquad SS = 38.4743$

(a) Write out the analysis of variance table. (b) Test for interaction and main effects ($\alpha = 0.01$). (c) Estimate the motor variance, tester variance, and interaction variance. (d) How many observations are required per combination to detect a tester variance of 0.30 with probability 0.75? Use the value found in (c) as an estimate of the interaction variance. Use the "within combinations" mean square as an estimate of σ^2.

30. A quality assurance engineer for an electronics components manufacturing firm intuitively feels that there is large variability among the 36 ovens used by this firm in testing the life of the various components. To determine whether or not he is correct he chooses a single type component and obtains the following data for the two temperatures normally used for life testing of this item. The component is operated in an oven until it fails. Three randomly selected ovens are used for the experiment.

	Life Components (minutes)		
Temperature	*Oven 1*	*Oven 2*	*Oven 3*
550°F	237	208	192
	254	178	186
	246	187	183
600°F	178	146	142
	179	145	125
	183	141	136

$SS_4 = 9{,}646.3333 \qquad SS_3 = 13{,}667.5$

$SS_1 = 837.3333 \qquad SS = 24{,}426$

(a) From the data given above, determine whether or not the engineer is correct in his belief that there is variability among the ovens. Use $\alpha = 0.01$. Test for interaction and temperature effects. (b) Estimate the oven variability.

31. There are three time-study engineers employed by a certain company. The chief industrial engineer for whom these men set standard times wants to determine whether or not all three of them essentially follow his prescribed methods in setting standards. Also he wishes to ascertain whether or not the standards set by these three men are influenced by any particular combinations of time-study engineers and machine operators. To obtain this information the chief engineer selects at random four of 25 turret lathe operators, and directs each of the three engineers to set three standard times (each independent of the others) for each operator, all standards being for the same job. The following data are obtained.

	Time Study Engineer		
Operator	1	2	3
1	2.95	2.07	2.18
	2.93	2.99	2.33
	2.42	2.53	2.21
2	2.58	2.59	2.15
	2.26	2.36	2.94
	2.05	2.78	2.55
3	2.47	2.38	2.72
	2.44	2.48	2.85
	2.52	2.82	2.73
4	2.66	2.39	2.67
	2.95	2.61	2.89
	2.27	2.01	2.75

Std. Time, in minutes

$SS_4 = 10.4404$ $\quad SS_3 = 5.4555$
$SS_1 = 43.2067$ $\quad SS = 79.3178$

(a) Test the hypothesis that the main effect due to time study engineers equals zero. (Use $\alpha = 0.05$.) (b) Are there interaction or operator effects ($\alpha = 0.05$)? (c) If the time study engineers differ, determine which differ. (d) What is the approximate probability of saying that the time study engineers do not differ when they differ by $\frac{1}{2}$ minute? Use the interaction mean square as an estimate of $\sigma^2 + n\sigma_\eta^2$.

32. The same experiment as described in Problem 21 is repeated three times, yielding the following results.

	Nozzles		
Operator	A	B	C
1	96.6	97.7	97.6
	96.0	96.0	97.5
	94.4	97.7	94.6
2	98.5	96.0	92.2
	97.2	96.0	96.8
	96.5	97.9	94.4
3	97.5	97.1	97.1
	96.0	97.4	95.4
	96.6	97.4	94.7
4	98.3	95.6	97.5
	97.6	96.2	95.2
	96.3	95.3	94.3
5	98.6	97.1	95.4
	97.2	96.0	97.4
	98.3	96.3	98.1

(a) Write out the analysis of variance table and test for the significance of inter-action, operator effect, and nozzle effect. Use $\alpha = 0.05$. What would be the physical meaning of interaction if any existed in this problem? (b) If nozzles differ, determine which nozzles differ. Obtain confidence intervals for those effects which differ. (c) What is the approximate probability of accepting the hypothesis of no nozzle effects when the maximum difference between nozzles is actually 2.4 cc? Use the interaction mean square as an estimate of $\sigma^2 + n\sigma_\eta^2$.

33. Five methods have been suggested for mixing concrete. Three batches were mixed with two specimens from each batch used with each method and tested for compression strength (in psi).

			Method		
Batch	1	2	3	4	5
A	4,250 4,260	4,110 4,120	4,390 4,380	4,020 4,040	3,880 3,890
B	4,260 4,250	4,110 4,100	4,380 4,370	4,030 4,020	3,870 3,860
C	4,180 4,160	4,020 4,020	4,300 4,200	3,950 3,940	3,780 3,770

(a) Test for interaction at the 5% level. (b) Do the methods have any effect on the strength of the concrete ($\alpha = 0.05$)? (c) If the methods do differ, determine Tukey confidence intervals to make statements about which ones differ.

34. A factory employs a large number of people who assemble electronic tubes. Managment is interested in whether there tends to be a variation in the average number of tubes assembled from worker to worker, and is particularly interested in detecting a variation between workers as large as one-and-one-half times an individual's daily variation. It is decided that 18 observations may be taken according to one of two schemes. (a) Six workers are chosen at random and ob-served on each of three days. (b) Three workers are chosen at random and observed on each of six days. Which scheme do you prefer? Why?

35. Two examiners are checked on the time they take to use a certain gauge. Duplicate readings are taken for each examiner. The results are:

Examiner I	*Examiner II*
a	c
b	d

where $a < b < c < d$. The value of F for the data above is:

$$\frac{2\left\{\left[\dfrac{a+b}{2} - \dfrac{a+b+c+d}{4}\right]^2 + \left[\dfrac{c+d}{2} - \dfrac{a+b+c+d}{4}\right]^2\right\}}{\dfrac{1}{2}\left\{\left[a - \dfrac{(a+b)}{2}\right]^2 + \left[b - \dfrac{(a+b)}{2}\right]^2 + \left[c - \dfrac{(c+d)}{2}\right]^2 + \left[d - \dfrac{(c+d)}{2}\right]^2\right\}}$$

$$= \frac{\left[\dfrac{(a+b)-(c+d)}{2}\right]^2 + \left[\dfrac{(c+d)-(a+b)}{2}\right]^2}{\left(\dfrac{a-b}{2}\right)^2 + \left(\dfrac{b-a}{2}\right)^2 + \left(\dfrac{c-d}{2}\right)^2 + \left(\dfrac{d-c}{2}\right)^2}$$

$$= \frac{[(a+b)-(c+d)]^2}{(a-b)^2 + (c-d)^2}.$$

Using the principle of randomization, enumerate all the possible values of F and accept or reject the hypothesis of equality of means at the $33\frac{1}{3}\%$ level of significance.

11

SOME FURTHER TECHNIQUES

OF DATA ANALYSIS

11.1 Introduction

Most of the text up to this point has dealt with data which came from measurements on a particular variable that was usually on a continuous scale. Examples are such things as dimensions, voltage, and Rockwell hardness. In these situations, the underlying distribution was generally assumed to be known, except possibly for parameters about which tests of hypotheses were to be made or estimates obtained. In many situations these assumptions could be justified by theoretical considerations such as the central limit theorem, or by noting that the test statistics or estimators were robust. A decision procedure is called robust if its performance is not sensitive to departures from the underlying assumptions. When such information was unavailable, non-parametric procedures were suggested. However, little has been said about verifying assumptions made about the underlying probability distribution, and this is one of the problems of concern in this chapter.

In many experiments it is not possible or useful to record measurements because the data available are expressed as the number of cases which fall into specified categories. For example, one may count the number of defective items produced by various machines, the number of errors made by several operators, the number of successful and unsuccessful satellite launches made by several launch devices, or the number of accidents as a function of shift. The analysis of this type of data is also considered in this chapter.

11.2 Qualitative Techniques for Determining the Form of a Distribution

The problem of determining the form of a distribution is probably one of the most important problems encountered in practice, but also one of the most difficult. As indicated above, the experimenter uses decision procedures or estimators that are robust whenever possible. However, there must be occasions when he has to check his assumptions about the form of the underlying distribution. Hopefully, there are data available for this purpose.

A method frequently used is to construct a histogram (see Sec. 1.2) or a sample cumulative distribution function (see Sec. 2.6)[1] and compare them with some anticipated density function (or probability distribution if the random variable is discrete) or cumulative distribution function. Even granted that this is a qualitative measure, it suffers greatly from an inability to distinguish between anomalies in shapes due to sampling fluctuations and theoretical considerations. For example, if a histogram is bimodal, an experimenter may be wary of assuming normality. However, the bimodal appearance may be due to sampling fluctuations, and could well represent data from a normal distribution.

An alternative procedure is to plot the data on "probability paper." By a suitable monotonic transformation of the vertical scale of the graph of a cumulative distribution function, it is possible to transform this curve so that it will plot as a straight line. Graph paper possessing this property is available for several distributions and it is common practice to plot data on the appropriate graph paper as a check on the form of the distribution from which the data were obtained. In particular, this is often done for "checking" on the assumption of normality. The procedure is:

(a) Label a sheet of normal probability paper so that the horizontal axis represents the observation scale, and the vertical axis represents a fraction between 0 and 1.[2]

(b) Based on a random sample, arrange the observations in ascending order, i.e., $x_{(1)} \leq x_{(2)} \leq \cdots \leq x_{(n)}$, and plot the points $(x_{(1)}, 1/2n)$, $(x_{(2)}, 3/2n)$, $\cdots$, $(x_{(i)}, (2i - 1)/2n)$, $\cdots$, $(x_{(n)}, (2n - 1)/2n)$ on the graph. [The choice of ordinate is somewhat arbitrary. An alternative plot is the points $(x_{(i)}, i/(n + 1))$].

(c) Fit a straight line to these points. This is usually done by eye, but the method of least squares can be used.

These data should plot as a straight line if the underlying distribution is normal since the actual CDF will plot as a straight line on this paper. The parameters of the normal distribution can be estimated from the graph. The estimate of the mean is the value of the abscissa (horizontal coordinate)

---

[1] If the data available are already grouped into intervals, a relative cumulative frequency function (see Sec. 1.2) can be used.

[2] Normal probability paper is constructed by plotting the fraction p at a vertical distance u above the horizontal axis where

$$p = \int_{-\infty}^{u} \frac{1}{\sqrt{2\pi}} e^{-z^2/2} \, dz.$$

where the fitted line intersects 0.50 on the vertical scale. The estimate of the standard deviation is the difference on the horizontal scale between the points where the fitted line passes through 0.8413 and 0.5 on the vertical scale.

As an example, 25 observations on the tensile strength of a rubber compound are obtained and the data are:

2,954.62	3,025.33	2,984.65	3,034.51	2,780.27
2,980.58	2,861.48	2,892.86	3,020.70	2,914.04
3,053.02	2,979.04	3,051.01	2,948.42	2,924.46
3,009.29	3,109.85	3,006.77	2,998.25	3,087.05
3,166.46	3,045.65	2,925.79	3,035.31	2,982.11

These data are plotted on normal probability paper $(x_{(i)}, (2i - 1)/2n)$ and are shown in Fig. 11.1. The data appear to be linear on this graph so that a line was drawn through the data by eye. If this is taken as evidence that the underlying distribution is normal, then an estimate of the mean and standard deviation can be obtained. The estimate of the mean is the value of the horizontal coordinate where the line intersects 0.50, i.e., 2,990 psi, and the estimate of the standard deviation is the difference on the horizontal scale between the points where the line passes through 0.8413 and 0.5 on the vertical scale, i.e., $(3,070 - 2,990) = 80$.

11.3 Quantitative Techniques for Determining the Form of a Distribution

The qualitative techniques presented in Sec. 11.2 suffer from the fact that no objective criteria were presented to judge whether the data fit the assumed distribution. This section is concerned with two statistical tests to be used for this purpose: the Kolmogorov-Smirnov test and the chi-square goodness of fit test.

11.3.1 The Kolmogorov-Smirnov Test

The Kolmogorov-Smirnov test is a very good test to use whenever the hypothesized distribution is completely specified; e.g., X is an exponentially distributed random variable with parameter equal to 10, or X is a normally distributed random variable with mean equal to 3 and standard deviation equal to 20. It should not be used when the form of the hypothesized distribution is specified, but one or more of the parameters remain unspecified; e.g., X is exponentially distributed, or X is normally distributed. The Kolmogorov-Smirnov procedure tests the hypothesis that the CDF, $F(x)$, is $F_0(x)$. A random sample of size n is drawn from a continuous distribution $F(x)$. Denote the sample cumulative distribution function by $F_n(x)$ (see Sec. 2.6).[1] The sample CDF, $F_n(x)$, is compared with the hypothesized CDF,

---

[1] The sample CDF can be defined simply as

$$F_n(x) = \frac{i}{n} \quad \text{for } X_{(i)} \leq x < X_{(i+1)}, \quad i = 0, \cdots, n,$$

and where $X_{(1)}$ is the random variable representing the smallest in the random sample, $X_{(2)}$ the second smallest, $\cdots$, $X_{(n)}$ the largest; $X_{(0)} = -\infty$ and $X_{(n+1)} = \infty$.

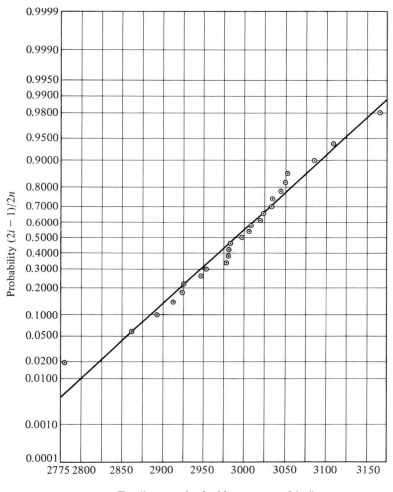

Tensile strength of rubber compound (psi)

Fig. 11.1. *Plot of tensile strength data on normal probability paper.*

$F_0(x)$. If $F_n(x)$ is "too far" from F_0, this is ample evidence that $F(x)$ is not $F_0(x)$. The "comparison" between $F_n(x)$ and $F_0(x)$ is the magnitude of the absolute value of their difference for each x, and "too far" is measured in terms of the maximum of these differences.

The Kolmogorov-Smirnov test statistic is then given by

$$D_n = \max_{\text{all } x} |F_n(x) - F_0(x)|.$$

When the hypothesis that $F_n(x) = F_0(x)$ is true, D_n has a distribution which is independent of $F_0(x)$. This distribution of D_n has been tabulated as a function of n and α (level of significance) when $F(x) = F_0(x)$ and excerpts are presented in Table 11.1. The hypothesis that $F(x) = F_0(x)$ is rejected at the α level of significance whenever $D_n(x) > d_{\alpha;n}$, where $d_{\alpha;n}$ are the values given in Table 11.1.

Table 11.1. Table of the Percentage Points for the Kolmogorov-Smirnov Test Statistic[1]

Sample Size (n)	Significance Level		
	10%	5%	1%
1	0.950	0.975	0.995
2	0.776	0.842	0.929
3	0.642	0.708	0.829
4	0.564	0.624	0.734
5	0.510	0.563	0.669
6	0.470	0.521	0.618
7	0.438	0.486	0.577
8	0.411	0.457	0.543
9	0.388	0.432	0.514
10	0.368	0.409	0.486
11	0.352	0.391	0.468
12	0.338	0.375	0.450
13	0.325	0.361	0.433
14	0.314	0.349	0.418
15	0.304	0.338	0.404
16	0.295	0.328	0.392
17	0.286	0.318	0.381
18	0.278	0.309	0.371
19	0.272	0.301	0.363
20	0.264	0.294	0.352
25	0.24	0.264	0.317
30	0.22	0.242	0.290
35	0.21	0.23	0.27
40		0.210	0.252
50		0.188	0.226
60		0.172	0.207
70		0.160	0.192
80		0.150	0.180
90		0.141	
100		0.134	
Approximate formula:	$\dfrac{1.22}{\sqrt{n}}$	$\dfrac{1.36}{\sqrt{n}}$	$\dfrac{1.63}{\sqrt{n}}$

[1] Abridged and reproduced by permission from "The Kolmogorov-Smirnov Test for Goodness of Fit" by F. J. Massey, Jr., *J. Amer. Stat. Assn.*, **46**, March, 1951. Corrections and additional entries are taken with permission from "Numerical Tabulation of the Distribution of Kolmogorov's Statistic for Finite Sample Size" by Z. W. Birnbaum, *J. Amer. Stat. Assn.*, **47**, September, 1952.

Although the description of the Kolmogorov-Smirnov test procedure is quite simple, carrying out the instructions is more difficult. In principle, the differences between $F_n(x)$ and $F_0(x)$ must be examined for all x. In fact, these differences need be examined only at the "jump points" of $F_n(x)$. The jump points occur at the observed values of the random variables. Typical jump points are shown in Fig. 11.2. At each jump point $x_{(i)}$ the two differences $|F_0(x_{(i)}) - F_n(x_{(i)})|$ and $|F_0(x_{(i)}) - F_n(x_{(i-1)})|$ must be obtained. For a random sample of size n, the maximum absolute deviation, D_n, must be among these $2n$ differences.

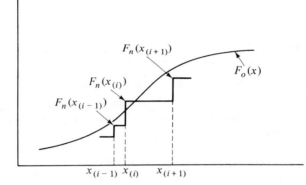

Fig. 11.2. *Comparison of $F_0(x)$ and $F_n(x)$ at typical jump points.*

The above test assumed that X was a continuous random variable. If X is discrete, the same procedure can be followed, but the actual level of significance achieved will be less than or equal to the nominal value associated with $d_{\alpha;n}$.

The previous discussion dealt with testing the hypothesis that $F(x) = F_0(x)$. Using the usual relationship between tests of hypotheses and confidence intervals, a two-sided $100(1 - \alpha)\%$ confidence interval on $F(x)$ can be found and is given by

$$F_n(x) \pm d_{\alpha;n}.$$

As an example, the Kolmogorov-Smirnov test will be used on the tensile strength data plotted on normal probability paper in Sec. 11.2 to test the hypothesis that the underlying distribution is normal with a mean equal to 3,000 psi and a standard deviation equal to 100. A comparison between the sample CDF, $F_n(x)$, and the hypothesized distribution $F_0(x)$ is shown in Fig. 11.3. At each "jump point" the differences $|F_0(x_{(i)}) - F_n(x_{(i)})|$ and $|F_0(x_{(i)}) - F_n(x_{(i-1)})|$ are obtained. The maximum absolute difference occurs at the jump point corresponding to the tensile strength of 3,053.02 ($x_{(i)} = x_{(22)}$), and the maximum absolute difference is $0.880 - 0.702 = 0.178$.

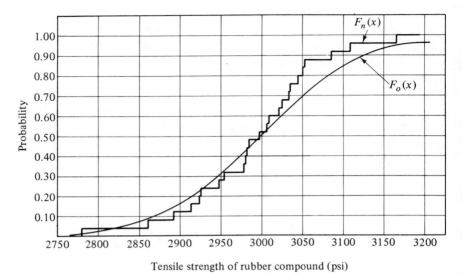

Fig. 11.3. *Comparison between the sample CDF, $F_n(x)$, of tensile strength data and the normal CDF, $F_0(x)$, with mean equal to 3,000 psi and standard deviation equal to 100.*

If a 5% level of significance is used, the critical value $d_{0.05; 25} = 0.264$ is obtained from Table 11.1. Since the maximum absolute difference between $F_0(x)$ and $F_n(x)$ is 0.178 which is less than 0.264, the hypothesis is accepted that $F_0(x)$ is normal with mean equal to 3,000 and standard deviation equal to 100.

11.3.2 THE CHI-SQUARE GOODNESS OF FIT TEST

An alternative test of the hypothesis that $F(x)$ equals $F_0(x)$ is given by the chi-square goodness of fit test. Although this test is not quite as good as the Kolmogorov-Smirnov test, it is applicable to a larger class of problems, i.e., it is applicable even when $F_0(x)$ is not completely specified. It can be used to test the hypothesis that a random variable X is normally distributed, exponentially distributed, Poisson distributed, etc., without specifying the parameter of these distributions. The chi-square goodness of fit test compares the observed frequency in an interval with the expected number in that interval if $F_0(x)$ is the CDF. In particular, assume that each outcome of a random variable falls into one and only one of the intervals; let $O_1, O_2, \cdots, O_k$ be the observed frequencies for each interval. Denote by $E_1, E_2, \cdots, E_k$ the expected number of observations falling into each interval if $F_0(x)$ is the CDF of the random variable X. If the hypothesis that $F(x) = F_0(x)$ is true, then

$$\chi^2 = \sum_{i=1}^{k} \frac{(O_i - E_i)^2}{E_i}$$

has an approximate chi-square distribution when the sample size is large.[1] Hence, the hypothesis that $F(x) = F_0(x)$ is accepted if $\chi^2 \leq \chi^2_{\alpha;\nu}$, where $\chi^2_{\alpha;\nu}$ is the α percentage point of the chi-square distribution with ν degrees of freedom and found in Appendix Table 2. The degrees of freedom associated with this chi-square random variable is $(k - 1)$ provided that $F_0(x)$ is completely specified. If only the form of $F_0(x)$ is specified without specifying the parameters, then the degrees of freedom are $(k - 1)$ minus the number of parameters that must be estimated from the data. For example, when testing the hypothesis that $F(x)$ is normally distributed, without specifying the mean and variance, these parameters must be estimated from the data, and the number of degrees of freedom becomes $(k - 1) - 2 = k - 3$. The method of maximum likelihood is a suitable method for estimating the unknown parameters. The above procedure does not indicate how to choose the number k. This choice is somewhat arbitrary, but a reasonable rule of thumb to follow so that the asymptotic theory will be applicable is to choose the number of intervals so that the expected number falling into each interval is at least 5, although deviations for some intervals may be tolerable. Furthermore, the intervals should be chosen so that the probability of an observation falling into each interval should be nearly equal when $F_0(x)$ is the true underlying distribution.

As an example, the chi-square goodness of fit test will be used on the tensile strength data presented in Sec. 11.2 (and subjected to the Kolmogorov-Smirnov test in Sec. 11.3.1). However, the hypothesis to be tested now will be that $F(x) = F_0(x)$, where $F_0(x)$ is normally distributed (as opposed to the previous section where $F_0(x)$ was specified to be normal with mean equal to 3000 and standard deviation equal to 100). Since μ and σ are not specified, they must be estimated from the data, at a "cost" of two degrees of freedom. The maximum likelihood estimators are $\hat{\mu} = \sum X_i/n$ and $\hat{\sigma} = \sqrt{\sum (X_i - \bar{X})^2/n}$, and the values obtained from the data are $\hat{\mu} = 2{,}990.86$ and $\hat{\sigma} = 80.01$. Since there are 25 observations and approximately 5 observations are required in each interval, five intervals, each having probability equal to 0.2 when $F(x) = F_0(x)$, will be chosen. The intervals are obtained from Appendix Table 1 and are shown in Fig. 11.4. The observed frequencies of the data falling into these intervals are $O_1 = 4; O_2 = 4; O_3 = 7; O_4 = 7; O_5 = 3$. Since the expected number E_i falling into each interval is 5, the χ^2 statistic takes on the value

$$\chi^2 = \frac{(4-5)^2}{5} + \frac{(4-5)^2}{5} + \frac{(7-5)^2}{5} + \frac{(7-5)^2}{5} + \frac{(3-5)^2}{5} = 2.8.$$

If a 5% level of significance is used, the percentage point of the chi-square distribution corresponding to $5 - 1 - 2 = 2$ degrees of freedom is 5.991. Since 2.8 is less than 5.991, the hypothesis that $F(x)$ is normal is accepted.

---

[1] Section 11.4 deals with a large variety of problems concerned with attribute data where this statistic is applicable.

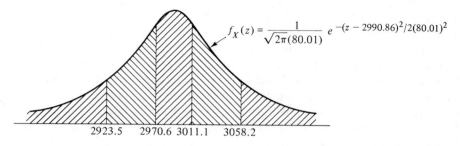

$$f_X(z) = \frac{1}{\sqrt{2\pi}(80.01)} \, e^{-(z - 2990.86)^2/2(80.01)^2}$$

2923.5   2970.6 3011.1   3058.2

Fig. 11.4. *Intervals containing equal areas for chi-square goodness of fit test.*

11.4 Chi-Square Tests

Enumeration or attribute data consist of the number of observations in a sample which fall into certain specified categories. The data are usually taken to test hypotheses about the true relative frequency of these categories. To be specific, assume that each sample observation must fall into one and only one of k categories; let $O_1, O_2, \cdots, O_k$ be the observed frequencies for each category. The experimenter is interested in testing hypotheses about the true relative frequencies.

Category	Observed Frequency	Theoretical Frequency
1	O_1	E_1
2	O_2	E_2
.	.	.
.	.	.
.	.	.
k	O_k	E_k

The test statistic is

$$\chi^2 = \sum_{i=1}^{k} \frac{(O_i - E_i)^2}{E_i}$$

which for large enough samples[1] has an approximate chi-square distribution. The number of degrees of freedom depends on how the data are used computing the E_i. A small value of χ^2 is associated with good agreement between observed and theoretical values; a large value tends to indicate discrepancy.

---

[1] The following rules of thumb can be used to assess adequacy of sample size: 1. If χ^2 has from 2 to 15 degrees of freedom, all E_i should be at least 2.5, although at least 5 is preferable. 2. If χ^2 is computed from a 2×2 table (which will be discussed in Sec. 11.4.4) all E_i should be at least 5; however, if all but one of them is at least 5, the remaining one may be as small as 1 with little distortion in significance level.

Whether the discrepancy is likely to arise by chance is decided by reference to a table of percentage points of the chi-square distributions, Appendix Table 2.

11.4.1 THE HYPOTHESIS COMPLETELY SPECIFIES THE THEORETICAL FREQUENCY

Assume that each of n observations can fall into one of k categories and let $p_1, p_2, \cdots, p_k$ be the true probabilities, i.e., p_i is the probability of a random observation falling in the ith category. In this type of problem, the theoretical frequencies are calculated from the formula

$$E_i = np_i$$

and the chi-square statistic has $k - 1$ degrees of freedom. Note that this is similar to the chi-square goodness of fit test discussed in Sec. 11.3.2.

For example, the total number of defective units in a day's production was tabulated by shifts. If there is no difference in quality between the shifts, the probability of any defect occurring in any shift i is $\frac{1}{3}$. There are 80 observations, so the theoretical frequency would be $80/3 = 26.67$.

	Observed Frequency	Theoretical Frequency
Shift 1	20	26.67
Shift 2	36	26.67
Shift 3	24	26.67
	80	

It is of interest to see whether the variation among shifts is due to chance or whether there is a real difference in the occurrence of defectives. That is, the hypothesis to be tested is that the true relative frequency of defectives is the same on all shifts, i.e., $p_1 = p_2 = p_3 = \frac{1}{3}$. The test statistic is

$$\chi^2 = \frac{(20 - 26.67)^2}{26.67} + \frac{(36 - 26.67)^2}{26.67} + \frac{(24 - 26.67)^2}{26.67}$$
$$= \frac{44.4889}{26.67} + \frac{87.0489}{26.67} + \frac{7.1289}{26.67}$$
$$= 1.6681 + 3.2639 + 0.26730$$
$$= 5.1993 < \chi^2_{0.05; 2} = 5.991.$$

Therefore, the hypothesis is accepted and it is concluded that the data do not indicate that the frequencies differ.

11.4.2 DICHOTOMOUS DATA

An interesting special case of completely specified frequency occurs when each observation can fall into one of two categories. Examples are the toss of a coin, which may be heads or tails, and the quality of an item of product, which may be defective or non-defective. Suppose that a coin is tossed 100 times with the following results:

Heads	40
Tails	60
Number of tosses	100

It is desired to see whether the coin is fair, i.e., whether $p_1 = p_2 = \frac{1}{2}$. The expected number of heads or tails is 50. By the method of the previous section, it follows that

$$\chi^2 = \frac{(10)^2}{50} + \frac{(10)^2}{50} = \frac{200}{50} = 4,$$

where χ^2 has one degree of freedom. The critical value is obtained from Appendix Table 2, i.e., $\chi^2_{0.05;1} = 3.841$, so the hypothesis is rejected at the 5% level.

Actually, it would not be necessary to apply a chi-square test to these data as an exact test is available. The number of heads (or the number of tails) is a binomial random variable (see Sec. 4.5). The probability of obtaining k successes in n trials is given by

$$P\{X = k\} = \binom{n}{k} p^k q^{n-k}$$

where p is the individual trial probability and $q = 1 - p$. An exact calculation of the probability of departing by 10 or more heads from the expected number of 50, for $n = 100$ and $p = \frac{1}{2}$, is given by

$$\sum_{k=0}^{40} \binom{100}{k} p^k q^{100-k} + \sum_{k=60}^{100} \binom{100}{k} p^k q^{100-k} = \frac{\sum_{k=0}^{40} \binom{100}{k} + \sum_{k=60}^{100} \binom{100}{k}}{2^{100}}.$$

This probability can be evaluated directly. However, for this n, the normal approximation can be used by noting that

$$\frac{(X/n) - p}{\sqrt{pq/n}}$$

is approximately normal with zero mean and unit standard deviation. The approximate probability of a deviation as large as 10 heads would be found finally by calculating the standard normal deviate corresponding to 40, which is -2, and the normal deviate corresponding to 60, which is $+2$, and calculating the area in a normal curve outside the interval -2 to $+2$. This probability is 0.0456, slightly less than 0.05, so the hypothesis of a fair coin would be rejected at the 0.05 level.

The fact that the result based on the normal approximation to the binomial and the result based on χ^2 are similar is no accident; actually, the two methods are equivalent. Chi-square in general is the sum of squares of standard normal deviates, and chi-square with one degree of freedom is just the square of a normal deviate.

In general, consider a sample n classified into one of two categories, and let $p_1 = p$ and $p_2 = 1 - p = q$.

Category	Observed Frequency	Theoretical Frequency
1	O_1	np
2	O_2	nq
Total	n	n

Then O_1 would have a binomial distribution and the quantity

$$\frac{(O_1/n) - p}{\sqrt{pq/n}} = \frac{O_1 - np}{\sqrt{npq}}$$

would be approximately normal. The chi-square statistic would be

$$\chi^2 = \frac{(O_1 - np)^2}{np} + \frac{(O_2 - nq)^2}{nq}.$$

Since $O_2 = n - O_1$ and $q = 1 - p$,

$$\chi^2 = \frac{(O_1 - np)^2}{npq}$$

or just the square of the normal variable above. This special case is considered in detail to give an indication of the nature of the chi-square approximation and why it is used.

11.4.3 Test of Independence in a Two-Way Table

Often frequency data are tabulated according to two criteria, with a view toward testing whether the criteria are associated. Consider the following analysis of the 157 machine breakdowns during a given quarter.

Number of Breakdowns

	A	B	C	D	Total per Shift
Shift 1	10	6	12	13	41
Shift 2	10	12	19	21	62
Shift 3	13	10	13	18	54
Total per machine	33	28	44	52	157

It is of interest to determine whether the same percentage of breakdowns occurs on each machine during each shift or whether there is some difference due perhaps to untrained operators or other factors peculiar to a given shift.

If the number of breakdowns is independent of shifts and machines, the probability of a breakdown occurring in the first shift and in the first machine can be estimated by multiplying the proportion of first shift breakdowns by the proportion of machine A breakdowns, i.e.,

$$p_{11} = \frac{41}{157} \times \frac{33}{157} = 0.05489.$$

If there are 157 breakdowns, the expected number of breakdowns on this shift and machine is estimated as

$$E_{11} = 157 \times p_{11} = 8.6177.$$

Similarly, for the third shift and second machine

$$p_{32} = \frac{54}{157} \times \frac{28}{157} = 0.06134,$$

and

$$E_{32} = 157 \times p_{32} = 9.630.$$

This is done for all categories and

$$\chi^2 = \sum_{i=1}^{3} \sum_{j=1}^{4} \frac{(O_{ij} - E_{ij})^2}{E_{ij}}$$
$$= \frac{(10 - 8.6177)^2}{8.6177} + \frac{(6 - 7.3120)^2}{7.3120} + \cdots + \frac{(18 - 17.885)^2}{17.885}$$
$$= 2.02.$$

This is to be compared with the percentage point of χ^2 for $(3 - 1)(4 - 1)$ $= 6$ degrees of freedom, i.e., $\chi^2_{0.05;6} = 12.6$. Hence, the hypothesis is accepted and it is concluded that the data do not indicate different percentages of breakdowns on each machine during each shift.

In general, for an r by s table,

$$\chi^2 = \sum_{i=1}^{r} \sum_{j=1}^{s} \frac{(O_{ij} - E_{ij})^2}{E_{ij}},$$

which has an approximate chi-square distribution with $(r - 1)(s - 1)$ degrees of freedom.

11.4.4 COMPUTING FORM FOR TEST OF INDEPENDENCE IN A 2×2 TABLE

An important special case of the test of independence arises when both criteria of classification have two categories; the data may be represented in a 2×2 table. The data may be shown in the following way:

	First Criteria		
Second Criteria	a	b	$a + b$
	c	d	$c + d$
	$a + c$	$b + d$	

In this case the chi-square statistic, which has one degree of freedom, may be written

$$\chi^2 = \frac{(ad - bc)^2(a + b + c + d)}{(a + b)(a + c)(b + d)(c + d)}.$$

As an example, the following data are used to determine whether the same percentage of breakdowns occur on each machine using material from each supplier.

Number of Machine Breakdowns
Source of Raw Materials

Machine	Supplier A	Supplier B	
I	4	9	13
II	15	3	18
Total	19	12	31

$$\chi^2 = \frac{(12 - 135)^2(31)}{(13)(19)(12)(18)} = 8.8$$

$$\geq \chi^2_{0.05;1} = 3.84.$$

Therefore, the hypothesis is rejected that the percentage of breakdowns is the same for each machine using material from each supplier.

11.5 Comparison of Two Percentages

A problem which in many cases is equivalent to the test of independence in a 2×2 table is the problem of testing equality of two percentages. Suppose, following Wallis,[1] these data on two fire control devices are presented:

	Hits	Misses	
Old	3	197	200
New	4	196	200
			400

It is desired to test the hypothesis that the percentage of hits is the same, and rejection is desired when the new is superior, i.e., has a larger percentage of hits. If p_E denotes the observed proportion for the new experimental method, p_S the observed proportion for the standard, and n the number of trials with each method, then

$$U = \frac{\sqrt{n}(p_E - p_S)}{\sqrt{(p_E + p_S)\left(1 - \frac{p_E + p_S}{2}\right)}}$$

---

[1] Eisenhart, Hastay, Wallis, *Techniques of Statistical Analysis*, McGraw-Hill Book Company, Inc., New York, 1947.

has approximately the normal distribution with zero mean and unit standard deviation. Hence, the hypothesis is rejected when

$$U > K_\alpha.$$

For the data above

$$p_S = 0.015$$
$$p_E = 0.020$$

and

$$U = \frac{\sqrt{200}(0.005)}{\sqrt{(0.020 + 0.015)\left(1 - \dfrac{0.035}{2}\right)}} = 0.3795 < K_{0.05} = 1.645.$$

Thus, the hypothesis is accepted and it is concluded that the data do not reveal that the new device is better than the old one.

In terms of the notation of the last section,

$$U = \frac{\sqrt{a + b + c + d}\,(ad - bc)}{\sqrt{(a + b)(a + c)(b + d)(c + d)}}.$$

If rejection is desired whenever the proportions differ, the test procedure should call for rejection when

$$|U| > K_{\alpha/2}$$

or when $U^2 > K_{\alpha/2}^2$. This is exactly the same as analyzing the data by the method used in Sec. 11.4.4.

11.6 Confidence Intervals for Proportion

Exact confidence intervals for a proportion were not included in Chapter 8, Estimation, since most of the test procedures discussed up to that point depend on the normality of the observations. Yet this topic is of considerable importance in applied statistics. One may want to estimate the proportion of people who will vote for a particular candidate, buy a new detergent, prefer movies to television, etc. In manufacturing, estimation of the proportion of defective items or the proportion of rejects to be expected is frequently required. In Sec. 8.2.5, a "process control" example was presented where the probability of a unit being defective, p, was considered to be the parameter of a Bernoulli random variable. Alternatively, if items are formed into lots, and the lot size is large, p is a good approximation to the proportion of defective items in the lot. If a random sample of size n of Bernoulli random variables is taken from the process (or the lot), the number of defective units X, in the sample has a binomial distribution with parameters n and p (see, Sec. 4.5), i.e.,

$$P\{X = k\} = \binom{n}{k}p^k q^{n-k}, \qquad \text{for } k = 0, 1, \cdots, n.$$

The parameter p can be viewed as the parameter of a Bernoulli random variable or, equivalently, as a parameter of the induced binomial distribution. In either case, a "good" point estimator is given by $\hat{p} = X/n$. In Sec. 8.3.8, an approximate confidence interval estimate was presented for large n, using the asymptotic properties of maximum likelihood estimators. The $100(1 - \alpha)\%$ approximate confidence interval was given by

$$\hat{p} \pm K_{\alpha/2} \sqrt{\frac{\hat{p}(1 - \hat{p})}{n}},$$

where $K_{\alpha/2}$ is the $\alpha/2$ percentage point of the normal distribution.

For small n, these results may be inadequate and an exact confidence interval for p is desirable. Such an interval can be obtained as follows: Let the lower confidence limits for p be denoted by $\underline{p}$ and the upper by $\bar{p}$. It is possible to obtain exact expressions for $\underline{p}$ and $\bar{p}$ in terms of the percentage points of the F distribution given in Appendix Table 4. In particular, if x defective items are observed in a random sample of size n, a $100(1 - \alpha)$ confidence interval is given by

$$\bar{p} = \frac{(x + 1)F_{\alpha/2;\, 2(x+1),\, 2(n-x)}}{(n - x) + (x + 1)F_{\alpha/2;\, 2(x+1),\, 2(n-x)}}$$

$$\underline{p} = \frac{x}{x + (n - x + 1)F_{\alpha/2;\, 2(n-x+1),\, 2x}}.$$

For example, suppose that 4 defects are observed in a sample of 25. The point estimate is $\hat{p} = 0.16$. To find a 95% confidence interval for p with $\alpha \doteq 0.05$, $n = 25$, $x = 4$,

$$F_{0.025;\, 10,\, 42} = 2.37$$

$$F_{0.025;\, 44,\, 8} = 3.82.$$

The interval is then given by

$$\bar{p} = \frac{5(2.37)}{21 + 5(2.37)} = 0.361$$

$$\underline{p} = \frac{4}{4 + 22(3.82)} = 0.045.$$

PROBLEMS

1. In the example of Sec. 7.2.3, plot the data for the new supplier on normal probability paper.

2. In the example of Sec. 7.2.3, plot the data for the old supplier on normal probability paper.

3. Using the data in Problem 1, perform the Kolmogorov-Smirnov test at the 5% level of significance to determine whether the data can be assumed to come from a normal distribution with mean 550 and variance 100 .

4. Using the data in Problem 2, perform the Kolmogorov-Smirnov test at the 1% level of significance to determine whether the data can be assumed to come from a normal distribution with mean 550 and variance 100.

5. Using the data in Problem 1, perform the chi-square goodness of fit test at the 5% level of significance to determine whether the data can be assumed to come from a normal distribution. Note that the required calculations for obtaining the maximum likelihood estimates have been performed in Sec. 7.2.3.

6. Using the data in Problem 2, perform the chi-square goodness of fit test at the 1% level of significance to determine whether the data can be assumed to come from a normal distribution. Note that the required calculations for obtaining the maximum likelihood estimates have been performed in Sec. 7.2.3.

7. The failures in a week's around-the-clock run are distributed over four shift groups as follows:

Shift *A*	10
Shift *B*	9
Shift *C*	7
Shift *D*	14

Is one justified in saying that Shift *D* has too many failures compared with the other shifts? (Choose $\alpha = 0.05$.)

8. In a game of craps, a suspicious player tallies the results of 100 throws by a particular opponent as follows:

Result of Throw	Observed Frequency
11	12
7	18
other	70
	100

Would you say his suspicions are justified? (Choose $\alpha = 0.05$.)

9. Two inspectors are testing pumps. It is suspected that one of the men was departing from the standard test procedure with a resulting bias toward reporting an undue number of failures. A record was kept of the same number of tests made by the two men in the same period of time.

Frequency of Pump Failures

Operator *A*	158
Operator *B*	118

Do the operators appear to differ? (Choose $\alpha = 0.01$.)

10. The past output of a machine indicates that an output of 500 pieces would be graded as follows:

Expected Output

Top grade	200
High grade	150
Medium grade . .	100
Low grade.	50
	500

A new machine is designed and built to do the same job. A sample of 500 pieces from the new machine is graded as follows:

Top grade	231
High grade	120
Medium grade ..	82
Low grade......	67
	500

Can the difference in the pattern of output be ascribed to chance? (Choose $\alpha = 0.01$.)

11. In a mass production process, a sample of 200 was taken from each day's production and the number of defective items was calculated:

	Number of Defects
Monday	12
Tuesday	16
Wednesday	8
Thursday	16
Friday	10

Test the hypothesis that the proportion of defects is constant from day to day. (Choose $\alpha = 0.05$.)

12. An experiment on rockets yields the following data on the characteristics of lateral deflection and ranges.

	Lateral Deflection		
Range	Left	Normal	Right
0–1199	12	15	14
1200–1799	9	6	11
1800–2699	16	17	8

Test the hypothesis that deflection and range are independent. (Choose $\alpha = 0.01$.)

13. The table below shows the number of defective and acceptable items in samples both before and after the introduction of a modification intended to improve the process of manufacturing. The proportion of defectives has dropped and it is desired to determine if this change is significant at the 0.01 level.

	Defective	Acceptable
Before	25	217
After	8	92

14. A study is made on the failures of vacuum tubes. There are four types of failures and two blocks of tubes in the equipment. Test the hypothesis that the type of failure is distributed independently of the location of the tube at the 0.05 level.

	Type of Failure			
	A	B	C	D
Top block	75	10	15	20
Bottom block	40	30	10	17

15. In a study of tire failures during a road test of wearing strength, the following results were obtained.

| | Number of Failures | | |
	Front	Rear	Total
Left	115	65	180
Right	125	95	220
Total	240	160	400

Test, at the 1 % level, the hypothesis that left-right wear is independent of front-rear wear (See Acheson J. Duncan, "Chi-Square Tests of Independence and the Comparison of Percentages," *Industrial Quality Control*, June, 1955, p. 9.)

16. The result of testing 164 fuses at two temperature levels is:

| | Fuses Tested | |
Result of Test	Temp. 1	Temp. 2
Successes	111	92
Failures	16	9

Is there a significant difference at the 5 % level in the proportion of failures at the two levels?

17. The grade distribution of three sections, taught by different instructors, of a course based on this book at Stanford University is:

Instructor	A	B	C	Other	Total
L	27	34	36	10	107
B	18	29	27	8	82
T	17	19	22	7	65

Do the instructors differ in the percentages of grades? (Choose $\alpha = 0.01$.)

18. Fabric is often inspected and graded into three classifications: *A, B, C*. In a study of four looms, the following results were obtained:

| | Number of Pieces of Fabric in | | |
Loom	Grade A	B	C
1	191	22	18
2	184	26	15
3	157	32	11
4	170	14	8

Is the output homogeneous or does it depend on the loom? (Choose $\alpha = 0.05$.)

19. A vaccine supposed to reduce absenteeism caused by flu was given to 167 individuals in a plant; 214 people were untreated. The results were:

	Caught Flu	Well	Total
Treated....	17	150	167
Untreated..	38	176	214

Was the vaccine worthwhile? (Choose $\alpha = 0.05$.)

20. Grades in an English course and a statistics course taken simultaneously were as follows for a sample of students:

| | English Grade | | | |
Statistics Grade	A	B	C	Other
A	20	10	17	8
B	17	16	18	7
C	19	4	15	12
Other	12	8	12	23

Are the grades in statistics and English related? (Choose $\alpha = 0.05$.)

21. In a sample of 275 items from a process, eight defects were found. Find a 99% confidence interval for the proportion defective in the lot.

22. In a sample of 712 college students, 414 were found to be in favor of selling beer on the campus. Find a 95% confidence interval for the number in favor. Would you be sure that a majority favored it?

23. In a sample of 400 pieces of cloth, 306 were found to be grade A. Compare the exact and the normal approximation methods for finding 95% confidence intervals for the true proportion of grade A cloth.

24. In the test of a vaccine, 27 out of 1,000 people tested were found to have an allergic reaction. Find a 95% confidence interval for the true proportion of people allergic.

25. The proportion of defectives in a random sample of 200 items checked on a go/no-go gauge is 9%. What are the 95% confidence limits for the proportion defective for the total population?

26. In 100 throws of dice the number 11 was found to appear 14 times. Does the 95% confidence interval for the true probability include the probability of obtaining 11, if the dice are unbiased?

12

STATISTICAL QUALITY CONTROL:
CONTROL CHARTS

12.1 Introduction

The previous chapters have emphasized that there always exists some inherent, uncontrollable source of variation associated with any process or experiment. Although attempts are made to insure that this natural variation is as small as possible, it cannot be eliminated completely. In applying this to the quality of manufactured items, this natural inherent variability is often referred to as a "stable system (or pattern) of chance causes",[1] and is viewed as being an acceptable source of variation. However, any variation in excess of this natural pattern is unacceptable and its detection and correction is required. These variations outside the stable pattern are known as *assignable causes* of quality variation. A process operating in the absence of any assignable causes of erratic fluctuations is said to be in *statistical control*.

To the manufacturer, the primary purpose of a control chart is to provide a basis for *action*. The introduction of a control chart aids in determining the capabilities of the production process. Action is taken when these estimated capabilities are unsatisfactory in relation to the design specifications. Furthermore, once the process capabilities have been determined, and are satisfactory, action is taken only when the control chart indicates that the process has fallen out of control, e.g., assignable causes of variation have entered.

There are several types of control charts that can be constructed. If data are obtained for a quality characteristic that can be measured and expressed

---

[1] For example, see E. L. Grant, *Statistical Quality Control*, 3rd ed., McGraw-Hill Book Co., Inc., New York, 1964.

in numbers, control charts are generally required for measures of central tendency and variability since the quality of a product can often be summarized in terms of these two quantities. An $\bar{X}$ chart is frequently used for "controlling" central tendency, whereas an R chart (or σ chart) is frequently used for "controlling" variability. If data are obtained for a quality characteristic by classifying an item simply as defective or non-defective, a control chart is required for "controlling" the fraction defective, and is frequently accomplished by means of a p chart. Finally, if *data* are obtained for a quality characteristic by counting the number of defects appearing on a unit, a control chart is required for "controlling" the average number of defects per unit, and is frequently accomplished by means of a c chart.

12.2 An Overview of Control Charts

The essential feature of the control chart method is the drawing of inferences about the production process on the basis of samples drawn from the production line. The success of the technique depends on grouping observations under consideration into subgroups or samples, within which a stable system of chance causes is operating, and between which the variations may be due to assignable causes whose presence is suspected or considered possible. Such a grouping is called a rational subgroup. Order of production is one of the more commonly used bases for obtaining rational subgroups. If items are coming from more than one source, the source may be a basis for rational subgrouping. The size of the subgroup or sample usually is not less than 4. In industry, 5 seems to be the most common.[1] It is preferable that all samples be of equal size.

The concept of a control chart is simple. Each subgroup, within which a stable system of chance causes is operating, can be interpreted as a random sample. The data obtained for a quality characteristic from each subgroup is then graphed on a chart (vertical axis), generally as a function of time (horizontal axis). "Control limits" are obtained for these data and shown on this chart. These limits should contain "almost all" the plotted points, provided that the process is in control. As long as points fall within these limits, the process is assumed to be in control and no action is required. As soon as a point falls outside, this is evidence that the process is no longer in control and corrective action is called for. This procedure is similar to testing the hypothesis that the process is in control, with each subgroup playing the role of a random sample. However, it differs from a usual test in that the procedure is dynamic, i.e., each subgroup (random sample) provides a test of the hypothesis so that repeated tests are being made.

---

[1] When characteristics are classified into two groups, those containing defects and those not containing defects, and no other measurement is recorded, the sample size is usually much larger.

12.3 Control Chart for Variables: $\bar{X}$ Charts

The notation in this chapter and in the ensuing chapter differs from the notation used throughout the book. This is done so that the notation in this chapter on control charts coincides with that recommended by the American Society for Quality Control. The basic changes are in the symbols for the parameters of the normal distribution. In the previous sections, the symbols μ and σ represented the mean and standard deviation of the normal distribution. In this chapter these symbols are replaced by $\bar{X}'$ and σ', respectively. The "prime" always represents a parameter of a probability distribution. $\bar{X}$ still denotes the sample average and is used as an estimate of the mean of the normal distribution. However, the symbol σ is now used as an estimate of the standard deviation of the normal distribution, whereas in the previous chapters the symbol actually represented this parameter.

12.3.1 STATISTICAL CONCEPTS

In Chapter 3 it was shown that if the measurable characteristic of an item is normally distributed, with mean $\bar{X}'$ and standard deviation σ', it is possible to find the probability that it will lie in a fixed interval by referring to Appendix Table 1. Furthermore, if this measurable characteristic is denoted by X, the probability that X will fall within the interval $[\bar{X}' - 3\sigma', \bar{X}' + 3\sigma']$ is 0.9973. In other words, if a plot such as that in Fig. 12.1 is made, on the average only 27 observations out of 10,000 will fall outside the above interval, *provided the distribution of the variable is normal with mean $\bar{X}'$ and standard deviation σ'*. If this is the case, it follows that the distribution of averages of subgroups of size n is also normal with mean $\bar{X}'$ and standard deviation $\sigma'_{\bar{x}} = \sigma'/\sqrt{n}$. Denoting the n observations by $X_1, X_2, X_3, \ldots, X_n$, the average of these n observations, $\bar{X}$, is defined as

$$\bar{X} = \frac{X_1 + X_2 + \cdots + X_n}{n}.$$

If, instead of plotting individual values as in Fig. 12.1, averages of samples of size n are plotted, it is expected that on the average only 27 in 10,000 values of these averages will fall outside the interval

$$\bar{X}' \pm 3\sigma'_{\bar{x}} = \bar{X}' \pm \frac{3\sigma'}{\sqrt{n}}.$$

The plot in Fig. 12.2 is called a control chart for $\bar{X}$. Here $\bar{X}' + 3\sigma'/\sqrt{n}$ is referred to as the upper control limit (UCL) while $\bar{X}' - 3\sigma'/\sqrt{n}$ is referred to as the lower control limit (LCL).

In this discussion it has always been assumed that the underlying distribution is normal. Although in actual practice many distributions of observations are "nearly" normally distributed, many others do not resemble a normal distribution at all. However, if sample averages are considered as

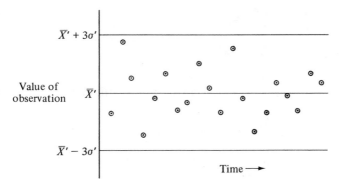

Fig. 12.1. *Plot of individual observations.*

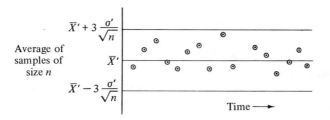

Fig. 12.2. *Control chart for $\bar{X}$.*

the variable plotted on the control chart, the central limit theorem (see Chapter 3) enables one to use the properties of the normal distribution even if the form of the underlying distribution is unknown (when n is large). It is evident, then, that the central limit theorem is one of the keys to the success of the control chart for $\bar{X}$.

12.3.2 ESTIMATE OF $\bar{X}'$

In Fig. 12.2, a control chart for $\bar{X}$ is shown, where the control limits are drawn in as functions of $\bar{X}'$ and σ'. In most practical applications $\bar{X}'$ and σ' are not known, and consequently estimates of these parameters must be obtained. It is desirable that these estimates be based on at least 25 subgroups of n observations. Naturally, the larger the number of subgroups, the better the estimates of the parameters of the distribution, provided the process is in control.

Suppose that observations are to be obtained from k subgroups, each of size n. Denoting the averages of the subgroups by $\bar{X}_1, \bar{X}_2, \ldots, \bar{X}_k$, the best estimator of $\bar{X}'$, the mean of the probability distribution, is

$$\bar{\bar{X}} = \frac{\bar{X}_1 + \bar{X}_2 + \cdots + \bar{X}_k}{k},$$

where $\bar{X}$ is also the average of all the nk observations. Furthermore, $\bar{\bar{X}}$ is an unbiased estimator of $\bar{X}'$.

12.3.3 ESTIMATE OF σ' BY $\bar{\sigma}$

As was pointed out in the previous section, it is usually necessary to estimate σ', the standard deviation of the distribution. Define, for each subgroup,

$$
\sigma = \sqrt{\frac{(X_1 - \bar{X})^2 + (X_2 - \bar{X})^2 + \cdots + (X_n - \bar{X})^2}{n}}
$$
$$
= \sqrt{\frac{X_1^2 + X_2^2 + \cdots + X_n^2 - n\bar{X}^2}{n}}.
$$

Also define $\bar{\sigma} = (\sigma_1 + \sigma_2 + \cdots + \sigma_k)/k$, where $\sigma_1, \sigma_2, \ldots, \sigma_k$ are the values of σ computed from the $1, 2, \ldots, k$ subgroups, respectively. An unbiased estimator of σ' is given by $\bar{\sigma}/c_2$, where values of c_2 for different n can be found in Table 12.1. The quantity c_2 is determined such that $E(\bar{\sigma}/c_2) = \sigma'$ or $E(\bar{\sigma})/\sigma' = c_2$.

To summarize, the estimated central line for the $\bar{X}$ control chart is $\bar{\bar{X}}$, and the estimated control limits are $\bar{\bar{X}} \pm 3\bar{\sigma}/(\sqrt{n}\,c_2)$. Values of $3/(\sqrt{n}\,c_2) = A_1$ are given in Table 12.1 so that the estimated control limits can be written as $\bar{\bar{X}} \pm A_1\bar{\sigma}$.

Formulas for Central Lines and Control Limits

Statistic	Standards Given		Analysis of Past Data	
	Central Line	Limits	Central Line	Limits
Average, using σ'	$\bar{X}'$	$\bar{X}' \pm A\sigma$	$\bar{\bar{X}}$	$\bar{\bar{X}} \pm A_1\bar{\sigma}$
Average, using R	—	—	$\bar{\bar{X}}$	$\bar{\bar{X}} \pm A_2\bar{R}$
Standard deviation	$c_2\sigma'$	$B_1\sigma', B_2\sigma'$	$\bar{\sigma}$	$B_3\bar{\sigma}, B_4\bar{\sigma}$
Range	$d_2\sigma'$	$D_1\sigma', D_2\sigma'$	$\bar{R}$	$D_3\bar{R}, D_4\bar{R}$

12.3.4 ESTIMATE OF σ' BY $\bar{R}$

Define, as the range (R) of a subgroup of n observations, the difference between the largest and smallest values. Let $\bar{R}$ be the average of the ranges of the k subgroups, i.e.,

$$
\bar{R} = \frac{R_1 + R_2 + \cdots + R_k}{k}.
$$

Another unbiased estimator of σ' is $\bar{R}/d_2$. Values of d_2 for different values of n can also be found in Table 12.1. The quantity d_2 is determined such that $E(\bar{R}/d_2) = \sigma'$ or $E(\bar{R})/\sigma' = d_2$.

Table 12.1. Factors for Computing Control Chart Lines[1]

| | Chart for Averages | | | | | Chart for Standard Deviations | | | | | | Chart for Ranges | | | | |
| | Factors for Control Limits | | | Factors for Central Line | | Factors for Control Limits | | | | Factors for Central Line | | | Factors for Control Limits | | | |
Number of Observations in Sample, n	A	A_1	A_2	c_2	$1/c_2$	B_1	B_2	B_3	B_4	d_2	$1/d_2$	d_3	D_1	D_2	D_3	D_4
2	2.121	3.760	1.880	0.5642	1.7725	0	1.843	0	3.267	1.128	0.8865	0.853	0	3.686	0	3.276
3	1.732	2.394	1.023	0.7236	1.3820	0	1.858	0	2.568	1.693	0.5907	0.888	0	4.358	0	2.575
4	1.501	1.880	0.729	0.7979	1.2533	0	1.808	0	2.266	2.059	0.4857	0.880	0	4.698	0	2.282
5	1.342	1.596	0.577	0.8407	1.1894	0	1.756	0	2.089	2.326	0.4299	0.864	0	4.918	0	2.115
6	1.225	1.410	0.483	0.8686	1.1512	0.026	1.711	0.030	1.970	2.534	0.3946	0.848	0	5.078	0	2.004
7	1.134	1.277	0.419	0.8882	1.1259	0.105	1.672	0.118	1.882	2.704	0.3698	0.833	0.205	5.203	0.076	1.924
8	1.061	1.175	0.373	0.9027	1.1078	0.167	1.638	0.185	1.815	2.847	0.3512	0.820	0.387	5.307	0.136	1.864
9	1.000	1.094	0.337	0.9139	1.0942	0.219	1.609	0.239	1.761	2.970	0.3367	0.808	0.546	5.394	0.184	1.816
10	0.949	1.028	0.308	0.9227	1.0837	0.262	1.584	0.284	1.716	3.078	0.3249	0.797	0.687	5.469	0.223	1.777
11	0.905	0.973	0.285	0.9300	1.0753	0.299	1.561	0.321	1.679	3.173	0.3152	0.787	0.812	5.534	0.256	1.744
12	0.866	0.925	0.266	0.9359	1.0684	0.331	1.541	0.354	1.646	3.258	0.3069	0.778	0.924	5.592	0.284	1.719
13	0.832	0.884	0.249	0.9410	1.0627	0.359	1.523	0.382	1.618	3.336	0.2998	0.770	1.026	5.646	0.308	1.692
14	0.802	0.848	0.235	0.9453	1.0579	0.384	1.507	0.406	1.594	3.407	0.2935	0.762	1.121	5.693	0.329	1.671
15	0.775	0.816	0.223	0.9490	1.0537	0.406	1.492	0.428	1.572	3.472	0.2880	0.755	1.207	5.737	0.348	1.652
16	0.750	0.788	0.212	0.9523	1.0501	0.427	1.478	0.448	1.552	3.532	0.2831	0.749	1.285	5.779	0.364	1.636
17	0.728	0.762	0.203	0.9551	1.0470	0.445	1.465	0.466	1.534	3.588	0.2787	0.743	1.359	5.817	0.379	1.621
18	0.707	0.738	0.194	0.9576	1.0442	0.461	1.454	0.482	1.518	3.640	0.2747	0.738	1.426	5.854	0.392	1.608
19	0.688	0.717	0.187	0.9599	1.0418	0.477	1.443	0.497	1.503	3.689	0.2711	0.733	1.490	5.888	0.404	1.596
20	0.671	0.697	0.180	0.9619	1.0396	0.491	1.433	0.510	1.490	3.735	0.2677	0.729	1.548	5.922	0.414	1.586
21	0.655	0.679	0.173	0.9638	1.0376	0.504	1.424	0.523	1.477	3.778	0.2647	0.724	1.606	5.950	0.425	1.575
22	0.640	0.662	0.167	0.9655	1.0358	0.516	1.415	0.534	1.466	3.819	0.2618	0.720	1.659	5.979	0.434	1.566
23	0.626	0.647	0.162	0.9670	1.0342	0.527	1.407	0.545	1.455	3.858	0.2592	0.716	1.710	6.006	0.443	1.557
24	0.612	0.632	0.157	0.9684	1.0327	0.538	1.399	0.555	1.445	3.895	0.2567	0.712	1.759	6.031	0.452	1.548
25	0.600	0.619	0.153	0.9696	1.0313	0.548	1.392	0.565	1.435	3.931	0.2544	0.709	1.804	6.058	0.459	1.541
Over 25	$\dfrac{3}{\sqrt{n}}$	$\dfrac{3}{\sqrt{n}}$	—	—	—	‡	§	‡‡	§	—	—	—	—	—	—	—

[1] Reproduced by permission from *ASTM Manual on Quality Control of Materials*, American Society for Testing Materials, Philadelphia, Pa., 1951.

‡ $1 - \dfrac{3}{\sqrt{2n}}$ § $1 + \dfrac{3}{\sqrt{2n}}$

477

In estimating the control limits for $\bar{X}$ with $\bar{R}$ as an estimate of σ', the limits become $\bar{\bar{X}} \pm 3\bar{R}/(\sqrt{n}\,d_2)$. Values of $3/(\sqrt{n}\,d_2) = A_2$ are given in Table 12.1 so that the estimated control limits can be written as $\bar{\bar{X}} \pm A_2\bar{R}$.

It must be pointed out that although $\bar{R}$ is simpler to calculate than $\bar{\sigma}$, the estimator of σ' based on $\bar{\sigma}$ is a better estimator in the sense of having a smaller variance; i.e., it is more efficient.

12.3.5. STARTING A CONTROL CHART FOR $\bar{X}$

It has been indicated that observations should be classified into rational subgroups, and each subgroup should contain at least 4 observations. The determination of the minimum number of subgroups is a compromise between obtaining the guidance of the control chart as quickly as possible, and the desire for the guidance to be reliable. Usually, at least 25 subgroups should be chosen.

With the data at hand, trial control limits can be calculated. Estimates of $\bar{X}'$ and σ' can be obtained in the manner described previously. The trial control limits are then $\bar{\bar{X}} \pm A_1\bar{\sigma}$ or $\bar{\bar{X}} \pm A_2\bar{R}$, depending on which estimator of σ' is chosen. If σ' and/or $\bar{X}'$ are known, the control limits should be calculated, using the known values; e.g., if both are known, $\bar{X}' \pm 3\sigma'/\sqrt{n}$. Values of $3/\sqrt{n} = A$ are given in Table 12.1, so that the control limits can be written as $\bar{X}' \pm A\sigma'$.

Returning to the case where the parameters $\bar{X}'$ and σ' are unknown, and estimates are computed, some modifications may have to be made in the trial control limits. Lack of control is usually indicated by points falling outside these limits. If all the points fall within the limits, it is concluded that the process is in control. However, there is still no assurance that assignable causes of variation are absent, and that this is a constant cause system. It merely means that for practical purposes it pays to act as if no assignable causes of variation are present, while realizing that an error of judgment is quite possible.

The trial control limits serve the purpose of determining whether past operations are in control. To continue using these limits as a basis for action on future production may require a revision of the trial limits, especially if a lack of control is exhibited by points falling outside the trial control limits. If the process is not in control, all the points do not come from a stable distribution. However, it is evident that future control limits should be based on data from a controlled process. As a practical rule, then, those points falling outside the trial control limits are eliminated, and new trial control limits are computed, using the remaining points. This procedure may be continued until all points fall within the control limits. There is no theoretical justification for this, other than the fact that those points falling outside the trial control limits are more likely to belong to another probability distribution; i.e., they may be due to some assignable cause.

The control chart when used as a basis for action on future production may be set, using aimed-at values of $\bar{X}'$ and/or σ'. In this case, the control limits are modified by using the aimed-at values as if they were the known values in the computations. If this is done, and if in the future the control chart exhibits a lack of control, the trouble may be due to the process failing to be in control when using the aimed-at values, although it is in control at some other level. For example, suppose that the probability distribution of the measurable quantity of an item being produced has a mean, $\bar{X}'$, equal to 20. If the aimed-at value is $\bar{X}' = 25$, the control chart based on $\bar{X}' = 25$ will exhibit a lack of control. Suppose the mean can be shifted by a change in machine setting. The interpretation, then, is that there is an assignable cause present, namely, the machine setting, preventing the production process from operating at the aimed-at value. Yet, with the present machine setting, the process is in control at $\bar{X}' = 20$. The term "state of control" should be interpreted in this light. Aimed-at values of $\bar{X}'$ or σ' are often used when they can be achieved by a simple adjustment of the machine.

12.3.6 Relation between Natural Tolerance Limits and Specification Limits

After ascertaining at what level the process is in control by means of the control chart, it remains to determine whether or not the process can meet the specification limits set for the item. In Secs. 12.3.3 and 12.3.4 it was shown that σ' can be estimated from $\bar{\sigma}$ or $\bar{R}$; i.e., the estimator of σ' is $\bar{\sigma}/c_2$ or $\bar{R}/d_2$. Furthermore, $\bar{X}'$ can be estimated from $\bar{X}$. Setting $\bar{X}'$ and σ' equal to the values that the estimators take on and assuming that these are the correct values, it follows that almost all the individual items will lie between the known limits $\bar{X}' \pm 3\sigma'$, i.e., on the average, 9,973 out of 10,000, provided the distribution is normal. Under this assumption, if the natural tolerances are defined to mean those limits containing all but 27 in 10,000 values, $\bar{X}' \pm 3\sigma'$ will coincide with the natural tolerances. On the other hand, if the natural tolerances are defined to mean those limits containing all but 1 in 1,000, the known limits $\bar{X}' \pm 3.29\sigma'$ will coincide with the natural tolerances. Once the meaning of natural tolerances is established and $\bar{X}'$ and σ' (or their estimates) are determined, the natural tolerance limits can be computed. These natural tolerances are then compared with the specification limits. If the natural tolerances are not included within the specification limits, the process must be readjusted with respect to either $\bar{X}'$ or σ', or both, or the specification limits must be changed. If the natural tolerances are included well within the specification limits, this often signifies unrealistic specification limits, or increased labor and material costs resulting in a product which is "too good." After a period of time, a process in equilibrium usually has the tolerance limits coinciding with the specification limits or operating just within the specification limits.

As an example, suppose the designer indicates specification limits of 20 $\pm$ 2, and natural tolerance limits are defined as those limits excluding only 1 in 1,000. If $\bar{X}'$ is 20.40 and $\sigma' = 1$, it follows that on the average 999 out of 1,000 items will fall between $(\bar{X}' - 3.29\sigma', \bar{X}' + 3.29\sigma') = (17.11, 23.69)$. The specification limits are given as (18, 22). Consequently, the process cannot meet the specification limits, and either the process must be changed or the specification limits revised. Of course, in practical situations, $\bar{X}'$ and σ' are never really known and estimates are used. If the estimates are good, these procedures are adequate. However, if the estimates are from small samples, a technique such as that described in Chapter 8, using statistical tolerance limits, is in order.

12.3.7 Interpretation of Control Charts for $\bar{X}$

As long as the process is in control, almost all the values of $\bar{X}$ will fall within the control limits. In this case, no action need be taken. A point falling outside these limits is a signal to hunt for trouble. In some instances, production may be stopped until a source of trouble has been discovered. Occasionally, approximately 27 in 10,000 times, an error will be made, in that trouble will be sought even though nothing has gone wrong with the process. If something has gone wrong with the process, points will begin to fall outside the control limits. For example, a shift in the mean $\bar{X}'$ (σ' remaining constant) will result in points falling outside the control limits. This is usually indicated by points falling above or below the limits (but rarely both), depending on whether the shift is in the positive or negative direction. On the other hand, an increase in σ' ($\bar{X}'$ remaining constant) will result in points falling above *and* below the control limits. In addition to examining the process, when points fall outside the control limits, the control limits themselves should be re-examined, and perhaps brought up to date.

It has been pointed out that an error may be committed when action is taken after a point falls outside the control limits. There is a small probability that a point will fall outside the control limits even though the process is in control; i.e., if $3\sigma'_{\bar{x}}$ limits are used, the probability is 0.0027. This probability of falling outside the control limits when the process is in control is known as the probability of committing a *Type I error*. Similarly, if the process goes out of control, there is a probability greater than 0 that a point will fall within the control limits. This is known as the probability of a *Type II error*. These two errors are related. A decrease in one results in an increase in the other. An increase in one results in a decrease in the other. The use of $3\sigma'_{\bar{x}}$ limits implies a probability of a Type I error of 0.0027. If the process jumps out of control, and the new level is specified, the probability of a Type II error can be computed. If $2\sigma'_{\bar{x}}$ limits were used instead of $3\sigma'_{\bar{x}}$, the probability of a Type I error is increased to about 0.0455, but the probability of a Type II error is decreased.

These are the same concepts that were applied in the chapters on significance tests, i.e., Chapters 5, 6, and 7. The control chart is a procedure used to test the hypothesis that the process is in control. This procedure has an OC curve with its associated Type I and Type II errors, just as any other procedure.

For example, suppose $\bar{X}'$ is 25, and $\sigma' = 1$, and there are 4 observations in each subgroup. The control limits are then $\bar{X}' \pm A\sigma' = 25 \pm \frac{3}{2}$. The probability of a Type I error is 0.0027. If the mean shifts to 27, the probability of a point falling inside the control limits of $25 \pm \frac{3}{2}$ is 0.1587. On the other hand, if $2\sigma'_{\bar{x}}$ limits are used, thereby resulting in control limits of 25 ± 1, the probability of a Type I error is increased to 0.0455, whereas the probability of committing a Type II error is only 0.0228.

It is evident that Type I errors or Type II errors (but not both) can be made as small as is desirable at the expense of increasing the other type of error. It is also evident that using $3\sigma'_{\bar{x}}$ limits is conservative from the point of view of considering the Type I error. A point falling outside the control limits is almost sure evidence that the process is no longer in control. On the other hand, one cannot be assured that a small shift has not taken place even if the points fall within the control limits.

12.4 R Charts and σ Charts

12.4.1 STATISTICAL CONCEPTS

Although both R and σ do not have normal distributions, both of these functions of a sample are random variables. The probability distribution of σ is related to the chi-square distribution, and the distribution of R is approximately related to the chi-square distribution. Furthermore, it can be shown that almost all of the probability distribution of each is contained within the limits bounded by the expected value of the variable plus and minus three standard deviations (of the variable). Consequently, for control chart purposes, it is only necessary to calculate both the expected value and standard deviation of these random variables.

The expected value of σ is $c_2\sigma'$. The standard deviation of σ is

$$\sigma_\sigma = [2(n - 1) - 2nc_2^2]^{1/2}\frac{\sigma'}{\sqrt{2n}}.$$

The control limits are then $c_2\sigma' \pm 3\sigma_\sigma$, which can be written as

$$\sigma'\left(c_2 \pm \frac{3}{\sqrt{2n}}[2(n - 1) - 2nc_2^2]^{1/2}\right).$$

The factors

$$B_2 = c_2 + \frac{3}{\sqrt{2n}}[2(n-1) - 2nc_2^2]^{1/2}$$

and

$$B_1 = c_2 - \frac{3}{\sqrt{2n}}[2(n-1) - 2nc_2^2]^{1/2}$$

can be obtained from Table 12.1. Thus,

$$\text{UCL}_\sigma = B_2\sigma'; \qquad \text{LCL}_\sigma = B_1\sigma'.$$

When σ' is unknown, it can be estimated from $\bar{\sigma}/c_2$, so that the estimated control limits are written as

$$\bar{\sigma}\left(1 \pm \frac{3}{\sqrt{2n}\,c_2}[2(n-1) - 2nc_2^2]^{1/2}\right).$$

The factors

$$B_4 = 1 + \frac{3}{\sqrt{2n}\,c_2}[2(n-1) - 2nc_2^2]^{1/2}$$

and

$$B_3 = 1 - \frac{3}{\sqrt{2n}\,c_2}[2(n-1) - 2nc_2^2]^{1/2}$$

are also given in Table 12.1. The estimated control limits are then

$$\text{UCL}_\sigma = B_4\bar{\sigma}; \qquad \text{LCL}_\sigma = B_3\bar{\sigma}.$$

Note that the center line is $c_2\sigma'$ if σ' is known, and $\bar{\sigma}$ if σ' is unknown.

The control limits for R are obtained in a similar manner. The expected value of R is $d_2\sigma'$. The standard deviation of R can be expressed as $\sigma_R = d_3\sigma'$, where values of d_3 are given in Table 12.1. The control limits are then $d_2\sigma' \pm 3\sigma_R$, which can be written as $\sigma'(d_2 \pm 3d_3)$. The factors

$$D_2 = d_2 + 3d_3; \qquad D_1 = d_2 - 3d_3$$

can be obtained from Table 12.1. Thus,

$$\text{UCL}_R = D_2\sigma'; \qquad \text{LCL}_R = D_1\sigma'.$$

When σ' is unknown, it can be estimated from $\bar{R}/d_2$ so that the estimated control limits are written as

$$\bar{R}\left[1 \pm \frac{3d_3}{d_2}\right].$$

The factors

$$D_4 = 1 + \frac{3d_3}{d_2}; \qquad D_3 = 1 - \frac{3d_3}{d_2}$$

can be obtained from Table 12.1. The estimated control limits are then

$$\text{UCL}_R = D_4\bar{R}; \qquad \text{LCL}_R = D_3\bar{R}.$$

The center line is $d_2\sigma'$ if σ' is known, and $\bar{R}$ if σ' is unknown.

12.4.2 SETTING UP A CONTROL CHART FOR R OR σ

The range of each subgroup is obtained and $\bar{R}$ calculated from these values. The trial control limits are given by

$$\text{UCL}_R = D_4\bar{R}; \qquad \text{LCL}_R = D_3\bar{R}.$$

If a σ chart is desired, similar calculations are made in accordance with rules described above. If all the points fall inside the trial control limits, no modification is made unless it is desired to reduce the process dispersion. In this case, an aimed-at value of σ' should be used in the calculations. When the $\bar{R}$ (or σ) chart indicates a possible lack of control by points falling outside the trial control limits, it is desirable to estimate the value of σ' that might be attained if the dispersion were brought into control. A method of estimation is to eliminate those points out of control (only those above the UCL_R if points fall both above and below) and recompute the values of the control limits based only on the remaining observations. If more points fall out of control, the procedure is repeated.

The final revised values of $\bar{R}$, $\bar{\sigma}$, or σ', whichever is appropriate, may also be used to obtain new control limits for $\bar{X}$. This has the effect of tightening the limits on the $\bar{X}$ chart, making them consistent with a σ' that may be estimated from the revised $\bar{R}$ or $\bar{\sigma}$.

Control limits should be revised from time to time as additional data are accumulated.

12.5 Example of $\bar{X}$ and R Chart

The following are the $\bar{X}$ and R values for 20 subgroups of five readings. The specifications for this product characteristic are 0.4037 ± 0.0010. The

Table 12.2. Data on Dimensions

Subgroup	$\bar{X}$	R	Subgroup	$\bar{X}$	R	Subgroup	$\bar{X}$	R
1	34.0	4	8	32.6	10	15	33.8	7
2	31.6	4	9	33.8	19	16	31.6	5
3	30.8	2	10	37.8	6	17	33.0	5
4	33.0	3	11	35.8	4	18	28.2	3
5	35.0	5	12	38.4	4	19	31.8	9
6	32.2	2	13	34.0	14	20	35.6	6
7	33.0	5	14	35.0	7			
	$\sum \bar{X} = 671.0$;		$\bar{X} = 33.6$;		$\sum R = 124$;		$\bar{R} = 6.20$.	

values given are the last two figures of the dimension reading; i.e., 31.6 should be 0.40316. The trial control limits are computed from

$$\bar{\bar{X}} \pm A_2 \bar{R}; \qquad \text{UCL}_{\bar{x}} = 37.2; \qquad \text{UCL}_R = (2.115)(6.20)$$

$$\bar{\bar{X}} \pm (0.577)(6.20); \qquad \qquad \qquad = 13.1$$

$$\bar{\bar{X}} \pm 3.58. \qquad \text{LCL}_{\bar{x}} = 30.0; \qquad \text{LCL}_R = 0$$

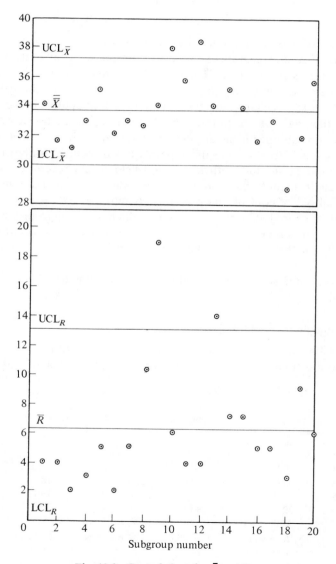

Fig. 12.3. *Control chart for $\bar{X}$ and R.*

The control charts for $\bar{X}$ and R with trial limits are shown in Fig. 12.3. Since some points fall outside the control limits, the process is assumed to be out of control. Eliminating these points, new control limits are computed:

$$\sum \bar{X} = 566.6; \qquad \bar{\bar{X}} = 33.3;$$

$$\sum R = 91; \qquad \bar{R} = 5.06;$$

$$\text{UCL}_{\bar{x}} = 33.3 + (0.577)(5.06) = 33.3 + 2.92 = 36.2;$$

$$\text{LCL}_{\bar{x}} = 33.3 - (0.577)(5.06) = 33.3 - 2.92 = 30.4;$$

$$\text{UCL}_R = 2.115(5.06) = 10.7; \quad \text{LCL}_R = 2.115(0) = 0.$$

None of the remaining points fall outside the limits. If it is now assumed that the process can be brought into control at this level, it is found that $\bar{X}' = 33.3$ and $\sigma' = 2.175$. The value of σ' is estimated from $\bar{R}/d_2$, i.e.,

$$\bar{R}/d_2 = 5.06/2.326 = 2.175.$$

Therefore,

$$(\bar{X}' - 3\sigma', \bar{X}' + 3\sigma') = (\bar{X}' - 6.525, \bar{X}' + 6.525) = (26.8, 39.8).$$

In terms of the actual data, $(\bar{X}' - 3\sigma', \bar{X}' + 3\sigma') = (0.4027, 0.4040)$. Thus, $(0.4027, 0.4040)$ are the estimated natural tolerance limits, where these limits are defined to include all but 27 in 10,000. The specifications are given as

$$(0.4037 - 0.0010, 0.4037 + 0.0010) = (0.4027, 0.4047).$$

This production process is then able to meet these specifications, even though the process is not centered at the nominal value of 0.4037, provided the process remains in control at the above level.

12.6 Control Chart for Fraction Defective

12.6.1 RELATION BETWEEN CONTROL CHARTS BASED ON VARIABLES DATA AND CHARTS BASED ON ATTRIBUTES DATA

The $\bar{X}$ and R charts are charts for variables, i.e., for quality characteristics that can be measured and expressed in numbers. However, many quality characteristics can be observed only as attributes, i.e., by classifying the item into one of two classes, usually defective or non-defective. Furthermore, with the existing techniques, an $\bar{X}$ and R chart can be used for only one measurable characteristic at a time. For example, if an item consists of 10,000 measurable characteristics, each characteristic is a candidate for an $\bar{X}$ and R chart. However, it would be impossible to have 10,000 such charts, and only the most important and troublesome would be charted. As an alternative to $\bar{X}$ and R charts, and as a substitute when characteristics are measured only by attributes, a control chart based on the fraction defective can be used. This is known as a p chart.[1] A p chart can be applied to quality characteristics that are actually observed as attributes, even though they may have been measured as variables. The cost of obtaining attribute data is usually less than that for obtaining variables data. The cost of computing and charting may also be less, since one p chart can apply to any number of characteristics.

12.6.2 STATISTICAL THEORY

Consider a random sample of n Bernoulli random variables, $X_1, X_2, \ldots,$ X_n, each with parameter p', where $p' = P\{X = 1\}$, i.e., p' represents the probability that an item is defective. Define the random variable D by

$$D = X_1 + X_2 + \cdots + X_n,$$

so that D represents the total number of defective items found in the sample. It was shown in Chapter 4 that the probability distribution of the total number of defective items found in the sample has a binomial distribution with parameters n and p', i.e.,

$$P_D(k) = P\{D = k\} = \binom{n}{k} p'^k (1 - p')^{n-k}, \text{ for } k = 0, 1, 2, \ldots, n.$$

The probability that the random sample will contain d or fewer defectives is just the CDF of the random variable D, i.e.,

$$F_D(d) = \sum_{k=0}^{d} \binom{n}{k} p'^k (1 - p')^{n-k} = \binom{n}{0} p'^0 (1 - p')^n + \binom{n}{1} p'^1 (1 - p')^{n-1}$$
$$+ \cdots + \binom{n}{d} p'^d (1 - p')^{n-d}.$$

Furthermore, the expected value of the total number of defectives in a sample of n is np', and the standard deviation is $\sqrt{np'(1 - p')}$.

If the fraction defective, p, is defined as the ratio of the number of defectives D to the total number of items in the sample n, i.e., D/n, the expected value of the fraction defective is p' and the standard deviation is $\sqrt{p'(1 - p')/n}$.

This theory can be applied to control charts. As a working rule, the control limits for the fraction defective, p, are recommended to be

$$\text{UCL}_p = p' + 3\sqrt{\frac{p'(1 - p')}{n}}; \qquad \text{LCL}_p = p' - 3\sqrt{\frac{p'(1 - p')}{n}}.$$

It is important to note that the actual probability of D/n falling within these

---

[1] Again, the notation is changed to coincide with that recommended by the American Society for Quality Control. In the previous chapters, the symbol p represents the parameter of a Bernoulli or a binomial distribution, i.e., p denotes the probability that an item is defective. In this chapter p' denotes this parameter and p denotes the fraction of defective items found in a sample (subgroup) of size n, i.e., p is an estimator of p'.

limits depends on the value of p', even if the process is in control.[1] However, almost all the D/n will fall within the given limits, provided the process is in control. Furthermore, the above limits result in a simple empirical rule which, for large n, lead to an accurate approximation. This follows from the fact that D/n is an average. Hence, the central limit theorem can be invoked for large n, thereby implying that D/n is approximately normally distributed.

The exact probability of a point falling within the control limits can be evaluated by using the expression

$$P\left\{p' - 3\sqrt{\frac{p'(1-p')}{n}} \leq \frac{D}{n} \leq p' + 3\sqrt{\frac{p'(1-p')}{n}}\right\}$$
$$= P\{np' - 3\sqrt{np'(1-p')} \leq D \leq np' + 3\sqrt{np'(1-p')}\}.$$

For a fixed p', the right-hand side of this equation is just the probability that a binomial random variable lies between two fixed constants, and can be obtained from the CDF of the binomial distribution.

If p' is not known, it is usually estimated from past data. The rules are similar to those used in estimating the parameters for the $\bar{X}$ chart. In this case, the estimator of p' is D_T/n_T, where D_T is the total number of defectives found in the n_T units of past data.

It often happens in control charts for fraction defective that the size of the subgroup varies. In this case, three possible solutions to the problem are as follows: (1) Compute control limits for every subgroup and show these fluctuating limits on the p chart. (2) Estimate the average subgroup size, and compute one set of limits for this average and draw them on the control chart. This method is approximate and is appropriate only when the subgroup sizes are not too variable. Points near the limits may have to be re-examined in accordance with (1). (3) Draw several sets of control limits on the chart corresponding to different subgroup sizes. This method is also approximate and is actually a cross between (1) and (2). Again, points falling near the limits should be re-examined in accordance with (1).

12.6.3 STARTING THE CONTROL CHART

The subgroup size is usually large compared with that used for $\bar{X}$ and R charts. The main reason for this is that if p' is very small, and n is small, the expected number of defectives in a subgroup will be very close to zero.

For each subgroup compute p, where

$$p = \frac{\text{number of defectives in the subgroup}}{\text{number inspected in the subgroup}}.$$

Whenever practicable, no fewer that 25 subgroups should be used to compute

---

[1] This is quite different from the $\bar{X}$ chart, where the probability of $\bar{X}$ falling between $\bar{X}' \pm 3\sigma'/\sqrt{n}$ is 0.9973 for any values of $\bar{X}'$ and σ', provided the process is in control.

trial control limits. These limits are obtained from the data by finding the average fraction defective $\bar{p}$, where

$$\bar{p} = \frac{\text{total number of defectives during period}}{\text{total number inspected during period}},$$

and then evaluating the expression

$$\text{UCL}_p = \bar{p} + \frac{3\sqrt{\bar{p}(1-\bar{p})}}{\sqrt{n}}; \qquad \text{LCL}_p = \bar{p} - \frac{3\sqrt{\bar{p}(1-\bar{p})}}{\sqrt{n}}.$$

Inferences about the existence of control or lack of control can be drawn in a manner similar to that described for the $\bar{X}$ and R chart.

12.6.4 CONTINUING THE p CHART

The preliminary plot reveals two facts, namely, whether or not the process is in an apparent state of control and the quality level. If the process appears to be in control, it may be in control at too high a level—i.e., the estimate of p' is higher than can be tolerated. In this case, the production process must be examined, and possible major changes made. On the other hand, the estimate of p' may indicate a good quality level, even though the process may appear to be out of control. In this case, assignable causes should be sought and eliminated. Of course, control limits can be determined on the basis of an aimed-at value of p', but an indicated lack of control must be interpreted with this in mind.

At first glance, points falling below the lower control limit may appear to be desirable. However, this may be attributed to a poor estimate of p', although it may possibly indicate a change for the better in the quality level. In this situation, tracking down the assignable cause may enable an improve-

Table 12.3. Screw Machine Data

Subgroup	D	p	Subgroup	D	p	Subgroup	D	p
1	1	0.02	9	1	0.02	18	0	0.00
2	2	0.04	10	0	0.00	19	0	0.00
3	5	0.10	11	0	0.00	20	1	0.02
4	6	0.12	12	1	0.02	21	1	0.02
5	3	0.06	13	0	0.00	22	0	0.00
6	5	0.10	14	1	0.02	23	0	0.00
7	2	0.04	15	0	0.00	24	1	0.02
8	1	0.02	16	2	0.04	25	0	0.00
			17	1	0.02			

$$\bar{p} = \frac{34}{1,250} = 0.0272,$$

$$\text{UCL}_p = 0.0272 + \frac{3\sqrt{(0.0272)(0.9728)}}{7.071} = 0.0963,$$

$$\text{LCL}_p = 0.$$

ment to be made in the production process. In either case, points falling below the lower control limit call for a re-examination of the control chart or the production process or both.

12.6.5 EXAMPLE

A sample of 50 pieces is drawn from the production of the last two hours from a single spindle automatic screw machine and each item is checked by go and no-go gauges for several possible sources of defectives. The number of defective items found in 25 such successive samples is given in Table 12.3.

The control chart for p with trial limits is shown in Fig. 12.4. Many points fall outside the control limits, and it appears that an assignable cause of variation was present when the first samples were taken that was not present when the later samples were taken. Recomputing the control limits starting with the seventh sample, one finds

$$\bar{p} = \frac{12}{950} = 0.0126,$$

$$\text{UCL}_p = 0.0126 + \frac{3\sqrt{(0.0126)(0.9874)}}{7.071} = 0.0599,$$

$$\text{LCL}_p = 0.$$

No points fall outside these limits; hence, they are used as the control limits for future production.

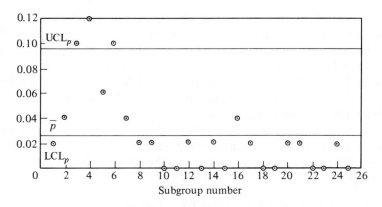

Fig. 12.4. *Control chart for p.*

12.7 Control Charts for Defects

12.7.1 DIFFERENCE BETWEEN A DEFECT AND A DEFECTIVE ITEM

An item is considered to be defective if it fails to conform to the specifications in any of the characteristics. Each characteristic that does not meet the specifications is a defect. An item is defective if it contains at least one defect.

The c chart is a control chart for defects per unit. The unit considered may be a single item, a group of items, part of an item, etc. The unit is examined and the number of defects found is recorded on the c chart. If for each unit there are numerous opportunities for defects to occur, and if the probability of a defect occurring in a particular spot is small, the statistical theory for the c chart is based on the Poisson distribution. Some examples where c charts are applicable are counting the number of defective rivets on an airplane wing, the number of imperfections in a piece of cloth, etc.

12.7.2 STATISTICAL THEORY

Let the random variable c denote the number of defects appearing on a unit.[1] If c is assumed to have a Poisson distribution, then the probability that the number of defects on a unit is equal to k is given by

$$P_c(k) = P\{c = k\} = \frac{c'^k e^{-c'}}{k!}, \quad \text{for } k = 0, 1, 2, \ldots$$

where c' is the *average* number of defects per unit. The standard deviation of this random variable is $\sqrt{c'}$.

As a working rule, control limits for the c chart are recommended to be

$$\text{UCL}_c = c' + 3\sqrt{c'}; \quad \text{LCL}_c = c' - 3\sqrt{c'}.$$

As in the case of the control chart for the fraction of defectives, these limits do not contain a fixed proportion of the distribution for all values of c' even if the process is in control; i.e., the probability of c falling between these limits depends on the value of c'. However, for practical purposes, it can be assumed that "almost all the values" will fall between the control limits, provided the process is in control. For large n, the random variable c is approximately normally distributed.

If c' is not known, it is usually estimated from past data. The rules are similar to those used in estimating the parameters for the $\bar{X}$ chart. In this case the estimator of c' is c_T/N_T, where c_T is the total number of defects found in N_T units.

12.7.3 STARTING AND CONTINUING THE c CHART

The discussion on starting and continuing the control chart for fraction defective given in Section 12.6.3 and 12.6.4 is pertinent to the c chart.

---

[1] To conform to standard quality control notation, a lower case c is used to denote the random variable.

12.7.4 EXAMPLE

The following table gives the number of missing rivets noted at final inspection on 25 airplanes. Read the columns downward and left to right.

Table 12.4. Data on Missing Rivets

10	14	21	15	12
17	7	13	11	18
16	14	10	12	6
20	19	11	8	10
10	16	25	30	8

$$\bar{c} = \frac{353}{25} = 14.12,$$
$$UCL_c = 14.12 + 3\sqrt{14.12} = 25.40,$$
$$LCL_c = 14.12 - 3\sqrt{14.12} = 2.84.$$

The control chart for c with trial limits is shown in Fig. 12.5. However, the twentieth plane falls outside the control limits, and it would be best

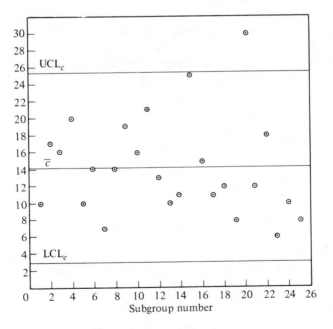

Fig. 12.5. *Control chart for c.*

to base future calculations on a $\bar{c}$ which did not involve this airplane. This results in

$$\bar{c} = \frac{323}{24} = 13.46,$$

$$\text{UCL}_c = 13.46 + 3\sqrt{13.46} = 24.47, \text{ and}$$

$$\text{LCL}_c = 13.46 - 3\sqrt{13.46} = 2.45.$$

The fifteenth airplane now falls outside the control limits, so that $\bar{c}$ is again recomputed without the results for this plane. The final starting control limits are given by

$$\bar{c} = \frac{298}{23} = 12.96,$$

$$\text{UCL}_c = 12.96 + 3\sqrt{12.96} = 23.76,$$

$$\text{LCL}_c = 12.96 - 3\sqrt{12.96} = 2.16.$$

12.8 Further Developments on Control Charts

All the control charts previously discussed belong to a class known as Shewhart control charts.[1] These charts, together with some modifications, have been widely used in industry. Among the modifications that appeared were control charts with "warning limits," usually in the form of two sigma limits, inside the control limits. Points falling within the control limits but outside the warning limits were signals that trouble could be brewing. Further modifications were concerned with taking action if a fixed number of consecutive points fell above or below the center line. However, these modifications tended to vitiate the important advantage of simplicity that the Shewhart control charts possess.

A major failing of Shewhart control charts is their inability to utilize all the information contained in the sequence of plotted points. Although the past data are used to update the limits on the charts, the plotted points are viewed independently. Even the suggested modifications use only the information contained in a fixed number of past points. Two alternative types of control charts have been introduced which tend to correct this deficiency. They will be referred to as the signed sequential rank control chart (SSRCC) and the cumulative sum control chart (CSCC), and both have "stopping rules" that incorporate the information contained in all the sample points. They will be illustrated for the case of controlling the mean, although the CSCC is a competitor for all the Shewhart control charts.

---

[1] The concept of control charts was originally published by W. A. Shewhart in *The Economic Control of Manufactured Product*, D. Van Nostrand, New York, 1931.

12.8.1 THE SIGNED SEQUENTIAL RANK CONTROL CHART

The signed sequential rank control chart was originally proposed by E. A. Parent, Jr.[1] Before describing the SSRCC procedure, it is desirable to introduce some further notation. As with a Shewhart $\bar{X}$ control chart, subgroups of size n are observed, and the sample means obtained. Suppose that sufficient data have been accumulated so that it can be assumed that $\bar{X}'$ is known, i.e., the process is in control at level $\bar{X}' = \bar{X}'_0$. Alternatively, $\bar{X}'_0$ can be viewed as the "target" value for the level of control of the mean. It will also be assumed that only a shift in $\bar{X}'$ is likely to occur (σ' remains constant). Since $\bar{X}'_0$ can now be considered to be a known number, form the random variable

$$Y = \bar{X} - \bar{X}'_0,$$

for each observed subgroup, i.e., the deviation from the process mean. Thus, a sequence of independent random variables $Y_1, Y_2, \ldots$ are being observed sequentially.

The SSRCC procedure can be defined in terms of the following steps:

1. For the kth observation Y_k (corresponding to the kth subgroup) find the rank of $|Y_k|$ relative to $|Y_1|, |Y_2|, \ldots, |Y_k|$. If it is the smallest, it receives rank 1; second smallest, rank 2; ... ; largest, rank k. Denote this rank by R_k. Ties should be assigned an average rank.
2. Examine the sign of Y_k ($+$ if $Y_k \geq 0$ and $-$ if $Y_k < 0$), and attach this sign to R_k/k. Denote the result by Z_k, i.e., Z_k is either R_k/k or $-R_k/k$. Thus, the sequence $Z_1, Z_2, Z_3, \ldots$ is obtained.
3. Form the cumulative sums of these Z's, i.e.,

$$S_1 = Z_1$$
$$S_2 = Z_1 + Z_2$$
$$S_3 = Z_1 + Z_2 + Z_3$$
$$\vdots$$
$$S_m = Z_1 + Z_2 + \ldots + Z_m.$$

4. The SSRCC procedure calls for stopping the process to investigate for possible trouble as soon as

$$|S_m| > a,$$

that is, as long as $-a \leq S_m \leq a$ the process is assumed to be in control and no action is required. The choice of a will be discussed later.

---

[1] E. A. Parent, Jr., "Signed Sequential Ranking Procedures." Technical Report No. 80, Department of Statistics, Stanford University, 1965.

5. Of course the sequence S_m can be plotted in a manner analogous to the graph of a Shewhart control chart with the limits $\pm a$ corresponding to the 3 sigma limits of the Shewhart chart.

Consider the following example: Suppose that it has been determined from past data in subgroups of size 4 that a process is in control with $\bar{X}'_0 = 10$. A SSRCC is to be kept and a is chosen to be 2.58. The data are actually obtained in a sequential order and presented in Table 12.5. The last point indicates that the process should be investigated since $2.63 > a = 2.58$. Some of the entries in Table 12.5 will be verified. For sample number 3, $\bar{X}_3$ is 9.66. Following the SSRCC procedure, $Y_3 = 9.66 - 10 = -0.34$. Hence, $|Y_3| = 0.34$. The rank, R_3, of $|Y_3| = 0.34$ relative to $|Y_1| = 1.14$, $|Y_2| = 0.91$, and $|Y_3| = 0.34$ (all the values to date) is 1 since it is the smallest. Attaching the sign of Y_3 (minus) to R_3, and dividing by 3, leads to

Table 12.5. Data for a Signed Sequential Rank Control Chart

Sample	$\bar{X}_k$	Y_k	$\|Y_k\|$	R_k	Z_k	$S_m = \sum_{k=1}^{m} Z_k$
1	8.86	-1.14	1.14	1	-1	-1
2	10.91	0.91	0.91	1	$\frac{1}{2}$	$-\frac{1}{2}$
3	9.66	-0.34	0.34	1	$-\frac{1}{3}$	$-\frac{5}{6}$
4	10.95	0.95	0.95	3	$\frac{3}{4}$	$-\frac{1}{12}$
5	11.23	1.23	1.23	5	$\frac{5}{5}$	$\frac{11}{12}$
6	11.50	1.50	1.50	6	$\frac{6}{6}$	$1\frac{11}{12}$
7	11.16	1.16	1.16	5	$\frac{5}{7}$	2.63

$Z_3 = -\frac{1}{3}$. Finally, $S_3 = Z_1 + Z_2 + Z_3 = -\frac{5}{6}$. Since $-2.58 \leq S_3 \leq 2.58$, another subgroup is to be examined. For sample number 7, $\bar{X}_7$ is 11.16. Again following the SSRCC procedure, $Y_7 = 11.16 - 10 = 1.16$. Hence, $|Y_7| = 1.16$. The rank, R_7, of $|Y_7| = 1.16$ relative to $|Y_1|, |Y_2|, \ldots, |Y_7|$ is 5 since it is the 5th smallest. Attaching the sign of Y_7 (plus) to R_7, and dividing by 7, leads to $Z_7 = \frac{5}{7}$. Finally, $S_7 = \sum_{k=1}^{7} Z_k = 2.63$. Since $S_7 > 2.58$, the process should be investigated since there is evidence that it may be out of control.

The choice of the value of the constant a requires some discussion. Recall that a Shewhart control chart can be interpreted as repeated tests of hypotheses about the mean of a distribution. This philosophy has some unfortunate consequences in that the usual interpretation of α and β is not very meaningful. It is evident that even for a process in control at a satisfactory level "eventually" a point will fall outside the control limits. Similarly, when the process is no longer in control, a point will "eventually" fall outside the control limits. The word "eventually" is the key to evaluating a control procedure. Good procedures will take a long time before they indicate that an adjustment is necessary when the process is in control (an incorrect action), and take a short time to call for action when the process is no longer in control. A measure of this "speed" is the average run length (ARL) before the control chart will indicate a lack of control. For a Shewhart control chart, the average run length is

$$ARL = 1/p$$

where p is the probability of a single point falling outside the control limits ($p = 0.0027$ for the usual 3 sigma limits when the process is in control so the ARL = 370). For a SSRCC it can be shown that the ARL, when the process is in control, is given approximately by

$$ARL = 3a^2.$$

In the example, a was chosen so that the ARL was equal to 20 when the process was in control, i.e., on the average 20 subgroups would be examined before the SSRCC indicated trouble when, in reality, the process was still in control. Unfortunately, there is no simple analytical expression available for determining the ARL when the process is not in control. However, some simulation work has been done by Kao[1] which reveals that the SSRCC is very good relative to the Shewhart control chart in detecting small shifts in the process mean (less than one standard deviation of $\bar{X}$). Moreover, the SSRCC procedure is non-parametric and, hence, does not require any assumption about normality (although for the ARL equation to hold, a symmetric distribution is necessary). The non-parametric feature has the added advantage that no assumption is required about the magnitude of the process standard deviation to actually carry out the control chart procedure. Of course, the speed with which the SSRCC detects a shift in the process mean does depend on the standard deviation (as well as the subgroup size). Finally, the mechanics of a SSRCC is also simple in that the calculations are of an elementary nature.

12.8.2 THE CUMULATIVE SUM CONTROL CHART

Another alternative to the Shewhart control chart that has a "stopping rule" which incorporates the information contained in all the sample points is the cumulative sum control chart (CSCC). Following the notation introduced for the SSRCC, the Y_k are formed from the mean of each subgroup of size n ($Y_k = \bar{X}_k - \bar{X}'$). The CSCC is a sequential plot of the points $\{m, \sum_{k=1}^{m} Y_k\}$ where m denotes the mth subgroup and $\sum_{k=1}^{m} Y_k$ denotes the cumulative sum up to and including the mth sampling. The process is assumed to be in control as long as these points "behave properly." These points are assessed to behave properly as follows. A separate V mask is constructed.

[1] E.P. Kao, "Comparison of Several Procedures for Process Control," Technical Report No. 84, Department of Statistics, Stanford University, 1965, revised.

Such a mask is shown in Fig. 12.6 and is clearly defined by specifying the angle θ and the distance d. The V mask is the shaded area in Fig. 12.6. The CSCC procedure calls for placing the V mask on the chart with the point 0 coinciding with the last point plotted, and such that the mask lies in a horizontal position. If any of the previously plotted points are covered by the mask (shaded area), the process is assumed to be out of control and corrective action is called for. If none of the previous points are covered, the next subgroup is observed. In Fig. 12.6 the process indicates a lack of control with the plot of the mth point since the $(m - 2)$ point is hidden by the mask.

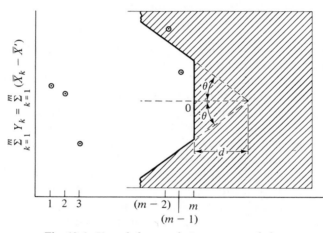

Fig. 12.6. *V mask for cumulative sum control chart.*

The choice of d and θ determines the performance of the CSCC. Furthermore, these values depend on the scale used. In a survey paper, Ewan[1] gave some good rules of thumb, based on his industrial experience, for these choices. Define the scale factor, w, as the vertical scale distance per horizontal plotting interval. It is assumed that the standard deviation of the sample mean, $\sigma'_{\bar{x}}$, is known or can be estimated from past experience. He suggests that this scale factor, w, should lie in the range $1\,\sigma'_{\bar{x}}$ to $2\,\sigma'_{\bar{x}}$, preferably near $2\,\sigma'_{\bar{x}}$. He recommends a $\frac{1}{10}$ inch horizontal interval as a satisfactory horizontal scale. Hence, the vertical scale should be approximately $\frac{1}{10}$ inch $= 2\,\sigma'_{\bar{x}}$, rounding down to obtain reasonable intervals. For example, if $\sigma'_{\bar{x}} = 1.6$ and the horizontal plotting interval is $\frac{1}{10}$ inch, then the vertical scale should be $\frac{1}{10}$ inch $= 3.2$ units. Since this may be an awkward scale, the vertical scale can be rounded to $\frac{1}{10}$ inch $= 3$ units [a scale factor of

---

[1] Ewan, W.D., "When and How to Use the Cu-Sum Charts," *Technometrics*, Vol. 5, No. 1, Feb., 1963.

$w = (3/1.6)\sigma'_{\bar{x}} = 1.9\sigma'_{\bar{x}}$]. Ewan also adopts the viewpoint of the importance of the average run length, and suggests using a CSCC, which has a resultant ARL similar to that of the Shewhart control chart when the process is in control. This recommended plan calls for choosing the parameters θ and d as follows:

$\tan \theta = 0.5 \, \sigma'_{\bar{x}}/w$, and

$\quad d = 10.0$ horizontal plotting intervals (or a distance, measured in terms of the vertical scale, equal to 10 w).

Thus, with a scale factor of $w = 2 \, \sigma'_{\bar{x}}$,

$\theta = $ arc tan $0.25 = 14°$, and

$\quad d = 10$ horizontal plotting intervals (or a distance, measured in terms of the vertical scale, equal to 20 $\sigma'_{\bar{x}}$).

Actually, the ARL for this procedure is approximately 500 when the process is in control whereas the ARL for the Shewhart control charts is 370. A CSCC whose ARL is 370 when the process is in control is given by

$\tan \theta = 0.5 \, \sigma'_{\bar{x}}/w$, and

$\quad d = 9.4$ horizontal plotting intervals (or a distance, measured in terms of the vertical scale, equal to 9.4 w).

Table 12.6 provides a means for choosing the CSCC parameters θ and d when the user is able to specify some of the properties that he desires from the test.[1] Let $L(\delta)$ be the ARL of the test when the process mean is $\pm\delta \, \sigma'_{\bar{x}}$ units from the target value. A good CSCC procedure has the following properties: (1) When there is a large shift in the process mean from the target value, $L(\delta)$ should be small. Recall that $L(\delta)$ is the expected time before an out-of-control signal appears when the process mean is $\pm\delta \, \sigma'_{\bar{x}}$ units from the target value. (2) When the process is in control, then the expected time before a false alarm is sounded, i.e., $L(0)$, should be large. For given values of δ and $L(0)$, Table 12.6 gives the values of $(w/\sigma'_{\bar{x}}) \tan \theta$ and d which minimize $L(\delta)$. The value of $L(\delta)$ for the minimizing values of $(w/\sigma'_{\bar{x}})$ and d is also given. Note that the test given in the table provides the minimum value of $L(\delta)$, assuming δ and $L(0)$ are fixed.

For example, if it is desired to detect a deviation of $\pm 0.5\sigma'_{\bar{x}}$ from the target value and to have an ARL of 400 when the process is in control at the target value, entering Table 12.6 results in a CSCC procedure given by $\theta = $ arctan $(0.28 \, \sigma'_{\bar{x}}/w)$ and $d = 27.3$. Furthermore, the ARL equals 29.0 when there is a shift in the process mean of $\pm 0.5 \, \sigma'_{\bar{x}}$ from the target value, and this is the smallest ARL for $\delta = 0.5$ and $L(0) = 400$. To carry the

---

[1] An excellent discussion of this material appears in C. S. Van Dobben De Bruyn, *Cumulative Sum Tests—Theory and Practice*, Hafner Publishing Co., New York, 1968.

example further, suppose it is decided that $L(0.5) = 29.0$ is too large when the process is out of control at $\pm 0.5\,\sigma'_{\bar{x}}$ units from the target value, and instead, the test must signal a lack of control within 20 time units (on the average). To get $L(0.5) \leq 20$ it is necessary to shift the column entry to the left. For $L(0) = 100$ a test with $L(0.5) = 19.0$ is found. The corresponding CSCC procedure is given by $\theta = \arctan\,(0.28\,\sigma'_{\bar{x}}/w)$ and $d = 18.2$. The penalty that one must pay for reducing $L(0.5)$ from 29.0 to 19.0 is to increase the expected number of false alarms from one every 400 time units to one every 100 time units.

The CSCC is particularly effective in terms of detecting shifts in $\bar{X}'$ for intermediate changes in $\bar{X}'$ in the range of $\pm 1\,\sigma'_{\bar{x}}$ to $\pm 2\,\sigma'_{\bar{x}}$.

Table 12.6. Optimal Cusum Tests

$L(0)$ = Expected Run Length when Process Is in Control

δ = Deviation from Target Value (in standard deviations)			50	100	200	300	400	500
	0.25	$(w/\sigma'_{\bar{x}}) \tan \theta$	0.125			0.195		0.248
		d	47.6			46.2		37.4
		$L(0.25)$	28.3			74.0		94.0
	0.50	$(w/\sigma'_{\bar{x}}) \tan \theta$	0.25	0.28	0.29	0.28	0.28	0.27
		d	17.5	18.2	21.4	24.7	27.3	29.6
		$L(0.5)$	15.8	19.0	24.0	26.7	29.0	30.0
	0.75	$(w/\sigma'_{\bar{x}}) \tan \theta$	0.375	0.375	0.375	0.375	0.375	0.375
		d	9.2	11.3	13.8	15.0	16.2	16.8
		$L(0.75)$	8.9	11.0	13.4	14.5	15.7	16.5
	1.0	$(w/\sigma'_{\bar{x}}) \tan \theta$	0.50	0.50	0.50	0.50	0.50	0.50
		d	5.7	6.9	8.2	9.0	9.6	10.0
		$L(1.0)$	6.1	7.4	8.7	9.4	10.0	10.5
	1.5	$(w/\sigma'_{\bar{x}}) \tan \theta$	0.75	0.75	0.75	0.75	0.75	0.75
		d	2.7	3.3	3.9	4.3	4.5	4.7
		$L(1.5)$	3.4	4.0	4.6	5.0	5.2	5.4
	2.0	$(w/\sigma'_{\bar{x}}) \tan \theta$	1.0	1.0	1.0	1.0	1.0	1.0
		d	1.5	1.9	2.2	2.4	2.5	2.7
		$L(2.0)$	2.26	2.63	2.96	3.15	3.3	3.4

PROBLEMS

1. Suppose probability limits are desired for $\bar{X}$ charts, and probability limits corresponding to a probability of 0.004 of a point falling outside the control limits (0.002 outside each limit) are to be used. Calculate the control limits for a sample of size 4.

2. Suppose that a process is in control at $\bar{X}' = 17$ and $\sigma' = 2$. A control chart for $\bar{X}$ based on a subgroup size of $n = 5$ is constructed, using 3-sigma limits. Suppose that a shift in the mean of 2.5 units occurs. What is the probability of the next $\bar{X}$ falling outside the control limits?

3. Samples of five items each are taken from a process at regular intervals. $\bar{X}$ and R are calculated for each sample for a certain quality characteristic, X. The sums of the $\bar{X}$ and R values for the first 25 samples are determined to be

$$\sum \bar{X} = 358.50, \qquad \sum R = 9.90.$$

(a) Compute the control limits for the $\bar{X}$ and R charts. (b) Assuming the process is in control at the level found in (a) (assume estimates are equal to the parameters being estimated), what are the 3-sigma natural tolerance limits of the process? (c) If the specification limits are 14.40 ± 0.45, what conclusions can you draw concerning the ability of the process to produce items within these specifications? (d) What percentage of the lot will fall outside the specification limits if the process is in control as in (b)?

4. The following are the $\bar{X}$ and R values for twenty subgroups of five readings. The specifications for this product characteristic are 0.4035 ± 0.0010. The values given are the last two figures of the dimension reading; i.e., 31.6 should be 0.40316.

Subgroup No.	$\bar{X}$	R	Subgroup No.	$\bar{X}$	R
1	34.0	4	11	35.8	4
2	31.6	4	12	35.8	4
3	30.8	2	13	34.0	14
4	33.0	3	14	35.0	4
5	35.0	5	15	33.8	7
6	32.2	2	16	31.6	5
7	33.0	5	17	33.0	5
8	32.6	13	18	33.2	3
9	33.8	19	19	31.8	9
10	35.8	6	20	35.6	6

$$\sum \bar{X} = 671.0 \qquad \bar{X} = 33.6$$
$$\sum R = 124 \qquad \bar{R} = 6.2$$
$$\bar{X} \text{ chart UCL} = 37.2$$
$$\bar{X} \text{ chart LCL} = 30.0$$
$$R \text{ chart UCL} = 13.1$$
$$R \text{ chart LCL} = 0$$

(a) Is this process probably in control? (b) If not, suggest values for UCL and LCL for use with succeeding subgroups. (c) If the process is in control with $\bar{X}'$ and σ' equal to these values used in obtaining the UCL and LCL values in (b), will this process be able to meet the specifications for the dimension? If not, what specification can you expect the process to meet?

5. Control charts for $\bar{X}$ and σ are maintained on the shear strength of spot welds. After 30 subgroups of 4, $\sum \bar{X} = 12,660$ and $\sum \sigma = 930$. Assume that the process is in control. (a) What are the $\bar{X}$ chart limits? (b) What are the σ chart limits? (c) Estimate the standard deviation for the process. (d) If the minimum specification for this weld is 400 pounds, what percentage of the welds does not meet the minimum specification?

6. Control charts for $\bar{X}$ and R are maintained on the breaking strength in pounds of a certain metal piece. The subgroup size is 5. The values of $\bar{X}'$ and σ' computed after a long period are 7.3 and 1.1, respectively. Compute the values of the 3-sigma limits for the $\bar{X}$ and R charts.

7. Control charts for $\bar{X}$ and R are maintained on resistors (in ohms). The subgroup size is 4. The values of $\bar{X}$ and R are computed for each subgroup. After 20 subgroups, $\sum \bar{X} = 8,620$, and $\sum R = 910$. (a) Compute the values of the 3-sigma limits for the $\bar{X}$ and R charts. (b) Estimate the value of σ' on the assumption that the process is in statistical control. (c) If the specification limits are 430 ± 30, what conclusions can you draw regarding the ability of the process to produce items within these specifications? (d) If $\bar{X}'$ is increased by 60, what is the probability of a subgroup average falling outside the control limits?

8. (a) Assume that probability limits rather than 3-sigma limits were to be used in Problem 7. Assuming that the value of σ' is as found in Problem 7, what would these limits be on the $\bar{X}$ chart if the probability of a point falling outside each limit is 0.001? (b) How many observations are required in Problem 7 such that $\bar{\bar{X}}$ is within five units of $\bar{X}'$ with probability greater than 0.99? Assume σ' is as found in Problem 7.

9. Given that the expected value of $\bar{\sigma} = 15$, what is the expected value of $\bar{R}$? Assume these values are based on subgroups of size 5.

10. The range of a subgroup of two items is 4. What is σ for this subgroup? Estimate σ', using both R and σ. Estimate the standard deviation for subgroup averages, i.e., $\sigma'_{\bar{x}}$.

11. A process is in control at $\bar{X}' = 17$ and $\sigma' = 3$. Subgroups of size 4 are being plotted for an $\bar{X}$ and σ chart. (a) What are the 3-sigma control limits for the $\bar{X}$ chart? (b) The $\bar{X}$ chart (with 3-sigma limits) is to be used as a sampling inspection plan. If a point falls outside the limits, the entire production period between this point and the previous point plotted is rejected. If the process is in control at the given level, what is the probability of rejecting one out of two production periods? (c) Suppose that $\bar{X}'$ was unknown but σ' was known to be equal to 3. How many observations are required so that $\bar{\bar{X}}$ differs from $\bar{X}'$ by no more than one unit with probability greater than 0.95? (d) What is the upper limit for the σ chart so that the probability of a point falling above the upper limit is 0.001? [*Hint*: Note that $n\sigma^2/(\sigma')^2$ has a χ^2 distribution with $(n - 1)$ degrees of freedom.]

12. A sample of 50 pieces is drawn from the production of the last two hours from a single spindle automatic screw machine and each item is checked by go and no-go gauges for several possible sources of defectiveness. The number of defective items found in 25 successive samples was (read each line left to right in order):

$$0, \quad 2, \quad 5, \quad 6, \quad 3, \quad 6, \quad 2, \quad 1, \quad 1, \quad 0, \quad 0, \quad 1, \quad 0,$$
$$1, \quad 0, \quad 2, \quad 1, \quad 0, \quad 0, \quad 1, \quad 1, \quad 0, \quad 0, \quad 1, \quad 1:$$
total, 35 defectives.

(a) Is this process in control? (b) If the consumer is willing to accept lots of this product that are 2.5% or less defective, will the process in the present condition produce an acceptable product? (c) From looking at the chart of this process, what do you think happened during the run?

13. The following data represent the number of defective transistors produced for a daily production of 2,800 units:

Day	Defectives
1	116
2	117
3	112
4	95
5	130
6	120
7	119
8	113

(a) Draw an appropriate control chart for these data. (b) What can you conclude?

14. Assume that a process is in control at $p' = 2\%$ and $n = 100$. (a) Calculate the 3-sigma control limits for a p chart. (b) What is the probability of a point falling above the upper control limit? (Use the Poisson approximation.) (c) If the process is in control at $p' = 2\%$, how many observations, N, are required so that $\bar{p} = T/N$, where T is the total number of defectives found in a sample of size N, is within ± 0.01 of p' with probability greater than 0.95? (Use the normal approximation.) (d) If there is no information about the value of p', how many observations are required so that $\bar{p}$ is within ± 0.01 of p' with probability greater than 0.95? (Use the normal approximation.)

15. The process average has been shown to be 0.04. Your control chart for fraction defective calls for taking daily samples of 800 items. What is the chance that, if the process average should suddenly shift to 0.08, you would catch the shift on the first sample taken after the shift?

16. In "work sampling," the fraction of time that an employee works is often estimated by taking a random sample and finding the ratio of the number of times he is working to the total number of tries, i.e., $p = D/N$. Thus, p is a binomial random variable. (a) Assuming the normal approximation, find the number of observations necessary so that the estimated value (p) differs from the true value (p') by no more than ± 0.03 with probability greater than 0.95. (*Hint*: Find the maximum value of the standard deviation of p.) (b) Suppose p' is estimated to be 0.95 in (a). Find the control limits for this "process" assuming that 100 observations are permitted.

17. Three-sigma control limits for control charts for percent of defectives are:

$$\text{UCL}_p = 6.20\%,$$
$$\text{LCL}_p = 0,$$

where the process average is 2% defective and the subgroup size is 100. Using the Poisson approximation, find: (a) the probability of a point falling outside these limits if the incoming quality is 2% defective, (b) the probability of a point falling outside these limits if the incoming quality shifts to 4% defective, (c) the probability of at least two points out of ten falling outside these limits if the incoming quality remains at 4% defective.

18. Over the past eight or nine months an assembly line has been producing 400 units of a product per day. Final inspection records show that, on an average, 20 units a day are defective. A survey of the daily reports shows that on an occasional day as many as 32 or 33 units are defective and that on occasional days as few as three or four are defective. (a) Do these records indicate statistical control of quality? (b) What help might be obtained, if any, from these records in improving quality performance at this line?

19. A control chart on the number of breakdowns in successive lengths of 10,000 feet each of rubber-covered wire leads to a value of $\bar{c} = 5.2$. (a) Assuming that the process is in control at this level, what are the 3-sigma control limits? (b) What is the probability of a point falling above the upper control limit?

20. Twenty-five 100-yard pieces of woolen goods were measured and the average number of defects per unit was found to be 3.8. Determine the 3-sigma control limits. What are the probability limits corresponding to a probability of 0.002 of a point falling outside the upper control limit?

21. Using the data in Problem 4, construct a SSRCC by using limits of $a = \pm11$. Assume $\bar{x}' = 33.6$. Does the process indicate a state of control? Assume the data are obtained sequentially.

22. Using the data in Problem 4, construct a CSCC by using the plan recommended in Sec. 12.8.2 (having an ARL of 500). Assume $\bar{X}' = 33.6$ and $\sigma'_{\bar{x}} = 1.2$ and $w = 2\,\sigma'_{\bar{x}} = 2.4$.

13

SAMPLING INSPECTION

BY ATTRIBUTES

13.1 The Problem of Sampling Inspection

13.1.1 INTRODUCTION

Sampling inspection is of two kinds, namely, lot-by-lot sampling inspection and continuous sampling inspection. In the former, items are formed into lots, a sample is drawn from the lot, and the lot is either accepted or rejected on the basis of the quality of the sample. This is most appropriate for acceptance inspection. In continuous sampling inspection, current inspection results are used to determine whether sampling inspection or screening inspection is to be used for the next articles to be inspected. Sampling plans are further classified, depending on whether the quality characteristics are measured and expressed in numbers, i.e., variables inspection, or whether articles are classified only as defective or non-defective, i.e., attributes inspection. Chapter 14 is concerned with sampling inspection by variables.

An alternative to sampling inspection is screening every item. The cost of such a scheme is prohibitive, with perfect quality rarely achieved. Still another advantage of sampling inspection schemes—if properly designed—is that they create more effective pressure for quality improvement and, therefore, result in the submission of better quality product for inspection.

Although the above factors were recognized somewhat prior to World War II, and a few sampling inspection schemes were then in use, it was not until the outbreak of hostilities that modern sampling inspection received its impetus. Good sampling plans replaced bad ones, sometimes at the expense of an increase in the total amount of inspection but almost always resulting indirectly in better quality being produced.

Any lot-by-lot sampling plan has as its primary purpose the acceptance of good lots and the rejection of bad lots. It is important to define what is meant by a good lot. Naturally, the consumer would like all of his accepted lots to be free of defectives. On the other hand, the manufacturer will usually consider this to be an unreasonable request since some defectives are bound to appear in the manufacturing process. If the manufacturer screens the lot a few times he may eliminate all the defectives, but at the prohibitive cost of screening. This cost will naturally be reflected in his price to the consumer. Ordinarily, the consumer can tolerate some defectives in his lot, provided the number is not too large. Consequently, the manufacturer and the consumer get together and agree on what constitutes good quality. If lots are submitted at this quality or better, the lot should be accepted; otherwise, rejected. Again this is an imposing task and can be accomplished only at the expense of screening. It is at this point that sampling inspection, with its corresponding advantage of reduced inspection costs, can be instituted. This advantage should not be minimized. Few manufacturers or consumers, whichever has to bear the cost of inspection, can stay in business very long if all lots are screened.

13.1.2 DRAWING THE SAMPLE

A decision must be made on what shall constitute a lot for acceptance purposes, for each lot must be identified. Each lot should represent, as nearly as possible, the output of one machine or process during one interval of time, so that all parts or products in the lot have been produced under essentially the same conditions. Wherever practicable, parts from different sources or different conditions should not be mixed into one lot. The power of the sampling plans to distinguish between good and bad lots is dependent on the variation from lot to lot, and provision should be made to maintain the identity of, and prevent the mixing of, the different lots.

A sample from each lot supplies the information on which the decision to accept or reject the lot is based. Therefore it is essential that the sample be drawn from each lot in a random manner. A sample is random if every piece in the lot has an equal chance of being selected.[1] This may be accomplished by using a table of random numbers, a deck of cards, or some such chance method, to insure the equal chance for each piece. Human attempts to randomize a sample without such aids often result in biased samples.

---

[1] This definition differs from that given in Chapter 2, which is inapplicable in situations where the probability distribution always depends on the results of the previous trials; if the size of the lot is finite, the probability of an item being defective depends on the outcome of previous trials.

13.2 Lot-by-Lot Sampling Inspection by Attributes

13.2.1 SINGLE SAMPLING PLANS

13.2.1.1 *Single Sampling.* A single sampling procedure can be characterized by the following: One sample of n items is drawn from a lot of N items; the lot is accepted if the number of defectives D in the sample does not exceed c. Here c is referred to as the acceptance number.

Certain risks must be taken if sampling inspection is to be used. A graph of these risks plotted as a function of the incoming lot quality (p') is known as an operating characteristic (OC) curve, and is illustrated in Fig. 13.1.

If quality is good, it is desirable to have the probability of acceptance, $L(p')$, high. On the other hand, if quality is bad, it is desirable to have the probability of acceptance small. Note that if p' is 0, the lot contains no defectives and, hence, the lot will always be accepted, i.e., $L(p') = 1$. If p' is 1, the entire lot is defective so that the sample contains all defectives, thereby insuring that the lot will always be rejected, i.e., $L(p') = 0$.

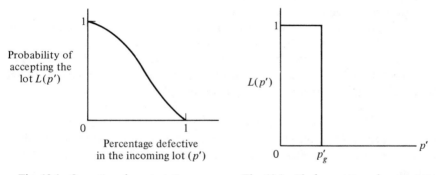

Fig. 13.1. *Operating characteristic curve.*

Fig. 13.2. *Ideal operating characteristic curve.*

Assume that a quality standard p'_g is established and that all lots better than this standard are considered to be "good" and all lots worse are considered to be "bad." An "ideal" OC curve would then be the form shown in Fig. 13.2 with all good lots accepted and all bad ones rejected. Obviously, no sampling plan can have such a curve. The degree of approximation to the *ideal curve* depends on n and c.

If c is held constant and n is increased, the slope becomes steeper. On the other hand, holding n constant and changing the acceptance number has the effect of shifting the OC curve to the left or right. These concepts are illustrated in Figs. 13.3 and 13.4.

If the lot size N is large compared with the sample size n, the OC curve is essentially independent of the lot size. In other words, if the OC curve of

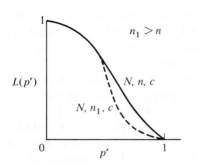

Fig. 13.3. *Effect on OC of varying n.*

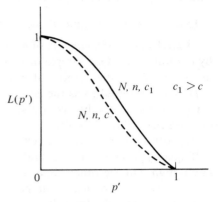

Fig. 13.4. *Effect on OC of varying c.*

a sampling plan, defined by a sample of size n, a lot of size N, and an acceptance number c, is compared with the OC curve of a sampling plan, defined by a sample of size n, a lot of size N_1 ($N_1 > N$), and an acceptance number c, the difference is negligible if the sample size is small compared with the lot size.[1] Since this is the situation in most industrial applications, this assumption is made throughout the rest of this section. The implications of this assumption are: The quality of each item in the sample is an independent Bernoulli random variable with parameter p'; furthermore, the sample drawn constitutes a random sample in the sense of the definition given in Chapter 2. As a result, a single sampling plan can now be defined by only two numbers, the sample size n and acceptance number c, and the probability distribution of the number of defectives will be given by the binomial distribution.

The single sampling procedure can be seen from another viewpoint. By drawing a sample of n, and looking at the number of defectives D present, the consumer is essentially estimating the fraction defective in the lot; i.e., $p = D/n$ is an estimate of the incoming quality p'. A lot is rejected if this estimate is too high, i.e., if $p > c/n = M$. Therefore, a procedure which calls for rejection of a lot if the number of defectives D is greater than c in a sample of size n is equivalent to rejecting a lot if the estimated fraction defective, $D/n = p$, is greater than $c/n = M$. This viewpoint will be useful later in the discussion of sampling inspection by variables.

The concept of an OC curve described above is identical with the concept presented in the chapters on significance tests. Sampling inspection can be viewed as testing the hypothesis that $p' \leq p'_g$ against the alternative that

---

[1] The notion of taking a sample which is a prescribed percentage of the lot is unsatisfactory. The protection obtained from a sampling plan depends on the absolute size of the sample rather than the relative size.

$p' > p'_g$. The decision procedure is to take a sample of size n and accept the lot if the number of defectives D is less than or equal to c. The choice of n and c depends on the OC curve desired.

13.2.1.2 *Choosing a Sampling Plan.* The consumer has at his disposal the choice of one of many OC curves. The one chosen should reflect his views as to the cost of making wrong decisions. By varying n and c, he can always find an OC curve which will pass through two preassigned points. In other words, by specifying the points $[p'_1, L(p'_1)]$ and $[p'_2, L(p'_2)]$ an n and a c can be found such that the OC curve passes through these two points. Before using sampling inspection, then, the consumer can locate two points p'_1 and p'_2 such that if the submitted quality is better than p'_1, he will accept the lot with probability greater than $L(p'_1) = 1 - \alpha$; if the submitted quality is worse than p'_2, he will accept the lot with probability less than $L(p'_2) = \beta$. Here α is known as the producer's risk and β is known as the consumer's risk. The long-run average of submitted quality is known as the process average.

13.2.1.3 *Calculation of OC Curves for Single Sampling Plans.* It was pointed out previously that a sampling plan is defined by two numbers: n, the sample size, and c, the acceptance number. Thus, the sampling procedure is to accept a lot if the number of defectives, D, in a sample of n does not exceed c. The OC curve is defined by the following procedure: If lot quality is submitted at p', the probability that the number of defective items, D, is equal to d in the sample of size n, i.e., $P\{D = d\}$, is given by

$$P\{D = d\} = P_D(d) = \binom{n}{d}(p')^d(1 - p')^{n-d}.$$

Since the lot is accepted if $D \leq c$, the probability of accepting the lot $L(p')$ is obtained from

$$L(p') = P\{D \leq c\} = \sum_{d=0}^{c} P_D\{d\} = P_D(0) + P_D(1) + \cdots + P_D(c)$$

$$= \sum_{d=0}^{c} \binom{n}{d}(p')^d(1 - p')^{n-d}$$

$$= (1 - p')^n + \binom{n}{1}(p')(1 - p')^{n-1} + \binom{n}{2}(p')^2(1 - p')^{n-2}$$

$$+ \cdots + \binom{n}{c}(p')^c(1 - p')^{n-c}.$$

Such calculations as the one above are often cumbersome, and an approximation is desirable. Approximate answers may be obtained very rapidly by the use of the Poisson distribution, which is tabulated in Appendix Table 5.[1] The Poisson approximation to the binomial is discussed in Chapter 4, Sec. 4.5.

---

[1] $L(p')$ is approximately the value read out of Appendix Table 5 with entries c and np'. The larger the n and smaller the p', the closer the approximate answer is to the true probability.

13.2.1.4 *Example.* What is the probability of accepting a lot whose incoming quality is 4.0% defective, using a sample of size 30 and an acceptance number $c = 1$?

$$L(0.04) = \sum_{d=0}^{1} \binom{30}{d}(0.04)^d(0.96)^{30-d}$$

$$= (0.96)^{30} + \frac{(30)!}{(29)!(1)!} 0.04(0.96)^{29} = 0.661.$$

Using the Poisson approximation,

$$np' = 30 \times 0.04 = 1.20; \qquad c = 1; \qquad L(0.04) = 0.663$$

which, in this case, is in good agreement with the correct value.

13.2.2 DOUBLE SAMPLING PLANS

13.2.2.1 *Double Sampling.* A double sampling procedure can be characterized by the following: A sample of n_1 items is drawn from a lot. The lot is accepted if there are no more than c_1 defective items. The lot is rejected if there are r_1 or more defective items. If there are more than c_1 but less than r_1 defective items, a second sample of size n_2 is drawn. The lot is accepted if there are no more than c_2 defectives in the combined sample of $n_1 + n_2$; the lot is rejected if there are more than c_2 defective items in the combined sample of $n_1 + n_2$. This procedure is summarized in Table 13.1.

Thus, a lot will be accepted on the first sample if the incoming quality is very good. Similarly, a lot will be rejected on the first sample if the incoming quality is very bad. If the lot is of intermediate quality, a second sample may be required. Double sampling plans have the psychological advantage of giving a second chance to doubtful lots. Furthermore, they can have the additional advantage of requiring fewer total inspections, on the average, than single sampling plans for any given quality protection. On the other side of the ledger, double sampling plans can have the following disadvantages: (1) more complex administrative duties; (2) inspectors will have variable inspection loads, depending on whether one or two samples are taken; (3) the maximum amount of inspection generally exceeds that for single sampling plans (which is constant).

Table 13.1. Double Sampling Plan

		Combined Samples		
Sample	Sample Size	Size	Acceptance Number	Rejection Number
First	n_1	n_1	c_1	r_1
Second	n_2	$n_1 + n_2$	c_2	$c_2 + 1$

*13.2.2.2 *OC Curves for Double Sampling Plans.* A lot will be accepted if and only if (1) the number of defectives D_1 in the first sample of n_1 does not exceed c_1, i.e., $D_1 \leq c_1$; (2) there are more than c_1 but less than r_1 defectives in the first sample and the total number of defectives, $D_1 + D_2$, in the combined sample of $n_1 + n_2$ does not exceed c_2, i.e., if $c_1 < D_1 < r_1$, then $D_1 + D_2 \leq c_2$.

Thus, for any incoming quality p', the probability of accepting the lot is

$$
\begin{aligned}
L(p') &= P\{D_1 \leq c_1, \text{ given } p'\} \\
&\quad + P\{D_1 + D_2 \leq c_2 \text{ and } c_1 < D_1 < r_1 \text{ given } p'\} \\
&= P\{D_1 \leq c_1 \mid p'\} + P\{D_1 + D_2 \leq c_2 \text{ and } c_1 < D_1 < r_1 \mid p'\},
\end{aligned}
$$

where the symbol | reads "given" and $P\{D_1 \leq c_1 \mid p'\}$ reads "the probability that D_1 is less than or equal to c_1, given p'." The OC curve is then given by the expression

$$
\begin{aligned}
L(p') &= \sum_{d_1=0}^{c_1} \binom{n_1}{d_1}(p')^{d_1}(1 - p')^{n_1 - d_1} \\
&\quad + \sum_{d_1=c_1+1}^{r_1-1} \sum_{d_2=0}^{c_2-d_1} \binom{n_1}{d_1}(p')^{d_1}(1 - p')^{n_1 - d_1} \binom{n_2}{d_2}(p')^{d_2}(1 - p')^{n_2 - d_2} \\
&= \sum_{d_1=0}^{c_1} \binom{n_1}{d_1}(p')^{d_1}(1 - p')^{n_1 - d_1} \\
&\quad + \sum_{d_1=c_1+1}^{r_1-1} \left[\binom{n_1}{d_1}(p')^{d_1}(1 - p')^{n_1 - d_1} \sum_{d_2=0}^{c_2-d_1} \binom{n_2}{d_2}(p')^{d_2}(1 - p')^{n_2 - d_2} \right].
\end{aligned}
$$

*13.2.2.3 *Example.* A large lot of $\frac{3}{4}$-inch screws is submitted for sampling inspection by means of double sampling. If the first sample of 20 contains no defectives, the lot is to be accepted. If it contains more than two defectives, the lot is to be rejected. Otherwise, a second sample of 40 is to be drawn, and the lot is to be accepted if the total number of defectives in both samples does not exceed two. This plan can be characterized by the following table. Find the probability of accepting the lot if $p' = 5\%$.

Sample	Sample Size	Combined Samples		
		Size	Acceptance Number	Rejection Number
First	20	20	0	3
Second	40	60	2	3

The lot will be accepted if (1) there is 0 defective in the first sample; (2) there is 1 defective in the first sample followed by 0 or 1 defectives in the second sample; (3) there are 2 defectives in the first sample followed by 0 defectives in the second sample. The probability of accepting the lot is then given by the sum of the probabilities of these three terms:

(1) $\sum_{d_1=0}^{0} \binom{n_1}{d_1}(p')^{d_1}(1-p')^{n_1-d_1} = (1-p')^{20} = (0.95)^{20} = 0.358;$

(2) $\binom{n_1}{1}(p')(1-p')^{n_1-1} \sum_{d_2=0}^{1} \binom{n_2}{d_2}(p')^{d_2}(1-p')^{n_2-d_2}$

$= \binom{n_1}{1}(p')(1-p')^{n_1-1}\left[(1-p')^{n_2} + \binom{n_2}{1}(p')(1-p')^{n_2-1}\right]$

$= \binom{20}{1}0.05(0.95)^{19}\left[(0.95)^{40} + \binom{40}{1}0.05(0.95)^{39}\right]$

$= 0.377[0.129 + 0.270] = 0.150;$

(3) $\binom{n_1}{2}(p')^2(1-p')^{n_1-2} \sum_{d_2=0}^{0} \binom{n_2}{d_2}(p')^{d_2}(1-p')^{n_2-d_2}$

$= \binom{n_1}{2}(p')^2(1-p')^{n_1-2}(1-p')^{n_2}$

$= \binom{20}{2}(0.05)^2(0.95)^{18}(0.95)^{40} = 0.189 \times 0.129 = 0.02438.$

Thus,

$$L(p') = (1) + (2) + (3) = 0.532.$$

Using the Poisson approximation, the following is obtained:

(1) $0.368;$

(2) $(0.736 - 0.368)(0.406) = 0.368 \times 0.406 = 0.149;$[1]

(3) $(0.920 - 0.736)(0.135) = 0.184 \times 0.135 = 0.025.$

Thus,

$$L(p') = (1) + (2) + (3) = 0.542.$$

13.2.3 MULTIPLE SAMPLING PLANS

A multiple sampling procedure is represented in Table 13.2.

A first sample of n_1 is drawn. The lot is accepted if there are no more than c_1 defectives; the lot is rejected if there are r_1 or more defectives. Otherwise, a second sample of n_2 is drawn. The lot is accepted if there are no more than c_2 defectives in the combined sample of $n_1 + n_2$. The lot is rejected if there are r_2 or more defectives in the combined sample of $n_1 + n_2$. The procedure is continued in accordance with the table above. If, by the end of the sixth sample, the lot is neither accepted nor rejected, a sample of n_7 is drawn. The lot is accepted if the number of defectives in the combined sample of

---

[1] The Poisson table is set up to give approximate values of $\sum_{d=0}^{c} \binom{n}{d}(p')^d(1-p')^{n-d}$. To get an individual term, say, $\binom{n}{c}(p')^c(1-p')^{n-c}$, one uses

$$\sum_{d=0}^{c} \binom{n}{d}(p')^d(1-p')^{n-d} - \sum_{d=0}^{c-1} \binom{n}{d}(p')^d(1-p')^{n-d} = \binom{n}{c}(p')^c(1-p')^{n-c}.$$

Table 13.2. Multiple Sampling Plan

| Sample | Sample Size | Combined Samples | | |
		Size	Acceptance Number	Rejection Number
First	n_1	n_1	c_1	r_1
Second	n_2	$n_1 + n_2$	c_2	r_2
Third	n_3	$n_1 + n_2 + n_3$	c_3	r_3
Fourth	n_4	$n_1 + n_2 + n_3 + n_4$	c_4	r_4
Fifth	n_5	$n_1 + n_2 + n_3 + n_4 + n_5$	c_5	r_5
Sixth	n_6	$n_1 + n_2 + n_3 + n_4 + n_5 + n_6$	c_6	r_6
Seventh	n_7	$n_1 + n_2 + n_3 + n_4 + n_5 + n_6 + n_7$	c_7	$c_7 + 1$

$n_1 + n_2 + n_3 + n_4 + n_5 + n_6 + n_7$ does not exceed c_7. Otherwise, the lot is rejected. Note that $c_1 < c_2 < \cdots < c_7$ and $c_i < r_i$, for all i. Of course, plans can be devised which permit any number of samples before a decision is reached. A multiple sampling plan will generally involve less inspection, on the average, than the corresponding single or double sampling plan guaranteeing the same protection. The advantage is quite important, since inspection costs are directly related to sample sizes. On the other hand, some of the disadvantages of multiple sampling plans are: (1) they usually require higher administrative costs than single or double sampling plans; (2) the variability of the inspection load introduces difficulties in scheduling inspection time; (3) higher caliber inspection personnel may be necessary to guarantee proper use of the plans; (4) adequate storage facilities must be provided for the lot while multiple sampling is being carried on.

13.2.4 CLASSIFICATION OF SAMPLING PLANS

Published tables of sampling plans have been classified into four categories, namely, by acceptable quality level (AQL), by lot tolerance percent defective (LTPD), by point of control, and by average outgoing quality limit (AOQL).

13.2.4.1 *Classification by* AQL. Acceptance procedures based on the acceptable quality level (AQL) generally make use of the process average to determine the sampling plan to be used. The AQL may be viewed as the highest percent defective that is acceptable as a process average. Under most sampling inspection procedures, a lot submitted at AQL quality will have a high probability of acceptance. The probability of acceptance, $1 - \alpha$, at the AQL is usually set near the 95% point, and α is known as the producer's risk. Thus, a producer has good protection against rejection of submitted lots from a process that is at the AQL or better. On the other hand, this type of classification does not specify anything about the protection the consumer has against the acceptance of a lot worse than the AQL. The most widely used published tables classified by the AQL are those of Military Standard 105D, which will be discussed in detail in Sec. 13.2.6.

13.2.4.2 *Classification by* LTPD. The lot tolerance percent defective (LTPD) usually refers to that incoming quality above which there is a small chance that a lot will be accepted. This probability is usually taken to be near the 10% point. Thus, a consumer inspecting lots submitted from a process that is at the LTPD or worse, has a small probability of accepting such lots. This probability is known as the consumer's risk. On the other hand, this type of classification does not specify anything about the protection the producer has against the rejection of lots better than the LTPD. The most widely used published tables classified by the LTPD are the Dodge-Romig tables which will be discussed in detail in Sec. 13.2.5.

13.2.4.3 *Classification by Point of Control.* The point of control is the lot quality for which the probability of acceptance is 0.50. The concept here is one of splitting the risk between producer and consumer. Lots submitted from processes whose quality is better than the point of control have a probability of acceptance that is higher than 50%. Lots submitted from processes whose quality is worse than the point of control have a probability of acceptance smaller than 50%.

A set of plans indexed according to the point of control was developed by H. Hamaker and is known as the Philips Standard Sampling System.[1] Plans are selected according to lot size and the point of control (0.25%, 0.5%, 1%, 2%, 3%, 5%, 7%, and 10%). Single sampling plans are used for lot sizes less than or equal to 1,000, whereas double sampling plans are called for when the lot size exceeds 1,000.

13.2.4.4 *Classification by* AOQL. The average outgoing quality limit (AOQL) does not refer to a point on the OC curve, but to the upper limit on outgoing quality that may be expected in the long run when all rejected lots are subjected to 100% inspection, with all defective articles removed and replaced by good articles. The average outgoing quality limit (AOQL) can be computed from the formulas

$$\text{AOQ} = p'L(p'); \qquad \text{AOQL} = \max_{p'} \text{AOQ}$$

where $\max_{p'} \text{AOQ}$ is the maximum AOQ over the range $p' = 0$ to $p' = 1$. The most widely used published tables classified by the AOQL are the Dodge-Romig tables which will be discussed in Sec. 13.2.5.

Figure 13.5 shows a graphic comparison of these four methods of indexing acceptance plans in relation to a stated quality standard. The four curves

---

[1] This standard has been published by N. V. Philips' Gloerlampenfabrieken of Eindhoven, Netherlands.

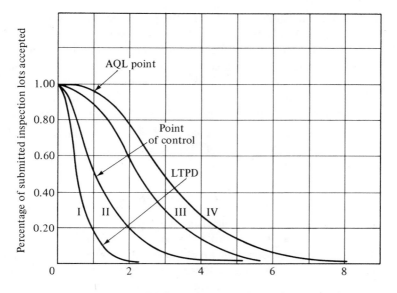

Percentage of defective items in submitted inspection lots

Fig. 13.5. *Single sampling classified on:*
I. 1% LTPD: n = *335;* c = *1.*
II. Point of control = *1%:* n = *150;* c = *1.*
III. 1% AOQL: n = *120;* c = *2.*
IV. AQL = *1%:* n = *125;* c = *3.*

shown in this figure are OC curves for single sampling plans that are indexed in published tables by the 1.0% defective figure for a lot size of 1,000. The most lenient plan (curve IV) is that classified by the AQL; in this plan with an AQL of 1.0%, three defectives are permitted in a sample of 125. The strictest plan (curve I) is the LTPD plan permitting one defective in a sample of 335. The point of control plan (curve II) permits 1 defective in a sample of 150; here the probability of acceptance of a 1.0% defective lot is 0.50. The 1.0% AOQL plan allows 2 defectives in a sample of 120; the corresponding OC curve (curve III) is intermediate between the AQL 1.0% and the point of control 1.0% plans.

13.2.5 DODGE-ROMIG TABLES

The most widely published tables of sampling plans classified by LTPD and AOQL are due to H.F. Dodge and H.G. Romig.[1] These tables originated in the Bell Telephone Laboratories and reflect many years of experience

---

[1] H. F. Dodge and H. G. Romig, *Sampling Inspection Tables*, 2nd ed., John Wiley & Sons, Inc., New York, 1959.

with acceptance sampling in the Bell Telephone System. Since the Bell Telephone System fulfills the role of both producer and consumer, these tables are concerned with the total cost of inspection, including both the cost of sampling inspection and the cost of screening rejected lots. In particular, when these tables are used properly, they provide plans which minimize the expected total amount of inspection. Four sets of plans are included, i.e., single and double sampling plans classified by LTPD and AOQL.

13.2.5.1 *Single Sampling Lot Tolerance Tables.* These plans are indexed according to the following lot tolerance percent defectives (LTPD) with the consumer's risk set equal to 0.10.

$$0.5\% \quad 1.0\% \quad 2.0\% \quad 3.0\%$$
$$4.0\% \quad 5.0\% \quad 7.0\% \quad 10.0\%.$$

Table 13.3 presents a typical table from the Dodge-Romig single sampling LTPD plans. Each plan in this table has the same LTPD. Furthermore, if rejected lots are screened, the table gives the AOQL associated with the sampling plan. For any given lot size there is a choice of six plans. A plan is chosen from a column headed by a range of possible process averages. If the true process average of the production process falls within this range, and if rejected lots are screened, the chosen plan will guarantee that the expected total amount of inspection, including sampling inspection and screening of rejected lots, will be smaller than that for any of the remaining five plans.[1] The process average is usually estimated from past data by taking the ratio of the total number of defectives to the total number inspected. If there are no past data to estimate the process average, the plan should be selected from the right-hand column of the table. This gives good lots a better chance of acceptance.

It must be emphasized that if screening of rejected lots does not take place, these plans still guarantee a stated LTPD.

13.2.5.2 *Double Sampling Lot Tolerance Tables.* These plans are indexed according to the same LTPD's as for the single sampling LTPD plans. A typical double sampling LTPD table is shown in Table 13.4. This table has the same features as the single sampling LTPD tables with the exception that it uses double sampling instead of single sampling. The first sample is always smaller than the corresponding single sampling plan, but if a second sample is required, the combined sample size exceeds the sample size of the single sampling plan. However, the average amount of sampling inspection is generally smaller with the double sampling plan. Since the second sample is not always taken (depending on the submitted quality), the process average is usually estimated from the first sample only.

---

[1] The average amount of inspection (AOI) is given by the formula

$$\text{AOI} = nL(p') + N[1 - L(p')].$$

Table 13.3. Example of Dodge-Romig Single Sampling Lot Tolerance Tables[1]

(Lot tolerance percent defective = 4.0%; consumer's risk = 0.10)

Lot Size	Process Average 0 to 0.04%			Process Average 0.05 to 0.40%			Process Average 0.41 to 0.80%			Process Average 0.81 to 1.20%			Process Average 1.21 to 1.60%			Process Average 1.61 to 2.00%		
	n	c	AOQL %	n	c	AOQL %	n	c	AOQL %	n	c	AOQL %	n	c	AOQL %	n	c	AOQL %
1–35	All	0	0	All	0	0	All	0	0	All	0	0	All	0	0	All	0	0
36–50	34	0	0.35	34	0	0.35	34	0	0.35	34	0	0.35	34	0	0.35	34	0	0.35
51–100	44	0	0.47	44	0	0.47	44	0	0.47	44	0	0.47	44	0	0.47	44	0	0.47
101–200	50	0	0.55	50	0	0.55	50	0	0.55	50	0	0.55	50	0	0.55	50	0	0.55
201–300	55	0	0.57	55	0	0.57	85	1	0.71	85	1	0.71	85	1	0.71	85	1	0.71
301–400	55	0	0.58	55	0	0.58	90	1	0.72	120	2	0.80	120	2	0.80	145	3	0.86
401–500	55	0	0.60	55	0	0.60	90	1	0.77	120	2	0.87	150	3	0.91	150	3	0.91
501–600	55	0	0.61	95	1	0.76	125	2	0.87	125	2	0.87	155	3	0.93	185	4	0.95
601–800	55	0	0.62	95	1	0.78	125	2	0.93	160	3	0.97	190	4	1.0	220	5	1.0
801–1,000	55	0	0.63	95	1	0.80	130	2	0.92	165	3	0.98	220	5	1.1	255	6	1.1
1,001–2,000	55	0	0.65	95	1	0.84	165	3	1.1	195	4	1.2	255	6	1.3	315	8	1.4
2,001–3,000	95	1	0.86	130	2	1.0	165	3	1.1	230	5	1.3	320	8	1.4	405	11	1.6
3,001–4,000	95	1	0.86	130	2	1.0	195	4	1.2	260	6	1.4	350	9	1.5	465	13	1.6
4,001–5,000	95	1	0.87	130	2	1.0	195	4	1.2	290	7	1.4	380	10	1.6	520	15	1.7
5,001–7,000	95	1	0.87	130	2	1.0	200	4	1.2	290	7	1.5	410	11	1.7	575	17	1.9
7,001–10,000	95	1	0.88	130	2	1.1	230	5	1.4	325	8	1.5	440	12	1.7	645	19	1.9
10,001–20,000	95	1	0.88	165	3	1.2	265	6	1.4	355	9	1.6	500	14	1.8	730	22	2.0
20,001–50,000	95	1	0.88	165	3	1.2	295	7	1.5	380	10	1.7	590	17	2.0	870	26	2.1
50,001–100,000	95	1	0.88	200	4	1.3	325	8	1.6	410	11	1.8	620	18	2.0	925	29	2.2

[1] From H. F. Dodge and H. Romig, *Sampling Inspection Tables*, 2nd ed., John Wiley & Sons, Inc., New York, 1959. Reprinted by permission of John Wiley & Sons, Inc.

Table 13.4. Example of Dodge-Romig Double Sampling Lot Tolerance Tables[1]
(Lot tolerance percent defective = 4.0%; consumer's risk = 0.10)

Lot Size	Process Average 0 to 0.04%						Process Average 0.05 to 0.40%						Process Average 0.41 to 0.80%					
	Trial 1		Trial 2			AOQL in %	Trial 1		Trial 2			AOQL in %	Trial 1		Trial 2			AOQL in %
	n_1	c_1	n_2	n_1+n_2	c_2		n_1	c_1	n_2	n_1+n_2	c_2		n_1	c_1	n_2	n_1+n_2	c_2	
1-35	All	0	—	—	—	0	All	0	—	—	—	0	All	0	—	—	—	0
36-50	34	0	—	—	—	0.35	34	0	—	—	—	0.35	34	0	—	—	—	0.35
51-75	40	0	—	—	—	0.43	40	0	—	—	—	0.43	40	0	—	—	—	0.43
76-100	50	0	25	75	1	0.46	50	0	25	75	1	0.46	50	0	25	75	1	0.46
101-150	55	0	30	85	1	0.55	55	0	30	85	1	0.55	55	0	30	85	1	0.55
151-200	60	0	30	90	1	0.64	60	0	30	90	1	0.64	60	0	30	90	1	0.64
201-300	60	0	35	95	1	0.70	60	0	35	95	1	0.70	60	0	65	125	2	0.75
301-400	65	0	35	100	1	0.71	65	0	35	100	1	0.71	65	0	65	130	2	0.80
401-500	65	0	40	105	1	0.73	65	0	70	135	2	0.83	65	0	70	135	2	0.83
501-600	65	0	40	105	1	0.74	65	0	75	140	2	0.85	65	0	100	165	3	0.93
601-800	65	0	40	105	1	0.75	65	0	75	140	2	0.87	65	0	110	175	3	0.97
801-1,000	70	0	40	110	1	0.76	70	0	75	145	2	0.90	70	0	105	175	3	0.98
1,001-2,000	70	0	40	110	1	0.78	70	0	80	150	2	0.94	70	0	145	215	4	1.1
2,001-3,000	70	0	80	150	2	0.95	70	0	115	185	3	1.1	70	0	180	250	5	1.2
3,001-4,000	70	0	80	150	2	0.96	70	0	115	185	3	1.1	110	1	175	285	6	1.3
4,001-5,000	70	0	80	150	2	0.97	70	0	115	185	3	1.1	115	1	170	285	6	1.3
5,001-7,000	70	0	80	150	2	0.98	70	0	115	185	3	1.1	115	1	205	320	7	1.4
7,001-10,000	70	0	80	150	2	0.98	70	0	150	220	4	1.2	115	1	205	320	7	1.4
10,001-20,000	70	0	80	150	2	0.98	70	0	150	220	4	1.2	115	1	235	350	8	1.5
20,001-50,000	70	0	80	150	2	0.99	70	0	150	220	4	1.2	115	1	270	385	9	1.6
50,001-100,000	70	0	80	150	2	0.99	70	0	185	255	5	1.3	115	1	300	415	10	1.7

Table 13.4 (continued). Example of Dodge-Romig Double Sampling Lot Tolerance Tables

Lot Size	Process Average 0.81 to 1.20%						Process Average 1.21 to 1.60%						Process Average 1.61 to 2.00%					
	Trial 1		Trial 2			AOQL in %	Trial 1		Trial 2			AOQL in %	Trial 1		Trial 2			AOQL in %
	n_1	c_1	n_2	$n_1 + n_2$	c_2		n_1	c_1	n_2	$n_1 + n_2$	c_2		n_1	c_1	n_2	$n_1 + n_2$	c_2	
1–35	All	0	—	—	—	0	All	0	—	—	—	0	All	0	—	—	—	0
36–50	34	0	—	—	—	0.35	34	0	—	—	—	0.35	34	0	—	—	—	0.35
51–75	40	0	—	—	—	0.43	40	0	—	—	—	0.43	40	0	—	—	—	0.43
76–100	50	0	25	75	1	0.46	50	0	25	75	1	0.46	50	0	25	75	1	0.46
101–150	55	0	30	85	1	0.55	55	0	30	85	1	0.55	55	0	30	85	1	0.55
151–200	60	0	55	115	2	0.68	60	0	55	115	2	0.68	60	0	55	115	2	0.68
201–300	60	0	65	125	2	0.75	60	0	90	150	3	0.84	60	0	90	150	3	0.84
301–400	65	0	95	160	3	0.86	65	0	95	160	3	0.86	65	0	120	185	4	0.92
401–500	65	0	100	165	3	0.92	65	0	130	195	4	0.96	105	1	140	245	6	1.0
501–600	65	0	135	200	4	1.0	105	1	145	250	6	1.1	105	1	175	280	7	1.1
601–800	65	0	140	205	4	1.0	105	1	185	290	7	1.2	105	1	210	315	8	1.2
801–1,000	110	1	155	265	6	1.2	110	1	210	320	8	1.2	145	2	230	375	10	1.3
1,001–2,000	110	1	195	305	7	1.3	150	2	240	390	10	1.5	180	3	295	475	13	1.6
2,001–3,000	110	1	260	370	9	1.4	185	3	305	490	13	1.6	220	4	410	630	18	1.7
3,001–4,000	150	2	255	405	10	1.5	185	3	340	525	14	1.6	285	6	465	750	22	1.8
4,001–5,000	150	2	285	435	11	1.6	185	3	395	580	16	1.7	285	6	520	805	24	1.9
5,001–7,000	150	2	320	470	12	1.6	185	3	435	620	17	1.7	320	7	585	905	27	2.0
7,001–10,000	150	2	325	475	12	1.7	220	4	460	680	19	1.9	320	7	645	965	29	2.1
10,001–20,000	150	2	355	505	13	1.7	220	4	495	715	20	1.9	350	8	790	1,140	35	2.2
20,001–50,000	150	2	420	570	15	1.7	255	5	575	830	24	2.0	385	9	895	1,280	40	2.3
50,001–100,000	150	2	450	600	16	1.8	255	5	665	920	27	2.1	415	10	985	1,400	44	2.4

[1] From H. F. Dodge and H. Romig, *Sampling Inspection Tables*, 2nd ed., John Wiley & Sons, Inc., New York, 1959. Reprinted by permission of John Wiley & Sons, Inc.

13.2.5.3 *Single Sampling* AOQL *Tables.* These plans are indexed according to the following average outgoing quality limits (AOQL).

0.1%	1.5%	4.0%
0.25%	2.0%	5.0%
0.5%	2.5%	7.0%
0.75%	3.0%	10.0%
1.0%		

A typical single sampling AOQL table is shown in Table 13.5. All the sampling plans in this table have the same AOQL, assuming all rejected lots are screened. Furthermore, the table gives the LTPD values for each plan. For any given lot size there is a choice of six plans. A plan is chosen from a column headed by a range of possible process averages. If the true process average falls within this range, and if rejected lots are screened, the chosen plan will guarantee that the expected total amount of inspection, including sampling inspection and screening of rejected lots, will be smaller than that for any of the remaining five plans.

13.2.5.4 *Double Sampling* AOQL *Tables.* These plans are indexed according to the same AOQL's as for the single sampling AOQL plans. A typical double sampling AOQL table is shown in Table 13.6. This table has the same features as the single sampling AOQL tables, except that it uses double sampling instead of single sampling.

13.2.6 MILITARY STANDARD 105D

13.2.6.1 *History.* There have been various sets of published sampling tables, using the AQL as an index. The first set of tables was developed in 1943 for Army Ordnance and, with some changes and extensions, became the Army Service Forces tables, used during the final years of World War II.

An administration manual, *Standard Sampling Inspection Procedures,* was issued by the Navy in October, 1945. This manual, including tables and OC (operating characteristic) curves for all the acceptance sampling plans given, had been prepared by the Statistical Research Group of Columbia University during World War II. Although the tables and procedures in this manual were similar in many respects to those of the Army Service Forces, sequential (multiple) sampling plans were included in addition to single and double sampling plans, and there were several other important points of difference. The actual tables from this manual were issued by the Navy Department in April, 1946, as Appendix X to *General Specifications for Inspection of Material.* After the unification of the armed services, Appendix X was adopted by the Department of Defense in February, 1949, as JAN-STD-105.

In 1949, a book entitled *Sampling Inspection,* edited by H.A. Freeman, M. Friedman, F. Mosteller, and W.A. Wallis,[1] was published. This book was

---

[1] H. A. Freeman, M. Friedman, F. Mosteller, and W. A. Wallis, *Sampling Inspection,* McGraw-Hill Book Company, Inc., New York, 1949.

Table 13.5. Example of Dodge-Romig Single Sampling AOQL Tables[1]

(Average outgoing quality limit = 1.0%)

Lot Size	Process Average 0 to 0.02%			Process Average 0.03 to 0.20%			Process Average 0.21 to 0.40%			Process Average 0.41 to 0.60%			Process Average 0.61 to 0.80%			Process Average 0.81 to 1.00%		
	n	c	p_t %	n	c	p_t %	n	c	p_t %	n	c	p_t %	n	c	p_t %	n	c	p_t %
1–25	All	0	—	All	0	—	All	0	—	All	0	—	All	0	—	All	0	—
26–50	22	0	7.7	22	0	7.7	22	0	7.7	22	0	7.7	22	0	7.7	22	0	7.7
51–100	27	0	7.1	27	0	7.1	27	0	7.1	27	0	7.1	27	0	7.1	27	0	7.1
101–200	32	0	6.4	32	0	6.4	32	0	6.4	32	0	6.4	32	0	6.4	32	0	6.4
201–300	33	0	6.3	33	0	6.3	33	0	6.3	33	0	6.3	33	0	6.3	65	1	5.0
301–400	34	0	6.1	34	0	6.1	34	0	6.1	70	1	4.6	70	1	4.6	70	1	4.6
401–500	35	0	6.1	35	0	6.1	35	0	6.1	70	1	4.7	70	1	4.7	70	1	4.7
501–600	35	0	6.1	35	0	6.1	75	1	4.4	75	1	4.4	75	1	4.4	75	1	4.4
601–800	35	0	6.2	35	0	6.2	75	1	4.4	75	1	4.4	75	1	4.4	120	2	4.2
801–1,000	35	0	6.3	35	0	6.3	80	1	4.4	80	1	4.4	120	2	4.3	120	2	4.3
1,001–2,000	36	0	6.2	80	1	4.5	80	1	4.5	130	2	4.0	130	2	4.0	180	3	3.7
2,001–3,000	36	0	6.2	80	1	4.6	80	1	4.6	130	2	4.0	185	3	3.6	235	4	3.3
3,001–4,000	36	0	6.2	80	1	4.7	135	2	3.9	135	2	3.9	185	3	3.6	295	5	3.1
4,001–5,000	36	0	6.2	85	1	4.6	135	2	3.9	190	3	3.5	245	4	3.2	300	5	3.1
5,001–7,000	37	0	6.1	85	1	4.6	135	2	3.9	190	3	3.5	305	5	3.0	420	7	2.8
7,001–10,000	37	0	6.2	85	1	4.6	135	2	3.9	245	4	3.2	310	5	3.0	430	7	2.7
10,001–20,000	85	1	4.6	135	2	3.9	195	3	3.4	250	4	3.2	435	7	2.7	635	10	2.4
20,001–50,000	85	1	4.6	135	2	3.9	255	4	3.1	380	6	2.8	575	9	2.5	990	15	2.1
50,001–100,000	85	1	4.6	135	2	3.9	255	4	3.1	445	7	2.6	790	12	2.3	1,520	22	1.9

[1] From H. F. Dodge and H. Romig, *Sampling Inspection Tables*, 2nd ed., John Wiley & Sons, Inc., New York, 1959. Reprinted by permission of John Wiley & Sons, Inc.

Table 13.6. Example of Dodge-Romig Double Sampling AOQL Tables[1]

(Average outgoing quality limit = 1.0%)

Lot Size	Process Average 0 to 0.02%						Process Average 0.03 to 0.20%						Process Average 0.21 to 0.40%					
	Trial 1		Trial 2			p_t %	Trial 1		Trial 2			p_t %	Trial 1		Trial 2			p_t %
	n_1	c_1	n_2	n_1+n_2	c_2		n_1	c_1	n_2	n_1+n_2	c_2		n_1	c_1	n_2	n_1+n_2	c_2	
1–25	All	0	—	—	—	—	All	0	—	—	—	—	All	0	—	—	—	—
26–50	22	0	—	—	—	7.7	22	0	—	—	—	7.7	22	0	—	—	—	7.7
51–100	33	0	17	50	1	6.9	33	0	17	50	1	6.9	33	0	17	50	1	6.9
101–200	43	0	22	65	1	5.8	43	0	22	65	1	5.8	43	0	22	65	1	5.8
201–300	47	0	28	75	1	5.5	47	0	28	75	1	5.5	47	0	28	75	1	5.5
301–400	49	0	31	80	1	5.4	49	0	31	80	1	5.4	55	0	60	115	2	4.8
401–500	50	0	30	80	1	5.4	50	0	30	80	1	5.4	55	0	65	120	2	4.7
501–600	50	0	30	80	1	5.4	50	0	30	80	1	5.4	60	0	65	125	2	4.6
601–800	50	0	35	85	1	5.3	60	0	70	130	2	4.5	60	0	70	130	2	4.5
801–1,000	55	0	30	85	1	5.2	60	0	75	135	2	4.4	60	0	75	135	2	4.4
1,001–2,000	55	0	35	90	1	5.1	65	0	75	140	2	4.3	75	0	120	195	3	3.8
2,001–3,000	65	0	80	145	2	4.2	65	0	80	145	2	4.2	75	0	125	200	3	3.7
3,001–4,000	70	0	80	150	2	4.1	70	0	80	150	2	4.1	80	0	175	255	4	3.5
4,001–5,000	70	0	80	150	2	4.1	70	0	80	150	2	4.1	80	0	180	260	4	3.4
5,001–7,000	70	0	80	150	2	4.1	75	0	125	200	3	3.7	80	0	180	260	4	3.4
7,001–10,000	70	0	80	150	2	4.1	80	0	125	205	3	3.6	85	0	180	265	4	3.3
10,001–20,000	70	0	80	150	2	4.1	80	0	130	210	3	3.6	90	0	230	320	5	3.2
20,001–50,000	75	0	80	155	2	4.0	80	0	135	215	3	3.6	95	0	300	395	6	2.9
50,001–100,000	75	0	80	155	2	4.0	85	0	180	265	4	3.3	170	1	380	550	8	2.6

Table 13.6 (continued). Example of Dodge-Romig Double Sampling AOQL Tables

Lot Size	Process Average 0.41 to 0.60%						Process Average 0.61 to 0.80%						Process Average 0.81 to 1.00%					
	Trial 1		Trial 2			p_t %	Trial 1		Trial 2			p_t %	Trial 1		Trial 2			p_t %
	n_1	c_1	n_2	$n_1 + n_2$	c_2		n_1	c_1	n_2	$n_1 + n_2$	c_2		n_1	c_1	n_2	$n_1 + n_2$	c_2	
1–25	All	0	—	—	—	—	All	0	—	—	—	—	All	0	—	—	—	—
26–50	22	0	—	—	—	7.7	22	0	—	—	—	7.7	22	0	—	—	—	7.7
51–100	33	0	17	50	1	6.9	33	0	17	50	1	6.9	33	0	17	50	1	6.9
101–200	43	0	22	65	1	5.8	43	0	22	65	1	5.8	47	0	43	90	2	5.4
201–300	55	0	50	105	2	4.9	55	0	50	105	2	4.9	55	0	50	105	2	4.9
301–400	55	0	60	115	2	4.8	55	0	60	115	2	4.8	60	0	80	140	3	4.5
401–500	55	0	65	120	2	4.7	60	0	95	155	3	4.3	60	0	95	155	3	4.3
501–600	60	0	65	125	2	4.6	65	0	100	165	3	4.2	65	0	100	165	3	4.2
601–800	65	0	105	170	3	4.1	65	0	105	170	3	4.1	70	0	140	210	4	3.9
801–1,000	65	0	110	175	3	4.0	70	0	150	220	4	3.8	125	1	180	305	6	3.5
1,001–2,000	80	0	165	245	4	3.7	135	1	200	335	6	3.3	140	1	245	385	7	3.2
2,001–3,000	80	0	170	250	4	3.6	150	1	265	415	7	3.0	215	2	355	570	10	2.8
3,001–4,000	85	0	220	305	5	3.3	160	1	330	490	8	2.8	225	2	455	680	12	2.7
4,001–5,000	145	1	225	370	6	3.1	225	2	375	600	10	2.7	240	2	595	835	14	2.5
5,001–7,000	155	1	285	440	7	2.9	235	2	440	675	11	2.6	310	3	665	975	16	2.4
7,001–10,000	165	1	355	520	8	2.7	250	2	585	835	13	2.4	385	4	785	1,170	19	2.3
10,001–20,000	175	1	415	590	9	2.6	325	3	655	980	15	2.3	520	6	980	1,500	24	2.2
20,001–50,000	250	2	490	740	11	2.4	340	3	910	1,250	19	2.2	610	7	1,410	2,020	32	2.1
50,001–100,000	275	2	700	975	14	2.2	420	4	1,050	1,470	22	2.1	770	9	1,850	2,620	41	2.0

[1] From H. F. Dodge and H. Romig, *Sampling Inspection Tables*, 2nd ed., John Wiley & Sons, Inc., New York, 1959. Reprinted by permission of John Wiley & Sons, Inc.

prepared by the Statistical Research Group of Columbia University, and with a few minor differences, the tables in *Sampling Inspection* are identical with JAN-STD-105.

JAN-STD-105 was superseded by MIL-STD-105A in 1950. Minor revisions resulted in MIL-STD-105B in 1958 and MIL-STD-105C in 1961. In April, 1963, a major revision, MIL-STD-105D appeared.[1] This new standard was developed by representatives from the military services of the U.S.A., the United Kingdom, and Canada, *and* with the cooperation of the American and European organizations for quality control.[2] Because of its international flavor, this standard will undoubtedly have great influence on acceptance procedures used in industry throughout the world.

MIL-STD-105D is one of a class of sampling systems. A sampling system is a collection of sampling plans together with rules for their operation. Its purpose is to (a) protect the consumer from accepting poor lots, (b) protect the producer from having good lots rejected, and (c) encourage the producer to submit good quality. This is generally accomplished by controlling the amount of inspection and the acceptance criteria according to specific rules of operation formulated in the sampling system.

The remaining subsections of Sec. 13.2.6 will be concerned with a discussion of this document.

13.2.6.2 *Classification of Defects.* Defects are grouped into three classes, i.e., critical, major, and minor. A critical defect "is a defect that judgment and experience indicate is likely to result in hazardous or unsafe conditions for individuals using, maintaining, or depending on the product; or a defect that judgment and experience indicate is likely to prevent performance of the tactical function of a major end item, such as a ship, aircraft, tank, missile, or space vehicle." A major defect "is a defect, other than critical, that is likely to result in failure, or to reduce materially the usability of the unit of product for its intended purpose." A minor defect "is a defect that is not likely to reduce materially the usability of the unit of product for its intended purpose, or is a departure from established standards having little bearing on the effective use or operation of the unit."

13.2.6.3 *Acceptable Quality Levels.* The acceptable quality level (AQL) "is the maximum percent defective (or the maximum number of defects per hundred units) that for purposes of sampling inspection, can be considered satisfactory as a process average." The distinction between a defect and a defective is readily seen if a defective item is defined as one containing one or more defects. If the percentage defective is relatively small, more than one

---

[1] MIL-STD-105D is available for private use and is for sale by the Superintendent of Documents, U. S. Government Printing Office, Washington, D. C., 20402. Price 40 cents.

[2] The international designation of this standard is ABC-STD-105. A denotes America, B denotes Britain, and C denotes Canada.

defect will occur on an item infrequently. Consequently, in this case, the mathematical theory using defects is essentially equivalent to using defectives. The value of the AQL's are

0.010				
0.015				
0.025	0.25	2.5	25.0	250.0
0.040	0.40	4.0	40.0	400.0
0.065	0.65	6.5	65.0	650.0
0.10	1.0	10.0	100.0	1,000.0
0.15	1.5	15.0	150.0	

Here AQL values of 10.0 or less are expressed either in percent defective or in defects per hundred units; those over 10.0 are expressed in defects per hundred units only. These points are approximately in geometric progression and multiples of 1, 1.5, 2.5, 4.0 and 6.5. The probabilities of acceptance of submitted lots having these AQL's range from about 0.88 for the smallest sample sizes to about 0.99 for the largest sample sizes. The particular AQL to be used is "designated in the contract or by the responsible authority." According to the standard "different AQL's may be designated for groups of defects considered collectively, or for individual defects. An AQL for a group of defects may be designated in addition to AQL's for individual defects, or subgroups, within that group." In summary then, in order to use the standard, AQL's are specified for groups of defects, together with a list of characteristics to be inspected for.

13.2.6.4 *Normal, Tightened, and Reduced Inspection.* The standard calls for using normal inspection at the start unless otherwise directed. Normal inspection is intended when the submitted quality is better than or equal to the AQL. Normal inspection is continued as long as neither tightened nor reduced inspection is called for.

Tightened inspection is intended when the submitted quality is worse than the AQL, and the standard provides rules for switching between normal and tightened inspection. Tightened inspection is instituted when 2 out of 5 consecutive lots have been rejected. Normal inspection is reinstated when 5 consecutive lots have been accepted. The standard contains an interesting rule for discontinuing inspection altogether. If 10 consecutive lots remain on tightened inspection, inspection under MIL-STD-105D is to cease pending action to improve the submitted quality.

Under tightened inspection, the producer's risk is increased, whereas the consumer's risk is decreased. In other words, the probability of accepting bad lots (as well as good lots) is decreased. In MIL-STD-105D this is generally accomplished by keeping the sample size fixed (the same as under normal inspection) while reducing the acceptance number. This is frequently accomplished by shifting the sampling plan one step to the next lower (left) AQL plan, e.g., if a 4% AQL plan is used under normal inspection, a 2.5% AQL plan is required under tightened inspection.

Since the OC curve depends on whether normal or tightened inspection is in effect, the switching rules lead to an over-all sampling plan with a composite operating characteristic curve. In other words, if lots are submitted with incoming quality p', and the rules for switching from normal to tightened inspection are used, what is the probability of accepting the lot? The composite OC curve always lies between the OC curves of the normal and tightened sampling plans, and for all practical purposes coincides with the OC curve for normal inspection when submitted quality is "good" and coincides with the OC curve for tightened inspection when submitted quality is "bad," i.e., the composite OC curve is steeper than each of its components—a very desirable situation.

Reduced inspection is intended when the submitted quality is substantially better than the AQL. It is not an essential part of the sampling system, but does provide an opportunity for substantial savings in inspection costs when the submitted quality is good. Reduced inspection is instituted, provided that all of the following conditions are satisfied:

1. The preceding 10 lots [or occasionally more to satisfy condition (2)] have been on normal inspection and none has been rejected on original inspection.
2. The total number of defectives (or defects) in the samples from the preceding 10 lots is equal to or less than a number tabled in the standard. This table occasionally requires results from more than 10 lots. These limits are approximately the two-sigma units based on the Poisson, i.e., $N(AQL) - 2\sqrt{N(AQL)}$, where N is the number of items looked at in the preceding 10 lots. If double or multiple sampling is in use, all samples inspected are to be included, not "first" samples only.
3. Production is at a steady rate.
4. Reduced inspection is considered desirable by the responsible authority.

Normal inspection is reinstated if any of the following occur on original inspection:

1. A lot is rejected.
2. A lot is considered acceptable, but the number of defective items is "too large." "Too large" is defined for each reduced sampling plan.
3. Production becomes irregular or delayed.
4. Other conditions warrant that normal inspection shall be instituted.

Under reduced inspection the sample size is decreased by approximately 40% of normal inspection. This is accomplished at the expense of increasing the consumer's risk and decreasing (slightly) the producer's risk.

The use of tightened and reduced inspection is an essential feature in the success of MIL-STD-105D. Associated with each normal inspection plan is a tightened inspection plan, together with switching rules. If the sampling system concept is to be valid, both normal and tightened inspection must be used because the threat of tightened inspection is often necessary to insure submission of quality better than or equal to the AQL. If quality is submitted close to but above the AQL, there is still a relatively large probability that it will be accepted under normal inspection. Although this appears to be harmful on the surface, in reality it may not be too damaging. In the first place, rejected lots cost the manufacturer a great deal of money. In fact, too many rejected lots can easily force a manufacturer out of business. For example, few manufacturers can tolerate even a 20% rejection rate for their lots. Second, if lots are constantly being submitted above the AQL, 2 out of 5 consecutive lots will be rejected, resulting in tightened inspection. Recall that for this case the composite OC curve is that of the tightened plan. This will cause even more of the manufacturer's lots to be rejected. Consequently, the manufacturer is actually forced to keep his submitted quality better than the agreed-upon level.

On the other hand, if the manufacturer continues to submit quality a great deal better than the AQL, the consumer is able to use reduced inspection, thereby resulting in a saving of inspection costs.

13.2.6.5 *Sampling Plans.* Sample sizes are designated by code letters from *A* to *S* (omitting *I* and *O*). The sample size code letter depends on the inspection level and the lot size. There are three inspection levels, I, II, and III, for general use. Unless otherwise specified, inspection level II is used. Four additional special levels, *S*-1, *S*-2, *S*-3, and *S*-4, are also given for use when relatively small sample sizes are necessary and large risks can or must be tolerated. The sample size code letter applicable to the specified inspection level and for lots of given size is obtained from Table 13.7 (Table I of MIL-STD-105D).

Since it has been pointed out that the OC curves are essentially independent of lot size (for large lots), it may appear incongruous that one of the entries to the table is lot size. However, a moment's reflection will reveal that to the inspection agency the acceptance of a bad lot is much more serious when the lot size is large than when it is small. Consequently, better protection in the form of steeper OC curves is desired for large lots.

By entering Table 13.7 (Table I of MIL-STD-105D), with the lot size and inspection level, a code letter is selected. This code letter, together with a designated AQL, results in a sampling plan if the type of sampling is specified, i.e., if single, double, or multiple sampling is called for. Alternatively, each code letter and AQL lead to three sampling plans: single, double, or multiple. The standard is constructed so that these plans have essentially equivalent OC curves. Hence, the choice of the type of sampling is based on criteria discussed in Secs. 13.2.2 and 13.2.3. Under double sampling, both sample sizes are chosen to be the same and, furthermore, are approximately equal

Table 13.7. Sample Size Code Letters
(Table I of MIL-STD-105D)

Lot or Batch Size			Special Inspection Levels				General Inspection Levels		
			S–1	S–2	S–3	S–4	I	II	III
2	to	8	A	A	A	A	A	A	B
9	to	15	A	A	A	A	A	B	C
16	to	25	A	A	B	B	B	C	D
26	to	50	A	B	B	C	C	D	E
51	to	90	B	B	C	C	C	E	F
91	to	150	B	B	C	D	D	F	G
151	to	280	B	C	D	E	E	G	H
281	to	500	B	C	D	E	F	H	J
501	to	1200	C	C	E	F	G	J	K
1201	to	3200	C	D	E	G	H	K	L
3201	to	10000	C	D	F	G	J	L	M
10001	to	35000	C	D	F	H	K	M	N
35001	to	150000	D	E	G	J	L	N	P
150001	to	500000	D	E	G	J	M	P	Q
500001	and	over	D	E	H	K	N	Q	R

to $\frac{5}{8}$ of the sample size of the corresponding single sampling plan. Under multiple sampling there are seven equal samples and each is approximately equal to $\frac{1}{4}$ of the sample size of the corresponding single sampling plan.

The master tables for single sampling plans, double sampling plans, and multiple sampling plans for normal inspection are presented in Table 13.8 (Table II-A of MIL-STD-105D), Table 13.9 (Table III-A of MIL-STD-105D, and Table 13.10 (Table IV-A of MIL-STD-105D), respectively. The standard also contains master tables for (a) single sampling plans for tightened and reduced inspection, (b) double sampling plans for tightened and reduced inspection, and (c) multiple sampling plans for tightened and reduced inspection. A table of limit numbers for reduced inspection (alluded to in Sec. 13.2.6.4) is included. For each single sampling plan under normal and tightened inspection, the AOQL is given. For each single sampling plan under normal inspection, the LTPD (called limiting quality in the standard) corresponding to 5% and 10% probabilities of acceptance are available. The LTPD corresponding to 10% is reproduced in Table 13.11 (Table VI-A of MIL-STD-105D). The standard also contains plots of the approximate average sample numbers for double and multiple sampling and compares them to the corresponding single sampling plan. Finally, a complete set of OC curves is included. Although these curves are not reproduced in this text, Table 13.11 contains those incoming qualities corresponding to probabilities of acceptance of $P_a = 0.10$, $P_a = 0.50$, and $P_a = 0.95$.

13.2.6.6 *Summary of the Procedure to be Followed in the Selection of a Sampling Plan from MIL-STD-105D.* Frequently, contracts exist between a producer and a consumer that call for sampling inspection to be performed under MIL-STD-105D. This implies that agreement exists on the lot size, choice of AQL, inspection level (usually level II), form of inspection (usually normal inspection), and type of sampling (single sampling is the most commonly used). For example, suppose that a contract exists between a supplier and consumer calling for delivery of lots of size 1000, with inspection to be performed under MIL-STD-105D. An AQL of 4% is specified. The contract calls for starting under normal inspection at inspection level II, and using a single sampling plan. With information on the lot size and inspection level, Table 13.7 (Table I of MIL-STD-105D) is entered, and the sample size code letter is obtained. For this example, *J* is the appropriate code letter. Using this code letter, the AQL (4%), information about the form of inspection (normal inspection), and type of sampling (single sampling), Table 13.8 (Table II-A of MIL-STD-105D) is entered, and the sampling plan, $n = 80$ and $c = 7$, is selected. Thus, a random sample of size 80 is to be drawn from each lot, and the lot accepted if the number of defective items in the sample is less than or equal to 7. Assuming that the benefits of a sampling system approach are to be enjoyed, then inspection under MIL-STD-105D requires that all the steps indicated in the standard are to be followed, including the provisions for tightened and reduced inspection.

13.2.7 DESIGNING YOUR OWN ATTRIBUTE PLAN

In many cases, it will not be desirable to select a plan according to the procedure of MIL-STD-105D, implying, as it does, a rather arbitrary relation between lot size and sample size. If the characteristics of an OC curve are specified in advance, it is possible to determine the sample size and acceptance number accordingly. This section deals with procedures for determining sampling plans when two points are specified, and with finding the OC curve of an arbitrary sampling plan.

13.2.7.1 *Computing the OC Curve of a Single Sampling Plan.* In Sec. 13.2.1.3 a method for computing the OC curve of a single sampling plan with sample size *n* and acceptance number *c* was presented. The OC curve may also be constructed from Table 13.12 (which is based on the Poisson approximation) by dividing each entry in the row for the given *c* by the sample size. The fraction defective p', for which the probability of acceptance is shown in the column heading, is obtained.

As an example, the OC curve for $n = 75$ and $c = 4$ will be found.

Table 13.8. Single Sampling Plans for Normal Inspection (Master Table)
(Table II-A of MIL-STD-105D)

Acceptable Quality Levels (Normal Inspection). Each cell shows Ac Re (Acceptance number, Rejection number).

Sample Size Code Letter	Sample Size	0.010	0.015	0.025	0.040	0.065	0.10	0.15	0.25	0.40	0.65	1.0	1.5
A	2	↓	↓	↓	↓	↓	↓	↓	↓	↓	↓	↓	↓
B	3	↓	↓	↓	↓	↓	↓	↓	↓	↓	↓	↓	0 1
C	5	↓	↓	↓	↓	↓	↓	↓	↓	↓	↓	0 1	1 2
D	8	↓	↓	↓	↓	↓	↓	↓	↓	↓	0 1	1 2	2 3
E	13	↓	↓	↓	↓	↓	↓	↓	↓	0 1	1 2	2 3	3 4
F	20	↓	↓	↓	↓	↓	↓	↓	0 1	1 2	2 3	3 4	5 6
G	32	↓	↓	↓	↓	↓	↓	0 1	1 2	2 3	3 4	5 6	7 8
H	50	↓	↓	↓	↓	↓	0 1	1 2	2 3	3 4	5 6	7 8	10 11
J	80	↓	↓	↓	↓	0 1	1 2	2 3	3 4	5 6	7 8	10 11	14 15
K	125	↓	↓	↓	0 1	1 2	2 3	3 4	5 6	7 8	10 11	14 15	21 22
L	200	↓	↓	0 1	1 2	2 3	3 4	5 6	7 8	10 11	14 15	21 22	↑
M	315	↓	0 1	1 2	2 3	3 4	5 6	7 8	10 11	14 15	21 22	↑	↑
N	500	0 1	1 2	2 3	3 4	5 6	7 8	10 11	14 15	21 22	↑	↑	↑
P	800	1 2	2 3	3 4	5 6	7 8	10 11	14 15	21 22	↑	↑	↑	↑
Q	1250	2 3	3 4	5 6	7 8	10 11	14 15	21 22	↑	↑	↑	↑	↑
R	2000	3 4	5 6	7 8	10 11	14 15	21 22	↑	↑	↑	↑	↑	↑

↓ = Use first sampling plan below arrow. If sample size equals or exceeds lot or batch size, do 100 per cent inspection.

↑ = Use first sampling plan above arrow.

Ac = Acceptance number.

Re = Rejection number.

Table 13.8 (continued). Single Sampling Plans for Normal Inspection (Master Table)
(Table II-A of MIL-STD-105D)

Acceptable Quality Levels (Normal Inspection)

Sample Size Code Letter	Sample Size	2.5 Ac	Re	4.0 Ac	Re	6.5 Ac	Re	10 Ac	Re	15 Ac	Re	25 Ac	Re	40 Ac	Re	65 Ac	Re	100 Ac	Re	150 Ac	Re	250 Ac	Re	400 Ac	Re	650 Ac	Re	1000 Ac	Re
A	2	↓		↓		↓		↓		0	1	1	2	2	3	3	4	5	6	7	8	10	11	14	15	21	22	30	31
B	3	↓		↓		↓		0	1	1	2	2	3	3	4	5	6	7	8	10	11	14	15	21	22	30	31	44	45
C	5	↓		↓		0	1	1	2	2	3	3	4	5	6	7	8	10	11	14	15	21	22	30	31	44	45	↑	
D	8	↓		0	1	1	2	2	3	3	4	5	6	7	8	10	11	14	15	21	22	30	31	44	45	↑		↑	
E	13	0	1	1	2	2	3	3	4	5	6	7	8	10	11	14	15	21	22	30	31	44	45	↑		↑		↑	
F	20	1	2	2	3	3	4	5	6	7	8	10	11	14	15	21	22	↑		↑		↑		↑		↑		↑	
G	32	2	3	3	4	5	6	7	8	10	11	14	15	21	22	↑		↑		↑		↑		↑		↑		↑	
H	50	3	4	5	6	7	8	10	11	14	15	21	22	↑		↑		↑		↑		↑		↑		↑		↑	
J	80	5	6	7	8	10	11	14	15	21	22	↑		↑		↑		↑		↑		↑		↑		↑		↑	
K	125	7	8	10	11	14	15	21	22	↑		↑		↑		↑		↑		↑		↑		↑		↑		↑	
L	200	10	11	14	15	21	22	↑		↑		↑		↑		↑		↑		↑		↑		↑		↑		↑	
M	315	14	15	21	22	↑		↑		↑		↑		↑		↑		↑		↑		↑		↑		↑		↑	
N	500	21	22	↑		↑		↑		↑		↑		↑		↑		↑		↑		↑		↑		↑		↑	
P	800	↑		↑		↑		↑		↑		↑		↑		↑		↑		↑		↑		↑		↑		↑	
Q	1250	↑		↑		↑		↑		↑		↑		↑		↑		↑		↑		↑		↑		↑		↑	
R	2000																												

↓ = Use first sampling plan below arrow.
↑ = Use first sampling plan above arrow.
Ac = Acceptance number.
Re = Rejection number.

Table 13.9. Double Sampling Plans for Normal Inspection (Master Table)
(Table III-A of MIL-STD-105D)

Sample size code letter	Sample	Sample size	Cumulative sample size	0.010		0.015		0.025		0.040		0.065		0.10		0.15		0.25		0.40		0.65		1.0	
				Ac	Re	Ac	Re	Ac	Re	Ac	Re	Ac	Re	Ac	Re	Ac	Re	Ac	Re	Ac	Re	Ac	Re	Ac	Re
A																									
B	First	2	2																						
	Second	2	4																						
C	First	3	3																						
	Second	3	6																						
D	First	5	5																					↓	
	Second	5	10																						
E	First	8	8																			↓		•	
	Second	8	16																						
F	First	13	13																	↓		•		↑	
	Second	13	26																						
G	First	20	20															↓		•		↑			
	Second	20	40																						
H	First	32	32													↓		•		↑				0	2
	Second	32	64																					1	2
J	First	50	50											↓		•		↑				0	2	0	3
	Second	50	100																			1	2	3	4
K	First	80	80									↓		•		↑				0	2	0	3	1	4
	Second	80	160																	1	2	3	4	4	5
L	First	125	125							↓		•		↑				0	2	0	3	1	4	2	5
	Second	125	250															1	2	3	4	4	5	6	7
M	First	200	200					↓		•		↑				0	2	0	3	1	4	2	5	3	7
	Second	200	400													1	2	3	4	4	5	6	7	8	9
N	First	315	315			↓		•		↑				0	2	0	3	1	4	2	5	3	7	5	9
	Second	315	630											1	2	3	4	4	5	6	7	8	9	12	13
P	First	500	500	↓		•		↑				0	2	0	3	1	4	2	5	3	7	5	9	7	11
	Second	500	1000									1	2	3	4	4	5	6	7	8	9	12	13	18	19
Q	First	800	800	•		↑				0	2	0	3	1	4	2	5	3	7	5	9	7	11	11	16
	Second	800	1600							1	2	3	4	4	5	6	7	8	9	12	13	18	19	26	27
R	First	1250	1250	↑		↑		0	2	0	3	1	4	2	5	3	7	5	9	7	11	11	16	↑	
	Second	1250	2500					1	2	3	4	4	5	6	7	8	9	12	13	18	19	26	27		

↓ = Use first sampling plan below arrow. If sample size equals or exceeds lot or batch size, do 100% inspection.

↑ = Use first sampling plan above arrow.

Ac = Acceptance number.

Re = Rejection number.

• = Use corresponding single sampling plan (or alternatively, use double sampling plan below, where available).

Acceptable Quality Levels (normal inspection)

.5 Re	2.5 Ac Re	4.0 Ac Re	6.5 Ac Re	10 Ac Re	15 Ac Re	25 Ac Re	40 Ac Re	65 Ac Re	100 Ac Re	150 Ac Re	250 Ac Re	400 Ac Re	650 Ac Re	1000 Ac Re	Sample size code letter
↓	↓	↓	•		↓	•	•	•	•	•	•	•	•	•	A
		•	↑	↓	0 2 / 1 2	0 3 / 3 4	1 4 / 4 5	2 5 / 6 7	3 7 / 8 9	5 9 / 12 13	7 11 / 18 19	11 16 / 26 27	17 22 / 37 38	25 31 / 56 57	B
↓	•	↑	↓	0 2 / 1 2	0 3 / 3 4	1 4 / 4 5	2 5 / 6 7	3 7 / 8 9	5 9 / 12 13	7 11 / 18 19	11 16 / 26 27	17 22 / 37 38	25 31 / 56 57	↑	C
•	↑	↓	0 2 / 1 2	0 3 / 3 4	1 4 / 4 5	2 5 / 6 7	3 7 / 8 9	5 9 / 12 13	7 11 / 18 19	11 16 / 26 27	17 22 / 37 38	25 31 / 56 57	↑		D
↑	↓	0 2 / 1 2	0 3 / 3 4	1 4 / 4 5	2 5 / 6 7	3 7 / 8 9	5 9 / 12 13	7 11 / 18 19	11 16 / 26 27	17 22 / 37 38	25 31 / 56 57	↑			E
↓	0 2 / 1 2	0 3 / 3 4	1 4 / 4 5	2 5 / 6 7	3 7 / 8 9	5 9 / 12 13	7 11 / 18 19	11 16 / 26 27	↑	↑	↑				F
2 / 2	0 3 / 3 4	1 4 / 4 5	2 5 / 6 7	3 7 / 8 9	5 9 / 12 13	7 11 / 18 19	11 16 / 26 27	↑							G
3 / 4	1 4 / 4 5	2 5 / 6 7	3 7 / 8 9	5 9 / 12 13	7 11 / 18 19	11 16 / 26 27	↑								H
4 / 5	2 5 / 6 7	3 7 / 8 9	5 9 / 12 13	7 11 / 18 19	11 16 / 26 27	↑									J
5 / 7	3 7 / 8 9	5 9 / 12 13	7 11 / 18 19	11 16 / 26 27	↑										K
7 / 9	5 9 / 12 13	7 11 / 18 19	11 16 / 26 27	↑											L
9 / 13	7 11 / 18 19	11 16 / 26 27	↑												M
11 / 19	11 16 / 26 27	↑													N
16 / 27	↑														P
↑															Q
															R

531

Table 13.10. Multiple Sampling Plans for Normal Inspection (Master Table)
(Table IV-A of MIL-STD-105D)

Sample size code letter	Sample	Sample size	Cumulative sample size	Acceptable Quality Levels (normal inspection)										
				0.010	0.015	0.025	0.040	0.065	0.10	0.15	0.25	0.40	0.65	1.0
				Ac Re	Ac Re	Ac Re	Ac Re	Ac Re	Ac Re	Ac Re	Ac Re	Ac Re	Ac Re	Ac Re
A														
B														
C														
D	First	2	2											
	Second	2	4											
	Third	2	6											
	Fourth	2	8											
	Fifth	2	10											
	Sixth	2	12											
	Seventh	2	14											↓
E	First	3	3											⋅
	Second	3	6											
	Third	3	9											
	Fourth	3	12											
	Fifth	3	15											
	Sixth	3	18									↓		
	Seventh	3	21											
F	First	5	5										⋅	↑
	Second	5	10											
	Third	5	15											
	Fourth	5	20											
	Fifth	5	25											
	Sixth	5	30								↓			
	Seventh	5	35											
G	First	8	8									⋅	↑	
	Second	8	16											
	Third	8	24											
	Fourth	8	32											
	Fifth	8	40											
	Sixth	8	48							↓				↓
	Seventh	8	56											
H	First	13	13									⋅	↑	″ 2
	Second	13	26											″ 2
	Third	13	39											0 2
	Fourth	13	52											0 3
	Fifth	13	65											1 3
	Sixth	13	78							↓			↓	1 3
	Seventh	13	91											2 3
J	First	20	20							⋅	↑		″ 2	″ 2
	Second	20	40										* 2	0 3
	Third	20	60										0 2	0 3
	Fourth	20	80										0 3	1 4
	Fifth	20	100										1 3	2 4
	Sixth	20	120	↓	↓	↓	↓	↓	↓		↑	↓	1 3	3 5
	Seventh	20	140										2 3	4 5

↓ = Use first sampling plan below arrow (refer to continuation of table on following pages, when necessary). If sample size equals or exceeds lot or batch size, do 100% inspection.

↑ = Use first sampling plan above arrow.

Ac = Acceptance number.

Re = Rejection number.

⋅ = Use corresponding single sampling plan (or alternatively, use multiple sampling plan below, where available).

†† = Use corresponding double sampling plan (or alternatively, use multiple sampling plan below, where available).

″ = Acceptance not permitted at this sample size.

532

Table 13.10 (continued). Multiple Sampling Plans for Normal Inspection (Master Table)
(Table IV-A of MIL-STD-105D)

	Acceptable Quality Levels (normal inspection) 2.5		4.0		6.5		10		15		25		40		65		100		150		250		400		650		1000		Sample size code letter
Re	Ac	Re	Ac	Re	Ac	Re	Ac	Re	Ac	Re	Ac	Re	Ac	Re	Ac	Re	Ac	Re	Ac	Re	Ac	Re	Ac	Re	Ac	Re	Ac	Re	
	↓		↓		•		↓		↓		•		•		•		•		•		•		•		•		•		A
			•		↑↓				††		††		††		††		††		††		††		††		††		††		B
	•		↑				††		††		††		††		††		††		††		††		††		††		↑		C
•	↑		↓		″	2	″	2	″	3	″	4	0	4	0	5	1	7	2	9	4	12	6	16	↑				D
					″	2	0	3	0	3	1	5	1	6	3	8	4	10	7	14	11	19	17	27					
					0	2	0	3	1	4	2	6	3	8	6	10	8	13	13	19	19	27	29	39					
					0	3	1	4	2	5	3	7	5	10	8	13	12	17	19	25	27	34	40	49					
					1	3	2	4	3	6	5	8	7	11	11	15	17	20	25	29	36	40	53	58					
					1	3	3	5	4	6	7	9	10	12	14	17	21	23	31	33	45	47	65	68					
					2	3	4	5	6	7	9	10	13	14	18	19	25	26	37	38	53	54	77	78					
	↓		″	2	″	2	″	3	″	4	0	4	0	5	1	7	2	9	4	12	6	16	↑						E
			″	2	0	3	0	3	1	5	1	6	3	8	4	10	7	14	11	19	17	27							
			0	2	0	3	1	4	2	6	3	8	6	10	8	13	13	19	19	27	29	39							
			0	3	1	4	2	5	3	7	5	10	8	13	12	17	19	25	40	34	40	49							
			1	3	2	4	3	6	5	8	7	11	11	15	17	20	25	29	36	40	53	58							
			1	3	3	5	4	6	7	9	10	12	14	17	21	23	31	33	45	47	65	68							
			2	3	4	5	6	7	9	10	13	14	18	19	25	26	37	38	53	54	77	78							
	″	2	″	2	″	3	″	4	0	4	0	5	1	7	2	9	↑		↑		↑								F
	″	2	0	3	0	3	1	5	1	6	3	8	4	10	7	14													
	0	2	0	3	1	4	2	6	3	8	6	10	8	13	13	19													
	0	3	1	4	2	5	3	7	5	10	8	13	12	17	19	25													
	1	3	2	4	3	6	5	8	7	11	11	15	17	20	25	29													
	1	3	3	5	4	6	7	9	10	12	14	17	21	23	31	33													
	2	3	4	5	6	7	9	10	13	14	18	19	25	26	37	38													
2	″	2	″	3	″	4	0	4	0	5	1	7	2	9	↑														G
2	0	3	0	3	1	5	1	6	3	8	4	10	7	14															
2	0	3	1	4	2	6	3	8	6	10	8	13	13	19															
3	1	4	2	5	3	7	5	10	8	13	12	17	19	25															
3	2	4	3	6	5	8	7	11	11	15	17	20	25	29															
3	3	5	4	6	7	9	10	12	14	17	21	23	31	33															
3	4	5	6	7	9	10	13	14	18	19	25	26	37	38															
2	″	3	″	4	0	4	0	5	1	7	2	9	↑																H
3	0	3	1	5	1	6	3	8	4	10	7	14																	
3	1	4	2	6	3	8	6	10	8	13	13	19																	
4	2	5	3	7	5	10	8	13	12	17	19	25																	
4	3	6	5	8	7	11	11	15	17	20	25	29																	
5	4	6	7	9	10	12	14	17	21	23	31	33																	
5	6	7	9	10	13	14	18	19	25	26	37	38																	
3	″	4	0	4	0	5	1	7	2	9	↑																		J
3	1	5	1	6	3	8	4	10	7	14																			
4	2	6	3	8	6	10	8	13	13	19																			
5	3	7	5	10	8	13	12	17	19	25																			
6	5	8	7	11	11	15	17	20	25	29																			
6	7	9	10	12	14	17	21	23	31	33																			
7	9	10	13	14	18	19	25	26	37	38																			

Table 13.10 (continued). **Multiple Sampling Plans for Normal Inspection (Master Table)**

Acceptable Quality Levels (normal inspection)

Sample size code letter	Sample	Sample size	Cumulative sample size	0.010		0.015		0.025		0.040		0.065		0.10		0.15		0.25		0.40		0.65		1.0
				Ac	Re	Ac	Re	Ac	Re	Ac	Re	Ac	Re	Ac	Re	Ac	Re	Ac	Re	Ac	Re	Ac	Re	Ac
K	First	32	32	↓		↓		↓		↓		↓		·		↑		↑		"	2	"	2	"
	Second	32	64																	"	2	0	3	0
	Third	32	96																	0	2	0	3	1
	Fourth	32	128																	0	3	1	4	2
	Fifth	32	160																	1	3	2	4	3
	Sixth	32	192																	1	3	3	5	4
	Seventh	32	224																	2	3	4	5	6
L	First	50	50	↓		↓		↓		↓		·		↑		↑		"	2	"	2	"	3	"
	Second	50	100															"	2	0	3	0	3	1
	Third	50	150															0	2	0	3	1	4	2
	Fourth	50	200															0	3	1	4	2	5	3
	Fifth	50	250															1	3	2	4	3	6	5
	Sixth	50	300															1	3	3	5	4	6	7
	Seventh	50	350															2	3	4	5	6	7	9
M	First	80	80	↓		↓		↓		·		↑		↑		"	2	"	2	"	3	"	4	0
	Second	80	160													"	2	0	3	0	3	1	5	1
	Third	80	240													0	2	0	3	1	4	2	6	3
	Fourth	80	320													0	3	1	4	2	5	3	7	5
	Fifth	80	400													1	3	2	4	3	6	5	8	7
	Sixth	80	480													1	3	3	5	4	6	7	9	10
	Seventh	80	560													2	3	4	5	6	7	9	10	13
N	First	125	125	↓		↓		·		↑		↑		"	2	"	2	"	3	"	4	0	4	0
	Second	125	250											"	2	0	3	0	3	1	5	1	6	3
	Third	125	375											0	2	0	3	1	4	2	6	3	8	6
	Fourth	125	500											0	3	1	4	2	5	3	7	5	10	8
	Fifth	125	625											1	3	2	4	3	6	5	8	7	11	11
	Sixth	125	750											1	3	3	5	4	6	7	9	10	12	14
	Seventh	125	875											2	3	4	5	6	7	9	10	13	14	18
P	First	200	200	↓		·		↑		↑		"	2	"	2	"	3	"	4	0	4	0	5	1
	Second	200	400									"	2	0	3	0	3	1	5	1	6	3	8	4
	Third	200	600									0	2	0	3	1	4	2	6	3	8	6	10	8
	Fourth	200	800									0	3	1	4	2	5	3	7	5	10	8	13	12
	Fifth	200	1,000									1	3	2	4	3	6	5	8	7	11	11	15	17
	Sixth	200	1,200									1	3	3	5	4	6	7	9	10	12	14	17	21
	Seventh	200	1,400									2	3	4	5	6	7	9	10	13	14	18	19	25
Q	First	315	315	·		↑		↑		"	2	"	2	"	3	"	4	0	4	0	5	1	7	2
	Second	315	630							"	2	0	3	0	3	1	5	1	6	3	8	4	10	7
	Third	315	945							0	2	0	3	1	4	2	6	3	8	6	10	8	13	13
	Fourth	315	1,260							0	3	1	4	2	5	3	7	5	10	8	13	12	17	19
	Fifth	315	1,575							1	3	2	4	3	6	5	8	7	11	11	15	17	20	25
	Sixth	315	1,890							1	3	3	5	4	6	7	9	10	12	14	17	21	23	31
	Seventh	315	2,205							2	3	4	5	6	7	9	10	13	14	18	19	25	26	37
R	First	500	500	↑		↑		"	2	"	2	"	3	"	4	0	4	0	5	1	7	2	9	↑
	Second	500	1,000					"	2	0	3	0	3	1	5	1	6	3	8	4	10	7	14	
	Third	500	1,500					0	2	0	3	1	4	2	6	3	8	6	10	8	13	13	19	
	Fourth	500	2,000					0	3	1	4	2	5	3	7	5	10	8	13	12	17	19	25	
	Fifth	500	2,500					1	3	2	4	3	6	5	8	7	11	11	15	17	20	25	29	
	Sixth	500	3,000					1	3	3	5	4	6	7	9	10	12	14	17	21	23	31	33	
	Seventh	500	3,500					2	3	4	5	6	7	9	10	13	14	18	19	25	26	37	38	

↓ = Use first sampling plan below arrow. If sample size equals or exceeds lot or batch size, do 100% inspection.
↑ = Use first sampling plan above arrow (refer to preceding page when necessary).
Ac = Acceptance number.
Re = Rejection number.
· = Use corresponding single sampling plan (or alternatively, use multiple plan below, where available).
" = Acceptance not permitted at this sample size.

534

						Acceptable Quality Levels (normal inspection)										Sample size code letter
.5	2.5		4.0		6.5		10		15	25	40	65	100	150	250	400	650	1000	
Re	Ac	Re	Ac	Re	Ac	Re	Ac	Re	Ac Re	Ac Re	Ac Re	Ac Re	Ac Re	Ac Re	Ac Re	Ac Re	Ac Re	

Code letter **K**:

.5 Re	2.5 Ac Re	4.0 Ac Re	6.5 Ac Re	10 Ac Re
4	0 4	0 5	1 7	2 9
5	1 6	3 8	4 10	7 14
6	3 8	6 10	8 13	13 19
7	5 10	8 13	12 17	19 25
8	7 11	11 15	17 20	25 29
9	10 12	14 17	21 23	31 33
10	13 14	18 19	25 26	37 38

(AQL 15 through 1000: ↑)

Code letter **L**:

.5 Re	2.5 Ac Re	4.0 Ac Re	6.5 Ac Re
4	0 5	1 7	2 9
6	3 8	4 10	7 14
8	6 10	8 13	13 19
10	8 13	12 17	19 25
11	11 15	17 20	25 29
12	14 17	21 23	31 33
14	18 19	25 26	37 38

(AQL 10: ↑)

Code letter **M**:

.5 Re	2.5 Ac Re	4.0 Ac Re
5	1 7	2 9
8	4 10	7 14
10	8 13	13 19
13	12 17	19 25
15	17 20	25 29
17	21 23	31 33
19	25 26	37 38

(AQL 6.5: ↑)

Code letter **N**:

.5 Re	2.5 Ac Re
7	2 9
10	7 14
13	13 19
17	19 25
20	25 29
23	31 33
26	37 38

(AQL 4.0: ↑)

Code letter **P**:

.5 Re
9
14
19
25
29
33
38

(AQL 2.5: ↑)

Code letter **Q**: (AQL .5: ↑)

Code letter **R**: (↑)

535

Table 13.11. Incoming Qualities (in percent defective) for which $P_a = 95\%$ (top entry), 50% (middle entry), 10% (bottom entry) for Normal Inspection, Single Sampling (Includes entries in Table VI-A of MIL-STD-105D)

Code letter	Sample size	0.010	0.015	0.025	0.040	0.065	0.10	0.15	0.25	0.40	0.65	1.0	1.5	2.5	4.0	6.5	10
A	2															2.5 29 68	
B	3														1.7 21 54		
C	5													1.0 13 37			7.6 31 59
D	8												0.64 8.3 25			2.6 20 41	11 32 54
E	13											0.39 5.2 16			2.8 13 27	6.6 20 36	11 28 44
F	20										0.26 3.4 11			1.8 8.3 18	4.2 13 25	7.1 18 30	14 28 42
G	32									0.16 2.1 6.9			1.1 5.2 12	2.6 8.3 16	4.4 11 20	8.5 18 27	13 24 34
H	50								0.10 1.4 4.5			0.71 3.3 7.6	1.7 5.3 10	2.8 7.3 13	5.3 11 18	8.2 15 22	13 21 29
J	80							0.064 0.86 2.8			0.44 2.1 4.8	1.0 3.3 6.5	1.7 4.6 8.2	3.3 7.1 11	5.1 9.6 14	7.9 13 19	12 18 24
K	125						0.041 0.55 1.8			0.28 1.3 3.1	0.65 2.1 4.3	1.1 2.9 5.4	2.1 4.5 7.4	3.2 6.1 9.4	4.9 8.5 12	7.4 12 16	12 17 23
L	200					0.026 0.35 1.2			0.18 0.84 2.0	0.41 1.3 2.7	0.68 1.8 3.3	1.3 2.8 4.6	2.0 3.8 5.9	3.1 5.3 7.7	4.6 7.3 10	7.5 11 14	
M	315				0.016 0.22 0.73			0.11 0.53 1.2	0.26 0.85 1.7	0.43 1.2 2.1	0.83 1.8 2.9	1.2 2.4 3.7	2.0 3.4 4.9	2.9 4.7 6.4	4.7 6.9 9.0		
N	500			0.010 0.14 0.46			0.071 0.34 0.78	0.16 0.54 1.1	0.27 0.73 1.3	0.52 1.1 1.9	0.80 1.3 2.4	1.2 2.1 3.1	1.9 2.9 4.0	3.0 4.3 5.6			
P	800		0.0064 0.087 0.29			0.044 0.21 0.49	0.10 0.33 0.67	0.17 0.46 0.84	0.33 0.71 1.2	0.50 0.96 1.5	0.77 1.3 1.9	1.2 1.8 2.5	1.9 2.7 3.5				
Q	1,250	0.0041 0.055 0.18		0.018	0.041	0.065 0.21 0.43	0.11 0.29 0.53	0.21 0.45 0.74	0.32 0.61 0.94	0.49 0.85 1.2	0.74 1.2 1.6	1.2 1.7 2.3					
R	2,000			0.018	0.041	0.068	0.13	0.20	0.31	0.46	0.75						

536

Dividing the numbers in the row $c = 4$ by 75 in Table 13.12, the following 13 points on the OC curve are found:

Probability of Acceptance	Fraction Defective	Probability of Acceptance	Fraction Defective
0.995	$\dfrac{1.078}{75} = 0.01437$	0.250	$\dfrac{6.274}{75} = 0.08365$
0.990	$\dfrac{1.279}{75} = 0.01705$	0.100	$\dfrac{7.994}{75} = 0.1066$
0.975	$\dfrac{1.623}{75} = 0.02164$	0.050	$\dfrac{9.154}{75} = 0.1221$
0.950	$\dfrac{1.970}{75} = 0.02627$	0.025	$\dfrac{10.242}{75} = 0.1372$
0.900	$\dfrac{2.433}{75} = 0.03244$	0.010	$\dfrac{11.605}{75} = 0.1547$
0.750	$\dfrac{3.369}{75} = 0.04492$	0.005	$\dfrac{12.594}{75} = 0.1679$
0.500	$\dfrac{4.671}{75} = 0.06228$		

13.2.7.2 *Finding a Sampling Plan Whose* OC *Curve Passes Through Two Points.* As pointed out in Sec. 13.2.1.2, it is often useful to specify two points on an OC curve, p_1' and p_2' such that $L(p_1') = 1 - \alpha$ and $L(p_2') = \beta$. Here p_1' is usually denoted as acceptable quality (quality for which acceptance is desired), and p_2' usually represents unacceptable quality (quality for which rejection is desired). Hence, α and β are usually taken as small numbers, $0.01, 0.05,$ or 0.10. For these values, sampling plans may be found in Table 13.13 (which is also based on the Poisson approximation). To construct a plan for a given p_1', α and p_2', β, calculate the ratio p_2'/p_1' and find the entry in the appropriate α, β column which is equal to or just less than the desired ratio. The acceptance number is read off directly and the sample size is determined by dividing np_1' by p_1'.

As an example, the sample size and acceptance number corresponding to $p_1' = 0.02, p_2' = 0.04, \alpha = 0.05,$ and $\beta = 0.05$ will be found as follows:

$$\frac{p_2'}{p_1'} = \frac{0.04}{0.02} = 2.$$

The value in Table 13.13 just less than 2 is 1.999. Hence,

$$c = 22, n = 15.719/0.02 = 786.$$

Because of the discreteness of n and c this plan does not pass through the desired points, but is very "close."

13.2.7.3 *Design of Item-by-Item Sequential Plans.* It is not possible to give simple formulas for the acceptance and rejection numbers in double or multiple sampling plans; however, simple formulas do exist for item-by-item sequential plans, i.e., plans in which the decision to accept, reject, or continue

Table 13.12. Values of np'_1 for Which the Probability of Acceptance of c or Fewer Defectives in a Sample of n is $P(A)$[1]

[To find the fraction defective p, corresponding to a probability of acceptance $P(A)$ in a single sampling plan with sample size n and acceptance number c, divide by n the entry in the row for the given c and the column for the given $P(A)$.]

c	P(A) = 0.995	P(A) = 0.990	P(A) = 0.975	P(A) = 0.950	P(A) = 0.900	P(A) = 0.750	P(A) = 0.500	P(A) = 0.250	P(A) = 0.100	P(A) = 0.050	P(A) = 0.025	P(A) = 0.010	P(A) = 0.005
0	0.00501	0.0101	0.0253	0.0513	0.105	0.288	0.693	1.386	2.303	2.996	3.689	4.605	5.298
1	0.103	0.149	0.242	0.355	0.532	0.961	1.678	2.693	3.890	4.744	5.572	6.638	7.430
2	0.338	0.436	0.619	0.818	1.102	1.727	2.674	3.920	5.322	6.296	7.224	8.406	9.274
3	0.672	0.823	1.090	1.366	1.745	2.535	3.672	5.109	6.681	7.754	8.768	10.045	10.978
4	1.078	1.279	1.623	1.970	2.433	3.369	4.671	6.274	7.994	9.154	10.242	11.605	12.594
5	1.537	1.785	2.202	2.613	3.152	4.219	5.670	7.423	9.275	10.513	11.668	13.108	14.150
6	2.037	2.330	2.814	3.286	3.895	5.083	6.670	8.558	10.532	11.842	13.060	14.571	15.660
7	2.571	2.906	3.454	3.981	4.656	5.956	7.669	9.684	11.771	13.148	14.422	16.000	17.134
8	3.132	3.507	4.115	4.695	5.432	6.838	8.669	10.802	12.995	14.434	15.763	17.403	18.578
9	3.717	4.130	4.795	5.426	6.221	7.726	9.669	11.914	14.206	15.705	17.085	18.783	19.998
10	4.321	4.771	5.491	6.169	7.021	8.620	10.668	13.020	15.407	16.962	18.390	20.145	21.398
11	4.943	5.428	6.201	6.924	7.829	9.519	11.668	14.121	16.598	18.208	19.682	21.490	22.779
12	5.580	6.099	6.922	7.690	8.646	10.422	12.668	15.217	17.782	19.442	20.962	22.821	24.145
13	6.231	6.782	7.654	8.464	9.470	11.329	13.668	16.310	18.958	20.668	22.230	24.139	25.496
14	6.893	7.477	8.396	9.246	10.300	12.239	14.668	17.400	20.128	21.886	23.490	25.446	26.836
15	7.566	8.181	9.144	10.035	11.135	13.152	15.668	18.486	21.292	23.098	24.741	26.743	28.166
16	8.249	8.895	9.902	10.831	11.976	14.068	16.668	19.570	22.452	24.302	25.984	28.031	29.484
17	8.942	9.616	10.666	11.633	12.822	14.986	17.668	20.652	23.606	25.500	27.220	29.310	30.792
18	9.644	10.346	11.438	12.442	13.672	15.907	18.668	21.731	24.756	26.692	28.448	30.581	32.092
19	10.353	11.082	12.216	13.254	14.525	16.830	19.668	22.808	25.902	27.879	29.671	31.845	33.383
20	11.069	11.825	12.999	14.072	15.383	17.755	20.668	23.883	27.045	29.062	30.888	33.103	34.668
21	11.791	12.574	13.787	14.894	16.244	18.682	21.668	24.956	28.184	30.241	32.102	34.355	35.947
22	12.520	13.329	14.580	15.719	17.108	19.610	22.668	26.028	29.320	31.416	33.309	35.601	37.219
23	13.255	14.088	15.377	16.548	17.975	20.540	23.668	27.098	30.453	32.586	34.512	36.841	38.485
24	13.995	14.853	16.178	17.382	18.844	21.471	24.668	28.167	31.584	33.752	35.710	38.077	39.745
25	14.740	15.623	16.984	18.218	19.717	22.404	25.667	29.234	32.711	34.916	36.905	39.308	41.000
26	15.490	16.397	17.793	19.058	20.592	23.338	26.667	30.300	33.836	36.077	38.096	40.535	42.252
27	16.245	17.175	18.606	19.900	21.469	24.273	27.667	31.365	34.959	37.234	39.284	41.757	43.497

Table 13.12 (Continued). Values of np_1' for Which the Probability of Acceptance of c or Fewer Defectives in a Sample of n is $P(A)$

n													
28	17.004	17.957	19.422	20.746	22.348	25.209	28.667	32.428	36.080	38.389	40.468	42.975	44.738
29	17.767	18.742	20.241	21.594	23.229	26.147	29.667	33.491	37.198	39.541	41.649	44.190	45.976
30	18.534	19.532	21.063	22.444	24.113	27.086	30.667	34.552	38.315	40.690	42.827	45.401	47.210
31	19.305	20.324	21.888	23.298	24.998	28.025	31.667	35.613	39.430	41.838	44.002	46.609	48.440
32	20.079	21.120	22.716	24.152	25.885	28.966	32.667	36.672	40.543	42.982	45.174	47.813	49.666
33	20.856	21.919	23.546	25.010	26.774	29.907	33.667	37.731	41.654	44.125	46.344	49.015	50.888
34	21.638	22.721	24.379	25.870	27.664	30.849	34.667	38.788	42.764	45.266	47.512	50.213	52.108
35	22.422	23.525	25.214	26.731	28.556	31.792	35.667	39.845	43.872	46.404	48.676	51.409	53.324
36	23.208	24.333	26.052	27.594	29.450	32.736	36.667	40.901	44.978	47.540	49.840	52.601	54.538
37	23.998	25.143	26.891	28.460	30.345	33.681	37.667	41.957	46.083	48.676	51.000	53.791	55.748
38	24.791	25.955	27.733	29.327	31.241	34.626	38.667	43.011	47.187	49.808	52.158	54.979	56.956
39	25.586	26.770	28.576	30.196	32.139	35.572	39.667	44.065	48.289	50.940	53.314	56.164	58.160
40	26.384	27.587	29.422	31.066	33.038	36.519	40.667	45.118	49.390	52.069	54.469	57.347	59.363
41	27.184	28.406	30.270	31.938	33.938	37.466	41.667	46.171	50.490	53.197	55.622	58.528	60.563
42	27.986	29.228	31.120	32.812	34.839	38.414	42.667	47.223	51.589	54.324	56.772	59.717	61.761
43	28.791	30.051	31.970	33.686	35.742	39.363	43.667	48.274	52.686	55.449	57.921	60.884	62.956
44	29.598	30.877	32.824	34.563	36.646	40.312	44.667	49.325	53.782	56.572	59.068	62.059	64.150
45	30.408	31.704	33.678	35.441	37.550	41.262	45.667	50.375	54.878	57.695	60.214	63.231	65.340
46	31.219	32.534	34.534	36.320	38.456	42.212	46.667	51.425	55.972	58.816	61.358	64.402	66.529
47	32.032	33.365	35.392	37.200	39.363	43.163	47.667	52.474	57.065	59.936	62.500	65.571	67.716
48	32.848	34.198	36.250	38.082	40.270	44.115	48.667	53.522	58.158	61.054	63.641	66.738	68.901
49	33.664	35.032	37.111	38.965	41.179	45.067	49.667	54.571	59.249	62.171	64.780	67.903	70.084

¹ Reprinted by permission from "Tables for Constructing and Computing the Operating Characteristics of Single-Sampling Plans," by J. M. Cameron, *Industrial Quality Control*, July, 1952, p. 38.

Table 13.13. Values of np'_1 and c for Constructing Single Sampling Plans Whose OC Curve Is Required to Pass through the Two Points $(p'_1, 1 - \alpha)$ and $(p'_2 \beta)$[1]

(Here p'_1 is the fraction defective for which the risk of rejection is to be α, and p'_2 is the fraction defective for which the risk of acceptance is to be β. To construct the plan, find the tabular value of p'_2/p'_1 in the column for the given α and β which is equal to or just less than the value of the ratio. The sample size is found by dividing the np'_1 corresponding to the selected ratio by p'_1. The acceptance number is the value of c corresponding to the selected value of the ratio.)

c	Values of p_2/p_1 for:			np_1	c	Values of p_2/p_1 for:			np_1
	$\alpha = 0.05$ $\beta = 0.10$	$\alpha = 0.05$ $\beta = 0.05$	$\alpha = 0.05$ $\beta = 0.01$			$\alpha = 0.01$ $\beta = 0.10$	$\alpha = 0.01$ $\beta = 0.05$	$\alpha = 0.01$ $\beta = 0.01$	
0	44.890	58.404	89.781	0.052	0	229.105	298.073	458.210	0.010
1	10.946	13.349	18.681	0.355	1	26.184	31.933	44.686	0.149
2	6.509	7.699	10.280	0.818	2	12.206	14.439	19.278	0.436
3	4.890	5.675	7.352	1.366	3	8.115	9.418	12.202	0.823
4	4.057	4.646	5.890	1.970	4	6.249	7.156	9.072	1.279
5	3.549	4.023	5.017	2.613	5	5.195	5.889	7.343	1.785
6	3.206	3.604	4.435	3.286	6	4.520	5.082	6.253	2.330
7	2.957	3.303	4.019	3.981	7	4.050	4.524	5.506	2.906
8	2.768	3.074	3.707	4.695	8	3.705	4.115	4.962	3.507
9	2.618	2.895	3.462	5.426	9	3.440	3.803	4.548	4.130
10	2.497	2.750	3.265	6.169	10	3.229	3.555	4.222	4.771
11	2.397	2.630	3.104	6.924	11	3.058	3.354	3.959	5.428
12	2.312	2.528	2.968	7.690	12	2.915	3.188	3.742	6.099
13	2.240	2.442	2.852	8.464	13	2.795	3.047	3.559	6.782
14	2.177	2.367	2.752	9.246	14	2.692	2.927	3.403	7.477
15	2.122	2.302	2.665	10.035	15	2.603	2.823	3.269	8.181
16	2.073	2.244	2.588	10.831	16	2.524	2.732	3.151	8.895
17	2.029	2.192	2.520	11.633	17	2.455	2.652	3.048	9.616
18	1.990	2.145	2.458	12.442	18	2.393	2.580	2.956	10.346
19	1.954	2.103	2.403	13.254	19	2.337	2.516	2.874	11.082

Table 13.13 (Continued). Values of np'_1 and c for Constructing Single Sampling Plans Whose OC Curve Is Required to Pass through the Two Points $(p'_1, 1 - \alpha)$ and $(p'_2\beta)$

				c				
1.922	2.065	2.352	14.072	20	2.287	2.458	2.799	11.825
1.892	2.030	2.307	14.894	21	2.241	2.405	2.733	12.574
1.865	1.999	2.265	15.719	22	2.200	2.357	2.671	13.329
1.840	1.969	2.226	16.548	23	2.162	2.313	2.615	14.088
1.817	1.942	2.191	17.382	24	2.126	2.272	2.564	14.853
1.795	1.917	2.158	18.218	25	2.094	2.235	2.516	15.623
1.775	1.893	2.127	19.058	26	2.064	2.200	2.472	16.397
1.757	1.871	2.098	19.900	27	2.035	2.168	2.431	17.175
1.739	1.850	2.071	20.746	28	2.009	2.138	2.393	17.957
1.723	1.831	2.046	21.594	29	1.985	2.110	2.358	18.742
1.707	1.813	2.023	22.444	30	1.962	2.083	2.324	19.532
1.692	1.796	2.001	23.298	31	1.940	2.059	2.293	20.324
1.679	1.780	1.980	24.152	32	1.920	2.035	2.264	21.120
1.665	1.764	1.960	25.010	33	1.900	2.013	2.236	21.919
1.653	1.750	1.941	25.870	34	1.882	1.992	2.210	22.721
1.641	1.736	1.923	26.731	35	1.865	1.973	2.185	23.525
1.630	1.723	1.906	27.594	36	1.848	1.954	2.162	24.333
1.619	1.710	1.890	28.460	37	1.833	1.936	2.139	25.143
1.609	1.698	1.875	29.327	38	1.818	1.920	2.118	25.955
1.599	1.687	1.860	30.196	39	1.804	1.903	2.098	26.770
1.590	1.676	1.846	31.066	40	1.790	1.887	2.079	27.587
1.581	1.666	1.833	31.938	41	1.777	1.873	2.060	28.406
1.572	1.656	1.820	32.812	42	1.765	1.859	2.043	29.228
1.564	1.646	1.807	33.686	43	1.753	1.845	2.026	30.051
1.556	1.637	1.796	34.563	44	1.742	1.832	2.010	30.877
1.548	1.628	1.784	35.441	45	1.731	1.820	1.994	31.704
1.541	1.619	1.773	36.320	46	1.720	1.808	1.980	32.534
1.534	1.611	1.763	37.200	47	1.710	1.796	1.965	33.365
1.527	1.603	1.752	38.082	48	1.701	1.785	1.952	34.198
1.521	1.596	1.743	38.965	49	1.691	1.775	1.938	35.032

[1] Reprinted by permission from "Tables for Constructing and Computing the Operating Characteristics of Single-Sampling Plans," by J. M. Cameron, *Industrial Quality Control*, July, 1952, p. 39.

sampling is made after each observation. The plan can be represented graphically as a pair of parallel lines (Fig. 13.6). The plan is determined by the acceptance line $a_m = -h_1 + ms$ and the rejection line $r_m = h_2 + ms$. To calculate h_1, h_2, and s, the following quantities are introduced:

$$b = \ln \frac{1 - \alpha}{\beta}; \qquad a = \ln \frac{1 - \beta}{\alpha}; \qquad g_1 = \ln \frac{p'_2}{p'_1}; \qquad g_2 = \ln \frac{1 - p'_1}{1 - p'_2}.$$

Then,

$$h_1 = \frac{b}{g_1 + g_2}; \qquad h_2 = \frac{a}{g_1 + g_2}; \qquad s = \frac{g_2}{g_1 + g_2}.$$

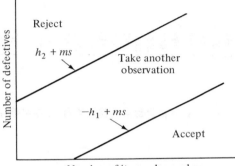

Fig. 13.6. *Graphic procedure for item-by-item sequential sampling plan.*

As an example, suppose $p'_1 = 0.01, p'_2 = 0.10, \alpha = 0.05, \beta = 0.20$; the following are calculated:

$$b = \ln \frac{0.95}{0.20} = 1.55815; \qquad a = \ln \frac{0.80}{0.05} = 2.77259;$$

$$g_1 = \ln \frac{0.10}{0.01} = 2.30258; \qquad g_2 = \ln \frac{0.99}{0.90} = 0.09531;$$

$$h_1 = 0.649796; \qquad h_2 = 1.156407; \qquad s = 0.039747.$$

This results in the equations

$$a_m = -0.649796 + 0.039747(m); \qquad r_m = 1.156407 + 0.039747(m).$$

Usually the sequential plan is applied in tabular rather than graphical form; a_m and r_m are computed for every n and listed in Table 13.14. Note that a_m is rounded down to the nearest integer and r_m is rounded up to the nearest integer.

Simple formulas may be found for the probability of acceptance and the average sample number (ASN) from five values of the fraction defective including the two points defining the plan.

Table 13.14. Tabular Procedure for Item-by-Item Sequential Sampling Plan

Number of Observations	Acceptance Number	Rejection Number	Observed Number of Defectives
m	a_m	r_m	d
1	...	...	
2	...	...	
3	...	2	
17	0	2	
22	0	3	
42	0	3	
47	1	4	
67	2	4	
72	2	5	
97	3	6	
117	4	6	
122	4	7	
143	5	7	
148	5	8	

$$p' = 0; \qquad L(p') = 1; \qquad \bar{n}_0 = \frac{h_1}{s}.$$

$$p' = p_1'; \qquad L(p') = 1 - \alpha; \qquad \bar{n}_{p'_1} = \frac{(1 - \alpha)h_1 - \alpha h_2}{s - p_1'}.$$

$$p' = s; \qquad L(p') = \frac{h_2}{h_1 + h_2}; \qquad \bar{n}_s = \frac{h_1 h_2}{s(1 - s)}.$$

$$p' = p_2'; \qquad L(p') = \beta; \qquad \bar{n}_{p'_2} = \frac{(1 - \beta)h_2 - \beta h_1}{p_2' - s}.$$

$$p' = 1; \qquad L(p') = 0; \qquad \bar{n}_1 = \frac{h_2}{1 - s}.$$

In the example above,

$$p' = 0; \qquad L(0) = 1; \qquad \bar{n}_0 = \frac{0.649796}{0.039747} = 16.348.$$

$$p' = 0.01; \qquad L(0.01) = 0.95;$$

$$\bar{n}_{0.01} = \frac{(0.95)(0.649796) - (0.05)(1.156407)}{0.029747} = 18.808.$$

$$p' = 0.039747; \qquad L(0.039747) = \frac{1.156407}{0.649796 + 1.156407} = 0.64024;$$

$$\bar{n}_{(0.039747)} = \frac{(0.649796)(1.156407)}{(0.039747)(0.960253)} = 19.688.$$

$$p' = 0.10; \qquad L(0.10) = 0.20;$$

$$\bar{n}_{0.10} = \frac{0.80(1.156407) - 0.20(0.649796)}{0.10 - 0.039747} = 13.197.$$

$$p' = 1.00; \qquad L(1.00) = 0; \qquad \bar{n}_{1.00} = \frac{1.156407}{0.960253} = 1.2043.$$

Usually, these five points may be used to make an adequate sketch of OC and ASN curves such as the ones in Figs. 13.7 and 13.8.

If more points are needed, they may be found by substituting values between 1 and ∞ for x in the equations

$$L(p') = \frac{x^{h_1 + h_2} - x^{h_1}}{x^{h_1 + h_2} - 1}; \qquad p' = \frac{x^s - 1}{x - 1}.$$

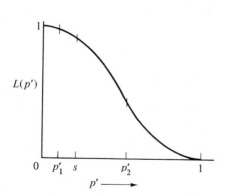

Fig. 13.7. *The OC curve of sequential plan sketched from five points.*

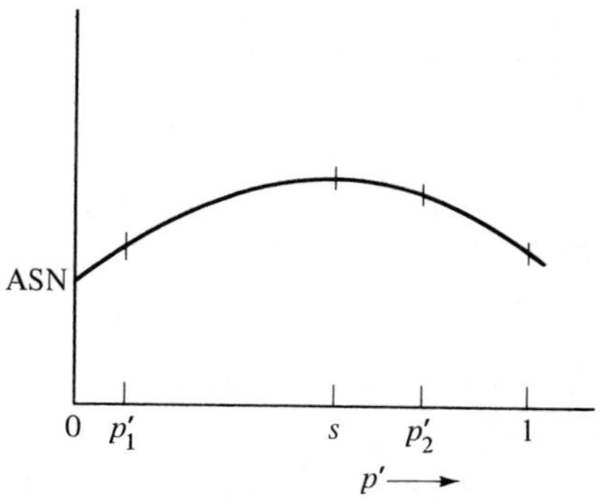

Fig. 13.8. *The ASN curve of sequential plan sketched from five points.*

The point $[p', L(p')]$ has the conjugate $[p'_c, L(p'_c)]$ given by

$$L(p'_c) = \frac{L(p')}{x^{h_1}}; \qquad p'_c = p'(x^{1-s}).$$

The expected number of observations is

$$\bar{n}_{p'} = \frac{L(p')(h_1 + h_2) - h_2}{s - p'}.$$

13.2.8 A BAYESIAN APPROACH TO SAMPLING INSPECTION

Another approach to the sampling inspection problem is to follow the Bayesian ideas introduced in Chapter 5. A lot containing N items is submitted for acceptance inspection. Assume that the lot contains M defectives (unknown). A random sample of size n is to be drawn and a decision about the acceptability of the lot is required. Instead of determining a sampling plan (n, c) by means of its OC curve, decision analysis can be used provided that the appropriate economic considerations are imposed and a prior probability distribution is available.

13.2.8.1 *Economic Structure.* Consider the following linear cost structure which is relevant to a decision to accept or reject a lot:

1. The cost of inspecting an item (I_1). This cost generally includes the inspection and testing costs, and is perhaps the simplest cost to evaluate.
2. Additional cost of correcting or replacing a defective item found during inspection (I_2). Determining the additional cost of inspecting a defective item depends upon its disposition. Frequently, a manufacturer is committed to supply a certain number of items so that defective units found must be replaced or repaired with a corresponding cost incurred.
3. Cost or value of accepting an item in the uninspected portion of an accepted lot (A_1). This cost is often small, or zero. It could be negative if each accepted unit is viewed as potential revenue.
4. Additional cost if an accepted item is defective (A_2). This cost is by far the most difficult to evaluate. Although A_1 is frequently zero or small, A_2 may be very large. It can include the cost of disassembly, reassembly, damage to other components, servicing, repairing, warranty, and the elusive cost of "good will."
5. Cost of rejecting an item in the uninspected portion of a rejected lot (R_1). Evaluation of R_1 depends on the disposition of rejected lots. Therefore, such costs as screening, scraping, salvaging, reworking, and discounting can be included.
6. Additional cost if a rejected item is defective (R_2). If a rejected lot is screened, then R_2 can be interpreted as the additional cost of repairing or replacing a defective item so that the screened lot will contain only "good" items.

In addition to some of the factors mentioned above it is evident that the interpretation and values of these costs are dependent on (a) the purpose of the sampling inspection, e.g., is it internal between departments, is it a producer's final inspection, or is it a consumer's receiving inspection; and (b) which party is responsible for bearing the inspection costs, i.e., the producer or consumer.

Whereas the aforementioned costs are arbitrary, it is reasonable to assume certain ordering relationships. In particular, if an item is non-defective, the preferred action should be acceptance; hence, $A_1 < R_1$. Similarly, if an item is defective, the preferred action should be rejection; hence, $R_1 + R_2 < A_1 + A_2$. From these two inequalities it follows that

$$A_2 - R_2 > R_1 - A_1 > 0.$$

Also, it is reasonable to assume that

$$I_1 \geq R_1 \quad \text{and} \quad I_2 \geq R_2.$$

13.2.8.2 *A Decision Analysis Model.* A process which generates the lot can be described as follows: The quality of each item produced is assumed to be independent, identically distributed Bernoulli random variables with parameter p'. The lot under study contains N of these units. In terms of the decision analysis formulation, this is a two-action problem, where action a denotes acceptance of the lot and action $\bar{a}$ denotes rejection of the lot. The number of possible states of nature p' is assumed to be infinite, $0 \leq p' \leq 1$. The loss function can be expressed in terms of the six cost factors alluded to in Sec. 13.2.8.1. If a random sample of size n is drawn, the cost of accepting the lot is given by

$$\text{cost of acceptance} = I_1 n + I_2 X + A_1(N - n) + A_2(M - X),$$

and the cost of rejecting the lot is given by

$$\text{cost of rejection} = I_1 n + I_2 X + R_1(N - n) + R_2(M - X),$$

where X is the number of defective items in the sample and M is the number of defective items in the lot. Note that these costs depend on the number of defective units found in the sample as well as the number of defective units in the lot. Such a loss function has not been encountered previously, so that the techniques described in Sec. 5.3.3 are not directly applicable. However, a modified loss function, independent of the sample outcome, will be exhibited.

Define a decision function d such that, when $X = x$,

$$d(x) = \begin{cases} a, & \text{if } x \leq c,[1] \\ \bar{a}, & \text{if } x > c, \end{cases}$$

---

[1] It can be shown rigorously that decision rules of this type are the only rules that need be considered.

where c is a constant to be determined. Letting $Y = M - X$ denote the number of defective items remaining in the lot (after the sample is drawn), the cost function using this decision rule is given by

$$I_1 n + I_2 X + A_1(N - n) + A_2 Y, \quad \text{if } X \leq c,$$
$$I_1 n + I_2 X + R_1(N - n) + R_2 Y, \quad \text{if } X > c.$$

For a given decision procedure d, the risk function $R(d, p')$ is just the expected value of the cost function, with the expectation taken with respect to the bivariate random variables (X, Y). It can be easily shown that X and Y are *independent* random variables each having binomial distributions with parameters n and p' and $(N - n)$ and p', respectively. (This result is somewhat intuitive because both X and Y are generated from different independent Bernoulli random variables.) Therefore, the risk function can be expressed as

$$R(d, p') = \sum_{k=0}^{c} \left\{ \sum_{j=0}^{N-n} [I_1 n + I_2 k + A_1(N - n) + A_2 j] \right.$$
$$\left. \times \binom{N-n}{j} p'^j (1 - p')^{N-n-j} \right\} \binom{n}{k} p'^k (1 - p')^{n-k}$$
$$+ \sum_{k=c+1}^{n} \left\{ \sum_{j=0}^{N-n} [I_1 n + I_2 k + R_1(N - n) + R_2 j] \right.$$
$$\left. \times \binom{N-n}{j} p'^j (1 - p')^{N-n-j} \right\} \binom{n}{k} p'^k (1 - p')^{n-k}.$$

Evaluating this expression leads to

$R(d, p')$
$$= \sum_{k=0}^{c} \{n(I_1 + I_2 p') + A_1(N - n) + A_2(N - n)p'\} \binom{n}{k} p'^k (1 - p')^{n-k}$$
$$+ \sum_{k=c+1}^{n} \{n(I_1 + I_2 p') + R_1(N - n) + R_2(N - n)p'\} \binom{n}{k} p'^k (1 - p')^{n-k}$$
$$= l(a, p') \sum_{k=0}^{c} \binom{n}{k} p'^k (1 - p')^{n-k} + l(\bar{a}, p') \sum_{k=c+1}^{n} \binom{n}{k} p'^k (1 - p')^{n-k}$$

where

$$l(a, p') = n(I_1 + I_2 p') + A_1(N - n) + A_2(N - n)p', \quad \text{and}$$
$$l(\bar{a}, p') = n(I_1 + I_2 p') + R_1(N - n) + R_2(N - n)p'.$$

For fixed n, this expression for the risk is of the form given in Sec. 5.3.2 so that the decision analysis framework of Sec. 5.3.3 can be utilized to find the Bayes procedure, i.e., for finding the optimal c (for fixed n). When the state of nature is p' the "loss" incurred if action a is taken (accepting the lot) is then $l(a, p')$, and the "loss" incurred if action $\bar{a}$ is taken (rejecting the lot) is $l(\bar{a}, p')$.

13.2.8.3 *Bayes Procedures.* Suppose that no sampling inspection is permitted and that $\pi_{p'}(z)$ is the density function of the prior distribution. The expected loss when the lot is accepted is given by

$$l(a) = \int_0^1 l(a, z)\pi_{p'}(z)\, dz = A_1 N + A_2 N\, E(p'),$$

and the expected loss when the lot is rejected is given by

$$l(\bar{a}) = \int_0^1 l(\bar{a}, z)\pi_{p'}(z)\, dz = R_1 N + R_2 N\, E(p').$$

Hence, the Bayes procedure calls for accepting the lot whenever

$$A_1 N + A_2 N\, E(p') \leq R_1 N + R_2 N\, E(p')$$

or equivalently, whenever

$$E(p') \leq \frac{R_1 - A_1}{A_2 - R_2} = p'_0.^1$$

Suppose sampling is now allowed and that n is fixed in advance. Denoting the density function of the posterior distribution of p', given that $X = x$ by $g_{p'}(z)$, it becomes necessary to compute the expected losses, $l_g(a)$ and $l_g(\bar{a})$, where the expectation is taken with respect to the posterior distribution, i.e.,

$$l_g(a) = \int_0^1 l(a, z)g_{p'}(z)\, dz$$
$$= n[I_1 + I_2 E_g(p')] + A_1(N - n) + A_2(N - n)E_g(p')$$

and

$$l_g(\bar{a}) = \int_0^1 l(\bar{a}, z)g_{p'}(z)\, dz$$
$$= n[I_1 + I_2 E_g(p')] + R_1(N - n) + R_2(N - n)E_g(p'),$$

where $E_g(p')$ is the expected value of p', given that x defectives were observed in the sample of size n.

Therefore, for fixed sample size n the Bayes procedure calls for accepting the lot whenever

$$n[I_1 + I_2 E_g(p')] + A_1(N - n) + A_2(N - n)E_g(p')$$
$$\leq n[I_1 + I_2 E_g(p')] + R_1(N - n) + R_2(N - n)E_g(p'),$$

or equivalently whenever

$$E_g(p') \leq \frac{R_1 - A_1}{A_2 - R_2} = p'_0.$$

If the prior distribution of p' is assumed to be a beta distribution with parameters $C > 0$ and $D > 0$, i.e.,

$$\pi_{p'}(z) = \frac{\Gamma(C + D)}{\Gamma(C)\Gamma(D)} z^{(C-1)}(1 - z)^{(D-1)}, \qquad \text{for } 0 \leq z \leq 1,$$

---

[1] The symbol p'_0 is often referred to as the breakeven quality. If the incoming quality p' were known exactly, the optimal procedure calls for accepting the lot if $p' \leq p'_0$, and rejecting the lot if $p' \geq p'_0$.

then the posterior distribution of p', given that $X = x$, also has a beta distribution (see Table 8.1) but with parameters $C + x > 0$ and $D + n - x > 0$, i.e.,

$$g_{p'|X=x}(z) = \frac{\Gamma(C + D + n)}{\Gamma(C + x)\Gamma(D + n - x)} z^{(C+x-1)}(1 - z)^{(D+n-x-1)},$$

for $0 \leq z \leq 1$.

Again referring to Table 8.1, it is seen that the expected value of p', given that x defectives were observed in the sample of size n, i.e., the expected value of p' taken with respect to the posterior distribution is given by

$$E_g(p') = \frac{C + x}{C + D + n}.$$

Hence, for this prior distribution, and for fixed sample size n, the Bayes procedure calls for accepting the lot whenever

$$\frac{C + x}{C + D + n} \leq p'_0,$$

which is equivalent to

$$x \leq p'_0(C + D + n) - C = c^*(n).$$

Thus, for fixed n, the Bayes procedure calls for accepting the lot whenever the number of defective items found in the sample is less than or equal to $c^*(n)$. In principle, this technique can be used to find the optimal sample size. For each n, $n = 0, 1, 2, \ldots, N$, an acceptance number $c^*(n)$ is determined. The Bayes risk, $B(n)$, is then evaluated, where

$$B(n) = \int_0^1 \left[l(a, z) \sum_{k=0}^{c^*(n)} \binom{n}{k} z^k (1 - z)^{n-k} \right.$$
$$\left. + l(\bar{a}, z) \sum_{k=c^*(n)+1}^{n} \binom{n}{k} z^k (1 - z)^{n-k} \right] \pi_{p'}(z) \, dz.$$

Thus, the sequence of pairs $(c^*(0), B(0))$, $(c^*(1), B(1))$, $(c^*(2), B(2))$, $\ldots$, $(c^*(N), B(N))$ are generated, and the optimal sampling inspection plan is obtained by choosing the n and $c^*(n)$ corresponding to the smallest $B(n)$. In principle, this appears to be a relatively simple set of computations. In fact, the number of required calculations may be quite large so that other techniques are used.[1]

---

[1] For computational purposes, it is generally easier to fix c^* and find the optimal n. Also, there exist asymptotic results for finding optimal sampling plans. Such results appear in D. Guthrie and M. V. Johns, "Bayes Acceptance Sampling Procedures for Large Lots," *Annals of Mathematical Statistics*, Vol. 30, 1959, 896–925; and A. Hald, "The Compound Hypergeometric Distribution and a System of Single Sampling Inspection Plans Based on Prior Distribution and Cost," *Technometrics*, Vol. 2, 1960, 275–340.

The following example is reported by Smith.[1] "It concerns a basically mechanical process of several steps each requiring great skill from the operators. Prior to assembly the component can be checked under high magnification to determine if it is suitable for assembly. Such inspection costs $0.187 per item. If an item is found to be defective it can be corrected for $0.235 each. Good items which are accepted involve no cost penalty. Defective items accepted are processed along with the acceptable items eventually being incorporated into an assembly. The avoidable costs of extra fabrication and lost parts are $12.44 per defective item. Rejection of a lot would mean subjecting each part to an inspection as above and a subsequent correction of defective parts." These data lead to the following cost structure:

$$I_1 = \$0.187 \qquad A_1 = 0 \qquad R_1 = \$0.187$$
$$I_2 = \$0.235 \qquad A_2 = \$12.44 \qquad R_2 = \$0.235.$$

The breakeven quality, p_0', is then given by

$$p_0' = \frac{R_1 - A_1}{A_2 - R_2} = \frac{0.187 - 0}{12.44 - 0.235} = 0.0153.$$

The lot size is specified as $N = 205$.

By examining past data, Smith concludes that the prior distribution of p' can be approximated by a beta distribution with parameters $C = 0.204$ and $D = 5.14$. The optimal plan is then given by $n = 40$ and $c = 0$, and the corresponding Bayes risk is $22.20 (the expected cost incurred without any sampling inspection is $40.20).

13.3 Continuous Sampling Inspection

13.3.1 INTRODUCTION

The purpose of this section on continuous sampling is to present the continuous sampling plans most commonly used in industry. These plans are used where the formation of inspection lots for lot-by-lot acceptance may be impractical or artificial, often the case for conveyor line production. The inspection is carried out by alternate sequences of consecutive item inspection (often called the 100% inspection) and sequences of production which are not inspected or from which sample items are inspected. In the plans discussed in this section, each item inspected is classified as defective or non-defective. Unless otherwise noted, it is assumed that all defective items found when sampling or when on 100% inspection are replaced by non-defective items.

---

[1] B. E. Smith, "Some Economic Aspects of Quality Control," Technical Report No. 53 (1961), Nonr 225 (53) (NR 042-002), Applied Mathematics and Statistics Laboratory, Stanford University, Stanford, California.

13.3.2 Dodge Continuous Sampling Plans

The simplest continuous sampling plan is the one proposed by Dodge in his pioneer paper in 1943.[1] This procedure (called CSP-1) is as follows: At the outset of inspection, inspect 100% of the units consecutively as produced and continue such inspection until i units in succession are found clear of defects. When this happens discontinue 100% inspection and inspect only a fraction f of the units, selecting one unit at random from each segment of $1/f$ items. If a single defective is found, revert immediately to 100% inspection of succeeding units and continue until again i units in succession are found clear of defects.

The objective of this plan and a wide class of generalizations of it is to provide assurance that the long-run percentage of defective units in the accepted product will be held within a prescribed limit, the average outgoing quality limit or AOQL. Evaluation of the statistical properties of the plan has been made under the assumption of control—qualities of the items are mutually independent Bernoulli random variables with constant parameter p'. Its statistical properties may be described by an average outgoing quality (AOQ) curve and by an average fraction inspected (AFI) curve. The statistical properties of most of the plans discussed in this section are described in the same way; AFI and AOQ curves are computed under the assumption of control and plans are classified by AOQL and sampling rate. The AOQL is defined as the worst average outgoing quality that will result from using a continuous sampling acceptance plan over a long-run period, regardless of the presented quality. The average outgoing quality (AOQ) will usually be better than the AOQL. The AOQL to be used for a given item is determined by the risk of accepting some defective items that the consumer is willing to take.

In his paper, Dodge presented equations and charts for determining the AOQL as functions of the parameters f and i, under the assumption that the process is in a state of statistical control. Figure 13.9 presents the necessary chart for the selection of the appropriate plan.[2]

Clearly an abrupt change between 100% inspection and partial inspection may sometimes be unnecessary. In the first place, even a production process which is at a satisfactory quality level will produce a certain number of defectives and eventually one of the sampled items will be defective. Furthermore, this abrupt change may lead to hardships in personnel assignments and, hence, in administration. In a very complicated and expensive item, such as an aircraft engine, this transition may require major readjustments. In a

---

[1] H. F. Dodge, "A Sampling Inspection Plan For Continuous Production," *Annals of Mathematical Statistics*, 1943, Vol. 14, pp. 264–279.

[2] If defective units are removed but not replaced, i should be increased by one to maintain the prescribed AOQL while holding f fixed.

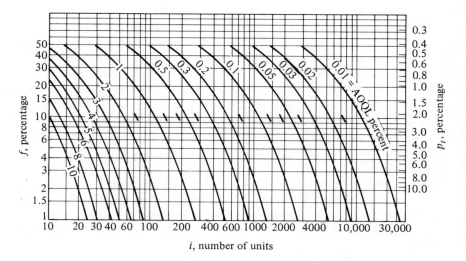

Fig. 13.9. *Curves for determining values of* f *and* i *for a given value of AOQL in Dodge's CSP-1 plan for continuous production.*
(Reprinted by permission from "A Sampling Inspection Plan for Continuous Production" by H. F. Dodge, *Annals of Mathematical Statistics*, September, 1943.)

later paper published in 1951, Dodge and Torrey propose two modifications of the plan (CSP-2 and CSP-3) which delay the beginning of 100% inspection and also add some protection against spotty quality.[1] In the CSP-2 plan inspection is begun, as in the CSP-1 plan, by inspecting 100% until i successive good items are found, after which partial inspection is introduced; the plan reverts to 100% inspection not on the basis of a single defective, but whenever two defectives occur spaced less than k units apart. The CSP-3 plan is similar to the CSP-2 plan, except that after a defective is found the next four units are inspected.

13.3.3 MULTI-LEVEL SAMPLING PLANS

A plan which allows for smooth transition between sampling and 100% inspection, which requires 100% inspection only when quality is quite inferior, and which allows for the amount of inspection to continue to reduce when quality is definitely good, is a multi-level sampling plan. This plan allows for any number of sampling levels subject to the provisions that transitions can occur only between adjacent levels. The origin of this plan is somewhat obscure. It seems to have been used by the Air Force for inspection of aircraft engines and was proposed by several people, including Joseph Greenwood of the Navy Bureau of Aeronautics.[2] A particular multi-level

---

[1] H. F. Dodge and M. N. Torrey, "Additional Continuous Sampling Plans," *Industrial Quality Control*, 1951, Vol. 7, pp. 7–12.

[2] J. A. Greenwood, "A Continuous Sampling Plan and Its Operating Characteristics," Bureau of Aeronautics, Navy Department, Washington, D.C., unpublished memorandum.

plan has been discussed in considerable detail by Lieberman and Solomon.[1] Their plan is called MLP and is as follows.

As with the Dodge plan, inspect 100% of the units consecutively as produced and continue until i units in succession are found clear of defects. When i units are found clear of defects, discontinue 100% inspection and inspect only a fraction f. If the next i units inspected are non-defective, then proceed to sampling at rate f^2. If a defective is found, revert to inspection at the next lowest level, etc. This plan can be used with any number of levels, but from two to four levels seem to be of the greatest practical interest. The first Dodge plan, CSP-1, is easily recognized as a special case containing only one sampling level. Curves of constant AOQL are given in Fig. 13.10 as a function of f and i for one-level and infinite-level plans. Figures 13.11, 13.12, and 13.13 present curves of constant AOQL for two-, three-, and four-level plans, respectively.

Several variations of the Lieberman-Solomon multi-level plans have appeared in the literature. The original Lieberman-Solomon plan suggested that sampling be reduced to the next lower level if a defective is found. That is, if sampling were at the rate f^k it would be reduced to the rate f^{k-1}. In a subsequent paper,[2] various modifications were considered, schemes which permit reversion to sampling at any rate when a defective is found or even back to 100% inspection.

One feature of all these plans is the requirement that, while sampling, one item is to be selected for inspection from a block of $1/f^j$ items. This type of

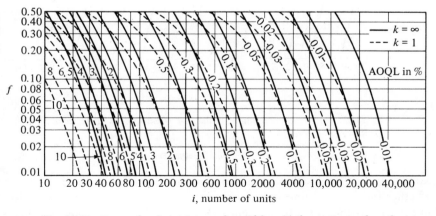

Fig. 13.10. *Curves for determining values of* f *and* i *for a given value of AOQL for one-level and for infinite-level plans.*

[1] G. J. Lieberman and H. Solomon, "Multi-Level Continuous Sampling Plans," *Annals of Mathematical Statistics*, 1955, Vol. 26, pp. 686–704.

[2] C. Derman, S. Littauer, and H. Solomon, "Tightened Multi-level Continuous Sampling Plans," *Annals of Mathematical Statistics*, 1957, Vol. 28, pp. 395–404.

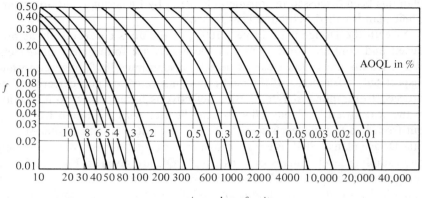

i, number of units

Fig. 13.11. *Curves for determining values of f and i for a given value of AOQL for two-level plans.*

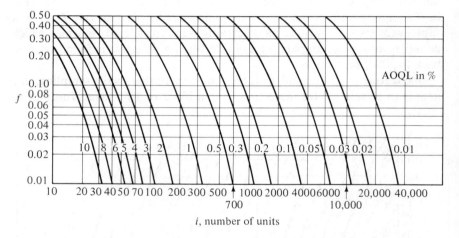

i, number of units

Fig. 13.12. *Curves for determining values of f and i for a given value of AOQL for three-level plans.*
(Reprinted by permission from "Multilevel Continuous Sampling Plans" by G. J. Lieberman and H. Solomon, *Annals of Mathematical Statistics*, December, 1955.)

sampling is called block sampling and assumes that each block is passed whether the sampled unit is defective or not. The AOQ function is derived under this assumption. If the item is found to be defective and the block contains a large number of items, the inspection department might indeed be reluctant to pass the remainder of the block without inspection, since finding a defective might be the result of a sudden shift in the process due to machine breakdown or a similar occurrence. Even if the block is small, as in the production of complex items, inspection might be initiated on one of

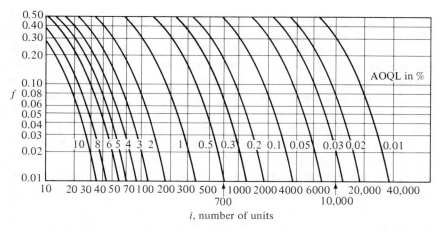

Fig. 13.13. *Curves for determining values of f and i for a given value of AOQL for four-level plans.*
(Reprinted by permission from "Multilevel Continuous Sampling Plans" by G. J. Lieberman and H. Solomon, *Annals of Mathematical Statistics*, December, 1955.)

the early items of the block before all the items are through the production line. Knowledge that one of the items has already been chosen for inspection and that the remainder are to be passed without inspection might prejudice the quality of the remaining items in the block.

An alternative scheme is to inspect every item with probability f and pass every item without inspection with probability $1 - f$. If the decision whether or not to inspect is made by coin tossing or some other random device after the item is off the production line, an operator will be unable to determine beforehand whether the item is to be inspected. Resnikoff[1] showed that the AOQ function for MLP with this modified type of sampling is identical with that obtained by Lieberman and Solomon for MLP with block sampling.

Another problem discussed by Resnikoff is that of truncation. When probability sampling is used instead of block sampling, there is a possibility of too long a string of uninspected items. With block sampling the inspection department is assured of having a "look" at the process within a specified production interval; with probability sampling no such assurance is possible. A natural method of overcoming the possibility of passing too many items without inspection is to truncate the sampling in some manner. For example, this might be done as follows:

Inspect each item with probability f^j when sampling at the jth level. It is mandatory, however, to inspect the $2/f^j$ item if $2/f^j - 1$ consecutive items

---

[1] G. J. Resnikoff, "Some Modifications of the Lieberman-Solomon Multi-Level Continuous Sampling Plan, MLP," *Technical Report No. 26*, Applied Mathematics and Statistics Laboratory, Stanford University, February 8, 1957.

have been passed without inspection. After inspecting the $2/f^j$ item resume the probability sampling.

Some numerical calculation reveals that for almost any reasonable truncation scheme the effect of truncation is negligible. The use of the AOQL contours already published by Lieberman and Solomon and Dodge is sufficiently close to those which might be computed with truncation, for practical purposes.

One of the most useful plans of this general type has already been adopted.[1] As in the Lieberman-Solomon plans, at the outset 100% of the product is inspected until a specified number of consecutive items, i, is found free of defects. At that time, the inspection is reduced to a fraction, f, of the items presented. If i successively inspected items are found free of defects, the inspection rate is reduced to a new fraction, f^2, of the items presented. This procedure is continued through the several levels for the plan. However, if at any level a defective item is found before i successively inspected items are passed, the plan provides a special procedure to determine if sampling should continue at the present rate or revert back to the previous sampling rate.

The special procedure, after a defective item has been found, consists of inspecting the next four consecutive items. If no defectives are found in the four consecutive items, sampling is resumed at the same rate as when the defective was found. If no more defectives are found before a total (including the four consecutive items) of i successively inspected items are determined to be acceptable, the sampling rate is reduced to the next smaller fraction (the next level). If a second defective is found, either in the four consecutive items or after resuming the sampling rate, revert to the former level (next larger sampling rate) and inspect four consecutive items, repeating the procedure just described. Thus, the sampling rate may be reduced or increased by specific steps as the quality of the presented items varies up and down. At the same time, the plans provide a check to differentiate between the random occurrence of a defective, which can occur even though the over-all quality level is acceptable, and inferior quality or spotty quality.

13.3.4 THE DODGE CSP-1 PLAN WITHOUT CONTROL

Up to this point the formulas for AOQ were derived on the assumption that the process is in control—a mathematical model for many processes which is rarely realized exactly. Several workers in the field have turned their attention to this problem. One approach is to examine the common procedures to see how they are affected by relaxing this condition.

---

[1] Inspection and Quality Control Handbook (Interim) H 106, Multi-Level Continuous Sampling Procedures and Tables for Inspection by Attributes, U. S. Department of Defense.

In 1953, Lieberman[1] showed that the Dodge procedure guarantees an AOQL whether or not the process is in a state of statistical control. In fact, without the assumption of control, and for a given f and i,

$$\text{AOQL} = \frac{(1/f) - 1}{(1/f) + i}.$$

Naturally this value of the AOQL is higher than the value given by the Dodge result for a fixed f and i. This is not to imply that the Dodge result is not useful. The AOQL is itself an upper bound, rarely achieved. The AOQ may be much less than the AOQL. Consequently, the AOQ for a process which is not in control may be less than the Dodge AOQL. In fact, this AOQL is achieved only for the pathological process which produces all defective items during partial inspection, and produces all non-defective items during 100% inspection. However, the result does point out that if the process behaves irregularly, the standard plans can lead to trouble.

13.3.5 WALD-WOLFOWITZ CONTINUOUS SAMPLING PLANS

Another approach has been to derive plans specifically designed to prescribe an AOQL. A major result in the field was obtained by Wald and Wolfowitz[2] in 1945 when they proposed a class of what might be called one-sided sequential plans. A particular type of this general class of plans is as follows:

Begin with partial inspection, choosing one at random from each successive segment of $1/f$ items. Let d_n be the total number of defectives found after the nth stage of partial inspection. Continue partial inspection as long as $d_n < h + sn$, where h and s are fixed constants.

If for some n, $d_n \geq h + sn$, terminate partial inspection and inspect h/s segments 100%. Repeat the procedure. Except for a slight approximation the AOQL for this plan is equal to $(1 - f)s$.

This one-sided sequential inspection plan is optimum from the point of view of cost of inspection (at least in the case of controlled production); it is not a plan which can be recommended for cases in which the excessive variability of the outgoing quality in finite batches of the product is an important factor. This becomes apparent when one considers the fact that for any material to be inspected 100%, the point

$$(d_j, j),$$

which represents the total number of defectives found after the jth stage of

[1] G. J. Lieberman, "A Note on Dodge's Continuous Sampling Plan," *Annals of Mathematical Statistics*, 1953, Vol. 24, pp. 480–484.

[2] A. Wald and J. Wolfowitz, "Sampling Inspection Plans for Continuous Production Which Insure a Prescribed Limit on the Outgoing Quality," *Annals of Mathematical Statistics*, 1945, Vol. 16, pp. 30–49.

partial inspection, must reach or exceed the line $y = h + sj$ for some sample size j. But if a few defective items are found during a long stretch of partial inspection, the point can wander so far away from the line that a great deal of unacceptable material can pass by before this fact is noted and the quality of the product improved by the 100% inspection. This situation can be remedied to some extent by employing a two-sided sequential inspection plan.

A two-sided sequential inspection plan is defined by two lines

$$y = h_2 + sj$$

and

$$y = -h_1 + sj$$

where h_1 and h_2 are positive constants and j represents the number of items inspected at the jth stage of partial inspection. In this plan, partial inspection continues as long as the point

$$(d_j, j)$$

lies between the two lines above. Partial inspection terminates when, for some $j = n$, either

$$d_n \leqq -h_1 + sn$$

or

$$d_n \geqq h_2 + sn.$$

In the former case, no 100% inspection is called for and the inspection procedure is simply repeated on new material. In the latter case, the inspection is resumed on new material only after h_2/s segments, i.e., h_2/fs items, have been inspected 100%.

Except for the action taken, the inspection procedure described above is identical with the sequential method of lot-by-lot acceptance inspection. Consequently, all the sequential theory can be employed in studying the plan.

13.3.6 GIRSHICK CONTINUOUS SAMPLING PLAN

One other type of sequential plan was presented by M.A. Girshick.[1] It is defined by the three constants m, N, and f. The plan operates as follows. The units of product in the production sequence are divided into segments of size $1/f$ items. The items are inspected in sequence and the number of defectives found, as well as the number of items examined, are cumulated. This procedure is continued until the cumulative number of defectives reaches m. At this point, the size of the sample n is compared with the integer N. If $n \geqq N$, the product which has passed through inspection is considered acceptable and the inspection procedure is repeated on the new incoming product. If, on the other hand, $n < N$, the following actions are taken:

---

[1] M. A. Girshick, "A Sequential Inspection Plan for Quality Control," *Technical Report No. 16*, Applied Mathematics and Statistics Laboratory, Stanford University, 1954.

(a) the next $N - n$ segments $[(N - n)1/f$ units] are inspected 100% and (b) after that, the inspection procedure is repeated. This procedure always guarantees, whether the process is in control or not, that the AOQL cannot exceed $(1 - f)m/N$.

Girshick also presented various operating characteristics of this plan, under control, such as the probability of inspection terminating with acceptance $(n \geq N)$, and discussed the biased and unbiased estimates of the process average.

13.3.7 PLANS WHICH PROVIDE FOR TERMINATION OF PRODUCTION

A criticism of all these continuous sampling plans, particularly those mentioned in the last section, is that they emphasize doing enough inspection to bring quality down to the AOQL but do not provide automatic penalties for poor quality. If the inspection is performed by a consumer or purchaser, he may do a very large amount of inspection to insure quality. Sometimes this may be avoided by administrative action. Dodge's original paper says:

"The inspection plan is most effective in practice if it is administered in such a way as to provide an incentive to clear up causes of trouble promptly. Such an incentive may be had by imposing a penalty on the operating or manufacturing department when defects are encountered. Normally no such penalty is imposed if both the sampling inspection and the 100% inspection are performed by this same person or group of persons; then the two costs merge. The inspector then merely serves as an agency for screening defects when quality goes bad. It is accordingly recommended that sampling inspection and 100% inspection operations be treated as two separate functions." With this possibility in mind, Army Ordnance[1] brought out a very comprehensive set of Dodge plans to use as a standard procedure for continuous sampling. In these plans, the government inspector performs all the sequences of partial inspection and the manufacturer is required to provide a screening crew to perform the 100% inspection. The government, in some cases, does 100% inspection but will charge the contractors for it.

The government inspector may verify the 100% inspection by inspecting the screened material and, of course, take fairly drastic action if any errors are found.

Another attack on this problem has been made by Rosenblatt and Weingarten,[2] who modify the Dodge plan by limiting the number of 100% inspection sequences an inspector is allowed to perform. In addition to f and i, their plan specifies a maximum allowable number of sequences of

---

[1] "Procedures and Tables for Continuous Sampling by Attributes," *Ordnance Inspection Handbook*, ORD-M608-11, August, 1954.

[2] H. Rosenblatt and H. Weingarten, "Sampling Procedures and Tables for Inspection on a Moving Line," *Continuous Sampling Plans*, NAVORD-Std 81.

consecutive item inspection beyond the first. If this number is exceeded at any time during a day's production, all inspection is stopped and the remaining items on the line are not accepted (although any product that has passed the inspector is accepted). The inspector informs the manufacturer as to which defects have occurred. He will not begin inspection again until the manufacturer locates the source of difficulty and gives him assurance that he has removed the cause.

A feature of the Navy Ordnance plan is that it comes close to satisfying the assumption of control. A state of control is rarely achieved over a long period of time. However, over a short period of time, like a day, the process may be in a state of control. The operating characteristics of this plan are calculated for a day's production.

This provision for shutting down the acceptance line seems to be desirable in a continuous sampling plan. That is, the plan should check on the product, do enough inspection to make quality good if there are any minor fluctuations, and if they are major to provide a basis for shutting down the line and looking for corrective action. The original Dodge plan and the Army Ordnance plan rely largely on the fact that if a product is generally bad or if it deteriorates, a large amount of the product will have to be inspected. However, the Navy Ordnance plan has the advantage that specific criteria for making this decision are provided. Other plans providing for termination of inspection have been introduced, using various mathematical models for the production process. However, there has been little industrial use of these procedures.

PROBLEMS

1. What is the probability of accepting a lot whose incoming quality is 8% defective, using a sample of size 5 and an acceptance number of 0? What is the probability when using an acceptance number of 1?

2. What is the probability of accepting a lot whose incoming quality is 10% defective, using a sample of size 4 and acceptance numbers of 0, 1, 2, and 3?

3. A lot containing 4,000 pieces is submitted for acceptance by a vendor. The consumer knows that this vendor's previous lots have averaged 2.0% defective. A sample of size 50 is to be drawn. If it is assumed that this lot is not significantly different from previous lots, i.e., it contains 2.0% defective, (a) what is the probability of the sample containing exactly three defectives by the binomial approximation? (Set up the equation.) (b) What is the probability of the sample containing two or more defectives by the Poisson approximation?

4. A product in lots is submitted for inspection.

$$n = 60 \qquad c = 3$$

Assuming the lot is very large compared to the sample sizes, determine the probability of acceptance if the submitted quality is 5% defective. Use the Poisson approximation.

5. A process is in control at a level such that $\bar{X}' = 100$ and $\sigma' = 3.3$. This quality characteristic is going to be accepted by an attributes sampling acceptance plan. The purchaser has indicated that it will accept practically all the lots which are 2% defective or less. The specification is 100 ± 10 for this characteristic. (a) What percent defective is being submitted? (b) If the producer were running a p chart ($n = 50$) as a quality assurance check on the in-process quality control function, what should the limits for this p chart be? Why? (c) Using the binomial approximation, determine the expression for the probability of a lot being accepted by the consumer if he is using the single sampling plan $n = 50$, $c = 1$. (d) Using the Poisson table, select a sampling plan that will accept 2% defective lots 95% of the time. Use $c = 3$.

6. The following double sampling plan is used in the acceptance inspection of a product:

First sample size	200	Second sample size	400
Acceptance number,		Cumulative sample size	600
first sample	2	Acceptance number,	
Rejection number,		combined samples	4
first sample	5	Rejection number,	
		combined samples	5

(a) If a lot that is 2% defective is submitted to this plan, what is the probability that it will be accepted? You may assume the Poisson distribution will give an answer of satisfactory accuracy. (b) If a great many lots that are 2% defective are submitted to this plan, what will be the average amount of inspection per lot? Consider only the inspection done by the receiver who does not screen rejected lots.

7. When complex items which are costly are tested by using a destructive test, a common acceptance procedure calls for taking a sample of size 1. If the item is good, the lot is accepted. If it is bad, a second sample of size 1 is obtained. If this item is also defective, the lot is rejected. If it is good, the lot is accepted. (a) Write down the expression for the OC curve for this plan, assuming that the submitted lot is large enough so that the selection of a defective article on the first sample does not make any appreciable change in the probability of getting a defective article on the second sample. (b) A single sampling procedure having the same OC curve as found in (a) is as follows: Take a sample of size $n = 2$. If the number of defectives is less than or equal to one, the lot is accepted. Otherwise, the lot is rejected. Prove this statement. (c) Compare the "average sample number" for the single and double sampling plans above.

8. A single test on a fuse costs more than the whole lot of 1,000. Alternative tests not proving satisfactory, it was decreed that not more than three tests of a fuse in actual performance could be made per lot. The following plan was prepared. Test two. If both function, accept the lot. If both fail, reject. Otherwise, test another. If it fails, reject; or if it functions, accept. (a) Assuming the binomial distribution is applicable, write down the expression for the OC curve. Express the final result in terms of q' only. (b) Suppose the single sampling plan $n = 3$, $c = 1$ were used instead of the one described above. Write down the expression for the OC curve. Express the final result in terms of q' only. (c) Compare the "average sample number" of the plans described in (a) and (b).

9. (a) Determine from the Dodge-Romig Single Sampling AOQL Table the single sampling plan for an AOQL of 1%, a process average of 0.5%, and lot size of 700. (b) What will be the total percentage of inspection, considering both sampling inspection and screening inspection of rejected lots, if the incoming lots are 0.5% defective? (c) What is the average outgoing quality (AOQ) if the incoming lots are 0.5% defective?

10. Determine the average outgoing quality for the single sampling plan $N = 100$, $n = 20$, $c = 1$, when submitted quality is 1.8% defective. Use Table 13.11 for finding the probability of acceptance.

11. A manufacturer is using a Dodge-Romig 4% LTPD single sampling plan on his own porduct. The lot size is 900 units and his process average is running at 1%. The consumer who is purchasing this product is using normal single sampling under MIL-STD-105D with an AQL of 6.5%. (a) What is the probability that a 6.5% defective lot will pass the manufacturer's sampling inspection and be rejected by the government? (b) Suppose the first five lots submitted by the vendor are exactly 10% defective. This is somewhat worse than the AQL of 6.5%. What is the approximate probability that a switch to tightened inspection will be called for?

12. Single sampling with an AQL of 1% is being used with acceptance criteria determined by MIL-STD-105D. The sample size code letter is F. (a) What is the probability of accepting a lot containing 2% defectives? (b) Suppose the first 5 lots submitted are each exactly 2% defective. What is the approximate probability that a switch to tightened inspection will be called for?

13. (a) In acceptance sampling under MIL-STD-105D, single sampling is to be used with inspection level I, an AQL of 4%, and a lot size of 200. What are the acceptance criteria under normal inspection? (b) What is the probability of accepting a lot containing 8 defectives (use Poisson tables)?

14. Billions of dollars worth of material are purchased under MIL-STD-105D. For any given lot, the probability of accepting a lot whose quality is at the next higher AQL is very good. Yet, most of the submitted product is at a quality level better than the AQL. Explain how MIL-STD-105D applies pressure to the producer to submit quality better than the AQL.

15. (a) In acceptance sampling under MIL-STD-105D single sampling is to be used with inspection level II, an AQL of 1%, and a lot size of 10,000. (a) Use the Poisson table to compute the approximate probability of acceptance of a 1% defective lot under normal inspection. (b) The comparable plan under tightened inspection corresponds to a 0.65% AQL plan (same sample size). What is the approximate probability of accepting a lot under tightened inspection when the incoming quality is 6.5%?

16. A product is submitted for acceptance inspection in lots of 25,000. Inspection is to be done in accordance with MIL-STD-105D single sampling. The AQL specified is 1%. Normal inspection at inspection level II is to be used. What is the probability of accepting a lot if the product is submitted at the AQL? What is the highest value of incoming quality permitting continuance of normal inspection with probability at least 0.90 based on the results of the immediately preceding 5 lots?

17. A product is submitted for acceptance inspection by attributes in lots of 10,000. Inspection is to be done in accordance with MIL-STD-105D single sampling. The AQL specified is 0.4%. Normal inspection at inspection level II is to be used. What is the probability of accepting a lot if the product is submitted at the AQL (use Poisson approximation to calculate the probability)? The comparable plan under tightened inspection corresponds to a 0.25% AQL plan (same sample size). What is the approximate probability of accepting a lot under tightened inspection when $p' = 0.4\%$?

18. In normal single sampling under MIL-STD-105D, the AQL is 2.5% and the lot size is 200. Inspection level II is to be used. What is the probability of accepting a lot submitted at the AQL?

19. Solve Problem 17, using double sampling.

20. Solve Problem 18, using double sampling.

21. Find a single sample attribute plan with the following properties (use Table 13.13): $p'_1 = 0.025$, $\alpha = 0.05$; $p'_2 = 0.08$, $\beta = 0.10$. Sketch the OC curve of the plan.

22. Suppose a single sampling plan is desired with $p'_1 = 0.01$, $\alpha = 0.01$, and $p'_2 = 0.03$. Find n if $\beta = 0.01$, $\beta = 0.05$, and $\beta = 0.10$. Use Table 13.13.

23. Find a single sampling plan with $p'_1 = 0.04$, $p'_2 = 0.10$, $\alpha = 0.05$, $\beta = 0.10$. Sketch the OC curve of the plan.

24. Find the boundaries for the graphical procedure of the item-by-item sequential plan for Problem 23. Sketch the sequential OC curve from five points. How does this OC curve compare with the OC curve of the single sampling plan determined by the same p'_1, α and p'_2, β?

25. Find a single sampling plan with properties $p'_1 = 0.01$, $\alpha = 0.05$, $p'_2 = 0.08$, $\beta = 0.10$ from Table 13.13. Find a plan in MIL-STD-105D which closely approximates this plan. Use Table 13.11 for the MIL-STD-105D plan.

26. Suppose the following two points on the OC curve are specified:

$$(p'_1 = 0.05, \ \alpha = 0.05) \quad \text{and} \quad (p'_2 = 0.10, \ \beta = 0.10).$$

Thus,

(1) $$P\{D \leq c\} = 0.95, \quad \text{where } p' = 0.05$$

(2) $$P\{D \leq c\} = 0.10, \quad \text{where } p' = 0.10.$$

If it is assumed that D is normally distributed with mean np' and standard deviation $\sqrt{np'q'}$, then

(3) $$P\{D \leq c\} = P\left\{\frac{D - np'}{\sqrt{np'q'}} \leq \frac{c - np'}{\sqrt{np'q'}}\right\}$$

$$= P\left\{N \leq \frac{c - np'}{\sqrt{np'q'}}\right\},$$

where N is normally distributed with mean 0 and standard deviation 1. Using the knowledge about equations (1), (2), and (3), two simultaneous equations can be set up which can be solved for the two unknowns n and c, thereby determining a sampling plan passing through two specified points. Find the values of n and c.

27. (a) Assuming the process is in control and the CSP-1 procedure is used, what is the necessary value of i if $f = \frac{1}{10}$ and the desired AOQL = 5%? (b) If control is not assumed, what is the required value of i?

28. (a) Assuming the process is in control and the CSP-1 procedure is used, what is the necessary value of i if $f = \frac{1}{25}$ and the desired AOQL = 1%? (b) If control is not assumed, what is the required value of i?

29. In Dodge's CSP-1 continuous sampling plan, it is desired to apply inspection to one piece out of every ten and to maintain an AOQL of 1%. (a) What should be the value of i, assuming the process is in control? (b) What should be the value of i, assuming the process is not in control? (c) If the process is in control at $p' = 2\%$ and sampling inspection is in effect, what is the probability of returning to 100% inspection on the next item looked at?

30. Assuming the process is in control and the MLP procedure is used with two levels, what is the necessary value of i if $f = \frac{1}{10}$ and the desired AOQL = 2%? If the process is in control at $p' = 4\%$ and sampling inspection is in effect at level III, what is the probability of returning to 100% inspection after the next two items are looked at?

31. Assuming the process is in control and the MLP procedure is used with three levels, what is the necessary value of i if $f = \frac{1}{10}$ and the desired AOQL = 4%? What is the value of i if $f = \frac{1}{2}, \frac{1}{3}$?

32. Assuming the process is in control and the MLP procedure is used with four levels, what is the necessary value of i if $f = \frac{1}{2}$ and the desired AOQL = 2%?

33. Using the Girshick continuous sampling procedure, what is the required value of N corresponding to $m = 2$, $f = \frac{1}{10}$ and AOQL = 4%?

34. Using the Girshick continuous sampling procedure, what is the AOQL corresponding to $m = 4$, $f = \frac{1}{10}$, and $N = 100$?

14

LOT-BY-LOT SAMPLING
INSPECTION BY VARIABLES

14.1 Introduction

In lot-by-lot acceptance sampling by attributes, each item of a sample drawn from a lot of manufactured items is classified simply as defective or non-defective. In single sampling, a random sample is drawn from the lot and the lot is either accepted or rejected, depending solely on the number of defectives in the sample; i.e., the lot is accepted if in a sample of size n, the number of defectives, D, does not exceed some preassigned constant, c; or, equivalently, if the estimate of the fraction defective, $p = D/n$, does not exceed $M = c/n$. Inspection procedures by variables are based on the outcomes of a quality characteristic, measured on a continuous scale, and the decision to accept or to reject the lot is a function of these measurements (as opposed to the number of defectives). Variables inspection is applicable whenever the testing of individual items involves measurement on a continuous scale and the form of the distribution is known. Since inspection by variables makes greater use of the information concerning the lot than does inspection by attributes, variables plans require smaller sample sizes than attributes plans furnishing the same protection. Variables sampling plans pertain to a single quality characteristic, and it is usually assumed that measurements of this quality characteristic are independent, identically distributed normal random variables having mean $\bar{X}'$ and standard deviation σ' either known or unknown. Such an assumption is made throughout this section on variables inspection.

Sampling inspection by variables is divided into three categories: known standard deviation plans, unknown standard deviation plans based on the sample standard deviation, and unknown standard deviation plans based on the average range. Known standard deviation plans are functions of the sample mean and the known standard deviation. Unknown standard deviation plans based on the sample standard deviation are functions of the sample mean and the sample standard deviation. Unknown standard deviation plans based on the average range are functions of the sample mean and the average range in sub-samples.

14.2 General Inspection Criteria

Associated with each inspection characteristic are the design specifications. If only an upper specification limit, U, is given, the item is considered defective if its measurement exceeds U; the percent defective in the entire lot, p'_U is represented by the shaded area in Fig. 14.1. If only a lower specification limit, L, is given, the item is considered defective if its measurement is smaller than L; the percent defective in the entire lot, p'_L, is represented by the shaded area in Fig. 14.2. Whenever both upper and lower limits are specified, the item is considered defective if its measurement either exceeds U or is smaller than L; the percent defective in the entire lot, $p'_U + p'_L$, is represented by the shaded area in Fig. 14.3. If the mathematical model is realistic, i.e., the assumption of a normal distribution is valid, the ratio of the total number

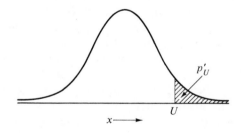

Fig. 14.1. *The shaded area represents the fraction defective in the lot, given an upper specification limit.*

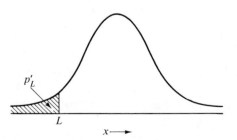

Fig. 14.2. *The shaded area represents the fraction defective in the lot, given a lower specification limit.*

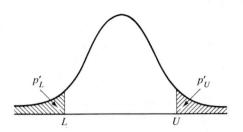

Fig. 14.3. *The shaded areas represent the fraction defective in the lot, given both upper and lower specification limits.*

of defective units in the lot to the lot size should approximate the area outside of the specification limits.

If the percent defective of a submitted lot is known, i.e., p'_L and/or p'_U are given (whichever is appropriate), sampling inspection is unnecessary to determine whether or not the lot is to be accepted. If the percent defective is sufficiently small, the lot is accepted; otherwise, it is rejected. Since such knowledge about the incoming quality is rare, a logical procedure is to *estimate* the percent defective from a sample, and to accept or reject the lot on the basis of this estimate. A sampling plan is then described as consisting of the sample size n; a method for estimating the percent defective with the estimate being denoted by p_U, p_L, or $p_U + p_L$, whichever is appropriate; and a maximum allowable percent defective, M. If only an upper specification limit, U, is given, the estimate of the percentage above this limit, p_U, is obtained from the sample of size n. If $p_U \leq M$, the lot is accepted. If only a lower specification limit, L, is given, the estimate of the percentage below this limit, p_L, is obtained from the sample of size n. If $p_L \leq M$, the lot is accepted. If a double specification limit is given, both p_U and p_L are computed. If $p_U + p_L \leq M$, the lot is accepted.

It is evident that these acceptance procedures are similar to the procedures used in attributes inspection, with p_L and/or p_U playing the role of D/n as the estimator of the incoming quality and M playing the role of c/n as the acceptance criterion. The OC curve is still a function of the incoming quality, p', and depends on the choice of n and M.[1]

As in attributes inspection, increasing the sample size, while holding M fixed, increases the steepness of the OC curve. Similarly, increasing M, while holding the sample size fixed, essentially shifts the OC curve to the right.

---

[1] p' denotes the incoming quality. If there is only an upper specification limit given, the symbol p'_U coincides with p'. If there is only a lower specification limit given, the symbol p'_L is interpreted as p'. If there are both upper and lower specification limits, p' is equivalent to $p'_U + p'_L$.

14.3 Estimates of the Percent Defective

14.3.1 ESTIMATE OF THE PERCENT DEFECTIVE WHEN THE STANDARD DEVIATION IS UNKNOWN BUT ESTIMATED BY THE SAMPLE STANDARD DEVIATION

As indicated in Chapter 8, a "good" estimator of a parameter satisfies certain requirements. If the property of unbiasedness is considered important, the "best" estimator of the percent defective in the lot is the minimum variance unbiased estimator; i.e., it is the most efficient estimator among all unbiased estimators. For an upper specification limit, U, the minimum variance unbiased estimator is a function of $Q_U = (U - \bar{X})/S$ and is denoted by p_U, where $\bar{X}$ is the sample mean and S is the sample standard deviation, i.e.,

$$S = \sqrt{\frac{\sum(X_i - \bar{X})^2}{n - 1}}.$$

For a lower specification limit, L, the minimum variance unbiased estimate is a function of $Q_L = (\bar{X} - L)/S$ and is denoted by p_L. The form of the functions, p_U and p_L, although rather complicated, is the same, and extensive tabulation exists.[1] A graph of the values that these estimators take on versus

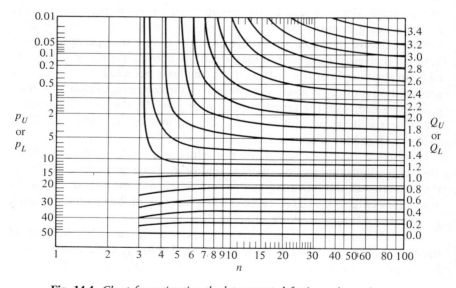

Fig. 14.4. *Chart for estimating the lot percent defective, using estimators based on the sample standard deviation.*
(Reprinted with permission from A. Duncan, *Quality Control and Industrial Statistics,* rev. ed., Richard D. Irwin, Inc., Homewood, Ill., p. 220.)

---

[1] These estimates are tabulated and are described in detail in an article by G.J. Lieberman and G.J. Resnikoff, "Sampling Plans For Inspection By Variables," *Journal*

Q_U or Q_L and n is given in Fig. 14.4. Therefore, in order to obtain the quantity p_U or p_L, Fig. 14.4 is entered with the sample size and the quality index Q_U or the quality index Q_L, whichever is appropriate; the required estimate, in percent, is read from the chart. For two-sided specification limits, the minimum variance unbiased estimate of the percent defective is given by $p_U + p_L$ where these quantities are also read from Fig. 14.4.

For an upper specification limit, the lot is accepted whenever $p_U \leq M$, where M is a preassigned constant. Alternatively, it can be shown that $p_U \leq M$ if, and only if, $(U - \bar{X})/S \geq k$, where k is a constant related to M. Hence, an equivalent procedure is to accept the lot if $(U - \bar{X})/S \geq k$.

For a lower specification limit, the lot is accepted whenever $p_L \leq M$. Alternatively, it can be shown that $p_L \leq M$ if, and only if, $(\bar{X} - L)/S \geq k$. Hence, an equivalent procedure is to accept the lot if $(\bar{X} - L)/S \geq k$.

Finally, for two-sided specification limits the lot is accepted whenever $p_U + p_L \leq M$. There is no equivalent k procedure for this situation.

14.3.2 ESTIMATE OF THE PERCENT DEFECTIVE WHEN THE STANDARD DEVIATION IS UNKNOWN BUT ESTIMATED BY THE AVERAGE RANGE

An estimator of the standard deviation σ' based on the average range has the advantage of computational ease over an estimator based on the sample standard deviation. However, it is known that the average range estimator is not as efficient as the sample standard deviation estimator. Moreover, there is no minimum variance unbiased estimator of p' which is a function of $\bar{X}$ and $\bar{R}$. An estimator has been found whose distribution is approximately equivalent to the distribution of the minimum variance unbiased estimator of p' based on $\bar{X}$ and S, but at the expense of an increased sample size. For an upper specification limit, U, this estimator of p'_U is a function of $Q_U = (U - \bar{X})c/\bar{R}$ and is denoted by p_U, where $\bar{X}$ is the sample mean, $\bar{R}$ is the average range of sub-group ranges (each subgroup usually consists of five measurements), and c is a constant depending on the sample size. For a lower specification limit, L, the estimator of p'_L is a function of $Q_L = (\bar{X} - L)c/\bar{R}$ and is denoted by p_L. The form of the functions, p_U and p_L, although rather complicated, is the same, and extensive tabulation exists.[1] A graph of the values that these estimators take on versus Q_U or Q_L and n is given in Fig. 14.5. For sample sizes of 3, 4, 5, and 7, $\bar{R}$ is taken to be the range of the sample. Values of c for various sample sizes is given in Table 14.1. Therefore, to obtain the quantity p_U or p_L Fig. 14.5 is entered with the sample size and the quality index Q_U or the quality index Q_L, whichever is appropriate; the required estimate, in percent, is then read from the chart. For two-sided

---

of the American Statistical Association, June, 1955, Vol. 50, pp. 457–516. They are also tabulated in MIL-STD-414, "Sampling Procedures and Tables for Inspection by Variables for Percent Defective," U. S. Government Printing Office.

[1] See footnote on tabulation of estimates based on the sample standard deviation.

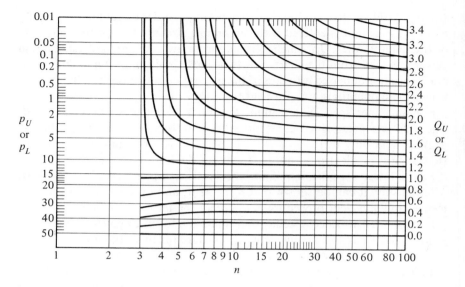

Fig. 14.5. *Chart for estimating the lot percent defective, using estimators based on the average range.*

(Reprinted with permission from A. Duncan, *Quality Control and Industrial Statistics*, rev. ed., Richard D. Irwin, Inc., Homewood, Ill., p. 227.)

Table 14.1. Table of c Factors

Sample Size	c Factor	Sample Size	c Factor
3	1.910	35	2.349
4	2.234	40	2.346
5	2.474	50	2.342
7	2.830	60	2.339
10	2.405	85	2.335
15	2.379	115	2.333
25	2.358	175	2.333
30	2.353	230	2.333

specification limits, the estimate of percent defective is given by $p_U + p_L$ where these quantities are also read from Fig. 14.5.

For an upper specification limit, the lot is accepted whenever $p_U \leq M$, where M is a preassigned constant. Alternatively, it can be shown that $p_U \leq M$ if, and only if, $(U - \bar{X})/\bar{R} \geq k$, where k is a constant related to M. Hence, an equivalent procedure is to accept the lot if $(U - \bar{X})/\bar{R} \geq k$.

For a lower specification limit, the lot is accepted whenever $p_L \leq M$. Alternatively, it can be shown that $p_L \leq M$ if, and only if, $(\bar{X} - L)/\bar{R} \geq k$. Hence, an equivalent procedure is to accept the lot if $(\bar{X} - L)/\bar{R} \geq k$.

Finally, for a two-sided specification limit, the lot is accepted whenever $p_U + p_L \leq M$. There is no equivalent k procedure for this situation.

Operating characteristic curves based on average range plans can be constructed which match OC curves based on sample standard deviation plans by suitably varying the sample size n and the maximum allowable percent defective M (or n and k if this procedure is used). However, the resulting sample size, using the range plans, will always be greater than or equal to the sample size using the sample standard deviation plans. If a sample standard deviation plan requires a small n, the comparable range plan requires a not too different sample size; this difference increases as the sample size of the sample standard deviation plan increases.

14.3.3 ESTIMATE OF THE PERCENT DEFECTIVE WHEN THE STANDARD DEVIATION IS KNOWN

When the standard deviation, σ', of the normal distribution is known, there exists a minimum variance unbiased estimator of p'. For an upper specification limit, U, the minimum variance unbiased estimator is a function of

$$Q_U = \left(\frac{U - \bar{X}}{\sigma'}\right)\sqrt{\frac{n}{n-1}}$$

and is given by p_U, where

$$p_U = \int_{Q_U}^{\infty} \frac{1}{\sqrt{2\pi}} e^{-z^2/2}\, dz.$$

It is evident that p_U is just the area of the standardized normal distribution above Q_U, and can be evaluated by using Appendix Table 1 (table of the percentage points of the normal distribution). For a lower specification limit, L, the minimum variance unbiased estimator is a function of

$$Q_L = \left(\frac{\bar{X} - L}{\sigma'}\right)\sqrt{\frac{n}{n-1}}$$

and is given by p_L, where

$$p_L = \int_{-\infty}^{-Q_L} \frac{1}{\sqrt{2\pi}} e^{-z^2/2}\, dz.$$

This integral is equivalent to the same integral between the limits Q_L and ∞ so that p_L is just the area of the standardized normal distribution above Q_L, and also can be evaluated by using Appendix Table 1. To obtain the quantities p_U and p_L, Appendix Table 1 is entered with the quality index Q_U or the quality index Q_L, whichever is appropriate. The required estimate is read from the table and converted to readings in percent by multiplying by one hundred. For two-sided specification limits, the minimum variance unbiased estimate of the percent defective is given by $p_U + p_L$.

For an upper specification limit, the lot is accepted whenever $p_U \leq M$, where M is a preassigned constant. Alternatively, it is evident from the expression for p_U that $p_U \leq M$ if, and only if, $(U - \bar{X})/\sigma' \geq k$, where k is a constant given by

$$k = \sqrt{\frac{n-1}{n}} K_M$$

and K_M is the M percentage point of the normal distribution. Hence, an equivalent procedure is to accept the lot if

$$(U - \bar{X})/\sigma' \geq k.$$

For a lower specification limit, the lot is accepted whenever $p_L \leq M$. Alternatively, it is evident from the expression for p_L that $p_L \leq M$ if, and only if, $(\bar{X} - L)/\sigma' \geq k$, where k is defined as above. Hence, an equivalent procedure is to accept the lot if

$$(\bar{X} - L)/\sigma' \geq k.$$

Finally, for a two-sided specification limit, the lot is accepted whenever $p_U + p_L \leq M$.

OC curves based on known standard deviation procedures can be constructed which match OC curves for unknown standard deviation procedures by suitably varying the sample size n and the maximum allowable percent defective M (or n and k if this procedure is used). The resulting sample size, using the known standard deviation procedure, will always be smaller than the sample size using the unknown standard deviation procedure.

Sampling plans based on a known standard deviation are easily designed. Suppose that such a plan is desired so that its OC curve passes through the two points $(p_1', 1 - \alpha)$ and (p_2', β). Consider a plan using an upper specification limit U. (The results are the same for a lower specification limit.) It has been shown that the lot is to be accepted whenever

$$(U - \bar{X})/\sigma' \geq k.$$

Hence, the requirement that the OC curve of the desired plan pass through the two prescribed points can be written as

$$P\{(U - \bar{X})/\sigma' \geq k \,|\, p' = p_1'\} = 1 - \alpha$$

and

$$P\{(U - \bar{X})/\sigma' \geq k \,|\, p' = p_2'\} = \beta.$$

These expressions are equivalent to

$$P\{\bar{X} + k\sigma' \leq U\}$$
$$= P\left\{ \frac{\sqrt{n}\,(\bar{X} - \mu)}{\sigma'} \leq \frac{\sqrt{n}\,(U - \mu)}{\sigma'} - \sqrt{n}\,k \,\Big|\, p' = p_1' \right\} = 1 - \alpha,$$

and

$$P\{\bar{X} + k\sigma' \leq U\}$$

$$= P\left\{\frac{\sqrt{n}(\bar{X} - \mu)}{\sigma'} \leq \frac{\sqrt{n}(U - \mu)}{\sigma'} - \sqrt{n}\,k \,|\, p' = p'_2\right\} = \beta.$$

Note that $N = \sqrt{n}(\bar{X} - \mu)/\sigma'$ is a normally distributed random variable having zero mean and unit variance; $(U - \mu)/\sigma' = K_{p'_1}$ (the p'_1 percentage point of the normal distribution) when $p' = p'_1$; and $(U - \mu)/\sigma' = K_{p'_2}$ (the p'_2 percentage point of the normal distribution) when $p' = p'_2$. It then follows that the conditions on the OC curve are equivalent to

$$P\{N \leq \sqrt{n}(K_{p'_1} - k)\} = 1 - \alpha$$

and

$$P\{N \leq \sqrt{n}(K_{p'_2} - k)\} = \beta,$$

so that

$$\sqrt{n}(K_{p'_1} - k) = K_\alpha$$

and

$$\sqrt{n}(K_{p'_2} - k) = K_{1-\beta} = -K_\beta.$$

Solving these equations simultaneously results in the equations for the sampling plan, i.e.,

$$k = \frac{K_{p'_2}K_\alpha + K_{p'_1}K_\beta}{K_\alpha + K_\beta}$$

$$n = \frac{(K_\alpha + K_\beta)^2}{(K_{p'_1} - K_{p'_2})^2}.^1$$

As an example, let $p'_1 = 0.04$, $\alpha = 0.05$, $p'_2 = 0.08$, and $\beta = 0.10$. From Appendix Table 1, the following are found:

$$K_{0.04} = 1.751; \qquad K_{0.05} = 1.645;$$
$$K_{0.08} = 1.405; \qquad K_{0.10} = 1.282.$$

The desired sampling plan is then given by

$$k = \frac{(1.405)(1.645) + (1.751)(1.282)}{(1.645 + 1.282)} = 1.557$$

$$n = \frac{(1.645 + 1.282)^2}{(1.751 - 1.405)^2} = 72.$$

---

[1] When σ' is unknown, and estimated by the sample standard deviation, S, similar results can be obtained for large n, using the asymptotic distribution of $\bar{X} + kS$. Since $\bar{X}$ and S are independent, it can be shown that $\bar{X} + kS$ is approximately normally distributed with mean equal to $\bar{X}' + k\sigma'$ and variance $\sigma'^2(1/n + k^2/2n)$. Using these properties, it follows that the expression for k is the same as when σ' is known and n is modified so that

$$n \approx \left(1 + \frac{k^2}{2}\right)\frac{(K_\alpha + K_\beta)^2}{(K_{p'_1} - K_{p'_2})^2}.$$

However, the user is cautioned that this is just an approximation and is reasonable only for large n.

The above derivation can also be used to construct an OC curve for a given sampling plan (n, k). It has been shown that when the incoming quality is p',

$$L(p') = P\{N \leq \sqrt{n}(K_{p'} - k)\},$$

where $L(p')$ is the probability of accepting the lot when the incoming quality is p', and N denotes a normally distributed random variable with zero mean and unit standard deviation. Thus, using $n = 72$ and $k = 1.557$ (the plan just obtained) and for $p' = 0.06$,

$$L(0.06) = P\{N \leq \sqrt{72}(1.555 - 1.557)\} = 0.493.$$

14.4 Comparison of Variables Procedures with M and k

It was indicated that single specification limit procedures based on estimates of the percent defective are equivalent to single specification limit procedures involving k. Any advantages or disadvantages of one over the other are independent of OC curve arguments since they both result in the same OC curve, provided M and k are chosen properly. The estimation procedure has several advantages over the k procedure. Like attributes plans, variables plans based on the estimation procedure can involve measurement of lot quality by the percent defective. The k procedure involves measurement of lot quality by the average and variability of the measurements. Thus, with the estimation procedure it is unnecessary to shift to attributes sampling whenever percent defective is the appropriate measure. Secondly, the estimation procedure has intuitive appeal because it is a logical one, whereas relating a measurement of lot quality based on the average and variability of the measurements to a practical criterion is difficult. In other words, rejecting a lot because the estimated percent defective is too large is understandable, whereas rejecting a lot because $(U - \bar{X})/S$ is too large is difficult to explain. Finally, the estimation procedure leaves the vendor and the consumer with an estimate of lot quality whether the lot is accepted or rejected. On the other hand the k procedure has the advantage of requiring one less step in carrying through the procedure. It does not require a chart (or table) look-up for estimate of the lot quality before deciding on acceptance or rejection.

All variables sampling procedures for two-sided specification limits suffer from the fact that the probability of accepting a submitted lot with given percent defective p' does not depend on p' alone but on the division of p' into two components, the percent lying above the upper specification limit, p'_U, and the percent lying below the lower specification limit, p'_L. For this reason a two-sided procedure does not have a unique OC curve but rather a band of curves, each curve within the band representing a possible division of p'. Using the double specification procedure based on the estimate of the lot quality, this band is so narrow as to be, for all practical purposes, a single curve. Since the OC curve for the single specification limit is contained within this narrow band (corresponding to all the defectives outside one limit and none outside the other), it is used as the OC curve for the double specification procedure. No procedure based on k has this desirable property.

14.5 The Military Standard for Inspection by Variables, MIL-STD-414

14.5.1 INTRODUCTION

In recent years there has been a recognition of the importance of inspection by variables for percent defective as a test procedure to evaluate product quality. Both government and industry prepared sets of standard sampling plans designed for their own needs. The use of sampling inspection by variables increased to such an extent that the need to prepare a military standard, to serve as an alternative to Military Standard 105D for inspection by attributes, became apparent. In the latter part of 1957, the first sampling system for inspection by variables, MIL-STD-414, was issued.

The variables standard is divided into four sections, namely:

Section A—General Description of Sampling Plans;
Section B—Variability Unknown—Standard Deviation Method;
Section C—Variability Unknown—Range Method;
Section D—Variability Known.

Section A is always used in conjunction with the other sections since it provides general concepts and definitions needed for sampling inspection by variables. Each of the Sections B, C, and D consists of three parts:

(1) Sampling Plans for the Single Specification Limit Case;
(2) Sampling Plans for the Double Specification Limit Case;
(3) Procedures for Estimation of Process Average and Criteria for Tightened and Reduced Inspection.

For the single specification limit case, the acceptability criterion is given in two forms, namely, the k procedure (form 1), e.g., accept the lot if $(U - \bar{X})/S \geq k$; and the estimation procedure based on M (form 2), e.g., accept the lot if $P_U \leq M$. Either of the forms may be used since they require the same sample size and have identical OC curves. Form 2 is required for the estimation of the process average, if such an estimate is desired. The estimation procedure (form 2) is required for double specification limits.

Like the attribute standard, MIL-STD-414 has provisions for normal, tightened, and reduced inspection, and the user is expected to use these features if he is to reap the full benefits of the sampling system. For each of the methods presented in Section B, C, and D, rules for switching between normal, tightened, and reduced inspection are provided. These rules usually depend on an estimate of the process average and an estimate of the percent defective in each lot. For example, tightened inspection is called for when the estimated process average computed from the preceding 10 lots is greater than the AQL, and whenever more than a preassigned number of these lots have estimates of the percent defective exceeding the AQL. The method of estimating the process average, determining the preassigned number alluded to above, etc., are discussed in detail in the standard.

Under tightened inspection the sample size remains the same as under normal inspection, but the acceptance criterion is generally moved to a sampling plan one step to the next lower (left) AQL plan; e.g., if a 4% AQL plan is used under normal inspection, a 2.5% AQL plan is required under tightened inspection.

Under reduced inspection substantial decreases in the sample size (compared to that under normal inspection) generally occur. A 40% decrease is a typical figure. This is accomplished at the expense of increasing the consumer's risk and decreasing (slightly) the producer's risk.

14.5.2 Section A—General Description of Sampling Plans

The variables standard has many features which are similar to the attributes standard. Defects are grouped into three classes, critical, major, and minor, with the definition of each similar to that in MIL-STD-105D. Like the attributes standard, the sampling plans are indexed by AQL, inspection level, and lot size. The definition of the AQL is different from that found in MIL-STD-105D. In MIL-STD-414 the acceptable quality level, AQL, is defined as "a nominal value expressed in terms of percent defective specified for a single quality characteristic."[1] The AQL values given, in percent, are

0.04	0.40	4.0
0.065	0.65	6.5
0.10	1.0	10.0
0.15	1.5	15.0
0.25	2.5	

The probabilities of acceptance of submitted lots having these AQL range from about 0.89 for the smallest sample sizes to about 0.99 for the largest sample sizes. The particular AQL value to be used for a single quality characteristic of a given product must be specified. In the case of a double specification limit, either an AQL value is specified for the total percent defective outside both upper and lower specification limits or two AQL values are specified, one for the upper limit and another for the lower limit.

Sample sizes are designated by code letters B to Q. The sample size code letter depends on the inspection level and the lot size. There are five inspection levels: I, II, III, IV, V. Unless otherwise specified, inspection level IV is used. The sample size code letter applicable to the specified inspection level and for lots of given size is obtained from Table 14.2 (Table A-2 of MIL-STD-414).

---

[1] This follows the definition appearing in many of the previous attribute standards, e.g., MIL-STD-105C.

Table 14.2. Sample Size Code Letters[1]
(Table A-2 of MIL-STD-414)

Lot Size			Inspection Levels				
			I	II	III	IV	V
3 to	8		B	B	B	B	C
9 to	15		B	B	B	B	D
16 to	25		B	B	B	C	E
26 to	40		B	B	B	D	F
41 to	65		B	B	C	E	G
66 to	110		B	B	D	F	H
111 to	180		B	C	E	G	I
181 to	300		B	D	F	H	J
301 to	500		C	E	G	I	K
501 to	800		D	F	H	J	L
801 to	1,300		E	G	I	K	L
1,301 to	3,200		F	H	J	L	M
3,201 to	8,000		G	I	L	M	N
8,001 to	22,000		H	J	M	N	O
22,001 to	110,000		I	K	N	O	P
110,001 to	550,000		I	K	O	P	Q
550,001 and over			I	K	P	Q	Q

[1] Sample size code letters given in body of table are applicable when the indicated inspection levels are to be used.

After the AQL, inspection level, and sample code letter are chosen, the appropriate sampling plan is chosen from a master table in Section B, C, or D of the standard, whichever is appropriate. It is interesting to note that the variables standard states that "unless otherwise specified, unknown variability, standard deviation method sampling plans and the acceptability criterion of form 2 (for the single specification limit case) shall be used." This refers to Section B and the estimation procedure.

Section A also contains all the OC curves. These OC curves represent the plans in Section B exactly. The OC curves of the corresponding plans in Sections C and D were matched as well as possible and are extremely close in most cases. Any discrepancies are due to the requirement of a constant sample size (multiple of five) for a code letter for the range plans, and integer values of the sample size for the known standard deviation plans. Hence, for a given lot size, inspection level, and AQL, the user may select a plan

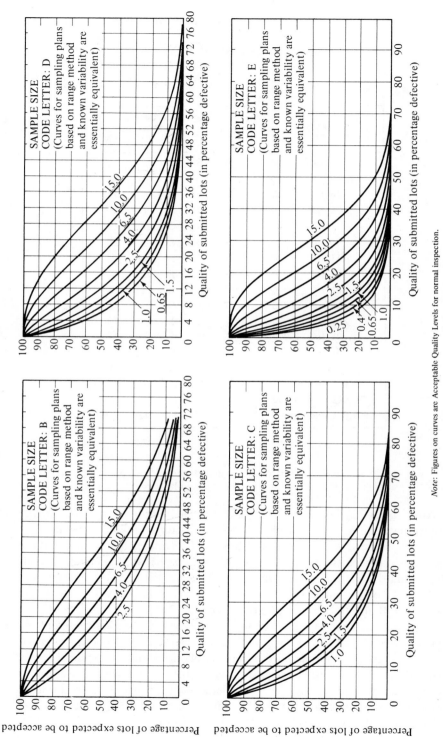

Note: Figures on curves are Acceptable Quality Levels for normal inspection.

Fig. 14.6. *The OC curves from MIL-STD-414.*

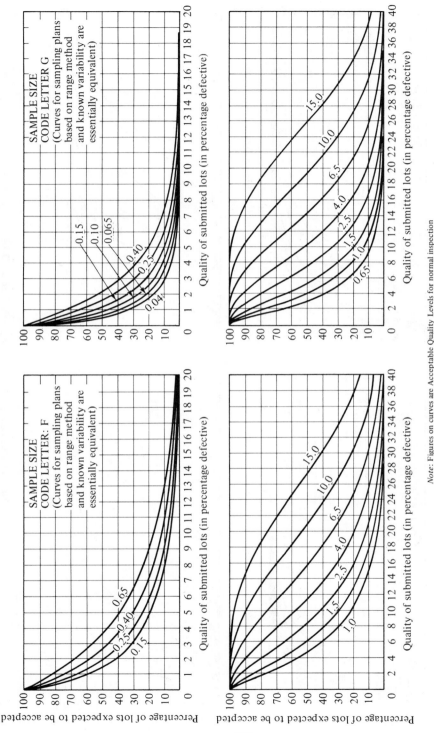

Note: Figures on curves are Acceptable Quality Levels for normal inspection

Fig. 14.6 (continued). *The OC curves from MIL-STD-414.*

579

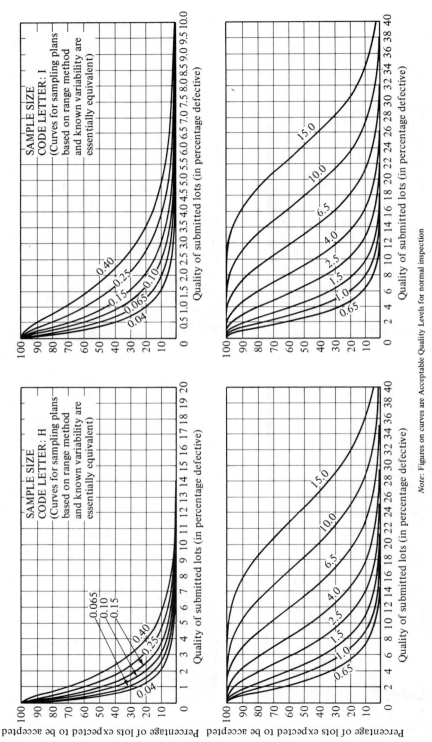

Fig. 14.6 (continued). *The OC curves from MIL-STD-414.*

Note: Figures on curves are Acceptable Quality Levels for normal inspection

580

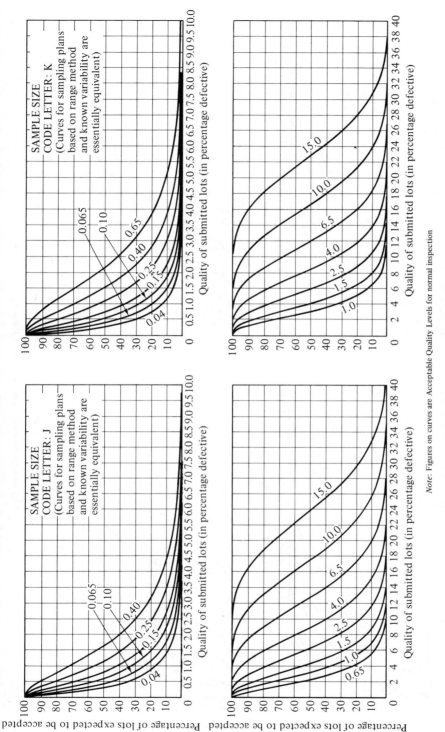

Note: Figures on curves are Acceptable Quality Levels for normal inspection

Fig. 14.6 (continued). *The OC curves from MIL-STD-414.*

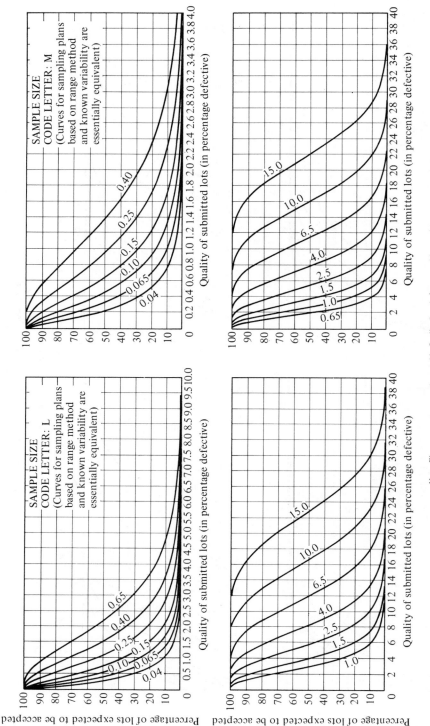

Note: Figures on curves are Acceptable Quality Levels for normal inspection

Fig. 14.6 (continued). *The OC curves from MIL-STD-414.*

582

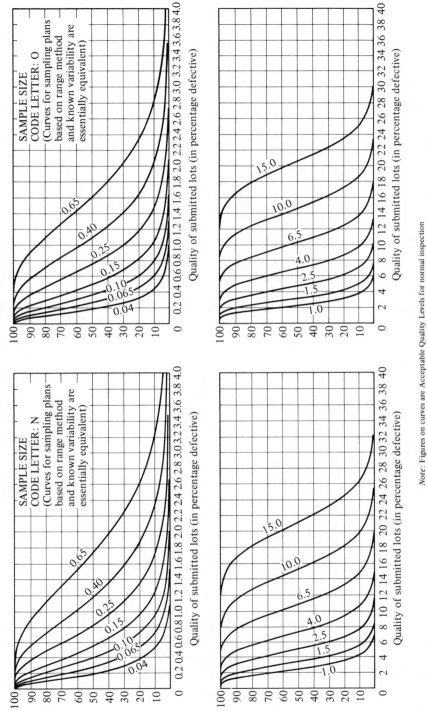

Note: Figures on curves are Acceptable Quality Levels for normal inspection

Fig. 14.6 (continued). *The OC curves from MIL-STD-414.*

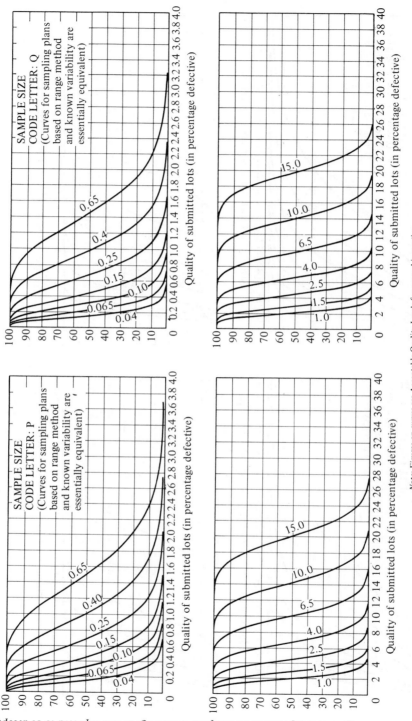

Note: Figures on curves are Acceptable Quality Levels for normal inspection.

Fig. 14.6 (continued). *The OC curves from MIL-STD-414.*

from any of Sections B, C, or D, and be assured of the same probability of accepting or rejecting material for any given quality. These OC curves are reproduced in Fig. 14.6.[1]

14.5.3 SECTION B—VARIABILITY UNKNOWN, STANDARD DEVIATION METHOD

When there is a single specification limit and form 1 is used (k procedure), the appropriate master sampling table for normal or tightened inspection is Table 14.3 (Table B-1 of MIL-STD-414). The sample size, n, and acceptance criterion, k, are read from this table. The lot is accepted if $(U - \bar{X})/S \geq k$ or $(\bar{X} - L)/S \geq k$, whichever is appropriate.

When there is a single specification limit or a double specification limit (with the AQL applying to both limits) and form 2 is used (estimation procedure), the appropriate master sampling table for normal or tightened inspection is Table 14.4 (Table B-3 of MIL-STD-414). The sample size, n, and the acceptance criterion, M, are read from this table. The quality indices $Q_U = (U - \bar{X})/S$ and/or $Q_L = (\bar{X} - L)/S$ are computed. Figure 14.4 (Table B-5 of MIL-STD-414) is entered with n and Q_U and/or Q_L, whichever is appropriate; the estimates p_U and/or p_L, in percent, are read from the chart. For an upper specification limit, the lot is accepted if $p_U \leq M$; for a lower specification limit, the lot is accepted if $p_L \leq M$; and for both upper and lower specification limits, the lot is accepted if $p_U + p_L \leq M$. Procedures are also given for two-sided specification limits when one AQL value is specified for the upper limit and another for the lower limit.

This section of the standard also contains a master table for reduced inspection, as well as the required tables to be used in conjunction with the switching rules.

14.5.4 SECTION C—VARIABILITY UNKNOWN, RANGE METHOD

When there is a single specification limit and form 1 is used (k procedure), the appropriate master sampling table for normal or tightened inspection is Table 14.5 (Table C-1 of MIL-STD-414). The sample size, n, and acceptance criterion, k, are read from this table. The lot is accepted if $(U - \bar{X})/\bar{R} \geq k$ or $(\bar{X} - L)/\bar{R} \geq k$, whichever is appropriate.

When there is a single specification limit or a double specification limit (with the AQL applying to both limits) and form 2 is used (estimation procedure), the appropriate master sampling table for normal or tightened inspection is Table 14.6 (Table C-3 of MIL-STD-414). The sample size, n, and the acceptance criterion, M, are read from these tables. The quality

---

[1] These OC curves are reproduced because the reader is unable to construct his own OC curves since the necessary distribution theory is beyond the scope of the text. This is not the situation for MIL-STD-105D where the OC curves are not reproduced.

Table 14.3. Master Table for Normal and Tightened Inspection for Plans Based on Variability Unknown, Standard Deviation Method (Single Specification Limit—Form 1; Table B-1 of MIL-STD-414)

Sample size code letter	Sample size	Acceptable quality levels (normal inspection)													
		0.04	0.065	0.10	0.15	0.25	0.40	0.65	1.00	1.50	2.50	4.00	6.50	10.00	15.00
		k	k	k	k	k	k	k	k	k	k	k	k	k	k
B	3							↓	↓	↓	1.12	0.958	0.765	0.566	0.341
C	4							↓	1.45	1.34	1.17	1.01	0.814	0.617	0.393
D	5						↓	1.65	1.53	1.40	1.24	1.07	0.874	0.675	0.455
E	7				↓	2.00	1.88	1.75	1.62	1.50	1.33	1.15	0.955	0.755	0.536
F	10	↓	↓	↓	2.24	2.11	1.98	1.84	1.72	1.58	1.41	1.23	1.03	0.828	0.611
G	15	2.64	2.53	2.42	2.32	2.20	2.06	1.91	1.79	1.65	1.47	1.30	1.09	0.886	0.664
H	20	2.69	2.58	2.47	2.36	2.24	2.11	1.96	1.82	1.69	1.51	1.33	1.12	0.917	0.695
I	25	2.72	2.61	2.50	2.40	2.26	2.14	1.98	1.85	1.72	1.53	1.35	1.14	0.936	0.712
J	30	2.73	2.61	2.51	2.41	2.28	2.15	2.00	1.86	1.73	1.55	1.36	1.15	0.946	0.723
K	35	2.77	2.65	2.54	2.45	2.31	2.18	2.03	1.89	1.76	1.57	1.39	1.18	0.969	0.745
L	40	2.77	2.66	2.55	2.44	2.31	2.18	2.03	1.89	1.76	1.58	1.39	1.18	0.971	0.746
M	50	2.83	2.71	2.60	2.50	2.35	2.22	2.08	1.93	1.80	1.61	1.42	1.21	1.00	0.774
N	75	2.90	2.77	2.66	2.55	2.41	2.27	2.12	1.98	1.84	1.65	1.46	1.24	1.03	0.804
O	100	2.92	2.80	2.69	2.58	2.43	2.29	2.14	2.00	1.86	1.67	1.48	1.26	1.05	0.819
P	150	2.96	2.84	2.73	2.61	2.47	2.33	2.18	2.03	1.89	1.70	1.51	1.29	1.07	0.841
Q	200	2.97	2.85	2.73	2.62	2.47	2.33	2.18	2.04	1.89	1.70	1.51	1.29	1.07	0.845
		0.065	0.10	0.15	0.25	0.40	0.65	1.00	1.50	2.50	4.00	6.50	10.00	15.00	
		Acceptable quality levels (tightened inspection)													

All AQL values are in percent defective.
Use first sampling plan below arrow, that is, both sample size as well as k value. When sample size equals or exceeds lot size, every item in the lot must be inspected.

Table 14.4. Master Table for Normal and Tightened Inspection for Plans Based on Variability Unknown, Standard Deviation Method
(Double Specification Limit and Form 2—Single Specification Limit; Table B-3 of MIL-STD-414)

Sample size code letter	Sample size	Acceptable quality levels (normal inspection)													
		0.04	0.065	0.10	0.15	0.25	0.40	0.65	1.00	1.50	2.50	4.00	6.50	10.00	15.00
		M	M	M	M	M	M	M	M	M	M	M	M	M	M
B	3								↓	↓	7.59	18.86	26.94	33.69	40.47
C	4								1.53	5.50	10.92	16.45	22.86	29.45	36.90
D	5	↓	↓	↓	↓	↓	↓	1.33	3.32	5.83	9.80	14.39	20.19	26.56	33.99
E	7	↓	↓	↓	↓	0.422	1.06	2.14	3.55	5.35	8.40	12.20	17.35	23.29	30.50
F	10	↓	↓	↓	0.349	0.716	1.30	2.17	3.26	4.77	7.29	10.54	15.17	20.74	27.57
G	15	0.099	0.186	0.312	0.503	0.818	1.31	2.11	3.05	4.31	6.56	9.46	13.71	18.94	25.61
H	20	0.135	0.228	0.365	0.544	0.846	1.29	2.05	2.95	4.09	6.17	8.92	12.99	18.03	24.53
I	25	0.155	0.250	0.380	0.551	0.877	1.29	2.00	2.86	3.97	5.97	8.63	12.57	17.51	23.97
J	30	0.179	0.280	0.413	0.581	0.879	1.29	1.98	2.83	3.91	5.86	8.47	12.36	17.24	23.58
K	35	0.170	0.264	0.388	0.535	0.847	1.23	1.87	2.68	3.70	5.57	8.10	11.87	16.65	22.91
L	40	0.179	0.275	0.401	0.566	0.873	1.26	1.88	2.71	3.72	5.58	8.09	11.85	16.61	22.86
M	50	0.163	0.250	0.363	0.503	0.789	1.17	1.71	2.49	3.45	5.20	7.61	11.23	15.87	22.00
N	75	0.147	0.228	0.330	0.467	0.720	1.07	1.60	2.29	3.20	4.87	7.15	10.63	15.13	21.11
O	100	0.145	0.220	0.317	0.447	0.689	1.02	1.53	2.20	3.07	4.69	6.91	10.32	14.75	20.66
P	150	0.134	0.203	0.293	0.413	0.638	0.949	1.43	2.05	2.89	4.43	6.57	9.88	14.20	20.02
Q	200	0.135	0.204	0.294	0.414	0.637	0.945	1.42	2.04	2.87	4.40	6.53	9.81	14.12	19.92
		0.065	0.10	0.15	0.25	0.40	0.65	1.00	1.50	2.50	4.00	6.50	10.00	15.00	
		Acceptable quality levels (tightened inspection)													

All AQL and table values are in percent defective.
Use first sampling plan below arrow, that is, both sample size as well as M value. When sample size equals or exceeds lot size, every item in the lot must be inspected.

Table 14.5. Master Table for Normal and Tightened Inspection for Plans Based on Variability Unknown, Range Method

(Single Specification Limit—Form 1; Table C-1 of MIL-STD-414)

Sample size code letter	Sample size	\[Acceptable quality levels (normal inspection)\] 0.04 k	0.065 k	0.10 k	0.15 k	0.25 k	0.40 k	0.65 k	1.00 k	1.50 k	2.50 k	4.00 k	6.50 k	10.00 k	15.00 k
B	3	↓	↓	↓	↓	↓	↓	↓	↓	↓	0.587	0.502	0.401	0.296	0.178
C	4	↓	↓	↓	↓	↓	↓	↓	0.651	0.598	0.525	0.450	0.364	0.276	0.176
D	5	↓	↓	↓	↓	↓	↓	0.663	0.614	0.565	0.498	0.431	0.352	0.272	0.184
E	7	↓	↓	↓	↓	0.702	0.659	0.613	0.569	0.525	0.465	0.405	0.336	0.266	0.189
F	10	↓	↓	↓	0.916	0.863	0.811	0.755	0.703	0.650	0.579	0.507	0.424	0.341	0.252
G	15	1.09	1.04	0.999	0.958	0.903	0.850	0.792	0.738	0.684	0.610	0.536	0.452	0.368	0.276
H	25	1.14	1.10	1.05	1.01	0.951	0.896	0.835	0.779	0.723	0.647	0.571	0.484	0.398	0.305
I	30	1.15	1.10	1.06	1.02	0.959	0.904	0.843	0.787	0.730	0.654	0.577	0.490	0.403	0.310
J	35	1.16	1.11	1.07	1.02	0.964	0.908	0.848	0.791	0.734	0.658	0.581	0.494	0.406	0.313
K	40	1.18	1.13	1.08	1.04	0.978	0.921	0.860	0.803	0.746	0.668	0.591	0.503	0.415	0.321
L	50	1.19	1.14	1.09	1.05	0.988	0.931	0.893	0.812	0.754	0.676	0.598	0.510	0.421	0.327
M	60	1.21	1.16	1.11	1.06	1.00	0.948	0.885	0.826	0.768	0.689	0.610	0.521	0.432	0.336
N	85	1.23	1.17	1.13	1.08	1.02	0.962	0.899	0.839	0.780	0.701	0.621	0.530	0.441	0.345
O	115	1.24	1.19	1.14	1.09	1.03	0.975	0.911	0.851	0.791	0.711	0.631	0.539	0.449	0.353
P	175	1.26	1.21	1.16	1.11	1.05	0.994	0.929	0.868	0.807	0.726	0.644	0.552	0.460	0.363
Q	230	1.27	1.21	1.16	1.12	1.06	0.996	0.931	0.870	0.809	0.728	0.646	0.553	0.462	0.364
		0.065	0.10	0.15	0.25	0.40	0.65	1.00	1.50	2.50	4.00	6.50	10.00	15.00	
		\[Acceptable quality levels (tightened inspection)\]													

All AQL values are in percent defective.

Use first sampling plan below arrow, that is, both sample size as well as k value. When sample size equals or exceeds lot size, every item in the lot must be inspected.

Table 14.6. Master Table for Normal and Tightened Inspection for Plans Based on Variability Unknown, Range Method
(Double Specification Limit and Form 2—Single Specification Limit; Table C-3 of MIL-STD-414)

Sample size code letter	Sample size	c factor	Acceptable quality levels (normal inspection)													
			0.04	0.065	0.10	0.15	0.25	0.40	0.65	1.00	1.50	2.50	4.00	6.50	10.00	15.00
			M	M	M	M	M	M	M	M	M	M	M	M	M	M
B	3	1.910								↓	↓	7.59	18.86	26.94	33.69	40.47
C	4	2.234							↓	1.53	5.50	10.92	16.45	22.86	29.45	36.90
D	5	2.474					↓	↓	1.42	3.44	5.93	9.90	14.47	20.27	26.59	33.95
E	7	2.830				↓	0.28	0.89	1.99	3.46	5.32	8.47	12.35	17.54	23.50	30.66
F	10	2.405	↓	↓	↓	0.23	0.58	1.14	2.05	3.23	4.77	7.42	10.79	15.49	21.06	27.90
G	15	2.379	0.061	0.136	0.253	0.430	0.786	1.30	2.10	3.11	4.44	6.76	9.76	14.09	19.30	25.92
H	25	2.358	0.125	0.214	0.336	0.506	0.827	1.27	1.95	2.82	3.96	5.98	8.65	12.59	17.48	23.79
I	30	2.353	0.147	0.240	0.366	0.537	0.856	1.29	1.96	2.81	3.92	5.88	8.50	12.36	17.19	23.42
J	35	2.349	0.165	0.261	0.391	0.564	0.883	1.33	1.98	2.82	3.90	5.85	8.42	12.24	17.03	23.21
K	40	2.346	0.160	0.252	0.375	0.539	0.842	1.25	1.88	2.69	3.73	5.61	8.11	11.84	16.55	22.38
L	50	2.342	0.153	0.261	0.381	0.542	0.838	1.25	1.60	2.63	3.64	5.47	7.91	11.57	16.20	22.26
M	60	2.339	0.158	0.244	0.356	0.504	0.781	1.16	1.74	2.47	3.44	5.17	7.54	11.10	15.64	21.63
N	85	2.335	0.156	0.242	0.350	0.493	0.755	1.12	1.67	2.37	3.30	4.97	7.27	10.73	15.17	21.05
O	115	2.333	0.153	0.230	0.333	0.468	0.718	1.06	1.58	2.25	3.14	4.76	6.99	10.37	14.74	20.57
P	175	2.333	0.139	0.210	0.303	0.427	0.655	0.972	1.46	2.08	2.93	4.47	6.60	9.89	14.15	19.88
Q	230	2.333	0.142	0.215	0.308	0.432	0.661	0.976	1.47	2.08	2.92	4.46	6.57	9.84	14.10	19.82
			0.065	0.10	0.15	0.25	0.40	0.65	1.00	1.50	2.50	4.00	6.50	10.00	15.00	
			Acceptable quality level (tightened inspection)													

All AQL and table values are in percent defective.
Use first sampling plan below arrow, that is, both sample size as well as M value. When sample size equals or exceeds lot size, every item in the lot must be inspected.

indices $Q_U = (U - \bar{X})c/\bar{R}$ and/or $Q_L = (\bar{X} - L)c/\bar{R}$ are computed. The factor c is obtained from Table 14.1 and is also given in the Master Table 14.6. Figure 14.5 (Table C-5 of MIL-STD-414) is entered with n and Q_U and/or Q_L, whichever is appropriate; the estimates p_U and/or p_L, in percent, are read from the table. For an upper specification limit, the lot is accepted if $p_U \leq M$; for a lower specification limit, the lot is accepted if $p_L \leq M$; and for both upper and lower specification limits, the lot is accepted if $p_U + p_L \leq M$.

This section of the standard also contains a master table for reduced inspection, as well as the required tables to be used in conjunction with the switching rules.

As noted earlier, the sample size required when using a range plan is always greater than or equal to the sample size required when using the sample standard deviation plan. For a sample standard deviation plan requiring a small n, the comparable range plan requires a not too different sample size.

14.5.5 SECTION D—VARIABILITY KNOWN

This section assumes that the standard deviation, σ', of the normal distribution is known.[1]

When there is a single specification limit and form 1 is used (k procedure), the appropriate master sampling table for normal or tightened inspection is Table 14.7 (Table D-1 of MIL-STD-414). The sample size, n, and acceptance criterion, k, are read from this table. The lot is accepted if $(U - \bar{X})/\sigma' \geq k$ or $(\bar{X} - L)/\sigma' \geq k$, whichever is appropriate.

When there is a single specification limit or a double specification limit (with the AQL applying to both limits) and form 2 is used (estimation procedure), the appropriate master sampling table for normal or tightened inspection is Table 14.8 (Table D-3 of MIL-STD-414). The sample size, n, and the acceptance criterion, M, are read from this table. The quality indices

$$Q_U = \left(\frac{U - \bar{X}}{\sigma'}\right)\sqrt{\frac{n}{n-1}}$$

and/or

$$Q_L = \left(\frac{\bar{X} - L}{\sigma'}\right)\sqrt{\frac{n}{n-1}}$$

are computed. The quantity $\sqrt{n/(n-1)}$ is denoted by v and is given in the master Table 14.8. Appendix Table 1 is entered with Q_U and/or Q_L, whichever is appropriate; the estimates p_U and/or p_L are read from the table and con-

---

[1] The known standard deviation is denoted by σ in MIL-STD-414. The symbol σ' is used in this section so that the notation conforms to the notation of the American Society for Quality Control.

verted to readings in percent by multiplying by one hundred. For an upper specification limit, the lot is accepted if $p_U \leq M$; for a lower specification limit, the lot is accepted if $p_L \leq M$; and for both upper and lower specification limits, the lot is accepted if $p_U + p_L \leq M$.

This section of the standard also contains a master table for reduced inspection, as well as the required tables to be used in conjunction with the switching rules.

It is evident that the sample size required when using a known standard deviation plan is always smaller than the sample size required when using a comparable sample standard deviation plan.

14.5.6 Example Using MIL-STD-414

The following example of a two-sided specification limits procedure, when the variability is unknown and the standard deviation method is used, is given in MIL-STD-414.

Example of Calculations. Variability Unknown—Standard Deviation Method (*One* AQL *Value for Both Upper and Lower Specification Limit Combined*)

The minimum temperature of operation for a certain device is specified as 180°F. The maximum temperature is 209°F. A lot of 40 items is submitted for inspection. Inspection Level IV, normal inspection with AQL = 1%, is to be used. From the appropriate tables it is seen that a sample of size 5 is required. Suppose the measurements obtained are as follows:

$$197°, \quad 188°, \quad 184°, \quad 205°, \quad \text{and} \quad 201°,$$

and compliance with the acceptability criterion is to be determined.

The lot meets the acceptability criterion, since $p = p_U + p_L$ is less than M.

Line	Information Needed	Value Obtained	Explanation
1	Sample size: n	5	
2	Sum of measurements: $\sum X$	975	
3	Sum of squared measurements: $\sum X^2$	190,435	
4	Correction factor (CF): $(\sum X)^2/n$	190,125	$(975)^2/5$
5	Corrected sum of squares (SS): $\sum X^2 - CF$	310	190,435—190,125
6	Variance (V): $SS/(n-1)$	77.5	310/4
7	Estimate of lot standard deviation (S): $\sqrt{V}$	8.81	$\sqrt{77.5}$
8	Sample mean ($\bar{X}$): $\sum X/n$	195	975/5
9	Upper specification limit: U	209	
10	Lower specification limit: L	180	
11	Quality index: $Q_U = (U - \bar{X})/S$	1.59	(209–195)/8.81
12	Quality index: $Q_L = (\bar{X} - L)/S$	1.70	(195–180)/8.81
13	Est. of lot % defective: above $U(p_U)$	2.19%	See Fig. 14.4.
14	Est. of lot % defective: below $L(p_L)$	0.66%	See Fig. 14.4.
15	Total est. % defective in lot (p): $p = p_U + p_L$	2.85%	2.19% + .66%
16	Max. allowable % defective: (M)	3.32%	See Table 14.4
17	Acceptability Criterion: Compare $p = p_U + p_L$ with M: 2.85% < 3.32%.		

Table 14.7. Master Table for Normal and Tightened Inspection for Plans Based on Variability Known
(Single Specification Limit—Form 1; Table D-1 of MIL-STD-414)

Acceptable quality levels (normal inspection)

Code letter	0.04		0.065		0.10		0.15		0.25		0.40		0.65	
	n	k	n	k	n	k	n	k	n	k	n	k	n	k
B														
C														
D													2	1.58
E									2	1.94	2	1.81	3	1.69
F							3	2.19	3	2.07	3	1.91	4	1.80
G	3	2.58	3	2.49	4	2.39	4	2.30	4	2.14	5	2.05	5	1.88
H	4	2.65	4	2.55	5	2.46	5	2.34	6	2.23	6	2.08	7	1.95
I	5	2.69	6	2.59	6	2.49	6	2.37	7	2.25	8	2.13	8	1.96
J	6	2.72	6	2.58	7	2.50	7	2.38	8	2.26	9	2.13	10	1.99
K	7	2.77	7	2.63	8	2.54	9	2.45	9	2.29	10	2.16	11	2.01
L	8	2.77	8	2.64	9	2.54	10	2.45	11	2.31	12	2.18	13	2.03
M	10	2.83	11	2.72	11	2.59	12	2.49	13	2.35	14	2.21	16	2.07
N	14	2.88	15	2.77	16	2.65	17	2.54	19	2.41	21	2.27	23	2.12
O	19	2.92	20	2.80	22	2.69	23	2.57	25	2.43	27	2.29	30	2.14
P	27	2.96	30	2.84	31	2.72	34	2.62	37	2.47	40	2.33	44	2.17
Q	37	2.97	40	2.85	42	2.73	45	2.62	49	2.48	54	2.34	59	2.18
	0.065		0.10		0.15		0.25		0.40		0.65		1.00	

Acceptable quality levels (tightened inspection)

All AQL values are in percent defective.

Use first sampling plan below arrow, that is, both sample size as well as k value. When sample size equals or exceeds lot size, every item in the lot must be inspected.

Table 14.7 (continued). Master Table for Normal and Tightened Inspection for Plans Based on Variability Known
(Single Specification Limit—Form 1; Table D-1 of MIL-STD-414)

Code letter	Acceptable quality levels (normal inspection)													
	1.00		1.50		2.50		4.00		6.50		10.00		15.00	
	n	k	n	k	n	k	n	k	n	k	n	k	n	k
B	↓		↓		↓		↓		↓		↓		↓	
C	2	1.36	2	1.25	2	1.09	2	0.936	3	0.755	3	0.573	4	0.344
D	2	1.42	2	1.33	3	1.17	3	1.01	3	0.825	4	0.641	4	0.429
E	3	1.56	3	1.44	4	1.28	4	1.11	5	0.919	5	0.728	6	0.515
F	4	1.69	4	1.53	5	1.39	5	1.20	6	0.991	7	0.797	8	0.584
G	6	1.78	6	1.62	7	1.45	8	1.28	9	1.07	11	0.877	12	0.649
H	7	1.80	8	1.68	9	1.49	10	1.31	12	1.11	14	0.906	16	0.685
I	9	1.83	10	1.70	11	1.51	13	1.34	15	1.13	17	0.924	20	0.706
J	11	1.86	12	1.72	13	1.53	15	1.35	18	1.15	21	0.942	24	0.719
K	12	1.88	14	1.75	15	1.56	18	1.38	20	1.17	24	0.964	27	0.737
L	14	1.89	15	1.75	18	1.57	20	1.38	23	1.17	27	0.965	31	0.741
M	17	1.93	19	1.79	22	1.61	25	1.42	29	1.21	33	0.995	38	0.770
N	25	1.97	28	1.84	32	1.65	36	1.46	42	1.24	49	1.03	56	0.803
O	33	2.00	36	1.86	42	1.67	48	1.48	55	1.26	64	1.05	75	0.819
P	49	2.03	54	1.89	61	1.69	70	1.51	82	1.29	95	1.07	111	0.841
Q	65	2.04	71	1.89	81	1.70	93	1.51	109	1.29	127	1.07	147	0.845
	1.50		2.50		4.00		6.50		10.00		15.00			
	Acceptable quality levels (tightened inspection)													

All AQL values are in percent defective.

Use first sampling plan below arrow, that is, both sample size as well as k value. When sample size equals or exceeds lot size, every item in the lot must be inspected.

Table 14.8. Master Table for Normal and Tightened Inspection for Plans Based on Variability Known
(Double Specification Limit and Form 2—Single Specification Limit; Table D-3 of MIL-STD-414)

Acceptable quality levels (normal inspection)

Code letter	0.04 n	0.04 M	0.04 v	0.065 n	0.065 M	0.065 v	0.10 n	0.10 M	0.10 v	0.15 n	0.15 M	0.15 v	0.25 n	0.25 M	0.25 v	0.40 n	0.40 M	0.40 v	0.65 n	0.65 M	0.65 v
B		→			→			→			→			→			→			→	
C		→			→			→			→			→			→			→	
D		→			→			→			→			→			→		2	1.28	1.414
E		→			→			→			→		2	0.310	1.414	2	0.510	1.414	3	1.94	1.225
F		→			→			→		3	0.369	1.225	3	0.568	1.225	3	0.959	1.225	4	1.88	1.155
G	3	0.079	1.225	3	0.114	1.225	4	0.290	1.155	4	0.399	1.155	4	0.681	1.155	5	1.09	1.118	5	1.76	1.118
H	4	0.111	1.155	4	0.161	1.155	5	0.296	1.118	5	0.445	1.118	6	0.721	1.095	6	1.14	1.095	7	1.75	1.080
I	5	0.130	1.118	6	0.230	1.095	6	0.321	1.095	6	0.478	1.095	7	0.756	1.080	8	1.14	1.069	8	1.80	1.069
J	6	0.145	1.095	6	0.234	1.095	7	0.343	1.080	7	0.507	1.080	8	0.791	1.069	9	1.18	1.061	10	1.79	1.054
K	7	0.141	1.080	7	0.226	1.080	8	0.330	1.069	9	0.469	1.061	9	0.760	1.061	10	1.14	1.054	11	1.73	1.049
L	8	0.153	1.069	8	0.243	1.069	9	0.351	1.061	10	0.494	1.054	11	0.768	1.049	12	1.15	1.045	13	1.74	1.041
M	10	0.141	1.054	11	0.217	1.049	11	0.326	1.049	12	0.461	1.045	13	0.721	1.041	14	1.08	1.038	16	1.62	1.033
N	14	0.138	1.038	15	0.211	1.035	16	0.308	1.033	17	0.438	1.031	19	0.673	1.027	21	1.00	1.025	23	1.51	1.023
O	19	0.134	1.027	20	0.207	1.026	22	0.296	1.024	23	0.423	1.023	25	0.655	1.021	27	0.980	1.019	30	1.47	1.017
P	27	0.129	1.019	30	0.193	1.017	31	0.283	1.017	34	0.397	1.015	37	0.615	1.014	40	0.921	1.013	44	1.39	1.012
Q	37	0.130	1.014	40	0.196	1.013	42	0.285	1.012	45	0.402	1.011	49	0.620	1.010	54	0.920	1.009	59	1.39	1.009
(tightened AQL)	0.065			0.10			0.15			0.25			0.40			0.65			1.00		

Acceptable quality levels (tightened inspection)

594

Table 14.8 (continued). Master Table for Normal and Tightened Inspection for Plans Based on Variability Known
(Double Specification Limit and Form 2—Single Specification Limit; Table D-3 of MIL-STD-414)

Acceptable quality levels (normal inspection)

Code letter	1.00 n	1.00 M	1.00 v	1.50 n	1.50 M	1.50 v	2.50 n	2.50 M	2.50 v	4.00 n	4.00 M	4.00 v	6.50 n	6.50 M	6.50 v	10.00 n	10.00 M	10.00 v	15.00 n	15.00 M	15.00 v
B	2	→2.73	1.414	2	→3.90	1.414	2	→6.11	1.414	2	→9.27	1.414	3	→17.74	1.225	3	→24.22	1.225	4	→33.67	1.225
C																					
D	2	2.23	1.414	2	3.00	1.414	3	7.56	1.225	3	10.79	1.225	3	15.60	1.225	4	22.97	1.155	4	31.01	1.155
E	3	2.76	1.225	3	3.85	1.225	4	6.99	1.155	4	9.97	1.155	5	15.21	1.118	5	20.80	1.118	6	28.64	1.095
F	4	2.58	1.155	4	3.87	1.155	5	6.05	1.118	5	8.92	1.118	6	13.89	1.095	7	19.46	1.080	8	26.64	1.069
G	6	2.57	1.095	6	3.77	1.095	7	5.83	1.080	8	8.62	1.069	9	12.88	1.061	11	17.88	1.049	12	24.88	1.045
H	7	2.62	1.080	8	3.68	1.069	9	5.68	1.061	10	8.43	1.054	12	12.35	1.045	14	17.36	1.038	16	23.96	1.033
I	9	2.59	1.061	10	3.63	1.054	11	5.60	1.049	13	8.13	1.041	15	12.04	1.035	17	17.05	1.031	20	23.43	1.026
J	11	2.57	1.049	12	3.61	1.045	13	5.58	1.041	15	8.13	1.035	18	11.88	1.029	21	16.71	1.025	24	23.13	1.022
K	12	2.49	1.045	14	3.43	1.038	15	5.34	1.035	18	7.72	1.029	20	11.57	1.026	24	16.23	1.022	27	22.63	1.019
L	14	2.51	1.038	15	3.54	1.035	18	5.29	1.029	20	7.80	1.026	23	11.56	1.023	27	16.27	1.019	31	22.57	1.017
M	17	2.35	1.031	19	3.28	1.027	22	4.98	1.024	25	7.34	1.021	29	10.93	1.018	33	15.61	1.016	38	21.77	1.013
N	25	2.19	1.021	28	3.05	1.018	32	4.68	1.016	36	6.95	1.014	42	10.40	1.012	49	14.87	1.010	56	20.90	1.009
O	33	2.12	1.016	36	2.99	1.014	42	4.55	1.012	48	6.75	1.011	55	10.17	1.009	64	14.58	1.008	75	20.48	1.007
P	49	2.00	1.010	54	2.82	1.009	61	4.35	1.008	70	6.48	1.007	82	9.76	1.006	95	14.09	1.005	111	19.90	1.005
Q	65	2.00	1.008	71	2.82	1.007	81	4.34	1.006	93	6.46	1.005	109	9.73	1.005	127	14.02	1.004	147	19.84	1.003
(tightened AQL)		1.50			2.50			4.00			6.50			10.00			15.00				

Acceptable quality levels (tightened inspection)

All AQL and table values are in percent defective.

Use first sampling plan below arrow, that is, both sample size as well as M value. When sample size equals or exceeds lot size, every item in the lot must be inspected.

PROBLEMS

1. Design a known standard deviation sampling plan whose OC curve passes through the points ($p'_1 = 0.025$, $1 - \alpha = 0.95$); ($p'_2 = 0.08$, $\beta = 0.10$).

2. Design a known standard deviation sampling plan whose OC curve passes through the points ($p'_1 = 0.025$, $1 - \alpha = 0.95$); ($p'_2 = 0.08$, $\beta = 0.05$).

3. Solve Problem 1 for an unknown standard deviation plan, but use the sample standard deviation to estimate σ'. Use the asymptotic results about the distribution of $\bar{X} + kS$.

4. The minimum acceptable value of the shear strength in pounds of spot welds is 300 pounds. A lot containing 800 such welds is submitted for inspection under MIL-STD-414. Inspection level IV, normal inspection with AQL = 4%, is to be used. The variability is unknown and the standard deviation method is to be used. (a) Using form 2, what are the acceptance criteria? (b) If $\bar{X} = 351.4$ and $S = 22.3$, would the lot be accepted?

5. The maximum specification for a certain device is 6.5. A lot containing 750 items is submitted for inspection under MIL-STD-414. Inspection level II, normal inspection with AQL = 2.5%, is to be used. The variability is unknown and the standard deviation method is to be used. (a) Using form 2, what are the acceptance criteria? (b) Suppose the data are 5.00, 5.88, 3.86, 4.13, 4.01, 3.92, 2.92, 5.82, 6.20, and 4.90. Should the lot be accepted? (c) What are the acceptance criteria if form 1 is used? ($S^2 = 1.1442$.)

6. The specification limits for a certain type of resistance are specified as 75 ± 5 ohms. A lot containing 300 items is submitted for inspection under MIL-STD-414. Inspection level III, normal inspection with AQL = 6.5%, is to be used. The variability is unknown and the standard deivation method is to be used. (a) Using form 2, what are the acceptance criteria? (b) Suppose the data are as follows: 74.3, 75.1, 71.9, 72.7, 75.8, 77.0, 77.3, 70.1, 73.8, and 75.0. Determine whether or not to accept the lot.

7. The minimum requirement on the content of a particular type of can is 80 ounces. A lot containing 700 cans is submitted for inspection under MIL-STD-414. Inspection level III, normal inspection with an AQL = 4%, is to be used. The variability is unknown and the standard deviation method is to be used. (a) Using form 2, what are the acceptance criteria? (b) Suppose the data are 81.4, 81.0, 81.0, 80.2, 82.1, 79.8, 80.3, 80.5, 80.8, 80.6, 80.3, 81.0, 80.5, 81.0, 82.1, 81.4, 80.6, 81.9, 80.8, and 80.7. Would you accept the lot?

8. The temperature specifications of a certain device are given as 195 degrees $\pm$ 15 degrees. A lot containing 5,000 items is submitted for inspection under MIL-STD-414. Inspection level V, normal inspection with an AQL = 0.65%, is to be used. The variability is unknown and the standard deviation method is to be used. (a) What are the acceptance criteria? (b) If $\bar{X} = 194.8$ degrees and $S = 6.1$ degrees, would the lot be accepted?

9. Solve Problem 4, assuming that the variability is unknown, but using the average range method. Assume that $\bar{R} = 49.7$ in part (b).

10. Solve Problem 5, assuming that the variability is unknown, but using the average range method. Assume that the observations presented are in order, reading from left to right.

11. Solve Problem 6, assuming that the variability is unknown, but using the average range method. Assume that the observations presented are in order, reading from left to right.

12. Can you solve Problem 7 by using the average range method with the data given? If not, assume that the average of the additional observations equals the average of the 20 observations given in part (b) of Problem 7. Also assume that the ranges of any additional subgroups equal the average range of the four subgroups of five given in part (b) of Problem 7.

13. Solve Problem 8, assuming that the variability is unknown, but using the average range method. Assume that $\bar{R} = 12.7$ in part (b).

14. Solve Problem 4, assuming that the variability is known and equal to $\sigma' = 22$.

15. Solve Problem 5, assuming that the variability is known and equal to $\sigma' = 1.0$. Use only the required number of observations in the order that they are presented, reading from left to right.

16. Solve Problem 6, assuming that the variability is known and equal to $\sigma' = 1.3$. Use only the required number of observations in the order that they are presented, reading from left to right.

17. Solve Problem 7, assuming that the variability is known and equal to $\sigma' = 0.80$. Use only the required number of observations in the order that they are presented, reading from left to right.

18. Solve Problem 8, assuming that the variability is known and equal to $\sigma' = 6$ degrees.

19. In Problem 15, verify that the value of k used in form 1 equals $K_M\sqrt{(n-1)/n}$.

20. In Problem 15, calculate exactly the probability of accepting the lot when the incoming quality is at the AQL.

21. In Problem 17, calculate exactly the probability of accepting the lot when the incoming quality is at the AQL.

APPENDIX

Table 1. Areas under the Normal Curve from K_α to ∞*

$$\int_{K_\alpha}^{\infty} \frac{1}{\sqrt{2\pi}}\, e^{-x^2/2}\, dx = \alpha$$

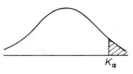

K_α	.00	.01	.02	.03	.04	.05	.06	.07	.08	.09
0.0	.5000	.4960	.4920	.4880	.4840	.4801	.4761	.4721	.4681	.4641
0.1	.4602	.4562	.4522	.4483	.4443	.4404	.4364	.4325	.4286	.4247
0.2	.4207	.4168	.4129	.4090	.4052	.4013	.3974	.3936	.3897	.3859
0.3	.3821	.3783	.3745	.3707	.3669	.3632	.3594	.3557	.3520	.3483
0.4	.3446	.3409	.3372	.3336	.3300	.3264	.3228	.3192	.3156	.3121
0.5	.3085	.3050	.3015	.2981	.2946	.2912	.2877	.2843	.2810	.2776
0.6	.2743	.2709	.2676	.2643	.2611	.2578	.2546	.2514	.2483	.2451
0.7	.2420	.2389	.2358	.2327	.2296	.2266	.2236	.2206	.2177	.2148
0.8	.2119	.2090	.2061	.2033	.2005	.1977	.1949	.1922	.1894	.1867
0.9	.1841	.1814	.1788	.1762	.1736	.1711	.1685	.1660	.1635	.1611
1.0	.1587	.1562	.1539	.1515	.1492	.1469	.1446	.1423	.1401	.1379
1.1	.1357	.1335	.1314	.1292	.1271	.1251	.1230	.1210	.1190	.1170
1.2	.1151	.1131	.1112	.1093	.1075	.1056	.1038	.1020	.1003	.0985
1.3	.0968	.0951	.0934	.0918	.0901	.0885	.0869	.0853	.0838	.0823
1.4	.0808	.0793	.0778	.0764	.0749	.0735	.0721	.0708	.0694	.0681
1.5	.0668	.0655	.0643	.0630	.0618	.0606	.0594	.0582	.0571	.0559
1.6	.0548	.0537	.0526	.0516	.0505	.0495	.0485	.0475	.0465	.0455
1.7	.0446	.0436	.0427	.0418	.0409	.0401	.0392	.0384	.0375	.0367
1.8	.0359	.0351	.0344	.0336	.0329	.0322	.0314	.0307	.0301	.0294
1.9	.0287	.0281	.0274	.0268	.0262	.0256	.0250	.0244	.0239	.0233
2.0	.0228	.0222	.0217	.0212	.0207	.0202	.0197	.0192	.0188	.0183
2.1	.0179	.0174	.0170	.0166	.0162	.0158	.0154	.0150	.0146	.0143
2.2	.0139	.0136	.0132	.0129	.0125	.0122	.0119	.0116	.0113	.0110
2.3	.0107	.0104	.0102	.00990	.00964	.00939	.00914	.00889	.00866	.00842
2.4	.00820	.00798	.00776	.00755	.00734	.00714	.00695	.00676	.00657	.00639
2.5	.00621	.00604	.00587	.00570	.00554	.00539	.00523	.00508	.00494	.00480
2.6	.00466	.00453	.00440	.00427	.00415	.00402	.00391	.00379	.00368	.00357
2.7	.00347	.00336	.00326	.00317	.00307	.00298	.00289	.00280	.00272	.00264
2.8	.00256	.00248	.00240	.00233	.00226	.00219	.00212	.00205	.00199	.00193
2.9	.00187	.00181	.00175	.00169	.00164	.00159	.00154	.00149	.00144	.00139

K_α	.0	.1	.2	.3	.4	.5	.6	.7	.8	.9
3	.00135	.0³968	.0³687	.0³483	.0³337	.0³233	.0³159	.0³108	.0⁴723	.0⁴481
4	.0⁴317	.0⁴207	.0⁴133	.0⁵854	.0⁵541	.0⁵340	.0⁵211	.0⁵130	.0⁶793	.0⁶479
5	.0⁶287	.0⁶170	.0⁷996	.0⁷579	.0⁷333	.0⁷190	.0⁷107	.0⁸599	.0⁸332	.0⁸182
6	.0⁹987	.0⁹530	.0⁹282	.0⁹149	.0¹⁰777	.0¹⁰402	.0¹⁰206	.0¹⁰104	.0¹¹523	.0¹¹260

† From *Tables of Areas in Two Tails and in One Tail of the Normal Curve*, by Frederick E. Croxton. Copyright 1949 by Prentice-Hall, Inc. Permission is given to reproduce this table provided credit is given to the author and the Prentice-Hall copyright line is included.

Table 2. Percentage Points of the χ^2 Distribution*

Table of $\chi^2_{\alpha;\nu}$—the 100 α percentage point of the χ^2 distribution for ν degrees of freedom

ν \ α	.995	.99	.975	.95	.90	.75	.50
1	.0⁴393	.0³157	.0³982	.00393	.0158	.102	.455
2	.0100	.0201	.0506	.103	.211	.575	1.386
3	.0717	.115	.216	.352	.584	1.213	2.366
4	.207	.297	.484	.711	1.064	1.923	3.357
5	.412	.554	.831	1.145	1.610	2.675	4.351
6	.676	.872	1.237	1.635	2.204	3.455	5.348
7	.989	1.239	1.690	2.167	2.833	4.255	6.346
6	1.344	1.646	2.180	2.733	3.490	5.071	7.344
9	1.735	2.088	2.700	3.325	4.168	5.899	8.343
10	2.156	2.558	3.247	3.940	4.865	6.737	9.342
11	2.603	3.053	3.816	4.575	5.578	7.584	10.341
12	3.074	3.571	4.404	5.226	6.304	8.438	11.340
13	3.565	4.107	5.009	5.892	7.042	9.299	12.340
14	4.075	4.660	5.629	6.571	7.790	10.165	13.339
15	4.601	5.229	6.262	7.261	8.547	11.036	14.339
16	5.142	5.812	6.908	7.962	9.312	11.912	15.338
17	5.697	6.408	7.564	8.672	10.085	12.792	16.338
18	6.265	7.015	8.231	9.390	10.865	13.675	17.338
19	6.844	7.633	8.907	10.117	11.651	14.562	18.338
20	7.434	8.260	9.591	10.851	12.443	15.452	19.337
21	8.034	8.897	10.283	11.591	13.240	16.344	20.337
22	8.643	9.542	10.982	12.338	14.041	17.240	21.337
23	9.260	10.196	11.688	13.091	14.848	18.137	22.337
24	9.886	10.856	12.401	13.848	15.659	19.037	22.337
25	10.520	11.524	13.120	14.611	16.473	19.939	24.337
26	11.160	12.198	13.844	15.379	17.292	20.843	25.336
27	11.808	12.879	14.573	16.151	18.114	21.749	26.336
28	12.461	13.565	15.308	16.928	18.939	22.657	27.336
29	13.121	14.256	16.047	17.708	19.768	23.567	28.336
30	13.787	14.953	16.791	18.493	20.599	24.478	29.336
40	20.707	22.164	24.433	26.509	29.051	33.660	39.335
50	27.991	29.707	32.357	34.764	37.689	42.942	49.335
60	35.535	37.485	40.482	43.188	46.459	52.294	59.335
70	43.275	45.442	48.758	51.739	55.329	61.698	69.334
80	51.172	53.540	57.153	60.391	64.278	71.145	79.334
90	59.196	61.754	65.647	69.126	73.291	80.625	89.334
100	67.328	70.065	74.222	77.929	82.358	90.133	99.334
K_α	−2.576	−2.326	−1.960	−1.645	−1.282	−0.6745	0.000

Table 2 (continued). Percentage Points of the χ^2 Distribution

Table of $\chi^2_{\alpha;\,v}$—the 100 α percentage point of the χ^2 distribution for v degrees of freedom

.25	.10	.05	.025	.01	.005	.001	α v
1.323	2.706	3.841	5.024	6.635	7.879	10.828	1
2.773	4.605	5.991	7.378	9.210	10.597	13.816	2
4.108	6.251	7.815	9.348	11.345	12.838	16.266	3
5.385	7.779	9.488	11.143	13.277	14.860	18.467	4
6.626	9.236	11.070	12.832	15.086	16.750	20.515	5
7.841	10.645	12.592	14.449	16.812	18.548	22.458	6
9.037	12.017	14.067	16.013	18.475	20.278	24.322	7
10.219	13.362	15.507	17.535	20.090	21.955	26.125	8
11.389	14.684	16.919	19.023	21.666	23.589	27.877	9
12.549	15.987	18.307	20.483	23.209	25.188	29.588	10
13.701	17.275	19.675	21.920	24.725	26.757	31.264	11
14.845	18.549	21.026	23.337	26.217	28.300	32.909	12
15.984	19.812	22.362	24.736	27.688	29.819	34.528	13
17.117	21.064	23.685	26.119	29.141	31.319	36.123	14
18.245	22.307	24.996	27.488	30.578	32.801	37.697	15
19.369	23.542	26.296	28.845	32.000	34.267	39.252	16
20.489	24.769	27.587	30.191	33.409	35.718	40.790	17
21.605	25.989	28.869	31.526	34.805	37.156	43.312	18
22.718	27.204	30.144	32.852	36.191	38.582	43.820	19
23.828	28.412	31.410	34.170	37.566	39.997	45.315	20
24.935	29.615	32.671	35.479	38.932	41.401	46.797	21
26.039	30.813	33.924	36.781	40.289	42.796	48.268	22
27.141	32.007	35.172	38.076	41.638	44.181	49.728	23
28.241	33.196	36.415	39.364	42.980	45.558	51.179	24
29.339	34.382	37.652	40.646	44.314	46.928	52.620	25
30.434	35.563	38.885	41.923	45.642	48.290	54.052	26
31.528	36.741	40.113	43.194	46.963	49.645	55.476	27
32.620	37.916	41.337	44.461	48.278	50.993	56.892	28
33.711	39.087	42.557	45.722	49.588	52.336	58.302	29
34.800	40.256	43.773	46.979	50.892	53.672	59.703	30
45.616	51.805	55.758	59.342	63.691	66.766	73.402	40
56.334	63.167	67.505	71.420	76.154	79.490	86.661	50
66.981	74.397	79.082	83.298	88.379	91.952	99.607	60
77.577	85.527	90.531	95.023	100.425	104.215	112.317	70
88.130	96.578	101.879	106.629	112.329	116.321	124.839	80
98.650	107.565	113.145	118.136	124.116	128.299	137.208	90
109.141	118.498	124.342	129.561	135.807	140.169	149.449	100
+0.6745	+1.282	+1.645	+1.960	+2.326	+2.576	+3.090	K_α

For $v > 100$ take

$$\chi^2 = v\left\{1 - \frac{2}{9v} + K_\alpha\sqrt{\frac{2}{9v}}\right\}^3 \quad \text{or} \quad \chi^2 = \frac{1}{2}\{K_\alpha + \sqrt{(2v - 1)}\}^2,$$

according to the degree of accuracy required. K_α is the standardized normal deviate corresponding to α, and is shown in the bottom line of the table.

Table 3. Percentage Points of the t Distribution*

Table of $t_{\alpha;\,\nu}$—the 100 α percentage point of the t distribution
for ν degrees of freedom

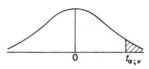

ν \ α	0.40	0.25	0.10	0.05	0.025	0.01	0.005	0.0025	0.001	0.0005
1	0.325	1.000	3.078	6.314	12.706	31.821	63.657	127.32	318.31	636.62
2	.289	0.816	1.886	2.920	4.303	6.965	9.925	14.089	23.326	31.598
3	.277	.765	1.638	2.353	3.182	4.541	5.841	7.453	10.213	12.924
4	.271	.741	1.533	2.132	2.776	3.747	4.604	5.598	7.173	8.610
5	0.267	0.727	1.476	2.015	2.571	3.365	4.032	4.773	5.893	6.869
6	.265	.718	1.440	1.943	2.447	3.143	3.707	4.317	5.208	5.959
7	.263	.711	1.415	1.895	2.365	2.998	3.499	4.029	4.785	5.408
8	.262	.706	1.397	1.860	2.306	2.896	3.355	3.833	4.501	5.041
9	.261	.703	1.383	1.833	2.262	2.821	3.250	3.690	4.297	4.781
10	0.260	0.700	1.372	1.812	2.228	2.764	3.169	3.581	4.144	4.587
11	.260	.697	1.363	1.796	2.201	2.718	3.106	3.497	4.025	4.437
12	.259	.695	1.356	1.782	2.179	2.681	3.055	3.428	3.930	4.318
13	.259	.694	1.350	1.771	2.160	2.650	3.012	3.372	3.852	4.221
14	.258	.692	1.345	1.761	2.145	2.624	2.977	3.326	3.787	4.140
15	0.258	0.691	1.341	1.753	2.131	2.602	2.947	3.286	3.733	4.073
16	.258	.690	1.337	1.746	2.120	2.583	2.921	3.252	3.686	4.015
17	.257	.689	1.333	1.740	2.110	2.567	2.898	3.222	3.646	3.965
18	.257	.688	1.330	1.734	2.101	2.552	2.878	3.197	3.610	3.922
19	.257	.688	1.328	1.729	2.093	2.539	2.861	3.174	3.579	3.883
20	0.257	0.687	1.325	1.725	2.086	2.528	2.845	3.153	3.552	3.850
21	.257	.686	1.323	1.721	2.080	2.518	2.831	3.135	3.527	3.819
22	.256	.686	1.321	1.717	2.074	2.508	2.819	3.119	3.505	3.792
23	.256	.685	1.319	1.714	2.069	2.500	2.807	3.104	3.485	3.767
24	.256	.685	1.318	1.711	2.064	2.492	2.797	3.091	3.467	3.745
25	0.256	0.684	1.316	1.708	2.060	2.485	2.787	3.078	3.450	3.725
26	.256	.684	1.315	1.706	2.056	2.479	2.779	3.067	3.435	3.707
27	.256	.684	1.314	1.703	2.052	2.473	2.771	3.057	3.421	3.690
28	.256	.683	1.313	1.701	2.048	2.467	2.763	3.047	3.408	3.674
29	.256	.683	1.311	1.699	2.045	2.462	2.756	3.038	3.396	3.659
30	0.256	0.683	1.310	1.697	2.042	2.457	2.750	3.030	3.385	3.646
40	.255	.681	1.303	1.684	2.021	2.423	2.704	2.971	3.307	3.551
60	.254	.679	1.296	1.671	2.000	2.390	2.660	2.915	3.232	3.460
120	.254	.677	1.289	1.658	1.980	2.358	2.617	2.860	3.160	3.373
∞	.253	.674	1.282	1.645	1.960	2.326	2.576	2.807	3.090	3.291

Table 4. 25 Percentage Points of the F Distribution*

Table of $F_{0.25;\,v_1,v_2}$

$v_2 \backslash v_1$	1	2	3	4	5	6	7	8	9	10	12	15	20	24	30	40	60	120	∞
1	5.83	7.50	8.20	8.58	8.82	8.98	9.10	9.19	9.26	9.32	9.41	9.49	9.58	9.63	9.67	9.71	9.76	9.80	9.85
2	2.57	3.00	3.15	3.23	3.28	3.31	3.34	3.35	3.37	3.38	3.39	3.41	3.43	3.43	3.44	3.45	3.46	3.47	3.48
3	2.02	2.28	2.36	2.39	2.41	2.42	2.43	2.44	2.44	2.44	2.45	2.46	2.46	2.46	2.47	2.47	2.47	2.47	2.47
4	1.81	2.00	2.05	2.06	2.07	2.08	2.08	2.08	2.08	2.08	2.08	2.08	2.08	2.08	2.08	2.08	2.08	2.08	2.08
5	1.69	1.85	1.88	1.89	1.89	1.89	1.89	1.89	1.89	1.89	1.89	1.89	1.88	1.88	1.88	1.88	1.87	1.87	1.87
6	1.62	1.76	1.78	1.79	1.79	1.78	1.78	1.78	1.77	1.77	1.77	1.76	1.76	1.75	1.75	1.75	1.74	1.74	1.74
7	1.57	1.70	1.72	1.72	1.71	1.71	1.70	1.70	1.70	1.69	1.68	1.68	1.67	1.67	1.66	1.66	1.65	1.65	1.65
8	1.54	1.66	1.67	1.66	1.66	1.65	1.64	1.64	1.63	1.63	1.62	1.62	1.61	1.60	1.60	1.59	1.59	1.58	1.58
9	1.51	1.62	1.63	1.63	1.62	1.61	1.60	1.60	1.59	1.59	1.58	1.57	1.56	1.56	1.55	1.54	1.54	1.53	1.53
10	1.49	1.60	1.60	1.59	1.59	1.58	1.57	1.56	1.56	1.55	1.54	1.53	1.52	1.52	1.51	1.51	1.50	1.49	1.48
11	1.47	1.58	1.58	1.57	1.56	1.55	1.54	1.53	1.53	1.52	1.51	1.50	1.49	1.49	1.48	1.47	1.47	1.46	1.45
12	1.46	1.56	1.56	1.55	1.54	1.53	1.52	1.51	1.51	1.50	1.49	1.48	1.47	1.46	1.45	1.45	1.44	1.43	1.42
13	1.45	1.55	1.55	1.53	1.52	1.51	1.50	1.49	1.49	1.48	1.47	1.46	1.45	1.44	1.43	1.42	1.42	1.41	1.40
14	1.44	1.53	1.53	1.52	1.51	1.50	1.49	1.48	1.47	1.46	1.45	1.44	1.43	1.42	1.41	1.41	1.40	1.39	1.38
15	1.43	1.52	1.52	1.51	1.49	1.48	1.47	1.46	1.46	1.45	1.44	1.43	1.41	1.41	1.40	1.39	1.38	1.37	1.36
16	1.42	1.51	1.51	1.50	1.48	1.47	1.46	1.45	1.44	1.44	1.43	1.41	1.40	1.39	1.38	1.37	1.36	1.35	1.34
17	1.42	1.51	1.50	1.49	1.47	1.46	1.45	1.44	1.43	1.43	1.41	1.40	1.39	1.38	1.37	1.36	1.35	1.34	1.33
18	1.41	1.50	1.49	1.48	1.46	1.45	1.44	1.43	1.42	1.42	1.40	1.39	1.38	1.37	1.36	1.35	1.34	1.33	1.32
19	1.41	1.49	1.49	1.47	1.46	1.44	1.43	1.42	1.41	1.41	1.40	1.38	1.37	1.36	1.35	1.34	1.33	1.32	1.30
20	1.40	1.49	1.48	1.47	1.45	1.44	1.43	1.42	1.41	1.40	1.39	1.37	1.36	1.35	1.34	1.33	1.32	1.31	1.29
21	1.40	1.48	1.48	1.46	1.44	1.43	1.42	1.41	1.40	1.39	1.38	1.37	1.35	1.34	1.33	1.32	1.31	1.30	1.28
22	1.40	1.48	1.47	1.45	1.44	1.42	1.41	1.40	1.39	1.39	1.37	1.36	1.34	1.33	1.32	1.31	1.30	1.29	1.28
23	1.39	1.47	1.47	1.45	1.43	1.42	1.41	1.40	1.39	1.38	1.37	1.35	1.34	1.33	1.32	1.31	1.30	1.28	1.27
24	1.39	1.47	1.46	1.44	1.43	1.41	1.40	1.39	1.38	1.38	1.36	1.35	1.33	1.32	1.31	1.30	1.29	1.28	1.26
25	1.39	1.47	1.46	1.44	1.42	1.41	1.40	1.39	1.38	1.37	1.36	1.34	1.33	1.32	1.31	1.29	1.28	1.27	1.25
26	1.38	1.46	1.45	1.44	1.42	1.41	1.39	1.38	1.37	1.37	1.35	1.34	1.32	1.31	1.30	1.29	1.28	1.26	1.25
27	1.38	1.46	1.45	1.43	1.42	1.40	1.39	1.38	1.37	1.36	1.35	1.33	1.32	1.31	1.30	1.28	1.27	1.26	1.24
28	1.38	1.46	1.45	1.43	1.41	1.40	1.39	1.38	1.37	1.36	1.34	1.33	1.31	1.30	1.29	1.28	1.27	1.25	1.24
29	1.38	1.45	1.45	1.43	1.41	1.40	1.38	1.37	1.36	1.35	1.34	1.32	1.31	1.30	1.29	1.27	1.26	1.25	1.23
30	1.38	1.45	1.44	1.42	1.41	1.39	1.38	1.37	1.36	1.35	1.34	1.32	1.30	1.29	1.28	1.27	1.26	1.24	1.23
40	1.36	1.44	1.42	1.40	1.39	1.37	1.36	1.35	1.34	1.33	1.31	1.30	1.28	1.26	1.25	1.24	1.22	1.21	1.19
60	1.35	1.42	1.41	1.38	1.37	1.35	1.33	1.32	1.31	1.30	1.29	1.27	1.25	1.24	1.22	1.21	1.19	1.17	1.15
120	1.34	1.40	1.39	1.37	1.35	1.33	1.31	1.30	1.29	1.28	1.26	1.24	1.22	1.21	1.19	1.18	1.16	1.13	1.10
∞	1.32	1.39	1.37	1.35	1.33	1.31	1.29	1.28	1.27	1.25	1.24	1.22	1.19	1.18	1.16	1.14	1.12	1.08	1.00

Degrees of freedom for the numerator (v_1)

Degrees of freedom for the denominator (v_2)

$F_{0.75;\,v_1,v_2} = 1/F_{0.25;\,v_2,v_1}$

Example: $P\{F > F_{0.25;\,9,\,15}\} = P\{F > 1.46\} = 0.25$.

Example: $F_{0.75;\,9,\,15} = 1/F_{0.25;\,15,\,9} = 1/1.57 = 0.637$.

*This table is reproduced from Table 18 of *Biometrika Tables for Statisticians*, Volume 1, 1962, by permission of the Biometrika Trustees.

Table 4 (continued). 10 Percentage Points of the F Distribution

Table of $F_{0.10;\,v_1,\,v_2}$

v_2 \ v_1	1	2	3	4	5	6	7	8	9	10	12	15	20	24	30	40	60	120	∞
1	39.86	49.50	53.59	55.83	57.24	58.20	58.91	59.44	59.86	60.19	60.71	61.22	61.74	62.00	62.26	62.53	62.79	63.06	63.33
2	8.53	9.00	9.16	9.24	9.29	9.33	9.35	9.37	9.38	9.39	9.41	9.42	9.44	9.45	9.46	9.47	9.47	9.48	9.49
3	5.54	5.46	5.39	5.34	5.31	5.28	5.27	5.25	5.24	5.23	5.22	5.20	5.18	5.18	5.17	5.16	5.15	5.14	5.13
4	4.54	4.32	4.19	4.11	4.05	4.01	3.98	3.95	3.94	3.92	3.90	3.87	3.84	3.83	3.82	3.80	3.79	3.78	3.76
5	4.06	3.78	3.62	3.52	3.45	3.40	3.37	3.34	3.32	3.30	3.27	3.24	3.21	3.19	3.17	3.16	3.14	3.12	3.10
6	3.78	3.46	3.29	3.18	3.11	3.05	3.01	2.98	2.96	2.94	2.90	2.87	2.84	2.82	2.80	2.78	2.76	2.74	2.72
7	3.59	3.26	3.07	2.96	2.88	2.83	2.78	2.75	2.72	2.70	2.67	2.63	2.59	2.58	2.56	2.54	2.51	2.49	2.47
8	3.46	3.11	2.92	2.81	2.73	2.67	2.62	2.59	2.56	2.54	2.50	2.46	2.42	2.40	2.38	2.36	2.34	2.32	2.29
9	3.36	3.01	2.81	2.69	2.61	2.55	2.51	2.47	2.44	2.42	2.38	2.34	2.30	2.28	2.25	2.23	2.21	2.18	2.16
10	3.29	2.92	2.73	2.61	2.52	2.46	2.41	2.38	2.35	2.32	2.28	2.24	2.20	2.18	2.16	2.13	2.11	2.08	2.06
11	3.23	2.86	2.66	2.54	2.45	2.39	2.34	2.30	2.27	2.25	2.21	2.17	2.12	2.10	2.08	2.05	2.03	2.00	1.97
12	3.18	2.81	2.61	2.48	2.39	2.33	2.28	2.24	2.21	2.19	2.15	2.10	2.06	2.04	2.01	1.99	1.96	1.93	1.90
13	3.14	2.76	2.56	2.43	2.35	2.28	2.23	2.20	2.16	2.14	2.10	2.05	2.01	1.98	1.96	1.93	1.90	1.88	1.85
14	3.10	2.73	2.52	2.39	2.31	2.24	2.19	2.15	2.12	2.10	2.05	2.01	1.96	1.94	1.91	1.89	1.86	1.83	1.80
15	3.07	2.70	2.49	2.36	2.27	2.21	2.16	2.12	2.09	2.06	2.02	1.97	1.92	1.90	1.87	1.85	1.82	1.79	1.76
16	3.05	2.67	2.46	2.33	2.24	2.18	2.13	2.09	2.06	2.03	1.99	1.94	1.89	1.87	1.84	1.81	1.78	1.75	1.72
17	3.03	2.64	2.44	2.31	2.22	2.15	2.10	2.06	2.03	2.00	1.96	1.91	1.86	1.84	1.81	1.78	1.75	1.72	1.69
18	3.01	2.62	2.42	2.29	2.20	2.13	2.08	2.04	2.00	1.98	1.93	1.89	1.84	1.81	1.78	1.75	1.72	1.69	1.66
19	2.99	2.61	2.40	2.27	2.18	2.11	2.06	2.02	1.98	1.96	1.91	1.86	1.81	1.79	1.76	1.73	1.70	1.67	1.63
20	2.97	2.59	2.38	2.25	2.16	2.09	2.04	2.00	1.96	1.94	1.89	1.84	1.79	1.77	1.74	1.71	1.68	1.64	1.61
21	2.96	2.57	2.36	2.23	2.14	2.08	2.02	1.98	1.95	1.92	1.87	1.83	1.78	1.75	1.72	1.69	1.66	1.62	1.59
22	2.95	2.56	2.35	2.22	2.13	2.06	2.01	1.97	1.93	1.90	1.86	1.81	1.76	1.73	1.70	1.67	1.64	1.60	1.57
23	2.94	2.55	2.34	2.21	2.11	2.05	1.99	1.95	1.92	1.89	1.84	1.80	1.74	1.72	1.69	1.66	1.62	1.59	1.55
24	2.93	2.54	2.33	2.19	2.10	2.04	1.98	1.94	1.91	1.88	1.83	1.78	1.73	1.70	1.67	1.64	1.61	1.57	1.53
25	2.92	2.53	2.32	2.18	2.09	2.02	1.97	1.93	1.89	1.87	1.82	1.77	1.72	1.69	1.66	1.63	1.59	1.56	1.52
26	2.91	2.52	2.31	2.17	2.08	2.01	1.96	1.92	1.88	1.86	1.81	1.76	1.71	1.68	1.65	1.61	1.58	1.54	1.50
27	2.90	2.51	2.30	2.17	2.07	2.00	1.95	1.91	1.87	1.85	1.80	1.75	1.70	1.67	1.64	1.60	1.57	1.53	1.49
28	2.89	2.50	2.29	2.16	2.06	2.00	1.94	1.90	1.87	1.84	1.79	1.74	1.69	1.66	1.63	1.59	1.56	1.52	1.48
29	2.89	2.50	2.28	2.15	2.06	1.99	1.93	1.89	1.86	1.83	1.78	1.73	1.68	1.65	1.62	1.58	1.55	1.51	1.47
30	2.88	2.49	2.28	2.14	2.05	1.98	1.93	1.88	1.85	1.82	1.77	1.72	1.67	1.64	1.61	1.57	1.54	1.50	1.46
40	2.84	2.44	2.23	2.09	2.00	1.93	1.87	1.83	1.79	1.76	1.71	1.66	1.61	1.57	1.54	1.51	1.47	1.42	1.38
60	2.79	2.39	2.18	2.04	1.95	1.87	1.82	1.77	1.74	1.71	1.66	1.60	1.54	1.51	1.48	1.44	1.40	1.35	1.29
120	2.75	2.35	2.13	1.99	1.90	1.82	1.77	1.72	1.68	1.65	1.60	1.55	1.48	1.45	1.41	1.37	1.32	1.26	1.19
∞	2.71	2.30	2.08	1.94	1.85	1.77	1.72	1.67	1.63	1.60	1.55	1.49	1.42	1.38	1.34	1.30	1.24	1.17	1.00

Degrees of freedom for the numerator (v_1)

Degrees of freedom for the denominator (v_2)

$$F_{0.90;\,v_1,\,v_2} = 1/F_{0.10;\,v_2,\,v_1}$$

Example: $P\{F > F_{0.10;\,9,\,15} = P\{F > 2.09\} = 0.10$.

Example: $F_{0.90;\,9,\,15} = 1/F_{0.10;\,15,\,9} = 1/2.34 = 0.427$.

605

Table 4 (continued). 5 Percentage Points of the F Distribution

Table of $F_{0.05;\,\nu_1,\,\nu_2}$

Degrees of freedom for the numerator (ν_1)

ν_2	1	2	3	4	5	6	7	8	9	10	12	15	20	24	30	40	60	120	∞
1	161.4	199.5	215.7	224.6	230.2	234.0	236.8	238.9	240.5	241.9	243.9	245.9	248.0	249.1	250.1	251.1	252.2	253.3	254.3
2	18.51	19.00	19.16	19.25	19.30	19.33	19.35	19.37	19.38	19.40	19.41	19.43	19.45	19.45	19.46	19.47	19.48	19.49	19.50
3	10.13	9.55	9.28	9.12	9.01	8.94	8.89	8.85	8.81	8.79	8.74	8.70	8.66	8.64	8.62	8.59	8.57	8.55	8.53
4	7.71	6.94	6.59	6.39	6.26	6.16	6.09	6.04	6.00	5.96	5.91	5.86	5.80	5.77	5.75	5.72	5.69	5.66	5.63
5	6.61	5.79	5.41	5.19	5.05	4.95	4.88	4.82	4.77	4.74	4.68	4.62	4.56	4.53	4.50	4.46	4.43	4.40	4.36
6	5.99	5.14	4.76	4.53	4.39	4.28	4.21	4.15	4.10	4.06	4.00	3.94	3.87	3.84	3.81	3.77	3.74	3.70	3.67
7	5.59	4.74	4.35	4.12	3.97	3.87	3.79	3.73	3.68	3.64	3.57	3.51	3.44	3.41	3.38	3.34	3.30	3.27	3.23
8	5.32	4.46	4.07	3.84	3.69	3.58	3.50	3.44	3.39	3.35	3.28	3.22	3.15	3.12	3.08	3.04	3.01	2.97	2.93
9	5.12	4.26	3.86	3.63	3.48	3.37	3.29	3.23	3.18	3.14	3.07	3.01	2.94	2.90	2.86	2.83	2.79	2.75	2.71
10	4.96	4.10	3.71	3.48	3.33	3.22	3.14	3.07	3.02	2.98	2.91	2.85	2.77	2.74	2.70	2.66	2.62	2.58	2.54
11	4.84	3.98	3.59	3.36	3.20	3.09	3.01	2.95	2.90	2.85	2.79	2.72	2.65	2.61	2.57	2.53	2.49	2.45	2.40
12	4.75	3.89	3.49	3.26	3.11	3.00	2.91	2.85	2.80	2.75	2.69	2.62	2.54	2.51	2.47	2.43	2.38	2.34	2.30
13	4.67	3.81	3.41	3.18	3.03	2.92	2.83	2.77	2.71	2.67	2.60	2.53	2.46	2.42	2.38	2.34	2.30	2.25	2.21
14	4.60	3.74	3.34	3.11	2.96	2.85	2.76	2.70	2.65	2.60	2.53	2.46	2.39	2.35	2.31	2.27	2.22	2.18	2.13
15	4.54	3.68	3.29	3.06	2.90	2.79	2.71	2.64	2.59	2.54	2.48	2.40	2.33	2.29	2.25	2.20	2.16	2.11	2.07
16	4.49	3.63	3.24	3.01	2.85	2.74	2.66	2.59	2.54	2.49	2.42	2.35	2.28	2.24	2.19	2.15	2.11	2.06	2.01
17	4.45	3.59	3.20	2.96	2.81	2.70	2.61	2.55	2.49	2.45	2.38	2.31	2.23	2.19	2.15	2.10	2.06	2.01	1.96
18	4.41	3.55	3.16	2.93	2.77	2.66	2.58	2.51	2.46	2.41	2.34	2.27	2.19	2.15	2.11	2.06	2.02	1.97	1.92
19	4.38	3.52	3.13	2.90	2.74	2.63	2.54	2.48	2.42	2.38	2.31	2.23	2.16	2.11	2.07	2.03	1.98	1.93	1.88
20	4.35	3.49	3.10	2.87	2.71	2.60	2.51	2.45	2.39	2.35	2.28	2.20	2.12	2.08	2.04	1.99	1.95	1.90	1.84
21	4.32	3.47	3.07	2.84	2.68	2.57	2.49	2.42	2.37	2.32	2.25	2.18	2.10	2.05	2.01	1.96	1.92	1.87	1.81
22	4.30	3.44	3.05	2.82	2.66	2.55	2.46	2.40	2.34	2.30	2.23	2.15	2.07	2.03	1.98	1.94	1.89	1.84	1.78
23	4.28	3.42	3.03	2.80	2.64	2.53	2.44	2.37	2.32	2.27	2.20	2.13	2.05	2.01	1.96	1.91	1.86	1.81	1.76
24	4.26	3.40	3.01	2.78	2.62	2.51	2.42	2.36	2.30	2.25	2.18	2.11	2.03	1.98	1.94	1.89	1.84	1.79	1.73
25	4.24	3.39	2.99	2.76	2.60	2.49	2.40	2.34	2.28	2.24	2.16	2.09	2.01	1.96	1.92	1.87	1.82	1.77	1.71
26	4.23	3.37	2.98	2.74	2.59	2.47	2.39	2.32	2.27	2.22	2.15	2.07	1.99	1.95	1.90	1.85	1.80	1.75	1.69
27	4.21	3.35	2.96	2.73	2.57	2.46	2.37	2.31	2.25	2.20	2.13	2.06	1.97	1.93	1.88	1.84	1.79	1.73	1.67
28	4.20	3.34	2.95	2.71	2.56	2.45	2.36	2.29	2.24	2.19	2.12	2.04	1.96	1.91	1.87	1.82	1.77	1.71	1.65
29	4.18	3.33	2.93	2.70	2.55	2.43	2.35	2.28	2.22	2.18	2.10	2.03	1.94	1.90	1.85	1.81	1.75	1.70	1.64
30	4.17	3.32	2.92	2.69	2.53	2.42	2.33	2.27	2.21	2.16	2.09	2.01	1.93	1.89	1.84	1.79	1.74	1.68	1.62
40	4.08	3.23	2.84	2.61	2.45	2.34	2.25	2.18	2.12	2.08	2.00	1.92	1.84	1.79	1.74	1.69	1.64	1.58	1.51
60	4.00	3.15	2.76	2.53	2.37	2.25	2.17	2.10	2.04	1.99	1.92	1.84	1.75	1.70	1.65	1.59	1.53	1.47	1.39
120	3.92	3.07	2.68	2.45	2.29	2.17	2.09	2.02	1.96	1.91	1.83	1.75	1.66	1.61	1.55	1.50	1.43	1.35	1.25
∞	3.84	3.00	2.60	2.37	2.21	2.10	2.01	1.94	1.88	1.83	1.75	1.67	1.57	1.52	1.46	1.39	1.32	1.22	1.00

Degrees of freedom for the denominator (ν_2)

$$F_{0.95;\,\nu_1,\,\nu_2} = 1/F_{0.05;\,\nu_2,\,\nu_1}$$

Example: $P\{F > F_{0.05;\,9,\,15} = P\{F > 2.59\} = 0.05.$

Example: $F_{0.95;\,9,\,15} = 1/F_{0.05;\,15,\,9} = 1/3.01 = 0.332.$

Table 4 (continued). 2.5 Percentage Points of the F Distribution

Table of $F_{0.025; \nu_1, \nu_2}$

ν_2 \ ν_1	1	2	3	4	5	6	7	8	9	10	12	15	20	24	30	40	60	120	∞
1	647.8	799.5	864.2	899.6	921.8	937.1	948.2	956.7	963.3	968.6	976.7	984.9	993.1	997.2	1001	1006	1010	1014	1018
2	38.51	39.00	39.17	39.25	39.30	39.33	39.36	39.37	39.39	39.40	39.41	39.43	39.45	39.46	39.46	39.47	39.48	39.49	39.50
3	17.44	16.04	15.44	15.10	14.88	14.73	14.62	14.54	14.47	14.42	14.34	14.25	14.17	14.12	14.08	14.04	13.99	13.95	13.90
4	12.22	10.65	9.98	9.60	9.36	9.20	9.07	8.98	8.90	8.84	8.75	8.66	8.56	8.51	8.46	8.41	8.36	8.31	8.26
5	10.01	8.43	7.76	7.39	7.15	6.98	6.85	6.76	6.68	6.62	6.52	6.43	6.33	6.28	6.23	6.18	6.12	6.07	6.02
6	8.81	7.26	6.60	6.23	5.99	5.82	5.70	5.60	5.52	5.46	5.37	5.27	5.17	5.12	5.07	5.01	4.96	4.90	4.85
7	8.07	6.54	5.89	5.52	5.29	5.12	4.99	4.90	4.82	4.76	4.67	4.57	4.47	4.42	4.36	4.31	4.25	4.20	4.14
8	7.57	6.06	5.42	5.05	4.82	4.65	4.53	4.43	4.36	4.30	4.20	4.10	4.00	3.95	3.89	3.84	3.78	3.73	3.67
9	7.21	5.71	5.08	4.72	4.48	4.32	4.20	4.10	4.03	3.96	3.87	3.77	3.67	3.61	3.56	3.51	3.45	3.39	3.33
10	6.94	5.46	4.83	4.47	4.24	4.07	3.95	3.85	3.78	3.72	3.62	3.52	3.42	3.37	3.31	3.26	3.20	3.14	3.08
11	6.72	5.26	4.63	4.28	4.04	3.88	3.76	3.66	3.59	3.53	3.43	3.33	3.23	3.17	3.12	3.06	3.00	2.94	2.88
12	6.55	5.10	4.47	4.12	3.89	3.73	3.61	3.51	3.44	3.37	3.28	3.18	3.07	3.02	2.96	2.91	2.85	2.79	2.72
13	6.41	4.97	4.35	4.00	3.77	3.60	3.48	3.39	3.31	3.25	3.15	3.05	2.95	2.89	2.84	2.78	2.72	2.66	2.60
14	6.30	4.86	4.24	3.89	3.66	3.50	3.38	3.29	3.21	3.15	3.05	2.95	2.84	2.79	2.73	2.67	2.61	2.55	2.49
15	6.20	4.77	4.15	3.80	3.58	3.41	3.29	3.20	3.12	3.06	2.96	2.86	2.76	2.70	2.64	2.59	2.52	2.46	2.40
16	6.12	4.69	4.08	3.73	3.50	3.34	3.22	3.12	3.05	2.99	2.89	2.79	2.68	2.63	2.57	2.51	2.45	2.38	2.32
17	6.04	4.62	4.01	3.66	3.44	3.28	3.16	3.06	2.98	2.92	2.82	2.72	2.62	2.56	2.50	2.44	2.38	2.32	2.25
18	5.98	4.56	3.95	3.61	3.38	3.22	3.10	3.01	2.93	2.87	2.77	2.67	2.56	2.50	2.44	2.38	2.32	2.26	2.19
19	5.92	4.51	3.90	3.56	3.33	3.17	3.05	2.96	2.88	2.82	2.72	2.62	2.51	2.45	2.39	2.33	2.27	2.20	2.13
20	5.87	4.46	3.86	3.51	3.29	3.13	3.01	2.91	2.84	2.77	2.68	2.57	2.46	2.41	2.35	2.29	2.22	2.16	2.09
21	5.83	4.42	3.82	3.48	3.25	3.09	2.97	2.87	2.80	2.73	2.64	2.53	2.42	2.37	2.31	2.25	2.18	2.11	2.04
22	5.79	4.38	3.78	3.44	3.22	3.05	2.93	2.84	2.76	2.70	2.60	2.50	2.39	2.33	2.27	2.21	2.14	2.08	2.00
23	5.75	4.35	3.75	3.41	3.18	3.02	2.90	2.81	2.73	2.67	2.57	2.47	2.36	2.30	2.24	2.18	2.11	2.04	1.97
24	5.72	4.32	3.72	3.38	3.15	2.99	2.87	2.78	2.70	2.64	2.54	2.44	2.33	2.27	2.21	2.15	2.08	2.01	1.94
25	5.69	4.29	3.69	3.35	3.13	2.97	2.85	2.75	2.68	2.61	2.51	2.41	2.30	2.24	2.18	2.12	2.05	1.98	1.91
26	5.66	4.27	3.67	3.33	3.10	2.94	2.82	2.73	2.65	2.59	2.49	2.39	2.28	2.22	2.16	2.09	2.03	1.95	1.88
27	5.63	4.24	3.65	3.31	3.08	2.92	2.80	2.71	2.63	2.57	2.47	2.36	2.25	2.19	2.13	2.07	2.00	1.93	1.85
28	5.61	4.22	3.63	3.29	3.06	2.90	2.78	2.69	2.61	2.55	2.45	2.34	2.23	2.17	2.11	2.05	1.98	1.91	1.83
29	5.59	4.20	3.61	3.27	3.04	2.88	2.76	2.67	2.59	2.53	2.43	2.32	2.21	2.15	2.09	2.03	1.96	1.89	1.81
30	5.57	4.18	3.59	3.25	3.03	2.87	2.75	2.65	2.57	2.51	2.41	2.31	2.20	2.14	2.07	2.01	1.94	1.87	1.79
40	5.42	4.05	3.46	3.13	2.90	2.74	2.62	2.53	2.45	2.39	2.29	2.18	2.07	2.01	1.94	1.88	1.80	1.72	1.64
60	5.29	3.93	3.34	3.01	2.79	2.63	2.51	2.41	2.33	2.27	2.17	2.06	1.94	1.88	1.82	1.74	1.67	1.58	1.48
120	5.15	3.80	3.23	2.89	2.67	2.52	2.39	2.30	2.22	2.16	2.05	1.94	1.82	1.76	1.69	1.61	1.53	1.43	1.31
∞	5.02	3.69	3.12	2.79	2.57	2.41	2.29	2.19	2.11	2.05	1.94	1.83	1.71	1.64	1.57	1.48	1.39	1.27	1.00

Degrees of freedom for the numerator (ν_1)

Degrees of freedom for the denominator (ν_2)

Example: $P\{F > F_{0.025;9,15}\} = P\{F > 3.12\} = 0.025$.

Example: $F_{0.975;9,15} = 1/F_{0.025;15,9} = 1/3.77 = 0.265$.

$$F_{0.975;\nu_1,\nu_2} = 1/F_{0.025;\nu_2,\nu_1}$$

Table 4 (continued). 1 Percentage Points of the F Distribution

Table of $F_{0.01;\nu_1,\nu_2}$

ν_2 \ ν_1	Degrees of freedom for the numerator (ν_1)																		
	1	2	3	4	5	6	7	8	9	10	12	15	20	24	30	40	60	120	∞
1	4052	4999.5	5403	5625	5764	5859	5928	5982	6022	6056	6106	6157	6209	6235	6261	6287	6313	6339	6366
2	98.50	99.00	99.17	99.25	99.30	99.33	99.36	99.37	99.39	99.40	99.42	99.43	99.45	99.46	99.47	99.47	99.48	99.49	99.50
3	34.12	30.82	29.46	28.71	28.24	27.91	27.67	27.49	27.35	27.23	27.05	26.87	26.69	26.60	26.50	26.41	26.32	26.22	26.13
4	21.20	18.00	16.69	15.98	15.52	15.21	14.98	14.80	14.66	14.55	14.37	14.20	14.02	13.93	13.84	13.75	13.65	13.56	13.46
5	16.26	13.27	12.06	11.39	10.97	10.67	10.46	10.29	10.16	10.05	9.89	9.72	9.55	9.47	9.38	9.29	9.20	9.11	9.02
6	13.75	10.92	9.78	9.15	8.75	8.47	8.26	8.10	7.98	7.87	7.72	7.56	7.40	7.31	7.23	7.14	7.06	6.97	6.88
7	12.25	9.55	8.45	7.85	7.46	7.19	6.99	6.84	6.72	6.62	6.47	6.31	6.16	6.07	5.99	5.91	5.82	5.74	5.65
8	11.26	8.65	7.59	7.01	6.63	6.37	6.18	6.03	5.91	5.81	5.67	5.52	5.36	5.28	5.20	5.12	5.03	4.95	4.86
9	10.56	8.02	6.99	6.42	6.06	5.80	5.61	5.47	5.35	5.26	5.11	4.96	4.81	4.73	4.65	4.57	4.48	4.40	4.31
10	10.04	7.56	6.55	5.99	5.64	5.39	5.20	5.06	4.94	4.85	4.71	4.56	4.41	4.33	4.25	4.17	4.08	4.00	3.91
11	9.65	7.21	6.22	5.67	5.32	5.07	4.89	4.74	4.63	4.54	4.40	4.25	4.10	4.02	3.94	3.86	3.78	3.69	3.60
12	9.33	6.93	5.95	5.41	5.06	4.82	4.64	4.50	4.39	4.30	4.16	4.01	3.86	3.78	3.70	3.62	3.54	3.45	3.36
13	9.07	6.70	5.74	5.21	4.86	4.62	4.44	4.30	4.19	4.10	3.96	3.82	3.66	3.59	3.51	3.43	3.34	3.25	3.17
14	8.86	6.51	5.56	5.04	4.69	4.46	4.28	4.14	4.03	3.94	3.80	3.66	3.51	3.43	3.35	3.27	3.18	3.09	3.00
15	8.68	6.36	5.42	4.89	4.56	4.32	4.14	4.00	3.89	3.80	3.67	3.52	3.37	3.29	3.21	3.13	3.05	2.96	2.87
16	8.53	6.23	5.29	4.77	4.44	4.20	4.03	3.89	3.78	3.69	3.55	3.41	3.26	3.18	3.10	3.02	2.93	2.84	2.75
17	8.40	6.11	5.18	4.67	4.34	4.10	3.93	3.79	3.68	3.59	3.46	3.31	3.16	3.08	3.00	2.92	2.83	2.75	2.65
18	8.29	6.01	5.09	4.58	4.25	4.01	3.84	3.71	3.60	3.51	3.37	3.23	3.08	3.00	2.92	2.84	2.75	2.66	2.57
19	8.18	5.93	5.01	4.50	4.17	3.94	3.77	3.63	3.52	3.43	3.30	3.15	3.00	2.92	2.84	2.76	2.67	2.58	2.49
20	8.10	5.85	4.94	4.43	4.10	3.87	3.70	3.56	3.46	3.37	3.23	3.09	2.94	2.86	2.78	2.69	2.61	2.52	2.42
21	8.02	5.78	4.87	4.37	4.04	3.81	3.64	3.51	3.40	3.31	3.17	3.03	2.88	2.80	2.72	2.64	2.55	2.46	2.36
22	7.95	5.72	4.82	4.31	3.99	3.76	3.59	3.45	3.35	3.26	3.12	2.98	2.83	2.75	2.67	2.58	2.50	2.40	2.31
23	7.88	5.66	4.76	4.26	3.94	3.71	3.54	3.41	3.30	3.21	3.07	2.93	2.78	2.70	2.62	2.54	2.45	2.35	2.26
24	7.82	5.61	4.72	4.22	3.90	3.67	3.50	3.36	3.26	3.17	3.03	2.89	2.74	2.66	2.58	2.49	2.40	2.31	2.21
25	7.77	5.57	4.68	4.18	3.85	3.63	3.46	3.32	3.22	3.13	2.99	2.85	2.70	2.62	2.54	2.45	2.36	2.27	2.17
26	7.72	5.53	4.64	4.14	3.82	3.59	3.42	3.29	3.18	3.09	2.96	2.81	2.66	2.58	2.50	2.42	2.33	2.23	2.13
27	7.68	5.49	4.60	4.11	3.78	3.56	3.39	3.26	3.15	3.06	2.93	2.78	2.63	2.55	2.47	2.38	2.29	2.20	2.10
28	7.64	5.45	4.57	4.07	3.75	3.53	3.36	3.23	3.12	3.03	2.90	2.75	2.60	2.52	2.44	2.35	2.26	2.17	2.06
29	7.60	5.42	4.54	4.04	3.73	3.50	3.33	3.20	3.09	3.00	2.87	2.73	2.57	2.49	2.41	2.33	2.23	2.14	2.03
30	7.56	5.39	4.51	4.02	3.70	3.47	3.30	3.17	3.07	2.98	2.84	2.70	2.55	2.47	2.39	2.30	2.21	2.11	2.01
40	7.31	5.18	4.31	3.83	3.51	3.29	3.12	2.99	2.89	2.80	2.66	2.52	2.37	2.29	2.20	2.11	2.02	1.92	1.80
60	7.08	4.98	4.13	3.65	3.34	3.12	2.95	2.82	2.72	2.63	2.50	2.35	2.20	2.12	2.03	1.94	1.84	1.73	1.60
120	6.85	4.79	3.95	3.48	3.17	2.96	2.79	2.66	2.56	2.47	2.34	2.19	2.03	1.95	1.86	1.76	1.66	1.53	1.38
∞	6.63	4.61	3.78	3.32	3.02	2.80	2.64	2.51	2.41	2.32	2.18	2.04	1.88	1.79	1.70	1.59	1.47	1.32	1.00

Degrees of freedom for the denominator (ν_2)

$F_{0.99;\nu_1,\nu_2} = 1/F_{0.01;\nu_2,\nu_1}$

Example: $P\{F > F_{0.01;9,15}\} = P\{F > 3.89\} = 0.01$.

Example: $F_{0.99;9,15} = 1/F_{0.01;15,9} = 1/4.96 = 0.202$.

608

Table 4 (continued). 0.5 Percentage Points of the F Distribution

Table of $F_{0.005; \nu_1, \nu_2}$

Degrees of freedom for the numerator (ν_1)

ν_2	1	2	3	4	5	6	7	8	9	10	12	15	20	24	30	40	60	120	∞
1	16211	20000	21615	22500	23056	23437	23715	23925	24091	24224	24426	24630	24836	24940	25044	25148	25253	25359	25465
2	198.5	199.0	199.2	199.2	199.3	199.3	199.4	199.4	199.4	199.4	199.4	199.4	199.4	199.5	199.5	199.5	199.5	199.5	199.5
3	55.55	49.80	47.47	46.19	45.39	44.84	44.43	44.13	43.88	43.69	43.39	43.08	42.78	42.62	42.47	42.31	42.15	41.99	41.83
4	31.33	26.28	24.26	23.15	22.46	21.97	21.62	21.35	21.14	20.97	20.70	20.44	20.17	20.03	19.89	19.75	19.61	19.47	19.32
5	22.78	18.31	16.53	15.56	14.94	14.51	14.20	13.96	13.77	13.62	13.38	13.15	12.90	12.78	12.66	12.53	12.40	12.27	12.14
6	18.63	14.54	12.92	12.03	11.46	11.07	10.79	10.57	10.39	10.25	10.03	9.81	9.59	9.47	9.36	9.24	9.12	9.00	8.88
7	16.24	12.40	10.88	10.05	9.52	9.16	8.89	8.68	8.51	8.38	8.18	7.97	7.75	7.65	7.53	7.42	7.31	7.19	7.08
8	14.69	11.04	9.60	8.81	8.30	7.95	7.69	7.50	7.34	7.21	7.01	6.81	6.61	6.50	6.40	6.29	6.18	6.06	5.95
9	13.61	10.11	8.72	7.96	7.47	7.13	6.88	6.69	6.54	6.42	6.23	6.03	5.83	5.73	5.62	5.52	5.41	5.30	5.19
10	12.83	9.43	8.08	7.34	6.87	6.54	6.30	6.12	5.97	5.85	5.66	5.47	5.27	5.17	5.07	4.97	4.86	4.75	4.64
11	12.23	8.91	7.60	6.88	6.42	6.10	5.86	5.68	5.54	5.42	5.24	5.05	4.86	4.76	4.65	4.55	4.44	4.34	4.23
12	11.75	8.51	7.23	6.52	6.07	5.76	5.52	5.35	5.20	5.09	4.91	4.72	4.53	4.43	4.33	4.23	4.12	4.01	3.90
13	11.37	8.19	6.93	6.23	5.79	5.48	5.25	5.08	4.94	4.82	4.64	4.46	4.27	4.17	4.07	3.97	3.87	3.76	3.65
14	11.06	7.92	6.68	6.00	5.56	5.26	5.03	4.86	4.72	4.60	4.43	4.25	4.06	3.96	3.86	3.76	3.66	3.55	3.44
15	10.80	7.70	6.48	5.80	5.37	5.07	4.85	4.67	4.54	4.42	4.25	4.07	3.88	3.79	3.69	3.58	3.48	3.37	3.26
16	10.58	7.51	6.30	5.64	5.21	4.91	4.69	4.52	4.38	4.27	4.10	3.92	3.73	3.64	3.54	3.44	3.33	3.22	3.11
17	10.38	7.35	6.16	5.50	5.07	4.78	4.56	4.39	4.25	4.14	3.97	3.79	3.61	3.51	3.41	3.31	3.21	3.10	2.98
18	10.22	7.21	6.03	5.37	4.96	4.66	4.44	4.28	4.14	4.03	3.86	3.68	3.50	3.40	3.30	3.20	3.10	2.99	2.87
19	10.07	7.09	5.92	5.27	4.85	4.56	4.34	4.18	4.04	3.93	3.76	3.59	3.40	3.31	3.21	3.11	3.00	2.89	2.78
20	9.94	6.99	5.82	5.17	4.76	4.47	4.26	4.09	3.96	3.85	3.68	3.50	3.32	3.22	3.12	3.02	2.92	2.81	2.69
21	9.83	6.89	5.73	5.09	4.68	4.39	4.18	4.01	3.88	3.77	3.60	3.43	3.24	3.15	3.05	2.95	2.84	2.73	2.61
22	9.73	6.81	5.65	5.02	4.61	4.32	4.11	3.94	3.81	3.70	3.54	3.36	3.18	3.08	2.98	2.88	2.77	2.66	2.55
23	9.63	6.73	5.58	4.95	4.54	4.26	4.05	3.88	3.75	3.64	3.47	3.30	3.12	3.02	2.92	2.82	2.71	2.60	2.48
24	9.55	6.66	5.52	4.89	4.49	4.20	3.99	3.83	3.69	3.59	3.42	3.25	3.06	2.97	2.87	2.77	2.66	2.55	2.43
25	9.48	6.60	5.46	4.84	4.43	4.15	3.94	3.78	3.64	3.54	3.37	3.20	3.01	2.92	2.82	2.72	2.61	2.50	2.38
26	9.41	6.54	5.41	4.79	4.38	4.10	3.89	3.73	3.60	3.49	3.33	3.15	2.97	2.87	2.77	2.67	2.56	2.45	2.33
27	9.34	6.49	5.36	4.74	4.34	4.06	3.85	3.69	3.56	3.45	3.28	3.11	2.93	2.83	2.73	2.63	2.52	2.41	2.29
28	9.28	6.44	5.32	4.70	4.30	4.02	3.81	3.65	3.52	3.41	3.25	3.07	2.89	2.79	2.69	2.59	2.48	2.37	2.25
29	9.23	6.40	5.28	4.66	4.26	3.98	3.77	3.61	3.48	3.38	3.21	3.04	2.86	2.76	2.66	2.56	2.45	2.33	2.21
30	9.18	6.35	5.24	4.62	4.23	3.95	3.74	3.58	3.45	3.34	3.18	3.01	2.82	2.73	2.63	2.52	2.42	2.30	2.18
40	8.83	6.07	4.98	4.37	3.99	3.71	3.51	3.35	3.22	3.12	2.95	2.78	2.60	2.50	2.40	2.30	2.18	2.06	1.93
60	8.49	5.79	4.73	4.14	3.76	3.49	3.29	3.13	3.01	2.90	2.74	2.57	2.39	2.29	2.19	2.08	1.96	1.83	1.69
120	8.18	5.54	4.50	3.92	3.55	3.28	3.09	2.93	2.81	2.71	2.54	2.37	2.19	2.09	1.98	1.87	1.75	1.61	1.43
∞	7.88	5.30	4.28	3.72	3.35	3.09	2.90	2.74	2.62	2.52	2.36	2.19	2.00	1.90	1.79	1.67	1.53	1.36	1.00

Degrees of freedom for the denominator (ν_2)

$$F_{0.995; \nu_1, \nu_2} = 1/F_{0.005; \nu_2, \nu_1}$$

Example: $P\{F > F_{0.005; 9, 15} = P\{F > 4.54\} = 0.005$.

Example: $F_{0.995; 9, 15} = 1/F_{0.005; 15, 9} = 1/6.03 = 0.166$.

Table 4 (continued). 0.1 Percentage Points of the F Distribution
Table of $F_{0.001; v_1, v_2}$

Degrees of freedom for the numerator (v_1)

v_2	1	2	3	4	5	6	7	8	9	10	12	15	20	24	30	40	60	120	∞
1	4053*	5000*	5404*	5625*	5764*	5859*	5929*	5981*	6023*	6056*	6107*	6158*	6209*	6235*	6261*	6287*	6313*	6340*	6366*
2	998.5	999.0	999.2	999.2	999.3	999.3	999.4	999.4	999.4	999.4	999.4	999.4	999.4	999.5	999.5	999.5	999.5	999.5	999.5
3	167.0	148.5	141.1	137.1	134.6	132.8	131.6	130.6	129.9	129.2	128.3	127.4	126.4	125.9	125.4	125.0	124.5	124.0	123.5
4	74.14	61.25	56.18	53.44	51.71	50.53	49.66	49.00	48.47	48.05	47.41	46.76	46.10	45.77	45.43	45.09	44.75	44.40	44.05
5	47.18	37.12	33.20	31.09	29.75	28.84	28.16	27.64	27.24	26.92	26.42	25.91	25.39	25.14	24.87	24.60	24.33	24.06	23.79
6	35.51	27.00	23.70	21.92	20.81	20.03	19.46	19.03	18.69	18.41	17.99	17.56	17.12	16.89	16.67	16.44	16.21	15.99	15.75
7	29.25	21.69	18.77	17.19	16.21	15.52	15.02	14.63	14.33	14.08	13.71	13.32	12.93	12.73	12.53	12.33	12.12	11.91	11.70
8	25.42	18.49	15.83	14.39	13.49	12.86	12.40	12.04	11.77	11.54	11.19	10.84	10.48	10.30	10.11	9.92	9.73	9.53	9.33
9	22.86	16.39	13.90	12.56	11.71	11.13	10.70	10.37	10.11	9.89	9.57	9.24	8.90	8.72	8.55	8.37	8.19	8.00	7.81
10	21.04	14.91	12.55	11.28	10.48	9.92	9.52	9.20	8.96	8.75	8.45	8.13	7.80	7.64	7.47	7.30	7.12	6.94	6.76
11	19.69	13.81	11.56	10.35	9.58	9.05	8.66	8.35	8.12	7.92	7.63	7.32	7.01	6.85	6.68	6.52	6.35	6.17	6.00
12	18.64	12.97	10.80	9.63	8.89	8.38	8.00	7.71	7.48	7.29	7.00	6.71	6.40	6.25	6.09	5.93	5.76	5.59	5.42
13	17.81	12.31	10.21	9.07	8.35	7.86	7.49	7.21	6.98	6.80	6.52	6.23	5.93	5.78	5.63	5.47	5.30	5.14	4.97
14	17.14	11.78	9.73	8.62	7.92	7.43	7.08	6.80	6.58	6.40	6.13	5.85	5.56	5.41	5.25	5.10	4.94	4.77	4.60
15	16.59	11.34	9.34	8.25	7.57	7.09	6.74	6.47	6.26	6.08	5.81	5.54	5.25	5.10	4.95	4.80	4.64	4.47	4.31
16	16.12	10.97	9.00	7.94	7.27	6.81	6.46	6.19	5.98	5.81	5.55	5.27	4.99	4.85	4.70	4.54	4.39	4.23	4.06
17	15.72	10.66	8.73	7.68	7.02	6.56	6.22	5.96	5.75	5.58	5.32	5.05	4.78	4.63	4.48	4.33	4.18	4.02	3.85
18	15.38	10.39	8.49	7.46	6.81	6.35	6.02	5.76	5.56	5.39	5.13	4.87	4.59	4.45	4.30	4.15	4.00	3.84	3.67
19	15.08	10.16	8.28	7.26	6.62	6.18	5.85	5.59	5.39	5.22	4.97	4.70	4.43	4.29	4.14	3.99	3.84	3.68	3.51
20	14.82	9.95	8.10	7.10	6.46	6.02	5.69	5.44	5.24	5.08	4.82	4.56	4.29	4.15	4.00	3.86	3.70	3.54	3.38
21	14.59	9.77	7.94	6.95	6.32	5.88	5.56	5.31	5.11	4.95	4.70	4.44	4.17	4.03	3.88	3.74	3.58	3.42	3.26
22	14.38	9.61	7.80	6.81	6.19	5.76	5.44	5.19	4.99	4.83	4.58	4.33	4.06	3.92	3.78	3.63	3.48	3.32	3.15
23	14.19	9.47	7.67	6.69	6.08	5.65	5.33	5.09	4.89	4.73	4.48	4.23	3.96	3.82	3.68	3.53	3.38	3.22	3.05
24	14.03	9.34	7.55	6.59	5.98	5.55	5.23	4.99	4.80	4.64	4.39	4.14	3.87	3.74	3.59	3.45	3.29	3.14	2.97
25	13.88	9.22	7.45	6.49	5.88	5.46	5.15	4.91	4.71	4.56	4.31	4.06	3.79	3.66	3.52	3.37	3.22	3.06	2.89
26	13.74	9.12	7.36	6.41	5.80	5.38	5.07	4.83	4.64	4.48	4.24	3.99	3.72	3.59	3.44	3.30	3.15	2.99	2.82
27	13.61	9.02	7.27	6.33	5.73	5.31	5.00	4.76	4.57	4.41	4.17	3.92	3.66	3.52	3.38	3.23	3.08	2.92	2.75
28	13.50	8.93	7.19	6.25	5.66	5.24	4.93	4.69	4.50	4.35	4.11	3.86	3.60	3.46	3.32	3.18	3.02	2.86	2.69
29	13.39	8.85	7.12	6.19	5.59	5.18	4.87	4.64	4.45	4.29	4.05	3.80	3.54	3.41	3.27	3.12	2.97	2.81	2.64
30	13.29	8.77	7.05	6.12	5.53	5.12	4.82	4.58	4.39	4.24	4.00	3.75	3.49	3.36	3.22	3.07	2.92	2.76	2.59
40	12.61	8.25	6.60	5.70	5.13	4.73	4.44	4.21	4.02	3.87	3.64	3.40	3.15	3.01	2.87	2.73	2.57	2.41	2.23
60	11.97	7.76	6.17	5.31	4.76	4.37	4.09	3.87	3.69	3.54	3.31	3.08	2.83	2.69	2.55	2.41	2.25	2.08	1.89
120	11.38	7.32	5.79	4.95	4.42	4.04	3.77	3.55	3.38	3.24	3.02	2.78	2.53	2.40	2.26	2.11	1.95	1.76	1.54
∞	10.83	6.91	5.42	4.62	4.10	3.74	3.47	3.27	3.10	2.96	2.74	2.51	2.27	2.13	1.99	1.84	1.66	1.45	1.00

Degrees of freedom for the denominator (v_2)

*Multiply these entries by 100.

Example: $P\{F > F_{0.001; 9, 15} = P\{F > 6.26\} = 0.001$.

Example: $F_{0.999; 9, 15} = 1/F_{0.001; 15, 9} = 1/9.24 = 0.108$.

$$F_{0.999; v_1, v_2} = 1/F_{0.001; v_2, v_1}$$

Table 5. Summation of Terms of the Poisson Distribution[1]

$$\left[1{,}000 \times \text{ probability of } c \text{ or less occurrences of a random variable that has an average number of occurrences equal to } c' \text{ or } np', \text{ i.e.,} \quad (1000) \sum_{d=0}^{c} (c')^d e^{-c'}/d! \right]$$

c' or np' c	0.01	0.02	0.03	0.04	0.05	0.06	0.07	0.08	0.09
0	990	980	970	961	951	942	932	923	914
1	1000	1000	1000	999	999	998	998	997	996
2				1000	1000	1000	1000	1000	1000

c' or np' c	0.10	0.15	0.20	0.25	0.30	0.35	0.40	0.45	0.50
0	905	861	819	779	741	705	670	638	607
1	995	990	982	974	963	951	938	925	910
2	1000	999	999	998	996	994	992	989	986
3		1000	1000	1000	1000	1000	999	999	998
4							1000	1000	1000

c' or np' c	0.55	0.60	0.65	0.70	0.75	0.80	0.85	0.90	0.95	1.00
0	577	549	522	497	472	449	427	407	387	368
1	894	878	861	844	827	809	791	772	754	736
2	982	977	972	966	959	953	945	937	929	920
3	998	997	996	994	993	991	989	987	984	981
4	1000	1000	999	999	999	999	998	998	997	996
5			1000	1000	1000	1000	1000	1000	1000	999
6										1000

c' or np' c	1.05	1.10	1.15	1.20	1.25	1.30	1.35	1.40	1.45	1.50
0	350	333	317	301	287	273	259	247	235	223
1	717	699	681	663	645	627	609	592	575	558
2	910	900	890	879	868	857	845	833	821	809
3	978	974	970	966	962	957	952	946	940	934
4	996	995	993	992	991	989	988	986	984	981
5	999	999	999	998	998	998	997	997	996	996
6	1000	1000	1000	1000	1000	1000	999	999	999	999
7							1000	1000	1000	1000

[1] Reproduced by permission from *Tables for Multiple-Server Queueing Systems Involving Erlang Distributions* by Frederick S. Hillier and Frederick D. Lo, Technical Report #14, NSF GK-2925, Department of Operations Research, Stanford University, December 28, 1971.

Table 5 (continued). Summation of Terms of the Poisson Distribution

c' or np' c	1.55	1.60	1.65	1.70	1.75	1.80	1.85	1.90	1.95	2.00
0	212	202	192	183	174	165	157	150	142	135
1	541	525	509	493	478	463	448	434	420	406
2	796	783	770	757	744	731	717	704	690	677
3	928	921	914	907	899	891	883	875	866	857
4	979	976	973	970	967	964	960	956	952	947
5	995	994	993	992	991	990	988	987	985	983
6	999	999	998	998	998	997	997	997	996	995
7	1000	1000	1000	1000	1000	999	999	999	999	999
8						1000	1000	1000	1000	1000

c' or np' c	2.10	2.20	2.30	2.40	2.50	2.60	2.70	2.80	2.90	3.00
0	122	111	100	091	082	074	067	061	055	050
1	380	355	331	308	287	267	249	231	215	199
2	650	623	596	570	544	518	494	469	446	423
3	839	819	799	779	758	736	714	692	670	647
4	938	928	916	904	891	877	863	848	832	815
5	980	975	970	964	958	951	943	935	926	916
6	994	993	991	988	986	983	979	976	971	966
7	999	998	997	997	996	995	993	992	9у0	988
8	1000	1000	999	999	999	999	998	998	997	996
9			1000	1000	1000	1000	999	999	999	999
10							1000	1000	1000	1000

c' or np' c	3.10	3.20	3.30	3.40	3.50	3.60	3.70	3.80	3.90	4.00
0	045	041	037	033	030	027	025	022	020	018
1	185	171	159	147	136	126	116	107	099	092
2	401	380	359	340	321	303	285	269	253	238
3	625	603	580	558	537	515	494	473	453	433
4	798	781	763	744	725	706	687	668	648	629
5	906	895	883	871	858	844	830	816	801	785
6	961	955	949	942	935	927	918	909	899	889
7	986	983	980	977	973	969	965	960	955	949
8	995	994	993	992	990	988	986	984	981	979
9	999	998	998	997	997	996	995	994	993	992
10	1000	1000	999	999	999	999	998	998	998	997
11			1000	1000	1000	1000	1000	999	999	999
12								1000	1000	1000

Table 5 (continued). Summation of Terms of the Poisson Distribution

c' or np' c	4.10	4.20	4.30	4.40	4.50	4.60	4.70	4.80	4.90	5.00
0	017	015	014	012	011	010	009	008	007	007
1	085	078	072	066	061	056	052	048	044	040
2	224	210	197	185	174	163	152	143	133	125
3	414	395	377	359	342	326	310	294	279	265
4	609	590	570	551	532	513	495	476	458	440
5	769	753	737	720	703	686	668	651	634	616
6	879	867	856	844	831	818	805	791	777	762
7	943	936	929	921	913	905	896	887	877	867
8	976	972	968	964	960	955	950	944	938	932
9	990	989	987	985	983	980	978	975	972	968
10	997	996	995	994	993	992	991	990	988	986
11	999	999	998	998	998	997	997	996	995	995
12	1000	1000	999	999	999	999	999	999	998	998
13			1000	1000	1000	1000	1000	1000	999	999
14									1000	1000

c' or np' c	5.10	5.20	5.30	5.40	5.50	5.60	5.70	5.80	5.90	6.00
0	006	006	005	005	004	004	003	003	003	002
1	037	034	031	029	027	024	022	021	019	017
2	116	109	102	095	088	082	077	072	067	062
3	251	238	225	213	202	191	180	170	160	151
4	423	406	390	373	358	342	327	313	299	285
5	598	581	563	546	529	512	495	478	462	446
6	747	732	717	702	686	670	654	638	622	606
7	856	845	833	822	809	797	784	771	758	744
8	925	918	911	903	894	886	877	867	857	847
9	964	960	956	951	946	941	935	929	923	916
10	984	982	980	977	975	972	969	965	961	957
11	994	993	992	990	989	988	986	984	982	980
12	998	997	997	996	996	995	994	993	992	991
13	999	999	999	999	998	998	998	997	997	996
14	1000	1000	1000	1000	999	999	999	999	999	999
15					1000	1000	1000	1000	1000	1000

613

Table 5 (continued). Summation of Terms of the Poisson Distribution

c' or np' c	6.10	6.20	6.30	6.40	6.50	6.60	6.70	6.80	6.90	7.00
0	002	002	002	002	002	001	001	001	001	001
1	016	015	013	012	011	010	009	009	008	007
2	058	054	050	046	043	040	037	034	032	030
3	143	134	126	119	112	105	099	093	087	082
4	272	259	247	235	224	213	202	192	182	173
5	430	414	399	384	369	355	341	327	314	301
6	590	574	558	542	527	511	495	480	465	450
7	730	716	702	687	673	658	643	628	614	599
8	837	826	815	803	792	780	767	755	742	729
9	909	902	894	886	877	869	860	850	840	830
10	953	949	944	939	933	927	921	915	908	901
11	978	975	972	969	966	963	959	955	951	947
12	990	989	987	986	984	982	980	978	976	973
13	996	995	995	994	993	992	991	990	989	987
14	998	998	998	997	997	997	996	996	995	994
15	999	999	999	999	999	999	998	998	998	998
16	1000	1000	1000	1000	1000	999	999	999	999	999
17						1000	1000	1000	1000	1000

c' or np' c	7.10	7.20	7.30	7.40	7.50	7.60	7.70	7.80	7.90	8.00
0	001	001	001	001	001	001	000	000	000	000
1	007	006	006	005	005	004	004	004	003	003
2	027	025	024	022	020	019	017	016	015	014
3	077	072	067	063	059	055	052	048	045	042
4	164	156	147	140	132	125	118	112	106	100
5	288	276	264	253	241	231	220	210	201	191
6	435	420	406	392	378	365	351	338	326	313
7	584	569	554	539	525	510	496	481	467	453
8	716	703	689	676	662	648	634	620	607	593
9	820	810	799	788	776	765	753	741	729	717
10	894	887	879	871	862	854	845	835	826	816
11	942	937	932	926	921	915	909	902	895	888
12	970	967	964	961	957	954	950	945	941	936
13	986	984	982	980	978	976	974	971	969	966
14	994	993	992	991	990	989	987	986	984	983
15	997	997	996	996	995	995	994	993	993	992
16	999	999	999	998	998	998	997	997	997	996
17	1000	1000	999	999	999	999	999	999	999	998
18			1000	1000	1000	1000	1000	1000	999	999
19									1000	1000

Table 5 (continued). Summation of Terms of the Poisson Distribution

c' or np' c	8.10	8.20	8.30	8.40	8.50	8.60	8.70	8.80	8.90	9.00
0	000	000	000	000	000	000	000	000	000	000
1	003	003	002	002	002	002	002	001	001	001
2	013	012	011	010	009	009	008	007	007	006
3	040	037	035	032	030	028	026	024	023	021
4	094	089	084	079	074	070	066	062	058	055
5	182	174	165	157	150	142	135	128	122	116
6	301	290	278	267	256	246	235	226	216	207
7	439	425	412	399	386	373	360	348	336	324
8	579	565	551	537	523	509	496	482	469	456
9	704	692	679	666	653	640	627	614	601	587
10	806	796	785	774	763	752	741	729	718	706
11	881	873	865	857	849	840	831	822	813	803
12	931	926	921	915	909	903	897	890	883	876
13	963	960	956	952	949	945	940	936	931	926
14	981	979	977	975	973	970	967	965	962	959
15	991	990	989	987	986	985	983	982	980	978
16	996	995	995	994	993	993	992	991	990	989
17	998	998	998	997	997	997	996	996	995	995
18	999	999	999	999	999	999	998	998	998	998
19	1000	1000	1000	1000	999	999	999	999	999	999
20					1000	1000	1000	1000	1000	1000

c' or np' c	9.10	9.20	9.30	9.40	9.50	9.60	9.70	9.80	9.90	10.00
0	000	000	000	000	000	000	000	000	000	000
1	001	001	001	001	001	001	001	001	001	000
2	006	005	005	005	004	004	004	003	003	003
3	020	018	017	016	015	014	013	012	011	010
4	052	049	046	043	040	038	035	033	031	029
5	110	104	099	093	089	084	079	075	071	067
6	198	189	181	173	165	157	150	143	137	130
7	312	301	290	279	269	258	248	239	229	220
8	443	430	417	404	392	380	368	356	344	333
9	574	561	548	535	522	509	496	483	471	458
10	694	682	670	658	645	633	621	608	596	583
11	793	783	773	763	752	741	730	719	708	697
12	868	861	853	845	836	828	819	810	801	792
13	921	916	910	904	898	892	885	879	872	864
14	955	952	948	944	940	936	931	927	922	917
15	976	974	972	969	967	964	961	958	955	951
16	988	987	985	984	982	981	979	977	975	973
17	994	993	993	992	991	990	989	988	987	986
18	997	997	997	996	996	995	995	994	993	993
19	999	999	998	998	998	998	998	997	997	997
20	999	999	999	999	999	999	999	999	999	998
21	1000	1000	1000	1000	1000	1000	1000	999	999	999
22								1000	1000	1000

Table 5 (continued). Summation of Terms of the Poisson Distribution

c′ or np′ c	10.5	11.0	11.5	12.0	12.5	13.0	13.5	14.0	14.5	15.0
0	000	000	000	000	000	000	000	000	000	000
1	000	000	000	000	000	000	000	000	000	000
2	002	001	001	001	000	000	000	000	000	000
3	007	005	003	002	002	001	001	000	000	000
4	021	015	011	008	005	004	003	002	001	001
5	050	038	028	020	015	011	008	006	004	003
6	102	079	060	046	035	026	019	014	010	008
7	179	143	114	090	070	054	041	032	024	018
8	279	232	191	155	125	100	079	062	048	037
9	397	341	289	242	201	166	135	109	088	070
10	521	460	402	347	297	252	211	176	145	118
11	639	579	520	462	406	353	304	260	220	185
12	742	689	633	576	519	463	409	358	311	268
13	825	781	733	682	628	573	518	464	413	363
14	888	854	815	772	725	675	623	570	518	466
15	932	907	878	844	806	764	718	669	619	568
16	960	944	924	899	869	835	798	756	711	664
17	978	968	954	937	916	890	861	827	790	749
18	988	982	974	963	948	930	908	883	853	819
19	994	991	986	979	969	957	942	923	901	875
20	997	995	992	988	983	975	965	952	936	917
21	999	998	996	994	991	986	980	971	960	947
22	999	999	998	997	995	992	989	983	976	967
23	1000	1000	999	999	998	996	994	991	986	981
24			1000	999	999	998	997	995	992	989
25				1000	999	999	998	997	996	994
26					1000	1000	999	999	998	997
27							1000	999	999	998
28								1000	999	999
29									1000	1000

Table 5 (continued). Summation of Terms of the Poisson Distribution

c' or np' / c	16	17	18	19	20	21	22	23	24	25
1	000	000	000	000	000	000	000	000	000	000
2	000	000	000	000	000	000	000	000	000	000
3	000	000	000	000	000	000	000	000	000	000
4	000	000	000	000	000	000	000	000	000	000
5	001	001	000	000	000	000	000	000	000	000
6	004	002	001	001	000	000	000	000	000	000
7	010	005	003	002	001	000	000	000	000	000
8	022	013	007	004	002	001	001	000	000	000
9	043	026	015	009	005	003	002	001	000	000
10	077	049	030	018	011	006	004	002	001	001
11	127	085	055	035	021	013	008	004	003	001
12	193	135	092	061	039	025	015	009	005	003
13	275	201	143	098	066	043	028	017	011	006
14	368	281	208	150	105	072	048	031	020	012
15	467	371	287	215	157	111	077	052	034	022
16	566	468	375	292	221	163	117	082	056	038
17	659	564	469	378	297	227	169	123	087	060
18	742	655	562	469	381	302	232	175	128	092
19	812	736	651	561	470	384	306	238	180	134
20	868	805	731	647	559	471	387	310	243	185
21	911	861	799	725	644	558	472	389	314	247
22	942	905	855	793	721	640	556	472	392	318
23	963	937	899	849	787	716	637	555	473	394
24	978	959	932	893	843	782	712	635	554	473
25	987	975	955	927	888	838	777	708	632	553
26	993	985	972	951	922	883	832	772	704	629
27	996	991	983	969	948	917	877	827	768	700
28	998	995	990	980	966	944	913	873	823	763
29	999	997	994	988	978	963	940	908	868	818
30	999	999	997	993	987	976	959	936	904	863
31	1000	999	998	996	992	985	973	956	932	900
32		1000	999	998	995	991	983	971	953	929
33			1000	999	997	994	989	981	969	950
34				999	999	997	994	988	979	966
35				1000	999	998	996	993	987	978
36					1000	999	998	996	992	985
37						999	999	997	995	991
38						1000	999	999	997	994
39							1000	999	998	997
40								1000	999	998
41									999	999
42									1000	999
43										1000

ANSWERS TO SELECTED PROBLEMS

Chapter 1

15. Mean is reduced by 500; standard deviation remains the same.
16. Both the mean and standard deviation are multiplied by 1/100.

Chapter 2

2. {Black, red, yellow, green}.
3. (a) $P_X(0) = \frac{1}{3}$, $P_X(1) = \frac{1}{2}$, $P_X(2) = \frac{1}{6}$.

(b) $F_X(b) = \begin{cases} 0 \text{ for } b < 0 \\ \frac{1}{3} \text{ for } 0 \le b < 1 \\ \frac{5}{6} \text{ for } 1 \le b < 2 \\ 1 \text{ for } b \ge 2. \end{cases}$

6. Let G denote a good item and B a bad item, {GGG, GGB, GBG, GBB, BGG, BGB, BBG, BBB}.
7. (a) $P_X(0) = \frac{1}{8}$, $P_X(1) = \frac{3}{8}$, $P_X(2) = \frac{3}{8}$, $P_X(3) = \frac{1}{8}$.

(b) $F_X(b) = \begin{cases} 0 \text{ for } b < 0 \\ \frac{1}{8} \text{ for } 0 \le b < 1 \\ \frac{1}{2} \text{ for } 1 \le b < 2 \\ \frac{7}{8} \text{ for } 2 \le b < 3 \\ 1 \text{ for } b \ge 3. \end{cases}$

8. $P_Y(3000) = \frac{1}{8}$, $P_Y(1750) = \frac{3}{8}$, $P_Y(500) = \frac{3}{8}$, $P_Y(-750) = \frac{1}{8}$.
10. (a) All values between zero and U. (b) $\Omega = \{\omega : 0 \le \omega \le 1\}$. The range of the random variable is all values between zero and one.
11. (a) $\Omega = \{\omega : 0 \le \omega \le 1,000\}$. (b) $Y = [6.5X - 500]$ cents.
13. (a) $d = \frac{1}{12}$. (b) $F_X(3.6) = \frac{5}{12}$. (c) $P\{3 \le X < 5\} = \frac{1}{3}$.
15. (a) $Y = (1,000) \cdot \min (X, 4)$.

(b) $P_Y(0) = P_Y(1,000) = P_Y(2,000) = P_Y(3,000) = \frac{1}{6}$, $P_Y(4,000) = \frac{1}{3}$.
17. (a) 1000.

(b) $F_L(b) = \begin{cases} 0 & \text{if } b < 1,000 \\ 1 - \dfrac{1,000}{b} & \text{if } b \ge 1,000. \end{cases}$

(c) $P\{L \ge 1,500\} = \frac{2}{3}$.
20. $a = 500$; $b = 776.4$.
34. $E(X) = 3\frac{1}{2}$, Var $(X) = 2\frac{7}{12}$.
40. $E(X) = 500$, Var $(X) = 250,000/6$.
46. $E(2X) = 1$, Var $(2X) = 1$.
62. (a) $P_{XZ}(j, 0) = \frac{1}{6}$ for $j = 0, 1, 2, 3, 4$; $P_{XZ}(5, 0) = 0$
  $P_{XZ}(j, 1) = 0$ for $j = 0, 1, 2, 3, 4$; $P_{XZ}(5, 1) = \frac{1}{6}$.

(b) $P_Z(0) = \frac{5}{6}$, $P_Z(1) = \frac{1}{6}$. (c) $E(Z) = \frac{1}{6}$.
64. $f_{X_L}(u) = \frac{2}{100} e^{-2u/100}$; $f_{X_U}(v) = \frac{2}{100} e^{-v/100}(1 - e^{-v/100})$.
74. No.
76. No.

82. $P_{Z|X=4}(0) = 1$, $P_{Z|X=4}(1) = 0$.

84. $f_{X_U|X_L=60}(v) = \begin{cases} \frac{1}{100}e^{60/100}e^{-v/100} & \text{for } v > 60 \\ 0 & \text{elsewhere.} \end{cases}$

Chapter 3

5. 0.01242.

6. 0.2514.

8. $\sigma \leq 0.0017$.

16. (a) 0.9082. (b) 0.9772.

18. (a) 0.9050. (b) approx 1.

21. $n = 5.38$ which is rounded to $n = 6$.

29. (a) 0.0228. (b) 0.0003 (approx).

Chapter 4

3. $n = [K_{\alpha/2}\,\sigma/b]^2$.

8. (a) 0.60 (approx). (b) 0.72 (approx). (c) 0.32 (approx).

14. 0.70.

27. (a) 0.15 (approx). (b) 0.985 (approx). (c) 0.99.

29. 0.91 (approx).

34. (a) 0.304. (b) 3.29. (c) 0.256.

35. (a) $\sigma_2/\sigma_1 = 3$. (b) 1/2.

42. (a) 0.512, 0.384, 0.096, 0.008. (b) 0.896.

46. (a) $n \geq 6807$. (b) $n \geq 6807$.

Chapter 5

3. Action a_1 is minimax.

5. Action a_1 is Bayes.

11. $R(d_3, 72) = 40{,}400$, $R(d_3, 75) = 55{,}200$.

14. (a) Action a_1 is Bayes. (b) Action a_1 is Bayes.

19. (a) $P\{D = d\,|\,p = 0.01\} = \dfrac{2!}{d!(2-d)!}(0.01)^d(0.99)^{2-d}$

$\quad\quad\ P\{D = d\,|\,p = 0.10\} = \dfrac{2!}{d!(2-d)!}(0.10)^d(0.90)^{2-d}$ $\Bigg\}$ for $d = 0, 1, 2$.

(b) $P\{p = 0.01\,|\,D = d\} = \dfrac{\frac{7}{8}(0.01)^d(0.99)^{2-d}}{\frac{7}{8}(0.01)^d(0.99)^{2-d} + \frac{1}{8}(0.10)^d(0.90)^{2-d}}$

$\quad\ P\{p = 0.10\,|\,D = d\} = \dfrac{\frac{1}{8}(0.10)^d(0.90)^{2-d}}{\frac{7}{8}(0.01)^d(0.99)^{2-d} + \frac{1}{8}(0.10)^d(0.90)^{2-d}}$.

(c) a_1 is Bayes if $d = 1$ or 2; a_2 is Bayes if $d = 0$.

23. $l(a_1) = \displaystyle\int_{p_0}^{1} r(p - p_0)(n + 1)(n)p(1 - p)^{n-1}\,dp$

$\quad\ l(a_2) = \displaystyle\int_{0}^{p_0} r(p_0 - p)(n + 1)(n)p(1 - p)^{n-1}\,dp$

Choose a_2 if $l(a_2) \leq l(a_1)$, which is actually the case.

26. 0.42.

27. $\alpha = 0.1359$, $\beta(73) = 0.769$.
29. $\alpha = 0.4997$, $\beta(73) = 0.2287$.

Chapter 6

1. (a) Accept H: $\mu \geq 180$. (b) 0.004.
7. (a) $n = 9$. (b) Reject H and conclude strength increased.
8. (a) $n = 3$. (b) Accept.
15. (a) Reject H: $\mu \leq 9.5$. (b) 0.15 (approx).
16. (a) Reject if $5(\bar{X} - 9)/S < -1.711$. (b) $\beta \approx 0.65$. (c) 0.05. (d) Don't buy.
25. (a) $n = 11$. (b) 0.321. (c) Accept H: $\sigma \leq 0.04$.
26. (a) $n = 11$. (b) Reject manufacturer's claim.

Chapter 7

1. (a) Accept H: $\mu_x = \mu_y$. (b) $n = 7$.
4. (a) Reject H: air-dried $(X) \leq$ kiln-dried (Y) if

$$U = \frac{\bar{X} - \bar{Y}}{\sqrt{(45)^2(1/n_x + 1/n_y)}} > 1.645.$$

(b) $n = 5$. (c) Reject H and use air-dried.
5. (a) Accept H: $\mu_1 \geq \mu_2$ and stay with X_1. (b) 1.5.
6. (a) $n = 7$. (b) Accept H: $\mu_x \leq \mu_y$.
8. (a) Accept H: $\mu_x = \mu_y$. (b) $n = 21$.
16. Reject H: $\mu_d = 0$ (vs A: $\mu_d > 0$), i.e., accept engineer's claim.
17. $n = 4$.
22. (a) $n = 9$. (b) Accept H: $\mu_T = \mu_C$ (vs A: $\mu_T < \mu_C$).
(c) Reject since $R = 54 < 56$.
24. Reject since all differences are of the same sign (critical value equals 1).
28. (a) Reject and conclude treatment has an effect ($r = 1 \leq r_c = 1$).
(b) Reject ($T = |-2| < 6 = T_c$).
39. (a) Reject H: $\sigma_Y = \sigma_X$ (vs A: $\sigma_Y > \sigma_X$) and conclude variability of inner > outer.
45. Accept H.

Chapter 8

1. T_2 is better than T_1 when $|\theta| < 4$.
2. (a) $E(X) = np$ so that $E(X/n) = p$. (b) $p(1 - p)/n$.
4. $\sum a_i = 1$.
6. (a) $\hat{\lambda} = \bar{D}$. (b) Yes. (c) λ/n. (d) Yes, since Var $(\hat{\lambda}) = \lambda/n$. (e) $\bar{D}$.
7. (a) $\bar{X}$. (b) $\bar{X}$. (c) $1/\bar{X}$. (d) $e^{-\tau_0/\bar{X}}$.
16. $(1 + n)/(\sum X_i + 1)$.
25. (2.0904, 2.2048).
32. (a) (0.7853, 0.7947). (b) (0.0076, 0.0146).
59. $(0.784 \leq \mu \leq 0.795)$ and $(0.00732 \leq \sigma \leq 0.0155)$.

Chapter 9

2. (a) $\tilde{Y} = -149,787 + 1,799x$. (b) Reject H: $A = -200,000$. (c) Accept
H: $B = 2,000$. (d) $(1,509 \leq B \leq 2,089)$, $(-175,087 \leq A \leq -124,487)$,
$(38,563 \leq E(Y \mid x = 105 \leq 39,653)$, $(509 \leq \sigma \leq 1,130)$. (e) $(37,618, 40,598)$.
12. (a) $\tilde{Y} = 0.712 + 18.53x$. (b) Accept H: $B = 20$. (c) 0.60 (approx).
(d) $(246.39, 273.87)$. (e) $(8.23, 10.03)$.

Chapter 10

1. (a)

Source	Sum of Squares	Degrees of Freedom	Mean Square
Between	0.009895	3	0.003298
Within	0.01296	16	0.00081
Total	0.022855	19	—

(b) Reject claim that there is no difference. (c) Accept H: $\mu_A = \mu_C$, using
regular t test. (d) A and D differ. (e) $c = 6$.
12. (a)

Source	Sum of Squares	Degrees of Freedom	Mean Square
Between cols.	1,468.5	3	489.5
Between rows	4,621.5	3	1,540.5
Residual	1,354.0	9	150.4
Total	7,444.0	15	—

There are row effects but no column effects. (b) A and B differ and A and
D differ. (c) At $\phi_{min} = 1.7$, P{accept} ≈ 0.35; at $\phi_{max} = 2.5$, P{accept} ≈ 0.06.

Chapter 11

7. No, since $\sum [(1/E_i)(0_i - E_i)^2] = 2.6 < \chi^2_{0.05;3} = 7.815$.
13. No change, $\chi^2 = 0.441 < \chi^2_{0.01;1} = 6.635$.
16. No $(\chi^2 = 0.774 < \chi^2_{0.05;1} = 3.841)$.
20. Yes.

Chapter 12

4. (a) No, 2 values of R are above the UCL. (b) $UCL_{\bar{x}} = 36.478$, $LCL_{\bar{x}} = 30.997$.
(c) Yes, $\bar{\bar{X}} \pm 3\sigma' = (27.6, 39.9)$.
18. (a) $UCL_p = 0.08269$, $LCL_p = 0.0173$; $3/400 = 0.0075$ so that process indicates out of control on lower side. (b) Find which days show lack of control
and try to find cause.
19. (a) $(0, 12.04)$. (b) 0.003.

Chapter 13

3. (a) $\binom{50}{3}(0.02)^3(0.97)^{47}$. (b) 0.264.

7. (a) $1 - p'^2$. (b) $1 - p'^2$. (c) ASN (Double) $= 1 + p'$; ASN (Single) $= 2$.

15. (a) $n = 200$, $c = 5$, 0.983. (b) $n = 200$, $c = 3$, 0.001.

23. $c = 10$, $n = 63$.

29. (a) 110 (approx). (b) 890. (c) 0.02.

33. $N = 45$.

Chapter 14

5. (a) Accept if $p_U < 7.29$, $n = 10$, $M = 7.29$.
  (b) $Q_U = 1.7164$; $p_U = 3.25 < 7.29$; accept. (c) Accept if $Q_U \geq 1.41$.

10. (a) Accept if $p_U < 7.42$, $n = 10$. (b) $p_U = 3.49$; accept.
  (c) Accept if $Q_U \geq 0.579$.

15. (a) Accept if $p_U < 6.05$, $n = 5$. (b) $p_U = 10.31$; reject.
  (c) Accept if $(6.5 - \bar{X})/1.07 \geq 1.39$.

INDEX

A

Acceptable quality level (AQL)
  acceptance procedures based on, 511-512,
    522-523
  definition of, 511-512, 576
  objection to classification by, 511
  relation of:
    to normal inspection, 523
    to reduced inspection, 524-525
    to tightened inspection, 523-524
  selection of, 523
  values in MIL-STD-105D, 523
Acceptance number, 505
Acceptance region, 165-166
Acceptance sampling (see Sampling inspection)
Action, 135-136
  criteria for choosing, 138-139
  dominated, 138
Action space, 135-136
  definition, 135
  example, 135-136
Addition theorem for chi-square distribution,
  110-111
Alternative to hypothesis:
  one-sided (see One-sided alternative)
  two-sided (see Two-sided alternative)
American Society for Quality Control, 474
Analysis of enumeration data (see
  Enumeration data)
Analysis of variance, 377-436
  comparison of means, table of factors k^*
    for, 387-388
  models and tests, summary of, 435
  one-way classification, 377-405
    computational procedure for, 381
    factors k^* for test of comparison of
      means, table of, 387-388

    fixed effects model, 377-380
      example using, 377-378, 380-381,
        395-396
      OC curve of, 391-394
      row effects, test for, 385
    random effects model, 380
      example using, 380, 381, 402-403
      OC curve for, 398-402
      row effects, test for, 396-397
    partition theorem applied to, 383-384
    procedure, heuristic justification of, 383
    randomization test, 403-405
  two-way classification, 405-436
    one observation per combination,
      405-420
      computational procedure, 409, 411
      fixed effects model, 405-408
        analysis of, 411-413
        confidence intervals for
          difference of two effects,
          412-413
        example, 405, 414-416
        OC curve for, 413-414
      interaction in, 407, 408, 409,
        416, 419
      mixed effects model, 409
        analysis of, 418-420
        confidence intervals for difference
          between two effects, 419-420
        example, 420
        OC curve for, 420
      random effects model, 408
        analysis of, 416-417
        example, 418
        OC curve for, 417-418
    n observations per combination, 421-436
      computational procedure, 423-424
      fixed effects model, 421
        analysis of, 424, 426-428

Analysis of variance, (cont.)
example, 421, 428-430
OC curve for 427-428
interaction, test for, 424, 430, 433
mixed effects model, 421
analysis of, 432-434
example, 434-436
OC curve of, 434
random effects model, 421
analysis of, 430-431
example, 432
OC curve of, 431-432
AOQ (see Average outgoing quality)
AOQL (see Average outgoing quality limit)
AQL (see Acceptable quality level)
Arbitrary origin, 9, 10
Assignable cause, 472
ASTM Manual on Quality Control of Materials, 477
Attribute data, 460
relation to variable data for control chart use, 485
Attributes, sampling inspection by, 503-560
classification of plans:
by AOQL, 512-513
by AQL, 511-512
comparison of, 513
by LTPD, 512
by point of control, 512
designing of plan, 527, 537-545
Dodge-Romig tables, 513-521
double sampling AOQL, 518, 520, 521
double sampling LTPD, 514, 516, 517
single sampling AOQL, 518, 519
single sampling LTPD, 514, 515
double sampling, 508-509
in MIL-STD-105D, 518-536
multiple sampling, 510-511
sequential sampling, 537, 542-545
single sampling, 505-508, 514, 518
Average fraction inspected (AFI), 551
Average outgoing quality (AOQ), 512, 551
Average outgoing quality limit (AOQL):
definition, 512-513
double sampling tables classified by, 518, 520, 521
single sampling tables classified by, 518, 519
Average run length, 495, 497-498
Average sample number (ASN) curve:
five points for sequential test, 542-544
general formula for sequential test, 543-544

B

Bayes
decision procedures, 152-155
estimators, 292-294
mean squared risk, 292
table of, 293
formula, 73
principle, 139-141
procedure, 141, 152-155
example, 141, 153-154
summary, 154-155

risk, 152-153
rule, 141, 153
Bayesian approach, 161-162
to sampling inspection, 545-550
Bayesian interval:
definition, 308
example, 308-309
estimates, 308-309
Beta distribution, 293
Bias, 242
Biased estimator, 283
Binomial:
coefficient, 123
distribution, 123-129, 486, 506
applied to control charts, 486
approximations to:
arc sine transformation, 128
normal, 127-128, 462, 467
Poisson, 129
expected value of, 125
mean of, 125
standard deviation of, 125, 126
tables of, 126-127
variance of, 125, 126
random variable, 123-126, 462, 466
Biometrika Tables for Statisticians, 601, 603, 604
Bivariate normal distribution, 350, 362
conditional mean, 350
conditional variance, 351
Bivariate probability distributions, 52-57
Bivariate random variable, 52
continuous, 53
discrete, 53
expected value of function of, 56
Bonferroni inequality, 305, 343
example, 307
Bush, S. H., 327, 328

C

c chart, 489-492
control limits for, 490
example of, 491-492
c factors, table of, 570
Cameron, J. M., 539, 541
Causes:
assignable, 472
chance, 472
Cell (class interval), 3
Central limit theorem, 99-101, 475
application to sample mean, 99, 100
assumptions of, 99
use in hypothesis testing, 192, 230, 231
Central lines for control charts:
factors for computing, 477
formulas for, 477
Central tendency, measures of, 7-8
Charts, control (see Control charts)
Chi-square distribution, 107-114
addition theorem for, 110-111
density function of, 107-108
range of, 109
examples of, 108, 109

expected value of, 109
mean of, 109
moments of, 109
parameter of, 108
percentage points of, 601-602
range of, 107
reproductive property of, 110, 119
standard deviation of, 109
variance of, 109
Chi-square goodness of fit test, 458-460
example, 459-460
test procedure, 458-459
Chi-square random variable, 107-110
definition, 108
degrees of freedom of, 108, 109
expected value of, 109
variance of, 109
Chi-square test using enumeration data:
assessing adequacy of sample size for, 460
hypothesis, 460
when hypothesis completely specifies
theoretical frequencies of
categories, 461-463
dichotomous data, 461-463
of independence, for two-way
classification, 463-465
in 2 by 2 table, 464-465
Chi-square test for variance of normal distribution, 207-217
OC curve:
of one-sided, 211-213
of two-sided, 209
Class interval, 3, 4, 5, 9, 10
end points of, 11
Clearance between components, problem
of, 92-93
Cochran's test for homogeneity of
variances, 263-265
Code letter, 525
Comparison of two percentages, 465-466
Conditional density function, 65
Conditional probability, 57-60
definition, 58
motivation, 59
reduced sample space, 59
Conditional probability distribution, 64
Confidence interval estimates, 294-295
approximate, 302, 304
average value of Y for a given x (straight
line), 338-340, 341
for CDF, 457
definition, 294-295
for difference between means of two
normal distributions with both standard
deviations known, 298-299
example of, 299
for difference between means of two
normal distributions with both standard
deviations unknown but equal, 300-301
example of, 301
for independent variable x of fitted straight
line associated with
observation on dependent variable Y,
340-342

for mean of a normal distribution
(known σ), 295-296
example of, 296
for mean of a normal distribution
(unknown σ), 296-297
example of, 297
probability statements about, 294, 295
for proportion, 466-467
exact, 467
normal approximations to, 467
for ratio of standard deviations of two
normal distributions, 301-302
example of, 302
significance tests, relationship to, 296
of slope and intercept of a straight line,
336-338, 341
for standard deviation of a normal distribution, 297-298
example of, 298
table summarizing procedures, 303
Confidence limits, 294
Consistent estimator, 284-285
Consumer's risk, 507
Continuous probability distribution (see Probability distribution,)
Continuous random variable (see Random
variable, continuous)
Continuous random variable, cumulative distribution function of (see Cumulative distribution function of continuous random variable)
Continuous sampling inspection:
Dodge plan (CSP-1), 551-557
without control, 556-557
Dodge-Torrey plan (CSP-2, CSP-3), 552
Girshick plan, 558-559
multi-level plans, 552-556
termination of production, plans providing
for, 559-560
Wald-Wolfowitz plans, 557-558
Control:
lack of, 478
point of, 512
to prevent bias, 242-243
state of, 478-479
statistical, 472
Control charts, 474-498
(for average (see X charts)
central lines:
factors for computing, 477
formulas for, 477
control limits:
factors for computing, 477
formulas for, 477
for varying subgroup sizes, 487
for defects (see c chart)
for fraction defective (see p charts)
interpretation of, 473, 478
overview, 473
for range (see R charts)
for standard deviation (see σ charts)
Control limits:
factors for computing, 477
formulas for, 477
and specification limits, 479

Correlated samples, 242-244
Correlation, 362-364
Correlation coefficient, 72, 362
  sample, formula for, 363
    relation to slope of fitted least
      squares line, 363
    test of null hypothesis of, 363
Countably infinite set of values, 23
Covariance, 71
Cramer-Rao inequality, 285
  example 286
Critical defect, 522, 576
Critical region, 165
Critical values, table of:
  for Sign test, 247
  for signed rank test, 250
  for test for two independent samples,
    253, 254
Croxton, F. E., 600
Cumulative distribution function, 27-33
  associated with a random variable, 29
  characterization of by moments, 47
  of continuous random variable, 37, 38, 40
    examples of, 41-42
    relation to density function, 37
  definition, 29
  of discrete random variable, 34
    examples of, 30, 33, 34, 37
  of exponential distribution, 42
  frequency interpretation, 32
  of maximum of two random variables, 71
  mean of, 47
  moments of, 47-48
    example of, 49-50
    first about the origin, 47
    j^{th} about the mean, 50
    j^{th} about the origin, 50
    mean or expected value, 47
    second about the mean, 48
    second about the origin, 47
    variance, 48
  of normal distribution, 76
  of Poisson distribution, 42, 129
  properties of, 31
  of rectangular distribution, 41
  relation to probability distribution, 34
  standard deviation of, 48
  variance of, 48
  of Weibull distribution, 71
Cumulative frequency table, 5, 6
Cumulative sum control charts, 495-498
  average run length, 497-498
  procedure, 495-496
  table of optimal tests, 498
  V mask, 496

D

Data analysis, 452-467
Davis, D. J., 2
Decision function, 149
Decision making, 13, 134-178
  examples, 135, 145-148, 158-161

with experimentation, 148-162
without experimentation, 134-148
Decision procedures, 148-149, 164
  Bayes, 152-155
  choice among, 149, 164
  definition, 148
  optimum, 170
Decisions, incorrect, 165-166
Defect, definition of, 489-490
Defective, definition of, 489-490, 522
Defectives:
  allowable fraction of, 86, 96
    choice of for linear combination of
      components, 89-90
Defects, classification of, 522
  control chart for (see c chart)
  critical, 552, 576
  major, 522, 576
  minor, 522, 576
Degrees of freedom:
  definition for chi-square destribution,
    108, 109
  definition for n variables, 383-384
  definition for t distribution, 115
Density function:
  definition for continuous random
    variable, 37
  exponential, 41-42
  extension to entire real line, 39-40
  of normal random variable, 75, 76, 77, 78-79
    points of inflection of, 77
    range of, 75
  probability interpretation, 40
  properties of, 40
  rectangular, 41
  relation to cumulative distribution
    function, 37
Derman, C., 553
Dichotomous data, 461-463
Discrete probability distribution, 33-37
Discrete random variable (see Random
  variable, discrete)
Discrete random variable, cumulative distribution function (see Cumulative distribution
  function of discrete random variable)
Dispersion tests, charts and tables to
  design, 208-214
Distribution:
  binomial, 123-129
  bivariate normal, 350, 362
  chi-square, 107-114
  of difference between two sample
    means, 118-119
  empirical, 2-6
  exponential:
    cumulative distribution function of, 42
    density function of, 41
  F, 120-123
  fraction cumulative frequency, 6
  of linear combinations of normal random
    variables, 82, 83
  normal, 75-101
  Poisson, 129
    cumulative distribution function of,
      37, 129

probability distribution of, 37, 129
probability (see Probability distribution)
rectangular:
    cumulative distribution function of, 41
    density function of, 41
    of sample mean, 84-86
        compared with distribution of normal
            random variable, 85-86
        expected value of, 57, 84
        standard deviation of, 84
        standardized, 85
        variance of, 84
        with unknown standard deviation,
            117-118
    of sample standard deviation (or sample
        variance), 111-114
    t, 114-119
    Weibull, cumulative distribution function
        of, 71
Distribution form, 452-460
    qualitative techniques, 452-454
    quantitative techniques, 454-460
        chi-square goodness of fit, 458-460
        Kolmogorov-Smirnov test, 454-458
Distribution-free tolerance limits, 310,
    315-316
Dodge, H. F., 513, 551
Dodge continuous sampling plan, 551-552
    without control, 556-557
Dodge-Romig tables, 513-521
Dominated action, 138
Double sampling:
    advantage of, 508
    characterization of, 508
    Dodge-Romig AOQL tables for, 518, 520,
        521
    Dodge-Romig lot tolerance tables for, 514,
        516, 517
    in MIL-STD-105D, 526, 530-531
    OC curves for, 509-510
        example of, 509-510

E

Efficient unbiased estimator, 285-286
    Cramer-Rao inequality, 285
    definition, 285
    example, 286
Eisenhart, C., 264, 311, 465
Empirical distribution, 2-6
Enumeration data, analysis of, 460-467
    dichotomous data, 461-463
        normal approximation to handle, 462
    discussion of, 460
    independence in two-way table, test
        of, 463-465
        computing form for 2 by 2, 464-465
        example of, 463-464, 465
    normal approximation for, 462, 467
        relation to chi-square, 462
    percentages, comparison of two, 465-466
    proportion, confidence intervals for, 467
        exact, 467
        normal approximations to, 467

test statistic, 460
theoretical frequencies, specified by
    hypothesis, 461
    example of, 461
Equilibrium, process in, 479
Error:
    experimental, 13
    theory of, 100
    type I, 165, 166, 173, 183-184, 480-481
        relation to type II error, 170, 480-481
    type II, 166, 173, 184, 480-481
        dependence upon mean of, 166
        relation to sample size, 170
        relation to type I error, 170, 480-481
Estimation, 279-316
    by Bayesian intervals, 308-309
    by confidence intervals, 294-308
        approximate, 302, 304
        of differences between means of two
            normal distributions with both
            standard deviations known, 298-299
        of difference between means of two
            normal distributions with both
            standard deviations unknown but
            equal, 300-301
        of mean of a normal distribution
            (known σ), 295-296
        of mean of a normal distribution
            (unknown σ), 296-297
        probability statements about, 294, 295
        of ratio of standard deviations of two
            normal distributions, 301-302
        relationship with significance tests, 296
        of slope and intercept of straight line,
            336-338, 341
        of standard deviation of normal
            distribution, 297-298
    point, 279-294
        consistent estimators, 284-285
        definition, 280
        efficient unbiased estimators, 285-286
        by method of Bayes, 292-294
        by method of maximum likelihood,
            286-290
        by method of moments, 290-291
        unbiased estimators, 283-284
    problem, statement of, 279-280
Estimators:
    comparison of, 280-283
        loss function, 280-281
        mean squared error risk, 281
        optimal, 283
        relative efficiency, 282
        risk function, 281
    definition, 280
Events, 18-19
    definition, 18
    disjoint, 19
    independent, 59
    intersection, 19
    mutually exclusive, 19
    union, 19
Ewan, W. D., 496
Expectation, 43-47, 56
Expected risk, 152-153

Expected value:
  of binomial distribution, 125
  of chi-square random variable, 109
  of constant, 51
  of constant times random variable, 51
  of continuous random variable, 43
    example of, 44-45
  of discrete random variable, 43
    example of, 43-44
  of F distribution, 121
  of function of bivariate random
    variable, 56-57
  of function of random variable, 45, 51-52
  of linear combination of normal random
    variables, 83
  of linear combination of two random
    variables, 57
  of non-linear functions of random
    variables, 97
  of normal random variable, 76, 78
    standardized, 84
  notation for, 43
  of Poisson distribution, 129
  of random variable, definition, 43
  relation to sample mean, 45
  of t distribution, 116
Experiment:
  relevant factors of, 17
  set of all possible outcomes of, 14-17
    probabilities associated with events, 25
    used in decision making, 19-20
Experimental error, 13
Exponential distribution:
  cumulative distribution function of, 41-42
  density function of, 41
  expectation, 45
  variance, 50

F

F distribution, 120-123
  degrees of freedom of, 120, 121
  density function of, 120, 121
    range of, 120
  examples of, 120
  expected value of, 121
  mean of, 121
  parameters of, 121
  percentage points of, 121-122, 604-610
  standard deviation of, 121
  variance of, 121
F random variable, 120-122
  definition, 120
  range of, 120
F test, OC curve of:
  one-sided, 259
  two-sided, 256, 257
F test of significance:
  equality of variances from two normal
    distributions, 254, 260
  for linearity in fitting straight line, 349-350
  slope and intercept of straight line, 346
Factorial experiment, 405
Ferris, C. D., 187, 201, 211, 212, 259

Fraction cumulative frequency function, 6
Fraction defective, control chart for
  (see p chart)
Fraction of defectives, 86, 96, 126
  control limits for, 486
Fraction of successful events, relation to
  binomial random variable, 126, 127, 128
Freeman, H. A., 244, 518
Frequency function:
  fraction cumulative, 6
  pathological, 8
Frequency interpretation of probability, 26
Frequency table, 3, 4
  cumulative, 5, 6
Friedman, M., 518
Functions of random variables (see Random
  variables, functions of)

G

Gamma distribution, 293
Gamma function, 108, 115
Girshick, M. A., 558
Girshick continuous sampling plan, 558-559
Grant, E. L., 472
Greenwood, J. A., 552
Grubbs, F. E., 187, 201, 211, 212, 259
Guthrie, D., 549

H

Hald, A., 127, 549
Hamaker, H., 512
Hartley, H. O., 387-388
Hastay, M. W., 264, 311, 465
Hillier, F. S., 611
Histogram, 3, 8
Homogeneity, tests for in analysis of variance
  (see Variance)
Homogeneity of variances, test for, 263-265
Hypothesis testing:
  notation for, 172-173

I

Independence in two-way table, test of
  463-465
Independent:
  events, 59
  random variables, 60-63, 362
    definition, 61, 62-63
  samples, non-parametric test for
    two, 251-254
    critical points for, 253, 254
    example of, 252
    hypothesis to be tested by, 251
    one-sided procedures, 252
    ties, treatment of, 252
Inspection:
  normal, 523, 575
  reduced, 524-525, 575-576
  tightened, 523-524, 575-576

Inspection levels, 525, 576
Intercept (see Straight line fitting)
Interference, between components of an
    assembly, 93

J

JAN-STD-105, 518, 522
Johns, M. V., 549
Johnson, R. H., 94, 97
Joint cumulative distribution function, 52
    associated with a bivariate random variable,
        52
    of continuous bivariate random variable, 54
    of discrete bivariate random variable, 53
    properties of, 53
Joint density function, 54
    properties of, 55
Joint probability distribution of discrete
    random variable, 54

K

Kao, E. P., 495
Known standard deviation:
    control chart lines for:
        R chart with, 477, 478, 482
        σ chart with, 477, 482
        $\bar{X}$ chart with, 477
    designing plans, 572-574
    interval estimation of:
        difference between means of two normal
            distributions with, 298-299
        mean of normal distribution with,
            295-296
    one-sided variable plans, acceptance
        criterion, 571-572
    significance test for:
        equality of means of two normal distri-
            butions with, 225-235
            decision rules:
                analytical determination of,
                    230-234
                tables and charts for determin-
                    ing, 226-230
            example of, 234-235
            OC curves for:
                analytical determination of,
                    232-234
                choice of, 225-226
                tables and charts for, 230
            one-sided procedures:
                acceptance region, 229-230,
                    231, 232
                OC curves for, abscissa scale, d,
                    of, 229, 231
                sample size, 229, 231, 232
                equivalent expression if
                    $n_x \neq ny$, 229
                summary of, 229-230
            table summarizing procedures, 231
            test statistic, 225, 226, 231

two-sided procedures:
    acceptance region for, 226,
        228, 231, 232
    OC curves for, abscissa scale, d,
        of, 227
    sample size, 227, 228, 232
        equivalent expression if
            $n_x \neq ny$, 227-228
    summary of, 228
    tables and charts for, 226-228
mean of normal distribution with,
    183-198
    decision rules, analytical determin-
        ation of, 192-197
    decision rules, tables and charts for
        determining, 184-192
    example of, 197-198
    OC curve for, choice of, 183-184
    OC curves for, discussion of, 192,
        196-197
        choice of level of significance,
            192
    one-sided procedures:
        acceptance region for, 189, 191,
            192-194
        OC curves for, abscissa scale, d,
            for, 189, 191
        sample size for, 189, 191, 193,
            194
            derivation of expression for,
                194-196
        summary of, 191-192
        tables and charts for, 189-191
    test statistic for, 185, 189, 191,
        192, 193
    two-sided procedures:
        acceptance region for, 185, 189,
            193
            example of, 185-186
        OC curves for, 186
            abscissa scale, d, for, 186
            standardized, 186
        sample size for, 188, 191, 193,
            194
        summary of, 188-189, 193
        tables and charts for, 184-188
two-sided variable plans, acceptance
    criterion, 572
Kolmogorov-Smirnov test, 454-458
    confidence interval on CDF, 457
    example, 457-458
    table of percentage points, 456
    test statistic, 455

L

LCN, 83, 84, 94, 106
Least squares:
    example of worksheet for fitting straight
        line by method of, 355-356, 359-360
    method of to estimate slope and intercept,
        331-335
    worksheet for fitting straight line by
        method of, 353-354

Level of significance, 165, 175
  effect on operating characteristic curve for
    fixed sample size, 170
Lieberman, G. J., 553, 557, 568
Limits:
  confidence (see Confidence limits)
  control (see Control limits)
  natural tolerance, 86, 479
    relation to specification limits, 86,
      479-480
  specification, 86, 479-480
    distribution of, with coincident
      natural tolerances, 86
    for linear combinations of components,
      87-89
    for non-linear functions of components,
      93-98
    relation to natural tolerance limits, 86,
      479-480
    standard deviation, determining
      maximum allowable, 86-87
      for linear combination of
        components, 88-89
      for non-linear functions of
        components, 95, 98
  tolerance (see Tolerance limits)
Linear combinations of normal random
    variables (see Normal random variables)
Linear regression, 331
Linear relationships, types of, 328-331
  degree of association, 330-331
  underlying physical relationship, 329-330
Linearity, ascertaining, 349-351
Littauer, S., 553
Lo, F. D., 611
Loss, 137
Loss function, 137-138
  definition, 137
  example, 137
  for Rockwell hardness example, 141-145
Lot, formation of for acceptance purposes,
    124, 504
Lot tolerance percent defective (LTPD):
  acceptance procedures based on, 512-514
  definition of, 512
  double sampling tables based on, 514,
    516, 517
  single sampling tables based on, 514, 515
LTPD (See Lot tolerance percent defective)

M

McCornack, R. L., 250
Major defect, 522, 576
Marginal density function, 55
Marginal probability distribution, 54
Mass analogy:
  to continuous probability distribution,
    38-39
  to discrete probability distribution, 34-35
  to mean of the data, 7
  to moments, 47
  to standard deviation of the data, 9

May, J. M., 387-388
Maximum likelihood:
  example, 287, 289
  method of, 286-290
  properties of, 290
Mean of the data, 7, 9-11
  computation of, for grouped data, 9-11
  computation of, for raw data, 7, 9-11
  mass analogy to, 7
Mean of distribution estimated by sample mean,
    475
Mean of the grouped data, 9-11
Mean of probability distribution, 47
  binomial distribution, 125
  chi-square distribution, 109
  confidence interval for, normal distribution:
    standard deviation known, 295-296
    standard deviation unknown, 296-297
  F distribution, 121
  normal distribution, 76, 78
    standardized, 84
  Poisson distribution, 129
  significance test for:
    equality of, both standard deviations
      known, 225-235
      decision rules:
        analytical determination of,
          230-234
        tables and charts for determining,
          226-230
      example of, 234-235
      OC curves for:
        analytical determination of,
          232-234
        choice of, 225-226
        tables and charts for, 230
      one-sided procedures:
        acceptance region, 229-230,
          231, 232
        OC curves for, abscissa scale, d,
          of, 229, 231
        sample size, 229, 231, 232
          equivalent expression if
            $n_x \neq ny$, 229
        summary of, 229-230
      table summarizing procedures, 231
      test statistic, 225, 226, 231
      two-sided procedures:
        acceptance region for, 226, 228,
          231, 232
        OC curves for, abscissa scale, d,
          of, 227
        sample size, 227, 228, 232
          equivalent expression if
            $n_x \neq ny$, 227-228
        summary of, 228
        tables and charts for, 226-228
    equality of, both standard deviations
      unknown but assumed equal, 235-240
      example of, 239-240
      OC curves for:
        choice of, 235
        tables and charts for, 238
      one-sided procedures:
        acceptance region, 238, 239

OC curves for, abscissa scale, *d*,
   of, 237, 239
sample size, 238, 239
summary of, 237-238
table summarizing procedures, 239
test statistic for, 236, 237, 239
   degrees of freedom, 236
two sample *t* tests:
   tables and charts for carrying out,
      235-239
two-sided procedures:
   acceptance region, 236, 237, 239
   OC curves for, abscissa scale, *d*,
      for, 236, 237, 239
   sample size, 236, 237, 239
   summary of, 237
   tables and charts for, 236-237
equality of, both standard deviations
   unknown but not necessarily equal,
   240-242
   example of, 241-242
   table summarizing procedures, 241
   test procedure for, 240-241
   test statistic, 241
      degrees of freedom of, 241
equality of, when observations are paired:
   assumptions of, 243
   example of, 244-246
   to prevent bias, 242:
      by control, 242-243
      by randomization, 242-243
   test procedure, 242-244
   test statistic for, 243
      degrees of freedom of, 244
specified value, known σ, 183-198
   decision rules:
      analytical determination of,
         192-197
      tables and charts for determin-
         ing, 184-192
   example of, 197-198
   OC curve for:
      choice of, 183-184
      discussion of, 192, 196-197
         choice of level of significance,
            192
   one-sided procedures:
      acceptance region for, 189, 191,
         192-194
      OC curves for, abscissa scale, *d*,
         for, 189, 191
      sample size for, 189, 191, 193,
         194
         derivation of expression for,
            194-196
      summary of, 191-192
      tables and charts for, 189-191
   test statistic for, 185, 189, 191, 192,
      193
   two-sided procedures:
      acceptance region for, 185, 189,
         193
         example of, 185-186
      OC curve for, 186
         abscissa scale, *d*, for, 186

standardized, 186
sample size for, 188, 191, 193,
   194
summary of, 188-189, 193
tables and charts for, 184-188
specified value, unknown σ, 198-207
   example of, 206-207
   OC curves for:
      choice of, 198-199
      tables and charts for, 205-206
      two points defining, 199
   one-sided procedures:
      acceptance region for, 202, 204
      OC curves for, abscissa scale, *d*,
         for, 202, 204, 205
      sample size, 203, 204, 205
      summary of, 204
      tables and charts for, 202-204
   t tests, tables and charts for, 199-206
   test statistic for, 198, 200, 205
      degrees of freedom of, 200
   two-sided procedures:
      acceptance region for, 200, 202,
         205
      OC curves for, abscissa scale, *d*,
         for, 200, 202, 205
      sample size for, 200, 202, 205
      tables and charts for, 199-202
t distribution, 116
Median, 7, 8
Military standard 414 (MIL-STD-414), 575-595
   example using, 595
   OC curves from, 578-584
   tables of, 586-589, 591-594
Military standard 105D (MIL-STD-105D), 518,
   522-527
   acceptable quality levels, 523
   classification of defects, 523
   code letters, 525
   inspection levels, 525
   normal inspection, 523
   reduced inspection, 524
   single sampling, 526, 528, 529
   summary of procedure, 527
   table for construction of OC curves, 536
   tightened inspection, 523-524
Minimax principle, 139, 152
   example, 139, 152
Minor defect, 522, 576
Mode, 8
Moments, 47-51
   examples of, 49-50
   first about the origin, 47
   jth about the mean, 50
   jth about the origin, 50
   mass analogy to, 47
   method, 290-291
   of normal distribution, 76, 78-79
   relations between, 48
   second about the mean, 48
   second about the origin, 47
   use to characterize probability distribu-
      tion, 47
      Tchebycheff's inequality, 51
Mosteller, F., 518

Multiple sampling, 510-511, 532-535
    advantages of, 510-511

N

National Bureau of Standards, 126
Natural tolerance limits, 86-91, 479
    relation to specification limits, 86,
        479-480
Nominal value, 86, 94
Non-linear functions of components:
    tolerance in, 93-99
Non-parametric tests:
    definition, 246
    sign test, 246-249
        asymptotic properties, 248
        critical values for, 247
        example of, 248
        hypothesis to be tested by, 247
        one-sided procedures, 249
        Pitman asymptotic efficiency, 248
        ties, treatment of, 248-249
    for two independent samples, 251-254
        asymptotic properties, 253
        critical points for, 253, 254
        example of, 252
        hypothesis to be tested by, 251
        one-sided procedures, 252
        Pitman asymptotic efficiency, 253
        ties, treatment of, 252
    Wilcoxon signed rank test, 249-251
        asymptotic properties, 251
        example of, 251
        hypothesis to be tested by, 249
        one-sided procedures, 251
        Pitman asymptotic efficiency, 251
        significance points for, 250
        ties, treatment of, 251
Non-symmetric operating characteristic curve,
    188
Normal deviate:
    definition, 79
    negative, 81
    use of, 185
Normal distribution, 75-101
    areas of, table, 79-82, 600
    assumption of in statistical analysis, 75,
        99-101, 474-475
    bivariate, 350, 362
    central limit theorem, 99-101
    comparison of two with different means,
        77
    comparison of two with different variances,
        78
    cumulative distribution function of, 76
    density function of, 75, 76, 77
        points of inflection of, 77
        range of, 75
    description, 75-76
    expected value of, 76, 78
    integral of, 79-82
    mean of, 76, 78
        estimation of, 106-107
    moments of, 76, 78-79

parameters of, 75, 76, 77
    estimation of (see Estimation)
    tests of (see Significance tests)
    percentage points of, 600
    relation to t distribution, 116
    reproductive property of, 110
    standard deviation of, 76, 77, 78
    standardized, 79
        compared with regular normal curve, 79
        normal deviate of, 79
        upper α percentage point, 79
    standardized normal random variable,
        79, 84
    tolerance limits for, 309-310
    variance of, 76, 78-79
Normal inspection, 523
Normal integral, tables of, 79-82
Normal probability paper, 453-454
    estimate of mean, 453-454
    estimate of standard deviation, 454
    example, 454, 455
    procedure, 453
Normal random variables:
    linear combinations of, 82-84
        distribution of, 82-84
        mean of, 83
        variance of, 83
    standardized, 79, 84
        distribution of, 84
        mean of, 84
        variance of, 84

O

Observations:
    notation for, 7
    use in experimentation, 1
OC curves (see Operating characteristic curves)
One-sided acceptance sampling tests:
    known-sigma variables plans, 571-574
    unknown-sigma variables plans:
        estimated by average range, 569-571
        estimated by sample standard devia-
            tion, 568-569
One-sided alternative:
    classification of, 178
    definition, 174
    examples, 174-178
One-sided procedures for tests of hypotheses,
    174-178
    definition, 174
    example, 174-177
    OC curve of, compared with one-sided pro-
        cedures, 174
    risks compared with two-sided proce-
        dures, 174
One-sided tests of hypotheses for comparison
    of two means, 228-230, 237-238, 240-241
One-sided tests of hypotheses for comparison
    of two variances:
    OC curves of, F tests, 259
One-sided tests of hypotheses for value of
    specified parameter, 189-192, 202-204,
    210-214

OC curves of:
  chi-square test, 211, 212, 213
  normal test, 190
  t test, 203
Operating characteristic curves (OC curves):
  abscissa scale:
  for test of hypothesis that mean of
    a normal distribution has
    specified value:
    known standard deviation, 186,
      189, 191
    unknown standard deviation, 200,
      202, 204
  for test of hypothesis that means of
    two normal distributions are equal:
    both standard deviations known,
      227, 231
    both standard deviations unknown
      but assumed equal, 236, 237,
      239
  for test of hypothesis that standard
    deviation of a normal distribution
    has a specified value, 207, 215
  for test of hypothesis that standard
    deviations of two normal distri-
    butions are equal, 254-255, 261
  for analysis of variance, one-way
  classification:
    fixed effects model, 391-394
    random effects model, 398-402
  for chi-square test:
    one-sided, 211, 212, 213
    two-sided, 209
  choice of:
    to test hypothesis that mean of normal
      distribution has specified value:
    known standard deviation, 183-184
    unknown standard deviation,
      198-199
    to test hypothesis that means of two
      normal distributions are equal:
    both standard deviations known,
      225-226
    both standard deviations unknown
      but assumed equal, 235
    to test hypothesis that standard
      deviation of normal distribution
      has specified value, 207-208
    to test hypothesis that standard
      deviations of two normal distri-
      butions are equal, 254-255
  comparison with risk function, 167-168
  comparison for several decision proce-
    dures, 168-169
  for fixed type I error, 169
  definition of, 165-167, 173, 505
  derivation of, 166-167, 173, 196, 215, 261
  determination of two points for signifi-
    cance tests, 171, 183-184, 198-199,
    207-208, 254-255
  for double sampling plans, 509
    example of, 509
  effect on varying acceptance number or
    level of significance, 171
  effect on varying sample size, 170, 505
  for *F*-test:
    one-sided, 259
    two-sided, 256, 257
  ideal, 171, 505
  limitation of use in significance testing,
    172
  non-symmetric, 188
  for normal test:
    one-sided, 190
    two-sided, 187, 188
  parameters of, 172
  relation between errors of types I and
    II, 170
  relation between sample size and type II
    error for, 170
  for sequential sampling, five point
    curve, 543-544
    general formula for, 543-544
  for single sampling, calculation by Poisson
    distribution, 507
    example of, 508
    table for construction given sample
      size and acceptance number, 538-539
    table for construction given two points
      on curve, 540-541
  symmetric, 185
  for *t* test:
    one-sided, 203
    two-sided, 201
Optimum procedure:
  definition, 170
  for significance tests, 185
Outcomes:
  set of all possible of an experiment, 14-17
  used in decision making, 19-20

P

p charts, 485-489
  control limits for, 486, 487
  example of, 489
  interpretation of, 488-489
  starting of, 487-488
Parallel axis theorem, 48
Parameter, 37, 41
  of chi-square distribution, 109
  definition, 136
  of *F* distribution, 121
  of normal distribution, 75, 76, 78
  of operating characteristic curves,
    172, 173
Parent, E. A., 493
Partition theorem, 383-385
Percentage points:
  for chi-square distribution, 601-602
  for *F* distribution, 604-610
  for normal distribution, 600
  for *t* distribution, 603
Percentages, comparison of two, 465-466
Permutation test, 403
Philips standard sampling system, 512
Pitman asymptotic efficiency, 248, 251, 253
Point of control, 512
Point estimation, 279-294

Point estimation. (cont.)
  consistent estimators, 284-285
  definition, 280
  efficient unbiased estimators, 285-286
  method of Bayes, 292-294
  method of maximum likelihood, 286-290
  method of moments, 290-291
  unbiased estimators, 283-284
Poisson distribution:
  convergence of binomial distribution to,
    129
  cumulative distribution function of,
    37, 129
  expectation, 44
  expected value of, 129
  expression for, 129, 490
  mean of, 129, 490
  probability distribution of, 37, 129
  relation to c chart, 490
  standard deviation of, 129, 490
  table of summation of terms of, 611-617
  use of in calculation of OC curves, 507
  variance of, 49, 129
Posterior distribution, 155
  calculation, 155-158
Prediction interval:
  for future observation on dependent
    variable Y of fitted straight
    line, 341, 342-345
  for k future observations on dependent
    variable Y of fitted straight
    line, 343
  for unspecified number of future obser-
    vations on dependent variable Y of
    fitted straight line, simultaneous
    tolerance intervals, 343, 344
Prior probability distribution, 139
Probability, 24-27
  definition, 25
  frequency interpretation, 26
  notation, 25
  properties of, 25-26
  subjective, 27
  Tchebycheff's inequality to estimate, 51
Probability distribution:
  of discrete random variables, 34, 35
  posterior, 155
  prior, 139
Probability paper, normal, 453-454
Producer's risk, 507
Properties of probability, 25
Proportion, confidence intervals for, 466-467
  exact, 467
  normal approximations to, 467

Q

Quality control, 472-498
Quality variation, 472

R

R charts, 481-483
  control limits:

for known sigma, 482
for unknown sigma, 482
example of, 483-485
setting up of, 483
Random sample, 60-63, 185
  definition, 63
Random sampling, 504
Random variable, 19-24, 52
  bivariate, 52
  chi-square, 107-110
  of chi-square distribution, density
    function of, 107, 109
  continuous, 23
    cumulative distribution function
      of, 37
      examples of, 40-42
    density function of, 37
      extension to entire real line,
        39-40
      properties of, 40
      relation to cumulative distri-
        bution function, 38
    expected value of, 43
      example of, 44
    mean of, 47
    standard deviation of, 48
      example of, 49-50
    variance of, 48
      example of, 49-50
  cumulative distribution function
    associated with, 29
  definition of, 20
  discrete, 23
    cumulative distribution function of,
      34, 35
      examples of, 29-30, 33, 37
    expected value of, 43
      example of, 43-44
    mean of, 47
    probability distribution of, 34
      examples of, 36, 37
    standard deviation of, 48
      example of, 49
    variance of, 48
      example of, 49
  examples of, 21-23
  functions of, 23, 97
    expected value of, 45, 51-52, 97
    properties of, 51-52, 97
    variance of, 51, 97
  independent, 61
  mean of, 47
  non-linear functions of, 97
    expected value, 97
    variance of, 97
  notation for, 20
Randomization:
  to prevent bias, 242-243
  tests in the analysis of variance,
    403-405
Range:
  definition of, 8, 476
  distribution of:
    expected value of, 482
    standard deviation of, 482

of F distribution, density function
    of, 120
of natural tolerances, 91
of normal distribution, density
    function of, 75
of t distribution, density
    function of, 115
use of as estimate of standard
    deviation, 482
Rational subgroups, 473
Rectangular distribution, 41
    cumulative distribution function of, 41
    density function of, 41
    of sample means from, 100
Reduced inspection, 524
Reduced sample space, 59
Region:
    of acceptance, 165, 166
    critical, 165
    of rejection, 165
Regression, 331
    linear, 331
Regret, 162
Rejection region, 165
Relevant factors, of an experiment, 21
Reproductive property, 110, 119
Resnikoff, G. J., 555, 568
Risk function, 150-152, 185
    comparison with OC curve, 167-168
    non-symmetric, 188
Romig, H. G., 126, 513
Rosenblatt. H., 559

S

Sample, drawing the, 504
Sample, random, 63, 504
Sample average (see Sample mean)
Sample correlation coefficient, 363
Sample cumulative distribution function, 32
Sample mean:
    control limits for, 474
    distribution of, 84-86, 106
        compared with distribution of normal
            random variable, 85-86
        difference between two, distribution
            of, 118-119
        expected value of, 84
        mean of, 84, 106
        standard deviation of, 84,106
        standardizing, 85
        from a rectangular distribution, 100
        from a triangular distribution, 100
        with unknown standard deviation,
            117-118
        variance of, 84, 106
    as estimator of mean μ of the distri-
        bution, 475
    independence of sample variance, 114
    relation to expected value, 45
    use in variables sampling inspection, 568,
        569, 570, 571, 572
Sample size:
    code letters for, 525, 576
    for control charts, 473

effect on operating characteristic curve
    for fixed level of significance, 170
effect of varying on OC curve, 505-506
relation to type II error, 170
Sample space, 14-17
Sample standard deviation:
    chi-square random variable related to,
        112
        degrees of freedom of, 112
    control limits for, 477, 481-482
    definition of, 107, 111
    distribution of, 111-114
        example of, 113
    as estimator of standard deviation, 476,
        481-482
    ratio of two, distribution of, 122-123
    use in variables sampling inspection,
        568
Sample variance:
    chi-square random variable related to,
        112
        degrees of freedom of, 112
    distribution of, 111-114
        example of, 113
    independence of sample mean, 114
    ratio of two, distribution of, 122-123
Sampling, double (see Double sampling)
Sampling, random, 504
Sampling, single (see Single sampling)
Sampling inspection, 124
Sampling inspection by attributes, 503-560
    advantage of, 503-504
    Bayesian approach, 545-550
        decision analysis model, 546-547
        economic structure, 545-546
        example, 550
    continuous (see Continuous sampling
        inspection)
    double (see Double sampling)
    problem of, 503-504
    sequential (see Sequential sampling)
    single (see Single sampling)
Sampling inspection by variables, 565-595
Sampling plan, choosing, 507
Scheffe, H., 385, 386
Screening inspection, 503-504
Sequential sampling:
    acceptance line for, 542
    ASN curve for, 5 points, 543-544
        general formula for, 543-544
    constants for construction of plan, 542
    example of, 542, 544
    graphic procedure for, 542
    OC curve for, 5 points, 543-544
        general formula for, 543-544
    rejection line for, 542
    tabular procedure, 543
Set of all possible outcomes of an experi-
    ment, 14-17
Shewhart, W. A., 492
σ charts, 481-483
    control limits for:
        for known sigma, 481-482
        for unknown sigma, 482
    setting up of, 483

Sign test, 246-249
  asymptotic properties, 248
  critical values for, 247
  example of, 248
  hypothesis to be tested by, 247
  one-sided procedures, 249
  Pitman asymptotic efficiency, 248
  ties, treatment of, 248-249
Signed sequential rank control charts,
  493-495
  average run length, 495
  example, 494
  procedure, 493
Significance, level of, 165, 175
  effect on operating characteristic curve
    for fixed sample size, 171
Significance points, table of, for Wilcoxon
  signed rank test, 250
Significance tests, 163-178, 183-278
  chi-square test, 207-217
  for coefficients of fitted straight line,
    345-346, 347
  for comparison of two means, 225-242
  on correlated samples, 242-251
    example of, 244-246
  for equality of means when observations
    are paired:
    assumptions of, 244
    example of, 244-246
    to prevent bias, 242
      by control, 242-243
      by randomization, 242-243
      test procedure, 242-244
      test statistic for, 243
        degrees of freedom of, 244
  for equality of two percentages, 465-466
  F test (see F test of significance)
  that mean of normal distribution has
    specified value:
    known standard deviation, 183-198
      decision rules:
        analytical determination of,
          192-197
        tables and charts for determin-
          ing, 184-192
      example of, 197-198
      OC curve for:
        choice of, 183-184
        discussion of, 192, 196-197
          choice of level of signifi-
            cance, 192
      one-sided procedures:
        acceptance region for, 189, 191
          192-194
        OC curves for, abscissa scale,
          d, for, 189, 191
        sample size for, 189, 191,
          193, 194
          derivation of expression
            for, 194-196
        summary of, 191-192
        tables and charts for, 189-191
      test statistic for, 185, 189, 191,
        192, 193

      two-sided procedures:
        acceptance region for, 185,
          189, 193
          example of, 185-186
        OC curve for, 186
          abscissa scale, d,
            for, 186
          standardized, 186
        sample size for, 188, 191,
          193, 194
        summary of, 188-189, 193
        tables and charts for, 184-188
    unknown standard deviation, 198-207
      example of, 206-207
      OC curves for:
        choice of, 198-199
        tables and charts for, 205-206
        two points defining, 199
      one-sided procedures:
        acceptance region for, 202, 204
        OC curves for, abscissa scale,
          d, for, 202, 204, 205
        sample size for, 203, 204, 205
        summary of, 204
        tables and charts for, 202-204
      t tests, tables and charts
        for, 199-206
      test statistic for, 198, 200, 205
        degrees of freedom of, 200
      Two-sided procedures:
        acceptance region for, 200,
          202, 205
        OC curves for, abscissa scale,
          d, for, 200, 202, 205
        sample size for, 200, 202, 205
        tables and charts for,
          199-202
  that means of two normal distributions
    are equal:
    both standard deviations known,
      225-235
      decision rules:
        analytical determination of,
          230-234
        tables and charts for deter-
          mining, 226-230
      example of, 234-235
      OC curves for:
        analytical determination of,
          232-234
        choice of, 225-226
        tables and charts for, 230
      one-sided procedures:
        acceptance region, 229-230,
          231, 232
        OC curves for, abscissa
          scale, d of, 229, 231
        sample size, 229, 231, 232
          equivalent expression if
            $n_x \neq n_y$, 229
        summary of, 229-230
      table summarizing procedures,
        231
      test statistic, 225, 226, 231

two-sided procedures:
  acceptance region for, 226,
    228, 231, 232
  OC curves for, abscissa
    scale, d, of, 227
  sample size, 227, 228, 232
    equivalent expression if
    $n_x \neq ny$, 227-228
  summary of, 228
  tables and charts for, 226-228
both standard deviations unknown but
  assumed equal, 235-240
  example of, 239-240
  OC curves for:
    choice of, 235
    tables and charts for, 238
  one-sided procedures:
    acceptance region, 238, 239
    OC curves for, abscissa scale, d,
      of, 237, 239
    sample size, 238, 239
    summary of, 237-238
  table summarizing procedures, 239
  test statistic for, 236, 237, 239
    degrees of freedom, 236
  two-sample t tests, tables and
    charts for carrying out, 235-239
  two-sided procedures:
    acceptance region, 236, 237, 239
    OC curves for, abscissa scale, d,
      for, 236, 237, 239
    sample size, 236, 237, 239
    summary of, 237
    tables and charts for, 236-237
both standard deviations unknown and not
  necessarily equal, 240-242
  example of, 241-242
  table summarizing procedures, 241
  test procedure for, 240-241
  test statistic, 241
    degrees of freedom of, 241
non-parametric (see Non-parametric tests)
OC curves of,
  for chi-square test:
    one-sided, 211, 212, 213
    two-sided, 209
  for F test:
    one-sided, 259
    two-sided, 256, 257
  for normal test:
    one-sided, 190
    two-sided, 187, 188
  for t test:
    one-sided, 203
    two-sided, 201
that standard deviation of a normal distribution has specified value, 207-217
  analytical treatment for chi-square
    tests, 215-216
  dispersion tests, charts and tables to
    design, 208-214
  example of, 216, 217
  OC curves for:
    abscissa scale, λ, for, 207, 215

    choice of, 207-208
    tables and charts for, 214
  one-sided procedures:
    acceptance region for, 210, 212,
      213, 214
    sample size, 210, 212, 213, 214
    summary of, 213-214
    tables and charts for, 210-212
  table summarizing procedures, 214
  test statistic, 207, 208, 214
    degrees of freedom of, 208
  two-sided procedures:
    acceptance region for, 208, 210, 214
    sample size for, 209, 210, 214, 216
    summary of, 209-210
    tables and charts for, 208-210
that standard deviations of two normal distributions are equal:
  example of, 262-263
  F tests:
    analytical treatment for, 260-263
    charts and tables for, 254-258
  OC curves for:
    choice of, 254-255
    tables and charts for, 258-260
  one-sided procedures:
    acceptance region, 257, 258, 260
    sample size, 258, 261-262
    summary of, 258
    tables and charts for, 257-258
  table summarizing procedures, 260
  test statistic for, 255, 260
    degrees of freedom of, 256
    sensitivity of, 258
  two-sided procedures:
    acceptance region, 255, 257
    sample size, 256, 257, 260, 262
    summary of, 256-257
    tables and charts for, 255-257
  summary table of, 193, 205, 214, 231, 239,
    241, 260
t test (see t test of significance)
for value of a specified parameter, 183-224
that variance of a normal distribution
  has a specified value, 207-217
that variances of two normal distributions are equal, 254-263
  example of, 262-263
Simultaneous confidence intervals, 304-308
  Bonferroni inequality, 305
    example of, 307
  definition, 304
Simultaneous tolerance intervals, 343, 344
Single sampling:
  characterization of, 505
  Dodge-Romig AOQL tables for, 518, 519
  Dodge-Romig lot tolerance tables for,
    514, 515
  in MIL-STD-105D, 528-529
  OC curves for, calculation of, 507
    example of, 508
    given sample size and acceptance
      number, table for constructing,
      538-539

Single sampling:, (cont.)
    passing through two points, table
      for constructing, 540-541
Slope (see Straight lines, fitting)
Smith, B. E., 550
Solomon, H., 553
Specification limits, 86, 479
    distribution of, with coincident
      natural tolerances, 86
    for linear combinations of components,
      89-91
    for non-linear functions of compo-
      nents, 93-98
    for one-sided tests, 566-567
    relation to natural tolerance limits, 86,
      479-480
    standard deviation, determining maximum
      allowable for, 86-87
      for linear combination of components,
        88
      for non-linear functions of compo-
        nents, 95, 98
    for two-sided tests, 566-567
Standard deviation, 48
    of binomial distribution, 125, 126
    of chi-square distribution, 109
    confidence interval for, 297-298
      ratio of two, 301-302
    of the data:
      computation for grouped data, 10
      computation of raw data, 8
      computational formula for, 8
      definition of, 8-9
      interpretation of, 9
      mass analogy to, 9
    distribution of sample, 111-114
    estimated by sample average range, 476,
      482, 569-571
    estimated by sample standard deviation,
      476, 481-482, 568-569
    of F distribution, 121
    of non-linear functions of random
      variables, 97
    of normal distribution, 76, 78-79
    of Poisson distribution, 129
    sample (see Sample standard deviation)
    significance test for, equality of
      two, 254-263
      example of, 262-263
      F tests:
        analytical treatment for,
          260-263
        charts and tables for, 254-258
        OC curves for:
          abscissa scale, λ, for 254-
            255, 261
          choice of, 254-255
          tables and charts for, 258-260
        one-sided procedures:
          acceptance region, 257, 258,
            260
          sample size, 258, 261-262
          summary of, 258
          tables and charts for, 257-258
        table summarizing procedures, 260

test statistic for, 255, 260
    degrees of freedom of, 256
    sensitivity of, 258
    two-sided procedures:
      acceptance region, 255, 257
      sample size, 256, 257, 260,
        262
      summary of, 256-257
      tables and charts for, 255-257
significance test for specified value,
    207-217
    analytical treatment for chi-
      square tests, 215-216
    dispersion tests, charts and
      tables to design, 208-214
    example of, 216-217
    OC curves for:
      abscissa scale, λ for, 207,
        215
      choice of, 207-208
      tables and charts for, 214
    one-sided procedures:
      acceptance region for, 210, 212,
        213, 214
      sample size, 210, 212, 213, 214
      summary of, 213-214
      tables and charts for, 210-212
    table summarizing procedures, 214
    test statistic, 207, 208, 214
      degrees of freedom of, 208
    two-sided procedures:
      acceptance region for, 208, 210,
        214
      sample size for, 209, 210, 214,
        216
      summary of, 209-210
      tables and charts for, 208-210
    of t distribution, 116
Standardized normal distribution(see Normal
    distribution of Normal random variable)
States of nature, 136-137
    definition, 136
    example of, 136
Statistical control, 472
Statistical quality control, 472-498
Statistical Research Group of Columbia
    University, 518
Statistical tolerance limits (see Tolerance
    limits)
Statistics, definition of science of, 1,
    134
Straight line, fitting, 325-364
    confidence interval estimate of average
      value of Y for a given x, 338-340, 341
      formulation of problem and results,
        338-339
      theory, 339-340
    confidence interval estimate of indepen-
      dent variable x associated with
      observation on dependent variable Y,
      340-342
    confidence interval estimates of slope and
      intercept, 336-338
      formulation of problem and results,
        336-337

theory, 337-338
confidence interval estimates when
  intercept is zero, 348
confidence interval and prediction
  interval estimates, table summarizing,
  341
examples of, 352-362
  worksheet for, 355-356, 359-360
general discussion, 325-328
intercept of, confidence interval estimate
  for, 337
intercept of, least squares estimate of,
  333, 334, 335
  distribution of, 336
  mean and variance of, 336
intercept of, tests of hypotheses about,
  345-346
least squares estimates of slope and
  intercept, 331-335
  estimated line, 332
    variance of, 333
    formulation of problem and re-
      sults, 331-333
  notation, 331
  properties of, 333, 334-335
  theory, 333-335
linear relationships, types of, 328-331
  degree of association, 330-331
  underlying physical relationship,
    329-330
linearity, ascertaining, 349-351
prediction interval for future observation
  on dependent variable Y, 342-345
  formulation of problem and results,
    342-343
  theory, 343-345
prediction interval for k future
  observations on dependent variable
  Y, 343
prediction interval for unspecified
  number of future observations on
  dependent variable Y, 343, 344
slope of:
  confidence interval estimate for,
    336, 338
    minimum of, 337
  estimation of when intercept is zero,
    347-349
  least squares estimate of, 332, 334
    distribution of, 336
    mean and variance of, 335, 336
  tests of hypotheses about, 345-346,
    347
table summarizing point, confidence
  interval, and prediction interval
  estimates, 341
table summarizing significance tests for
  slope and intercept, 347
tests of hypotheses about slope and
  intercept, 345-346, 347
transforming to a straight line, 351
variability about the line, estimate
  of, 337
variance stabilizing transformations,
  351

worksheet for by method of least squares,
  353-354
Straight line, transforming to, 351
Student's t distribution (see t distribution)
Subgroups, rational, 473
  size of, 473
Subjective probability, 27
Subjective probability distributions, 139
Symmetric operating characteristic curve,
  185

                    T

t distribution, 114-119
  application of to distribution of sample
    mean with standard deviation unknown,
    117-118
  degrees of freedom of, 115
  density function of, 115, 116
    range of, 115
  expected value of, 116
  mean of, 116
  percentage points of, 603
  relation to normal distribution, 116
  standard deviation of, 116
  variance of, 116
t random variable:
  definition, 114
  degrees of freedom of, 114, 115
  negative, 116
  range of, 115
t test, OC curve of:
  one-sided, 203
  two-sided, 201
t test of significance:
  for coefficients of fitted straight line,
    345-346
  for correlation coefficient, 363
  for equality of two means with standard
    deviations unknown:
    assumed equal, 235-240
    not necessarily equal, 240-242
  for mean with unknown standard deviation,
    198-207
  tables and charts for, 199-206
Table summarizing formulas for central lines
  and control limits, 476
Table summarizing procedures:
  for estimation by confidence intervals
    or points, 303
  for fitted straight lines, estimates by
    point, confidence interval and pre-
    diction interval, 341
  for test of hypothesis that mean of normal
    distribution has specified value:
    known standard deviation, 193
    unknown standard deviation, 205
  for test of hypothesis that means of two
    normal distributions are equal:
    both standard deviations known, 231
    both standard deviations unknown but
      assumed equal, 239
    both standard deviations unknown and
      not necessarily equal, 241

Table summarizing procedures:, (cont.)
    for test of hypothesis that standard
        deviation of a normal distribution
        has specified value, 214
    for test of hypothesis that standard
        deviations of two normal distributions
        are equal, 260
    for tests of hypotheses about slope and
        intercept of fitted straight line, 347
Taylor series expansion for non-linear
    functions of components, 94, 96
Tchebycheff's inequality, 51
Test statistic, 185
Tests of hypothesis, notation for, 172-173
Tests of significance (see Significance test)
Theorem of unconscious statistician, 46, 56
Tightened inspection, 523-524
Tolerance factors, table of, for normal
    distribution, 311-314
Tolerance intervals, simultaneous, 343, 344
Tolerance limits, 309-316
    definition, 309
    distribution-free, 310, 315-316
        one-sided:
            example of, 315-316
            sample size for, 315
        two-sided:
            example of, 315
            sample size for, 315
    for a normal distribution, 309-314
        one-sided, 310
            example of, 310
            table of factors, K, for,
                313-314
        two-sided, 309-310
            example of, 310
            table of factors, K, for,
                311-312
Tolerances, 86-93
    clearance between components, problem of,
        92-93
    in complex items, 93-98
    for linear combinations of components,
        88-91
    natural, 86-87
        range of, 91
    for non-linear functions of components,
        93-98
Torrey, M. N., 552
Triangular distribution of sample means, 100
Tukey, J., 385
Two-sided acceptance sampling tests:
    known standard deviation variables plans,
        572
    unknown standard deviation variables plans,
        569, 571
Two-sided alternative:
    definition, 174
    examples, 174
Two-sided procedures for tests of hypotheses,
    174-178
    acceptance region for, 174
    definition, 174
    OC curve of, compared with one-sided
        procedures, 174

    risks compared with one-sided procedures,
        174
Two-sided tests of hypotheses:
    OC curves of:
        chi-square test, 209
        F test, 256, 257
        normal test, 187, 188
        t test, 201
    for value of specified parameter, 193,
        205, 214
2 by 2 table, chi-square test of independence
    in, 464-465
Type I error, 166, 173, 184, 480-481
    relation to type II error, 170, 480-481
Type II error, 166, 173, 184, 480-481
    dependence upon mean of, 166-167

U

Unbiased estimator, 283-285
    of variance, 284
Unknown standard deviation:
    control chart lines for:
        R chart with, 477, 482-483
        σ chart with, 477, 482
        X chart with, 477
    designing approximate plans, 573
    interval estimation by confidence interval:
        difference between means of two normal
            distributions with, 300-301
        mean of normal distribution with,
            296-297
    one-sided variables plans:
        average range as estimate of standard
            deviation, acceptance criterion,
            570-571
        sample standard deviation as estimate
            of standard deviation, acceptance
            criterion, 569
    significance test for:
        equality of means of two normal distri-
            butions with, 235-242
            assumed equal:
                example of, 239-240
                OC curves for:
                    choice of, 235
                    tables and charts for, 238
                one-sided procedures:
                    acceptance region, 238, 239
                    OC curves for, abscissa scale,
                        d, of, 237, 239
                    sample size, 238, 239
                    summary of, 237-238
                    table summarizing procedures,
                        239
                test statistic for, 236, 237,
                    239
                    degrees of freedom, 236
                two-sample t tests:
                    tables and charts for
                        carrying out, 235-239
                two-sided procedures:
                    acceptance region, 236,
                        237, 239

OC curves for, abscissa scale,
d, for, 236, 237, 239
sample size, 236, 237, 239
summary of, 237
tables and charts for,
236-237
not necessarily equal:
example of, 241-242
table summarizing procedures,
241
test procedure for, 240-241
test statistic, 241
degrees of freedom of, 241
mean of normal distribution with,
198-207
example of, 206-207
OC curves for:
choice of, 198-199
tables and charts for, 205-
206
two points defining, 199
one-sided procedures:
acceptance region for, 202,
204
OC curves for, abscissa scale,
d, for, 202, 204, 205
sample size, 203, 204,
205
summary of, 204
tables and charts for,
202-204
t tests, tables and charts for,
199-206
test statistic for, 198, 200, 205
degrees of freedom of, 200
two-sided procedures:
acceptance region for, 200,
202, 205
OC curves for:
abscissa scale, d, for,
200, 202, 205
sample size for, 200, 202, 205
tables and charts for, 199-202
two-sided variables plans, acceptance
criterion, 569, 571

V

V mask, 496
Van Dobben De Bruyn, C. S., 497
Variability, inherent, 13
Variables, random (see Random variables)
Variables, sampling inspection by,
565-595
characterization of, 565-567
comparison of procedures, 574-575
designing plans, 572-574
one-sided test for:
known standard deviation, 571-574
unknown standard deviation, 568-571
percent defective, estimates of, 568-572
Variables data, relation to attribute data for

control chart use, 485-486
Variance, analysis of (see Analysis of variance)
Variance:
of binomial distribution, 125, 126
of chi-square random variable, 109
of constant, 52
of constant times random variable, 51
definition, 48
of F distribution, 121
of function of random variable, 51
homogeneity of, Cochran's test for, 263
of linear combination of independent
random variables, 63
of linear combinations of normal random
variables, 83
of non-linear functions of random
variables, 97
of normal distribution, 76, 78-79
standardized, 84
of Poisson distribution, 129
sample (see Sample variance, also Sample
standard deviation)
significance test for:
equality of two, 254-263
specified value (see also Standard
deviation), 207-217
of t distribution, 116
Variation, 2, 8, 13, 472
measures of, 8-9

W

Wallis, W. A., 264, 311, 465, 518
Wald, A., 557
Weaver, C. L., 187, 201, 211, 212, 259
Weibull distribution, cumulative distribution
function of, 71
Weingarten, H., 559
White, C., 253
Wilcoxon, F., 249, 251
Wilcoxon signed rank test, 249-251
example of, 251
hypothesis to be tested by, 249
one-sided procedures, 251
significance points for, 250
ties, treatment of, 251
Wolfowitz, J., 557

X-Y-Z

X charts, 474-481
central line for, 477
control limits for:
for known standard deviation, 474
for unknown standard deviation, 476,
477
example of, 483-485
interpretation of, 480-481
starting of, 478-479
Youden, W. J., 405, 421